Books by Dimitri Kececioglu, Ph.D., P.E.

Reliability Engineering Handbook Volume 1
Reliability Engineering Handbook Volume 2
The Reliability and Life Testing Handbook Volume 1
The Reliability and Life Testing Handbook Volume 2

RELIABILITY AND LIFE TESTING HANDBOOK

Volume I

Dimitri Kececioglu, Ph.D., P.E.

Department of Aerospace and Mechanical Engineering
The University of Arizona

PTR Prentice Hall
Englewood Cliffs, New Jersey 07632

Editorial/production supervision: *Brendan M. Stewart*
Buyer: *Mary Elizabeth McCartney*
Acquisitions editor: *Michael Hays*

ISBN 0-13-772377-6

Prentice-Hall International (UK) Limited, *London*
Prentice-Hall of Australia Pty. Limited, *Sydney*
Prentice-Hall Canada Inc., *Toronto*
Prentice-Hall Hispanoamericana, S.A., *Mexico*
Prentice Hall of India Private Limited, *New Delhi*
Prentice-Hall of Japan, Inc., *Tokyo*
Simon & Schuster Asia Pte. Ltd., *Singapore*
Editora Prentice-Hall do Brasil, Ltda., *Rio de Janeiro*

To my wonderful wife Lorene,
daughter Zoe,
and son John.

VOLUME 1

TABLE OF CONTENTS

PREFACE

NEED FOR RELIABILITY AND LIFE TESTING

Reliability and life testing is a very fast growing and very important field in consumer and capital goods industries, in space and defense industries, and in NASA and DoD agencies. Reliability and life testing provide the theoretical and practical tools whereby it can be ascertained that the probability and capability of parts, components, products, and systems to perform their required functions in specified environments for the desired function period without failure, or the desired or specified reliability and life have indeed been designed and manufactured in them.

This book is intended to be used as a second text in the reliability engineering instructional curricula by upper-level undergraduate and graduate students, and as a working book by the practicing reliability, product assurance and test engineers, or the initiates. Great emphasis has been placed on clarity of presentation and practicality of subject matter. Some knowledge of the mathematics of probability and statistics would be helpful but is not absolutely necessary, as the needed mathematical background and tools are presented in the text whenever deemed necessary.

It is hoped that efforts will be expended to attract students from all fields of engineering, as well as, from operations research, mathematics, statistics, probability, chemistry and physics into reliability engineering, because this field is interdisciplinary, the demand for qualified reliability and test engineers is great, and their supply is meager.

It is very essential that for any industry and technology to be competitive, in today's highly competitive world marketplace, all countries have to know the reliability of their products, have to be able to control it, and have to produce them at the optimum reliability level that yields the minimum life-cycle cost to the user.

It is hoped that, in the not too distant future, Deans of Colleges of Engineering will require at least one course in probability and statistics of all their students, so that students taking reliability engineering and testing courses would already have the necessary mathematical background. Furthermore most engineering courses today recognize the fact that nature and everything in it is probabilistic and statistical in nature; consequently, stochastic approaches have to be incorporated into

them. It is similarly hoped that engineering students will be required to take at least one course of "Reliability Engineering" or "Reliability and Life Testing" based on the contents of this handbook and of a previous book: "Reliability Engineering Handbook," by Dr. Dimitri B. Kececioglu, Prentice Hall, Inc., Englewood Cliffs, New Jersey 07632, Tel. (201)816–4108, Vol. 1, 720 pp. and Vol. 2, 568 pp., 1991, hereafter to be referred to as *REH*.

It is no good to design and build a product which meets the requirements of the performance specifications during the relatively short period of the input-output and efficiency-of-performance tests but fails shortly thereafter or before the required period of function or mission time is consummated. To optimally minimize such product failures, most engineers should be knowledgeable of "reliability engineering," and "reliability and life testing," This handbook and the *REH* are intended to fulfill this objective.

NEED FOR THIS BOOK

This book is the culmination of over thirty years of teaching by the author in the Reliability Engineering Master's Degree Program at The University of Arizona, Tucson, Arizona, which he initiated in 1963, of extensive consulting to over 80 companies and government agencies worldwide, of putting on over 290 institutes, short courses and seminars worldwide to train industry and government personnel. Over 10,000 students, and industry and government personnel have been exposed to the author's presentations on the subjects of this book and of the *REH*. It has been written to meet the needs of (1) the students attending the Reliability Engineering Program courses, and in particular the course "Reliability Testing", at The University of Arizona; (2) those attending the "Annual Reliability Engineering and Management Institute" and the "Annual Reliability Testing Institute," conceived, initiated and directed by the author; and (3) all present and future reliability, product assurance, test and quality assurance engineers.

These needs are the following:

1. Plan reliability tests and acquire the appropriate data.

2. Reduce the acquired data to useful information in terms of the most appropriate times-to-failure distributions and their parameters, failure rates, reliabilities, conditional reliabilities, probabilities of failure, and mean lives ($MTBF's$).

3. Compare the mean lives and the reliabilities of redesigned components and equipment to determine whether the redesign has improved them as intended, to decide which supplier of a component or equipment to prefer, and to determine if a change in the application and/or operation stresses affected the mean lives and reliabilities significantly enough.

4. Interpret and use correctly the reduced information.

5. Determine the confidence limits on the true mean lives and reliabilities at a desired confidence level.

6. Identify and use efficient, short-duration and small-sample-size reliability tests.

7. Devise and use burn-in tests.

8. Devise and use accelerated reliability and life tests.

9. Identify and use nonparametric tests when the underlying times-to-failure distribution is not known or cannot be determined within the time, manpower, monetary resources and test facilities available.

10. Use appropriate, and time and sample-size efficient, sequential tests to determine the mean life and the reliability of components and equipment with chosen consumer's and producer's risks.

11. Use appropriate, and even more time and sample-size efficient, Bayesian tests to determine the mean life and the reliability of components and equipment.

12. Provide reliability growth monitoring and evaluation techniques.

13. Determine how many components and equipment to test to quantify the designed-in reliability and mean life with the desired accuracy and confidence level.

14. Plan Environmental Stress Screens (ESS) to minimize, if not eliminate, shipping to customers and users products and systems with latent defects, and with manufacturing and packaging problems, so that they get relatively failure-free products and systems.

15. Conduct software debugging tests to reduct software errors by taking corrective actions, to determine the existing number of errors, to determine how long to debug the software to remove the remaining errors, and to determine the parameters of software models to quantify the software's reliability.

Numerous examples are worked out illustrating the applications of the subject matter discussed, and an abundance of problems is given to be used by teachers, students, and the practitioner for additional practice.

HOW TO USE THIS BOOK

The material in this handbook is intended to be used by reliability engineers, reliability and life test engineers, and product assurance engineers and practitioners; as well as a college textbook, preferably covered in two regular semesters. Every major section within each chapter has a statement of its objectives, and what the reader is expected to learn. Theoretical development is followed by illustrative examples. Each chapter has at least nine problems to be worked on by the student or the practitioner. All tables necessary to solve the problems are given in the appropriate chapter or in the Appendices at the end of this handbook, for the convenience of the user, so that he (she) will not have to hunt for them elsewhere. Each chapter is completely referenced, including the specific pages involved, for further in-depth study by the reader.

This book is an ideal sequel to the "Reliability Engineering Handbook," by Dr. Dimitri B. Kececioglu, published by Prentice Hall, Inc., Englewood Cliffs, N. J. 07632, Vol. 1, 720 pp. and Vol. 2, 568 pp., 1991. I recommend that those who teach reliability engineering courses start with this "Reliability Engineering Handbook", and then follow it by teaching a second course using this "Reliability and Life Testing Handbook."

In Volume 1, Chapter 1 establishes the objectives of this handbook, defines reliability testing, tells why it is needed, presents its applications and benefits, and discusses test planning, scheduling and documentation.

Chapter 2 discusses the type of reliability data that should be obtained, data acquisition forms, reliability data sources, the final format in which the reduced data should be presented or reported; and the "data acquisition, analysis, feedback and corrective active system."

Chapter 3 covers the five very important and basic analytical reliability engineering functions; namely, the probability density function, the reliability function, the conditional reliability function, the failure rate function, and the mean life function, with applications.

Chapter 4 determines the average failure rate and reliability estimates from raw data, and provides the methodology to calculate and plot the number failed, the average failure rate and the average reliability versus their time to failure in histogram form. The methodology covered provides a quick and preliminary way of determining the behavior of raw life data, their lifetime characteristics, their failure rates, and their reliability, with applications.

Chapter 5 presents details about the chi-square, the Student's t, and the F distributions, which are used frequently in reliability and life testing, their characteristics, their percentage points, and their applications.

Chapter 6 covers the determination of the $MTBF$, failure rate, and reliability of components and equipment functioning during their useful life, or when their failure rate remains constant; i.e., their times-to-failure distribution is the exponential, with applications.

Chapter 7 provides methods to determine at a desired confidence level the confidence limits on the life, failure rate, and reliability of components and equipment that have a constant failure rate; i.e., their times-to-failure distribution is the exponential. Also the test sample size and test duration are determined; as well as, the expected waiting time and the expected number of failures during such tests, their operating characteristic curves, the case when no failures occur, and the reliability when the test time is different than the mission time, with applications.

Chapter 8 provides methods for determining whether or not two or more sets of life data, from units with an exponential times-to-failure distribution, are significantly different. The objective is to see if a redesign has improved the reliability and the mean life of components and equipment, or a particular manufacturer supplies more reliable components or products than another, or a change in the application and operation stresses has changed the reliability and the mean life of components and products significantly, with applications.

Chapter 9 covers the normal times-to-failure distribution, reliability, conditional reliability, failure rate and mean life characteristics, with applications.

Chapter 10 provides methods for the determination of the confidence intervals on the mean life, life range, life limit, lower one-sided confidence limit on the reliability, and the tolerance limits on the life of normally distributed data, with applications.

Chapter 11 covers the lognormal times-to-failure distribution, reliability, conditional reliability, failure rate and mean life characteristics, with applications.

Chapter 12 covers the Weibull times-to-failure distribution, reliability, conditional reliability, failure rate and mean life characteristics; and the determination of its parameters using the probability plotting, least squares, matching moments and the maximum likelihood unbiased estimation methods, as well as the best linear unbiased and the best linear invariant estimation methods, with numerous illustrative examples.

Chapter 13 covers the determination of the confidence limits on the reliability, mean life, mission duration, mission range, and shape and scale parameters of Weibull distributed data, with numerous illustrative examples.

Chapter 14 provides a method for calculating explicitly any percentile rank to be used for probability plotting of data and for determining the probability of success and failure of any item in any sample, including three methods to determine the median ranks, with applications.

Chapter 15 covers unique methods of tests of comparison for Weibull distributed data at their mean life and at their 10% failed life, using graphical and nomograph methods; including the cases when the Weibull slopes are the same or different, and a method of constructing the nomograph involved, with many examples.

Chapter 16 covers the gamma times-to-failure distribution, reliability, conditional reliability, failure rate and mean life characteristics; as well as, the estimation of its parameters using probability plotting, with applications.

Chapter 17 covers the beta distribution, its reliability and failure rate characteristics, and the estimation of its parameters, with applications.

Chapter 18 covers in great detail numerous methods of parameter determination from all types of continuous raw data; namely, the least squares, the matching moments and the maximum likelihood methods with many illustrative examples. Additionally, four outliers tests are

presented and illustrated by examples, to determine whether or not all data actually belong to the population they should be representing.

Chapter 19 covers the chi-squared goodness-of-fit test and its applications, to see if the chosen distribution fits the raw data acceptably well.

Chapter 20 covers several Kolmogorov-Smirnov goodness-of-fit tests including the general, the refined and the modified, and their applications, to see if the chosen distribution fits the raw data acceptably well.

Chapter 21 covers the Anderson-Darling and Cramer-Von Mises goodness-of-fit tests, with numerous illustrative examples.

Chapters 19, 20 and 21 present six different goodness-of-fit tests, which are the most available anywhere.

Volume 2, Chapter 1 of this handbook covers the binomial distribution and its uses in the determination of the reliability and confidence limits of one-shot items or of binomial (Bernoulli) trials, including the methods of determining the exact binomial confidence limits on the true reliability.

Chapter 2 covers four unique methods of tests of comparison for the one-shot-items, or the binomial case; i.e., (1) the chi-square, (2) the binomial probability plotting, (3) the homology and (4) the normal approximation methods, and numerous applications are provided.

Chapter 3 covers a unique testing technique called the "Suspended Items Test." Suspended items are components or units that are withdrawn from the test prior to their failure or before the unit with the longest life fails. This technique is used to determine the life distribution and the reliability of data containing such suspended items. It is also used to analyze field and warranty data, to determine the life distribution and the current and projected reliability of products in the field or in the customers' hands. The confidence limits on the true reliability of these products are also determined. Test-time savings are quantified. This method also analyzes data obtained from samples where not all units in the test are tested to failure. Also covered are censored tests, the average waiting time for the *rth* failure in the nonreplacement and replacement cases for exponentially and Weibull distributed units, and the *EM* algorithm method of determining the parameters of mixed Weibull distributions from suspended data. Many illustrative problems are solved to illustrate the methodology.

Chapter 4 presents a very time efficient test called the "Sudden-

Death Test", which saves much test time, and at the same time provides good confidence limits on the true life and the reliability of components and products. Test-time savings are quantified and strategies for shorter test time plans are given. All are illustrated with many examples.

Chapter 5 covers unique nonparametric tests which enable the determination of the reliability and its confidence limits when the times-to-failure distribution of the tested components or units is not known, or cannot by determined conveniently. These are the Success-Run, CL-Rank, Percent-Surviving, Sign, Run, Randomness, Range, Binomial-Pearson, Binomial-Pearson Overload, and the Wilcoxon-White Rank-Sum tests.

Chapter 6 covers three methods of Accept-Reject testing for the exponential case, whereby the equipment are tested for a fixed test time and allowable number of failures. If during this test the failures are equal to or less than the allowable, then the equipment are accepted as having passed the mean life ($MTBF$) test. If more than the predetermined number of failures are observed then the equipment are rejected as not having met their mean life requirements with the prechosen consumer's and producer's risks. The optimum number of units to test with no failures, as well as with one or more failures is determined. Methods of determining the designed-in mean life, the minimum acceptable mean life, and the correct mean life confidence limits at the conclusion of the test are also presented.

Chapter 7 presents the efficient, Sequential Probability Ratio Test for the exponential case, whereby the mean life of components or equipment can be determined in a relatively short time, with prechosen consumer's and producer's risks. If the components or equipment tested have a high designed-in operational mean life; i.e., close to their design-to level, then the test *accepts* them as having met their requirements in a *relatively short test time*. If the components or equipment tested have a low designed-in, or close to their minimum acceptable operational mean life, then the test *rejects* them as not having met their requirements in a *relatively short test time*. The operating characteristic curve determination for this test is covered, as well as the expected test time, MIL-STD-781 and its test plans, and methods of determining the confidence limits on the true mean life. Numerous illustrative examples are given.

Chapter 8 covers the sequential test on the scale parameter of the

Weibull distribution.

Chapter 9 covers the test-time and sample-size efficient Accept-Reject Test for one-shot items, or the binomial test case, and gives numerous precalculated test plans to choose from and apply.

Chapter 10 covers the sample-size and cost-efficient Sequential Test for one-shot items, or the binomial test case. Procedures for constructing such nontruncated and truncated test plans are presented, with applications.

Chapter 11 covers the ultimate test type which requires the smallest test sample size and test duration of all tests covered in this book, namely the Bayesian *MTBF* and reliability demonstration tests. The inverted gamma prior distribution is used for the *MTBF*, and the beta prior distribution is used for the reliability. Using these priors and the results of a relatively small-sample-size and short-duration test of the current design, the *MTBF* or the reliability of the current design can be determined, as well as its Bayesian confidence limits, or its credible intervals. Methods of determining the prior distribution for a variety of prior test data are covered, as well as one-risk and two-risk tests with tables of such test plans, and a unique Bayesian test using a single prior belief of what reliability can be designed and manufactured into a new product, and with what probability. Charts are given to determine the accept and reject number of failures, without the necessity of calculating them, for this single prior belief test.

Chapter 12 covers the following, very important accelerated Life Tests: Arrhenius, Eyring, Inverse Power Law, Combination Arrhenius-Inverse Power Law, Generalized Eyring, Bazovsky, Temperature-Humidity, Weibull Stress-Life, Log-Log Stress-Life, Overload-Stress Reliability, Combined-Stress Percent-Life, Deterioration-Monitoring, Step-Stress, Distribution-Free Tolerance-Limits, Nonparametric, and Optimum Accelerated Life Test with Only Two Stress Levels of Testing. Recommendations on how to conduct these accelerated tests are made. Numerous examples are given illustrating the applications of these tests. Through these tests the mean life, the failure rate and the reliability of components and equipment are determined while they operate under higher stress levels than are required for normal use. From these results their mean life, failure rate and reliability are determined for stress levels of normal or derated use.

Chapter 13 covers another very important area of reliability and life tests, that of test sample size determination to establish the *MTBF*,

failure rate or reliability, with the desired accuracy and confidence level of components and equipment having normal, exponential and Weibull times-to-failure distributions. Minimum-cost test sample-size determination methods for single-risk and sequential tests in the exponential case are also covered. In the Weibull case, the required sample sizes for the determination of the shape parameter and of the mean life are determined. Trade-off relationships to reduce the required test sample size are also given.

Chapter 14 covers reliability growth monitoring techniques to determine if the test-analyze-fail-fix-test cycle is proceeding at a pace that assures the equipment will attain their $MTBF$ or reliability goal by the time the equipment will be subjected to an $MTBF$ or reliability demonstration test, or will reach the marketplace and the customers. The following growth models are covered: Duane $MTBF$ and failure rate, both cumulative and instantaneous; Gompertz reliability, unmodified and modified to get S-shaped reliability growth curves; failure discounting for reliability; and AMSAA.

Chapter 15 covers techniques for scientifically conducting burn-in tests on electronic and nonelectronic components and assemblies, to determine the required burn-in time, for both accelerated and unaccelerated burn-in tests, using mixed-population analysis. Also least cost burn-in test plans are presented and illustrated with examples.

Chapter 16 covers the latest recommended Environmental Stress Screening (ESS) techniques to precipitate latent defects introduced into the product during manufacturing and packaging, before any product is shipped. Temperature cycling and subjecting the product to random vibration provide effective ESS. Through ESS the customer receives a failure-free product, hence he is more satisfied with the product. Case histories depicting current practices and benefits of ESS are presented.

Chapter 17 covers quantified software reliability techniques. Current models for error correction and reliability prediction are developed, and methods for estimating the model parameters from test data are explained. The Linear, Exponential, Delayed S-shaped, Inflection S-shaped, and Hyper-exponential error models; the Shooman, Musa, Jelinski-Moranda, and Schick-Wolverton reliability models; and the Moments, Least Squares, and Maximum Likelihood parameter estimation methods are covered.

It is recommended that instructors teaching this course assign three to five problems every week for homework, the specific number depend-

ing on the degree of difficulty of the problems assigned, to be handed in within a week. After the homework is corrected and returned, it should be discussed in class and all subtleties in the solutions brought out.

ACKNOWLEDGEMENTS

The author would like to thank all his many colleagues and friends for making this book possible. Special thanks are due to: The Allis-Chalmers Manufacturing Company for starting him in the field of "Reliability Engineering" and "Reliability and Life Testing"; Mr. Igor Bazovsky for his personal support of the author's activities in these fields, for helping make the Annual Reliability Engineering and Management Institutes a success, and for helping lay the foundation for this book; Dr. Austin Bonis, Dr. Leslie W. Ball, Dr. Myron Lipow, and Dr. David K. Lloyd for their inspiration to write this book; to Dr. Harvey D. Christensen, former Head, Aerospace and Mechanical Engineering Department, for starting the instruction of Reliability and Maintainability Engineering courses; Dr. Walter J. Fahey, former Dean, College of Engineering; Dr. Lawrence B. Scott, Jr. and Dr. J. T. Chen, former Heads, Aerospace and Mechanical Engineering Department; Dr. Ernest T. Smerdon, present Dean, College of Engineering and Mines; and Dr. Pitu Mirchandani, present Head Systems and Industrial Engineering Department, all of The University of Arizona, for staunchly supporting the Reliability Engineering Program at this University; the many companies and government agencies he consulted for, who enabled him to gather the practical material included in this book and for inspiring him to arrive at the effective format of this book; the many reviewers for their suggestions and help; and the many outstanding graduate students who worked under the author and helped formulate and work out many of the examples and problems in it.

The author is deeply indebted to his untiring wife Lorene June Kececioglu, his highly accomplished daughter Zoe Diana Kececioglu/Draelos, M.D. in Dermatology, and his outstanding son Dr. John Dimitri Kececioglu, Ph.D. in Computer Science, who contributed Chapter 17 of Vol. 2 on Quantified Software Reliability Techniques.

The author is indebted greatly to his numerous undergraduate and graduate students who took his Reliability Engineering and Reliability and Life Testing courses at The University of Arizona and to over 95 graduate students who got their Master's Degree in Reliability Engi-

neering and their Ph.D. Degree with a Reliability Engineering minor under him, and who made many suggestions to improve the coverage in this book and to correct errors, and in particular to outstanding graduate students Dr. Dingjun Li, Dr. Siyuan Jiang, Mr. Phuong Hung Nguyen, Mr. James W. Coleman, Mr. Pantelis Vassiliou, Mr. Feng-Bin Sun, Mr. James Bartos and Mr. Andreas Stokas, who contributed to several chapters in this book and helped work out the examples and problems, and for keying in the manuscript in the LATEX language and making it camera ready, including all figures and tables.

The author is much indebted to the many secretaries who tirelessly typed the original manuscript, and in particular to Mrs. Florence Conant and Mrs. Dorothy A. Long. Mrs. Long contributed the lion's share with her superspeed typing and accurate work.

The Prentice Hall staff were very cooperative, and contributed much to the imaginative format and to the excellence of this book.

Dr. Dimitri B. Kececioglu
Tucson, Arizona

April 1992.

Chapter 1

OBJECTIVES, TYPES, SCHEDULING AND MANAGEMENT OF RELIABILITY AND LIFE TESTING

1.1 OBJECTIVES OF RELIABILITY AND LIFE TESTING

Reliability and life testing have the following objectives:

1. Determine if the performance of components, equipment, and systems, either under closely controlled and known stress conditions in a testing laboratory or under field use conditions, with or without corrective and preventive maintenance, and with known operating procedures, is within specifications for the desired function period, and if it is not, whether it is the result of a malfunction or of a failure which requires corrective action.

2. Determine the pattern of recurring failures, the causes of failure, the underlying times-to-failure distribution, and the associated stress levels.

3. Determine the failure rate, the mean life, and the reliability of components, equipment, and systems and their associated confidence limits at desired confidence levels.

1

4. Based on the results obtained, provide guidelines as to whether corrective actions should be taken and what these should be.

5. Provide reevaluation of the reliabilitywise performance of the units after corrective actions are taken to assure that these actions were the correct ones and as effective as intended.

6. Determine the growth in the mean life and/or the reliability of units during their research, engineering and development phase; and whether the growth rate is sufficient enough to meet the mean life and/or the reliability requirements of the specifications by the time the life and/or the reliability need to be demonstrated.

7. Provide a means to statistically and scientifically determine, with chosen risks, whether a redesign has indeed improved the failure rate, mean life, or reliability of components or equipment with the desired confidence, and if not, what corrective actions need to be taken.

8. Provide a statistical and scientific means to determine which one of two manufacturers of a component or equipment, or which design, should be preferred from the failure rate, mean life, or reliability point of view, all other factors being practically and economically the same, with the desired risk level.

9. Enable the selection of the best reliability test from those presented in this book from the point of view of the test time, risk levels, number of equipment, cost, and personnel available to demonstrate the specified or desired failure rate, mean life or reliability at the chosen confidence level.

10. Provide management with the reliability test results, as requested by them, and in the format that will convey the requested information in a very easily understood form throughout the complete evaluation of a component, equipment, or system to appreciate the value of the reliability testing efforts and make the right decisions.

It must be mentioned that the mathematics of analysis of the observed data are identical whether they come from a testing laboratory or from the field, and whether they are from electronic or nonelectronic components, or equipment or systems.

The reliability test objectives should be spelled out before reliability tests are conducted. The following test objectives should be kept in mind:

1. The objective should be stated in terms of expected results. The statement of the objectives establishes the criteria against which the results of the test will be measured.

2. The objective should be realistic. If the objective is unrealistic, the test will merely be an exercise in futility. However, since tests involve uncertainty and risk, objectives should not be cast in concrete; they should be flexible enough to be adapted to unforeseen developments.

3. A time period for completion should be included. This provides milestones by which progress can be measured, and forces an emphasis on results rather than on activity.

4. The results stipulated must be measurable. The objective should be stated in terms of the data that will result from the tests. Criteria for measuring the desired accuracy, precision, confidence intervals, risks, etc., should be included in the statement of the objectives.

To ensure the accomplishment of established objectives, to avoid being sidetracked by unexpected results, and to hold accountable those responsible for obtaining test data, all test results should be justified in light of the preset criteria for success. The process of justification involves the following steps:

1. Determine whether or not the objective for each test was accomplished. If a particular objective was not met, find out why.

2. Determine what was learned from the parameters measured and the data collected, and how these results contributed to meeting each objective.

3. Account for all positive or negative results caused by accidents or other unexpected developments.

4. Account for all discrepancies in the data.

5. Indicate the level of confidence placed in the test results.

6. If an objective was only partially met or was not met at all, make recommendations for future tests in terms of measurable objectives.

7. Designate specific individuals to be responsible for carrying out the recommended actions.

1.2   WHY RELIABILITY AND LIFE TESTING?

The needs for reliability and life testing as a function of stages of development of a component, equipment, or system are as follows:

1. In early stages the tests are conducted to determine whether a unit's configuration is feasible and will perform the intended functions successfully for the desired period of time.

2. Afterwards the tests are used to select from several possible configurations the one whose mean life, failure rate, and reliability; maintainability; performance under varying operating conditions; and whose development, production, and operating costs provide the best combination.

3. Once the configuration is essentially established, additional reliability tests are conducted to improve on its mean life, failure rate, reliability, maintainability, performance, and cost after comprehensive Reliability Design Reviews are conducted which lead to the reduction of the number of components, the rearrangement of components and subsubsystems for greater reliability and accessibility, the proper derating of components, the use of standard parts with established failure rates and reliabilities, etc.

4. Later qualifying tests need to be conducted to see whether the performance and reliability requirements can be met, after all of the improvements in the previous step are implemented.

5. Finally, the components, subsubsystems, equipment, subsystems, and systems are subjected to more severe and diverse application and operation stresses to prove that the inherent design is good enough to withstand more severe conditions than the rated, and thus allow less costly methods to be used in handling and transporting them, and permit more diversified applications of these units.

1.3   TEST TYPES, THEIR APPLICATIONS AND BENEFITS

It must be recognized that Reliability and Life Tests at the component level are different than at the system level. A component has a more precisely definable function in a system. A system is a major functional unit composed of a multitude of interacting components.

A component has to be and is much more reliable than the system in which it is functioning; consequently, to determine its failure rate,

mean life or reliability a large sample size is required, a large number of tests need to be conducted over a wide range of conditions that include burn-in tests and simulate system environments, and over a relatively long period of time, to determine the best range of its use in the system it is destined for and to demonstrate its reliability.

The system, having a lower reliability requirement, is easier to test for its reliability as more failures are expected to occur within a relatively shorter test time. Nevertheless, it must be realized that systems are bulkier, more costly, and fewer would be available for reliability testing. System testing should concentrate on determining whether the components, subsubsystems and subsystems are functioning reliably enough in the system. System testing should not eliminate component testing of course, but should help pinpoint faulty components so that they may be replaced by better ones, or modified and/or redesigned to improve their reliability. The overall purpose is to obtain failure rate, mean life, and reliability information as efficiently as possible and in an optimum manner so that the components' and systems' reliability can be attained successfully in the shortest possible time.

The tests should anticipate the end operation and be planned accordingly, otherwise it will be found too late that the components, equipment, or systems are not performing (particularly reliabilitywise) their intended functions and achieving their mission objectives. Such tests should reveal the equipment's weaknesses, behavior characteristics, and modes of failure. To accomplish this an appropriate combination of tests is recommended to be conducted. These tests are the following [1, pp. 349-353; 2, pp. 419-464; 3, pp. 233-365]: (1) Burn-in. (2) Time or life. (3) Event or one-shot. (4) Accelerated or overstress. (5) Environmental.

Burn-in tests are intended to minimize, if not eliminate, defective substandard components from going into the next level of assembly. A combination of high temperature, temperature cycling, and nonthermal stresses such as voltage, wattage, mechanical loads or stresses, etc., are applied to such units as they come out of production. Failure analysis needs to be conducted to ascertain the causes of defects creeping into the production of these components. Subsequent implementation of appropriate corrective actions should minimize if not eliminate the production of such defective components.

In time or life tests the times to failure, the modes of failure, and the mechanisms of failure are determined. At the component level the following are determined: 1. Which part of the component failed. 2. How did the component fail; i.e., by early, chance, wear-out, catastrophic, drift, or degradation type failure, by erratic performance, or other? 3. What failure mechanisms were involved; i.e., were the reasons for the failure inadequate design; poor materials, manufacturing,

quality control, testing, packaging, shipping, installation, startup, or poor operation; user abuse; misapplication; or other? Such testing generates failures which are analyzed. These analyses help find answers to the previous questions when operating time is the important performance parameter of the component. The data enable the determination of the times-to-failure distribution, which in turn enables the determination of the mean life, failure rate, and reliability of the components. These in turn tell us how long the burn-in process should last, how long the useful life will last, and when components should be replaced preventively because they have reached their wear-out life. At the equipment level such testing determines the equipment's weaknesses, behavior characteristics, and modes of failure. Time testing is best suited when the components and equipment operate continuously or experience a large number of cycles wherein the transient conditions of starting and stopping are not more severe than the accumulation of time.

Event tests are used more effectively when the starting and stopping operations are much more destructive than the mere accumulation of operating time. One-shot items also fall into this category of tests. In this case the test duration for each event has to be equal to at least the mission duration specified.

Accelerated or overstress tests determine the ability of components, equipment, or systems to withstand stresses much higher than would be expected under actual operating conditions. Failure rates, mean lives, or reliabilities are obtained at four or five accelerated stress levels, and are then extrapolated to those at actual operating, or derated, stress levels or conditions. Extreme care must be exercised in extrapolating the results, particularly when overstress tests precipitate failures in modes that do not occur at actual operating stress levels.

Environmental tests are performed over a wide range of environments stradling the ambient, and usually at less severe environmental levels than those required in qualification tests, for example. In addition, emphasis is placed on the sequence and pattern of test environments and stresses.

Four major reliability tests types are:

1. Development.

2. Qualification.

3. Demonstration.

4. Quality Assurance.

Some or most of the previously discussed tests are used in these four major types of tests to a lesser or greater degree. The basic

characteristics of these tests are given in Tables 1.1 through 1.4 [4, pp. 4-2 through 4-6].

1.4   TEST PLANNING, SCHEDULING AND DOCUMENTATION

Reliability tests may cost a lot in test facilities, test fixtures, hardware to be tested, and numerous test engineers and technical personnel. It is very important, therefore, that they are properly planned so that these resources are available when needed. Such a plan should give the required test procedures, environments, test points, test schedules, and test manpower, with inputs from reliability, maintainability, quality assurance, and design engineers. The recommended steps in overall test planning are summarized in Table 1.5 [4, p. 4-7]. The information categories that should be included in a test plan are described in Table 1.6 [4, p. 4-8].

The reliability test schedule should be complete, developed early enough in the program, and be flexible to allow the incorporation of needed changes which become necessary as the program progresses. Figures 1.1 through 1.4 give samples of test schedules and support plans, which can provide the readers ideas for their situations. A preliminary classification should be made for all proposed tests based on those given in Tables 1.1 through 1.4. Test plans should be recorded in spreadsheet format.

Test documentation needs are quite large. The information categories that should be included in a test plan, generated during and at the conclusion of the tests, are given in Table 1.5 [4, p. 4-8].

TABLE 1.1 - Management aspects of development tests.

1. Purpose of Tests

   To determine physical realizability, to determine functional capabilities, and to establish the basic design.

2. General Description

   Development tests are usually informal exploratory tests designed to provide fundamental research and development information about a basic design. Nominal environmental levels are used unless the test is oriented specifically to check for effects at environmental extremes. Sample sizes are limited, but the general principles of good experimental and statistical design should be followed.

3. Examples of Specific Types of Tests

3.1 Design-Evaluation Tests	3.2 Fatigue Tests
3.3 Environmental Tests	3.4 Functional Tests
3.5 Breadboard Tests	3.6 Critical-Weakness Tests
3.7 Compatibility Tests	3.8 Growth Tests
3.9 Accelerated Tests	

4. Test Scheduling

   Not usually specified formally. Design and engineering group establishes schedules to meet design and development objectives. Such schedules must conform to development program milestones.

5. Test Items

   Basic materials, off-the-shelf parts and assemblies, breadboard models, prototype hardware.

6. Test Documentation

   Engineering test reports and analyses. Performance, failure, reliability and maintainability information to be documented for later use in prediction, growth, evaluation, and test tasks.

7. Test Follow-up Action

   Determination of design feasibility or need for redesign. Implementation of test information in further design work. Approval, modification, or disapproval of design, materials, and parts.

8. Reliability/Maintainability Provisions

   Proposed materials and designs to yield acceptable reliability and maintainability performance are tested on limited samples. Material-fatigue tests, packaging tests, component-interaction tests, accelerated environmental tests, etc., are examples. All reliability and maintainability data should be fully documented for future use in prediction, assessment, and later testing activities.

TABLE 1.2 - Management aspects of qualification tests.

1. Purpose of Tests

   To demonstrate that the equipment or specified components, assemblies, and packages meet specified performance requirements under stated environmental conditions, before full production begins.

2. General Description

   Qualification tests are formal tests conducted according to procedures specified in the development contract. Sample size is small, and thus inferential analysis is limited.

3. Examples of Specific Types of Tests

3.1 Preproduction Tests	3.2 Environmental Tests
3.3 Functional Tests	3.4 Compatibility Tests
3.5 Safety-margin Tests	3.6 Continuity Tests
3.7 Quality Tests	3.8 Burn-in Tests
3.9 Stress Screening Tests	3.10 Accelerated Tests

4. Test Scheduling

   Normally contract specified. Should be performed before production release.

5. Test Items

   Pilot-line items produced, to the extent possible, under normal production methods.

6. Test Documentation

   Detailed test requirements and procedures. Test results fully documented, including analyses and conclusions concerning design qualification.

7. Test Follow-up Action

   Approval of design or implementation of recommended changes to correct deficiencies. Design approval permits production release.

8. Reliability/Maintainability Provisions

   Limited reliability and maintainability assessments may be specified during design qualification tests, such as a short continuous-operation test or tests of failure diagnostic routines. Primary applications are limited, however, to quality testing of parts and processes.

TABLE 1.3 - Management aspects of demonstration tests.

1. Purpose of Tests

   To demonstrate formally that operational requirements in terms of effectiveness parameters such as reliability, maintainability, and design capability are achieved.

2. General Description

   Demonstration tests are performed on the major end items, often at the highest system level, under realistic operational and environmental conditions. Rules are specified for classifying failures, performing repairs, allowing design changes, etc. Time is an inherent test parameter. The test design is usually directed towards providing a specified confidence level for making an appropriate decision.

3. Examples of Specific Types of Tests

3.1 Reliability Demonstration	3.2 Maintainability Demonstration
3.3 Availability Demonstration	3.4 Life Tests
3.5 Longevity Tests	3.6 Environmental Screening Tests

4. Test Scheduling

   Demonstration test schedules are normally contract-specified. They generally occur before full-scale production but after initial production when test samples are available.

5. Test Items

   Production hardware at major end-item level.

6. Test Documentation

   Contract-specified procedures or clauses requiring contractor to submit complete test plan. Test results fully documented, including analyses and conclusions concerning the meeting of contract goals.

7. Test Follow-up Actions

   Acceptance or rejection of equipment with respect to reliability, maintainability, and effectiveness goals. Failure to pass demonstration tests will require appropriate design and assurance efforts on the part of the contractor.

8. Reliability/Maintainability Provisions

   Demonstration tests are specifically designed to test for reliability, maintainability, and associated parameters at the equipment level. Demonstration tests may be continued throughout the production cycle on samples of equipment.

TABLE 1.4 - Management aspects of quality assurance tests.

1. Purpose of Tests

   Quality assurance tests are performed on samples of incoming or out-going products to assure that materials, parts, processes, and the final product meet the established performance and quality levels.

2. General Description

   Quality assurance tests, performed during the production phase, include two basic types: (1) Acceptance tests on samples of items, to accept or reject a lot; and (2) quality control tests on processes and machines, to ensure that the final product will be satisfactory. The tests are usually designed on a statistical basis to meet specified risk levels.

3. Examples of Specific Types of Tests

   3.1 Percent-defective tests 3.2 Parts-screening tests
   3.3 Production-control tests 3.4 Part-lot acceptance tests
   3.5 Incoming-inspection tests 3.6 Storage tests
   3.7 Machine-wear tests 3.8 Continuous-sampling tests
   3.9 Burn-in tests

4. Test Scheduling

   Quality assurance tests are scheduled throughout the production phase, on either a lot-by-lot basis or on a continuous basis, depending on the circumstances. Scheduling of tests can depend on past performance of the contractor.

5. Test Items

   Incoming material, machines that process the material, and production end items at all levels.

6. Test Follow-up Action

   Acceptance or rejection of processes or production lot. Rework of rejected lots may be provided for. Many plans tighten risk levels of poor producers, or relax levels if good quality is maintained.

7. Reliability/Maintainability Provisions

   An initial reliability acceptance test on go/no-go items may be considered to be a normal quality assurance test. Time tests for testing mean life may be scheduled periodically but may not be as extensive as the initial demonstration tests. Maintainability usually is tested only indirectly.

TABLE 1.5 - Steps in overall test planning.

1. Determine test requirements and objectives.

2. Review existing data to determine if any existing requirements can be met without tests.

3. Review a preliminary list of planned tests to determine whether economies can be realized by combining individual test requirements.

4. Determine the necessary tests.

5. Allocate time, funds, and effort to perform these tests.

6. Develop test specifications at an appropriate level, or make reference to applicable sections of the system specification to provide direction for later development of test specifications.

7. Assign responsibility for test conduct, monitoring, analysis, and integration.

8. Develop review and approval policies for test-reporting procedures and forms.

9. Develop procedures for maintaining test-status information throughout the entire program.

TABLE 1.6– Information categories for a developmental testing plan.

Information Category	Description
Equipment identification	Detailed description of equipment to be tested and necessary auxiliary equipment.
Quantity	Number of test specimens to be built or purchased.
Date	Dates of delivery of test specimens and equipment, dates testing is to commence and conclude.
Test duration	Expected time tests are to continue on an item.
Test type and purpose	Example: Test to failure, nondestructive test, life test, longevity test, qualification test, etc.
Environments	Stresses to be imposed and cycling rates, parameters to be monitored, applicable specifications.
Test procedures	Applicable specifications, frequency and type of monitoring, definitions of failure or satisfactory operation, repair actions to be allowed.
Test location and facility	Place(s) where tests are to be performed, facility requirements, tools; support, test and safety equipment; calibration support requirements; spares and consumables; test monitoring equipment.
Cost of tests	Cost of test specimens and special test equipment needed, total cost of tests.
Reporting	Frequency of interim reports, types of analyses to be reported, data forms to be employed and their distribution, test completion and issuance of final test report, distribution of reports.
Responsibilities and test team	Organization, quantity and training of test team, specific personnel obligations for preparation and design of test plans and procedures, procurement of test specimens and test equipment, operation of tests, analysis of test data and results, preparation of interim and final reports, participating agencies and their degree of participation.
Reliability requirements	Statement of reliability level to which equipment will be tested, minimum acceptable and designed-to reliabilities or $MTBF's$, confidence level, the consumer's and producer's risks, major and minor failures, acceptable performance limits.
Maintainability requirements	Statement of maintainability level to which equipment will be tested, preventive and corrective maintenance to be performed.

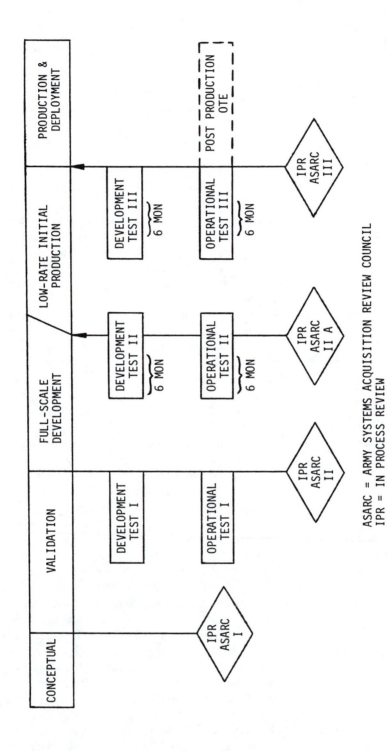

ASARC = ARMY SYSTEMS ACQUISITION REVIEW COUNCIL
IPR = IN PROCESS REVIEW

Fig. 1.1– Test schedule and support plan.

14

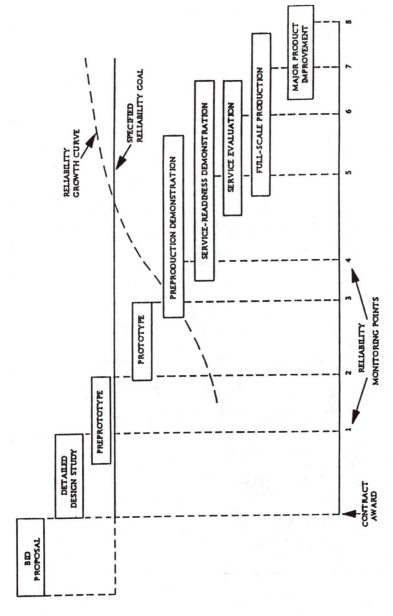

Fig. 1.2– Reliability growth tests and their schedule.

15

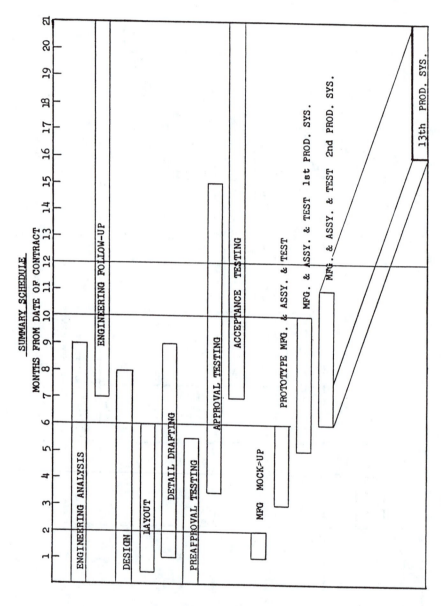

Fig. 1.3– Summary schedule of contract execution and tests.

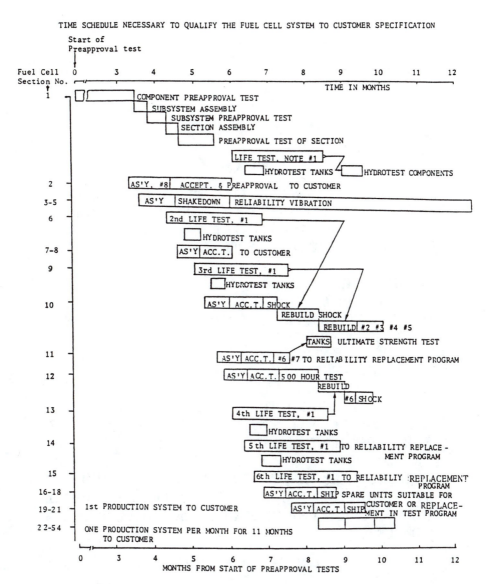

Fig. 1.4(a) – A comprehensive test schedule for a fuel cell system.

NOTES

#1 COMPONENTS FOR LIFE TEST AND TANKS FOR HYDROTEST MAKE ENOUGH COMPONENTS FOR ONE (1) SYSTEM.

#2 LOW TEMPERATURE TESTS.

#3 ALTITUDE TEST.

#4 HUMIDITY TEST.

#5 SALT SPRAY.

#6 VIBRATION TEST, RADIO INTERFERENCE TEST, NOISE TEST.

#7 ACCELERATION.

#8 AS'Y = ASSEMBLY.

*NOTHING INCLUDED FOR DEVELOPMENT TESTS OR PREAPPROVAL TESTS AS NECESSARY TO ASSURE THE DESIGNED COMPONENT WILL PERFORM AS EXPECTED. NO MANUFACTURING, QUALITY CONTROL OR INSPECTION TIME INCLUDED. NO TEST ANALYSIS TIME OR REPORT TIME INCLUDED. NO INCLUSION FOR PROCUREMENT OF MATERIALS TO TEST.

Fig. 1.4(b)– Notes to Fig. 1.4 (a).

PROBLEMS

1-1. List briefly five of the most important, in your opinion, objectives of reliability and life testing.

1-2. Spell out briefly the four specific reliability test objectives.

1-3. As all reliability and life tests should be justified in light of preset criteria, list briefly the four most important, in your opinion, steps in justifying these tests.

1-4. List briefly the five needs for reliability and life testing.

1-5. What are burn-in tests for?

1-6. What are time or life tests for?

1-7. What are event or one-shot tests for?

1-8. What are accelerated or overstress tests for?

1-9. What are environmental tests for?

1-10. Summarize in less than 50 words the eight management aspects of development tests.

1-11. Summarize in less than 50 words the management aspects of qualification tests.

1-12. Summarize in less than 50 words the management aspects of demonstration tests.

1-13. Summarize in less than 50 words the management aspects of quality assurance tests.

1-14. List briefly the nine steps in overall test planning.

1-15. Summarize in less than 100 words the information categories for a developmental testing plan.

1-16. What are key points that stand out in the test schedule and support plan of Fig. 1.1?

1-17. What are the key points that stand out in the reliability growth tests and their schedule as given in Fig. 1.2?

1-18. What are the key points that stand out in the summary schedule of contract execution and tests as given in Fig. 1.3.

1-19. What are the key points that stand out in the comprehensive test schedule for a fuel cell system given in Fig. 1.4(a).

REFERENCES

1. Lloyd, David K. and Lipow, Myron, *Reliability: Management, Methods, and Mathematics,* Lloyd and Lipow, 201 Calle Miramar, Redondo Beach, California 90277, 589 pp., 1977.

2. Mann, Nancy R., Schaefer, Ray E., and Singpurwalla, Nozer D., *Methods for Statistical Analysis of Reliability and Life Data,* John Wiley & Sons, Inc., New York, 564 pp., 1974.

3. Kapur, K.C. and Lamberson, L.R., *Reliability in Engineering Design,* John Wiley & Sons, Inc., New York, 586 pp., 1977.

4. Department of the Army, *Engineering Design Handbook–Development Guide for Reliability–Part four: Reliability Measurement,* Headquarters, United States Army Materiel Command, 5001 Eisenhower Avenue, Alexandria, VA 22333, 286 pp., Section 4, 7 January 1976.

Chapter 2

RELIABILITY AND LIFE DATA-THEIR ACQUISITION AND PROCESSING

2.1   OBJECTIVES

The objectives of this chapter are to (1) identify the data needed to determine the times-to-failure distribution, mean life, failure rate, and the reliability with their associated confidence limits, (2) provide sample reliability data acquisition forms, (3) present the final format in which the data should be prepared, (4) provide examples of ways for reducing and presenting these data, and (5) to discuss available reliability data sources for immediate use or for comparing their values with those obtained from in-house tests and analyses.

2.2   RELIABILITY AND LIFE DATA NEEDED

The reliability and life data needed should, as a minimum, provide the following information:

1. What component(s) failed?

2. When and after how many hours, miles, cycles, actuations, rounds, etc. did it fail?

3. How did it fail?

4. Why did it fail?

21

5. Is it a primary or a secondary failure?

6. What was the application stress level at failure?

7. What was the operation stress level at failure?

8. What corrective action was taken and why?

9. What was done to the failed component(s)?

10. How long it took to actively restore the equipment to a satisfactory functioning state?

11. What was the cost of the corrective action taken in parts, labor, downtime and other?

12. What was the administrative and logistic time expended in restoring the equipment to successful function?

13. What recommendations and remarks were made by the personnel assigned to the task?

14. Report number and initiation date, work completion date, signature and date of personnel assigned to the task, signature and date of approval of report, signature and date report was received by the processing organization, and signature and date report was data-processed.

Additional information should be added to meet the specific needs of the particular components, equipment, and systems involved; and to elaborate on the fourteen items listed previously as the case dictates.

2.3   RELIABILITY AND LIFE DATA ACQUISITION FORMS

Every organization or company has its own reliability and life data acquisition forms. Figure 2.1 provides the front side and Fig. 2.2 the back side (the form is in four copies) of a reliability report form. Figure 2.3 gives the front side of the GIDEP (Government-Industry Data Exchange Program) [1] form in which the background of the failure rate or reliability data should be reported to GIDEP by the supplier of such information. Figure 2.4 gives the GIDEP tabular failure rate data summary form. Figure 2.5(a) gives a typical, simple reliability report form. Figure 2.5(b) is the failure report form of the Fluid Systems Division of Allied-Signal Aerospace Company, Tempe, Arizona.

PRODUCT RELIABILITY FORM

1. PROJECT NAME _____
2. THIS REPORT COVERS ☐ REWORK ACTION ☐ TROUBLE ☐ FAILURE

ORIGINATOR

3. SYSTEM / SUBSYSTEM NAME	P / N	S / N	4. TIMER / COUNTER READINGS		5. DATE		
			HOURS	CYCLES	MO.	DAY	YR.

6. REPORTING FACILITY | 7. REPORTED BY | ORG. CODE

8. TROUBLE / FAILURE OBSERVED WAS (TEST CONDITIONS, SYMPTOMS, SEVERITY, ETC. INCLUDING "IS" AND "SHOULD BE" DATA)

9. TROUBLE / FAILURE OCCURRED DURING
(A) ☐ ASSIGNED USE (B) ☐ SCHEDULED MAINTENANCE (C) ☐ UNSCHEDULED MAINTENANCE (D) ☐ PRE-FLIGHT OR CURSORY CHECK

10. CORRECTIVE ACTION TAKEN
(A) ☐ Replaced Item (Complete 11 - 13)
(B) ☐ Repaired Item (Complete 14 - 28)

11. ITEM REPLACED NAME	P / N	S / N	12. ITEM INSTALLED NAME	P / N	S / N	13. DISPOSITION OF REPLACED ITEM

VERIFICATION

14. WAS REPORTED TROUBLE / FAILURE VERIFIED ?
(A) ☐ YES (B) ☐ NO (CHECK 17 B. AND EXPLAIN)

15. FAILED ITEM NAME	P / N	S / N	MANUFACTURER
16. NEXT HIGHER ASSY: NAME	P / N	S / N	MANUFACTURER

17. (A) ☐ REWORK REQUIRED IS (B) ☐ REWORK IS NOT REQUIRED BECAUSE | 18. VENDOR REWORK / ANALYSIS IS REQUIRED ☐

19. DISPOSITION OF ITEM IF NOT REWORKED | 20. AUTHORIZATION BY | ORG. CODE

REWORK & RETEST

21. LIST BELOW ALL PARTS REPLACED OR ADJUSTED

(A) CIRCUIT SYMBOL	(B) PART #	(C) S/N	(D) MANUFACTURER	(E) DEFECT	(G) REWORK ACTION TAKEN	(H) APPROVED PARTS YES	NO

22. OTHER REWORK ACTION

23. REWORKED BY ORG. CODE	24. RETESTED BY ORG. CODE	25. RETEST RESULTS APPROVAL DATE

26. RETEST RESULTS: REPAIRED ITEM IS ACCEPTABLE FOR
☐ UNRESTRICTED USE ☐ RESTRICTED USE ☐ NOT ACCEPTABLE (EXPLAIN AT 29) | 27. DISPOSITION OF ITEM AFTER RETEST | 28. Q.C.

ENGINEERING

29. CAUSE OF FAILURE - COMMENTS

30. BASIC CAUSE OF TROUBLE / FAILURE | 31. SIGNATURE | ORG. CODE | DATE

32. CORRECTIVE ENGINEERING ACTION TAKEN (IF NONE. JUSTIFY) - CONCLUSIONS

REF. DOCUMENT IMPLEMENTING CORRECTIVE ACTION (EO. ECR. ETC) IS NO. | 33. SIGNATURE | ORG. CODE | DATE

Fig. 2.1– A product reliability report form. Front side.

RELIABILITY ANALYSIS AND FOLLOW UP

1. CAUSE OF DIFFICULTY

- ☐ DESIGN
- ☐ MFG. PROCESS
- ☐ OPERATING TIME
- ☐ TEST EQUIPMENT
- ☐ TEST PROCEDURE
- ☐ HUMAN ERROR
- ☐ WORKMANSHIP

- ☐ WIRING ERROR
- ☐ ROUGH HANDLING
- ☐ ENVIRONMENTAL EFFECT(S)
- ☐ OTHER ITEM FAILURE _____ SPECIFY
- ☐ DEFECTIVE PARTS
- ☐ UNKNOWN
- ☐ OTHER _____ SPECIFY

2. EFFECT ON

(ITEM NAME)

- ☐ INOPERATIVE
- ☐ BADLY DEGRADED
- ☐ SLIGHTLY DEGRADED
- ☐ NUISANCE

3. CORRECTIVE ACTION STATUS　　DATE

- ☐ IN EFFECT _____
- ☐ OTHER INVESTIGATION _____
- ☐ SCHEDULED _____
- ☐ INTERIM FIX _____
- ☐ NOT REQUIRED _____
- ☐ ACTION PREVIOUSLY TAKEN _____

C. A. R. NO. _____

4. REPLACED / REPAIRED ITEM OPERATIONAL USAGE

HOURS	CYCLES

5. RELATED CASES

_____
_____
_____

6. CLASS OF FAILURE

FAIL CLASS	TEST TYPE	TEST ENV.	FAIL CAUSE	RELEV.

PARTS ANALYSIS

7. CIRCUIT SYMBOL	8. RESULTS	9. ANALYZED BY	ORG. CODE	DATE	10. FAILURE CODE
			| |		
			| |		
			| |		
			| |		

11. COMMENTS

12. RELIABILITY ANALYSIS BY	ORG. CODE	DATE
	| | |	MO.　DAY　YR.

APPROVALS:

13. RR CLOSED

VENDOR / CUSTOMER (IF APPLICABLE)　　ENGINEERING (IF APPLICABLE)　　RELIABILITY

Fig. 2.2– A product reliability report form. Back side.

MICROFILM ACCESS NUMBER	GOVERNMENT–INDUSTRY DATA EXCHANGE PROGRAM **RELIABILITY–MAINTAINABILITY DATA SUMMARY** *Please Type All Information – See Instructions On Reverse*	1 OF

1. ACTIVITY	CITY/STATE	NAME	TELEPHONE
2. CODE	3. CHECK ONE ☐ FAILURE RATES ☐ REPLACEMENT RATES	4. REPORT NUMBER	5. DATE
6. PROGRAM OR SYSTEM	7. OBSERVED ENVIRONMENT	8. OBSERVATION PERIOD (Mo/Yr) START FINISH	9. UNIT OF MEASURE
10. ELECTRICAL STRESS CAT % OF	11. TEMPERATURE CAT °C	12. MECHANICAL STRESS CAT TYPE	13. SCREENING CLASS SPEC NO.

LINE	14. ITEM NOMENCLATURE	15. APPLICATION
1		
2		
3		
4		
5		
6		

LINE	16. VENDOR & H4 CODE	17. ITEM NUMBER	18. OTHER ITEM NUMBER(S)	19. NO. OF ITEMS
1				
2				
3				
4				
5				
6				

LINE	20. DURATION	21. NO. FAILURES	22. FAILURE RATE	23. MEAN REPAIR TIME	25. PREDOMINANT FAILURE MODE(S)
1					
2					
3					
4					
5					
6					

25. ADDITIONAL COMMENTS AND INFORMATION
SIGNATURE

DD FORM 1 NOV 78 2001

REPRODUCTION OR DISPLAY OF THIS MATERIAL FOR SALES OR PUBLICITY PURPOSES IS PROHIBITED

Fig. 2.3– GIDEP Reliability-Maintainability Data Summary form.

TABULAR FAILURE RATE DATA SUMMARY
11ND-FMSAEG-8800/10 (REV. 2-66)

"BACKGROUND INFORMATION ON FAILURE RATE DATA" Should also be provided.

(1) PART/COMPONENT COMPLETE NOMENCLATURE (Identification including MIL Spec. No. if applicable)	(2) VENDOR AND VENDOR PART NO., MIL. STD. PART NO., OR FEDERAL STOCK NUMBER	(3) OBSERVATION ENVIRONMENT	(4) INTENDED INSTALLATION (End Use) ENVIRONMENT	(5) PARTS X HOURS IN MILLIONS t	(6) NUMBER OF FAILURES f	(7) FAILURE RATE λ f/t	(8) PART/ COMPONENT POPULATION

NOTE: (a) The following coding is used for radiation type:

NUCLEAR	ELECTROMAGNETIC
A - Fast	D - Gamma
B - Slow	E - X-Ray
C - Thermal	

Fig. 2.4(a)– GIDEP's tabular failure rate data summary
form, left half.

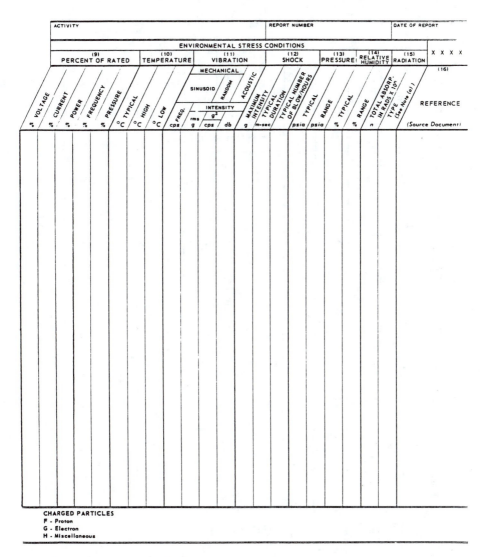

Fig. 2.4(b)– GIDEP's tabular failure rate data summary
form, right half.

1. REPORT NO.	2. INITIAL REPORT NO.	3. REPORTING ACTIVITY	4. MISSILE TMS.	5. MISSILE SERIAL NO.
6. FAILED ITEM PART NO.	7. FAILED ITEM B/N	8. FAILED ITEM NAME	9. FAILED ITEM MFR.	10. F/I REF. DESIG.
11. NEXT ASSY. PART NO.	12. NEXT ASSY. NAME	13. NEXT ASSY. MFR	14. NEXT ASSY. REF. DES.	15. SYSTEM NO.
16. FAILURE CODE	17. F/I MFR CODE	18. SUBSTITUTE PART NO.	19. REPLACEMENT S/N	20. DATE OF FAILURE.

21.1 OPERATIONAL USAGE — Hour Minutes Seconds | .2 CYCLES | .3 MONTHS | .4 MILES

22. FAILURE DISCOVERED DURING
.1 BENCH TEST   .5 CHECKOUT
.2 INSPECTION   .6 MAINTENANCE
.3 STORAGE   .7 MFR TEST
.4 SHIPPING   .8 OPERATION

23. REASON FOR REPORT
.1 FAILED ITEM
.2 T.O. DIRECT
.3 TIME EXPIRED
.4 OTHER

24. REPAIR OR DISPOSITION ACTION
.1 REPAIRED IN PLACE   .5 CONDEMNED
.2 REP. REINSTALLED   .6 HELD FOR REP
.3 ADJUSTED   .7 DEPOT REP.
.4 ELIMINATED   .8 FAILURE ANALYSIS

25. REPLACEMENT
.1 IDENTICAL PART
.2 SUBSTITUTE PART
.3 NONE NEEDED
.4 NOT AVAILABLE

26.1 DESCRIPTION OF TROUBLE

26.2 DISPOSITION

26.3 TEST CONDITION CODE	26.4 ENVIRONMENT CODE	26.5 SYSTEM AFFECTED	27. REPORTED BY:

FAILURE AND CONSUMPTION REPORT

Fig. 2.5(a)– A simple reliability report form.

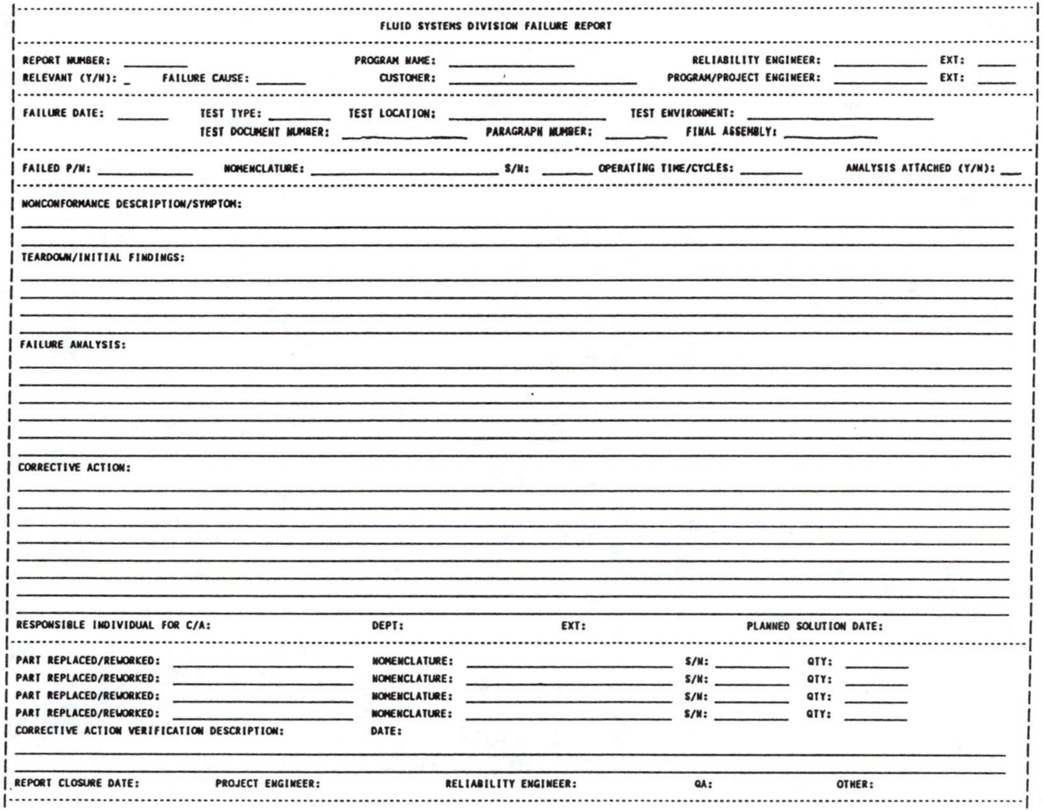
FLUID SYSTEMS DIVISION FAILURE REPORT

REPORT NUMBER: ________ PROGRAM NAME: ________ RELIABILITY ENGINEER: ________ EXT: ____
RELEVANT (Y/N): _ FAILURE CAUSE: ______ CUSTOMER: ________ PROGRAM/PROJECT ENGINEER: ________ EXT: ____

FAILURE DATE: ______ TEST TYPE: ______ TEST LOCATION: ________ TEST ENVIRONMENT: ________
TEST DOCUMENT NUMBER: ________ PARAGRAPH NUMBER: ______ FINAL ASSEMBLY: ________

FAILED P/N: ________ NOMENCLATURE: ________ S/N: ______ OPERATING TIME/CYCLES: ______ ANALYSIS ATTACHED (Y/N): __

NONCONFORMANCE DESCRIPTION/SYMPTOM:

TEARDOWN/INITIAL FINDINGS:

FAILURE ANALYSIS:

CORRECTIVE ACTION:

RESPONSIBLE INDIVIDUAL FOR C/A: DEPT: EXT: PLANNED SOLUTION DATE:

PART REPLACED/REWORKED: ________ NOMENCLATURE: ________ S/N: ______ QTY: ______
PART REPLACED/REWORKED: ________ NOMENCLATURE: ________ S/N: ______ QTY: ______
PART REPLACED/REWORKED: ________ NOMENCLATURE: ________ S/N: ______ QTY: ______
PART REPLACED/REWORKED: ________ NOMENCLATURE: ________ S/N: ______ QTY: ______
CORRECTIVE ACTION VERIFICATION DESCRIPTION: DATE:

REPORT CLOSURE DATE: PROJECT ENGINEER: RELIABILITY ENGINEER: QA: OTHER:

Fig. 2.5(b)– Failure report form of the Fluid Systems Division of Allied-Signal Aerospace Company, Tempe, Arizona.

TABLE 2.1– Ungrouped times-to-failure data.

Failure order number, j	Times to failure , T_j, hr
1	125
2	1,260
3	2,080
4	2,825
5	3,550
6	4,670

2.4   FINAL FORMAT OF DATA

The recommended format in which the partially processed data should be presented would be tabular of either individual times-to-failure data if there are few such data, or of grouped times-to-failure data if there is an abundance of such data. Table 2.1 provides a format for a few data and Table 2.2 for an abundance of data. Observe that the times-to-failure are tabulated in increasing value. From this stage on a variety of techniques can be used to further reduce and analyze the information appearing in these two tables. These techniques are the subject of the rest of this book.

2.5   FINAL REDUCED DATA-RESULTS

The final reduced data format and results may be presented, or reported, in at least one or more of the following ways:

1. Mean life estimate; e.g., $\hat{\bar{T}} = 980$ hr with seven units tested and five failures observed during 4,900 unit hours of operation.

2. Mean-time-to-failure estimate for nonrepairable components or equipment; e.g., $\hat{\bar{T}} = 583.33$ hr with seven units tested and six failures observed during 3,500 unit hours of operation.

3. Mean-time-between-failures estimate, $MTBF$, $\hat{m}$, or $\hat{\bar{T}}$; e.g., $MTBF = 675.6$ hr for repairable components or equipment with nine units tested and nine failures observed during 6,080 unit hours of operation.

TABLE 2.2–Grouped times-to-failure data.

Life in 100 hr starting at 200 hr	Number of failures, N_f
0-10	300
10-20	200
20-30	140
30-40	90
40-50	60
50-60	40
over 60	70
	$N = 900$

4. Failure rate (mean) estimate, $\hat{\bar{\lambda}} = 1/675.6 = 1,480$ failures per million unit hours of operation with nine units tested and nine failures observed during 6,080 unit hours of operation.

5. Mean life estimate and the two-sided confidence limits, m_{L2} and m_{U2}, at a prescribed confidence level, CL, and

$$P(m_{L2} \leq m \leq m_{U2}) = CL;$$

e.g., the mean life estimate $\hat{m} = 675.6$ hr, the lower two-sided confidence limit on the true mean life is $m_{L2} = 385.7$ hr, the upper two-sided confidence limit on the true mean life is $m_{U2} = 1,477.3$ hr at a confidence level of 95%, and the confidence interval for this sample at a CL of 95% is

$$(375.7 \text{ hr} ; 1,477.3 \text{ hr})$$

with nine units tested and nine failures observed during 6,080 unit hours of operation.

6. Mean life estimate, $\hat{m}$, and the lower one-sided confidence limit, m_{L1}, on the true mean life at a prescribed confidence level, and

$$P(m \leq m_{L1}) = CL;$$

e.g., $\hat{m} = 675.6$ hr and $m_{L1} = 421.2$ hr at a confidence level of 95% with nine units tested and nine failures observed during

6,080 unit hours of operation, and the confidence interval for this sample at a CL of 95% is

$$(421.2 \text{ hr} ; \infty).$$

7. Reliability (average or point) estimate, $\hat{\bar{R}}$; e.g., $\hat{\bar{R}} = 0.998887$ for a 2-hr mission.

8. Reliability (average or point) estimate, the two-sided confidence limits, R_{L2} and R_{U2}, on the true reliability at a prescribed confidence level, and

$$P(R_{L2} \leq R \leq R_{U2}) = CL;$$

e.g., $\hat{\bar{R}} = 0.99887$ for a 2-hr mission, $R_{L2} = 0.997661$ and $R_{U2} = 0.999561$, and the confidence interval on the true reliability at CL=90% is

$$(0.997661 ; 0.999561).$$

9. Reliability (average or point) estimate, $\hat{\bar{R}}$, the lower one-sided confidence limit, R_{L1}, on the true reliability at a prescribed confidence level,

$$P(R \geq R_{L1}) = CL;$$

e.g., $\hat{\bar{R}} = 0.99887$ for a 2-hr mission, and the confidence interval at CL=90% is

$$(0.997936 ; 1).$$

10. The name of the times-to-failure distribution found to fit the data best and the estimate of the parameters thereof; e.g., the Weibull distribution fits the times-to-failure data obtained from tests (or from field data) best with a location parameter estimate of $\hat{\gamma} = 250$ hr, a shape parameter estimate of $\hat{\beta} = 2.3$, and a scale parameter estimate of $\hat{\eta} = 3,550$ hr with eight units tested and five failures observed.

11. The probability plot of the data on the most appropriate probability plotting paper, the parameters of the distribution, the unreliability and reliability of the units from which the data were obtained for a specific period of operation, and the conditional reliability for the next mission. For examples, see Chapters 6, 9, 11, 12 and 16.

12. The improvement in the mean life, if any, due to redesign; e.g., the specific redesign has indeed improved the $MTBF$ from 150 hr to 750 hr at the 10% significance level or with 90% confidence.

13. The $MTBF$ of the product of Manufacturer A is better than that of Manufacturer B; e.g., Manufacturer A's product has a better $MTBF$ ($\hat{m}_A = 1,250$ hr) than that of Manufacturer B ($\hat{m}_B = 1,050$ hr) at the 5% significance level and the difference in the $MTBF's$ is not by chance alone.

14. The improvement in the average reliability, if any, due to redesign; e.g., the specific redesign has improved the reliability from $\hat{R} = 0.94$ to $\hat{R} = 0.96$ at the 5% significance level.

15. The components of Manufacturer A have a better average reliability than those of Manufacturer B; e.g., Manufacturer A's components have a better average reliability ($\hat{R}_A = 0.95$) than that of Manufacturer B ($\hat{R}_B = 0.93$) at the 5% significance level, and the difference in the two reliabilities is more than would occur by chance alone.

16. The units are accepted or rejected with prechosen consumer's (β) and producer's (α) risks, and specified minimum acceptable $MTBF$, m_{L1}, and design-to $MTBF$, m_{U1}; i.e., the units have passed the test (sequential) and thus have been accepted as having an $MTBF$ at least equal to the minimum acceptable, (200 hr), and the risk that they have an $MTBF$ lower than m_{L1} does not exceed the prechosen value of $\beta = 5\%$ in the long run. Another version would be that the units have not passed their test (sequential) and thus have been rejected, but the probability that they will be rejected, though they may have an $MTBF$ as high as the design-to, or $m_{U1} = 600$ hr, will be no greater than $\alpha = 5\%$ in the long run.

17. The equipment are accepted or rejected with prechosen β and α risks, and specified minimum acceptable reliability, R_{L1}, (binomial sequential test) by having completed a predetermined number of events successfully out of those undertaken up to that time; e.g., the equipment have passed the test (sequential) and thus have been accepted as having a reliability at least equal to the minimum acceptable reliability of $R_{L1} = 0.900$ and the risk that they have a reliability less than R_{L1} will not exceed the prechosen value of $\beta = 10\%$, by exhibiting only 4 failed trials out of 97 . Another version would be that the equipment have not passed their test (sequential) thus have been rejected,

and the risk that they will be rejected, though they may have a designed-in reliability as high as $R_{U1} = 0.950$, is no greater than the prechosen value of $\alpha = 5\%$ in the long run, by experiencing six failures out of 43 trials.

18. The equipment is being accepted (fixed failures accept/reject test) as having exhibited a specified $MTBF$ because the number of failures that occurred during a prechosen unit hours of test time, T_a, did not exceed a predetermined number; e.g., the equipment have been accepted because they demonstrated an $MTBF$ of at least 200 hr during 1,000 unit hours of test time, by observing no more than eight failures and the probability of passing this test was 95%.

19. The equipment is being accepted (fixed test time accept/reject test) as having experienced no more than a prechosen number of failures in a prefixed unit hours of test time and prechosen producer's and consumer's risks; e.g., the equipment has been accepted because no more than 12 failures occurred during 50,000 unit hours of test time with consumer's and producer's risks not exceeding 5%, and with a minimum acceptable $MTBF$ of 2,500 hr and a designed-to $MTBF$ of 6,500 hr.

20. The equipment is being accepted (binomial accept/reject test) as exhibiting no more than the allowable number of failed trials in a prechosen total number of trials; e.g., the equipment is accepted because, to demonstrate a minimum acceptable reliability of $R_{L1} = 0.950$ with a consumer's risk of $\beta = 10\%$ and a producer's risk of $\alpha = 5\%$, no trials failed in a total of 45 trials, having had designed into it a reliability of $R_{U1} = 0.999$.

21. The hardware has previously exhibited an inverted gamma mean life distribution (prior) with a scale parameter of α and a shape parameter of β. It is now accepted (Bayesian) after only a limited unit hours of operation and the associated number of failures. Furthermore, the current confidence level at which m_{L1} is demonstrated is determined as well, after the prior is updated (posterior) using the new test results to obtain the currently valid $MTBF$ distribution which is also inverted gamma distributed; e.g., the desired hardware is known to have a prior $MTBF$ distribution which is well represented by the inverted gamma distribution with parameters $\alpha = 500$ and $\beta = 2$. Units from the new production are tested for 1,330 unit hours and two failures are observed. The posterior $MTBF$ distribution with the incorporation of the new information into the prior yields another inverted

gamma distribution with parameters $\alpha = 1,830$ and $\beta = 4$, which yield the facts that the actual probability (confidence level) with which the desired $MTBF$ of 200 hr has been demonstrated is 98%, whereas for the prior it was only 71%, and the mean and mode of the posterior $MTBF$ distribution are 610 hr and 366 hr, respectively, whereas those of the prior were 500 hr and 167 hr, respectively.

22. The hardware has previously demonstrated a beta distribution (prior) for its reliability with parameters α and β. It is now accepted (Bayesian) after a few trials on the new production units based on the updated distribution (posterior) which is also beta distributed; e.g., the desired hardware has demonstrated in the past a beta reliability distribution with parameters $\alpha = 4$ and $\beta = 4$. After 20 trials on units of the new production, during which 5 trials fail, the posterior is also beta distributed with $\alpha = 19$ and $\beta = 9$; whereas, while the actual current confidence level with which a minimum acceptable reliability of $R_{L1} = 0.60$ was demonstrated by the prior was only 27%, that demonstrated by the posterior is 79%. Furthermore, whereas the mean and mode of the prior β distribution were 0.50 and 0.50, respectively, those of the posterior are 0.67 and 0.68, respectively.

The previous situations are some of the most frequently occurring ones. The subsequent chapters in this handbook cover these and many more situations. They have been presented here to give management, reliability and life test engineers, and product assurance engineers an idea as to what to expect to find and report through reliability and life testing.

2.6   RELIABILITY DATA SOURCES

All available in-house and externally available reliability data should be used in the process of predicting the failure rate and/or the reliability of components, equipment, subsystems and systems. These should generally include as a minimum those listed in the upper left corner of Fig 2.6. Important data sources are GIDEP [3], MIL-HDBK-217 [4], Non-Electronic Equipment Reliability Data [5], Non-Electronic Part Reliability Data [6], Missile Materiel Reliability Prediction Handbook - Parts Count Prediction [7] for storage (nonoperating) failure rates, RADC [8] and BELL CORE [9]. It is urged that such data, or reports containing such data, be made available to GIDEP so that they get disseminated to all who have use for this very valuable information.

This avoids duplication of testing, and provides means for judiciously comparing in-house generated data and results with those of others.

2.7   RELIABILITY DATA ANALYSIS, FEEDBACK, AND CORRECTIVE ACTION SYSTEM

A very important reliability engineering and testing function is the establishment of the "Reliability Data Sources, Acquisition, Analysis, Feedback, and Corrective Action System" and its implementation. Figure 2.6 gives a flow chart for this system. Major objectives are to assure that all in-plant and field failures are properly recorded on special product reliability forms, such as the ones given in Figs. 2.1 through 2.5, excessive failures are investigated, the responsibility for these failures in terms of engineering, manufacturing quality control, inspection, sales, field service, purchasing, suppliers and customers is identified, and corrective actions are initiated through the responsible departments. Furthermore, reliability engineering and testing should institute control actions to insure that these corrective actions are implemented and followed through to ascertain that the corrective actions were indeed the right ones. The lessons learned from such corrective actions should then be documented in an engineering retention log for immediate and/or future use by all concerned. Without this function reliability and test engineers cannot complete their responsibility concerning a product's improvement and reliable function. As mentioned in the previous section, all available data sources, such as those given in Fig. 2.6 and in [3] through [9], should be used.

The data processing may be done either by hand or by machine, the deciding criterion being the volume to be handled. Typical data processing systems are shown in Figs. 2.7 and 2.8. A diligent effort should be made to tabulate the data in a format which will expedite data analysis. This tabulation should contain as a minimum the following information from product reliability forms, such as those given earlier, for each part, assembly, and product model:

1. Part assembly or serial number.

2. Total failures, including primary and secondary failures.

3. Cost of parts per failure.

4. Cost of labor per failure (cost or manhours).

5. Customer downtime per failure.

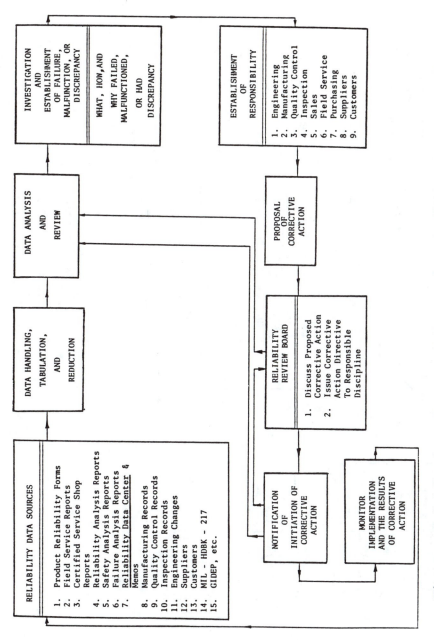

Fig. 2.6– Reliability data sources, acquisition, analysis, feedback and corrective action system.

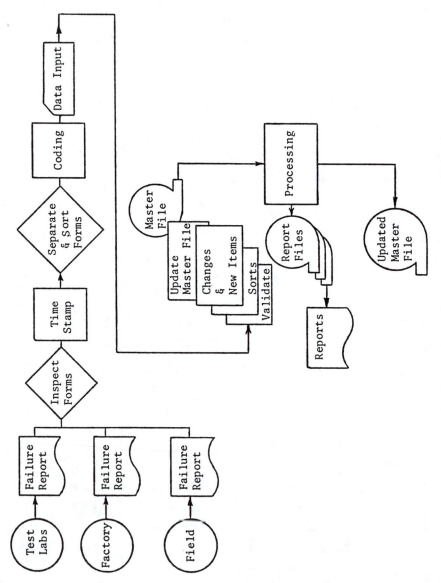

Fig. 2.7– Typical reliability data processing system.

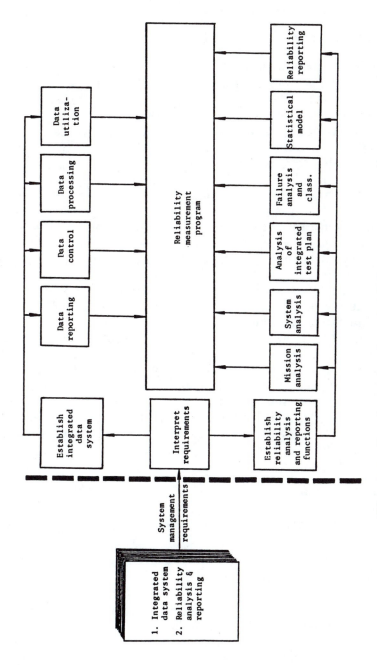

Fig. 2.8– Development of a reliability measurement program.

6. Number of failures for each category under the "Recommended Area for Corrective Action" on reliability report forms.

7. The usage factor for the model of machine.

The reliability report form should be completed at the time of failure in the test laboratory by test personnel and in the field at the time of failure by salesmen, servicepeople, product department personnel, or certified service shop personnel. More than one of the preceding should report the same failure whenever possible. Anyone who might receive news of a failure by phone or letter should also have copies of the form available to note down as many of the facts as possible.

After the reliability report form is completed, it should be distributed to appropriate personnel given in the form's instructions; namely, assigned personnel in the reliability data center. The supervisor should route them to the reliability data processing section for editing and as the case dictates to other personnel who should see the report. All reports should be filed in a central file of the reliability data center. Often a satisfactory investigation of a failure cannot be made at once. In this case a reliability report form should be made out and then more forms should be completed as additional information is obtained indicating the number of the previous reports issued.

The reliability report form could take care of all failure reporting; however, other information could be communicated from the field by the use of field service reports. Many field service reports may be issued, subsequently a task force should scan and fill out product reliability forms from such field reports.

The incoming reliability report forms and field service reports should be edited to ensure that the data are extracted correctly for calculation and tabulation. All reports should be classified under one of the following headings:

1. Product Failure.

2. Assembly Failure.

3. Part Failure.

4. No Failure.

The first three of the above categories should be subdivided into relevant and nonrelevant failures as follows:

1. Relevant failures.

1.1 Adjustment (maintenance).

1.2 Design defect.

1.3 Manufacturing defect.

1.4 Misapplication.

1.5 Part defect.

1.6 Part wear.

1.7 Workmanship.

1.8 Unknown.

1.9 Other.

2. Nonrelevant failures.

2.1 Usable

2.1.1 Drawing error.
2.1.2 Installation error.
2.1.3 Operator error.

2.2 Nonusable.

2.2.1 Accidental damage.
2.2.2 "Acts of God" circumstances, such as earthquakes, storms, floods, fires, etc.
2.2.3 Failure of another part.
2.2.4 Scheduled replacement.
2.2.5 Operator adjustment.

All relevant failures should be used in failure rate calculations. These may be broken down further into early, chance (constant failure rate) and wear-out failures.

The nonrelevant may be classified into two subheadings: usable and nonusable. The failures which are grouped under nonrelevant-usable, can be used to calculate human error failure rates. High failure rates of this type would indicate that something about the device induces human errors. Such a condition should spark corrective action activity. The data classified as nonrelevant, nonusable will not be used.

For those parts designated as critical by data analysis the following information would be required for each failure:

1. Subassembly name.

2. Part name and number.

3. Description of failure.

4. Previous corrective action(s).

5. Previous engineering change(s).

Data analysis includes as a minimum, the determination of failure rates, times-to-failure distributions and their parameters, costs of primary and secondary failures, repair costs, and customer downtime expenses.

The reliability review board may designate parts and subsystems in one or more of the following categories as critical, and consequently as requiring corrective action:

1. High failure rate.

2. High primary failure cost.

3. High secondary failure cost.

4. Many such identical parts in use in the same product.

5. High labor cost.

6. Excessive downtime for repair and restoration.

A reliability request for investigation of the critical components should be issued by the board as the first step toward corrective action. The basic questions that need to be answered during the investigation are: What component failed? How did it fail? Why did it fail? The investigators should examine all data, original data sources, past corrective action measures, and engineering changes. They should contact people involved in all pertinent disciplines affecting the product's performance for discussion of, and leads to, effective and expedient corrective action recommendations.

The objective of the investigation will be to establish corrective action responsibility and prepare a proposal containing recommendations for corrective actions. The cost of corrective actions and its effect on reliability should be the foremost factors influencing the proposed corrective actions. Responsibility may be established in any of the following areas where the recommended corrective actions shall be implemented:

1. Engineering.

2. Manufacturing.

3. Quality control.

4. Inspection.

5. Sales.

6. Service.

7. Purchasing.

8. Suppliers.

9. Customers.

The results of the investigation should be summarized in a reliability investigation report. Such a report should as a minimum address the following:

1. The problem under investigation should be well defined.

2. The contacts, discussions and efforts of the task force should be summarized.

3. Prior corrective actions in the experience retention log should be noted and evaluated.

4. Engineering changes affecting the problem should be evaluated.

5. Several alternative corrective actions should be presented if possible, and their cost and effect on product reliability should be considered for each.

The problems to be investigated will generally fall into one of the following general categories:

1. Misapplication of parts within the end product.

2. Unreliable parts used in the end product.

3. Inadequate or incorrect manufacturing procedures.

4. Human factors not properly considered in design.

5. Inadequate procedures for use of the end product.

The reliability review board should issue a corrective action directive. The chairman of the reliability review board should, in turn, be advised by the responsible party of the date the prescribed corrective action has been initiated through a corrective action notice.

The reliability review board should consist of responsible personnel from each department who come into contact with the product involved. Its responsibility may include the initiation, preparation, distribution and dissemination of the following:

1. Reliability report forms.

2. Procedures for collecting and processing these forms.

3. Reliability investigation request forms and reports.

4. Method of filing forms and documents.

5. Corrective action directive forms and directives.

6. Reliability investigation reports.

7. Reliability review board agenda.

8. Experience retention log.

9. Reliability review board minutes.

Further responsibilities may include the following:

1. Organize and conduct reliability design reviews.

2. Follow through on corrective action directives.

3. Establish reliability task forces as needed for data handling, analysis, corrective action investigations and preparation of reliability checklists.

4. Formulate reliability policies.

5. Maintain a reliability education program and library.

6. Check off engineering drawings.

7. Monitor all engineering changes.

The reliability task force may consist of working groups serving the reliability review board. The sizes, memberships and purposes would be determined by the reliability review board.

Some of the tasks that may be undertaken by reliability task forces are:

1. Perform corrective action investigations for the reliability review board.

2. Submit corrective actions to the reliability review board.

3. Perform special, complicated calculations and analyses of product reliability, as well as of $MTBF$'s and failure rates, and compare them with their goals.

4. Perform reliability check-offs of engineering drawings. The reliability group in a company could prepare periodically such reports as:

4.1 Reliability status.

4.2 Failure summary.

4.3 Historic test results.

4.4 Failure status.

4.5 Hardware summary.

4.6 Failure analysis follow-up.

4.7 Failure rate compendium.

These reports can be used for maintaining a test history; for establishing corrective actions; for failure rate, $MTBF$ and reliability improvement; for estimating spares and logistic requirements; for establishing a reliability experience retention log; for reliability design reviews; and for other uses.

The reliability status report presents estimates of the reliability of each component, equipment and subsystem; pinpoints reliability problem areas; and discusses possible corrective actions. The failure summary report presents a complete record of all failures occurring on a particular program, equipment, or hardware. The historic test result reports provide contents for a historic test result file containing records of all reliability tests and their results. The failure status reports are used to maintain an historical record of all failures and form the basis of the failure summary report. The hardware summary report contains an entry for each hardware tested, consisting of the total test time, failures accumulated, and the failure rate. The failure analysis follow-up reports are used to list every item which requires further action and its status. The failure rate compendia are a compilation and summary of the hardware test results contained in the historic test results file and the associated failure rates. The data from all projects should be summarized by hardware groupings to provide a reference document for failure rates. These may then be used to predict the failure rate of new systems, to make better design and management decisions, and to provide an insight into the causes of failures.

PROBLEMS

2-1. Summarize the reliability and life data needed as a minimum in less than 75 words.

2-2. Prepare a Product Reliability Form for a car.

2-3. Prepare a Product Reliability Form for a television set.

2-4. Prepare a Product Reliability Form for a microwave oven.

2-5. What are the key points that stand out in Fig. 2.2?

2-6. What are the key points that stand out in Fig. 2.3?

2-7. What are the key points that stand out in Fig. 2.5(a)?

2-8. What are the key points that stand out in Fig. 2.5(b)?

2-9. Give five good examples of final reduced data and results.

2-10. What are the four best reliability data sources?

2-11. What are the key points in the reliability data sources, acquisition, analysis, feedback and corrective action system of Fig. 2.6?

2-12. How would you classify the incoming reliability report form?

2-13. How would you subdivide the relevant failures?

2-14. How would you subdivide the nonrelevant failures?

2-15. How would you identify the critical parts and subsystems for corrective action?

2-16. What should a reliability investigation report address as a minimum?

2-17. List the general categories of the failure problems that should be investigated.

2-18. What are the responsibilities of the reliability review board?

2-19. What should be the composition of a reliability task force?

2-20. What are some of the tasks of the reliability task force?

REFERENCES

1. GIDEP, Government-Industry Data Exchange Program, Code 30G, GIDEP Operations Center, Naval Warfare Assessment Center, Corona, Calif. 91720. Telephone: (714) 273-4677; Autovon: 933-4677.

2. Grant, W. Ireson and Combs, C. F., Ed., *Handbook of Reliability Engineering and Management*, McGraw-Hill Book Co., New York, 656 pp., 1988.

3. GIDEP, Government-Industry Data Exchange Program, "Reliability-Maintainability Data Bank," GIDEP Operations Center, Naval Warfare Assessment Center, Corona, Calif. 91720.

4. Commander, Rome Air Development Center, "MIL-HDBK-217: Military Standardization Handbook - Reliability Prediction of Electronic Equipment, " RBRS, Griffiss Air Force Base, New York 13441.

5. Yurkowsky, William, Hughes Aircraft Company, *Non-electronic Reliability Notebook*, RADC-TR-69-458, Final Report to Rome Air Development Center, Air Force Systems Command, Griffiss Air Force Base, New York, March 1970.

6. Reliability Analysis Center, "Non-electronic Part Reliability Data, 1978," No. NPRD-1, RADC/RBRAC, Griffiss Air Force Base, New York, 250 pp., 1978.

7. Army Missile Command, "Missile Materiel Reliability Prediction Handbook, Parts Count Prediction," LC-77-1, RDE Lab, contact Leslie R. Conger, Redstone Arsenal, Alabama, 290 pp., October 1977.

8. Rome Air Development Center, RADC/RBRAC, Griffiss Air Force Base, New York 13441 (315–330–4151).

9. BELLCORE, 290 West Mt. Pleasant Ave., Room 4D–110, Livingston, N.J. 07039 (1–800–521–2673).

Chapter 3

FIVE VERY IMPORTANT ANALYTICAL FUNCTIONS IN RELIABILITY AND LIFE TESTING

3.1   THE FIVE FUNCTIONS

Five of the most important functions in reliability and life testing are the following: (1) The failure probability density function. (2) The reliability function. (3) The conditional reliability function. (4) The failure rate function. (5) The mean life function. Given these five functions, most reliability and life testing problems may be solved.

The failure probability density function enables the determination of the number of failures occurring over a period of time referred to the original, total population. The reliability function enables the determination of the probability of success of any unit in undertaking a mission of a prescribed duration or operating for the desired function period. The conditional reliability function enables the determination of the probability of success of any unit in undertaking a new mission given the unit already survived the previous mission. The failure rate function enables the determination of the number of failures occurring per unit time referred to the size of the population existing at the beginning of the period for which the failure rate is to be calculated. The mean life provides the average time of operation to a failure.

49

3.2 THE FAILURE PROBABILITY DENSITY FUNCTION

The work of a reliability and life test engineer starts with observations of times to failure, cycles to failure, revolutions to failure, actuations to failure, miles to failure, number of failed operations, and the like, for parts, components, equipment, products, or systems (all of which will be referred to as *unit* in this book), while they operate in specified application and operation environments according to their application specifications. Such observations are called data, and are subject to central tendencies and deviations from their mean value caused by *variations* in raw materials, manufacturing tools and processes, workmanship, quality control, and outgoing quality levels; in assembly, inspection, testing, packaging, shipping, startup and operation practices; in application and operation environments; and by experimental and human errors that prevail for the units whose reliabilitywise performance is being monitored.

The engineer and the manager want to know what such data, like those given in Table 3.1 which at first glance looks like a mass of numbers, is telling about the output and performance characteristics of the unit. They want to know how many of these observations fall in particular ranges of values, if there is a crowding of similar observations in a particular region, where they are less frequent, if there is a wide variability in the values of the observations, and if there are peculiar trends in their frequency of occurrence in particular regions or points. The best scientific way of obtaining such information is through the reduction of these data into more meaningful, condensed forms, namely their probability density function, *pdf*, $f(T)$, from which statements such as their mean, median, mode, and their standard deviation, the name of their *pdf*, the parameters of their *pdf*, their skewness and kurtosis, their goodness of fit, etc. can be made. This handbook presents the techniques for determining all of these.

For the step-by-step procedure for determining the probability density function, or distribution of such data for ungrouped and grouped, as well as their frequency histograms and polygons after their tally and frequency table is prepared, their relative frequency histogram and their frequency distribution function, the reader is referred to [1, Vol. 1, Chapter 4].

If it is assumed that the data in Table 3.1 follow the normal distri-

TABLE 3.1– Times-to-failure data for 225 components in multiples of 100 hr.

433	441	440	446	446	445	444	433	442	436	438	440	444
435	447	441	431	433	432	439	434	437	438	436	439	428
427	440	436	428	442	431	427	434	441	437	432	436	442
442	439	434	440	447	437	435	430	434	435	432	445	430
431	435	432	444	441	437	451	437	433	438	435	446	442
437	436	427	440	430	436	427	438	427	433	439	435	431
429	434	437	440	429	433	430	433	435	432	453	432	444
434	440	436	440	437	445	440	442	426	437	439	435	437
437	440	432	441	430	430	435	440	427	432	432	441	441
428	430	432	435	434	430	435	440	434	439	440	451	437
438	438	432	439	433	431	447	428	430	431	435	429	445
442	428	437	435	440	441	435	436	441	436	444	446	435
436	439	436	433	435	431	431	435	433	443	436	443	436
445	432	448	428	445	442	444	438	443	439	431	438	429
430	436	432	435	446	431	443	434	439	431	437	433	435
439	443	436	439	448	454	449	430	440	444	428	430	438
445	433	429	440	434	434	443	437	436	435	442	442	434
438	437	441	440									

bution, then their frequency distribution function $f(T)'$, would be

$$f(T)' = \frac{Nw}{\sigma_T\sqrt{2\pi}}e^{-\frac{1}{2}(\frac{T-\overline{T}}{\sigma_T})^2}, \tag{3.1}$$

where

$f(T)'$ = probability density, or a measure of the relative frequency of occurrence of a specific value of time to failure, T,

N = number of data values; e.g., in Table 3.1, $N = 225$,

w = class interval width, or data range of each interval into which the data have been grouped, in this case, $w = 300$ hr,

σ_T = standard deviation of the data, calculated from

$$\sigma_T = \left[\sum_{i=1}^{N=225} (T_i - \overline{T})^2/N - 1\right]^{1/2}, \tag{3.2}$$

in this case $\sigma_T = 564$ hr,

T_i = individual data values,

and

$\overline{T}$ = mean of the data, calculated from

$$\overline{T} = \sum_{i=1}^{N=225} T_i/N. \tag{3.3}$$

In this case $\overline{T} = 43,679$ hr.

Then the frequency distribution function for the data of Table 3.1, from Eq. (3.1), is

$$f(T)' = \frac{225(300)}{564\sqrt{2\pi}} e^{-\frac{1}{2}(\frac{T-43,679}{564})^2}. \tag{3.4}$$

This is the equation which when plotted superposed on the frequency histogram of the data, it represents this histogram.

Their probability density function is

$$f(T) = \frac{1}{\sigma_T\sqrt{2\pi}} e^{-\frac{1}{2}(\frac{T-\overline{T}}{\sigma_T})^2}, \tag{3.5}$$

which in this case is

$$f(T) = \frac{1}{564\sqrt{2\pi}} e^{-\frac{1}{2}(\frac{T-43,679}{564})^2}. \tag{3.6}$$

Figure 3.1 provides the plot of Eq. (3.6). It may be seen that the most frequently observed value is the mean value, $\overline{T} = 43,679$ hr, and not very many values are observed in the left and right tails.

The *pdf* starts at $T = -\infty$ at the height of zero, it increases as T increases, reaches a peak at $T = \overline{T}$, it decreases thereafter until at $T = +\infty$, $f(T = +\infty) = 0$. This is the typical behavior of the distribution of the times to wear-out failures. The normal *pdf* may be used to represent the distribution of the times to failure when the failure mode is the wear-out failure mode. For more details about the

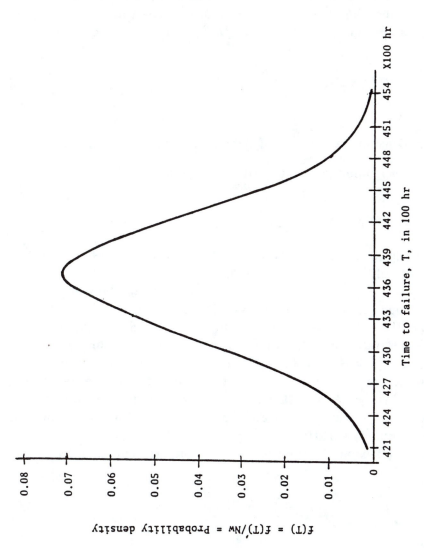

Fig. 3.1— The fitted normal distribution, or probability density function, for the data in Table 3.1.

normal *pdf* see Chapters 9 and 10 in this book, and Chapter 7 in [1, Vol. 1].

When the failure data represent times to failure occurring during the useful life of identical units, or when units fail due to chance failures only, or when they have a constant failure rate during their operating time, the appropriate distribution, $f(T)$, to use is the exponential, or

$$f(T) = \lambda e^{-\lambda T}, \tag{3.7}$$

where

$\lambda =$ constant failure rate of the units.

The corresponding frequency distribution function, $f(T)'$, is

$$f(T)' = N w e^{-\lambda T}. \tag{3.8}$$

This is the equation which, when plotted superposed on the frequency histogram of the data, represents this histogram.

Figure 3.2 provides the plot of Eq. (3.7). It may be seen that the most frequently observed time to failure is $T = 0$, amazingly enough, because the *pdf* has the highest value at $T = 0$! So the *pdf* starts at the highest value of $f(T = 0) = \lambda$ and it decreases thereafter until at $T = +\infty$, $f(T = +\infty) = 0$. With this distribution 63.2% of such units fail by the time each one operates for a period equal to their mean life, $\overline{T}$, given by $\overline{T} = \frac{1}{\lambda}$, and that is a large proportion! For more details about the exponential *pdf* see Chapters 6, 7 and 8 in this book and Chapter 5 in [1, Vol. 1].

A more flexible distribution to use is the Weibull, given by

$$f(T) = \frac{\beta}{\eta} \left(\frac{T - \gamma}{\eta} \right)^{\beta - 1} e^{-(\frac{T-\gamma}{\eta})^{\beta}}, \tag{3.9}$$

where

$\beta =$ shape parameter,

$\eta =$ scale parameter,

and

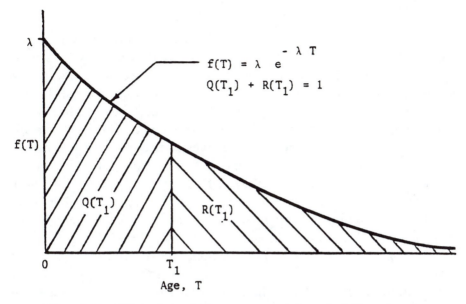

Fig. 3.2 – The plot of the exponential probability density
function, and of the areas which give unreliability
and reliability.

γ = location parameter.

Figure 3.3 provides the plot of Eq. (3.9) for various values of β. This is a three-parameter distribution, thereby its flexibility in assuming a variety of shapes thus representing the behavior of a large variety of life data. When $\gamma \leq 0$ and $0 < \beta < 1$, as in Fig. 3.3 (a), it represents times-to-failure data of units failing due to early life causes, namely due to defects introduced by improper materials used, out-of-control production processes, by improper packaging, transportation, storage and startup, etc. The *pdf* starts at $T = \gamma$, in this case at $T = 0$, at the height of $+\infty$, because $f(T = \gamma$, or $0) = +\infty$, it drops down sharply thereafter and at $T = +\infty$, $f(T = +\infty) = 0$. It may be seen that most failures occur near $T = \gamma$ or 0, as the case may be, or when these units are first turned on or shortly thereafter.

When $\beta = 1$ the Weibull *pdf* represents chance failures occurring during their useful life, after burn-in and stress screening, as depicted in Fig. 3.3 (b). Then the Weibull *pdf*, as a special case, becomes the two-parameter exponential distribution, or

$$f(T) = \frac{1}{\eta} e^{-\frac{1}{\eta}(T-\gamma)},\tag{3.10}$$

where

$$\frac{1}{\eta} = \lambda,$$

and

$\eta = \frac{1}{\lambda} = m$, or the mean life when the failure rate is constant.

When $\beta > 1$ then the Weibull *pdf* represents the life characteristic of wear-out failures, as depicted in Fig. 3.3 (c). In this case the *pdf* starts at the height of zero at $T = \gamma$ or 0, increases thereafter, reaches a peak, as in the case of the normal *pdf*, and decreases thereafter until at $T = \infty$, $f(T = \infty) = 0$. Details about the Weibull distribution are given in Chapters 13, 14 and 16 in this book, and in Chapter 6 in [1, Vol. 1].

It must be pointed out that the time to failure, T, is called a continuous random variable. A random variable is continuous when the

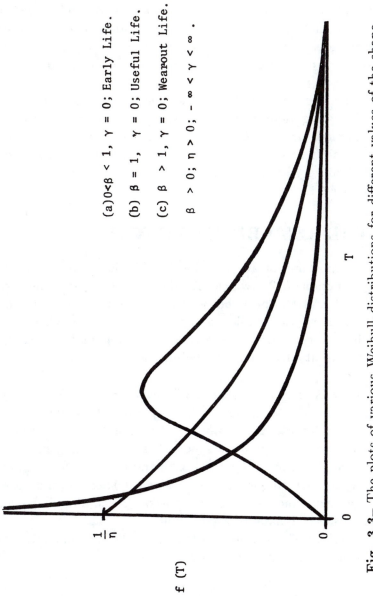

(a) $0 < \beta < 1$, $\gamma = 0$; Early Life.

(b) $\beta = 1$, $\gamma = 0$; Useful Life.

(c) $\beta > 1$, $\gamma = 0$; Wearout Life.

$\beta > 0$; $\eta > 0$; $-\infty < \gamma < \infty$.

Fig. 3.3– The plots of various Weibull distributions for different values of the shape parameter, β.

set of values it can assume can be placed in a one-to-one correspondence with the set of *all numbers* within a continuous range of $-\infty$ to $+\infty$, as opposed to, a random variable which is discrete. A random variable is discrete when the set of values it can assume can be placed in a one-to-one correspondence with all possible positive integers. The discussions in this and in most subsequent chapters deal with continuous random variables.

It may be seen that the first of the five very important in reliability and life testing analytical functions, namely the failure probability density function, or the *pdf*, tells us a lot about the behavior of life, or times-to-failure, data. Furthermore once the *pdf* is determined the remaining four very important analytical functions can be determined, as discussed next.

3.3   THE RELIABILITY FUNCTION

The second most important function in reliability engineering is the reliability function, $R(T)$, which provides the relationship between the age of a unit and the probability that the unit survives up to that age while starting the mission at age zero. The reliability function enables the determination of the conditional reliability function, the probability density function, the failure rate function and the mean life function.

Reliability is the (1) *conditional probability*, at a given (2) *confidence level*, that the equipment will (3) *perform* their intended functions satisfactorily or *without failure*; i.e., within specified performance limits, at a given (4) *age*, for a specified length of time, function period or (5) *mission time*, when used in the manner and for the purpose intended while operating under the specified *application* and *operation* environments with their associated (6) *stress levels*. The six key points in the definition of reliability are in italics.

The reliability of any unit for a mission of T_1 duration, starting the mission at age zero, can be quantified from

$$R(T_1) = P(T \geq T_1) = \int_{T_1}^{\infty} f(T)dT, \tag{3.11}$$

because reliability is the probability that the time to failure is equal to, or greater than, the mission duration. Then the unit cannot fail before

the mission is completed, because it has operated for a time equal to or longer than the mission duration, and the probability of not failing before the mission is completed is the reliability of the unit for that mission. As probabilities are areas under a *pdf*, it must be the area to the left of T, as depicted in Fig. 3.4.

For the normal *pdf* the reliability for a mission of T_1 duration, starting the mission at zero, is given by

$$R(T_1) = \int_{T_1}^{\infty} \frac{1}{\sigma_T \sqrt{2\pi}} e^{-\frac{1}{2}(\frac{T-\overline{T}}{\sigma_T})^2}, \qquad (3.12)$$

or

$$R(T_1) = \int_{z(T_1)}^{\infty} \phi(z) dz,$$

where

$$z(T_1) = \frac{T_1 - \overline{T}}{\sigma_T},$$

and $R(T_1)$ is obtained from area tables of the standardized normal *pdf* given in Appendix C. Figure 3.5 provides the plot of the reliability for units whose times-to-failure data are given in Table 3.1.

For the exponential *pdf* case the reliability for a mission of T duration, starting the mission at age zero, using Eq. (3.11), is given by

$$R(T_1) = \int_{T_1}^{\infty} \lambda e^{-\lambda T} dT = e^{-\lambda T_1}. \qquad (3.13)$$

See Fig. 6.6 in Chapter 6 for the behavior of this function.

For the Weibull case

$$R(T_1) = \int_{T_1}^{\infty} \frac{\beta}{\eta} \left(\frac{T-\gamma}{\eta}\right)^{\beta-1} e^{-\left(\frac{T-\gamma}{\eta}\right)^{\beta}} dT,$$

or

$$R(T_1) = e^{-\left(\frac{T-\gamma}{\eta}\right)^{\beta}}. \qquad (3.14)$$

See Fig. 12.5 in Chapter 12 for the behavior of this function.

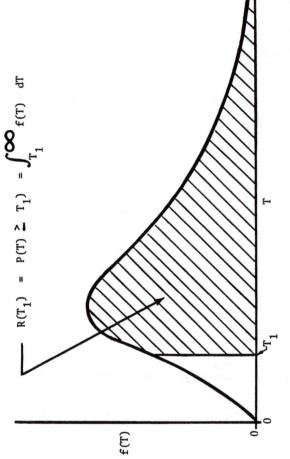

Fig. 3.4– The cross-hatched area under the *pdf* gives the reliability of a unit for a mission of T_1 duration starting the mission at age zero.

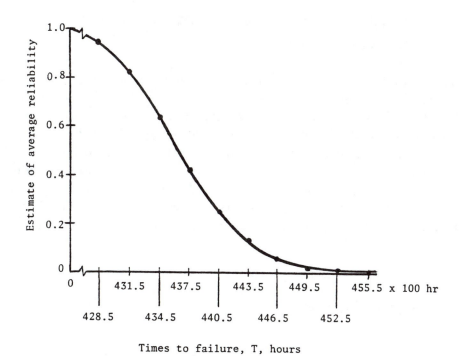

Fig. **3.5**– The reliability of units whose times-to-failure data are given in Table 3.1.

3.4   THE CONDITIONAL RELIABILITY FUNCTION

The third most important function in reliability and life testing is the conditional reliability function. If a unit has already accumulated T hours of operation and its reliability for a new mission of t hours duration thereafter is desired, this reliability is a conditional reliability; i.e., conditional to having survived the first T hours to be able to undertake the new mission. The time domains involved are shown in Fig. 3.6. The probability of surviving $T + t$ hours of operation, or $R(T + t)$, is equal to the probability of surviving T hours, or $R(T)$, and then of surviving an additional t hours, or $R(T, t)$. Mathematically

$$R(T + t) = R(T) \times R(T, t).$$

Solving for $R(T, t)$ gives

$$R(T, t) = \frac{R(T + t)}{R(T)}, \tag{3.15}$$

where $R(T, t)$ stands for the reliability of a new mission of t hours duration, having already operated successfully for T hours.

For the normal *pdf* case presented earlier where $\overline{T} = 43{,}679$ hr and $\sigma_T = 562$ hr the reliability for a mission of 300 hr duration, after such a unit had already accumulated 42,850 hr of operation is obtained as follows:

$$
\begin{aligned}
R(T, t) &= \frac{R(T + t)}{R(T)}, \\[2mm]
&= \frac{\int_{T+t}^{\infty} f(T)\, dT}{\int_{T}^{\infty} f(T)\, dT},
\end{aligned}
$$

or

$$
= \frac{\int_{42,850+300}^{\infty} f(T)\, dT}{\int_{42,850}^{\infty} f(T)\, dT}
$$

$$R(T = 42{,}850 \text{ hr}, t = 300 \text{ hr}) = \frac{\int_{-0.9412}^{\infty} \phi(z)\, dz}{\int_{-1.473}^{\infty} \phi(z)\, dz}, \tag{3.16}$$

where

Age or hours of operation

Fig. 3.6– Time domains for conditional reliability.

$$f(T) = \frac{1}{562\sqrt{2\pi}} e^{-\frac{1}{2}[\frac{T-43,679}{562}]^2}.$$

Substitution of the integral values in Eq. (3.16) gives the conditional reliability of

$$R(T = 42,850 \text{ hr}, t = 300 \text{ hr}) = \frac{0.8267}{0.9296} = 0.8893.$$

For the exponential case

$$R(T, t) = \frac{e^{-\lambda(T+t)}}{e^{-\lambda(T)}} = \frac{e^{-\lambda T} \times e^{-\lambda t}}{e^{-\lambda T}},$$

or

$$R(T, t) = e^{-\lambda t} = R(t). \tag{3.17}$$

Equation 3.17 says that when the unit's failure rate is constant over the period $(T + t)$, the reliability for a new mission is independent of

the age of the unit at the beginning of the mission, but is a function of its failure rate and the duration of the new mission only. For the nonexponential case, however, this does not hold.

For the Weibull case, for example, Eq. (3.15) becomes

$$R(T,t) = \frac{e^{-(\frac{T+t-\gamma}{\eta})^\beta}}{e^{-(\frac{T-\gamma}{\eta})^\beta}},$$

or

$$R(T,t) = e^{-[(\frac{T+t-\gamma}{\eta})^\beta - (\frac{T-\gamma}{\eta})^\beta]}. \tag{3.18}$$

Consequently, the reliability for a new mission, having already accumulated T hours of operation, or starting the new mission at age T, is not independent of the age of the unit when $\beta \neq 1$. It may be seen from Eqs. (3.16) and (3.18) that the reliability for a new mission following the previous mission is dependent on both the mission duration, t, and the age, T, at the beginning of the new mission. These are the cases when the failure rate of the unit is not constant.

3.5   THE FAILURE RATE FUNCTION

The fourth most important function in reliability and life testing is the failure rate function, $\lambda(T)$, which provides the relationship between the age of a unit and the failure frequency, or the number of failures occurring per unit time at age T. The failure rate function enables the determination of the reliability bath-tub curve, among other things.

For any age T, the average failure rate estimate, $\hat{\bar{\lambda}}(T)$, at that age for a homogeneous sample of identical units which are being subjected to a reliability test, or whose reliabilitywise performance is being monitored, while functioning in the same application and operation environment, may be calculated from

$$\hat{\bar{\lambda}}(T) = \frac{N_F(\Delta T)}{N_B(T) \times \Delta T}, \tag{3.19}$$

where

$N_F(\Delta T) =$ number of units failing in age increment ΔT, or in time period from age T to age $T + \Delta T$,

$N_B(T)$ = number of units in the test, or under observation, at the *beginning* (by definition) of the age increment ΔT or at age T,

and

ΔT = age increment during which the $N_F(\Delta T)$ units fail.

Another way to calculate the average failure rate estimate is to use Eq. (4.2), which provides a more accurate estimate. If $N_B(T)$ is large and ΔT is small Eq. (3.19) gives accurate enough results, otherwise Eq. (4.2) should be used. See Chapter 4, Section 4.1 and Chapter 6, Section 6.1 for a comparison of the average failure rate estimates obtained from Eqs. (3.19) and (4.2).

If in Eq. (3.19) we set $\Delta T \to 0$ we obtain the instantaneous failure rate, $\lambda(T)$, also called the hazard rate, or force of mortality, or

$$\lambda(T) = \frac{1}{N_B(T)} \frac{d[N_F(T)]}{dT}. \tag{3.20}$$

Equation (3.20) may be rewritten as

$$\lambda(T) = \frac{1}{N_B(T)} \frac{N}{N} \frac{d[N_F(T)]}{dT},$$

and a rearrangement gives

$$\lambda(T) = \frac{1}{N_B(T)/N} \left\{ \frac{1}{N} \frac{d[N_F(T)]}{dT} \right\}. \tag{3.21}$$

The first part of Eq. (3.21) is $1/R(T)$ and the second part is $f(T)$; consequently,

$$\lambda(T) = \frac{f(T)}{R(T)}. \tag{3.22}$$

Equation (3.22) gives the instantaneous failure rate function; and says that, to determine the instantaneous failure rate at age T, all we have to do is to divide the probability density at age T with the reliability for a mission of T duration, starting the mission at age 0.

If the *pdf* of identical units tested is the normal, then Eq. (3.22) yields

$$\lambda(T) = \frac{\frac{1}{\sigma_T\sqrt{2\pi}}e^{-\frac{1}{2}(\frac{T-\overline{T}}{\sigma_T})^2}}{\int_T^{\infty}\frac{1}{\sigma_T\sqrt{2\pi}}e^{-\frac{1}{2}(\frac{T-\overline{T}}{\sigma_T})^2}dT},$$

which results in an increasing failure rate with increasing age; consequently, $\lambda(T)$ is age dependent during wear-out.

The behavior of $f(T)$, $R(T)$ and $\lambda(T)$ for the units whose times-to-failure data are given in Table 3.1, is shown plotted in Fig. 3.7.

If the times to failure of identical units are represented by the exponential distribution, then Eq. (3.22) yields

$$\lambda(T) = \frac{\lambda e^{\lambda T}}{e^{\lambda T}} = \lambda. \tag{3.23}$$

Equation (3.23) says that the instantaneous failure rate of exponential units is independent of their age and is constant at a value equal to the useful life failure rate.

If the times to failure of identical units follow the Weibull distribution, then Eq. (3.22) yields

$$\lambda(T) = \frac{\frac{\beta}{\eta}(\frac{T-\gamma}{\eta})^{\beta-1}e^{-(\frac{T-\gamma}{\eta})^\beta}}{e^{-(\frac{T-\gamma}{\eta})^\beta}}, \tag{3.24}$$

or

$$\lambda(T) = \frac{\beta}{\eta}(\frac{T-\gamma}{\eta})^{\beta-1}. \tag{3.25}$$

It may be seen that the failure rate for Weibull units is not constant and varies with their age, when $\beta \neq 1$.

If $0 < \beta < 1$ the failure rate decreases with age, if $\beta = 1$ it is constant with and if $\beta > 1$ increases with age.

Table 3.2 provides the various units for failure rate currently in use and their conversion factors.

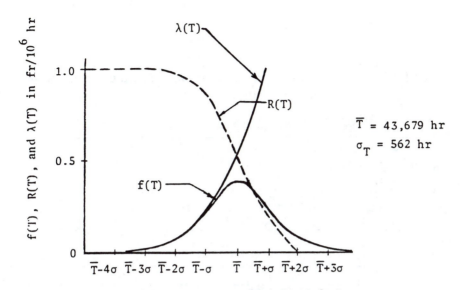

Fig. 3.7– Plots of $f(T)$, $R(T)$, and $\lambda(T)$ versus age, T, for the data in Table 3.1.

TABLE 3.2– Conversion factors for failure rate measurement units.

When you know	You can find	If you multiply by *
Failures per million hours, $fr/10^6$ hr	Failures per hour, fr/hr	10^{-6}
"	$\%/1,000$	10^{-1}
"	$\%fr/1,000$ hr	10^{-1}
"	Percent failures per 1,000 hr	10^{-1}
"	*FITS***	10^3
"	*BITS****	10^4
"	PPM hr, parts failing per million device-hours	1
$\%/1,000$	$fr/10^6$ hr	10
"	fr/hr	10^{-5}
"	FITS**	10^4
"	BITS***	10^5
"	PPM hr	10
FITS**	$fr/10^6$ hr	10^{-3}
"	$\%/1,000$ hr	10^{-4}
BITS***	$fr/10^6$ hr	10^{-4}
"	$\%/1,000$hr	10^{-5}

Example: If you know that the failure rate is 10 $fr/10^6$ hr, you can find the failure rate in % fr/1,000 hr if you multiply 10 by 10^{-1}, or $10 \times 10^{-1}=1$; consequently the failure rate is 1% fr/1,000 hr.

** 1 FIT = 1 $fr/10^9$ hr.

*** 1 BIT = 1 $fr/10^{10}$ hr.

3.6   THE MEAN LIFE FUNCTION

The fifth very important analytical function in reliability and life testing is the mean life function. The mean life is the average, or the
expected time to failure of identical units operating under identical
application and operation environment stresses. The mean life, $\overline{T}$, is
given by

$$\overline{T} = m = \int_{\gamma}^{\infty} T f(T) \, dT, \tag{3.26}$$

where

$$-\infty < \gamma < \infty,$$

$$-\infty < T < \infty,$$

and

$$f(T) \geq 0 \text{ for } T \geq \gamma, \text{ otherwise } 0.$$

Equation (3.26) is illustrated in Fig. 3.8 for a specific value of $T = T_1$.
A computationally easier way of determining the mean life is

$$\overline{T} = m = \gamma + \int_{\gamma}^{\infty} R(T) \, dT. \tag{3.27}$$

Equation (3.27) is obtained as follows: Consider first the case when
$\gamma = 0$. If T is a positive random variable, then

$$\overline{T} = \int_{0}^{\infty} T f(T) \, dT.$$

In the above integral, T can be written as

$$T = \int_{0}^{T} du,$$

therefore

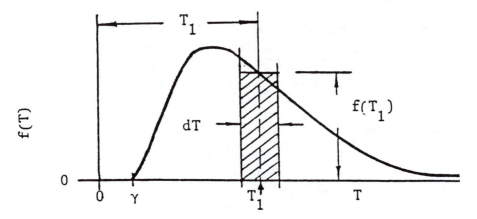

Fig. 3.8– Illustration of the first raw moment, or of the
mean life equation, or of Eq. (3.26).

$$\overline{T} = \int_0^\infty \int_0^T f(T)\,du\,dT. \tag{3.28}$$

The domain of this double integral is the area on the $u - T$ plane
bounded by the T-axis, the $u - T$ line and $T = +\infty$ as shown in Fig.
3.9. In Eq. (3.28), $f(T)$ is integrated from 0 to T with respect to u
first, then from 0 to ∞ with respect to T. By interchanging the order
of the integrals, $f(T)$ can be integrated first with respect to T, then be
integrated again with respect to u. In this case, to keep the integration
in the same domain, the limits of the inner integral would be from u
to ∞, and the limits of the second integral would still be from 0 to ∞;
i.e.,

$$\overline{T} = \int_0^\infty \int_u^\infty f(T)\,dT\,du. \tag{3.28'}$$

Since

$$\int_u^\infty f(T)\,dT = R(u),$$

then

$$\overline{T} = \int_0^\infty R(u)\,du = \int_0^\infty R(T)\,dT.$$

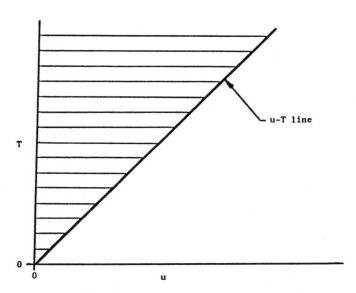

Fig 3.9– Illustration of the areas that yield $\overline{T}$ using Eq. (3.28) or (3.28'), and of the limits of the double integrals.

In the case when $\gamma > 0$, then

$$\overline{T} = \int_0^\infty R(u)\,du,$$

$$= \int_0^\gamma R(u)\,du + \int_\gamma^\infty R(u)\,du,$$

$$= \int_0^\gamma du + \int_\gamma^\infty R(u)\,du,$$

or

$$\overline{T} = \gamma + \int_\gamma^\infty R(u)\,du = \gamma + \int_\gamma^\infty R(T)\,dT.$$

In the normal *pdf* case Eq. (3.26) yields as the mean life of its units the value of $\overline{T}$, obtained from

$$\overline{T} = \sum_{i=1}^N T_i/N.$$

For the useful life, or the exponential *pdf* case, the mean life, using Eq. (3.26), is

$$\overline{T} = \int_0^\infty T\lambda e^{-\lambda T}\, dT = \frac{1}{\lambda}. \tag{3.29}$$

In the two-parameter exponential *pdf* case Eq. (3.27) yields

$$\overline{T} = \gamma + \int_\gamma^\infty e^{-\lambda(T-\gamma)}\, dT = \gamma + \frac{1}{\lambda}, \quad \gamma \neq 0. \tag{3.30}$$

For the Weibull case Eq. (3.27) yields [1, Vol. 1, pp. 151-152]

$$\overline{T} = \gamma + \eta\, \Gamma(\frac{1}{\beta} + 1), \tag{3.31}$$

where $\Gamma(\frac{1}{\beta}+1)$ is the value of the gamma function evaluated at $(\frac{1}{\beta}+1)$. $\Gamma(\frac{1}{\beta} + 1)$ may be obtained from Table 13.2.

The mean life is also called mean time between failures, m, $MTBF$, or mean time to failure, $MTTF$. $MTBF$ is used for units which are repairable, such that they can be reused after each failure and repair. Such a sequence of times of operation until the next failure gives rise to a mean time between failures. Consequently, $MTBF$ means that for a relatively large sequence of failures followed by repair, which restores the unit to the same time-to-failure distribution as before the first failure, a failure will occur on the average after every m hours of operation.

If the repair action does not restore the failed units to the same *pdf* as before the first first failure, then the *pdf* of the units will change after each repair and Eqs. (3.26) and (3.27) cannot then yield the correct mean life using $f(T)$ which is usually determined from the time to the first failure. One solution would be to determine the *pdf* of the times to the second failure, measured from the time the repair of the first failure is completed. Then we can talk about the mean time to the second failure from the first, and similarly the mean time to the third failure from the second, and so on.

For exponential units with repair, the $MTBF$, as determined by Eq. (3.26) from the times to the first failure, means that the failure rate of the units and their $f(T)$ has not been altered at all by the repair actions. For exponential units with repair, but with their $f(T)$ determined from the times to all failures from the previous failure and

completion of its repair, the *MTBF*, as determined by Eq. (3.26), means that the failure rate of the units, or their $f(T)$, has not been altered significantly by the repair actions.

MTTF is used when the units are not repairable or recoverable, or are one–shot items, hence it is used when there is only one period of operation until failure for each such unit, which is the first and only failure possible. The mean of such times to failure is the *MTTF*, and more precisely it is the mean time to the first failure only. Then it is best to use the term *MTTFF*, or the mean time to first failure.

3.7 ADDITIONAL COVERAGE

For more detailed coverage of the subject matter of this chapter, it is recommended that Chapters 4, 5 and 6 of [1, Vol. 1] be consulted.

PROBLEMS

3-1. Which are the five most important functions in reliability and life testing?

3-2. Which are the pertinent observations of interest in reliability and life testing?

3-3. Given in Table 3.3 are data for 110 miles to failure of one brand of passenger car tires in 100 miles; i.e., the first number in Table 3.3, 390 means 39,000 miles to failure. Do the following:

(1) Prepare a tally table and a frequency table.

(2) Determine the class midvalue, the class starting and end values, and the class lower and upper bounds.

(3) Plot the frequency histogram, and the relative frequency polygon.

(4) Plot in by eye the frequency function curve which in your judgment represents the data best.

(5) Plot the cumulative frequency polygon.

(6) Plot the relative cumulative frequency polygon.

(7) Plot in by eye the cumulative frequency function curve which in your judgment represents the data best.

TABLE 3.3 – Data of miles to failure, in 100 miles, for 110 of one brand of passenger car tires for Problems 3-3 and 3-4.

390	393	395	405	420	376	381	381	383	401
380	387	395	397	407	377	383	387	390	393
393	395	403	405	414	376	388	395	397	400
387	400	400	403	410	391	392	394	397	405
390	391	395	401	405	379	391	393	394	410
390	397	400	406	428	380	382	389	391	399
375	383	392	395	404	387	390	398	400	408
390	395	395	397	403	382	399	401	406	406
390	395	395	400	410	381	390	394	397	399
387	389	398	401	415	372	378	396	400	405
387	389	391	391	400	376	380	391	406	412

Identify all plots carefully and completely. Label all abscissas and ordinates, and their scales.

3-4. For the data in Problem 3-3, do the following:

(1) Find the mean and standard deviation.

(2) Assuming the data follow the normal distribution, write the following functions:

(2.1) Frequency distribution per class interval width.

(2.2) Frequency distribution per unit class interval width.

(2.3) Probability density.

(3) Plot the functions determined previously. Identify all plots carefully and completely. Label all abscissas and ordinates, and their scales.

3-5. Given the information of Table 3.4, do the following:

(1) Calculate $\bar{\bar{\lambda}}$, the average estimate of the average failure rates calculated for each age group.

(2) Write down the probability density function.

TABLE 3.4 – Field data for a population exhibiting an exponentially distributed failure characteristic (chance failures) for Problem 3.5.

Life in 100 hr, starting at 200 hr	Number of failures, N_F	Failure rate, $\hat{\lambda}$
0 – 10	300	
10 – 20	200	
20 – 30	140	
30 – 40	90	
40 – 50	60	
50 – 60	40	
Over 60	70	
	$N_T = 900$	

(3) Write down the failure rate function.

(4) Write down the reliability function.

(5) Write down the conditional reliability function.

(6) Write down the mean life function and calculate the mean life for these units.

(7) Determine the reliability of these components for a mission of 100 hr, starting the mission at the age of 200 hr.

(8) Determine the reliability of these components for a mission of 100 hr, starting the mission at the age of 3,000 hr.

(9) How many of 1,000 such identical components will fail by the end of the mission in Case 7?

(10) How many of 1,000 such identical components will fail by the end of the mission in Case 8?

(11) Plot the number-of-failures versus the time-to-failure histogram, and draw a smooth curve representing the histogram.

(12) Plot the failure-rate versus the time-to-failure histogram, and draw a smooth curve representing the histogram.

TABLE 3.5 – Data of hours to failure obtained by testing to failure all 112 complex amplifiers for Problems 3-6 and 3-7.

490.5	493.5	495.5	505.5	520.5	476.5	481.5	481.5
480.5	487.5	495.5	497.5	507.5	477.5	483.5	487.5
493.5	495.5	503.5	505.5	514.5	476.5	488.5	495.5
487.5	500.5	500.5	503.5	510.5	491.5	492.5	494.5
490.5	491.5	495.5	501.5	505.5	479.5	491.5	493.5
490.5	497.5	500.5	506.5	528.5	480.5	482.5	489.5
475.5	483.5	492.5	495.5	504.5	487.5	490.5	498.5
490.5	495.5	495.5	497.5	503.5	482.5	499.5	501.5
490.5	495.5	495.5	500.5	510.5	481.5	490.5	494.5
487.5	489.5	498.5	501.5	515.5	472.5	478.5	496.5
487.5	489.5	491.5	491.5	500.5	476.5	480.5	491.5
483.5	490.5	497.5	494.5	491.5	500.5	506.5	497.5
500.5	506.5	501.5	493.5	500.5	505.5	510.5	499.5
508.5	506.5	499.5	505.5	512.5	497.5	505.5	497.5

(13) Draw the reliability versus the time-to-failure (same as mission time if the mission starts at age zero) histogram and draw a smooth curve representing the histogram.

(14) Write down the frequency distribution function per class interval width, prepare a table of T and $f(T)'$ values, and plot them on the same figure as in Case 11.

(15) Prepare a table of T and $R(T)$ values, and plot them on the same figure as in Case 13.

3-6. Given the times-to-failure data in Table 3.5, do the following:

(1) Prepare a tally table and a frequency table.

(2) Determine the class midvalue, the class starting and end values, and the class lower and upper bounds.

(3) Plot the frequency histogram and the relative frequency polygon.

(4) Plot in by eye the frequency function curve which in your judgment represents the data best.

(5) Plot the cumulative frequency polygon.

(6) Plot the relative cumulative frequency polygon.

(7) Plot in by eye the cumulative frequency function curve which in your judgment represents the data best.

Identify all plots carefully and completely. Label all abscissas and ordinates, and their scales.

3-7. For the data in Problem 3-6, do the following:

(1) Find the mean and standard deviation.

(2) Assuming the data follow the normal distribution, write the following functions:

    (2.1) Frequency distribution per class interval width.

    (2.2) Frequency distribution per unit class interval width.

    (2.3) Probability density.

(3) Plot the functions determined previously. Identify all plots carefully and completely. Label all abscissas and ordinates, and their scales.

3-8. Given the data in Table 3.6, do the following:

(1) Calculate $\bar{\lambda}$, the average estimate of the average failure rates calculated for each age group.

(2) Write down the probability density function.

(3) Write down the failure rate function.

(4) Write down the reliability function.

(5) Write down the conditional reliability function.

(6) Write down the mean life function and calculate the mean life for these units.

(7) Determine the reliability of these components for a mission of 100 hr, starting the mission at the age of 200 hr.

(8) Determine the reliability of these components for a mission of 100 hr, starting the mission at the age of 3,000 hr.

TABLE 3.6 – Field data for a population exhibiting an exponentially distributed failure characteristic (chance failures) for Problem 3.8.

Life in 100 hr	Number of failures, N_F	Failure rate, $\hat{\bar{\lambda}}$
0 – 100	165	
100 – 200	110	
200 – 300	77	
300 – 400	50	
400 – 500	34	
500 – 600	27	
600 – 700	21	
700 – 800	16	
	$N_T = 500$	

(9) How many of 1,000 such identical components will fail by the end of the mission in Case 7?

(10) How many of 1,000 such identical components will fail by the end of the mission in Case 8?

(11) Plot the number-of-failures versus the time-to-failure histogram, and draw a smooth curve representing the histogram.

(12) Plot the failure-rate versus the time-to-failure histogram, and draw a smooth curve representing the histogram.

(13) Draw the reliability versus the time-to-failure (same as mission time if the mission starts at age zero) histogram and draw a smooth curve representing the histogram.

(14) Write down the frequency distribution function per class interval width, prepare a table of T and $f(T)$ values, and plot them on the same figure as in Case 11.

(15) Prepare a table of T and $R(T)$ values, and plot them on the same figure as in Case 13.

3-9. Given the *pdf*

$$f(T) = 0.005e^{-0.005T},$$

do the following:

(1) Write down the reliability function.
(2) Write down the unreliability function.
(3) Write down the failure rate function.
(4) Write down the conditional reliability function.
(5) Plot all four previous functions on the same figure and identify the scales on the separate ordinate scales for each function.

3-10. Given the reliability function

$$R(T) = e^{-(\frac{T-150}{500})^{2.0}},$$

do the following:

(1) Write down the corresponding *pdf*.
(2) Give the name of the corresponding *pdf*.
(3) Write down the failure rate function.
(4) Write down the unreliability function.
(5) Plot the $f(T)$, $\lambda(T)$, $R(T)$ and the $\phi(T)$ on the same figure and identify the scales on the separate ordinate scales for each function.

3-11. Given a normal *pdf* with $\overline{T} = 7,780$ hr and $\sigma_T = 95$ hr, find the following:

(1) The reliability for a mission of 300 hr, starting the mission at the age of 7,850 hr.
(2) The reliability for a new mission of 300 hr, starting the mission at the age of 8,150 hr.
(3) The number of units which should be on hand and ready to start a mission at the age of 7,850 hr when 50 are needed to complete a mission at the age of 8,450 hr.

3-12. Given is the reliability function

$$R(T) = \int_{T}^{\infty} \frac{1}{T\sigma_{T'}\sqrt{2}} e^{-\frac{1}{2}(\frac{T'-\overline{T'}}{\sigma_{T'}})^2} dT.$$

Do the following:

(1) Write down the corresponding *pdf*.

(2) Give the name of the corresponding *pdf*.

(3) Write down the corresponding failure rate function.

(4) Write down the corresponding unreliability function.

3-13. Given is the *pdf*

$$f(T) = \frac{1}{\eta \Gamma(\beta)} \left(\frac{T}{\eta} \right)^{\beta - 1} e^{-\frac{T}{\eta}}.$$

Write down the following corresponding functions:

(1) The reliability function.

(2) The unreliability function.

(3) The failure rate function.

(4) The conditional reliability function.

(5) Plot the $f(T)$, $\lambda(T)$ and $R(T)$ on the same figure and identify the scales on the separate ordinate scales for each function.

REFERENCE

1. Kececioglu, Dimitri B., *Reliability Engineering Handbook*, Prentice Hall, Inc., Englewood Cliffs, N. J., Vol. 1, 720 pp. and Vol. 2, 568 pp., 1991.

Chapter 4

FAILURE FREQUENCY, FAILURE RATE AND RELIABILITY DETERMINATION FROM FIELD DATA

4.1 SIMPLE DATA ANALYSIS

If there is an abundance of time-to-failure data; e.g., 35 or more, and preferably 100 or more, then a quick, simple and easy data analysis procedure would be the following:

1. Group the data by prechosen life intervals, such as 10-hr intervals, as in Table 4.1. When grouping make sure that values of times to failure up to but not equal to the end value of each interval are counted in that interval. Values equal to the end value and up to the end of the next interval but not including the end value of this interval, are counted as belonging to this next interval.

2. Count how many of the failures have occurred in each interval, like 91 failures in the first 10 hr of operation in Table 4.1, 45 failures during the second 10 hr of operation, etc., until all units are accounted for, or 230 units. In this case the test was stopped after the first 205 units failed; therefore, we do not know when the remaining 25 units will fail had the test been continued until all 230 units failed. Consequently, we do not know the ΔT to enable the calculation of the failure rate. As eventually all units will fail $N_S \to 0$ and $R \to 0$.

81

TABLE 4.1 – Field data for 230 identical units exhibiting early failure characteristics.

1	2	3		4	5
	Number of	Failure rate		No. of units sur-	
Life,	failures,	$\bar{\lambda} = \dfrac{N_F(\Delta T)}{N_B(T)\times\Delta T}$,	$\bar{\lambda} = \dfrac{N_F(\Delta T)}{[N_B(T)-N_F(\Delta T)]\Delta T+\frac{1}{2}[N_F(\Delta T)](\Delta T)}$,	viving period,	Reliability,
hr	N_F	fr/hr	fr/hr	N_S	$\hat{R}=N_S/N$
0–10	91	$\dfrac{91}{230\times10}=0.0396$	$\dfrac{91}{(230-91)\times10+\frac{1}{2}(91\times10)}=0.0493$	$230-91=139$	$\dfrac{139}{230}=0.604$
10–20	45	$\dfrac{45}{139\times10}=0.0324$	$\dfrac{45}{(139-45)\times10+\frac{1}{2}(45\times10)}=0.0386$	$139-45=94$	$\dfrac{94}{230}=0.409$
20–30	26	$\dfrac{26}{94\times10}=0.0277$	$\dfrac{26}{(94-26)\times10+\frac{1}{2}(26\times10)}=0.0321$	$94-26=68$	$\dfrac{68}{230}=0.296$
30–40	16	$\dfrac{16}{68\times10}=0.0235$	$\dfrac{16}{(68-16)\times10+\frac{1}{2}(16\times10)}=0.0267$	$68-16=52$	$\dfrac{52}{230}=0.226$
40–50	10	$\dfrac{10}{52\times10}=0.0192$	$\dfrac{10}{(52-10)\times10+\frac{1}{2}(10\times10)}=0.0213$	$52-10=42$	$\dfrac{42}{230}=0.183$
50–60	7	$\dfrac{7}{42\times10}=0.0167$	$\dfrac{7}{(42-7)\times10+\frac{1}{2}(7\times10)}=0.0182$	$42-7=35$	$\dfrac{35}{230}=0.152$
60–70	5	$\dfrac{5}{35\times10}=0.0143$	$\dfrac{5}{(35-5)\times10+\frac{1}{2}(5\times10)}=0.0154$	$35-5=30$	$\dfrac{30}{230}=0.130$
70–80	3	$\dfrac{3}{30\times10}=0.0100$	$\dfrac{3}{(30-3)\times10+\frac{1}{2}(3\times10)}=0.0105$	$30-3=27$	$\dfrac{27}{230}=0.117$
80–90	2	$\dfrac{2}{27\times10}=0.0074$	$\dfrac{2}{(27-2)\times10+\frac{1}{2}(2\times10)}=0.0077$	$27-2=25$	$\dfrac{25}{230}=0.109$
Over 90	25	—	—	$25-25=0$	$\dfrac{0}{230}=0.000$
	N=230				

3. With the information in Columns 1 and 2 of Table 4.1 construct a *number of failures histogram*, as shown in Fig. 4.1.

4. Draw in a smooth curve through the top midpoints of each histogram bar, if possible, as shown in Fig. 4.1, called the *failure frequency curve*. A study of the curve shows a relatively sharp drop in the number of failures with increasing operating time or time to failure, indicative of early type failures being exhibited by these units. Also an estimate of the burn-in time may be obtained, assuming that these 230 units are the ones suspected of failing due to early causes only. In this case an estimate of the burn-in time might be somewhat over 100 hr.

5. Calculate the failure rate estimate, $\hat{\lambda}$, for each life interval from

$$\hat{\lambda} = \frac{N_F(\Delta T)}{N_B(T) \times \Delta T},\tag{4.1}$$

where

$N_F(\Delta T)$ = number of units failing in age increment ΔT, or in time period from age T to age $T + \Delta T$,

$N_B(T)$ = number of units in the test, or being observed, at the beginning of the age increment ΔT, or at age T,

and

ΔT = age increment during which the $N_F(\Delta T)$ units fail.

A refinement to Eq. (4.1) is

$$\hat{\lambda} = \frac{N_F(\Delta T)}{[N_B(T) - N_F(\Delta T)]\Delta T + \frac{1}{2}[N_F(\Delta T)](\Delta T)},\tag{4.2}$$

which will provide a more accurate estimate of $\hat{\lambda}$, because in the denominator of Eq. (4.1) it is assumed that all failed units operate ΔT hr, which is not true. Consequently, Eq. (4.1) provides not as good a $\hat{\lambda}$ estimate as Eq. (4.2), where it is assumed that the $[N_B(T) - N_F(\Delta T)]$ units operate during the ΔT interval without failing, as is the case, and the units that fail, fail uniformly

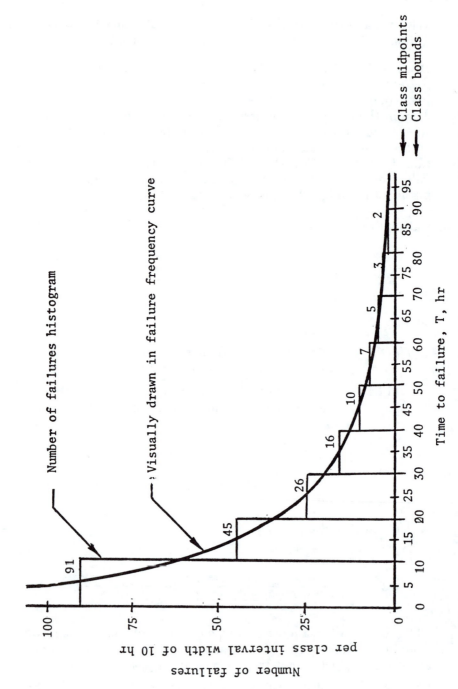

Fig. 4.1— Number of failures versus time-to-failure relationship for early failures.

84

throughout the ΔT period thus effectively each operating $\frac{1}{2}\Delta T$ hr before failing.

Equation (4.1) provides acceptable accuracy when $N_F(\Delta T)$ and ΔT are both small and $N_B(T)$ is comparatively large. In Table 4.1 the $\hat{\lambda}$ values obtained using both equations are given for comparative purposes. It may be seen that, for the data of Table 4.1, using Eq. (4.2) would be more desirable, even though the calculations would be somewhat lengthier.

6. Plot the failure rate histogram, as shown in Fig. 4.2.

7. Draw in a smooth curve through the top midpoints of each histogram bar, if possible, as shown in Fig. 4.2, called the *failure rate curve*. In this figure two curves have been drawn, one using Eq. (4.1) and another using Eq. (4.2). It may be seen that Eq. (4.2) yields higher failure rates; however, these are closer to the actual; consequently, Eq. (4.2) should be preferred.

8. Calculate the reliability estimate, $\hat{\bar{R}}(T)$, for each life interval from

$$\hat{\bar{R}}(T) = \frac{N_S(T)}{N},\tag{4.3}$$

where

$\hat{\bar{R}}(T)$ = average estimate of the reliability for a period of operation from age zero to age T at the end of each life interval,

$N_S(T)$ = number of units surviving up to the end of each life interval, at the end of which the age of the units is T; e.g., in Table 4.1, for the first interval, $N_S(T) = 139$,

and

N = total number of units being tested, or being monitored.

9. Plot the reliability histogram, as shown in Fig. 4.3.

10. Draw in a smooth curve that favors the top right side corners of the histogram bars, because the reliability values calculated are for a period of operation from age zero to an age equal to that at the end of each life interval, as shown in Fig. 4.3. For example, in Table 4.1, the reliability value of 0.296 is for a period of operation from age zero to an age of 30 hr for all of the 230 units put to test, or being monitored, etc.

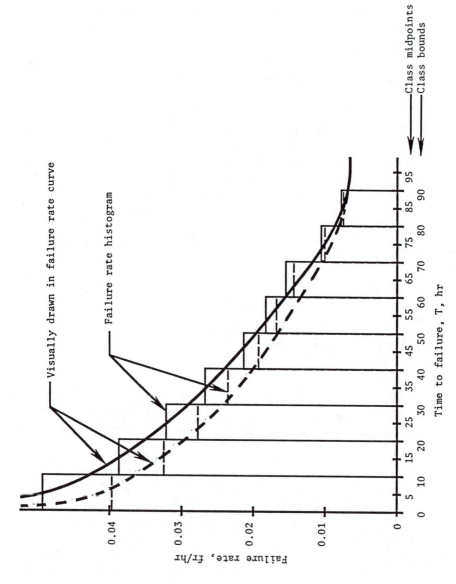

Fig 4.2– Failure rate versus time-to-failure relationship for early failures.

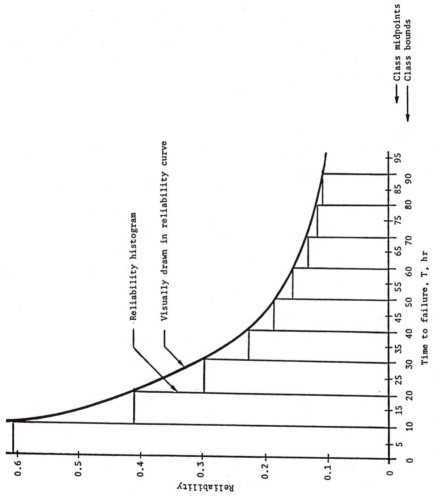

Fig. 4.3– Reliability versus time-to-failure relationship for early failures.

87

4.2   EARLY FAILURE CHARACTERISTICS

From the analysis of the data, of the calculated values of $\hat{\bar{\lambda}}$ and $\hat{\bar{R}}$, in Table 4.1, and of the results in Figs. 4.1, 4.2 and 4.3, the following observations may be made:

1. In Fig. 4.1 the number of units failing with increasing operating time is decreasing sharply, indicative of early-type failure characteristics. This is so because the weaker units fail first leaving the stronger ones that have lower failure rates to operate longer.

2. A burn-in time of somewhat longer than 100 hr appears to be the case.

3. In Fig. 4.2 the failure rate decreases with operating time, again indicating that early-type failures are occurring.

4. In Fig. 4.3 the reliability is decreasing sharply with operating time, again indicating that early-type failures are occurring.

4.3   USEFUL LIFE CHARACTERISTICS

The data in Columns 1 and 2 of Table 4.2 are analyzed the same way as those in Table 4.1. The same $\hat{\bar{\lambda}}$ and $\hat{\bar{R}}$ equations are used to calculate the values in Columns 3 and 5. Figures 4.4, 4.5 and 4.6 are plotted the same way as Figs. 4.1, 4.2 and 4.3.

From the analysis of the data, the values of $\hat{\bar{\lambda}}$ and $\hat{\bar{R}}$ in Table 4.2, and of the results in Figs. 4.4, 4.5 and 4.6, the following observations may be made:

1. In Fig. 4.4 it may be seen that the number of units failing decreases with operating time but not as sharply as for the early-type failures in Fig. 4.1.

2. In Fig. 4.5 it may be seen that the average failure rate, $\hat{\lambda}$, for each life increment of 1,000 hr, is essentially constant in value, with an overall failure rate of

$$\bar{\hat{\bar{\lambda}}} = \sum_{i=1}^{6} \hat{\bar{\lambda}}_i/6 = 0.000415 \text{ fr/hr.}$$

The deviation of the actual $\hat{\bar{\lambda}}_i$ from the global average $\bar{\hat{\bar{\lambda}}}$ is $\pm.0015\%$, which is within accepted engineering accuracy. Consequently, it

TABLE 4.2 – Field data for 900 identical units exhibiting chance failure characteristics.

1	2	3		4	5
Life in 100 hr, starting at 200 hr	Number of failures, N_F	Failure rate		No. of units surviving period, N_S	Reliability, $\hat{R} = N_S/N$
		$\bar{\hat{\lambda}} = \dfrac{N_F(\Delta T)}{N_B(T)\times\Delta T}$, fr/hr	$\bar{\hat{\lambda}} = \dfrac{N_F(\Delta T)}{[N_B(T)-N_F(\Delta T)]\Delta T+\frac{1}{2}[N_F(\Delta T)](\Delta T)}$, fr/hr		
0–10	300	$\dfrac{300}{900\times10\times100} = 0.000333$	$\dfrac{300}{(900-300)\times10\times100+\frac{1}{2}(300\times10\times100)} = 0.000400$	$900 - 300 = 600$	$\frac{600}{900} = 0.667$
10–20	200	$\dfrac{200}{600\times10\times100} = 0.000333$	$\dfrac{200}{(600-200)\times10\times100+\frac{1}{2}(200\times10\times100)} = 0.000400$	$600 - 200 = 400$	$\frac{400}{900} = 0.444$
20–30	140	$\dfrac{140}{400\times10\times100} = 0.000350$	$\dfrac{140}{(400-140)\times10\times100+\frac{1}{2}(140\times10\times100)} = 0.000424$	$400 - 140 = 260$	$\frac{260}{900} = 0.289$
30–40	90	$\dfrac{90}{260\times10\times100} = 0.000346$	$\dfrac{90}{(260-90)\times10\times100+\frac{1}{2}(90\times10\times100)} = 0.000419$	$260 - 90 = 170$	$\frac{170}{900} = 0.189$
40–50	60	$\dfrac{60}{170\times10\times100} = 0.000353$	$\dfrac{60}{(170-60)\times10\times100+\frac{1}{2}(60\times10\times100)} = 0.000429$	$170 - 60 = 110$	$\frac{110}{900} = 0.122$
50–60	38	$\dfrac{38}{110\times10\times100} = 0.000345$	$\dfrac{38}{(110-38)\times10\times100+\frac{1}{2}(38\times10\times100)} = 0.000418$	$110 - 38 = 72$	$\frac{72}{900} = 0.080$
Over 60	72	—	—	$72 - 72 = 0$	$\frac{0}{900} = 0.000$
	N=900	$\bar{\hat{\lambda}} = 0.000343$ fr/hr	$\bar{\hat{\lambda}} = 0.000415$ fr/hr		

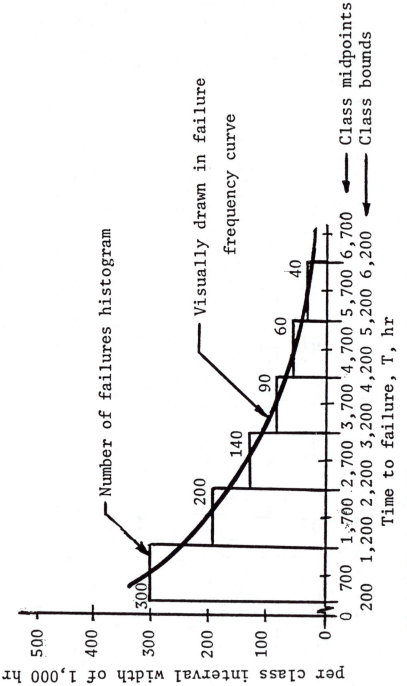

Fig 4.4– Number of failures versus time-to-failure relationship for chance failures.

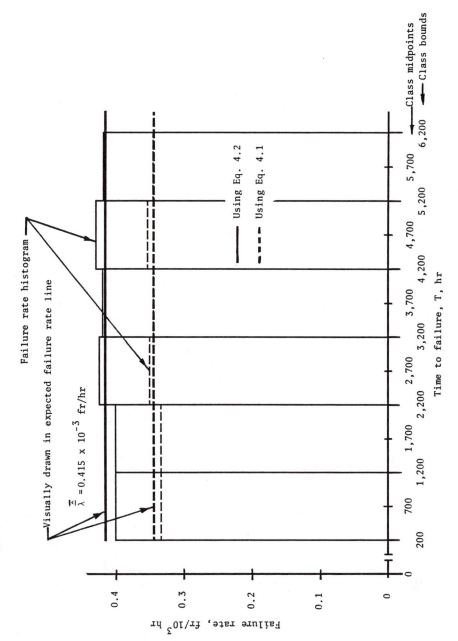

Fig. 4.5— Failure rate versus time-to-failure relationship for chance failures.

91

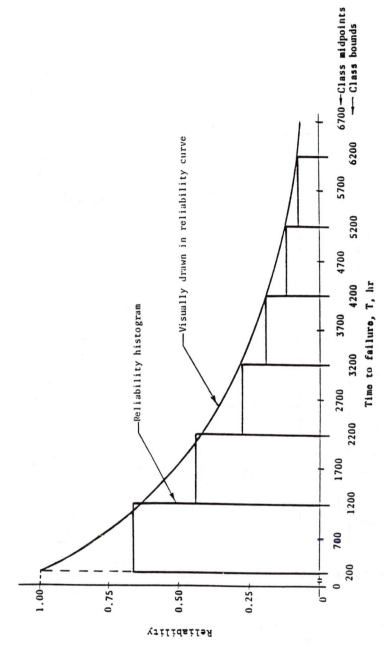

Fig. 4.6— Reliability versus time-to-failure relationship for chance failures.

may be concluded that these units are exhibiting chance failures, and that their times-to-failure *pdf* is the two-parameter exponential with parameters $\gamma = 200$ hr and $\lambda = 0.000415$ fr/hr.

3. In Fig. 4.6 it may be seen that the reliability does not drop as sharply as in Fig. 4.3 for early type failures.

4.4 WEAR-OUT LIFE CHARACTERISTICS

The data in Columns 1 and 2 of Table 4.3 are analyzed the same way as those in Table 4.1. The same $\hat{\bar{\lambda}}$ and $\hat{\bar{R}}$ equations are used to calculate the values in Columns 3 and 5. Figures 4.7, 4.8 and 4.9 are plotted the same way as Figs. 4.1, 4.2 and 4.3.

From the analysis of the data, the values of $\hat{\bar{\lambda}}$ and $\hat{\bar{R}}$ in Table 4.3, and Figs. 4.7, 4.8 and 4.9, the following observations may be made:

1. In Fig. 4.7 the number failing starts low, increases with operating time, reaches a peak at about 7,750 hr, and decreases thereafter. As T becomes large, such as after about 8,100 hr of operation, the number failing goes to essentially zero. This strongly indicates that wear-out failures are occurring.

2. In Fig. 4.8 the failure rate starts very low and increases steadily, again indicating that wear-out failures are occurring.

3. In Fig. 4.9 the reliability starts at the level of 1.00, stays at this level until about 7,550 hr of operation at which time a significant number of wear-out failures start to occur, decreases sharply thereafter as wear-out sets in, and finally it becomes essentially zero after about 8,100 hr. The reliability curve of units exhibiting wear-out failures, typically has a knee, as seen in Fig. 4.9.

4.5 MORE SOPHISTICATED DATA ANALYSES

It must be born in mind that this chapter presented very simple data analysis techniques. If there are a lot of data and a quick insight at what they are telling us is needed, then these techniques may prove to be adequate. However more sophisticated data analysis techniques are available, and are covered in the later chapters of this book. The reader may also wish to consult [1].

TABLE 4.3 – Field data for 900 identical units exhibiting wear-out failure characteristics.

1	2	3		4	5
		Failure rate		No. of units sur-	Reliability,
Life in 100 hr, starting at 7,500 hr	Number of failures, N_F	$\bar{\lambda} = \dfrac{N_F(\Delta T)}{N_B(T)\times \Delta T}$, fr/hr	$\bar{\lambda} = \dfrac{N_F(\Delta T)}{[N_B(T)-N_F(\Delta T)]\Delta T+\frac{1}{2}[N_F(\Delta T)](\Delta T)}$, fr/hr	viving period, N_S	$\hat{R} = N_S/N$
0–1	40	$\dfrac{40}{900\times100} = 0.00044$	$\dfrac{40}{(900-40)\times100+\frac{1}{2}(40\times100)} = 0.00046$	$900 - 40 = 860$	$\dfrac{860}{900} = 0.955$
1–2	210	$\dfrac{210}{860\times100} = 0.00244$	$\dfrac{210}{(860-210)\times100+\frac{1}{2}(210\times100)} = 0.00278$	$860 - 210 = 650$	$\dfrac{650}{900} = 0.772$
2–3	300	$\dfrac{300}{650\times100} = 0.00462$	$\dfrac{300}{(650-300)\times100+\frac{1}{2}(300\times100)} = 0.00600$	$650 - 300 = 350$	$\dfrac{350}{900} = 0.389$
3–4	250	$\dfrac{250}{350\times100} = 0.00714$	$\dfrac{250}{(350-250)\times100+\frac{1}{2}(250\times100)} = 0.01111$	$350 - 250 = 100$	$\dfrac{100}{900} = 0.111$
4–5	80	$\dfrac{80}{100\times100} = 0.00800$	$\dfrac{80}{(100-80)\times100+\frac{1}{2}(80\times100)} = 0.01333$	$100 - 80 = 20$	$\dfrac{20}{900} = 0.022$
5–6	20	$\dfrac{20}{20\times100} = 0.01000$	$\dfrac{20}{(20-20)\times100+\frac{1}{2}(20\times100)} = 0.02000$	$20 - 20 = 0$	$\dfrac{0}{900} = 0.000$
	N=900				

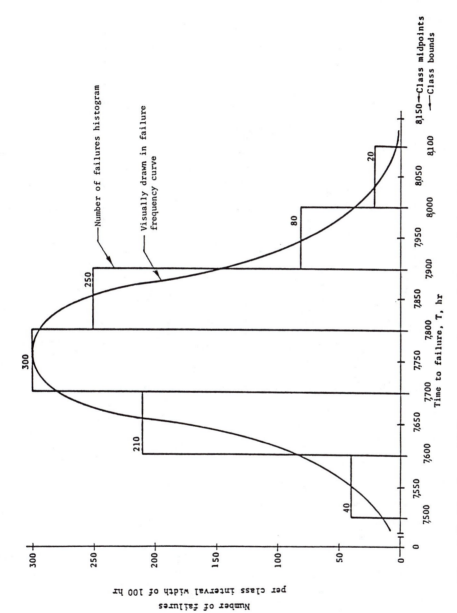

Fig. 4.7– Number of failures versus time-to-failure relationship for wear-out failures.

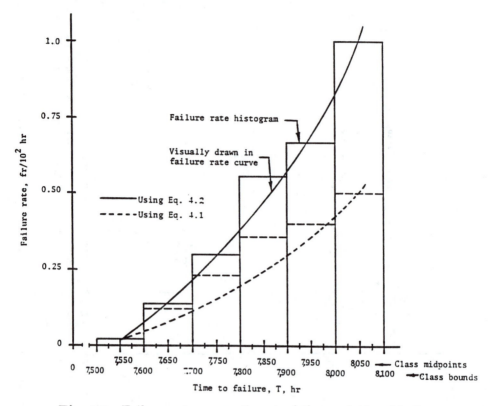

Fig 4.8– Failure rate versus time-to-failure relationship for
wear-out failures.

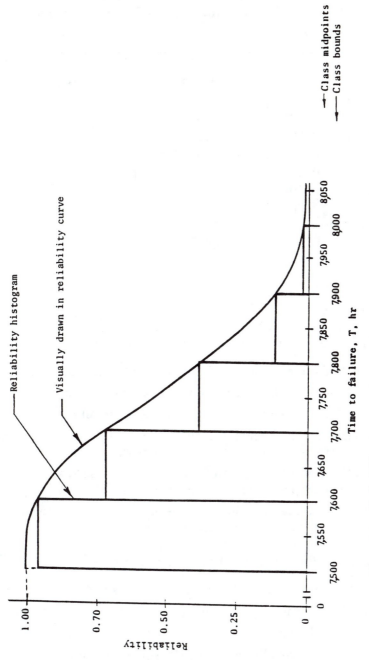

Fig. 4.9– Reliability versus time-to-failure relationship for wear-out failures.

Table 4.4 – Field data for a population exhibiting early failure characteristics for Problem 4-2.

Life hr,	Number of failures, N_F	Failure rate, $\hat{\bar{\lambda}}$
0–10	65	
10–20	22	
20–30	13	
30–40	8	
40–50	4	
50–60	1	
Over 60	787	
	$N_T = 900$	

PROBLEMS

4-1. Using the data in Table 4.3, do the following:

(1) Plot the number of failures versus the times to failure.

(2) Plot the failure rate versus the times to failure.

(3) Plot the reliability versus the times to failure.

(4) Discuss the behavior of the results in Case 1, after a smooth curve representing the plot is drawn.

(5) Discuss the behavior of the results in Case 2, after a smooth curve representing the plot is drawn.

(6) Discuss the behavior of the results in Case 3, after a smooth curve representing the plot is drawn.

4-2. Using the data in Table 4.4, do the following:

(1) Prepare a table similar to that in Table 4.1.

(2) Plot the number of failures versus the times to failure.

(3) Plot the failure rate versus the times to failure.

(4) Plot the reliability versus the times to failure.

(5) Discuss the behavior of the results in Case 2, after a smooth curve representing the plot is drawn.

Table 4.5 – Field data for a population exhibiting chance failure characteristics for Problem 4-3.

Life in hr, starting at 100 hr	Number of failures, N_F	Failure rate, $\hat{\lambda}$
0–20	47	
20–40	44	
40–60	42	
60–80	39	
80–100	38	
100–120	35	
Over 120	655	
	$N_T = 900$	

(6) Discuss the behavior of the results in Case 3, after a smooth curve representing the plot is drawn.

(7) Discuss the behavior of the results in Case 4, after a smooth curve representing the plot is drawn.

4-3. Using the data in Table 4.5, do the following:

(1) Prepare a table similar to that in Table 4.1.

(2) Plot the number of failures versus the times to failure.

(3) Plot the failure rate versus the times to failure.

(4) Plot the reliability versus the times to failure.

(5) Discuss the behavior of the results in Case 2, after a smooth curve representing the plot is drawn.

(6) Discuss the behavior of the results in Case 3, after a smooth curve representing the plot is drawn.

(7) Discuss the behavior of the results in Case 4, after a smooth curve representing the plot is drawn.

Table 4.6 – Times-to-failure data for 100 components in multiples of 10 hr for Problems 4-4 and 4-5.

439	434	440	447	437	435	430	434	435	432
435	432	444	441	437	451	437	433	438	435
436	427	440	430	436	427	438	427	433	439
434	437	440	429	433	430	438	435	432	453
440	436	440	437	445	440	442	426	437	439
440	432	441	430	430	435	440	427	432	432
430	432	435	434	430	435	440	434	439	440
438	432	439	433	431	447	428	430	431	435
428	437	435	440	441	435	436	441	436	444
439	436	433	435	431	431	435	433	443	436

4-4. Using the data in Table 4.6, do the following:

   (1) Prepare a tally table and a frequency table.

   (2) Determine the class midvalue, the class starting and end values, and the class lower and upper bounds.

   (3) Plot the frequency histogram, and the relative frequency polygon.

   (4) Plot in by eye the frequency function curve which in your judgment represents the data best.

   (5) Plot the cumulative frequency polygon.

   (6) Plot the relative cumulative frequency polygon.

   (7) Plot in by eye the cumulative frequency function curve which in your judgment represents the data best.

Identify all plots carefully and completely. Label all abscissas and ordinates, and their scales.

4-5. For the data in Table 4.6 do the following:

   (1) Find the mean and standard deviation.

   (2) Assuming the data follow the normal distribution, write the following functions:

   (2.1) Frequency distribution per class interval width.

   (2.2) Frequency distribution per unit class interval width.

   (2.3) Probability density.

Table 4.7 – **Field data for a population exhibiting an exponentially distributed failure characteristic for Problems 4-6 and 4-7.**

Life in hr, starting at 200 hr	Number of failures, N_F
0–100	30
100–200	22
200–300	15
300–400	10
400–500	7
500–600	5
Over 600	11
	$N_T = 100$

(3) Plot the functions determined previously. Identify all plots carefully and completely. Label all abscissas and ordinates, and their scales.

4-6. Using the data in Table 4.7, do the following:

(1) Prepare a table similar to that in Table 4.1.

(2) Plot the number of failures versus the times to failure.

(3) Plot the failure rate versus the times to failure.

(4) Plot the reliability versus the times to failure.

(5) Discuss the behavior of the results in Case 2, after a smooth curve representing the plot is drawn.

(6) Discuss the behavior of the results in Case 3, after a smooth curve representing the plot is drawn.

(7) Discuss the behavior of the results in Case 4, after a smooth curve representing the plot is drawn.

4-7. Given is the information of Table 4.7. Do the following:

(1) Calculate $\bar{\bar{\lambda}}$, the average estimate of the average failure rates calculated for each age group.

(2) Write down the probability density function.

(3) Write down the failure rate function.

Table 4.8 – **Field data for a population exhibiting an exponentially distributed failure characteristic for Problems 4-8 and 4-9.**

Life in hr, starting at 250 hr	Number of failures, N_F
0–1,000	70
1,000–2,000	43
2,000–3,000	30
3,000–4,000	21
4,000–5,000	13
5,000–6,000	8
Over 6,000	15
	$N_T = 200$

(4) Write down the reliability function.

(5) Write down the conditional reliability function.

(6) Write down the mean life function and calculate the mean life for these units.

(7) Determine the reliability of these components for a mission of 100 hr, starting the mission at the age of 200 hr.

(8) Determine the reliability of these components for a mission of 100 hr, starting the mission at the age of 3,000 hr.

(9) How many of 1,000 such identical components will fail by the end of the mission in Case 7?

(10) How many of 1,000 such identical components will fail by the end of the mission in Case 8?

4-8. Field data were obtained and grouped as given in Table 4.8. Do the following:

(1) Prepare a table similar to that in Table 4.1.

(2) Plot the number of failures versus the time-to-failure histogram, and draw in the best fitting curve in your judgement.

(3) Plot the failure rate versus the time-to-failure histogram, and draw in the best fitting curve in your judgement.

(4) Plot the reliability versus the mission time histogram and draw in the best fitting curve in your judgement.

(5) Discuss the behavior of the results in Case 2.

(6) Discuss the behavior of the results in Case 3.

(7) Discuss the behavior of the results in Case 4.

4-9. Given is the information in Table 4.8. Do the following:

(1) Calculate $\bar{\bar{\lambda}}$, the average estimate of the average failure rates calculated for each age group.

(2) Write down the probability density function.

(3) Write down the failure rate function.

(4) Write down the reliability function.

(5) Write down the conditional reliability function.

(6) Write down the mean life function and calculate the mean life for these units.

(7) Determine the reliability of these components for a mission of 100 hr, starting the mission at the age of 200 hr.

(8) Determine the reliability of these components for a mission of 100 hr, starting the mission at the age of 3,000 hr.

(9) How many of 1,000 such identical components will fail by the end of the mission in Case 7?

(10) How many of 1,000 such identical components will fail by the end of the mission in Case 8?

4-10. Using the results of Problem 4-9, do the following:

(1) Calculate, tabulate and plot the frequency distribution function per class interval width on the previous histogram. Label the axes.

(2) Calculate, tabulate and plot the frequency distribution per unit class interval width.

(3) Calculate, tabulate and plot the probability density function.

(4) Calculate, tabulate and plot the standardized probability density function.

4-11. Given is the information in Table 4.9. Do the following:

(1) Calculate $\bar{\bar{\lambda}}$, the average estimate of the average failure rates calculated for each age group.

Table 4.9 – Field data for a population exhibiting an exponentially distributed failure characteristic for Problem 4-11.

Life in hr, starting at 200 hr	Number of failures, N_F
0–1,000	300
1,000–2,000	200
2,000–3,000	140
3,000–4,000	90
4,000–5,000	60
5,000–6,000	40
Over 6,000	70
	$N_T = 900$

(2) Write down the probability density function.

(3) Write down the failure rate function.

(4) Write down the reliability function.

(5) Write down the conditional reliability function.

(6) Write down the mean life function and calculate the mean life for these units.

(7) Determine the reliability of these components for a mission of 100 hr, starting the mission at the age of 200 hr.

(8) Determine the reliability of these components for a mission of 100 hr, starting the mission at the age of 3,000 hr.

(9) How many of 1,000 such identical components will fail by the end of the mission in Case 7?

(10) How many of 1,000 such identical components will fail by the end of the mission in Case 8?

REFERENCE

1. Kececioglu, Dimitri B., *Reliability Engineering Handbook*, Prentice Hall, Inc., Englewood Cliffs, New Jersey, Vol. 1, 720 pp. and Vol. 2, 568 pp., 1991.

Chapter 5

CHI-SQUARE, STUDENT'S t AND F DISTRIBUTIONS

5.1 THE CHI-SQUARE DISTRIBUTION

5.1.1 CHI-SQUARE DISTRIBUTION CHARACTERISTICS

The chi-square, χ^2, distribution is widely used in statistics. Its *pdf* is given by [1, p.108]

$$f(\chi^2) = \frac{1}{2^{\frac{\nu}{2}}\Gamma(\frac{\nu}{2})}\chi^{2\frac{\nu}{2}-1}e^{-\frac{\chi^2}{2}}, \quad \chi^2 \geq 0, \tag{5.1}$$

where

ν = only parameter, called degrees of freedom,

and

$\Gamma(n)$ = gamma function.

Figure 5.1 shows how the shape of the χ^2 *pdf* changes as ν increases.

Some of the χ^2 *pdf's* characteristics are the following:

1. The mean, $\overline{\chi^2}$, of the χ^2 *pdf* is

$$\overline{\chi^2} = \nu. \tag{5.2}$$

2. The standard deviation, σ_{χ^2}, of the χ^2 *pdf* is

$$\sigma_{\chi^2} = (2\nu)^{\frac{1}{2}}. \tag{5.3}$$

105

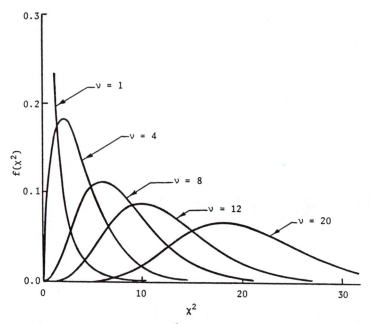

Fig.5.1 –Plot of the χ^2 *pdf* for various values of ν.

3. The coefficient of skewness, α_3, of the χ^2 *pdf* is given by [2, p. 4-53]

$$\alpha_3 = 2 \left(\frac{2}{\nu} \right)^{\frac{1}{2}}. \tag{5.4}$$

4. The coefficient of kurtosis, α_4, of the χ^2 *pdf* is given by

$$\alpha_4 = 3 + \frac{12}{\nu}. \tag{5.5}$$

5. The χ^2 *pdf* has a single mode at

$$\chi^2 = \nu - 2, \text{ for } \nu \geq 2. \tag{5.6}$$

6. The additivity of χ^2 random variables:

If χ_1^2 and χ_2^2 are independent, χ^2 random variables with ν_1 and ν_2 degrees of freedom, respectively, the sum of these random variables; i.e.,

$$\chi^2 = \chi_1^2 + \chi_2^2 \tag{5.7}$$

also has a χ^2 distribution with $\nu = \nu_1 + \nu_2$ degrees of freedom [1].

7. The relation to the normal distribution:

Let $X_1, X_2, ..., X_n$ be n independent, normally distributed random variables, each with zero mean and unit variance. Then the random variable

$$\chi^2 = X_1^2 + X_2^2 + ... + X_n^2, \quad n = 1, 2, ..., \tag{5.8}$$

has a χ^2 distribution with $\nu = n$ degrees of freedom [3].

If $X_1, X_2, ..., X_n$ are normally distributed random variables with means $\mu_1, \mu_2, ..., \mu_n$, and variances $\sigma_1^2, \sigma_2^2, ..., \sigma_n^2$, respectively, then the random variable

$$\chi^2 = \sum_{i=1}^{n} \left(\frac{X_i - \mu_i}{\sigma_i}\right)^2 \tag{5.9}$$

has a χ^2 distribution with $\nu = n$ degrees of freedom.

8. Let $X_1, X_2, ..., X_n$ be independent, normally distributed random variables, each having mean μ and variance σ^2. Then the random variable $(n-1)S^2/\sigma^2$ is χ^2 distributed with $\nu = n - 1$ degrees of freedom, where S^2, the sample variance, is given by

$$S^2 = \frac{1}{n-1} \sum_{i=1}^{n} (X_i - \bar{X})^2, \tag{5.10}$$

and $\bar{X}$, the sample mean, is given by

$$\bar{X} = \frac{1}{n} \sum_{i=1}^{n} X_i. \tag{5.11}$$

9. As ν tends to infinity, the χ^2 *pdf* slowly tends to a normal distribution with mean ν and variance 2ν. A good approximation [4] of the χ^2 *pdf* can be obtained by transforming χ^2 to

$$Y = \left(\frac{\chi^2}{\nu}\right)^{\frac{1}{3}}. \tag{5.12}$$

This new variable is asymptotically, normally distributed with mean

$$\mu_Y = 1 - \frac{2}{9\nu}, \tag{5.13}$$

and variance

$$\sigma_Y^2 = \frac{2}{9\nu}. \tag{5.14}$$

Also

$$V = (2\chi^2)^{\frac{1}{2}} \tag{5.15}$$

is asymptotically normally distributed as ν tends to infinity, with mean

$$\mu_V = (2\nu - 1)^{\frac{1}{2}}, \tag{5.16}$$

and variance

$$\sigma_V^2 = 1. \tag{5.17}$$

This approximation may be used to extend the χ^2 tables to larger values of ν than are tabulated.

10. The χ^2 *pdf* is a special case of the gamma *pdf*, when the scale parameter, η, of the gamma *pdf* is 2 and the shape parameter, β, is a multiple of $\frac{1}{2}$; i.e., $\beta = \frac{1}{2}\nu$.

11. The exponential *pdf* with failure rate $\lambda = \frac{1}{2}$ is a χ^2 *pdf* with $\nu = 2$ degrees of freedom. By the additivity property of χ^2 random variables, the sum of k exponential random variables with $\lambda = \frac{1}{2}$ is a χ^2 *pdf* with $\nu = 2k$ degrees of freedom.

5.1.2   THE CUMULATIVE DISTRIBUTION FUNCTION AND THE PERCENTAGE POINT OF THE CHI-SQUARE DISTRIBUTION

The cumulative distribution function of the χ^2 *pdf* is given by

$$F(\chi^2) = \int_0^{\chi^2} \frac{1}{2^{\frac{\nu}{2}}\Gamma(\frac{\nu}{2})} \tau^{\frac{\nu}{2}-1} e^{-\frac{\tau}{2}} d\tau. \tag{5.18}$$

Let $t = \frac{\tau}{2}$, then $dt = \frac{1}{2}d\tau$ and $\tau = 2t$, consequently

$$F(\chi^2) = \int_0^{\frac{\chi^2}{2}} \frac{1}{\Gamma(\frac{\nu}{2})} t^{\frac{\nu}{2}-1} e^{-t} dt,$$

or

$$F(\chi^2) = \frac{1}{\Gamma(\frac{\nu}{2})} I(\frac{\chi^2}{2}; \frac{\nu}{2}), \tag{5.19}$$

where

$$I(u; n) = \int_0^u t^{n-1} e^{-t} dt$$

is the incomplete gamma function.

Let α be the significance level, then the percentage point $\chi^2_{\alpha;\nu}$ is defined by

$$P(\chi^2 \geq \chi^2_{\alpha;\nu}) = \int_{\chi^2_{\alpha;\nu}}^{\infty} \frac{1}{2^{\frac{\nu}{2}} \Gamma(\frac{\nu}{2})} \tau^{\frac{\nu}{2}-1} e^{-\frac{\tau}{2}} d\tau = \alpha. \tag{5.20}$$

The values of the percentage points, $\chi^2_{\alpha;\nu} = \chi^2_{\delta;\nu}$, are tabulated for various values of ν and α, and are given in Appendix D. From Appendix D, the value of $\chi^2_{\alpha;\nu}$, given α and ν, can be found. For example, for $\delta = \alpha = 0.10$ and $\nu = 4$, $\chi^2_{0.10;4} = 7.779$; i.e.,

$$P(\chi^2 \geq \chi^2_{0.10;4}) = P(\chi^2 \geq 7.779) = 0.10.$$

If the value of ν exceeds 100, the value of $\chi^2_{\alpha;\nu}$ can be calculated using its normal approximation. From Eqs. (5.12) through (5.14),

$$z = \frac{Y - \mu_Y}{\sigma_Y} = \frac{(\frac{\chi^2}{\nu})^{\frac{1}{3}} - (1 - \frac{2}{9\nu})}{(\frac{2}{9\nu})^{\frac{1}{2}}} \sim N(0, 1). \tag{5.21}$$

Let z_α be the percentage point of the standardized normal distribution, then

$$\frac{(\frac{\chi^2_{\alpha;\nu}}{\nu})^{\frac{1}{3}} - (1 - \frac{2}{9\nu})}{(\frac{2}{9\nu})^{\frac{1}{2}}} = z_\alpha. \tag{5.22}$$

Solving for $\chi^2_{\alpha;\nu}$ yields

$$\chi^2_{\alpha;\nu} = \nu[1 - \frac{2}{9\nu} + z_\alpha(\frac{2}{9\nu})^{\frac{1}{2}}]^3. \tag{5.23}$$

From Eqs. (5.15) through (5.17),

$$z = \frac{(2\chi^2)^{\frac{1}{2}} - (2\nu - 1)^{\frac{1}{2}}}{1} \sim N(0, 1), \tag{5.24}$$

and

$$(2\chi^2_{\alpha;\nu})^{\frac{1}{2}} - (2\nu - 1)^{\frac{1}{2}} = z_\alpha.$$

Solving for $\chi^2_{\alpha;\nu}$ yields

$$\chi^2_{\alpha;\nu} = \frac{1}{2}[z_\alpha + (2\nu - 1)^{\frac{1}{2}}]^2. \qquad (5.25)$$

For example, for $\alpha = 0.10$ and $\nu = 100$, from Appendix D, $z_{0.10} = 1.282$. From Eq. (5.23)

$$\chi^2_{0.10;100} = 100\left\{1 - \frac{2}{9(100)} + 1.282\left[\frac{2}{9(100)}\right]^{\frac{1}{2}}\right\}^3,$$

or

$$\chi^2_{0.10;100} = 118.4999,$$

and from Eq. (5.25),

$$\chi^2_{0.10;100} = \frac{1}{2}\{1.282 + [(2)(100) - 1]^{\frac{1}{2}}\}^2,$$

or

$$\chi^2_{0.10;100} = 118.4066.$$

These values are close to the exact value of $\chi^2_{0.10;100} = 118.498$.

5.1.3   APPLICATIONS OF THE CHI-SQUARE DISTRIBUTION

5.1.3.1   ESTIMATE OF THE STANDARD DEVIATION OF THE NORMAL DISTRIBUTION

Let $X_1, X_2, ..., X_n$ be a random sample of size n, from a normal distribution with mean μ and variance σ^2. Let $\bar{X}$ and S^2 be the sample mean and variance, defined by Eqs. (5.11) and (5.10), respectively. As discussed in Section 5.1.1, the random variable $(n-1)S^2/\sigma^2$ is χ^2 distributed with $\nu = n - 1$ degrees of freedom; i.e.,

$$\frac{(n-1)S^2}{\sigma^2} \sim \chi^2(n-1). \qquad (5.26)$$

Therefore the two-sided confidence interval of σ^2 can be determined as follows:

Given a significance level α and a sample of size n, the sample mean and sample variance are given by Eqs. (5.11) and (5.10), then the following probability statement may be made:

$$P\left[\chi^2_{1-\frac{\alpha}{2};\nu} \le \frac{(n-1)S^2}{\sigma^2} \le \chi^2_{\frac{\alpha}{2};\nu}\right] = 1 - \alpha,$$

$$P\left[\frac{1}{\chi^2_{1-\frac{\alpha}{2};\nu}} \ge \frac{\sigma^2}{(n-1)S^2} \ge \frac{1}{\chi^2_{\frac{\alpha}{2};\nu}}\right] = 1 - \alpha,$$

or

$$P\left[\frac{(n-1)S^2}{\chi^2_{\frac{\alpha}{2};\nu}} \le \sigma^2 \le \frac{(n-1)S^2}{\chi^2_{1-\frac{\alpha}{2};\nu}}\right] = 1 - \alpha. \tag{5.27}$$

For example, if $\alpha = 0.10$ and $n = 10$, then $\nu = 10 - 1 = 9$, and from Appendix D

$$\chi^2_{1-\frac{\alpha}{2};\nu} = \chi^2_{1-0.05;9} = \chi^2_{0.95;9} = 3.325,$$

and

$$\chi^2_{\frac{\alpha}{2};\nu} = \chi^2_{0.05;9} = 16.919.$$

Given observations from this random sample, the sample variance, S^2, can be calculated by Eq. (5.10). Let's assume $S^2 = 5$, then, from Eq. (5.27), the upper confidence limit on σ^2 is

$$\sigma^2_{U2} = \frac{(n-1)S^2}{\chi^2_{1-\frac{\alpha}{2};\nu}} = \frac{(9)(5)}{3.325} = 13.534,$$

and the lower confidence limit on σ^2 is

$$\sigma^2_{L2} = \frac{(n-1)S^2}{\chi^2_{\frac{\alpha}{2};\nu}} = \frac{(9)(5)}{16.919} = 2.660.$$

Therefore, the confidence interval on σ^2 is

$$(2.660 \; ; 13.534).$$

The one-sided, confidence interval on σ^2 can be determined from

$$P\left[\frac{(n-1)S^2}{\sigma^2} \le \chi^2_{\alpha;\nu}\right] = 1 - \alpha,$$

and

$$P\left[\chi^2_{1-\alpha;\nu} \leq \frac{(n-1)S^2}{\sigma^2}\right] = 1 - \alpha.$$

Therefore, the lower, one-sided confidence interval is

$$P\left[\frac{(n-1)S^2}{\chi^2_{\alpha;\nu}} \leq \sigma^2\right] = 1 - \alpha, \tag{5.28}$$

and the upper, one-sided confidence interval is

$$P\left[\sigma^2 \leq \frac{(n-1)S^2}{\chi^2_{1-\alpha;\nu}}\right] = 1 - \alpha. \tag{5.29}$$

For example, if $\alpha = 0.10$, $n = 10$ and $S^2 = 5$, then from Appendix D,

$$\chi^2_{\alpha;\nu} = \chi^2_{0.10;(10-1)} = 14.684,$$

and

$$\chi^2_{1-\alpha;\nu} = \chi^2_{0.90;(10-1)} = 4.168.$$

Consequently, the upper, one-sided confidence limit is

$$\sigma^2_{U1} = \frac{(n-1)S^2}{\chi^2_{1-\alpha;\nu}} = \frac{(10-1)(5)}{4.168} = 10.797,$$

and the lower, one-sided confidence limit is

$$\sigma^2_{L1} = \frac{(n-1)S^2}{\chi^2_{\alpha;\nu}} = \frac{(10-1)(5)}{14.684} = 3.065.$$

5.1.3.2 GOODNESS-OF-FIT TEST

The chi-squared goodness-of-fit test is widely used to test the hypothesis that $F(x)$ equals $F_0(x)$; i.e.,

$$H_0: \quad F(x) = F_0(x) \quad \text{for all } x, \tag{5.30}$$

where $F_0(x)$ is the postulated distribution.

Let the range of the random variable X be divided into m successive, adjoining intervals $[a_{i-1}, a_i)$, $i = 1, 2, ..., m$. Let O_i denote the number of observations in the ith interval, where

$$N = \sum_{i=1}^{m} O_i$$

is the total number of observations.

For a specified distribution $F_0(x)$, the probability p_i of obtaining *one observation* in the *ith* interval can be calculated for $i = 1, 2, ..., m$, as

$$p_i = F_0(a_i) - F_0(a_{i-1}), \tag{5.31}$$

where

$$\sum_{i=1}^{m} p_i = 1.$$

The probability of obtaining O_i specific observations in the *ith* interval is $p_i^{O_i}$. The number of ways in which N objects can be divided into m groups of size O_i, $i = 1, 2, ..., m$, is given by

$$N!/(\prod_{i=1}^{m} O_i!).$$

Therefore given $F_0(x)$, the probability of obtaining the grouped sample is

$$P(O_1, O_2, ..., O_m; p_1, p_2, ..., p_m) = N! \prod_{i=1}^{m} \frac{p_i^{O_i}}{O_i!}. \tag{5.32}$$

This is a multinomial model, since it is also obtained as the general term of the multinomial expansion $(p_1 + p_2 + ... + p_m)^N$.

If $m = 2$, Eq. (5.32) becomes the well-known binomial model; i.e.,

$$P(O_1, O_2; p_1, p_2) = N! \frac{p_1^{O_1} p_2^{O_2}}{O_1! O_2!}.$$

Since $O_1 + O_2 = N$ and $p_1 + p_2 = 1$, the binomial model is usually written as

$$B(x; p) = \binom{N}{x} p^x (1 - p)^{N-x}, \tag{5.33}$$

where

$$x = O_1, \quad \text{and} \quad p = p_1.$$

Now consider the standardized version of the binomial variable O_1

$$\chi = \frac{O_1 - Np_1}{[Np_1(1 - p_1)]^{\frac{1}{2}}}. \tag{5.34}$$

From the central limit theorem, it may be seen that χ asymptotically approaches the standardized normal distribution, as $N \to \infty$, since O_1 is the sum of N Bernoulli trials. Then, from Eq. (5.8), for the case of $n = 1$, the square of Eq. (5.34) is chi-square distributed with $\nu = 1$ degree of freedom. Thus, the statistic

$$\chi^2 = \frac{(O_1 - Np_1)^2}{Np_1(1 - p_1)} = \frac{(O_1 - Np_1)^2}{Np_1} + \frac{(O_1 - Np_1)^2}{N(1 - p_1)},$$

$$= \frac{(O_1 - Np_1)^2}{Np_1} + \frac{[N - O_2 - N(1 - p_2)]^2}{Np_2},$$

$$= \frac{(O_1 - Np_1)^2}{Np_1} + \frac{(O_1 - Np_2)^2}{Np_2},$$

or

$$\chi^2 = \sum_{i=1}^{2} \frac{(O_i - Np_i)^2}{Np_i}, \tag{5.35}$$

has the limiting distribution of chi-square with $\nu = 2 - 1$ degrees of freedom, where $\nu = m - 1 = 2 - 1$ is due to the presence of the linear constraint $O_1 + O_2 = N$ on the observable frequencies O_1 and O_2.

Similarly it can be shown that standardized multinomial variables are asymptotically distributed according to the multivariate standardized normal model, so that their sum of squares

$$\chi^2 = \sum_{i=1}^{m} \frac{(O_i - Np_i)^2}{Np_i} \tag{5.36}$$

is chi-square distributed, with $\nu = m - 1$ degrees of freedom [5]. The difference between m and ν is due to the linear constraint

$$\sum_{i=1}^{m} O_i = N$$

on the observable frequencies O_i. Eq. (5.36) can be rewritten as

$$\chi_o^2 = \sum_{i=1}^{m} \frac{(O_i - E_i)^2}{E_i}, \tag{5.37}$$

where

O_i = observed frequency in the *ith* interval,

and

E_i = expected frequency in the *ith* interval.

Table 5.1 – Observed frequencies, expected frequencies and the observed χ^2 value for testing the honesty of a die for Example 5–1.

Number of face	Observed frequencies, O_i	Expected frequencies, E_i	$\frac{(O_i - E_i)^2}{E_i}$
1	13	20	2.45
2	28	20	3.20
3	16	20	0.80
4	10	20	5.00
5	32	20	7.20
6	21	20	0.05
	120	120	$\chi_o^2 = 18.70$

Appendix D gives the percentage points for various values of α and ν. Thus, probabilistic statements can be found for the statistic χ^2. Conversely, given a level of significance α, a critical value, χ_c^2, can be determined from these tables, to check if the observed frequencies O_i contradict the expected frequencies $N p_i$ which are calculated from the postulated model $F_0(x)$. It is not difficult to see that the smaller the observed χ^2 value the better the fit of the postulated model. The procedure of the test is illustrated by the next example.

EXAMPLE 5–1

A die is tossed 120 times, and the results are listed in Column 2 of Table 5.1. Using the chi-squared goodness-of-fit test, test the hypothesis that the die is honest.

SOLUTION TO EXAMPLE 5–1

Since each face of an honest die is expected to show one sixth of the time, in this example, the expected number of frequencies for each

face should be $(1/6)(120) = 20$. From Eq. (5.37) the observed χ^2 value is calculated to be

$$\chi_o^2 = 18.70.$$

Based on the level of significance of $\alpha = 0.05$, the critical χ^2 value with $\nu = 6 - 1 = 5$ degrees of freedom, from Appendix D, is

$$\chi_c^2 = \chi_{0.05;5}^2 = 11.070,$$

or

$$P\{\chi_{\nu=5}^2 \geq \chi_{0.05;5}^2 = 11.070\} = 0.05.$$

But

$$\chi_o^2 > \chi_c^2,$$

which means that the observed results come from an honest die with a probability of less than 0.05; i.e.,

$$P\{\chi_{\nu=5}^2 \geq 18.70\} < 0.05.$$

The conclusion is that we can not accept the hypothesis that the die is honest, or we reject that the die is honest at a significance level of $\alpha = 0.05$.

It must be emphasized that since the statistic χ_o^2 given by Eq. (5.37) has a limiting chi-square distribution, for high accuracy, the sample size has to be large. The recommended smallest sample size is 20 [4].

5.2   THE STUDENT'S t DISTRIBUTION

5.2.1   STUDENT'S t DISTRIBUTION CHARACTERISTICS

The Student's t distribution's *pdf* is given by [1, p. 115],

$$f(t) = \frac{\Gamma(\frac{\nu+1}{2})}{(\pi\nu)^{\frac{1}{2}}\Gamma(\frac{\nu}{2})}(1 + \frac{t^2}{\nu})^{-\frac{(\nu+1)}{2}}, \tag{5.38}$$

where

ν = degrees of freedom, the parameter of the Student's t distribution,

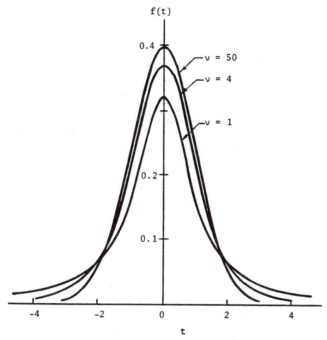

Fig. 5.2 –Plot of the Student's t *pdf* for various values of ν.

and

$\quad \Gamma(n) = $ gamma function.

Figure 5.2 shows how the peakedness of the Student's t *pdf* changes as the value of ν changes.

Some of the specific characteristics of the Student's t *pdf* are the following:

1. The Student's t *pdf* is symmetrical about $t = 0$, therefore, its mean, $\bar{t}$, median, $\check{t}$, and mode, t, are equal to 0.

2. The standard deviation, σ_t, of the Student's t *pdf* is

$$\sigma_t = (\frac{\nu}{\nu - 2})^{\frac{1}{2}}, \quad \nu > 2. \tag{5.39}$$

3. The coefficient of skewness, α_3, of the Student's t *pdf* is

$$\alpha_3 = 0. \tag{5.40}$$

4. The coefficient of kurtosis, α_4, of the Student's t pdf is given by [2, p. 4-55]

$$\alpha_4 = \frac{3(\nu - 2)}{\nu - 4}, \quad \nu > 4. \tag{5.41}$$

5. As the degrees of freedom, ν, tends to infinity, t becomes asymptotically normally distributed, with zero mean and unit variance. Therefore the normal approximation can be used to evaluate the characteristics of the Student's t distribution when ν is large, $\nu \geq 30$ say.

6. If X is a normally distributed random variable with zero mean and unit variance, and χ^2 is a random variable independent of X and has a chi-square distribution with ν degrees of freedom, then the random variable

$$t = \frac{X}{(\chi^2/\nu)^{\frac{1}{2}}} \tag{5.42}$$

is Student's t distributed with ν degrees of freedom. The usefulness of such a property in making decisions about the mean of a normal distribution when the variance is unknown is discussed in Section 5.2.3.

7. If $X_1, X_2, ..., X_{n_x}$ is a random sample of n_x independent normally distributed random variables, each having a mean, μ_x, and a variance, σ^2; if $Y_1, Y_2, ..., Y_{n_y}$ is a random sample of n_y independent normally distributed random variables, each having a mean, μ_y, and a variance, σ^2, which is the same as that of the Xs; and if all of the Xs and Ys are independent, then the random variable

$$t = \frac{(\bar{X} - \bar{Y}) - (\mu_x - \mu_y)}{[\frac{(n_x-1)S_x^2+(n_y-1)S_y^2}{n_x+n_y-2}]^{\frac{1}{2}}(\frac{1}{n_x} + \frac{1}{n_y})^{\frac{1}{2}}}, \tag{5.43}$$

is Student's t distributed with $n_x + n_y - 2$ degrees of freedom, where

$$\bar{X} = \frac{1}{n_x}\sum_{i=1}^{n_x}X_i \quad \text{and} \quad \bar{Y} = \frac{1}{n_y}\sum_{i=1}^{n_y}Y_i$$

are the sample means, and

$$S_x^2 = \frac{1}{n_x - 1}\sum_{i=1}^{n_x}(X_i - \bar{X})^2 \quad \text{and} \quad S_y^2 = \frac{1}{n_y - 1}\sum_{i=1}^{n_y}(Y_i - \bar{Y})^2$$

are the sample variances. Note that Eq. (5.43) does not contain σ [1, p. 119].

5.2.2 THE CUMULATIVE DISTRIBUTION FUNCTION AND THE PERCENTAGE POINT OF THE STUDENT'S t DISTRIBUTION

The cumulative distribution function of the Student's t variable is given by

$$F(t) = \frac{\Gamma(\frac{\nu+1}{2})}{(\pi\nu)^{\frac{1}{2}}\Gamma(\frac{\nu}{2})} \int_{-\infty}^{t} (1 + \frac{\tau^2}{\nu})^{-\frac{\nu+1}{2}} d\tau. \tag{5.44}$$

Let α be the significance level, then the percentage point, $t_{\alpha;\nu}$, is defined by

$$P(t \geq t_{\alpha;\nu}) = \frac{\Gamma(\frac{\nu+1}{2})}{(\pi\nu)^{\frac{1}{2}}\Gamma(\frac{\nu}{2})} \int_{t_{\alpha;\nu}}^{\infty} (1 + \frac{\tau^2}{\nu})^{-\frac{\nu+1}{2}} d\tau = \alpha. \tag{5.45}$$

The values of the percentage points, $t_{\alpha;\nu}$, have been tabulated for various values of ν and α, and are given in Appendix E. For example, if $\alpha = 0.05$ and $\nu = 10$, then $t_{0.05;10} = 1.812$; i.e.,

$$P(t \geq t_{0.05;10}) = P(t \geq 1.812) = 0.05.$$

Since the Student's t distribution is symmetrical about zero, the percentage points with $\alpha > 0.50$ can be found from

$$t_{\alpha;\nu} = -t_{1-\alpha;\nu}.$$

For example, if $\alpha = 0.95$ and $\nu = 10$, then $t_{0.95;10} = -1.812$; i.e.,

$$P(t \geq t_{0.95;\nu}) = P(t \geq -1.812) = 0.95.$$

If the degrees of freedom, ν, exceed 30, the value of $t_{\alpha;\nu}$ can be calculated using its normal approximation. From Section 5.1.2

$$t \sim N(0, 1), \tag{5.46}$$

as $\nu \to \infty$. Let z_α be the percentage point of the standardized normal distribution, then

$$t_{\alpha;\nu} \cong z_\alpha, \quad \text{for } \nu \geq 30. \tag{5.47}$$

For example if $\alpha = 0.05$ and $\nu = 120$ then $z_{0.05} = 1.645$ and $t_{0.05;120} = 1.658$.

5.2.3   SOME APPLICATIONS OF THE STUDENT'S t DISTRIBUTION

5.2.3.1   INTERVAL ESTIMATE OF THE MEAN OF THE NORMAL DISTRIBUTION WITH UNKNOWN STANDARD DEVIATION

Let $X_1, X_2, ..., X_n$ be independent normally distributed random variables, each having a mean, μ, and a variance, σ^2. Let $\bar{X}$ and S^2 be the sample mean and sample variance, defined by Eqs. (5.11) and (5.10), respectively. Then the random variable $(\bar{X} - \mu)\sqrt{n}/\sigma$ is normally distributed with zero mean and unit variance, and from Section 5.1.1, the random variable $(n-1)S^2/\sigma^2$ has a chi-square distribution with $n-1$ degrees of freedom. Furthermore, $\bar{X}$ and S^2 are independent random variables [1, p. 117]. From Eq. (5.42)

$$t = \frac{(\frac{\bar{X}-\mu}{\sigma/\sqrt{n}})}{[(\frac{(n-1)S^2}{\sigma^2})/(n-1)]^{\frac{1}{2}}} = \frac{(\bar{X}-\mu)}{S/n^{\frac{1}{2}}} \tag{5.48}$$

has a Student's t distribution with $(n-1)$ degrees of freedom. This random variable can be used to find the confidence interval of the mean of a normal random variable.

EXAMPLE 5–2

Seven observations are obtained from a normal distribution, and the sample mean and the sample variance are calculated, yielding $\bar{X} = 25.5$ and $S^2 = 17.6$. Find the two-sided confidence interval of the mean, μ, at a confidence level of 95%.

SOLUTION TO EXAMPLE 5–2

From Eq. (5.48)

$$P(t_{1-\frac{\alpha}{2};\nu} \leq \frac{\bar{X}-\mu}{S/\sqrt{n}} \leq t_{\frac{\alpha}{2};\nu}) = 1 - \alpha,$$

$$P(\frac{t_{1-\frac{\alpha}{2};\nu} \cdot S}{\sqrt{n}} - \bar{X} \leq -\mu \leq \frac{t_{\frac{\alpha}{2};\nu} \cdot S}{\sqrt{n}} - \bar{X}) = 1 - \alpha,$$

or

$$P(\bar{X} - \frac{t_{\frac{\alpha}{2};\nu} \cdot S}{\sqrt{n}} \leq \mu \leq \bar{X} - \frac{t_{1-\frac{\alpha}{2};\nu} \cdot S}{\sqrt{n}}) = 1 - \alpha. \tag{5.49}$$

In this example

$$\bar{X} = 25.5, \quad S^2 = 17.6 \text{ or } S = 4.195.$$

From Appendix E, with $\nu = n - 1 = 7 - 1 = 6$,

$$t_{\frac{\alpha}{2};\nu} = t_{0.025;6} = 2.447,$$

and

$$t_{1-\frac{\alpha}{2};\nu} = -t_{0.025;6} = -2.447.$$

Therefore, the lower, two-sided confidence limit on μ is given by

$$\mu_{L2} = \bar{X} - \frac{t_{\frac{\alpha}{2};\nu} \cdot S}{\sqrt{n}} = 25.5 - \frac{(2.447)(4.195)}{\sqrt{7}},$$

or

$$\mu_{L2} = 21.620.$$

The upper, two-sided confidence limit on μ is given by

$$\mu_{U2} = \bar{X} - \frac{t_{1-\frac{\alpha}{2};\nu} \cdot S}{\sqrt{n}} = 25.5 - \frac{(-2.447)(4.195)}{\sqrt{7}},$$

or

$$\mu_{U2} = 29.380.$$

Therefore, the confidence interval of the mean at a confidence level of 0.95 is

$$(21.620 \; ; 29.380).$$

5.2.3.2 COMPARISON OF TWO TREATMENTS

Experimentation is often concerned with comparing two treatments. Suppose a random sample $X_1, X_2, ..., X_{n_x}$, of n_x independent, normally distributed random variables, each having mean, μ_x, and variance, σ^2, is drawn using the standard treatment; a random sample $Y_1, Y_2, ..., Y_{n_y}$, of n_y independent, normally distributed random variables, each having mean, μ_y, and variance, σ^2(same as for the $X's$), is drawn using the new treatment; and all the $X's$ and $Y's$ are independent. Then the

comparison of these treatments is equivalent to a test of the hypothesis that μ_x is equal to μ_y; i.e.,

$$H_0 : \quad \mu_x = \mu_y. \tag{5.50}$$

Under this hypothesis, from Eq. (5.43), the random variable

$$t = \frac{\bar{X} - \bar{Y}}{[\frac{(n_x-1)S_x^2+(n_y-1)S_y^2}{n_x+n_y-2}]^{\frac{1}{2}}[\frac{1}{n_x} + \frac{1}{n_y}]^{\frac{1}{2}}}, \tag{5.51}$$

has a Student's t distribution with $\nu = n_x + n_t - 2$ degrees of freedom. Therefore, at a significance level, α, the hypothesis is accepted, if the observation of the statistic given by Eq. (5.51) falls in the acceptance region; i.e.,

$$(t_{1-\frac{\alpha}{2};n_x+n_y-2} ; \ t_{\frac{\alpha}{2};n_x+n_y-2}),$$

or

$$(-t_{\frac{\alpha}{2};n_x+n_y-2} ; \ t_{\frac{\alpha}{2};n_x+n_y-2}). \tag{5.52}$$

The hypothesis is rejected if the observed statistic, t, lies outside the above interval.

EXAMPLE 5–3

Strength tests on two types of materials used in the fabrication of printed circuit boards give the following results in pounds per square inch:

Type X : 138, 127, 134, 125.

Type Y : 134, 137, 135, 140, 130, 134.

Is there a significant difference between these two types of materials at a level of significance of $\alpha = 0.05$? Assume the variances of the strengths of these two types of materials are equal.

SOLUTION TO EXAMPLE 5–3

Assume that the strengths of these two types of materials are normally distributed. From the comments on Eq. (5.43)

$$\bar{X} = \frac{1}{4}\sum_{i=1}^{4} X_i = 131,$$

$$\bar{Y} = \frac{1}{6}\sum_{i=1}^{6} Y_i = 135,$$

$$S_x^2 = \frac{1}{4-1}\sum_{i=1}^{4}(X_i - \bar{X})^2 = 36.7,$$

and

$$S_y^2 = \frac{1}{6-1}\sum_{i=1}^{6}(Y_i - \bar{Y})^2 = 11.2.$$

Then the observed value of t_0, from Eq. (5.51), is

$$t_0 = \frac{131 - 135}{[\frac{(4-1)(36.7)+(6-1)(11.2)}{4+6-2}]^{\frac{1}{2}}[\frac{1}{4} + \frac{1}{6}]^{\frac{1}{2}}},$$

or

$$t_0 = -1.360.$$

Since the statistic t given by Eq. (5.51) has a Student's t distribution with $\nu = n_x + n_y - 2 = 4 + 6 - 2 = 8$ degrees of freedom, the acceptance region given by Eq. (5.52) can be determined as follows: From Appendix E,

$$t_{\frac{\alpha}{2};\nu} = t_{0.025;8} = 2.306.$$

Hence the acceptance region is

$$(-2.306 \; ; \; 2.306).$$

The observed value, t_0, is in the acceptance region; therefore, there is no significant difference between these two types of materials at a significance level of $\alpha = 0.05$.

5.3 THE *F* DISTRIBUTION

5.3.1 *F* DISTRIBUTION CHARACTERISTICS

The F distribution's *pdf* is given by [1, p. 120]

$$f(F) = \frac{\Gamma(\frac{\nu_1+\nu_2}{2})\nu_1^{\frac{\nu_1}{2}}\nu_2^{\frac{\nu_2}{2}}}{\Gamma(\frac{\nu_1}{2})\Gamma(\frac{\nu_2}{2})} \frac{F^{\frac{\nu_1}{2}-1}}{(\nu_2 + \nu_1 F)^{\frac{\nu_1+\nu_2}{2}}}, \quad F \geq 0, \quad (5.53)$$

where

ν_1 = first degree of freedom,

ν_2 = second degree of freedom,

and

$\Gamma(n)$ = gamma function.

Figures 5.3 and 5.4 show how the shape of the F distribution changes with ν_1 and ν_2.

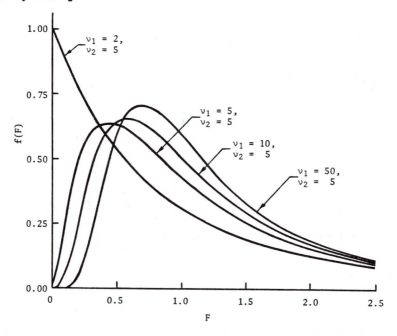

Fig. 5.3 –Plot of the F *pdf* for $\nu_2 = 5$ and various values of ν_1.

Some of the characteristics of the F distribution are the following:

1. The F distribution is skewed to the right as indicated in Figs. 5.3 and 5.4, and the *pdf* has a single mode when $\nu_2 \geq 2$.

2. The mean, $\bar{F}$, of the F distribution is given by [2]

$$\bar{F} = \frac{\nu_2}{\nu_2 - 2}, \quad \nu_2 > 2. \tag{5.54}$$

3. The mode, $\tilde{F}$, of the F distribution is given by

$$F = \frac{\nu_2(\nu_1 - 2)}{\nu_1(\nu_2 + 2)}, \quad \nu_1 \geq 2. \tag{5.55}$$

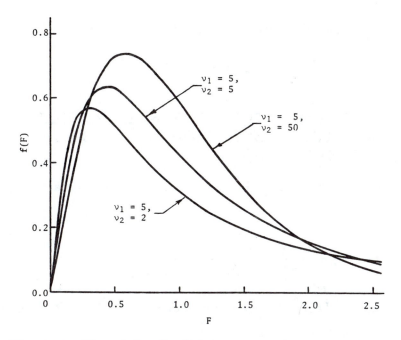

Fig. 5.4 –Plot of the F *pdf* for $\nu_1 = 5$ and various values of ν_2.

4. The standard deviation, σ_F, of the F distribution is given by

$$\sigma_F = [\frac{2\nu_2^2(\nu_2 + \nu_1 - 2)}{\nu_1(\nu_2 - 2)^2(\nu_2 - 4)}]^{\frac{1}{2}}, \quad \nu_2 > 4. \tag{5.56}$$

5. The coefficient of skewness, α_3, of the F distribution is given by

$$\alpha_3 = \frac{2\nu_2 + \nu_1 - 2}{\nu_2 - 6}[\frac{8(\nu_2 - 4)}{\nu_1 + \nu_2 - 2}]^{\frac{1}{2}}, \quad \nu_2 > 6. \tag{5.57}$$

6. If χ_1^2 and χ_2^2 are independent random variables which are chi-square distributed with ν_1 and ν_2 degrees of freedom, respectively, then the random variable

$$F = \frac{\chi_1^2/\nu_1}{\chi_2^2/\nu_2} \tag{5.58}$$

has an F distribution with ν_1 and ν_2 degrees of freedom.

7. If a random variable F has an F distribution with ν_1 and ν_2 degrees of freedom, then the random variable

$$X = \frac{\nu_2}{\nu_2 + \nu_1 F} \tag{5.59}$$

is beta distributed with parameters

$$\alpha = \frac{\nu_2}{2} - 1 \quad \text{and} \quad \beta = \frac{\nu_1}{2} - 1. \tag{5.60}$$

5.3.2   THE CUMULATIVE DISTRIBUTION FUNCTION AND THE PERCENTAGE POINT OF THE F DISTRIBUTION

The cumulative distribution function of the F random variable is given by

$$F(F) = \frac{\Gamma(\frac{\nu_1+\nu_2}{2})\nu_1^{\frac{\nu_1}{2}}\nu_2^{\frac{\nu_2}{2}}}{\Gamma(\frac{\nu_1}{2})\Gamma(\frac{\nu_2}{2})} \int_0^F \frac{t^{\frac{\nu_1}{2}-1}}{(\nu_2 + \nu_1 t)^{\frac{\nu_1+\nu_2}{2}}} dt. \tag{5.61}$$

For a fixed significance level, α, or confidence level, $1 - \alpha$, the percentage point, $F_{1-\alpha;\nu_1,\nu_2}$, is defined by

$$P(F \le F_{1-\alpha;\nu_1;\nu_2}) = \frac{\Gamma(\frac{\nu_1+\nu_2}{2})\nu_1^{\frac{\nu_1}{2}}\nu_2^{\frac{\nu_2}{2}}}{\Gamma(\frac{\nu_1}{2})\Gamma(\frac{\nu_2}{2})} \int_0^{F_{1-\alpha;\nu_1;\nu_2}} \frac{t^{\frac{\nu_1}{2}-1}}{(\nu_2 + \nu_1 t)^{\frac{\nu_1+\nu_2}{2}}} dt,$$

$$= 1 - \alpha. \tag{5.62}$$

Since $F_{1-\alpha;\nu_1;\nu_2}$ is a function of α, ν_1 and ν_2, a three-way table is needed to tabulate the value of $F_{1-\alpha;\nu_1;\nu_2}$ corresponding to different values of $(1 - \alpha)$, ν_1 and ν_2. Appendix F gives the values of $F_{1-\alpha;\nu_1;\nu_2}$ with $1 - \alpha \ge 0.50$ and various values of ν_1 and ν_2. For the cases where $\alpha < 0.50$ the following identity can be used:

$$F_{\alpha;\nu_1;\nu_2} = \frac{1}{F_{1-\alpha;\nu_2;\nu_1}}. \tag{5.63}$$

For example, the percentage point for $\alpha = 0.10$, $\nu_1 = 6$ and $\nu_2 = 10$, or $F_{0.10;6;10}$ can be obtained as follows:

$$F_{0.10;6;10} = \frac{1}{F_{0.90;10;6}} = \frac{1}{2.9369} = 0.3405.$$

5.3.3 SOME APPLICATIONS OF THE *F* DISTRIBUTION

5.3.3.1 TEST OF THE HYPOTHESIS THAT THE STANDARD DEVIATIONS OF TWO NORMAL DISTRIBUTIONS ARE EQUAL

Let $X_1, X_2, ..., X_{n_x}$ denote a random sample of size n_x, which is drawn from a normal population with unknown mean, μ_x, and unknown standard deviation, σ_x. Let $Y_1, Y_2, ..., Y_{n_y}$ denote a random sample of size n_y, which is drawn from a normal population with unknown mean, μ_y, and unknown standard deviation, σ_y. Furthermore, let all the Xs and Ys be independent. The procedure of testing the hypothesis that σ_x equals σ_y; i.e.,

$$H_0 : \quad \sigma_x = \sigma_y \tag{5.64}$$

can be determined as follows:
From Section 5.1.1,

$$\frac{(n_x - 1)S_x^2}{\sigma_x^2} \quad \text{and} \quad \frac{(n_y - 1)S_y^2}{\sigma_y^2}$$

have chi-square distributions with $\nu_x = n_x - 1$ and $\nu_y = n_y - 1$ degrees of freedom, respectively, where

$$S_x^2 = \frac{1}{(n_x - 1)} \sum_{i=1}^{n_x} (X_i - \bar{X})^2, \tag{5.65}$$

and

$$S_y^2 = \frac{1}{(n_y - 1)} \sum_{i=1}^{n_y} (Y_i - \bar{Y})^2, \tag{5.66}$$

and $\bar{X}$ and $\bar{Y}$ are the sample means of the two random samples, respectively.
Therefore, from Eq. (5.58),

$$\frac{[\frac{(n_x-1)S_x^2}{\sigma_x^2}]/(n_x - 1)}{[\frac{(n_y-1)S_y^2}{\sigma_y^2}]/(n_y - 1)} = \frac{S_x^2}{S_y^2} \frac{\sigma_y^2}{\sigma_x^2} \tag{5.67}$$

has an F distribution with $(n_x - 1)$ and $(n_y - 1)$ degrees of freedom. Under the hypothesis given by Eq. (5.64), S_x^2/S_y^2 has an F distribution with $(n_x - 1)$ and $(n_y - 1)$ degrees of freedom. If the value taken on by

the test statistic, as a result of experimentation, falls in the acceptance region, the hypothesis that $\sigma_x = \sigma_y$ is accepted, otherwise it is rejected.

If α denotes the level of significance, the probability statement

$$P(F_{\frac{\alpha}{2};n_x-1;n_y-1} \leq \frac{S_x^2}{S_y^2} \leq F_{1-\frac{\alpha}{2};n_x-1;n_y-1}) = 1 - \alpha \qquad (5.68)$$

can be used to determine the acceptance region for a two-sided test procedure; i.e.,

$$(F_{\frac{\alpha}{2};n_x-1;n_y-1} \; ; \; F_{1-\frac{\alpha}{2};n_x-1;n_y-1}),$$

and

$$(\frac{1}{F_{\frac{\alpha}{2};n_y-1;n_x-1}} \; ; \; F_{1-\frac{\alpha}{2};n_x-1;n_y-1}). \qquad (5.69)$$

EXAMPLE 5–4

Two types of materials used in the fabrication of printed circuit boards are tested, and the results given in Example 5–3 are obtained. Is there a difference between the two variances of these two types of materials at a level of significance of 0.05?

SOLUTION TO EXAMPLE 5–4

Assuming the strength of these two types of materials are normally distributed, from Example 5–3, the sample means and variances are

$$\bar{X} = 131, \quad \bar{Y} = 135, \quad S_x^2 = 36.7, \quad S_y^2 = 11.2.$$

Then the observed value of the statistic $(S_x^2/S_y^2)_0$ is

$$(S_x^2/S_y^2)_0 = \frac{36.7}{11.2} = 3.274.$$

From Appendix F,

$$F_{1-\frac{\alpha}{2};n_x-1;n_y-1} = F_{0.975;3;5} = 7.7636,$$

and

$$\frac{1}{F_{\frac{\alpha}{2};n_y-1;n_x-1}} = \frac{1}{F_{0.975;5;3}} = \frac{1}{14.885} = 0.0672.$$

Therefore, the acceptance region is

$$(0.0672 \; ; \; 7.7636),$$

and the observed value of $(S_x^2/S_y^2)_0 = 3.274$ falls in this region. Consequently, the conclusion is that the hypothesis $\sigma_x = \sigma_y$ is accepted at the significance level of 0.05.

5.3.3.2 TESTS OF COMPARISON ON THE EXPONENTIAL *PDF'S MTBF*

Another application of the F distribution is the test of comparison for the mean life of exponential units. This application is covered in detail in Chapter 8 and is illustrated by examples.

PROBLEMS

5-1. Given the sum of two independent χ^2 random variables with parameters ν_1 and ν_2, respectively, do the following:

  (1) Name the *pdf* of this sum.

  (2) Write down the *pdf* of this sum in terms of ν_1 and ν_2.

5-2. Given the sum of n independent, standardized normally distributed random variables, do the following:

  (1) Name the *pdf* of this sum.

  (2) Write down the *pdf* of this sum in terms of its parameter.

5-3. Given the sum of $x_1, x_2, ..., x_n$ independent non-standardized normal *pdfs*, do the following:

  (1) Name the *pdf* of this sum.

  (2) Write down the *pdf* of this sum in terms of the parameters of the individual non-standardized normal *pdfs*.

5-4. Choose a gamma *pdf* such that the χ^2 *pdf* is its special case. Write down explicitly each one of these two *pdfs* with numerically chosen parameters.

5-5. Show by an example that the exponential *pdf* with failure rate $\lambda = \frac{1}{2}$ is a χ^2 *pdf* with $\nu = 2$ degrees of freedom.

5-6. Show by an example that the sum of k exponential random variables with $\lambda = \frac{1}{2}$ is a χ^2 *pdf* with $\nu = 2k$ degrees of freedom.

5-7. Given the χ^2 *pdf* with $\nu = 30$, find the χ^2 value such that $P(\chi^2 > \chi^2_{0.05;30}) = 0.05$, as follows:

   (1) Using Appendix D.
   (2) Using Eq. (5.23).
   (3) Using Eq. (5.25).
   (4) Comparatively discuss the previous three results.

5-8. Find the following confidence intervals on the standard deviation of a random sample of size $n = 100$, drawn from a normal *pdf*, at a confidence level of 95%, when the sample's mean life is 25,000 hr and the standard deviation is 500 hr:

   (1) Two-sided.
   (2) One-sided lower.
   (3) One-sided upper.

5-9. Test data yielded an observed $\chi^2_o = 5.679$. At a significance level of $\alpha = 5\%$ would a three-parameter Weibull distribution represent these data well if the data were divided into nine intervals?

5-10. Given a Student's t *pdf* with degrees of freedom $\nu = 40$, do the following:

   (1) Find its mean, median and mode.
   (2) Find its variance.
   (3) Find its coefficient of variation, and interpret this result.
   (4) Derive its coefficient of skewness.
   (5) Derive its coefficient of kurtosis, and find it for this problem.

5-11. Given 20 observations from a normal *pdf* which yield $\bar{T} = 1,500$ hr and $\sigma_T = 100$ hr, find the following at a confidence level of 90%:

   (1) The two-sided confidence interval on the mean.
   (2) The one-sided, lower confidence interval on the mean life.
   (3) The one-sided, upper confidence interval on the mean life.

5-12. A sample of 20 units obtained from Manufacturer A has a normal *pdf* of lives. The life test results yield: $\bar{X} = 1,260$ hr and $\sigma_X = 16$ hr. Another sample of 30 units is drawn from Manufacturer B's production of identical units. Assuming their lives also follow a normal distribution, the test determined lives of these units yield

$\bar{Y} = 1,095$ hr and $\sigma_Y = 12$ hr. Is there a significant difference in the lives of these units obtained from these two manufacturers at a significance level of $\alpha = 0.10$?

5-13. Given an F *pdf* with $\nu_1 = 20$ and $\nu_2 = 120$, find the following for this *pdf*:

    (1) The mean.

    (2) The mode.

    (3) The standard deviation.

    (4) The coefficient of skewness.

    (5) The coefficient of kurtosis.

5-14. If the random variable χ_1^2 has 10 degrees of freedom and the random variable χ_2^2 has 20 degrees of freedom, write down the F *pdf* of the random variable

$$F = \frac{\chi_1^2/\nu_1}{\chi_2^2/\nu_2}.$$

5-15. Given a random variable that is F distributed with $\nu_1 = 10$ and $\nu_2 = 20$, write down the distribution of

$$X = \frac{\nu_2}{\nu_2 + \nu_1 F}.$$

5-16. Find $F_{0.05;10;20}$ and $F_{0.95;20;10}$.

5-17. Ultimate strength tests of a sample of 30 specimens of a special steel yield the normal *pdf* parameters of $\bar{X} = 175,000$ psi and $\sigma_X = 3,000$ psi. A change in the composition of this steel yields the normal *pdf* parameters for the ultimate strength of $\bar{Y} = 195,000$ psi and $\sigma_Y = 2,000$ psi, for a sample of 35 specimens. Is there a significant difference in the ultimate strength of these two steels? Which one would you recommend from its ultimate strength point of view?

REFERENCES

1. Bowker, A.H. and Lieberman, G.J., *Engineering Statistics*, Prentice Hall, Inc., Englewood Cliffs, N.J., 641 pp., 1972.

2. Ireson, W.G., *Reliability Handbook*, McGraw-Hill, Inc., New York, 720 pp., 1966.

3. Hahn, G.J. and Shapiro, S.S., *Statistical Models in Engineering*, John Wiley & Sons, Inc., New York, 355 pp., 1967.

4. Bury, K.V., *Statistical Models in Applied Science*, John Wiley & Sons, Inc., New York, 625 pp., 1975.

5. Lancaster, H.O., *The Chi-square Distribution*, John Wiley & Sons, Inc., New York, 356 pp., 1969.

Chapter 6

THE EXPONENTIAL DISTRIBUTION

6.1 EXPONENTIAL DISTRIBUTION CHARACTERISTICS

The exponential distribution is a very commonly used distribution in reliability and life testing, because it represents the times-to-failure distribution of components, equipment and systems exhibiting a constant failure rate characteristic with operating time. This chapter is an abbreviated version of Chapter 5 in [1], where extensive and detailed coverage of the exponential distribution is presented. The reader desiring more details should study Chapter 5 of [1]. Failures which result in a constant failure rate characteristic are called chance failures. Consequently, the times-to-failure distribution of chance failures is the exponential.

The *single-parameter exponential pdf* is

$$f(T) = \lambda e^{-\lambda T} = \frac{1}{m} e^{-\frac{1}{m}T}, \quad T \geq 0, \quad \lambda > 0, \quad m > 0, \tag{6.1}$$

where

λ = constant failure rate, in failures per unit of measurement period, e.g., failures per hour, per million hours, per million cycles, per million miles, per million actuations, per million rounds, etc.,

$\lambda = \frac{1}{m}$,

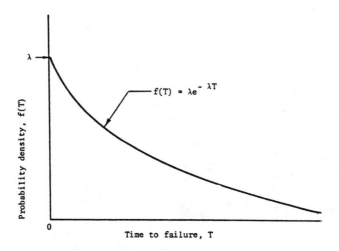

Fig. 6.1– Plot of the single-parameter exponential times-to-failure distribution.

m = mean time between failures, or mean time to a failure,

e = 2.718281828,

and

T = operating time, life, or age, in hours, cycles, miles, actuations, rounds, etc.

This distribution requires the knowledge of only one parameter, λ, for its application. Figure 6.1 illustrates Eq. (6.1).

The mean, $\bar{T}$, is

$$\bar{T} = m = \frac{1}{\lambda}, \tag{6.2}$$

the median, $\check{T}$, is

$$\check{T} = \frac{1}{\lambda}0.693 = 0.693\bar{T} = 0.693m, \tag{6.3}$$

the mode, $\tilde{T}$, is

$$\tilde{T} = 0, \tag{6.4}$$

the standard deviation, σ_T, is

$$\sigma_T = \frac{1}{\lambda} = m, \tag{6.5}$$

the coefficient of skewness, α_3, is

$$\alpha_3 = 2, \tag{6.6}$$

and the coefficient of kurtosis, α_4, is

$$\alpha_4 = 9. \tag{6.7}$$

The *two-parameter exponential pdf* is

$$f(T) = \lambda e^{-\lambda(T-\gamma)}, \quad f(T) \geq 0, \quad \lambda > 0, \quad T \geq \gamma, \tag{6.8}$$

where γ is the location parameter. This *pdf* applies to a sample or a population, which is put into operation at $T = 0$ and no failures occur until γ period of operation, but thereafter failures do occur and they are chance failures only. Figure 6.2 illustrates Eq. (6.8). The location parameter, γ, shifts the beginning of the distribution by a distance of γ to the right of the origin, signifying that the chance failures start to occur only after γ hours of operation and cannot occur before.

The mean of this *pdf* is

$$\bar{T} = \gamma + \frac{1}{\lambda} = m, \tag{6.9}$$

the median is

$$\check{T} = \gamma + \frac{1}{\lambda}0.693, \tag{6.10}$$

the mode is

$$\tilde{T} = \gamma, \tag{6.11}$$

the standard deviation is

$$\sigma_T = \frac{1}{\lambda}, \tag{6.12}$$

the coefficient of skewness is

$$\alpha_3 = 2, \tag{6.13}$$

and the coefficient of kurtosis is

$$\alpha_4 = 9. \tag{6.14}$$

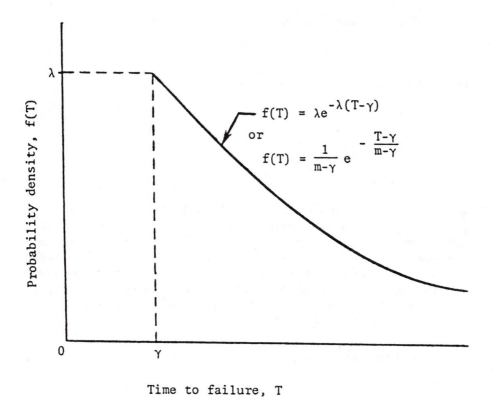

Fig. 6.2 --The two-parameter exponential distribution.

EXAMPLE 6–1

Using the grouped data in Table 6.1, do the following:

1. Fill in the table completely.

2. Find the mean failure rate.

3. Plot the number of failures versus the time-to-failure histogram and draw in the visually best fitting curve in your judgement.

4. Plot the failure-rate histogram and draw in the visually best fitting curve in your judgement.

5. Plot the reliability histogram and draw in the visually best fitting curve in your judgement.

6. Write down the frequency distribution function per class interval width.

7. Write down the frequency distribution function per unit class interval width.

8. Write down the probability density function.

9. Determine the mean life.

SOLUTIONS TO EXAMPLE 6–1

1. The average failure rate estimate for each group, given in Table 6.1 maybe calculated using Eq. (4.2), or

$$\hat{\bar{\lambda}}_i(T_i) = \frac{N_{F_i}(\Delta T)}{[N_B(T_i) - N_{F_i}(\Delta T)] \cdot \Delta T + \frac{1}{2}[N_{F_i}(\Delta T)] \cdot \Delta T}.$$

For the first group, for example, $N_{F_i}(\Delta T) = 75$, $N_B = 225$, and $\Delta T = 100$ hr. Then

$$\hat{\bar{\lambda}}_1(T_1) = \frac{75}{(225 - 75)100 + \frac{1}{2}(75)100} = 0.00400 \text{ fr/hr},$$

or

$$\hat{\bar{\lambda}}_1(T_1) = 4.00 \text{ fr/1,000 hr}.$$

The remaining average failure rate estimates are calculated similarly, and are given in Table 6.2. Note that such a calculation is not possible for the "Over 600" group, because its ΔT is not known.

TABLE 6.1– **Field data for a population exhibiting an exponentially distributed failure characteristic.**

1	2	3	4	5
Life starting at 300 hr, hr, $\Delta T = 100$ hr	Number of failures, N_F	Failure rate, $\hat{\bar{\lambda}}$, fr/1,000 hr	Number of units surviving period, N_S	Reliability, $\hat{\bar{R}} = \frac{N_S}{N_T}$
0–100	75			
100–200	50			
200–300	35			
300–400	22			
400–500	15			
500–600	10			
Over 600	18			
	$N_T = 225$			

2. The overall mean failure rate, $\hat{\bar{\lambda}}$, using the results given in Table 6.2, Column 3, is calculated from

$$\hat{\bar{\bar{\lambda}}} = \frac{1}{n} \sum_{i=1}^{n} \hat{\bar{\lambda}}_i,$$

where

$\hat{\bar{\lambda}}_i$ = average estimate of the failure rate for each group given in Column 3 of Table 6.2,

and

n = number of $\hat{\bar{\lambda}}_i$ values available, in this case six.

Substitution yields the overall mean failure rate of

$$\hat{\bar{\bar{\lambda}}} = \frac{4.00 + 4.00 + 4.24 + 4.07 + 4.23 + 4.35}{6},$$

or

$$\hat{\bar{\bar{\lambda}}} = 4.15 \text{ fr}/10^3 \text{ hr}.$$

TABLE 6.2– Field data for a population exhibiting an exponentially distributed failure characteristic.

1	2	3	4	5
Life starting at 300 hr, hr	Number of failures, N_F	Failure rate, $\hat{\bar{\lambda}}$, fr/1,000 hr	Number of units surviving period, N_S	Reliability, $\hat{\bar{R}} = \dfrac{N_S}{N_T}$
0–100	75	$\dfrac{75}{(225-75)\times100+\frac{1}{2}(75\times100)} = 4.00$	$N_S = 225 - 75,$ $N_S = 150.$	$R = \frac{150}{225},$ $R = 0.667.$
100–200	50	$\dfrac{50}{(150-50)\times100+\frac{1}{2}(50\times100)} = 4.00$	$N_S = 150 - 50,$ $N_S = 100.$	$R = \frac{100}{225},$ $R = 0.444.$
200–300	35	$\dfrac{35}{(100-35)\times100+\frac{1}{2}(35\times100)} = 4.24$	$N_S = 100 - 35,$ $N_S = 65.$	$R = \frac{65}{225},$ $R = 0.289.$
300–400	22	$\dfrac{22}{(65-22)\times100+\frac{1}{2}(22\times100)} = 4.07$	$N_S = 65 - 22,$ $N_S = 43.$	$R = \frac{43}{225},$ $R = 0.191.$
400–500	15	$\dfrac{15}{(43-15)\times100+\frac{1}{2}(15\times100)} = 4.23$	$N_S = 43 - 15,$ $N_S = 28.$	$R = \frac{28}{225},$ $R = 0.124.$
500–600	10	$\dfrac{10}{(28-10)\times100+\frac{1}{2}(10\times100)} = 4.35$	$N_S = 28 - 10,$ $N_S = 18.$	$R = \frac{18}{225},$ $R = 0.080.$
Over 600	18	—	$N_S = 18 - 18,$ $N_S = 0.$	$R = 0.$
	$N_T = 225$	$\lambda = \hat{\bar{\lambda}} = 4.15$ fr/1,000 hr		

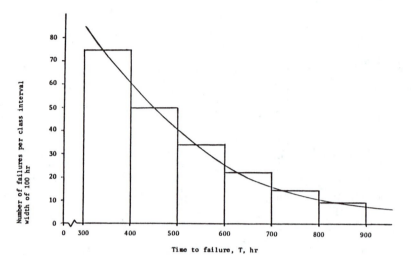

Fig. 6.3– Number of failures per class interval width, w, of 100 hr versus time-to-failure histogram and curve for Example 6–1.

3. The number-of-failures versus the time-to-failure histogram is given in Fig. 6.3, with the visually best fitting curve drawn in.

4. The failure rate histogram is given in Fig. 6.4, together with the visually best fitting curve.

5. The reliability histogram together with the visually best fitting curve is given in Fig. 6.5.

6. The frequency distribution function per class interval width, w, is

$$f(T)' = N w \lambda e^{-\lambda(T-\gamma)}, \tag{6.15}$$

where

N = 225, the number of times-to-failure observed,

w = 100 hr, the class interval width,

λ = 0.00415 fr/hr,

γ = 300 hr, the location parameter before which time no chance failures can occur,

and

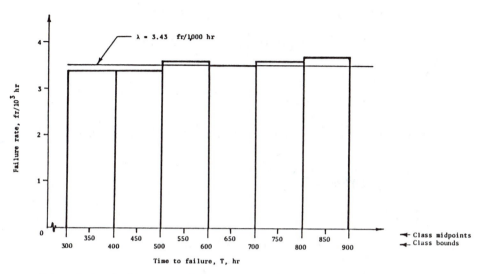

Fig. 6.4 –Failure rate histogram and curve for Example 6–1.

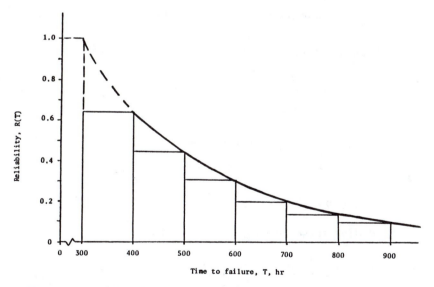

Fig. 6.5 –Reliability histogram and curve for Example 6–1.

$$T \ = \ \text{time to failure, hr.}$$

Substitution of these values into Eq. (6.15) yields

$$f(T)' = (225)(100)(0.00415) \ e^{-0.00415(T-300)},$$

or

$$f(T)' = 93.375 \ e^{-0.00415(T-300)}.$$

7. The frequency distribution function per unit class interval width is

$$f(T)'' = \frac{f(T)'}{w}, \tag{6.16}$$

where

$$w \ = \ 100 \ \text{hr, the class interval width;}$$

therefore,

$$f(T)'' = \frac{93.375 \ e^{-0.00415(T-300)}}{100},$$

or

$$f(T)'' = 0.93375 \ e^{-0.00415(T-300)}.$$

8. The probability density function is

$$f(T) = \frac{f(T)''}{N}, \tag{6.16'}$$

where

$$f(T)'' = 0.93375 \ e^{-0.00415(T-300)},$$

and

$$N = 225.$$

Substituting these into Eqs. (6.16) and (6.16') yields

$$f(T) = \frac{0.93375 \ e^{-0.00415(T-300)}}{225},$$

or

$$f(T) = 0.00415 \ e^{-0.00415(T-300)}.$$

9. The mean life of these units, from Eq. (6.9), is

$$m = \bar{T} = \gamma + \frac{1}{\lambda} = 300 + \frac{1}{0.00415},$$

or

$$m = 540.96 \text{ hr.}$$

This is the mean life measured from the time these units came out of production!

6.2 EXPONENTIAL RELIABILITY CHARACTERISTICS

The *one-parameter exponential reliability function* is

$$R(T) = e^{-\lambda T} = e^{-\frac{T}{m}}, \tag{6.17}$$

and is illustrated in Fig. 6.6. This function is the compliment of the exponential cumulative distribution function, or

$$R(T) = 1 - Q(T) = 1 - \int_0^T f(T) \, dT,$$

and

$$R(T) = 1 - \int_0^T \lambda e^{-\lambda T} \, dT = e^{-\lambda T}.$$

The reliability for a mission duration of $T = m = 1/\lambda$, or of one *MTBF* duration, is always equal to 0.3679 or 36.79%. This means that the reliability of a mission which is as long as one *MTBF* is relatively very low and is not recommended, because only 36.8% of the missions will be completed successfully, or of the equipment undertaking such a mission only 36.8% will survive their mission.

The *conditional reliability function* is

$$R(T,t) = \frac{R(T+t)}{R(T)} = \frac{e^{-\lambda(T+t)}}{e^{-\lambda T}} = e^{-\lambda t},$$

which says that the reliability for a mission of t duration undertaken after the component or equipment has already accumulated T hours of operation from age zero, is only a function of the mission duration and not a function of the age at the beginning and at the end of the

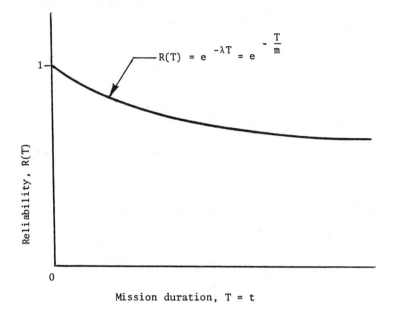

Fig. 6.6 –The one-parameter exponential reliability function.

mission, provided during the mission the failure rate remains constant at λ.

The *reliable life*, T_R, is the *mission duration* that guarantees a *desired reliability goal* as long as the units remain in their useful life until the end of the mission of T_R duration. For the single-parameter case, it is given by

$$R(T_R) = e^{-\lambda T_R},$$

$$\log_e[R(T_R)] = -\lambda T_R,$$

or

$$T_R = -\frac{\log_e[R(T_R)]}{\lambda}, \tag{6.18}$$

where

$$T_R = reliable\ life.$$

The *two-parameter exponential reliability function* is

$$R(T) = e^{-\lambda(T-\gamma)} = e^{-\frac{T-\gamma}{m-\gamma}}, \tag{6.17'}$$

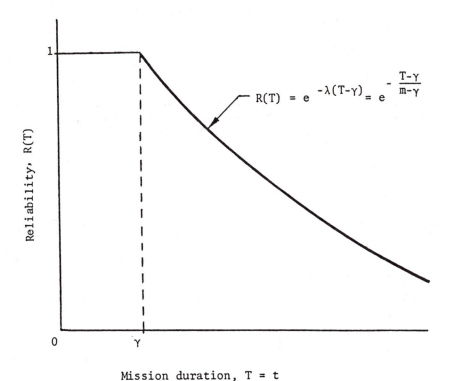

Mission duration, T = t

Fig. 6.7 –The two-parameter exponential reliability function.

and is illustrated in Fig. 6.7. The reliability for a mission of $T = m$ duration is always equal to 0.3679 or 36.79%.

The *conditional reliability function* for the two-parameter case is

$$R(T, t) = \frac{R(T + t)}{R(T)} = \frac{e^{-\lambda(T+t-\gamma)}}{e^{-\lambda(T-\gamma)}} = e^{-\lambda t},$$

and the reliable life, T_R, that guarantees a desired reliability goal, $R(T_R)$, is given by

$$T_R = \gamma - \frac{\log_e[R(T_R)]}{\lambda}, \tag{6.19}$$

when starting the operating period at age zero, as long as the units remain in their useful life until the end of their reliable life.

The mission duration, T_R, that guarantees a desired reliability goal, $R(T_R)$, after age T is given by

$$T_R = \begin{cases} -\frac{\log_e[R(T_R)]}{\lambda}, & \text{if } T \geq \gamma; \\ (\gamma - T) - \frac{\log_e[R(T_R)]}{\lambda}, & \text{if } T < \gamma, \end{cases} \qquad (6.19')$$

where T is the age of the unit. The first equation is used when starting the mission at any age, T, after age γ, and the second equation is used when starting the mission at any age, T, before age γ.

EXAMPLE 6–2

A component's useful life chance failure rate is 0.5 fr/10^6 hr. Do the following:

1. Determine its reliability function and plot it.

2. Determine the probability density function and plot it on the same plot as that of Case 1.

3. Determine the failure rate function and plot it on the same plot as that of Case 1.

SOLUTIONS TO EXAMPLE 6–2

1. The reliability function is

$$R(T) = e^{-\lambda T},$$

where $\lambda = 0.5$ fr/10^6 hr or 5×10^{-7} fr/hr; therefore

$$R(T) = e^{-5 \times 10^{-7} T},$$

where T is in hours. This function, for various values of mission time $t = T$, is shown plotted in Fig. 6.8 from the results in Table 6.3.

2. The probability density at time T, or $f(T)$, calculated from

$$f(T) = 5 \times 10^{-7} e^{-5 \times 10^{-7} T},$$

is given in Table 6.3 and is plotted in Fig. 6.8, along with the failure rate.

TABLE 6.3– Data for plotting the reliability function for Example 6–2.

Time to failure, hr	Reliability, $R(T)$	Probability density, $f(T)$	Failure rate, $\lambda(T)$, fr/10^6 hr
100	0.99995	4.99975×10^{-7}	0.5
200	0.99990	4.99950×10^{-7}	0.5
500	0.99975	4.99875×10^{-7}	0.5
1,000	0.99950	4.99750×10^{-7}	0.5
2,000	0.99900	4.99500×10^{-7}	0.5
5,000	0.99750	4.98752×10^{-7}	0.5
100,000	0.95123	4.97506×10^{-7}	0.5

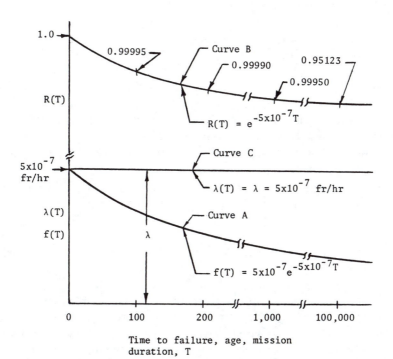

Fig. 6.8– The plot of the exponential reliability, probability density and failure rate functions for Example 6–2.

3. The failure rate function is

$$\lambda(T) = \frac{f(T)}{R(T)} = \frac{\lambda e^{-\lambda T}}{e^{-\lambda T}} = \lambda = 0.5 \text{ fr}/10^6 \text{ hr.}$$

This failure rate is also plotted in Fig. 6.8.

6.3   EXPONENTIAL FAILURE RATE AND MEAN-TIME-BETWEEN-FAILURES CHARACTERISTICS

The exponential failure rate function is

$$\lambda(T) = \frac{f(T)}{R(T)} = \frac{\lambda e^{-\lambda(T-\gamma)}}{e^{-\lambda(T-\gamma)}} = \lambda, \tag{6.20}$$

which supports the fundamental premise that the exponential distribution represents the case when the failure rate is constant with age at the value of λ. It is illustrated in Fig. 6.9 for the one-parameter exponential case and in Fig. 6.10 for the two-parameter exponential case.

It must be pointed out that the mean time between failures, or the mean life, for the exponential distribution, is given by

$$\bar{T} = \int_0^\infty T\lambda e^{-\lambda T}\,dT = \frac{1}{\lambda} = m = MTBF, \tag{6.21}$$

which says that the mean life is given by the reciprocal of the constant failure rate. Under no other situation the reciprocal of the failure rate gives the $MTBF$ or vice versa. It is true only when we are looking at identical units of whatever simplicity or complexity which exhibit a failure rate characteristic which is constant with age, life, or operating time for the period of concern and usage. To obtain, therefore, the mean life, or $MTBF$, for the chance failures and useful life situations all that is necessary to do is to determine the failure rate first and then take its reciprocal, but only when $\gamma=0$. If $\gamma \neq 0$, then the mean is obtained from Eq. (6.9).

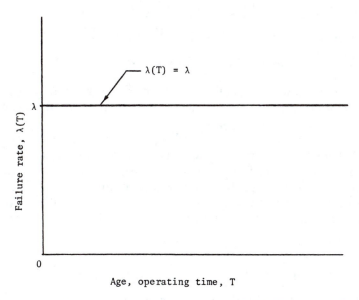

Fig. 6.9 –The one-parameter exponential failure rate function.

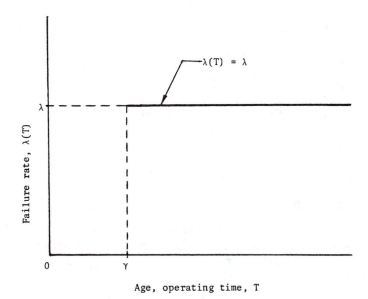

Fig. 6.10 –The two-parameter exponential failure rate function.

6.4 DETERMINATION OF THE EXPONENTIAL FAILURE RATE AND *MTBF* FROM INDIVIDUAL TIME-TO-FAILURE DATA

When relatively few exponential units are on test, or in the field, or in use, the following procedures are recommended to determine the *MTBF* estimate $\hat{m}$.

If N units are tested, k units fail and are replaced to operate until they fail again and are replaced again to operate until the test is terminated; and n units fail at times T_i and are not replaced, and the test is terminated at the rth failure at which time t_d test clock hours have elapsed; then the accumulated unit-hours of operation, T_a, of the failed and nonfailed units are given by

$$T_a = \sum_{i=1}^{n} T_i + (N - n)t_d, \tag{6.22}$$

and the total number of failures, r, is given by

$$r = k + n. \tag{6.23}$$

Consequently $\hat{m}$ is given by

$$\hat{m} = \frac{T_a}{r} = \frac{\sum\limits_{i=1}^{n} T_i + (N - n)t_d}{k + n}. \tag{6.24}$$

The total number of units participating in this test is

$$N' = N + k, \tag{6.25}$$

assuming that the replacements are new units.

EXAMPLE 6–3

Consider the results of the reliability test given in Fig. 6.11 where nine units are tested, six units fail and are replaced immediately, and three units fail and are not replaced. Determine $\hat{m}$ and the number of units that participated in this test when $t_d = 700$ hr.

SOLUTION TO EXAMPLE 6–3

From Fig. 6.11, $N = 9$, $k = 6$, $n = 3$, $r = k + n = 6 + 3 = 9$, $T_1 = 530$ hr, $T_2 = 650$ hr and $T_3 = 700$ hr. Substitution of these

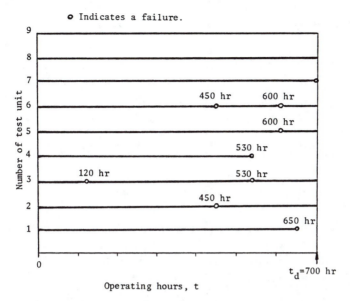

Fig. 6.11-- Hours-to-failure results obtained from a reliability test of nine units where Units 2, 3, 5, and 6 are replaced when they fail and Units 1 and 4 are not replaced when they fail in a failure-terminated, mixed replacement and nonreplacement test, for Example 6-3.

quantities into Eq. (6.24) yields

$$\hat{m} = \frac{(530 + 650 + 700) + (9 - 3)700}{6 + 3} = \frac{6,080}{9},$$

or

$$\hat{m} = 675.6 \text{ hr.}$$

In this test $N' = N + k = 9 + 6 = 15$ units participated, assuming the replacements were new units.

6.5 DETERMINATION OF THE EXPONENTIAL FAILURE RATE AND *MTBF* BY PROBABILITY PLOTTING

Another quick and efficient method of determining $\hat{m}$ and $\hat{\lambda}$ for the exponential *pdf* is the probability plotting method. This is based on

the fact that if the logarithm of

$$R(T) = e^{-\lambda T} = e^{-\frac{T}{m}} \tag{6.26}$$

is taken, then

$$\log_e R(T) = -\lambda T. \tag{6.27}$$

If we set $\log_e R(T) = y$ and $T = x$ then Eq. (6.27) becomes a straight line, or

$$y = -\lambda x.$$

It may be seen that if $\log_e R(T)$ is plotted versus T, a straight line will result. A straight line will also result when a semilogarithmic plotting paper, like that given in Fig. 6.12, is used, where $R(T)$ is plotted along the logarithmic ordinate scale and T is plotted along the linear abscissa scale. If the data plotted on such a paper fall acceptably well on a straight line, as is the case in Fig. 6.13, then the exponential distribution represents the data well.

From this plot $\hat{m}$ may be determined by entering the plot at the $R = 36.8\%$ level, going to the straight line fitted to the plotted data, and then dropping down vertically from the intersection and reading the value of T along the abscissa. This value is $\hat{m}$, and $\hat{\lambda} = 1/\hat{m}$. This is so because for $T = \hat{m}$ the right side of Eq. (6.26) becomes

$$R = e^{-\frac{\hat{m}}{\hat{m}}} = e^{-1} = 0.368, \quad \text{or} \quad 36.8\%.$$

For probability plotting the times-to-failure data are ranked; i.e., they are listed in increasing value. This has been done in Table 6.4, Column 2 and Table 6.5, Column 2. To be able to plot the data on the exponential probability plotting paper the plotting position for $R(T)$ above the corresponding value of T is needed. The best estimate of $R(T)$, $\hat{R}(T)$, is obtained from

$$\hat{R}(T) = 1 - \hat{Q}(T) = 1 - MR, \tag{6.28}$$

where

$\hat{Q}(T)$ = estimate of the unreliability,

and

MR = median rank obtained from the Median Rank tables in Appendix A.

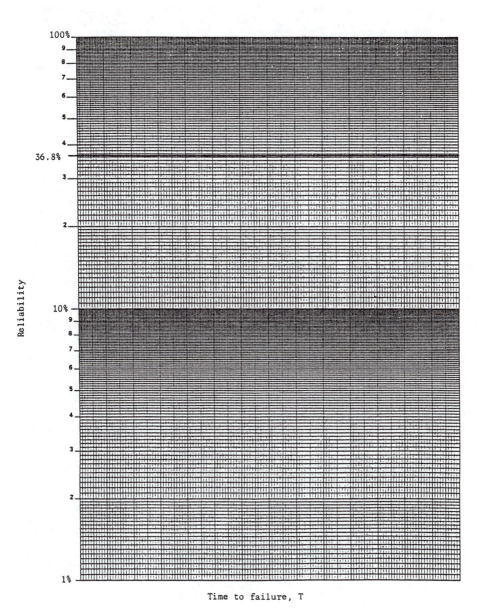

Fig. 6.12– Semilogarithmic plotting paper for the exponential distribution.

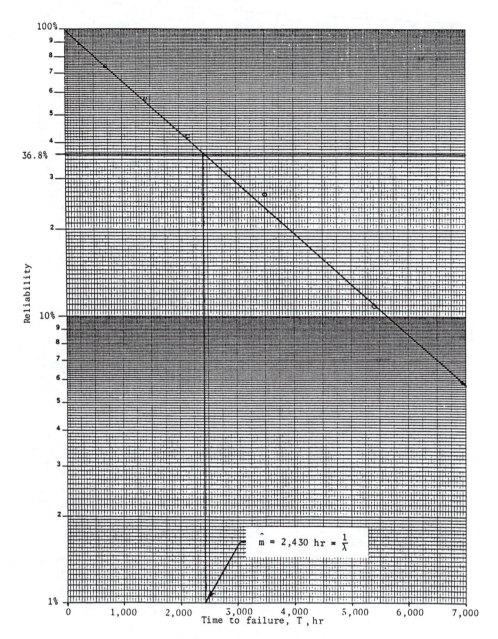

Fig. 6.13– Probability plot of the exponential data given in Tables 6.4 and 6.5.

TABLE 6.4– Ungrouped exponential time-to-failure data.

Failure order number, j	Time to failure, T_j, hr
1	250
2	710
3	1,400
4	2,150
5	3,400
6	5,400

The median rank provides the best, or the 50% confidence level, estimate of $Q(T)$.

For ungrouped data plot the $(1 - MR)$ value of each time to failure directly above the corresponding time to failure value. The MR values are given in the Median Rank tables *for the sample size tested* in the column headed by the value 50.0. For grouped data plot the $(1 - MR)$ value of each class, based on the cumulative number of failures which have occurred until the end of the group being considered, above the endpoint of the class interval, because reliability is determined for the end of the operating period. This procedure also enables the Kolmogorov-Smirnov goodness-of-fit test to be applied. See Chapter 20. This requires that the expected and observed values of $Q(T)$ at the endpoint of each class be compared. Finally, additional goodness-of-fit tests may be applied to determine whether the exponential is indeed the best distribution to use for the data being analyzed. See Chapters 19 and 20.

The reliability estimates listed in Column 3 of Table 6.5 may be obtained directly from the Median Rank tables for the sample size tested, in this case the sample size is six, by reading the values given in the column headed by 50.0 *from the bottom up*, because the two MR values equidistant from the middle value in this column add up to 100, or 100%. For example in the Sample Size 6 section of the Median Rank tables the two middle values 42.14 and 57.86 add up to 100.00, and the first and last values 10.91 and 89.09 add up to 100.00, consequently, $100.00 - 10.91 = 89.09$. But $R(T)\% = 100\% - Q(T)\%$ and $\hat{Q}(T)\% = MR\%$. Consequently the bottom-up values give directly the reliability estimate being sought.

TABLE 6.5– Ungrouped exponential time–to–failure data, median ranks, and reliability estimates for the data in Table 6.4.

1	2	3
Failure order number, j	Time to failure, T_j, hr	$\hat{R}(T_j) = 1 - MR_j$, Reliability estimate, %
1	250	89.09
2	710	73.56
3	1,400	57.86
4	2,150	42.14
5	3,400	26.44
6	5,400	10.91

6.6   A BETTER ESTIMATE OF RELIABILITY

If all units in a test are exponential, they are all tested to failure, and the mission time is less than the unit-hours accumulated by all units in the test, then a better estimate of the reliability, $\hat{R}(T)$, is obtained from [2]

$$R(t) = (1 - \frac{t}{T_a})^{N-1}, \quad t < T_a, \tag{6.29}$$

where

t = mission time, hr,

T_a = unit hours of operation accumulated by all units in the test, hr,

and

N = number of units in the test, with all N units tested to failure.

EXAMPLE 6–4

The exponential time-to-failure data given in Table 6.5 are obtained from six units tested to failure. Do the following:

1. Find the reliability estimates, $\hat{R}(T_j)$.

2. Plot the $\hat{R}(T_j)$ versus T_j, and draw the best possible line through these points.

3. Find $\hat{\lambda}$ and $\hat{m}$.

4. Write down the *pdf*, $f(T)$.

5. Write down the reliability function, $R(T)$.

6. Find the reliability for a mission of 100 hr from age zero.

7. Find the reliability for an additional mission of 100 hr duration.

8. Find the duration of the next mission to assure a reliability of 0.980.

9. Find the best reliability estimate using Eq. (6.29) for a mission of 100 hr, and compare it with the result in Case 6.

SOLUTIONS TO EXAMPLE 6–4

1. The reliability estimates are tabulated in Column 3 of Table 6.5. These values may also be obtained directly by reading the values for $N=6$ and 50.0 column values from bottom up in Appendix A, and entering them into Column 3 from the top down.

2. Plot the $\hat{R}(T_j)$ versus the T_j values, as shown in Fig. 6.13, and draw the best straight line through these points. It appears that the exponential distribution fits the data quite well.

3. To find $\hat{\lambda}$, enter the ordinate with $R = 36.8\%$, go to the fitted straight line, drop down vertically at the intersection, and read off $T=2,430$ hr, as shown. Then $\hat{m}=2,430$ hr and $\hat{\lambda} = 1/\hat{m} = 1/2,430 = 0.0004115$ fr/hr.

   To find $\hat{\lambda}$ another way, determine the slope of the fitted straight line and take the absolute value of this slope. The slope of the straight line in Fig. 6.13 may be calculated from

   $$\text{slope} = \frac{\Delta y}{\Delta x} = \frac{y_2 - y_1}{x_2 - x_1},$$

   using the $\hat{R}(T = 2,430 \text{ hr}) = 0.3679$ and the $\hat{R}(T = 0 \text{ hr}) = 1$ points, or

   $$\text{slope} = \frac{\log_e(R = 0.3679) - \log_e(R = 1)}{(T = 2,430) - 0} = \frac{-1.000 - 0}{2,430}.$$

or

$$\text{slope} = -0.0004115.$$

Therefore,

$$\hat{\lambda} = |\text{slope}| = |-0.0004115| = 0.0004115 \ \text{fr/hr},$$

or

$$\hat{\lambda} = 411.5 \ \text{fr}/10^6 \ \text{hr}.$$

Consequently,

$$\hat{m} = \frac{1}{\hat{\lambda}} = \frac{1}{0.0004115},$$

or

$$\hat{m} = 2,430 \ \text{hr}.$$

4. The *pdf* is

$$f(T) = \lambda e^{-\lambda T},$$

or

$$f(T) = 0.0004115 \ e^{-0.0004115T}.$$

5. The reliability function is

$$R(T) = e^{-\lambda T},$$

or

$$R(T) = e^{-0.0004115T}.$$

6. For a mission of 100 hr, using Fig. 6.13, $R=0.96$. Using the previous equation

$$R(T = 100 \ \text{hr}) = e^{-0.0004115 \times 100} = e^{-0.04115},$$

or

$$R(T = 100 \ \text{hr}) = 0.95969.$$

7. The reliability for an additional mission of 100 hr is given by

$$R(T,t) = \frac{R(T+t)}{R(T)},$$

$$R(T = 100 \text{ hr}, t = 100 \text{ hr}) = \frac{R(100+100)}{R(100)},$$

$$R(T,t) = \frac{e^{-0.0004115(100+100)}}{e^{-0.0004115(100)}},$$

$$R(T,t) = e^{-0.0004115 \times 100} = e^{-\lambda t} = R(t),$$

$$R(T,t) = e^{-0.04115},$$

or

$$R(T = 100 \text{ hr}, t = 100 \text{ hr}) = 0.95969.$$

Consequently, when the failure rate is constant the reliability for each successive mission of the same duration has the same value, as may be seen by comparing the results of Cases 6 and 7.

8. To assure a reliability of 0.980 with 50% confidence, the mission duration, from Eq. (6.18), should not exceed

$$T_R = -\frac{\log_e(R = 0.980)}{\lambda} = -\frac{0.020203}{0.0004115},$$

or

$$T_R = 49.1 \text{ hr}.$$

9. In Eq. (6.29) $t=100$ hr and $T_a=13,310$ hr, therefore $t < T_a$ and Eq. (6.29) is applicable. Therefore,

$$R(t = 100 \text{ hr}) = (1 - \frac{100}{13,310})^{6-1} = (1 - 0.00751)^5,$$

or

$$R(t = 100 \text{ hr}) = 0.96301.$$

This compares with $R(T = 100 \text{ hr}) = 0.95969$ obtained in Case 6.

EXAMPLE 6–5

Given are the following time-to-failure data obtained from seven units put to a reliability test:

Failure order number	Time to failure, hr
1	450
2	760
3	1,200
4	1,590
5	2,210

Using the *exponential probability plotting paper*, do the following:

1. Find the numerical value of the estimates of the mean life and of the location parameter of the *pdf* of these units.

2. Write down the probability density function for the time to failure of these units.

3. Write down the associated reliability function for these units.

4. Determine graphically the reliability for a mission of 500 hr duration, and also analytically using the results of Case 1.

5. Find the mission duration for a reliability of 90%. Find it from the probability plot and also analytically using the results of Case 3.

6. Find the median life.

7. Find the modal life.

8. Find the standard deviation of the times to failure.

SOLUTIONS TO EXAMPLE 6–5

1. From the Median Rank table for $N = 7$, the following table is prepared:

Failure order number	Time to failure, T, hr	$\hat{R}(T)$, %
1	450	90.572
2	760	77.151
3	1,120	63.588
4	1,590	50.000
5	2,210	36.412

Plotting the $\hat{R}(T)$ values versus the T values in Fig. 6.14 yields the estimate of the mean, or

$$\hat{\bar{T}} = 2,170 \text{ hr},$$

and of the location parameter, or

$$\hat{\gamma} = 290 \text{ hr}$$

from the intercept of the fitted straight line at the top of the plot and the corresponding $\hat{\gamma}$ value read off along the x-axis.

2. The *pdf* is

$$f(T) = \lambda e^{-\lambda(T-\gamma)},$$

where

$$\hat{\lambda} = \frac{1}{\hat{\bar{T}} - \hat{\gamma}} = \frac{1}{2,170 - 290} = \frac{1}{1,880} = 0.00053191 \text{ fr/hr.}$$

Then the *pdf* is

$$f(T) = 0.00053191 e^{-0.00053191(T-290)}.$$

3. The reliability function is

$$R(T) = e^{-\lambda(T-\gamma)} \text{ for } T \geq \gamma,$$

or

$$R(T) = e^{-0.00053191(T-290)} \text{ for } T \geq 290 \text{ hr.}$$

4. Graphically

$$R(500 \text{ hr}) = 88.5\%.$$

Analytically

$$R(500 \text{ hr}) = e^{-0.00053191(500-290)} = e^{-0.1117011} = 0.89431153,$$

or

$$R(500 \text{ hr}) = 89.43\%.$$

The results are relatively close to each other.

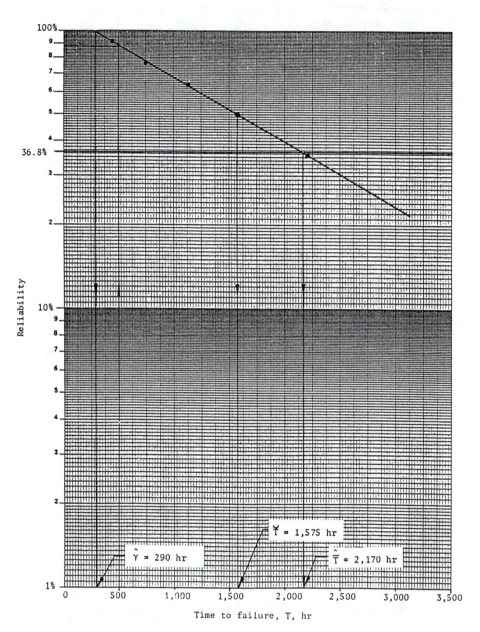

Fig. 6.14 –Probability plot of the data for Example 6–5.

5. The mission duration for a reliability of 90% is $T_R = 465$ hr, *graphically*, and

$$T_R = -\frac{\ln R(T_R)}{\lambda} + \gamma,$$

or

$$T_R = -\frac{\ln(0.90)}{0.00053191} + 290 = 488.1 \text{ hr, } \textit{analytically.}$$

6. From Fig. 6.14, the median life is obtained by entering the ordinate at the 50% level and dropping down vertically from the intersection with the fitted straight line and reading off the corresponding abscissa value. Then

$$\check{T} = 1,575 \text{ hr.}$$

Analytically

$$R(\check{T}) = 0.50 = e^{-(\frac{\check{T}-\gamma}{\bar{T}-\gamma})},$$
$$\check{T} = -(\log_e 0.50)(\bar{T} - \gamma) + \gamma,$$
$$\check{T} = -(\log_e 0.50)(2,170 - 290) + 290,$$

or

$$\check{T} = 1,593 \text{ hr.}$$

7. The mode of the time to failure is

$$\tilde{T} = \gamma = 290 \text{ hr.}$$

8. The standard deviation is

$$\sigma_T = \frac{1}{\lambda} = \frac{1}{0.00053191} = 1,880 \text{ hr.}$$

Also

$$\sigma_T = \bar{T} - \gamma = 2,170 - 290 = 1,880 \text{ hr.}$$

PROBLEMS

6-1. A valve has a useful life failure rate of 50 fr/10^6 hr.

(1) What is its reliability during a mission of 16 hr?

(2) If there were 10,000 such valves being used, how many shall survive a 16-hr mission?

(3) How many shall survive a 10-hr mission?

(4) How many will fail from $t = 10$ hr to $t = 16$ hr?

6-2. Three hundred airplanes fly 40 hr a month each. There are three identical electrohydraulic actuators in the flight control system of each airplane.

(1) If all of these actuators experience three malfunctions per month and have a constant failure rate, what is their *MTBF* if the malfunctioning actuators are replaced immediately?

(2) What is their failure rate in failures per million actuator-hours?

6-3. A certain radar system has an *MTBF* of 200 hr.

(1) What is the reliability for an operating time of $2\frac{1}{2}$ hr?

(2) What is the reliability for an operating time of 10 hr?

(3) What would be the maximum allowable time of operation if a reliability of 0.99 is required?

6-4. A valve has been designed for a mean time between failures of 2,000 hr.

(1) What is the failure rate in failure per million hours and in % failure/1,000 hr?

(2) What is the reliability for a 10-hr mission and for a 24-hr mission?

(3) What improvement in *MTBF* is necessary if a reliability requirement for a 24-hr mission of 0.995 must be met?

6-5. Given is a component with a mean time between failures of 100,000 hr.

(1) What is the component's reliability for a mission time of one hour?

(2) What is the reliability for a mission time of one hour for 10 components in series?

(3) What is the significance of the slope of the exponential reliability curve at $t = 0$?

(4) What is the reliability for a mission of 10 hr from an age of 1,000 hr and also from an age of 2,000 hr during its useful life?

(5) What is the reliability for an age of 2,010 hr starting new and debugged and staying within the useful life?

6-6. A component's useful life failure rate is 20 fr/10^6 hr.

(1) Write down its failure density function.

(2) Write down its failure rate function.

(3) Write down its reliability function for mission t.

(4) Write down its reliability function for a mission t, starting the mission at age T.

(5) Sketch these four functions comparatively.

6-7. A pump shaft has a failure rate of 1.50 fr/10^6 hr, and a mechanical shaft seal has a failure rate of 5.00 fr/10^6 hr. There are 10,000 pumps in the field every year, each operating 8,000 hr on the average per year, during the pump's useful life. How many shafts and seals will be replaced per year for all pumps?

(1) Solve it by assuming immediate replacement.

(2) Solve it by assuming a decaying population.

6-8. Given is a test sample of 100 units from the same production.

(1) What is the reliability and failure rate if, after one hour of testing, six units fail?

(2) What is the reliability and failure rate at each failure period if all that failed above are replaced and one more fails after the first hour, two fail after two hours, and three fail after 10 hours? All times are from zero time and all failed items are replaced.

(3) Find the reliability and failure rate in Case 2 if the failed units are not replaced.

(4) Discuss comparatively all results.

6-9. A centrifugal pump has one shaft with a failure rate of 2.5 hr/10^6 hr, and two mechanical shaft seals one near each end of the shaft. Each seal has a failure rate of 10 fr/10^6 hr. Assume that when the shaft fails it does not cause other pump components to fail.

TABLE 6.6 – Field data for a population exhibiting an exponentially distributed failure characteristic for Problem 6–10.

Life starting at 200 hr, hr	Number of failures, N_F	Failure rate, $\hat{\bar{\lambda}}$, fr/10^6 hr	Number of units surviving period, N_S	Reliability, $\hat{\bar{R}} = \frac{N_S}{N_T}$
0 – 100	30			
100 – 200	22			
200 – 300	15			
300 – 400	10			
400 – 500	7			
500 – 600	5			
over 600	11			
	$N_T = 100$			

There will be 10,000 pumps in the field next year and each one will operate 5,000 hr on the average per year during its useful life.

How many shafts and seals will be required as spares, on the average, for next year?

(1) Solve it by assuming immediate replacement.

(2) Solve it by assuming a decaying population.

6-10. Field data were obtained and grouped as given in Table 6.6, Columns 1 and 2. Do the following:

(1) Fill in the table completely to obtain the failure rate and the reliability of this equipment.

(2) Plot the failures versus the time-to-failure histogram and draw in the best fitting curve in your judgement.

(3) Plot the failure rate versus the time-to-failure histogram and draw in the best fitting curve in your judgement.

(4) Plot the reliability versus the mission-time histogram and draw in the best fitting curve in your judgement.

6-11. Ten pumps are put to an operational test of 1, 500 clock hours. During the test, two pumps fail after operating 400 hr, a third pump experiences an apparent indication of some abnormal performance after 750 hr of operation, but as it is not certain whether or not a reliability failure had occurred, testing on it is continued until the end of the reliability test. A failure analysis shows that this *was* a reliability failure. A fourth pump fails after 850 hr and is replaced by a new one. A fifth pump fails after 1,250 hr of testing. The malfunction causing the failure is corrected and the pump is put back into operation immediately. What is the failure rate estimate of such pumps? (All hours given are from the beginning of the test.) Assume these pumps have essentially a constant failure rate.

6-12. Ten pumps are put to a life test of 1,000 clock hours. During the test, two pumps fail after operating 400 hr, and are removed from the test, a third malfunctions after 350 hr but is kept in the test. A failure analysis shows that this was a reliability failure. A fourth pump fails after 350 hr and is replaced by a new one. A fifth pump malfunctions after 250 hr of test, the malfunction is corrected, and the pump is put back into operation immediately.

(1) What is the mean life estimate of these pumps?

(2) What is the failure rate of these pumps in $fr/10^6$ hr?

6-13. One hundred components, which come from a burned-in and de-bugged population, are subjected to a useful-life reliability test of 2,000 clock hours duration. The following test results are logged: Two failures occur after 500 clock hours; five failures occur after 800 clock hours; and three failures occur after 1,500 clock hours. All hours given here are from the beginning of the test.

(1) What is these components' *MTBF*, if this were a replacement test?

(2) What is these components' *MTBF*, if this were a nonreplacement test?

6-14. Given are the following time-to-failure data obtained from seven units put to a reliability test:

Failure order number	Time to failure, hr
1	150
2	460
3	820
4	1,290
5	1,910

Using the *exponential probability plotting paper,* do the following:

(1) Find the numerical value of the estimate of the mean time between failures from the probability plot, and the failure rate.

(2) Write down the probability density function for the times to failure for these units.

(3) Write down the reliability function for these units.

(4) Determine the reliability for a mission of 300 hr duration from the probability plot.

(5) Determine the reliability for a mission of 300 hr duration using the equation of Case 3.

(6) What should the mission duration be for a reliability of 95%?

6-15. Fifteen items were put to a 1,000-hr life test and a total of ten failures were observed after the following hours of operation were completed: 50, 110, 200, 270, 350, 450, 550, 650, 850, and 1,000.

(1) What is the *two-parameter exponential distribution* which fits these data, using *probability plotting?*.

(2) What should the mission duration be for a reliability of 98.5%, starting the mission at age zero?

(2.1) As determined from the probability plot.

(2.2) As determined by calculation.

(3) What should the mission duration be for a reliability of 95% starting the mission at the age of 250 hr?

6-16. Given are the following time-to-failure data obtained from nine units put to a reliability test:

Failure order number	Time to failure, hr
1	320
2	630
3	995
4	1,450
5	2,090

Using the *exponential probability plotting paper,* do the following:

(1) Find the numerical value of the estimate of the mean time between failures from the probability plot, and the failure rate.

(2) Write down the probability density function for the times to failure for these units.

(3) Write down the reliability function for these units.

(4) Determine the reliability for a mission of 500 hr duration from the probability plot.

(5) What should the mission duration be, in hours, for a reliability of 99%? Find it from the probability plot and also analytically using the results of Case 3.

6-17. Given are the following times-to-failure data obtained from *seven* units put to a reliability test:

Failure order number	Time to failure, hr
1	250
2	560
3	920
4	1,390
5	2,010

Using the *exponential probability plotting paper,* do the following:

(1) Find the numerical value of the estimates of the mean life and of the location parameter of the *pdf* for these units.

(2) Write down the probability density function for the times to failure for these units.

(3) Write down the associated reliability function for these units.

(4) Determine graphically the reliability for a mission of 500 hr duration, and also analytically using the results of Case 1.

(5) What should the mission duration be, in hours, for a reliability of 99%? Find it from the probability plot and also analytically using the results of Case 3.

6-18. Fifteen items were put to a 1,800-hr life test and a total of ten failures were observed after the following hours of operation were completed: 800, 860, 950, 1,020, 1,100, 1,200, 1,300, 1,400, 1,500, and 1,800 hr.

(1) What are the two parameters of the exponential distribution which fits these data, found using probability plotting?

(2) What should the mission duration be for a reliability of 98.5%, starting the mission at age zero?

(2.1) As determined from the probability plot.

(2.2) As determined by calculation using the two parameters found previously.

(3) What should the mission duration be for a reliability of 95% starting the mission at the age of 850 hr?

6-19. Given are the times-to-failure data of Table 5.4, where the sample size is seven. Do the following:

Failure order number	Time to failure, hr
1	480
2	790
3	1,070
4	1,640
5	2,010

(1) Determine the probability plot of these data.

(2) Find $\hat{m}$ and $\hat{\gamma}$.

(3) Find $\hat{\lambda}$.

(4) Write down the $f(T)$.

(5) Write down the $R(T)$.

(6) Find the reliability for a mission of 100 hr starting the mission at the age of 500 hr.

(7) Find the reliability for an additional mission of 100 hr.

(8) Find the duration of the next mission to assure a reliability of 0.980.

6-20. Given are the following times-to-failure data obtained from *nine* units put to a reliability test:

Failure order number	Time to failure, hr
1	155
2	450
3	830
4	1,280
5	1,910

Using the *exponential probability plotting paper*, do the following:

(1) Find the numerical value of the estimate of the mean time between failures graphically, assuming $\gamma = 0$.

(2) Write down the probability density function for the times to failure for these units.

(3) Write down the associated reliability function for these units.

(4) Determine graphically the reliability for a mission of 200 hr duration.

(5) What should the mission duration be, in hours, for a reliability of 99%?

6-21. Given are the following times-to-failure data obtained from *seven* units put to a reliability test:

Failure order number	Time to failure, hr
1	270
2	585
3	940
4	1,410
5	2,030

Using the *exponential probability plotting paper*, do the following:

(1) Find the numerical value of the estimates of the mean life and of the location parameter of the *pdf* for these units.

(2) Write down the probability density function for the times to failure for these units.

(3) Write down the associated reliability function for these units.

(4) Determine graphically the reliability for a mission of 500 hr duration, and also analytically using the results of Case 1.

(5) What should the mission duration be, in hours, for a reliability of 99%? Find it from the probability plot and also analytically using the results of Case 3.

REFERENCES

1. Kececioglu, Dimitri B., *Reliability Engineering Handbook*, Prentice Hall, Inc., Englewood Cliffs, New Jersey, Vol. 1, 720 pp. and Vol. 2, 568 pp., 1991.

2. Pugh, E. L., "The Best Estimate of Reliability in the Exponential Case," Operations Research, Vol. 11, pp. 57–61, 1963.

Chapter 7

$MTBF$ CONFIDENCE LIMITS, TEST TIME AND OC CURVES FOR THE EXPONENTIAL CASE

7.1   CONFIDENCE LEVEL AND CONFIDENCE LIMITS

The estimate of the mean time between failures, $\hat{m}$, obtained from tests is the expected, or the average value, or the mean life for that sample. Two or more samples from the same designated production line, tested under identical conditions, will give data from which different estimates of m, the true mean life, will be obtained. If it is claimed that the first estimate is the m of this equipment and we are asked to conduct another such test, we would be much embarrassed if it yields an $\hat{m}$ which turns out to be lower than that of the first estimate. To minimize the probability of being so embarrassed we utilize the concept of *confidence level*, such that the probability of our being proven wrong, or the risk α, that the second $\hat{m}$ turns out to be lower than the first, in the long run, is kept at a prechosen low level. The confidence level is designated by $CL = 1 - \alpha$, and the risk level by α.

Confidence level is equal to the proportion of the samples (from whose test results their individual $\hat{m}'s$ are calculated) whose confidence limits contain the true m.

This is illustrated in Fig. 7.1, where the lower, two-sided confidence limit on m, m_{L2}, and the upper, two-sided confidence limit on m, m_{U2}, for each sample is plotted, as well as the true $MTBF$ of this equipment, m. Were we to have a very large number of such tested

samples, and had counted those samples whose confidence limits on their m contained the true m (as is the case for samples numbered 1, 2, 4, 5, 7 and 8) and divided this number by the total number (a very large number) of such samples tested, then the resulting number would give us the confidence level, $CL = 1 - \alpha$, or

$$CL \quad = \quad \lim_{N_T \to \infty} \frac{N_{SC}}{N_T}, \qquad (7.1)$$

where

$$N_{SC} \quad = \quad \text{number of samples whose confidence limits on } \hat{m}$$
$$\text{contain the true } m,$$

and

$$N_T \quad = \quad \text{total number of samples tested.}$$

The number of samples whose confidence limits do not contain the true m (as is the case for samples numbered 3 and 6) divided by the total number of samples tested would give us the risk level, or α.

The case for the lower, one-sided confidence limit is illustrated in Fig. 7.2 where the true m is contained within the confidence limits of samples numbered 1, 2, 3, 4, 5 and 8, and not contained within the confidence limits of those numbered 6 and 7. The upper confidence limit is ∞.

It should be noted that statistics associated with a specified confidence level cannot be determined unless test data are available. This statement is made because a frequent question asked is "What is the confidence limit on the mean life at $100(1 - \alpha)\%$ confidence level for a unit?" but no data are given! Consequently, no confidence limit can be determined because we don't have the necessary information, such as the accumulated device hours of test time, T_a, and the total number of failures, r, for the exponential case.

The methodologies for determining the two-sided and one-sided confidence limits on m for units having an exponential time-to-failure distribution are presented next.

7.2   DETERMINATION OF THE CONFIDENCE LIMITS ON m

Once the estimate of m, $\hat{m}$, is calculated, the next step is to determine the confidence limits on m from a knowledge of $\hat{m}$, the number of failures, r, and the confidence level, $CL = 1 - \alpha$.

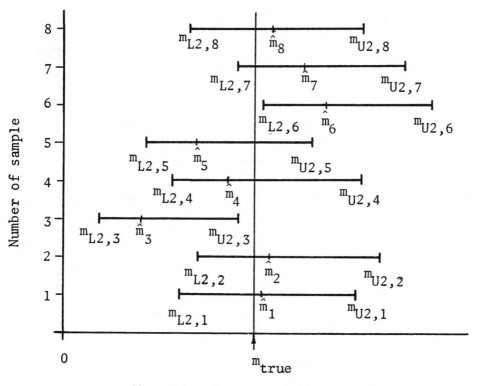

Fig. 7.1 – Two-sided confidence limits on the mean time between failures of identical units, m_{L2} and m_{U2}, and their true m.

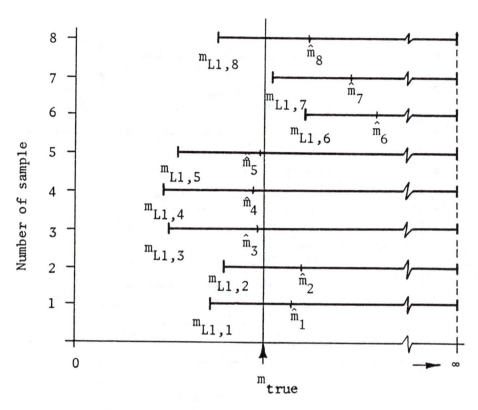

Mean-time-between-failures, m

Fig. 7.2– Lower, one-sided confidence limit on the mean time between failures of identical units, m_{L1}, and their true m.

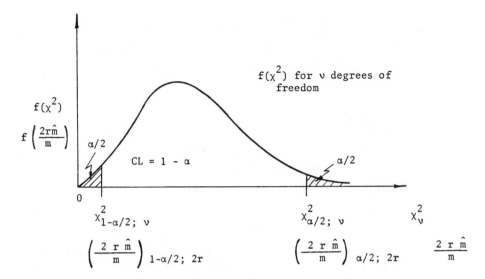

Fig. 7.3 − Distribution of $2r\hat{m}/m$, which is χ^2 distributed, and the associated $(1-\alpha/2)$ and $\alpha/2$ percentage points. These percentages are those corresponding to the areas under the *pdf* to the right of these points for a failure terminated test.

7.2.1   TWO-SIDED CONFIDENCE LIMITS ON m IN A FAILURE TERMINATED TEST

Epstein and Sobel [1] have shown that the random variable defined by $2r\hat{m}/m$ is χ^2 distributed with $2r$ degrees of freedom when the test is *failure terminated* at the *rth* failure. See Appendix 7C. Consequently, for two-sided confidence limits, the following probabilistic equation can be written, making use of the $(1-\alpha/2)$ and $\alpha/2$ percentage point values of the χ^2 distribution:

$$P(\chi^2_{1-\alpha/2;2r} \leq \frac{2r\hat{m}}{m} \leq \chi^2_{\alpha/2;2r}) = CL = 1 - \alpha. \qquad (7.2)$$

These percentage points are illustrated in Fig. 7.3. It must be noted that the χ^2 values should be so determined that the area under the χ^2 *pdf* to the *right* of them is equal to $(1-\alpha/2)$ for the left side of the inequality and $\alpha/2$ for the right side.

Equation (7.2) can be rearranged to yield

$$P(\frac{2r\hat{m}}{\chi^2_{\alpha/2;2r}} \leq m \leq \frac{2r\hat{m}}{\chi^2_{1-\alpha/2;2r}}) = 1 - \alpha, \tag{7.3}$$

from which it may be seen that the lower, two-sided confidence limit on m, m_{L2}, is

$$m_{L2} = \frac{2r\hat{m}}{\chi^2_{\alpha/2;2r}}, \tag{7.4}$$

and the upper, two-sided limit, m_{U2}, is

$$m_{U2} = \frac{2r\hat{m}}{\chi^2_{1-\alpha/2;2r}}. \tag{7.5}$$

It must be observed that in calculating m_{L2} the $\alpha/2$ percentage point is used and not the $(1 - \alpha/2)$, and in calculating m_{U2} the $(1 - \alpha/2)$ percentage point is used and not the $\alpha/2$, because m is in the denominator in Eq. (7.2) and solving for the confidence limits, as is done in Eq. (7.3), transposes the χ^2 percentage points appearing in Eq. (7.2).

It must be recalled that

$$\hat{m} = \frac{T_a}{r}.$$

Consequently, substituting this into Eq. (7.4) yields

$$m_{L2} = \frac{2T_a}{\chi^2_{\alpha/2;2r}}, \tag{7.6}$$

and into Eq. (7.5) yields

$$m_{U2} = \frac{2T_a}{\chi^2_{1-\alpha/2;2r}}. \tag{7.7}$$

Appendix D may be used to find the percentage point values of the χ^2 *pdf*. The χ^2 values given in the body of the table are those values *to the right* of which the area under the χ^2 *pdf* is equal to the desired probability; i.e., $\alpha/2$ for Eq. (7.6) and $(1 - \alpha/2)$ for Eq. (7.7).

EXAMPLE 7–1

In a failure terminated test 6,080 operating unit hours are accumulated and nine failures occur. This is the case of Example 6–3 in Chapter 6. Find the two-sided confidence limits on the mean time between failures at a 95% confidence level based on this sample.

SOLUTION TO EXAMPLE 7-1

In this example $T_a = 6,080$ hr, $r = 9$, $2r = 18$, $CL = 1 - \alpha = 0.95$, $\alpha = 0.05$, $\alpha/2 = 0.025$ and $(1 - \alpha/2) = 0.975$. From Eq. (7.6) the lower, two-sided confidence limit on the mean time between failures, based on this sample, is

$$m_{L2} = \frac{2 \times 6,080}{\chi^2_{0.025;18}}.$$

Entering Appendix D with $\delta = 0.025$ and degrees of freedom $\nu = 18$, we find

$$\chi^2_{0.025;18} = 31.526.$$

Then,

$$m_{L2} = \frac{12,160}{31.526} = 385.7 \text{ hr.}$$

From Eq. (7.7) the upper, two-sided confidence limit on the mean time between failures, based on this sample, is

$$m_{U2} = \frac{2 \times 6,080}{\chi^2_{0.975;18}}.$$

Entering Appendix D with $\delta = 0.975$ and $\nu = 18$, we find

$$\chi^2_{0.975;18} = 8.231.$$

Then,

$$m_{U2} = \frac{12,160}{8.231} = 1,477.3 \text{ hr.}$$

Table 7.1 [2] may also be used to evaluate Eqs. (7.6) and (7.7). It gives the factor $2/\chi^2_{\delta;\nu}$ for one-sided and two-sided confidence limits on m, at six two-sided and six one-sided confidence levels. Multiplying the appropriate factor obtained from this table by the observed T_a, or by $r\hat{m}$, gives the confidence limit on m. The next example illustrates the use of Table 7.1.

EXAMPLE 7-2

Work out Example 7-1 using Table 7.1.

TABLE 7.1– Factors for calculation of mean life confidence intervals from test data

$$[\text{Factors} = 2/\chi^2(\delta,\nu)].$$

(Assumption of exponential distribution)

Confidence	99% Two-sided 99.5% One-sided	98% Two-sided 99% One-sided	95% Two-sided 97.5% One-sided	90% Two-sided 95% One-sided	80% Two-sided 90% One-sided	60% Two-sided 80% One-sided

Degrees of freedom, ν	Lower limit						Upper limit					
2	0.185	0.217	0.272	0.333	0.433	0.619	4.47	9.462	19.388	39.580	100.000	200.000
4	0.135	0.151	0.180	0.210	0.257	0.334	1.21	1.882	2.826	4.102	6.667	10.000
6	0.108	0.119	0.139	0.159	0.188	0.234	0.652	0.909	1.221	1.613	2.308	3.007
8	0.0909	0.100	0.114	0.129	0.150	0.181	0.437	0.573	0.733	0.921	1.212	1.481
10	0.0800	0.0857	0.0976	0.109	0.125	0.149	0.324	0.411	0.508	0.600	0.789	0.909
12	0.0702	0.0759	0.0856	0.0952	0.107	0.126	0.256	0.317	0.383	0.454	0.555	0.645
14	0.0635	0.0690	0.0765	0.0843	0.0948	0.109	0.211	0.257	0.305	0.355	0.431	0.500
16	0.0588	0.0625	0.0693	0.0760	0.0848	0.0976	0.179	0.215	0.251	0.290	0.345	0.385
18	0.0536	0.0571	0.0633	0.0693	0.0769	0.0878	0.156	0.184	0.213	0.243	0.286	0.322
20	0.0500	0.0531	0.0585	0.0635	0.0703	0.0799	0.137	0.158	0.184	0.208	0.242	0.270
22	0.0465	0.0495	0.0543	0.0589	0.0648	0.0732	0.123	0.142	0.162	0.182	0.208	0.232
24	0.0439	0.0463	0.0507	0.0548	0.0601	0.0676	0.111	0.128	0.144	0.161	0.185	0.200
26	0.0417	0.0438	0.0476	0.0513	0.0561	0.0629	0.101	0.116	0.130	0.144	0.164	0.178
28	0.0392	0.0413	0.0449	0.0483	0.0527	0.0588	0.0927	0.106	0.118	0.131	0.147	0.161
30	0.0373	0.0393	0.0425	0.0456	0.0496	0.0551	0.0856	0.0971	0.108	0.119	0.133	0.145
32	0.0355	0.0374	0.0404	0.0433	0.0469	0.0519	0.0795	0.0899	0.0997	0.109	0.122	0.131
34	0.0339	0.0357	0.0385	0.0411	0.0445	0.0491	0.0742	0.0834	0.0925	0.101	0.113	0.122
36	0.0325	0.0342	0.0367	0.0392	0.0423	0.0466	0.0696	0.0781	0.0899	0.0939	0.104	0.111
38	0.0311	0.0327	0.0351	0.0375	0.0404	0.0443	0.0656	0.0732	0.0804	0.0874	0.0971	0.103
40	0.0299	0.0314	0.0337	0.0359	0.0386	0.0423	0.0619	0.0689	0.0756	0.0820	0.0901	0.097

To use: Multiply values shown by total test hours to get upper and lower confidence limits in hours.

Note: $\nu = 2$, except for the lower limit on tests truncated at a fixed time and where $r < n$. In such cases, use $\nu = 2(r+1)$.

SOLUTION TO EXAMPLE 7–2

Entering Table 7.1 with degrees of freedom $\nu = 18$ and the 95% two-sided columns, read off Lower Limit, $k_1 = 0.0633$ and Upper Limit, $k_2 = 0.243$. Then, from Eq. (7.6),

$$m_{L2} = \frac{2T_a}{\chi^2_{\alpha/2;2r}} = k_1 T_a,$$

where

$$T_a = 6,080 \text{ hr.}$$

Then,

$$m_{L2} = 0.0633 \times 6,080 = 384.7 \text{ hr,}$$

and from Eq. (7.7)

$$m_{U2} = \frac{2T_a}{\chi^2_{1-\alpha/2;2r}} = k_2 T_a,$$

or, for this sample,

$$m_{U2} = 0.243 \times 6,080 = 1,477.4 \text{ hr,}$$

with practically identical results as in Example 7–1.

For large values of the degrees of freedom, say more than 40, the following relationship [3] may be used:

$$\chi^2_{\delta;\nu} \cong \frac{1}{2}[(2\nu - 1)^{\frac{1}{2}} + z_\delta]^2, \tag{7.8}$$

where

$z_\delta = $ standardized normal distribution variate's value such that the area under this distribution to the right of this z_δ value is equal to δ. Table 7.2 gives the frequently used values of z_δ.

For another approximation formula of the χ^2 distribution's percentage points, see Eq. (5.23).

EXAMPLE 7–3

One hundred units are reliability tested, during which test 20 failures occur. The test is terminated when the twentieth failure occurs, and the failed and nonfailed units operate for a total of 110,000 unit hours. Find the estimates of the two-sided confidence limits on m, based on this sample, at the 90% confidence level.

TABLE 7.2– Values of z_δ to use in Eq. (7.8) for various δ probabilities.

δ	0.995	0.990	0.980	0.975	0.950
z_δ	-2.576	-2.326	-2.054	-1.960	-1.645
δ	0.900	0.800	0.750	0.700	0.500
z_δ	-0.1282	-0.8416	-0.6745	-0.5244	0
δ	0.300	0.250	0.200	0.100	0.050
z_δ	+0.5244	+0.6745	+0.8416	+1.282	+1.645
δ	0.025	0.020	0.010	0.005	0.001
z_δ	+1.960	+2.054	+2.326	+2.576	+3.090

SOLUTION TO EXAMPLE 7–3

In this case $CL = 1 - \alpha = 0.90$, $\alpha = 0.10$, $\delta = \alpha/2 = 0.05$, and $\nu = 2r = 2 \times 20 = 40$. From Eq. (7.6) and based on this sample,

$$m_{L2} = \frac{2T_a}{\chi^2_{\alpha/2;2r}} = \frac{2 \times 110,000}{\chi^2_{0.05;40}}.$$

The value of $\chi^2_{0.05;40}$ can be found from Eq. (7.8) using Table 7.2 where $z_{0.05} = +1.645$, or

$$\chi^2_{0.05;40} \cong \frac{1}{2}[(2 \times 40 - 1)^{\frac{1}{2}} + 1.645]^2,$$

or

$$\chi^2_{0.05;40} \cong 55.474.$$

This compares with the value from Table 7.1 of $2/\chi^2_{0.05;40} = 0.0359$ or $\chi^2_{0.05;40} = 55.710$. Consequently, using Table 7.2,

$$m_{L2} = \frac{220,000}{55.474} = 3,965.8 \text{ hr},$$

whereas using Table 7.1

$$m_{L2} = \frac{220,000}{55.710} = 3,949.0 \text{ hr},$$

or a difference of less than 0.5%. This difference decreases as r increases beyond 20.

From Eq. (7.7), where $(1 - \alpha/2) = 1 - 0.05 = 0.95$,

$$m_{U2} = \frac{2T_a}{\chi^2_{1-\alpha/2;40}} = \frac{2 \times 110,000}{\chi^2_{0.95;40}}.$$

$\chi^2_{0.95;40}$ can be found from Eq. (7.8) and Table 7.2, where $z_{0.95} = -1.645$.

Then,

$$\chi^2_{0.95;40} \cong \frac{1}{2}[(2 \times 40 - 1)^{\frac{1}{2}} - 1.645]^2,$$

or

$$\chi^2_{0.95;40} \cong 26.232.$$

This compares with the value from Table 7.1 of $2/\chi^2_{0.95;40} = 0.0756$, or $\chi^2_{0.95;40} = 26.455$. Consequently, using Table 7.2,

$$m_{U2} = \frac{220,000}{26.232} = 8,386.7 \text{ hr},$$

whereas, using Table 7.1,

$$m_{U2} = \frac{220,000}{26.455} = 8,316.0 \text{ hr},$$

or a difference of less than 1.0%.

7.2.2   IMPORTANT STATEMENT ABOUT THE CONFIDENCE INTERVAL

It should be noted here that before the data are observed, the confidence interval, $(m_{L2}; m_{U2})$, is a function of the random variables $(T_1, T_2,...,T_n)$. Then the following probability statement may be made:

$$P(m_{L2} < m < m_{U2}) = 1 - \alpha. \tag{7.9}$$

In other words the probability that the true mean time between failures, m, is contained in the random interval $(m_{L2}; m_{U2})$ is equal to $(1 - \alpha)$.

However, after the data are observed and the specific values of m_{L2} and m_{U2} are calculated from the test results of a finite sample, then the interval will either contain the true value of m or it will not. Therefore, it is not proper to attach a probability statement, such as that of Eq. (7.9), using the values of m_{L2} and m_{U2} calculated from a particular set of observed data. The appropriate statement would be that we do not know if the calculated interval from a specific sample contains the true value of m or not, but if many repeated samples of the same size were taken containing units from the same population, then the calculated confidence limits $(m_{L2i}; m_{U2i})$ in $100(1 - \alpha)\%$ of the samples will contain the true mean life m, as the number of samples tested approaches infinity theoretically or very large practically.

EXAMPLE 7–4

Write the associated statement for the results of Example 7–3.

SOLUTION TO EXAMPLE 7–4

Using the m_{L2} and m_{U2} values found in Example 7–3, it can be stated that, at the 90% confidence level, the estimated confidence interval on m based on this sample, using Table 7.2, is

$$(3, 965.8 \text{ hr} \; ; \; 8, 386.7 \text{ hr}).$$

7.3   ONE-SIDED CONFIDENCE LIMITS ON m IN A FAILURE TERMINATED TEST

The lower, one-sided confidence limit on m in a *failure terminated test* is obtained from Eq. (7.6) by replacing $\alpha/2$ by α, or

$$m_{L1} = \frac{2T_a}{\chi^2_{\alpha;2r}}, \qquad (7.10)$$

with the upper confidence limit on m being ∞.

The upper, one-sided confidence limit on m in a failure terminated test is obtained from Eq. (7.7) by replacing $(1 - \alpha/2)$ by $(1 - \alpha)$, or

$$m_{U1} = \frac{2T_a}{\chi^2_{1-\alpha;2r}}, \qquad (7.11)$$

with the lower confidence limit on m being zero.

Equations (7.10) and (7.11) may be evaluated by using the $\chi^2_{\delta;\nu}$ values in Appendix D, the $\frac{2}{\chi^2_{\delta;\nu}}$ values in Table 7.1, or Eq. (7.8).

The following probabilistic statements may be made using the results of Eqs. (7.10) and (7.11):

$$P(m \geq m_{L1}) = \text{Confidence level} = 1 - \alpha, \qquad (7.12)$$

and

$$P(m \leq m_{U1}) = \text{Confidence level} = 1 - \alpha. \qquad (7.12')$$

In other words, the probability that the true m is included in the intervals bounded by m_{L1} and ∞, or by 0 and m_{U1}, as the number of such samples tested goes to infinity theoretically, or is very large practically, is equal to the confidence level.

EXAMPLE 7–5

1. Find the lower, one-sided confidence limit on m based on the sample of Example 7–1.

2. Write a probabilistic statement for the result found in the previous case.

3. Find the upper, one-sided confidence limit on m based on the sample of Example 7–1.

4. Write the associated statement for the result found in the previous case.

SOLUTIONS TO EXAMPLE 7–5

1. In Example 7–1, $T_a = 6,080$ unit hr, $r = 9$, and $CL = 0.95$. Therefore,

$$\hat{m} = \frac{T_a}{r} = \frac{6,080}{9},$$

or

$$\hat{m} = 675.6 \text{ hr.}$$

The lower, one-sided confidence limit on m, based on this sample, is

$$m_{L1} = \frac{2T_a}{\chi^2_{\alpha;2r}} = \frac{2 \times 6,080}{\chi^2_{0.05;2\times9}} = \frac{12,160}{\chi^2_{0.05;18}} = \frac{12,160}{28.869},$$

or

$$m_{L1} = 421.2 \text{ hr,}$$

using Appendix D to obtain the $\chi^2_{0.05;18}$ value.

2. It can be stated that, based on this sample and at the 95% confidence level, the estimate of the one-sided, lower confidence interval is

$$(421.2 \text{ hr} \; ; \; \infty).$$

3. The upper, one-sided confidence limit on m, based on this sample, is

$$m_{U1} = \frac{2T_a}{\chi^2_{1-\alpha;2r}} = \frac{12,160}{\chi^2_{0.95;18}} = \frac{12,160}{9.390},$$

or

$$m_{U1} = 1,295.0 \text{ hr}.$$

4. It can be stated that, based on this sample and at the 95% confidence level, the estimate of the upper, one-sided confidence interval is

$$(0 \; ; \; 1,295.0 \text{ hr}).$$

7.4   TWO-SIDED AND ONE-SIDED CONFIDENCE LIMITS ON m IN A TIME TERMINATED TEST

When the test is time terminated only the approximate confidence limits on m can be determined. See Appendix 7B.3. The lower and upper, two-sided confidence limits, respectively, on m are given by

$$m_{L2} \cong \frac{2r\hat{m}}{\chi^2_{\alpha/2;2r+2}} = \frac{2T_a}{\chi^2_{\alpha/2;2r+2}}, \tag{7.13}$$

and

$$m_{U2} \cong \frac{2r\hat{m}}{\chi^2_{1-\alpha/2;2r}} = \frac{2T_a}{\chi^2_{1-\alpha/2;2r}}, \tag{7.14}$$

The corresponding probabilistic statement for the true m is

$$P(m_{L2} \leq m \leq m_{U2}) = \text{ Confidence level} = 1 - \alpha. \tag{7.15}$$

The lower and upper, one-sided confidence limits, respectively, on m are given by

$$m_{L1} \cong \frac{2r\hat{m}}{\chi^2_{\alpha;2r+2}} = \frac{2T_a}{\chi^2_{\alpha;2r+2}}, \qquad (7.16)$$

and

$$m_{U1} \cong \frac{2r\hat{m}}{\chi^2_{1-\alpha;2r}} = \frac{2T_a}{\chi^2_{1-\alpha;2r}}. \qquad (7.17)$$

The corresponding probabilistic statements for the true m are

$$P(m \geq m_{L1}) = \text{Confidence level} = 1 - \alpha, \qquad (7.18)$$

and

$$P(m \leq m_{U1}) = \text{Confidence level} = 1 - \alpha. \qquad (7.19)$$

It may be observed that the lower limits for the time terminated cases are determined with χ^2 values obtained from a χ^2 distribution with $(2r + 2)$ degrees of freedom, whereas for the failure terminated cases with $(2r)$ degrees of freedom. The upper limit equations are identical for both the failure terminated and the time terminated test cases, with the χ^2 values obtained from a χ^2 distribution with $(2r)$ degrees of freedom.

These confidence limits on m may be calculated using the appropriate $\chi^2_{\delta;\nu}$ values given in Appendix D, the $2/\chi^2_{\delta;\nu}$ values given in Table 7.1, or the values obtained from Eq. (7.8).

Figure 7.4 presents a graphical technique for determining the confidence limits on m for time terminated tests when the number of failures are known. It is based on Eqs. (7.13) and (7.14) such that for the lower, two-sided limits on m

$$m_{L2} = \frac{2r\hat{m}}{\chi^2_{\alpha/2;2r+2}} = k_{L2}\hat{m},$$

where

$$k_{L2} = \frac{2r}{\chi^2_{\alpha/2;2r+2}},$$

which can be calculated once the number of failures, r, and the confidence level, $1 - \alpha$, are known. For example, for $r = 5$ and $CL = 1 - \alpha = 0.90, \alpha = 0.10$, and $\alpha/2 = 0.05$,

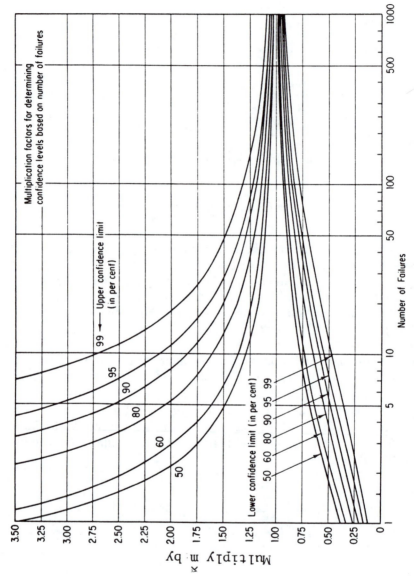

Fig. 7.4– Multiplication factors for $\hat{m}$ to determine the upper and lower confidence limits versus number of failures for time terminated tests.

188

$$k_{L2} = \frac{2 \times 5}{\chi^2_{0.05;2 \times 5 + 2}} = \frac{10}{\chi^2_{0.05;12}} = \frac{10}{21.026},$$

or

$$k_{L2} = 0.476.$$

Consequently, in Fig. 7.4 the intersection of the number-of-failures-of-5 line with the lower, 90% confidence limit curve should occur at 0.476, and it does. Therefore, to find m_{L2} enter Fig. 7.4 with $CL = 1 - \alpha$ and r, find k_{L2} and multiply $\hat{m}$ by this k_{L2} to find m_{L2}.

Similarly for the upper, two-sided confidence limit on m

$$m_{U2} = \frac{2r\hat{m}}{\chi^2_{1-\alpha/2;2r}} = k_{U2}\hat{m},$$

where

$$k_{U2} = \frac{2r}{\chi^2_{1-\alpha/2;2r}},$$

which can be calculated once the number of failures, r, and the confidence level, $1 - \alpha$, are known. For example, for $r = 5$ and $CL = 1 - \alpha = 0.90, \alpha = 0.10, \alpha/2 = 0.05$ and $(1 - \alpha/2) = 0.95$,

$$k_{U2} = \frac{2 \times 5}{\chi^2_{0.95;2 \times 5}} = \frac{10}{\chi^2_{0.95;10}} = \frac{10}{3.940},$$

or

$$k_{U2} = 2.54.$$

This is the value in Fig. 7.4 corresponding to the intersection of the $r = 5$ line and $CL = 1 - \alpha = 90\%$ upper confidence limit curve. Consequently, enter Fig. 7.4 with r and $CL = 1 - \alpha$, read the k_{U2} value off the ordinate, and multiply $\hat{m}$ by this k_{U2} to find m_{U2}.

For a failure terminated test find m_{U2} using Fig. 7.4 the same way as was described previously. To find m_{L2} for a failure terminated test, from Fig. 7.4, enter with one more failure than the observed failures.

EXAMPLE 7–6

Twenty units are subjected to a reliability test of 500 clock hr duration. During this test five failures occur at $T_1 = 110$ hr, $T_2 = 180$ hr, $T_3 = 300$ hr, $T_4 = 410$ hr and $T_5 = 480$ hr. The failed units are not replaced. Do the following:

1. Estimate the mean time between failures, $\hat{m}$.

2. Estimate the two-sided confidence limits on m, based on this sample, at a 90% confidence level.

3. Write the associated probabilistic statement concerning the results obtained in Case 2.

4. Estimate the lower, one-sided confidence limit on m, based on this sample, at a confidence level of 90%.

5. Write the associated probabilistic statement concerning the result obtained in Case 4.

6. Estimate the upper, one-sided confidence limit on m, based on this sample, at a confidence level of 90%.

7. Write the associated probabilistic statement concerning the result obtained in Case 6.

SOLUTIONS TO EXAMPLE 7–6

This is a time terminated test because the last failure occurred at $T_5 = 480$ hr, whereas the test's duration was $t_d = 500$ clock hr, and $N = 20, r = 5, CL = 1 - \alpha = 0.90; \alpha = 0.10, \alpha/2 = 0.05$ and $(1 - \alpha/2) = 0.95$.

1. The mean time between failures estimate, $\hat{m}$, is given by (See Appendix 7C.2)

$$\hat{m} = \frac{T_a}{r} = \frac{\sum\limits_{i=1}^{r} T_i + (N - r)t_d}{r},$$

$$\hat{m} = \frac{110 + 180 + 300 + 410 + 480 + (20 - 5)500}{5} = \frac{8,980}{5},$$

or

$$\hat{m} = 1,796 \text{ hr.}$$

2. The two-sided confidence limits on m, based on this sample and calculated from Eqs. (7.13) and (7.14), are

$$m_{L2} = \frac{2T_a}{\chi^2_{\alpha/2;2r+2}} = \frac{2 \times 8,980}{\chi^2_{0.05;2\times5+2}},$$

$$m_{L2} = \frac{17,960}{\chi^2_{0.05;12}} = \frac{17,960}{21.026},$$

or

$$m_{L2} = 854.2 \text{ hr},$$

and

$$m_{U2} = \frac{2T_a}{\chi^2_{1-\alpha/2;2r}} = \frac{2 \times 8,980}{\chi^2_{0.95;2\times5}},$$

$$m_{U2} = \frac{17,960}{\chi^2_{0.95;10}} = \frac{17,960}{3.940},$$

or

$$m_{U2} = 4,558.4 \text{ hr}.$$

3. It can be stated that, based on this sample, the estimate of the 90% confidence interval on m is

$$(854.2 \text{ hr} ; 4,558.4 \text{ hr}).$$

4. The lower, one-sided confidence limit on m, based on this sample and calculated from Eq. (7.16), is

$$m_{L1} = \frac{2T_a}{\chi^2_{\alpha;2r+2}} = \frac{2 \times 8,980}{\chi^2_{0.10;2\times5+2}},$$

$$m_{L1} = \frac{17,960}{\chi^2_{0.10;12}} = \frac{17,960}{18.549},$$

or

$$m_{L1} = 968.2 \text{ hr}.$$

5. It can be stated that, based on this sample, the estimate of the lower one-sided confidence interval for m is

$$(968.2 \text{ hr} \; ; \; \infty).$$

6. The upper, one-sided confidence limit on m, calculated from Eq. (7.17) and based on this sample, is

$$m_{U1} = \frac{2T_a}{\chi^2_{1-\alpha;2r}} = \frac{2 \times 8,980}{\chi^2_{0.90;2\times5}},$$

$$m_{U1} = \frac{17,960}{\chi^2_{0.90;10}} = \frac{17,960}{4.865},$$

or

$$m_{U1} = 3,691.7 \text{ hr.}$$

7. It can be stated that, based on this sample, the estimate of the upper one-sided confidence interval on m is

$$(0 \; ; \; 3,691.7 \text{ hr}).$$

7.5 DETERMINATION OF THE CONFIDENCE LIMITS ON THE RELIABILITY FOR THE EXPONENTIAL CASE

Given the confidence limits on m, the confidence limits on the associated reliability for the exponential case, when the failure rate is constant, and for a mission of t duration, may be calculated from

$$R(t) = e^{-\frac{t}{m}}, \tag{7.20}$$

using the appropriate confidence limit on m.

The two-sided confidence limits on the reliability may be obtained using m_{L2} and m_{U2}, depending on whether the test is failure or time terminated. Then

$$R_{L2}(t) = e^{-\frac{t}{m_{L2}}} \tag{7.21}$$

for the lower confidence limit, and

$$R_{U2}(t) = e^{-\frac{t}{m_{U2}}}, \tag{7.22}$$

for the upper confidence limit.

The corresponding probabilistic statement for the true reliability is

$$P(R_{L2} \leq R \leq R_{U2}) = \text{Confidence level} = 1 - \alpha,$$

or

$$P(e^{-\frac{t}{m_{L2}}} \leq R \leq e^{-\frac{t}{m_{U2}}}) = \text{Confidence level} = 1 - \alpha, \qquad (7.23)$$

whereas the estimate of the expected, or average, reliability is given by

$$\hat{\bar{R}}(t) = e^{-\frac{t}{\hat{m}}}. \qquad (7.24)$$

The lower, one-sided confidence limit on the reliability is given by

$$R_{L1} = e^{-\frac{t}{m_{L1}}}, \qquad (7.25)$$

and the upper, one-sided confidence limit by

$$R_{U1} = e^{-\frac{t}{m_{U1}}}. \qquad (7.26)$$

The following probabilistic statements may then be made for the true reliability using the results of Eqs. (7.25) and (7.26):

$$P(R \geq R_{L1}) \quad = \text{Confidence level} = 1 - \alpha,$$

or

$$P(R \geq e^{-\frac{t}{m_{L1}}}) = \text{Confidence level} = 1 - \alpha, \qquad (7.27)$$

and

$$P(R \leq R_{U1}) \quad = \text{Confidence level} = 1 - \alpha,$$

or

$$P(R \leq e^{-\frac{t}{m_{U1}}}) = \text{Confidence level} = 1 - \alpha. \qquad (7.28)$$

EXAMPLE 7–7

Using the results of the confidence limits on m found in Example 7–6, find the estimate of the 90% confidence level limits on the associated reliabilities for a mission of two hours duration.

SOLUTIONS TO EXAMPLE 7–7

1. The lower and upper, two-sided confidence limits on the reliability, for Case 2 in Example 7–6 for a mission of two hours duration and at a confidence level of 90%, calculated from Eqs. (7.21) and (7.22) and based on this sample, are

$$R_{L2}(t = 2 \text{ hr}) = e^{-\frac{t}{m_{L2}}} = e^{-\frac{2}{854.2}} = 0.997661,$$

and

$$R_{U2}(t = 2 \text{ hr}) = e^{-\frac{t}{m_{U2}}} = e^{-\frac{2}{4,558.4}} = 0.999561.$$

2. The lower and upper, one-sided confidence limits on the reliability for Cases 4 and 6 in Example 7–6 for a mission of two hours duration, calculated from Eqs. (7.25) and (7.26) and based on this sample, are

$$R_{L1}(t = 2 \text{ hr}) = e^{-\frac{t}{m_{L1}}} = e^{-\frac{2}{968.2}} = 0.997936,$$

and

$$R_{U1}(t = 2 \text{ hr}) = e^{-\frac{t}{m_{U1}}} = e^{-\frac{2}{3,691.7}} = 0.999458.$$

7.6 *MTBF* AND RELIABILITY DETERMINATION WHEN NO FAILURES ARE OBSERVED IN A TEST

If in a test no failures are observed, one might be tempted to say

$$\hat{m} = \frac{T_a}{r} = \frac{T_a}{0} = \infty,$$

which of course cannot be correct, because from a finite sample and a short test time, during which no failures occur, we cannot say this equipment has an infinite m and a corresponding reliability of unity, because a failure could have occurred any time after the termination of this test, had it been continued. One can, however, calculate the lower, one-sided confidence limit on m with zero failures, $m_{L1(0)}$, from Eq. (7.16) by setting $r = 0$, or

$$m_{L1(0)} = \frac{2T_a}{\chi^2_{\alpha;2\times0+2}}.$$

Therefore,

$$m_{L1(0)} = \frac{2T_a}{\chi^2_{\alpha;2}}. \tag{7.29}$$

The lower, one-sided confidence limit on the associated reliability is then given by

$$R_{L1(0)}(t) = e^{-\frac{t}{m_{L1(0)}}}. \tag{7.30}$$

EXAMPLE 7–8

Assuming that in Example 7–6 no failures occur, do the following:

1. Determine the lower, one-sided confidence limit on m, based on this sample, at a 90% confidence level.

2. Write down the statement associated with the result of Case 1.

3. Determine the lower, one-sided confidence limit on R, based on this sample, at a 90% confidence level for a mission of two hours duration.

4. Write down the statement associated with the result of Case 2.

SOLUTIONS TO EXAMPLE 7–8

1. Given are $CL = 1 - \alpha = 0.90, \alpha = 0.10, r = 0$, and $T_a = 20 \times 500 = 10,000$ hr. Consequently, from Eq. (7.29), the lower, one-sided confidence limit on m, based on this sample, is

$$m_{L1(0)} = \frac{2 \times 10,000}{\chi^2_{0.10;2}} = \frac{20,000}{4.605},$$

or

$$m_{L1(0)} = 4,343.1 \text{ hr.}$$

2. It can be stated that, based on this sample, the estimate of the lower, one-sided, 90% confidence interval on m is

   $(4,343.1 \text{ hr} ; \infty)$.

3. The lower, one-sided confidence limit on the reliability of these units, based on this sample, is

   $$R_{L1(0)}(t = 2 \text{ hr}) = e^{-\frac{2}{4,343.1}} = 0.999540.$$

4. It can be stated that, based on this sample, the estimate of the lower, one-sided, 90% confidence interval on R is

   $(0.999540 ; 1.0)$.

7.7 COMPARISON OF THE LOWER, ONE-SIDED CONFIDENCE LIMIT ON m WHEN ONE FAILURE OCCURS IN A FAILURE TERMINATED TEST WITH THE CASE WHEN NO FAILURES OCCUR IN A TIME TERMINATED TEST

When a test is terminated after only one failure occurs then $r = 1$. Consequently, Eq. (7.10) for the lower, one-sided confidence limit on m becomes

$$m_{L1} = \frac{2T_a}{\chi^2_{\alpha;2}}. \tag{7.31}$$

When no failures occur in a test then $r = 0$. Consequently, Eq. (7.16) for the lower, one-sided confidence limit on m, for a time terminated test, that this case is, becomes

$$m_{L1(0)} = \frac{2T_a}{\chi^2_{\alpha;2}}. \tag{7.32}$$

It may be seen from Eqs. (7.31) and (7.32) that

$m_{L1} = m_{L1(0)}.$

Consequently, the two cases yield the same value of m_{L1}. The implications are that the Eq. (7.16) approximation of m_{L1} may presuppose

that conservatively a failure could have very likely occurred an instant after the test was time terminated such that the failures would be $(r + 1)$ in Eq. (7.10), or

$$m_{L1} = \frac{2T_a}{\chi^2_{\alpha;2(r+1)}},$$

or

$$m_{L1} = \frac{2T_a}{\chi^2_{\alpha;2r+2}},$$

hence the same as Eq. (7.16).

7.8 EFFECT OF CONFIDENCE LEVEL, SAMPLE SIZE, AND OCCURRENCE OF FAILURES ON THE CONFIDENCE LIMITS ON m

A study of Table 7.1 and Fig. 7.4 will indicate that as the number of failures observed in a test increases the range between the lower and upper confidence limits on m, for the same confidence level, or the confidence interval, decreases; i.e., these two limits approach each other, and in the limit they approach $\hat{m}$ which in turn approaches the true m. This should be so because, for a given m, a larger number of failures will result when either a larger number of units are used in the test, or the test duration is much longer, or both; consequently, a better estimate of m is obtained whose limits are closer to each other and in the limit they tend to the true m.

It may also be seen that as the confidence level is increased, for the same number of failures, the range between the lower and upper confidence limits, or the confidence interval, increases because a larger percentage of the sample's $\hat{m}$s need to be contained within the confidence limits.

It is also of interest to observe from Eqs. (7.6), (7.10), (7.13) and (7.16) that for the same T_a and r, in both tests (failure and time terminated), the failure terminated test gives a higher value for the lower confidence limit on m than the time terminated one. It appears then that a failure terminated test should be preferred over a time terminated one. This opinion may be reversed, however, when sequential tests are involved.

7.9 THE SKEWNESS OF THE DISTRIBUTION OF THE MEAN TIME BETWEEN FAILURES

It is also of interest to find out if the *pdf* of $\hat{m}$ is skewed and in which direction. This may be determined by comparing the median value of m, $\check{m}$, with its expected value, $\hat{m}$. The median time-to-failure value for a *failure terminated test* can be obtained by taking $CL = 1 - \alpha = 0.50$, or $\alpha = 0.50$, in Eq. (7.6); therefore,

$$\check{m} = \frac{2T_a}{\chi^2_{0.50;2r}}. \tag{7.33}$$

Let us assume the following results were obtained in a failure terminated test:

$T_a = 9,000$ hr and $r = 8$.

From Eq. (7.33)

$$\check{m} = \frac{2 \times 9,000}{\chi^2_{0.50;2\times8}} = \frac{18,000}{\chi^2_{0.50;16}},$$

$$\check{m} = \frac{18,000}{18.338},$$

or

$$\check{m} = 1,173.6 \text{ hr}.$$

This compares with $\hat{m} = 9,000/8 = 1,125.0$ hr; which implies that the distribution of m is skewed to the left because $\check{m} > \hat{m}$. This behavior prevails for all practical values of T_a and r.

7.10 TEST DURATION AND SAMPLE SIZE FOR VARIOUS TEST TYPES AND CONFIDENCE LEVELS

It is recommended that the equipment's *MTBF* goal, m_G, be taken to be equal to m_{L1}, such that having established by test that $m_{L1} = m_G$ we are *CL%* sure that the equipment's actual *MTBF* will be at least equal to and greater than m_G, in the long run. Next we want to know what T_a, r and N should be to assure that the condition $m_{L1} = m_G$ is met at the chosen confidence level, *CL*.

If in a test *no failures* occur, what should T_a be to demonstrate that we have met the m_G requirement with a *CL* of 90%?

From Eq. (7.29)

$$m_{L1} = m_G = \frac{2T_a}{\chi^2_{\alpha;2}}.$$

Therefore,

$$T_a = m_G(\frac{1}{2}\chi^2_{\alpha;2}). \tag{7.34}$$

For a $CL = 1 - \alpha = 0.90$ and $\alpha = 0.10, \chi^2_{0.10;2} = 4.605$. Then,

$$T_a = m_G(\frac{1}{2} \times 4.605),$$

or

$$T_a = 2.3025m_G. \tag{7.35}$$

Consequently, to demonstrate that the *MTBF* goal is met with 90% confidence, one must test for unit hours equal to 2.3025 times the *MTBF* goal, and not for unit hours equal to just the *MTBF* goal!

For a 95% confidence level

$$T_a = m_G(\frac{1}{2}\chi^2_{0.05;2}) = m_G(\frac{1}{2} \times 5.991),$$

or

$$T_a = 2.9955m_G. \tag{7.36}$$

Consequently, we must test for unit hours equal to almost three times the *MTBF* goal, to demonstrate an m_G with 95% confidence.

If N units are used in the test; then, for a 90% confidence level the clock hours of test time, t_d, will be

$$T_a = 2.3025m_G = Nt_d, \tag{7.37}$$

or

$$t_d = \frac{2.3025m_G}{N}, \tag{7.38}$$

and, for a 95% confidence level,

$$t_d = \frac{2.9955m_G}{N}. \tag{7.39}$$

Equations (7.38) and (7.39) give the *minimum* test duration possible to demonstrate that the equipment has an *MTBF* of at least m_G with a confidence of 90% and 95%, respectively.

For $CL = 95\%$ and $N = 1$,

$$t_d \cong 3m_G.$$

If $N = 2$,

$$t_d \cong 1.5m_G.$$

If $N = 3$,

$$t_d \cong m_G.$$

If $N = 6$,

$$t_d \cong \frac{1}{2}m_G,$$

and so on. Similar values may be found for $CL = 90\%$. Thus the effect of testing more units on reducing the test duration may be determined.

If t_d is prescribed; e.g., based on delivery schedule, then the minimum test sample size will have to be

$$N = \frac{2.3025m_G}{t_d} \tag{7.40}$$

for $CL = 90\%$, and

$$N = \frac{2.9955m_G}{t_d} \tag{7.41}$$

for $CL = 95\%$.

If *one failure* occurs, or is allowed, in the test, then $r = 1$ and

$$m_{L1} = m_G = \frac{2T_a}{\chi^2_{\alpha;2r+2}} = \frac{2T_a}{\chi^2_{\alpha;4}},$$

and

$$T_a = m_G\left(\frac{1}{2}\chi^2_{\alpha;4}\right). \tag{7.42}$$

For $CL = 90\%$

$$T_a = m_G\left(\frac{1}{2}\chi^2_{0.10;4}\right) = m_G\left(\frac{1}{2} \times 7.779\right),$$

or

$$T_a = 3.8895m_G. \tag{7.43}$$

For $CL = 95\%$

$$T_a = m_G(\frac{1}{2}\chi^2_{0.05;4}) = T_a = m_G(\frac{1}{2} \times 9.488),$$

or

$$T_a = 4.744m_G. \tag{7.44}$$

Equations (7.43) and (7.44) give the unit hours of operation that have to be accumulated by all units in the test to demonstrate m_G with a CL of 90% and 95%, respectively, *with one failure.*

If the failure occurs at T_1, how many additional unit hours of operation, T_{ad}, should be accumulated, without an additional failure, to demonstrate m_G? This may be found from

$$T_{ad} = T_a - NT_1. \tag{7.45}$$

If it is a *nonreplacement test,* then the additional clock hours of test time, t_{ad}, *without an additional failure,* will be

$$t_{ad} = \frac{T_{ad}}{N - 1}, \tag{7.46}$$

and the total clock hours of test time will have to be

$$t_d = T_1 + t_{ad} = T_1 + \frac{T_{ad}}{N - 1} = T_1 + \frac{T_a - NT_1}{N - 1},$$

or

$$t_d = \frac{T_a - T_1}{N - 1}. \tag{7.47}$$

If the failed unit is *replaced* immediately, then

$$t_{ad} = \frac{T_{ad}}{N}, \tag{7.48}$$

and

$$t_d = T_1 + \frac{T_a d}{N} = T_1 + \frac{T_a - NT_1}{N},$$

or

$$t_d = \frac{T_a}{N}. \tag{7.49}$$

For $CL = 90\%$ and a *nonreplacement test*

$$t_d = \frac{3.8895m_G - T_1}{N - 1}, \tag{7.50}$$

and in an *immediate replacement test*

$$t_d = \frac{3.8895 m_G}{N}. \tag{7.51}$$

For $CL = 95\%$ and a *nonreplacement test*

$$t_d = \frac{4.744 m_G - T_1}{N - 1}, \tag{7.52}$$

and in an *immediate replacement test*

$$t_d = \frac{4.744 m_G}{N}. \tag{7.53}$$

If a *second failure* occurs, or is allowed, in the test, then $r = 2$, and

$$
\begin{aligned}
m_{L1} \;=\; m_G &= \frac{2T_a}{\chi^2_{\alpha;2r+2}}, \\
&= \frac{2T_a}{\chi^2_{\alpha;(2\times2+2)}}, \\
&= \frac{2T_a}{\chi^2_{\alpha;6}},
\end{aligned}
$$

and

$$T_a = m_G(\tfrac{1}{2}\chi^2_{\alpha;6}). \tag{7.54}$$

For $CL = 90\%$,

$$T_a = m_G(\tfrac{1}{2}\chi^2_{0.10;6}) = m_G(\tfrac{1}{2} \times 10.645),$$

or

$$T_a = 5.323 m_G. \tag{7.55}$$

For $CL = 95\%$,

$$T_a = m_G(\tfrac{1}{2}\chi^2_{0.05;6}) = m_G(\tfrac{1}{2} \times 12.592),$$

or

$$T_a = 6.296m_G. \tag{7.56}$$

Equations (7.55) and (7.56) give the unit hours of operation that have to be accumulated by all units in the test to demonstrate m_G at the confidence levels of 90% and 95%, respectively, *with two failures*.

If the first failure occurs at T_1 with the second at $T_2 > T_1$, how many additional unit hours of operation, T_{ad}, have to be accumulated, without an additional failure, to demonstrate m_G? For a *nonreplacement test* this is found from

$$T_{ad} = T_a - [NT_1 + (N-1)(T_2 - T_1)],$$

or

$$T_{ad} = T_a - [T_1 + (N-1)T_2]. \tag{7.57}$$

The additional clock hours, without an additional failure, will then be

$$t_{ad} = \frac{T_{ad}}{N-2} = \frac{T_a - [T_1 + (N-1)T_2]}{N-2}, \tag{7.58}$$

and the total clock hours of test time will then be

$$t_d = T_2 + t_{ad} = T_2 + \frac{T_a - [T_1 + (N-1)T_2]}{N-2},$$

or

$$t_d = \frac{T_a - (T_1 + T_2)}{N-2}. \tag{7.59}$$

Then, for $CL = 90\%$,

$$t_d = \frac{5.323m_G - (T_1 + T_2)}{N-2}, \tag{7.60}$$

and for $CL = 95\%$,

$$t_d = \frac{6.296m_G - (T_1 + T_2)}{N-2}. \tag{7.61}$$

If the test is an *immediate replacement test*; then, after the second failure,

$$T_{ad} = T_a - [NT_1 + N(T_2 - T_1)],$$

or

$$T_{ad} = T_a - NT_2. \tag{7.62}$$

Then,

$$t_{ad} = \frac{T_{ad}}{N} = \frac{T_a - NT_2}{N}, \tag{7.63}$$

and

$$t_d = T_2 + t_{ad} = T_2 + \frac{T_a - NT_2}{N}, $$

or

$$t_d = \frac{T_a}{N}. \tag{7.64}$$

For $CL = 90\%$,

$$t_d = \frac{5.323m_G}{N}, \tag{7.65}$$

and for $CL = 95\%$,

$$t_d = \frac{6.296m_G}{N}. \tag{7.66}$$

Similar equations may be derived for additional failures, for the failure terminated test case, and for other confidence levels.

EXAMPLE 7–9

Ten identical exponential units are in a nonreplacement reliability test. Two failures occur: The first at $T_1 = 300$ hr and the second at $T_2 = 600$ hr. At a 95% confidence level, determine the following:

1. What should the accumulated test time in unit hours be, to meet a target *MTBF* of $m_G = 1,000$ hr, if the test is time terminated?

2. What should the accumulated test time in unit hours be, to meet the same target *MTBF*, if the test is failure terminated?

3. What will the clock hours of test time be in Case 1?

4. What will be the clock hours of test time in Case 2?

5. Which test type requires the longer clock hours of test time?

6. If the test were allowed to run for a period of time which is 25% longer than the required unit hours, and during this period no additional failures occur, with what confidence level would the m_G have been demonstrated?

SOLUTIONS TO EXAMPLE 7–9

1. From Eq. (7.56), for two failures, the accumulated unit hours of test time in a time terminated test should be

$$T_a = 6.296 m_G = 6.296 \times 1,000,$$

or

$$T_a = 6,296 \text{ unit hr}$$

for $CL = 95\%$.

2. In a failure terminated test with two failures

$$m_{L1} = m_G = \frac{2T_a}{\chi^2_{\alpha;2r}},$$

or

$$T_a = m_G(\frac{1}{2}\chi^2_{\alpha;2r}),$$

and for $CL = 95\%$ and $r = 2$

$$T_a = m_G(\frac{1}{2}\chi^2_{0.05;4}) = m_G(\frac{1}{2} \times 9.488),$$

or

$$T_a = 4.744 m_G.$$

For this case

$$T_a = 4.744 \times 1,000,$$

or

$$T_a = 4,744 \text{ unit hr.}$$

This time is substantially less than that for the time terminated test case.

3. In Case 1, which is a time terminated, nonreplacement test, the clock hours of test duration, from Eq. (7.61), will be

$$t_d = \frac{6.296 m_G - (T_1 + T_2)}{N - 2},$$

or

$$t_d = \frac{6.296 \times 1,000 - (300 + 600)}{10 - 2} = \frac{5,396}{8},$$

or

$$t_d = 674.5 \text{ clock hr.}$$

4. In Case 2, which is a failure terminated, nonreplacement test, the clock hours of test duration will be

$$t_d = T_2,$$

or

$$t_d = 600 \text{ hr,}$$

because the test has to be terminated when the second failure occurs, this being a failure terminated test. A caution is in order here! One should check, in such a test, to ascertain that at least the required hours, T_a, have been accumulated to prove that the target *MTBF* has been attained, if not exceeded. In this case, the actual unit hours accumulated, $T_{a,a}$ is

$$T_{a,a} = 300 + 600 + (10 - 2)600,$$

or

$$T_{a,a} = 5,700 \text{ unit hr.}$$

The required hours, from Case 2, are

$$T_a = 4,744 \text{ unit hr.}$$

Consequently, since

$$T_{a,a} > T_a,$$

we can conclude that we have proven, and even exceeded, our *MTBF* goal.

5. From the answers in Cases 3 and 4 it may be seen that the time terminated test requires the longer clock hours of test time.

6. This is a time terminated case and

$$T_a' = 1.25T_a.$$

From Case 1, $T_a = 6,296$ unit hr. Therefore,

$$m_G = \frac{2T_a'}{\chi^2_{\alpha;6}} = \frac{2 \times 1.25T_a}{\chi^2_{\alpha;6}} = \frac{2.5T_a}{\chi^2_{\alpha;6}},$$

then

$$\chi^2_{\alpha;6} = \frac{2.5T_a}{m_G} = \frac{2.5 \times 6,296}{1,000},$$

or

$$\chi^2_{\alpha;6} = 15.740.$$

From Appendix D, $\alpha \cong 0.015$. Therefore, m_G has been demonstrated with a $CL \cong 98.5\%$.

7.11 EXPECTED WAITING TIMES FOR RELIABILITY TESTS IN THE EXPONENTIAL CASE

7.11.1 INTRODUCTION

The expected waiting time, $E(t)$, is defined as the average elapsed time from the start of the test to the time the decision is reached as to product acceptability.

In this section four types of tests are discussed:

1. Failure terminated with replacement.

2. Failure terminated without replacement.

3. Time terminated with replacement.

4. Time terminated without replacement.

7.11.2 EXPECTED WAITING TIME FOR A FAILURE TERMINATED RELIABILITY TEST WITH REPLACEMENT

Assume that the true mean life of the tested product is m, and the sample size is N. In this case, the expected waiting time, $E(t)$, is equal to the expectation of the *rth* order statistic, or

$$E(t) = E(T_{r,N}), \tag{7.67}$$

where $T_{r,N}$ is the *rth* order statistic, or

$$T_{r,N} = T_{1,N} + (T_{2,N} - T_{1,N}) + \ldots + (T_{r,N} - T_{r-1,N}), \qquad (7.68)$$

and the

$$T_{1,N}, (T_{2,N} - T_{1,N}), \ldots, (T_{r,N} - T_{r-1,N})$$

are independent, and identically exponentially distributed, with mean life m/N [5, pp. 6-8; 6, pp. 310-311]. See Appendix 7A for the proof. Then,

$$E(T_{r,N}) = E(T_{1,N}) + E(T_{2,N} - T_{1,N}) + \cdots + E(T_{r,N} - T_{r-1,N}),$$

where

$$E(T_{i,N}) = \frac{m}{N},$$

or

$$E(T_{r,N}) = \frac{rm}{N}.$$

Substituting this result into Eq. (7.67) yields

$$E(t) = \frac{rm}{N}. \qquad (7.69)$$

EXAMPLE 7-10

Twelve units are placed on a test, the failed units are replaced, and the test is terminated after the sixth failure. Calculate the expected waiting time when $m = 1,000$ hr.

SOLUTION TO EXAMPLE 7-10

From the given and Eq. (7.69),

$$E(t) = \frac{6 \times 1,000}{12} = 500 \text{ hr.}$$

7.11.3 EXPECTED WAITING TIME FOR A FAILURE TERMINATED RELIABILITY TEST WITHOUT REPLACEMENT

In this case, following the same reasoning as in Section 7.11.2,

$$E(t) = E(T_{r,N}),$$
(7.70)

and

$$T_{r,N} = T_{1,N} + (T_{2,N} - T_{1,N}) + ... + (T_{r,N} - T_{r-1,N}).$$
(7.71)

Let

$$W_1 = T_{1,N}, W_2 = T_{2,N} - T_{1,N}, ..., W_r = T_{r,N} - T_{r-1,N}.$$

It has been proven [6, pp. 304-309; 7, pp. 100-101] that
  1. $W_1, \cdots, W_r$ are mutually independent, and

  2. $W_i \sim EXP[m/(N - i + 1)]$. See Appendix 7C.1.

Therefore, for any $1 \le r \le N$,

$$E(T_{r,N}) = m \left[\sum_{i=1}^{r} \frac{1}{N - i + 1}\right].$$
(7.72)

Substituting Eq. (7.72) into Eq. (7.70) yields

$$E(t) = m \left[\sum_{i=1}^{r} \frac{1}{N - i + 1}\right].$$
(7.73)

EXAMPLE 7–11

Rework Example 7–10 for the nonreplacement case.

SOLUTION TO EXAMPLE 7–11

From the given and Eq. (7.73),

$$
\begin{aligned}
E(t) &= 1,000 \left[\sum_{i=1}^{6} \frac{1}{12 - i + 1}\right], \\
&= 1,000 \left(\frac{1}{12} + \frac{1}{11} + \frac{1}{10} + \frac{1}{9} + \frac{1}{8} + \frac{1}{7}\right), \\
&= 1,000 (0.617),
\end{aligned}
$$

or

$$E(t) = 617 \text{ hr.}$$

Comparing Example 7–10 and Example 7–11 results, it may be seen that the expected waiting time is longer in the nonreplacement case than in the replacement test case.

7.11.4   EXPECTED WAITING TIME FOR A TIME TERMINATED RELIABILITY TEST WITH REPLACEMENT

In this case the number of failures follows the Poisson distribution with parameter $\bar{r} = Nt_d/m$, that is

$$P(r = k) = \frac{\bar{r}^k e^{-\bar{r}}}{k!}, k = 0, 1, 2, ..., c,$$

and

$$P(r \geq c + 1) = 1 - \sum_{k=0}^{c} \frac{\bar{r}^k e^{-\bar{r}}}{k!}.$$

Then, the expected waiting time is given by [8]

$$E(t) = t_d \sum_{k=0}^{c} P(r = k) + P(r \geq c + 1)[E(T_{c+1,N}|r = c + 1)]. \quad (7.74)$$

Let $E(T_{c+1,N})$ be the unconditional expected waiting time to get the $(c+1)th$ failure; then,

$$
\begin{aligned}
E(T_{c+1,N}) &= \sum_{k=0}^{c} P(r = k)[E(T_{c+1,N}|r = k) \\
&+ P(r \geq c + 1)[E(T_{c+1,N}|r = c + 1)].
\end{aligned}
\quad (7.75)
$$

This gives

$$P(r \geq c + 1)[E(T_{c+1,N}|r = c + 1)] = E(T_{c+1,N})$$

$$- \sum_{k=0}^{c} P(r = k)[E(T_{c+1,N}|r = k)]. \quad (7.76)$$

Substituting Eq. (7.76) into Eq. (7.74) yields

$$E(t) = E(T_{c+1,N}) + \sum_{k=0}^{c} P(r = k)[t_d - E(T_{c+1,N}|r = k)]. \quad (7.77)$$

Since the underlying distribution is the exponential,

$$E(T_{c+1,N}|r = k) = t_d + E(T_{c+1-k,N}), k = 0, 1, 2, \cdots, c. \quad (7.78)$$

Furthermore, from Eqs. (7.67) and (7.69) and for $r = c + 1 - k$,

$$E(T_{c+1-k,N}) = \frac{(c+1-k)m}{N}.$$ (7.79)

Substituting Eqs. (7.78) and (7.79) into Eq. (7.77) yields

$$E(t) = E(T_{c+1,N}) + \sum_{k=0}^{c} P(r = k)[t_d - t_d - \frac{c+1-k}{N}m],$$

or

$$E(t) = \frac{c+1}{N}m - \frac{c+1}{N}m\sum_{k=0}^{c} P(r = k) + \frac{m}{N}\sum_{k=0}^{c} kP(r = k).$$ (7.80)

Since,

$$P(r = k) = \frac{\bar{r}^k e^{-\bar{r}}}{k!};$$

then, Eq. (7.80) becomes

$$E(t) = \frac{m}{N}\Big[(c+1)\Big(1 - \sum_{k=0}^{c} \frac{\bar{r}^k e^{-\bar{r}}}{k!}\Big) + \bar{r}\sum_{k=0}^{c-1} \frac{\bar{r}^k e^{-\bar{r}}}{k!}\Big],$$

or, from Eq. (7.89),

$$E(t) = \frac{m}{N}E(r).$$ (7.81)

EXAMPLE 7–12

The test plan is as follows: $N = 16$, the allowable number of failures is $c = 4$ and $t_d = 500$ hr. Calculate the expected waiting time for $m = 5,000$ hr if this were a replacement test.

SOLUTION TO EXAMPLE 7–12

From Example 7–14 of Section 7.12,

$$E(r) = 1.59.$$

Substituting this value into Eq. (7.81) gives

$$E(t) = \frac{5,000}{16}(1.59),$$

or

$$E(t) = 496.9 \text{ hr}.$$

7.11.5 EXPECTED WAITING TIME FOR A TIME TERMINATED RELIABILITY TEST WITHOUT REPLACEMENT

In this case the number of failures follows the binomial distribution; i.e.,

$$P(r = k) = \binom{N}{r} q^k (1 - q)^{N-k}, \quad k = 0, 1, 2, \cdots, c,$$

and

$$P(r \geq c + 1) = 1 - \sum_{k=0}^{c} \binom{N}{r} q^k (1 - q)^{N-k},$$

where

$$q = 1 - e^{-t_d/m},$$

and

$$m = \text{ true mean life of the tested product.}$$

Thus, analogous to Eq. (7.77),

$$E(t) = E(T_{c+1,N}) + \sum_{k=0}^{c} P(r = k)[t_d - E(T_{c+1,N}|r = k)], \quad (7.82)$$

and analogous to Eq. (7.78),

$$E(T_{c+1,N}|r = k) = t_d + E(T_{c+1-k,N-k}), k = 0, 1, 2, \cdots, c, \quad (7.83)$$

where $E(T_{c+1-k})$ is the unconditional expected waiting time to get the $(c + 1 - k)th$ failure in a random sample of size $(N - k)$.

From Eq. (7.72)

$$
\begin{aligned}
E(T_{c+1-k,N-k}) &= m \sum_{i=1}^{c+1-k} \frac{1}{N - k - i + 1}, \\
&= m \Big(\frac{1}{N - k} + \frac{1}{N - k - 1} + \cdots + \frac{1}{N - c} \Big). \quad (7.84)
\end{aligned}
$$

Rewriting Eq. (7.84) yields

$$
\begin{aligned}
E(T_{c+1-k,N-k}) &= m \Big(\frac{1}{N} + \frac{1}{N - 1} + \cdots + \frac{1}{N - c} \Big) \\
&\quad - m \Big(\frac{1}{N} + \frac{1}{N - 1} + \cdots + \frac{1}{N - k + 1} \Big),
\end{aligned}
$$

or

$$E(T_{c+1-k,N-k}) = E(T_{c+1,N}) - E(T_{k,N}). \qquad (7.85)$$

Substituting Eqs. (7.83) and (7.85) into Eq. (7.82) leads to

$$E(t) = E(T_{c+1,N}) + \sum_{k=0}^{c} P(r = k)[E(T_{k,N})]$$

$$- E(T_{c+1,N}) \sum_{k=0}^{c} P(r = k), \qquad (7.86)$$

or

$$E(t) = \left[1 - \sum_{k=0}^{c} P(r = k)\right] E(T_{c+1,N}) + \sum_{k=0}^{c} P(r = k)[E(T_{k,N})].$$

Therefore,

$$E(t) = \sum_{k=1}^{c} P(r = k)[E(T_{k,N})] + P(r \geq c + 1)[E(T_{c+1,N})], \quad (7.87)$$

where $E(T_{k,N})$ is given by Eq. (7.72).

EXAMPLE 7–13

Rework Example 7–12 for the nonreplacement test case.

SOLUTION TO EXAMPLE 7–13

Since $N = 16, c = 4, t_d = 500$ hr and $m = 5,000$ hr,

$$q = 1 - e^{-500/5,000} = 0.095.$$

To calculate $E(t)$, use the procedure of Table 7.3. Consequently,

$$E(t) = 497.67 \text{ hr}.$$

Comparing this result with that obtained for the replacement case, it may be seen that $E(t)$ is somewhat shorter in the replacement test case than in the nonreplacement test case.

7.12   EXPECTED NUMBER OF FAILURES FOR RELIABILITY TESTS IN THE EXPONENTIAL CASE

7.12.1   INTRODUCTION

The expected number of failures, $E(r)$, is defined as the average number of failures that will have occurred at the time the decision as to

TABLE 7.3–Procedure for calculating the expected test time for the nonreplacement case of Example 7–13.

1	2	3	4	5	6
k	$\binom{16}{k}0.095^k(0.905)^{16-k}$	m	$\frac{1}{N-k+1}$	$\sum \frac{1}{N-k+1}$	$E(t)$, 2 × 3 × 5
0	0.2025				
1	0.3401	5,000	0.0625	0.0625	106.28
2	0.2677	5,000	0.0667	0.1292	172.93
3	0.1312	5,000	0.0714	0.2006	131.59
4	0.0448	5,000	0.0769	0.2775	62.16
≥ 5	0.0137*	5,000	0.0833	0.3608	24.71
*$P(r \geq 5) = 1 - \sum\limits_{k=0}^{4} P(r = k) = 0.0137$				$\sum = 497.67$	

product acceptability is reached. For failure terminated reliability test plans, this number of failures is known in advance of the test, but for the time terminated reliability test plans, this number cannot be predetermined. Therefore, the latter cases need to be discussed.

7.12.2   THE EXPECTED NUMBER OF FAILURES IN A TIME TERMINATED RELIABILITY TEST WITH REPLACEMENT

The test plan is given by three numbers: (1) Sample size, N. (2) Test time, t_d. (3) Maximum allowed failure number, c. The number of failures, r, may have any value of $0, 1, 2, \cdots$, up to $c + 1$; since testing will be terminated even if the test time is less than t_d, but the $(c+1)th$ failure has already occurred.

If the true mean life of the tested product is m, then the actual failure number, r, follows the Poisson distribution with parameter $\bar{r} = Nt_d/m$. The expected number of failures is then given by [8]

$$E(r) = \sum_{k=0}^{c} k \cdot \frac{\bar{r}^k e^{-\bar{r}}}{k!} + (c+1)\left(1 - \sum_{k=0}^{c} \frac{\bar{r}^k e^{-\bar{r}}}{k!}\right). \qquad (7.88)$$

Rearranging Eq. (7.88) yields

$$E(r) = \bar{r} \sum_{k=0}^{c-1} \frac{\bar{r}^k e^{-\bar{r}}}{k!} + (c+1)\left(1 - \sum_{k=0}^{c} \frac{\bar{r}^k e^{-\bar{r}}}{k!}\right). \tag{7.89}$$

EXAMPLE 7–14

The test plan is as follows: $N = 16, c = 4$ and $t_d = 500$ hr. Find the expected number of failures if the true mean life of the tested product is 5,000 hr.

SOLUTION TO EXAMPLE 7–14

From the given,

$$\bar{r} = \frac{16 \times 500}{5,000} = 1.6.$$

Substitution of the values of $\bar{r}$ and c into Eq. (7.89) yields

$$E(r) = 1.6 \sum_{k=0}^{3} \frac{1.6^k e^{-1.6}}{k!} + 5\left(1 - \sum_{k=0}^{4} \frac{1.6^k e^{-1.6}}{k!}\right).$$

Entering the tables of the cumulative Poisson distribution yields

$$\sum_{k=0}^{3} \frac{1.6^k e^{-1.6}}{k!} = 0.921,$$

and

$$\sum_{k=0}^{4} \frac{1.6^k e^{-1.6}}{k!} = 0.976.$$

Therefore,

$$E(r) = 1.6 \times 0.921 + 5(1 - 0.976),$$

or

$$E(r) = 1.59.$$

7.12.3 THE EXPECTED NUMBER OF FAILURES IN A TIME TERMINATED RELIABILITY TEST WITHOUT REPLACEMENT

In this case, as discussed in Section 7.11.5, the number of failures, r, follows the binomial distribution with parameter $q = 1 - e^{-t_d/m}$. Thus the expected number of failures is given by [1]

$$E(r) = \sum_{k=0}^{c} k \cdot \binom{N}{k} q^k (1-q)^{N-k}$$

$$+ (c+1)\left[1 - \sum_{k=0}^{c} \binom{N}{k} q^k (1-q)^{N-k}\right]. \qquad (7.90)$$

But,

$$\sum_{k=0}^{c} k \binom{N}{k} q^k (1-q)^{N-k}$$

$$= \sum_{k=0}^{c} k \frac{N!}{k!(N-k)!} q^k (1-q)^{N-k},$$

$$= Nq \sum_{k=1}^{c} \frac{(N-1)!}{(k-1)!(N-k)!} q^{k-1} (1-q)^{N-k},$$

$$= Nq \sum_{j=0}^{c-1} \frac{(N-1)!}{j!(N-1-j)!} q^{j} (1-q)^{N-1-j},$$

$$= Nq \sum_{j=0}^{c-1} \binom{N-1}{j} q^{j} (1-q)^{N-1-j}, \qquad (7.91)$$

where in Eq. (7.91) the needed transformations are $N(N-1)! = N!$ and $k/k! = 1/(k-1)!$ Therefore, Eq. (7.91) becomes

$$E(r) = Nq \sum_{k=0}^{c-1} \binom{N-1}{k} q^k (1-q)^{N-1-k}$$

$$+ (c+1)[1 - \sum_{k=0}^{c} \binom{N}{k} q^k (1-q)^{N-k}]. \qquad (7.92)$$

EXAMPLE 7–15

Rework Example 7–14 for the nonreplacement test case.

SOLUTION TO EXAMPLE 7–15

From the given,

$$q = 1 - e^{-500/5,000} = 0.095.$$

Substituting the values of q, N, c and t_d into Eq. (7.92) yields

$$E(r) = 16(0.095) \sum_{k=0}^{3} \binom{15}{k} 0.095^k 0.905^{15-k}$$

$$+ 5[1 - \sum_{k=0}^{4} \binom{16}{k} 0.095^k 0.905^{16-k}].$$

Entering the tables of the cumulative binomial distribution yields

$$\sum_{k=0}^{3} \binom{15}{k} 0.095^k 0.905^{15-k} = 0.9526,$$

and

$$\sum_{k=0}^{4} \binom{16}{k} 0.095^k 0.905^{16-k} = 0.9862.$$

Therefore,

$$E(r) = 16(0.095)(0.9526) + 5(1 - 0.9862),$$

or

$$E(r) = 1.52.$$

7.13 OPERATING CHARACTERISTIC CURVES FOR RELIABILITY TESTS IN THE EXPONENTIAL CASE

7.13.1 INTRODUCTION

In this section, the formulas for calculating the operating characteristic, OC, curves for reliability tests in the exponential case are given.

The OC curve of a test is the plot of the probability of passing the test (acceptance probability) plotted along the y axis versus any

chosen test parameter, in this case the true or designed-in mean life for a given test plan, along the x axis. Each different test plan has its own different OC curve. Four situations are considered:

1. Failure terminated test with replacement.
2. Failure terminated test without replacement.
3. Time terminated test with replacement.
4. Time terminated test without replacement.

7.13.2 FAILURE TERMINATED TEST WITH REPLACEMENT

The test procedure is as follows: Place N items on test simultaneously, replace failed items until r failures occur, then terminate the test. Calculate $\hat{m}$ using

$$\hat{m} = \frac{T_a}{r}, \tag{7.93}$$

where

$$
\begin{aligned}
\hat{m} &= \text{estimate of the mean life,} \\
r &= \text{termination number of failures,} \\
T_a &= Nt_d, \text{ the accumulated unit-hours of test time,} \\
N &= \text{sample size,}
\end{aligned}
$$
and
$$
t_d = \text{time when the } rth \text{ failure occurs.}
$$

Accept the product if $\hat{m} \geq m_G$, otherwise reject.

The customer may require that if the true $MTBF$ of the product is m_L, it should be rejected with a high probability of $(1 - \beta)$, then

$$P(\text{Accept}|m = m_L) = P(\hat{m} \geq m_G|m = m_L) = \beta,$$

or

$$P(\frac{2r\hat{m}}{m_L} \geq \frac{2rm_G}{m_L}|m = m_L) = \beta,$$

which gives

$$m_G = m_L = \frac{2r\hat{m}}{\chi^2_{\beta;2r}}.$$

In this case, $P(A)$ for the OC curve is given by

$$P(A) = P(\hat{m} \geq m_G|m). \tag{7.94}$$

Since $2r\hat{m}/m$ is χ^2 distributed with $2r$ degrees of freedom, Eq. (7.94) can be written as

$$P(A) = P[\chi^2(2r) \geq \frac{2rm_G}{m}]. \tag{7.95}$$

Thus the OC curve can be calculated using Eq. (7.95) and χ^2 tables.

EXAMPLE 7–16

Twelve units are placed on test, the failed units are replaced and the test is terminated after the sixth failure.

If $m_G = 600$ hr, calculate the $P(A)$ and draw the OC curve for this test plan.

SOLUTION TO EXAMPLE 7–16

For $m = 2,000$ hr, from Eq. (7.95),

$$P(A) = P[\chi^2(2r) \geq \frac{2 \times 6 \times 600}{2,000}],$$

or

$$P(A) = P[\chi^2(12) \geq 3.6].$$

Entering χ^2 tables yields

$$P(A) = 0.990.$$

Additional values have been similarly calculated for different values of m, and are given in Table 7.4. The OC curve for this test plan is shown in Fig. 7.5, Curve I.

TABLE 7.4–Values of $P(A)$ for Examples 7–16, 7–18 and 7–19.

m, hr	P(A)		
	Example 7–16	Example 7–18	Example 7–19
2,000	0.990	0.949	0.968
1,750	0.981	0.918	0.949
1,500	0.962	0.868	0.914
1,250	0.930	0.781	0.851
1,000	0.840	0.629	0.731
750	0.652	0.384	0.509
500	0.275	0.100	0.180
200	0.0005	0.00002	0.0002

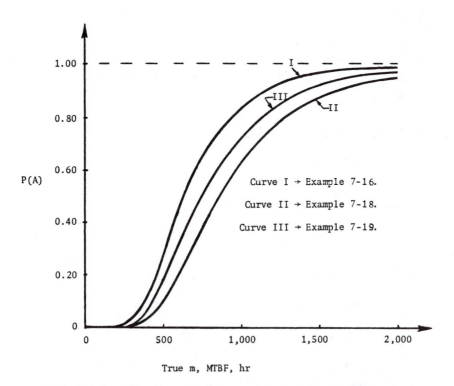

Fig. 7.5– OC curves for Examples 7–16, 7–18 and 7–19.

7.13.3 FAILURE TERMINATED TEST WITHOUT REPLACEMENT

The test procedure and the $P(A)$ formula for this test is the same as that of the failure terminated test with replacement, except that the failed units are not replaced. Then the accumulated unit hours of test, T_a, are given by

$$T_a = \sum_{i=1}^{r} T_i + (N - r)t_d,$$

where T_i is the time when the ith failure occurs.

EXAMPLE 7–17

In a test $N = 10, r = 5, m_G = 500$ hr, and the true designed-in mean life, m, is 1,500 hr. Calculate the producer's risk, α, for this test.

SOLUTION TO EXAMPLE 7–17

From Eq. (7.95)

$$P(A) = P[\chi^2(10) \geq \frac{2 \times 5 \times 500}{1,500}],$$

or

$$P(\chi^2(10) \geq 3.33).$$

Entering χ^2 tables yields

$$P(A) = 0.97;$$

therefore, the producer's risk is

$$\alpha = 1.00 - P(A) = 1.00 - 0.97 = 0.03, \text{ or } 3\%.$$

7.13.4 TIME TERMINATED TEST WITH REPLACEMENT

The test procedure is as follows: Place N units on test, replace the failed units, and terminate the test at a preassigned time, t_d, in clock hours.

Accept the product if $r \leq a$,

Reject the product if $r > a$,

where

r = actual failure number during t_d,

and

a = allowed maximum number of failures during t_d test clock hours.

Let the test plan be denoted by (N, a, t_d). If the true mean life of the tested units is m, and N such units are tested, then the number of failures, r, until time t_d, follows the Poisson distribution (POI) with parameter Nt_d/m, that is

$$r \sim POI(\frac{Nt_d}{m}).$$

Then, the acceptance probability for this test plan is

$$P(A) = P(r \leq a|m) = \sum_{k=0}^{a} \frac{\bar{r}^k e^{-\bar{r}}}{k!}, \tag{7.96}$$

where

$\bar{r} = Nt_d/m$ = expected number of failures.

Equation (7.96) is now used to calculate the OC curve.

EXAMPLE 7–18

If $N = 16, a = 4$ and $t_d = 250$ hr, calculate $P(A)$ and draw the OC curve.

SOLUTION TO EXAMPLE 7–18

For $m = 2,000$ hr,

$$\bar{r} = \frac{16 \times 250}{2,000} = 2.0.$$

From Eq. (7.96),

$$P(A) = \sum_{k=0}^{4} \frac{2.0^k e^{-2.0}}{k!}.$$

Entering the cumulative Poisson distribution tables yields

$$P(A) = 0.949.$$

Additional values were calculated similarly, for different values of m and are given in Table 7.4. The OC curve for this test plan is shown in Fig. 7.5, Curve II.

7.13.5 TIME TERMINATED TEST WITHOUT REPLACEMENT

The test procedure is the same as that given in the previous case, except that the failed units are not replaced. The accept and reject conditions are also the same. If the true mean life of the tested units is m, and N of them are tested, then the probability that they will fail before t_d is

$$q = 1 - e^{-t_d/m}.$$

Thus, the number of failures that will occur among the N units in the test, before time t_d, follows the binomial distribution, or

$$P(r = k) = \binom{N}{k} q^k (1 - q)^{N-k}.$$

Therefore, the probability that the number of failures before time t_d does not exceed a is given by

$$P(A) = \sum_{k=0}^{a} \binom{N}{k} q^k (1 - q)^{N-k}, \qquad (7.97)$$

which is the equation for calculating the OC curve for this test.

EXAMPLE 7–19

Rework Example 7–18 for the nonreplacement case.

SOLUTION TO EXAMPLE 7–19

In this case, for $m = 2,000$ hr,

$$q = 1 - e^{-250/2,000} = 0.1175.$$

Substituting this value into Eq. (7.97) yields

$$P(A) = \sum_{k=0}^{4} \binom{16}{k} 0.1175^k (1 - 0.1175)^{16-k}.$$

Entering the tables of the cumulative binomial distribution, with the values of N and q, gives

$$P(A) = 0.968.$$

Additional values have been calculated and are given in Table 7.4. The OC curve for this test plan is shown in Fig. 7.5, Curve III.

7.14 BINOMIAL RELIABILITY DEMONSTRATION TEST WHEN THE TEST TIME IS DIFFERENT THAN THE MISSION TIME, FOR THE EXPONENTIAL CASE

7.14.1 APPLICABILITY

This test is applicable when the units are exponential and it is required that their reliability for a mission of t duration be demonstrated but the available test duration is less than t. In particular the lower, one-sided confidence limit on the reliability of these units needs to be determined at a specified confidence level, and the test sample size and the number of allowable failures is desired to be established [4, pp. 303-307].

7.14.2 METHODOLOGY

A random sample of N units is tested, preferably simultaneously, to save time in reaching a decision, for time T not necessarily equal to the required mission duration, t. At the end of time T the number of failures are recorded, and if the number of failures, r, does not exceed the allowable then the specified R_{L1} is demonstrated at a specified CL. The test plan consists of determining N and r when t is the required mission duration.

Given R_{L1}, m_{L1} is obtained from

$$R_{L1}(t) = e^{-\frac{t}{m_{L1}}}, \tag{7.98}$$

from which

$$m_{L1} = -\frac{t}{log_e R_{L1}}. \tag{7.99}$$

The corresponding R_{L1} for a test time of $T < t$ would then be obtained from

$$R_{L1}(T < t) = e^{-\frac{T}{m_{L1}}}. \tag{7.100}$$

Now from the figures in Appendix M, choosing the one for the specified confidence level and entering across the top with the value of $R_{L1}(T)$ given by Eq. (7.100), the various combinations of N and r which satisfy the requirements of the reliability to be demonstrated are obtained.

The figures in Appendix M have been developed using

$$\sum_{j=0}^{r} \binom{N}{j} R_{L1}^{N-j}(1 - R_{L1})^{j} = \beta = 1 - CL. \qquad (7.101)$$

EXAMPLE 7–20

An electronic device is required to operate within its performance specifications for 100 hr. The specified minimum acceptable reliability is 0.990 at a 90% confidence level. If only 60 hr of test time were available, find three desirable test plans to demonstrate the specified reliability.

SOLUTION TO EXAMPLE 7–20

From Eq. (7.99),

$$m_{L1} = -\frac{100}{log_e 0.990} = 9,950 \text{ hr.}$$

From Eq. (7.100),

$$R_{L1}(T = 60 \text{ hours}) = e^{-\frac{60}{9,950}} = 0.9940.$$

Entering Fig. M.3, in Appendix M, with $R_{L1} = 0.9940$ across the top, one may find the following most desirable test plans:

1. $N = 383; r = 0$.

2. $N = 645; r = 1$.

3. $N = 885; r = 2$.

Their desirability is based on the fact that they require the smaller test sample sizes; consequently, the shorter expected test duration and lower test cost.

PROBLEMS

7-1. Onehundred identical units are subjected to a reliability test of 2,000 clock hr with the following results: Two failures occur at 500 clock hr, five failures at 800 clock hr, and three failures at 1,500 clock hr. Determine the following:

(1) The $\hat{m}$ of these units when the failed units are replaced.

(2) The $\hat{m}$ when they are not replaced.

(3) The one-sided confidence limits on m, based on this sample, at a 90% confidence level, when the failed units are not replaced.

(4) The one-sided, 90% confidence limits on the reliability for a 10 hr mission when the failed units are not replaced, based on this sample.

(5) The two-sided limits on m, based on this sample, at a 90% confidence level when the failed units are not replaced.

(6) The two-sided, 90% confidence limits on the reliability for a 10 hr mission with no replacements, based on this sample.

(7) The lower, one-sided, 90% confidence limit on m, based on this sample, if no failures occur.

(8) The lower, one-sided, 90% confidence limit on the reliability for a 10 hr mission, based on this sample, if no failures occur.

7-2. If in Problem 7–1 the last one of the three failures occurs at 2,000 hr, at which time the test is terminated, and the failed units are not replaced, determine the following at a confidence level of 90% and write down the associated probabilistic statements (The mission duration is 10 hr):

(1) $\hat{m}$.

(2) m_{L1} and m_{U1}.

(3) R_{L1} and R_{U1}.

(4) m_{L2} and m_{U2}.

(5) R_{L2} and R_{U2}.

7-3. Twenty exponential units are tested for 500 hr (clock hours of straight test time) and the following data are obtained:

Failures occur at 110, 180, 300 and 410 hr, and these units are not replaced.

Additional failures occur at 290, 340, 460 and 480 hr, and these failed units are replaced with units from the same production lot.

Do the following:

(1) Determine the two-sided, 80% confidence level limits on the *MTBF*, based on this sample.

(2) Write a probabilistic statement about the confidence interval in Case 1.

(3) Determine the two-sided, 80% confidence level limits on the reliability demonstrated by these units for a one-hour mission.

(4) Write a probabilistic statement about the confidence interval in Case 3.

(5) Determine the lower, one-sided, 80% confidence level limit on the *MTBF*, based on this sample.

(6) Write a probabilistic statement about the confidence interval in Case 5.

(7) Determine the lower, one-sided, 80% confidence level limit on the reliability demonstrated by these units for a one-hour mission.

(8) Write a probabilistic statement about the confidence interval in Case 7.

(9) Determine the lower, one-sided, 90% confidence level limit on the *MTBF*, based on this sample, and write the associated probabilistic statement.

(10) Determine the lower, one-sided 90% confidence level limit on the reliability for a one-hour mission, based on this sample, and write the associated probabilistic statement.

(11) Assuming that the failure which had occurred at 480 hr now occurs at 500 hr, at which time the test is terminated, repeat Case 1.

(12) Repeat Case 2 for the situation in Case 11.

(13) Repeat Case 3 for the situation in Case 11.

(14) Repeat Case 4 for the situation in Case 11.

(15) Repeat Case 5 for the situation in Case 11.

(16) Repeat Case 6 for the situation in Case 11.

(17) Repeat Case 7 for the situation in Case 11.

(18) Repeat Case 8 for the situation in Case 11.

(19) Repeat Case 9 for the situation in Case 11.

(20) Repeat Case 10 for the situation in Case 11.

7-4. Twenty exponential units are tested for 1,500 hr (clock hours of straight test time) and the following data are obtained:

Failures occur at 1,290; 1,340; 1,460 and 1,480 hr, and these failed units are replaced with units from the same production lot.

Do the following:

(1) Determine the two-sided, 90% confidence level limits on the *MTBF*, based on this sample.

(2) Write a probabilistic statement about the confidence interval in Case 1.

(3) Determine the two-sided, 90% confidence level limits on the reliability demonstrated by these units for a mission of 10 hours.

(4) Write a probabilistic statement about the confidence interval in Case 3.

(5) Determine the lower, one-sided, 95% confidence level limit on the *MTBF*, based on this sample.

(6) Write a probabilistic statement about the confidence interval in Case 5.

(7) Determine the lower, one-sided, 95% confidence level limit on the reliability demonstrated by these units for a mission of 10 hours.

(8) Write a probabilistic statement about the confidence interval in Case 7.

(9) Assuming that the failure which had occurred at 1,480 hr now occurs at 1,500 hr at which time the test is terminated, repeat Case 1.

(10) Repeat Case 2 for the situation in Case 9.

(11) Repeat Case 5 for the situation in Case 9.

(12) Repeat Case 6 for the situation in Case 9.

(13) Repeat Case 7 for the situation in Case 9.

(14) Repeat Case 8 for the situation in Case 9.

7-5. Six identical exponential units are in a reliability test. Three failures occur at the following times: $T_1 = 450$ hr, $T_2 = 555$ hr and $T_3 = 610$ hr. At a 90% confidence level, determine the following:

(1) What should the accumulated test time in unit hours be to meet the *MTBF* goal of 1,500 hr if the test is time terminated?

(2) What should the accumulated test time in unit hours be to meet the same *MTBF* goal if the test is failure terminated?

(3) What should the clock hours of test time be in Case 1 in a nonreplacement test?

(4) What should the clock hours of test time be in Case 2 in a nonreplacement test, and when should the third failure occur to meet the *MTBF* goal at a 95% confidence level?

(5) Which test type will require the longer test time in clock hours, the time terminated or the failure terminated test?

(6) If the test were allowed to run for 15% more unit hours than required, during which period no additional failures occur, with what confidence level would the *MTBF* goal have been demonstrated?

(7) If the test were allowed to run for 15% more unit hours than required, during which period no additional failures occur, what *MTBF* would have been demonstrated at a 95% confidence level?

7-6. Twenty exponential units from the same production are reliability tested with *immediate replacement* of the failed units. Failures occur at the following times:

$T_1 = 350$ hr, $T_2 = 700$ hr, $T_3 = 1,050$ hr and $T_4 = 1,100$ hr.

At a 95% confidence level do the following:

(1) Find the clock hours in this failure terminated test.

(2) Find the unit hours of test time required to prove that an *MTBF* goal of 3,000 hr has been met in a failure terminated test, assuming four failures were allowed in this test.

(3) Find the clock hours of test time required in Case 2 to prove the *MTBF* goal.

(4) Given the target *MTBF* of 3,000 hr, determine the actual confidence level at which this *MTBF* has been proven based on the results of this test.

(5) Find the actual minimum *MTBF* proven by this test at the 95% confidence level.

(6) Determine the *total* number of units that would have been needed in this test to meet the *MTBF* goal without any additional failures.

7-7. Ten exponential units are placed in a 600 hr test. Three units fail at 190, 455 and 580 hr and are not replaced; and two units fail at 430 and 575 hr and are replaced.

(1) What is the *MTBF* estimate of these units?

(2) What lower and upper, two-sided, 95% confidence level limits can be put on the true *MTBF*, based on this sample?

(3) In practice, the one-sided, lower confidence limit is very useful. What is the one-sided, 95% confidence level, lower confidence limit of the true *MTBF*, based on this sample?

(4) Find the confidence limits on the reliability for Case 2, for a mission of 100 hr.

(5) Find the confidence limit on the reliability for Case 3, for a mission of 100 hr.

7-8. Ten exponential units are placed in a 1,000-hr test. Three units fail at 140, 390 and 580 hr and are not replaced; and two units fail at 300 and 650 hr and are replaced.

(1) What is the *MTBF* estimate of these units?

(2) What lower and upper, two-sided, 95% confidence level limits can be put on the true *MTBF*, based on this sample?

(3) In practice, the one-sided, lower confidence limit is very useful. What is the one-sided, 95% confidence level, lower confidence limit of the true *MTBF* , based on this sample?

(4) Find the confidence limits on the reliability for Case 2.

(5) Find the confidence limit on the reliability for Case 3.

7-9. Fifteen identical exponential units are reliability tested in a non-replacement test. Three failures occur as follows: The first at 300 hr, the second at 600 hr and the third at 900 hr. At a 95% confidence level determine the following:

(1) What should the test time in unit hours be to meet a target *MTBF* of at least 1,500 hr if the test is terminated soon after the third failure occurs? (A time terminated test!)

(2) What should the test time in unit hours be to meet a target *MTBF* of at least 1,500 hr if the test is failure terminated?

(3) What should the straight test time in Case 1 be in clock hours? Explain your answer in relation to the time to failure of the third unit.

(4) What should the straight test time in Case 2 be in clock hours? Explain your answer in relation to the time to failure of the third unit.

(5) Which one will require the longest straight test time in clock hours, the time or failure terminated test?

(6) If the test were allowed to run for 25% more unit hr than required, during which period no additional failures occurred, at what confidence level would the target *MTBF* be established?

7-10. Nine units are placed in a 100-hr test. Three units fail at 14, 39 and 58 hr, and are not replaced.

(1) What is the estimated *MTBF* of these units?

(2) What lower and upper, two-sided, 95% confidence level limits can be put on the *MTBF*, based on this sample?

(3) In practice, the one-sided lower confidence limit is very useful. What is the one-sided, 95% confidence level, lower confidence limit on the *MTBF*, based on this sample?

7-11. Nine units are placed in a 100-hr test. Three units fail at 14, 39, and 58 hr and are not replaced.

(1) What is the estimated *MTBF* of these units?

(2) What lower and upper, two-sided, 90% confidence level limits can be put on the estimated *MTBF*, based on this sample?

(3) In practice, the one-sided lower confidence limit is very useful. What is the one-sided, 90% confidence level, lower confidence limit for the estimated *MTBF*, based on this sample?

7-12. A semiconductor manufacturer provides to its customers the RELIABILITY DATA given in Table 7.5.

Do the following:

(1) Determine the 90% confidence level failure rate at 25° C and 20 mA, based on this sample.

(2) Determine the 90% confidence level failure rate at 85° C and 85% relative humidity, based on this sample.

(3) Determine the 90% confidence level failure rate under air-to-air, -40° C to $+ 85^\circ$ C for 1,000 cycles, based on this sample.

(4) Comparatively discuss the results in Cases 1, 2, and 3.

7-13. Five thousand of a certain type of a device have performed for a total of 5,200,000 hr with 52 failures. What minimum reliability can be assured, with 95% confidence that future devices will operate 5,000 hr without failure, if these devices have a constant failure rate and the test is failure terminated?

7-14. In conducting a test of a certain device, one failure was experienced in 10,000 hr of testing. The required *MTBF* is 5,000 hr. With what confidence can we state that the *MTBF* equals or exceeds the required *MTBF* of 5,000 hr, if these devices have a constant failure rate and the test is failure terminated?

TABLE 7.5 – Reliability test data for OPTO displays.

RELIABILITY DATA

PRODUCT LINE: OPTO   SUBGROUP: Displays

DESCRIPTION: Results below are from Reliability audit testing performed on
                       0.43 inch (10.9 mm) seven (7) segment displays. These data are
                       representative of all displays made from the same process.

FAILURE CRITERIA: 1. Light output degrades in excess of 50% of initial value.
                       Digits to maintain minimum 2.0:1 matching.

                   2. Electrical parameter V_F, V_R and I_R exceed data sheet
                       limits.

                   3. Electrical opens, shorts-functional.

TEST DESCRIPTION: Operating Life – Continuous operation with forced current a 20
                                    mA per segment, 25°C. Duration, 1000 hours.

DEVICE	DATE CODE	168	500	RESULTS 1000 HOURS	LOT I.D.
5082-7650	8351	0/38	0/38	0/38	840189
5082-7663	8401	0/38	0/38	0/38	840220
5082-7750	8401	0/38	0/38	0/38	840320
5082-7751	8402	0/38	0/38	0/38	840218
5082-7673	8403	0/38	0/38	0/38	840219
5082-7671	8405	0/38	0/38	0/38	840263
5082-7673	8410	0/38	0/38	0/38	840321
5082-7666	8410	0/38	0/38	0/38	840322

TABLE 7.5 – Continued.

RELIABILITY DATA

TEST DESCRIPTION: <u>Temperature Humidity Non-Biased</u> – Storage at 85°C and 85% RH. Duration, 1000 hours.

DEVICE	DATE CODE	168	500	RESULTS 1000 HOURS	LOT I.D.
5082-7651	8322	0/38	0/38	0/38	03009
5082-7671	8323	0/38	0/38	0/38	03008
5082-7650	8351	0/38	0/38	0/38	840189
5082-7663	8401	0/38	0/38	0/38	840220
5082-7750	8401	0/38	0/38	0/38	840320
5082-7751	8402	0/38	0/38	0/38	840218
5082-7673	8403	0/38	0/38	0/38	840219
5082-7671	8405	0/38	0/38	0/38	840263
5082-7673	8410	0/38	0/38	0/38	840321
5082-7666	8410	0/38	0/38	0/38	840322

TEST DESCRIPTION: <u>Temperature Cycle</u> – Air-to-Air, –40°C to +85°C. Duration, 1000 cycles.

DEVICE	DATE CODE	500	RESULTS 1000 CYCLES	LOT I.D.
5082-7660	8321	0/38	0/38	03010
5082-7651	8322	0/38	1/38	03009
5082-7671	8323	0/38	1/38	03008
5082-7650	8351	0/38	0/38	840189
5082-7663	8401	0/38	0/38	840220
5082-7750	8401	0/38	0/38	840320
5082-7751	8402	0/38	0/38	840218
5082-7673	8403	0/38	0/38	840219
5082-7671	8405	0/38	1/38	840263
5082-7673	8410	3/38	1/35	840321
5082-7666	8410	0/38	0/38	840322

7-15. It is specified that a gun system, when installed on a specific aircraft, shall provide a minimum reliability, at a confidence level of 90%, of 95% of firing a one-second burst (45 rounds) without a stoppage. A stoppage occurs when gun firing is interrupted and cannot be resumed by use of the remote gun system control.

The gun system was tested extensively and during 53,588 rounds fired, two chargeable stoppages occurred.

Determine the following:

(1) Has the gun system demonstrated its reliability goal? Show this by calculation and state your assumptions explicitly.

(2) How many stoppages should have been observed for the gun system not to have met its reliability goal?

Assume this is a time terminated test.

7-16. Prove that

$$m = \frac{\sum_{i=1}^{r} T_i + (N - r)T_r}{r}$$

is the maximum likelihood estimate of the mean of an exponentially failing, single-parameter population of parts from which N units were drawn and submitted to a life test. During the life test r of the N units failed at $T_1, T_2, \cdots, T_r$ and $N - r$ units survived up to test termination at T_r, without failing. (Hint: Use the likelihood function of the sample remembering that $N - r$ units survive the test).

7-17. Twenty units are tested for a test duration of 1,000 hr. The allowable number of failures is four.

(1) If this were a replacement test calculate and plot the OC curve.

(2) If this were a nonreplacement test calculate and plot the OC curve.

(3) Comparatively discuss the previous two curves.

7-18. In an airborne environment 4,610 of a certain type of capacitor perform for a total of 25,027,000 hr with 57 failures. What minimum reliability can be assured, with 95% confidence, that future production items of this type will operate in an airborne environment for 3,600 hr without failure?

7-19. Ninety differential generators have performed in an airborne environment for a total of 1,544,000 generator hours with 46 failures. What operating or mission time, under the same environment, can be assured when the minimum reliability has been specified as 95% at the confidence level of 99%?

7-20. A total of five units of a certain electronic device are given, for which it is required that an *MTBF* of 253 hr be demonstrated at a confidence level of 80%. All items are tested simultaneously to failure. What cumulative test time is required with zero failures, and from one up to and including five failures, to demonstrate that the true *MTBF* is equal to or greater than the required *MTBF* of 253 hr?

7-21. In conducting a test of 10 equipment, what minimum *MTBF* can be calculated with 95% confidence, if the test is terminated after the third failure and the cumulative test time on all 10 devices is 2,400 hr at that instant?

7-22. If the test time available for testing one unit is only 1,500 hr, how many failures can be tolerated in a test to establish that the true *MTBF* is at least 300 hr, at a confidence level of 90%?

7-23. In conducting the test of a certain device, zero failures were experienced in 10,000 hr of testing. The required *MTBF* is 4,000 hr. With what confidence can we state that the true *MTBF* equals or exceeds the required *MTBF* of 4,000 hr?

7-24. A turboalternator drive in an aircraft is composed of the components give in Table 7.6, where all components function reliabilitywise in series.

Determine the following:

(1) The reliability for a mission of 10 hr.

(2) The mean life confidence limits of these drives, at a 90% confidence level, if the performance of these drives is monitored and in 185,400 hr of actual flight time a total of 238 failures occur.

(3) Compare the results of Case 2 with the average mean life.

(4) The reliability confidence limits at a 90% confidence level for a mission of 10 hr, based on this sample.

(5) Compare the results of Case 4 with the average reliability.

TABLE 7.6–Component reliability analysis data for the turboalternator drive of Problem 17-24.

Number	Component	Quantity	Basic failure rate ($\times 10^6$)
1	Ball Bearing	19	2.20
2	Microswitch	2	5.49
3	Packing and Gaskets	14	1.00
4	Oil Filter	5	1.45
5	Oil Cooler	1	53.62
6	Bellow	2	5.00
7	Actuators	2	129.00
8	Transfer Valve	1	22.00
9	Electrical Connector	16	0.48
10	Solenoid	1	9.50
11	Roller Bearing	1	2.20
12	Turbine Wheel	1	20.00
13	Turbine Seal	2	25.00
14	Oil Nozzle	6	1.00
15	Relief Valve	1	19.60
16	Tachometer Generator	1	46.29
17	Oil Pump	1	53.39
18	Pressure Regulator Valve	1	70.00
19	By-pass Valve	1	48.90
20	Relief Valve	1	19.60
21	Gear Train	1	9.33
22	Clutch	1	5.70
23	Start Oil Pump	1	103.40
24	Oil Line Fittings	62	3.00
25	Oil Tank	1	10.00
26	Electric Cable	4	0.20
27	Solenoid Valve	1	22.00
28	Pressure Regulator	1	58.70
29	Check Valve	1	14.10
30	Fluid Flow Restrictor	1	14.10
31	Thermal Relay	1	20.00
32	Thermal Relay Socket	1	3.27
33	Interference Filter	1	16.18
34	Armature Relay	1	50.00
35	Pressure Switch	1	15.01

TABLE 7.6– Continued.

Number	Component	Quantity	Basic failure rate ($\times 10^6$)
36	Transformer	5	1.90
37	Resistor	30	0.76
38	Variable Resistor	7	1.71
39	Wire Wound Resistor	1	2.34
40	Electrolytic Capacitor	2	0.60
41	Capacitor	12	0.47
42	Capacitor	5	1.33
43	Crystal Diode	30	0.30
44	Magnetic Amplifier	1	22.70

7-25. Six identical exponential units are in a reliability test. Three failures occur at the following times: $T_1 = 950$ hr, $T_2 = 1,555$ hr, $T_3 = 1,610$ hr. At a 95% confidence level, determine the following:

   (1) What should the accumulated test time in unit hours be, to meet the *MTBF* goal of 2,000 hr if the test is time terminated?

   (2) What should the accumulated test time in unit hours be, to meet the same *MTBF* goal if the test is failure terminated?

   (3) What should the clock hours of test time be in Case 1 in a nonreplacement test?

   (4) What should the clock hours of test time be in Case 2 in a nonreplacement test and when should the third failure occur to meet the *MTBF* goal at a 95% confidence level?

   (5) Which test will require the longer test time in clock hours, the time terminated or the failure terminated test?

7-26. Twenty units are tested and the failed units are replaced. The test is terminated at the third failure. The designed-in *MTBF* is 3,000 hr.

   (1) Determine the expected test duration.

   (2) Determine the expected test duration if this were a nonreplacement test.

7-27. Twenty units are tested and the failed units are replaced. The test duration is 1,000 hr, the allowable number of failures is three, and the designed-in *MTBF* is 3,000 hr.

   (1) Determine the expected test duration.

   (2) Determine the expected test duration if this were a nonreplacement test.

7-28. Thirty units are tested and the failed units are replaced. Four failures are allowed in a time terminated test duration of 1,000 hr. To demonstrate a designed-in *MTBF* of 3,000 hr, determine the following:

   (1) The expected number of failures.

   (2) The expected number of failures if this were a nonreplacement test.

REFERENCES

1. Epstein, B., and Sobel, M.,"Life Testing," *Journal of the American Statistical Association*, Vol. 48, pp. 486-502, Sept. 1953.

2. *RADC Reliability Notebook,* Rome Air Development Center, Air Force Systems Command, Griffiss Air Force Base, New York 13440, Section 5, PB 161894-2, obtainable from U.S. Department of Commerce, Office of Technical Services, Washington, D.C., December 31, 1961.

3. Hald, A., and Sinkbreak, S. A., "A Table of Percentage Points of the χ^2 distribution," *Skandinavisk*, pp. 168-175, 1950.

4. Lloyd, David K., and Lipow, Myron, *Reliability: Management, Methods, and Mathematics*, Second Edition, David K. Lloyd and Myron Lipow, Redondo Beach, Calif., 589 pp., 1977.

5. Sinha, S.K. and Kale, B.K., *Life Testing and Reliability Estimation*, John Wiley & Sons, New York, 196 pp., 1980.

6. Bain, Lee J., *Statistical Analysis of Reliability and Life-Testing Models*, Marcel Dekker, Inc., New York, 450 pp., 1978.

7. Mann, N.R., Schafer, R.E. and Singpurwalla, N.D., *Methods for Statistical Analysis of Reliability and Life Data*, John Wiley & Sons, New York, 564 pp., 1974.

8. Epstein, B., "Truncated Life Tests in the Exponential Case," *Annals of Mathematical Statistics*, Vol. 25, pp. 555-564, 1954.

APPENDIX 7A
PROOF OF $E(T_{r,N} - T_{r-1,N}) = \frac{m}{N}$.

Assume the failure time, T, is exponentially distributed with *pdf*

$$f(T) = \frac{1}{m}e^{-\frac{T}{m}}, T \geq 0, \tag{7A.1}$$

and N units are put in a replacement test. Let $T_{1;N} < T_{2;N} < \cdots < T_{r;N}$ be the ordered failure times in the test, and define W_i as

$$W_1 = T_{1;N}, W_2 = T_{2;N} - T_{1;N}, \cdots, W_r = T_{r;N} - T_{r-1;N}.$$

Then the following apply:

1. The $W_1, W_2, \cdots, W_N$, are independent and identically distributed (iid), and

2. $W_i \sim EXP\left(\frac{m}{N}\right)$.

The proof of these follows:

1. The first item follows from the facts that the exponential distribution has the unique property of being "memoryless" and the units that fail are immediately replaced by identical units, hence the independence of the W_i. Then, $W_1 = T_1$ is distributed as the first order statistic, T_1, in a sample of N units from an exponential distribution given by Eq. (7A.1). As, after the first failure occurs, the failed unit is replaced by a new one and because of the "memoryless" property of the exponential distribution, $W_2 = T_{2;N} - T_{1;N}$ is also distributed as the first order statistic T_1 in a sample of N units from an exponential distribution given by Eq. (7A.1), and similarly the $W_3, W_4, \cdots, W_N$. Consequently the $W_1, W_2, \cdots, W_n$, are independent and identically distributed.

2. The second item follows from the fact that,

$$
\begin{aligned}
P(W_1 < w) &= P(T_1 < w), \\
&= 1 - P(T_1 \geq w), \\
&= 1 - P[(T_1 \geq w) \cap (T_2 \geq w) \cap \cdots \cap (T_N \geq w)],
\end{aligned}
$$

or

$$P(W_1 < w) = 1 - [1 - F(w)]^N.$$

Since

$$F(w) = 1 - e^{-\frac{w}{m}},$$

then

$$P(W_1 < w) = 1 - e^{-\frac{N}{m}w}. \tag{7A.2}$$

Taking the derivative of Eq. (7A.2) yields the *pdf* of W_i as

$$f_{W_1}(w) = \frac{N}{m}e^{-\frac{N}{m}w} = \frac{1}{\frac{m}{N}}e^{-\frac{w}{\frac{m}{N}}}.$$

Thus,

$$W_1 \sim EXP(\frac{m}{N}).$$

As the W_i, $i = 1, 2, \cdots, N$ are identically distributed; then,

$$W_i \sim EXP(\frac{m}{N}), \text{ for } i = 1, 2, \cdots, N.$$

APPENDIX 7B

DERIVATION OF THE DISTRIBUTION OF $\hat{m}$ AND OF THE CONFIDENCE LIMITS ON THE $MTBF$

7B.1 FAILURE TERMINATED TEST WITHOUT REPLACEMENT

1. THE DISTRIBUTION OF $\hat{m}$

Let

$$Y_i = (N - i + 1)(T_i - T_{i-1}), \quad i = 1, 2, \cdots, r;$$

then,

$$T_a = \sum_{i=1}^{r} Y_i,$$

and

$$\hat{m} = \frac{\sum_{i=1}^{r} Y_i}{r}. \tag{7B.1}$$

It is known by Property 1 in Appendix 7C.1 that the $Y_i's$ are iid exponential variables with *pdf*

$$f(y_i) = \frac{1}{m}e^{-y_i/m}.$$

Also, by Property 3a in Appendix 7C.1,

$$T_a = \sum_{i=1}^{r} Y_i \sim gamma(m, r),$$

or

$$r\hat{m} \sim gamma(m, r).$$

Finally, by Property 3b in Appendix 7C.1, it can be seen that

$$\frac{2r\hat{m}}{m} \sim \chi^2(2r). \tag{7B.2}$$

2. CONFIDENCE LIMITS ON THE *MTBF*

By Eq. (7B.2), for given α,

$$P(\chi^2_{1-\frac{\alpha}{2};2r} \le \frac{2r\hat{m}}{m} \le \chi^2_{\frac{\alpha}{2};2r}) = 1 - \alpha$$

should hold, or

$$P(\frac{2r\hat{m}}{\chi^2_{\frac{\alpha}{2};2r}} \le m \le \frac{2r\hat{m}}{\chi^2_{1-\frac{\alpha}{2};2r}}) = 1 - \alpha.$$

Therefore, the two-sided, $100(1-\alpha)\%$ confidence interval on m is given by

$$(\frac{2r\hat{m}}{\chi^2_{\frac{\alpha}{2};2r}} \; ; \; \frac{2r\hat{m}}{\chi^2_{1-\frac{\alpha}{2};2r}}). \tag{7B.3}$$

Since

$$\hat{m} = \frac{T_a}{r},$$

substituting $\hat{m}$ with T_a/r in Eq. (7B.3) yields

$$(\frac{2T_a}{\chi^2_{\frac{\alpha}{2};2r}} \; ; \; \frac{2T_a}{\chi^2_{1-\frac{\alpha}{2};2r}}). \tag{7B.4}$$

7B.2 FAILURE TERMINATED TEST WITH REPLACEMENT

It can be seen that

$$T_r = T_1 + (T_2 - T_1) + \cdots + (T_r - T_{r-1}).$$

By Property 2 of exponential variables given in Appendix 7C.1 and Appendix 7A, the

$$W_i = T_i - T_{i-1}, \quad i = 1, 2, \cdots, r$$

are exponentially distributed with $MTBF = \frac{m}{N}$; therefore, by Property 3a in Appendix 7C.1,

$$T_r \sim gamma(\frac{m}{N}, r).$$

Thus, by Property 3b in Appendix 7C.1,

$$\frac{2NT_r}{m} \sim \chi^2(2r).$$

But, with replacement,

$$T_r = \frac{r\hat{m}}{N};$$

therefore,

$$\frac{2r\hat{m}}{m} \sim \chi^2(2r), \tag{7B.5}$$

which may be seen to be identical with Eq. (7B.2), and therefore the confidence limits are also given by Eq. (7B.4).

7B.3 THE APPROXIMATE CONFIDENCE LIMITS ON m IN THE TIME TERMINATED TEST WITH REPLACEMENT

In this case, the observation in the test is the number of failures, r, within the test interval $[0, t_d]$. Since the time to failure, T, is exponentially distributed, then r is Poisson distributed. Let $\bar{N}_F$ be the average number of failures in $[0, t_d]$, then

$$P(r = k) = \frac{\bar{N}_F^k e^{-\bar{N}_F}}{k!}, \quad k = 0, 1, 2, \cdots,$$

where

$$\bar{N}_F = \frac{T_a}{m}, \tag{7B.6}$$

and

$$T_a = N t_d.$$

From the relationship between $\bar{N}_F$ and m given by Eq. (7B.6), it may be seen that if the confidence limits on $\bar{N}_F$ can be established, then the confidence limits on m can be obtained using Eq. (7B.6).

The confidence limits on $\bar{N}_F$ can be obtained by solving the equations

$$\sum_{k=0}^{r} \frac{\bar{N}_{FU}^k e^{-\bar{N}_{FU}}}{k!} = \frac{\alpha}{2}, \tag{7B.7}$$

and

$$\sum_{k=r}^{\infty} \frac{\bar{N}_{FL}^k e^{-\bar{N}_{FL}}}{k!} = \frac{\alpha}{2}, \tag{7B.8}$$

where

$$\bar{N}_{FU} = \text{upper confidence limit of failures,}$$

and

$$\bar{N}_{FL} = \text{lower confidence limit of failures.}$$

It may be seen that due to the discreteness of the Poisson distribution, the *confidence limits* obtained by solving these two equations would *not be the exact limits.*

It is known that the following relationship exists between the χ^2 distribution and the Poisson distribution (See Appendix 1A in Vol. 2 of this Handbook):

If

$$Y \sim \chi^2(2r),$$

where r is a positive integer; then,

$$F_Y(a) = P(Y \le a) = 1 - \sum_{j=0}^{r-1} e^{-\frac{a}{2}} \frac{(\frac{a}{2})^j}{j!},$$

or

$$F_Y(2a) = P(Y \le 2a) = 1 - \sum_{j=0}^{r-1} e^{-a} \frac{a^j}{j!}.$$

Therefore, Eq. (7B.7) can be written as

$$P(Y_1 \leq 2\bar{N}_{FU2}) = 1 - \sum_{k=0}^{r} e^{-\bar{N}_{FU2}} \frac{\bar{N}_{FU2}^k}{k!} = 1 - \frac{\alpha}{2},$$

where

$$Y_1 \sim \chi^2[2(r+1)].$$

Thus,

$$2\bar{N}_{FU2} = \chi^2_{\frac{\alpha}{2};2(r+1)},$$

or

$$\bar{N}_{FU2} = \frac{\chi^2_{\frac{\alpha}{2};2(r+1)}}{2}. \tag{7B.9}$$

Since

$$\bar{N}_F = \frac{T_a}{m};$$

then,

$$m_{L2} = \frac{2T_a}{\chi^2_{\frac{\alpha}{2};2(r+1)}}. \tag{7B.10}$$

Similarly, Eq. (7B.8) can be written as

$$P(Y_2 \leq 2\bar{N}_{FL2}) = 1 - \sum_{k=0}^{r-1} e^{-\bar{N}_{FL2}} \frac{\bar{N}_{FL2}^k}{k!} = \frac{\alpha}{2},$$

where

$$Y_2 \sim \chi^2(2r).$$

Thus,

$$2\bar{N}_{FL2} = \chi^2_{1-\frac{\alpha}{2};2r},$$

or

$$m_{U2} = \frac{2T_a}{\chi^2_{1-\frac{\alpha}{2};2r}}. \tag{7B.11}$$

Therefore, the approximate $100(1-\alpha)\%$ confidence interval on m is given by

$$\left(\frac{2T_a}{\chi^2_{\frac{\alpha}{2};2(r+1)}} \; ; \; \frac{2T_a}{\chi^2_{1-\frac{\alpha}{2};2r}}\right). \tag{7B.12}$$

APPENDIX 7C

7C.1 SOME PROPERTIES OF EXPONENTIALLY DISTRIBUTED VARIABLES

In this section, three important properties of exponential variables are presented to carry on the analysis of the distribution of the Maximum Likelihood Estimator (MLE) of the $MTBF$, m.

1. If a complete exponential life test is conducted *without replacement*, and the ordered failure times are

$$T_1, T_2, \cdots, T_N;$$

then, the

$$Z_i = T_i - T_{i-1}, \quad i = 1, 2, 3, \cdots, N, \quad T(0) = 0,$$

have the *pdf* [1]

$$f(z_i) = \frac{N - i + 1}{m} exp\left(\frac{N - i + 1}{m} z_i\right).$$

Proof: The joint distribution of $T_1, T_2, \cdots, T_N$ is given by [2]

$$G(T_1, T_2, \cdots, T_N \mid m) = \frac{N!}{m^N} exp\left(\frac{-\sum\limits_{i=1}^{m} T_i}{m}\right).$$

Let

$$
\begin{aligned}
Y_1 &= NT_1, \\
Y_2 &= (N - 1)(T_2 - T_1), \\
&\;\;\vdots \\
Y_k &= (N - k + 1)(T_k - T_{k-1}), k = 1, 2, \cdots, N.
\end{aligned}
$$

Since [2, pp. 138-143]

$$\frac{\partial(Y_1, Y_2, \cdots, Y_N)}{\partial(T_1, T_2, \cdots, T_N)} = \begin{vmatrix} \frac{\partial Y_1}{\partial Z_1} & \frac{\partial Y_1}{\partial Z_2} & \cdots & \frac{\partial Y_1}{\partial Z_N} \\ \frac{\partial Y_2}{\partial Z_1} & \frac{\partial Y_2}{\partial Z_2} & \cdots & \frac{\partial Y_2}{\partial Z_N} \\ \vdots & \vdots & \cdots & \vdots \\ \frac{\partial Y_N}{\partial Z_1} & \frac{\partial Y_N}{\partial Z_2} & \cdots & \frac{\partial Y_N}{\partial Z_N} \end{vmatrix},$$

$$= \begin{vmatrix} N & 0 & \cdots & 0 \\ -(N-1) & (N-1) & \cdots & 0 \\ 0 & -(N-2) & \cdots & 0 \\ \vdots & \vdots & \vdots & \vdots \\ 0 & 0 & \cdots & 1 \end{vmatrix},$$

$$= N!,$$

the Jacobian is given by

$$|J| = \frac{1}{N!}.$$

Also, since

$$\sum_{i=1}^{N} Y_i = \sum_{i=1}^{N} T_i,$$

the joint distribution of $Y_1, Y_2, \cdots, Y_N$ is

$$f(Y_1, Y_2, \cdots, Y_N) = \frac{1}{m^N} exp \left(-\frac{\sum_{i=1}^{N} Y_i}{m} \right),$$

which means that the Y_i's are independent and have a common *pdf* of

$$f(y_i) = \frac{1}{m} exp \left(-\frac{y_i}{m} \right).$$

Since

$$Y_i = (N - i + 1)Z_i;$$

then,

$$\begin{aligned} F_{Z_i}(z_i) = P(Z_i \le z_i) &= P(\frac{Y_i}{N - i + 1} \le z_i), \\ &= P(Y_i \le (N - i + 1)z_i), \\ &= 1 - exp \left(-\frac{N - i + 1}{m} z_i \right). \end{aligned}$$

Taking the derivative of this equation yields

$$f(z_i) = \frac{N - i + 1}{m} exp \left(-\frac{N - i + 1}{m} z_i \right).$$

2. If a complete exponential life test *with replacement* is conducted, and the ordered failure times are

$$T_1, T_2, \cdots, T_N;$$

then, the

$$Z_i = T_i - T_{i-1}, \quad i = 1, 2, 3, \cdots, N, \quad T(0) = 0,$$

are independently and identically distributed with *pdf* [3]

$$g(z_i) = \frac{N}{m} exp\left(-\frac{N}{m} z_i\right).$$

Proof: From the no-memory property of the exponential distribution, and also because the failed units are replaced immediately, it may be seen that the Z_i's are independent, and they are the smallest observations in a sample of size N, therefore they are also identically distributed. Then the cumulative distribution of Z_i, is given by

$$
\begin{aligned}
G(z_1) &= P(Z_1 \leq z_1) = 1 - P(Z_1 > z_1), \\
&= 1 - [1 - F(z_1)]^N,
\end{aligned}
$$

where

$$F(z_1) = 1 - e^{-z_1/m}.$$

Therefore,

$$G(z_1) = 1 - e^{-\frac{N}{m} z_1},$$

or

$$g(z_1) = \frac{N}{m} e^{-\frac{N}{m} z_1}.$$

3. If the T_i, $(i = 1, 2, \cdots, N)$ are independently and identically distributed exponential variables, then [2]

$$a. \quad Z = \sum_{i=1}^{N} T_i \sim gamma(m, N),$$

and

$$b. \quad \frac{2Z}{m} = \frac{\sum_{i=1}^{N} T_i}{m} = \frac{2\frac{r}{r} \sum_{i=1}^{N} T_i}{m} = \frac{2r\hat{m}}{m} \sim \chi^2(2N).$$

Proof a: If X is a non-negative random variable with *pdf* $f(x)$; then, the Moment Generating Function (MGF) of X is defined as

$$
\begin{aligned}
M_X(t) &= E(e^{tX}), \\
&= \int_0^\infty e^{tx} f(x)\, dx,
\end{aligned}
$$

and for the exponential distribution its MGF is given by

$$
\begin{aligned}
M_X(t) &= \int_0^\infty e^{tT} f(T)\, dT, \\
&= \int_0^\infty e^{tT} \frac{1}{m} e^{-\frac{T}{m}}\, dT, \\
&= \int_0^\infty e^{-\frac{(1-mt)T}{m}}\, d\frac{T}{m}.
\end{aligned}
$$

Let $u = \frac{T}{m}$; then,

$$
\begin{aligned}
M_X(t) &= \int_0^\infty e^{(1-mt)u}\, du, \\
&= \frac{1}{1 - mt}.
\end{aligned}
$$

The *pdf* of the gamma distribution is given by

$$
f(T) = \frac{1}{\eta \Gamma(\beta)} \left(\frac{T}{\eta}\right)^{\beta-1} e^{-\frac{T}{\eta}}.
$$

If β is a positive integer; then,

$$
f(T) = \frac{1}{\eta(\beta-1)!} \left(\frac{T}{\eta}\right)^{\beta-1} e^{-\frac{T}{\eta}},
$$

and its MGF is given by

$$
\begin{aligned}
M_G(t) &= \int_0^\infty e^{tT} \frac{1}{\eta(\beta-1)!} \left(\frac{T}{\eta}\right)^{\beta-1} e^{-\frac{T}{\eta}}\, dT, \\
&= \int_0^\infty \frac{1}{\eta(\beta-1)!} \left(\frac{T}{\eta}\right)^{\beta-1} e^{-\frac{(1-t\eta)T}{\eta}}\, dT, \\
&= \int_0^\infty \frac{1}{(\beta-1)!} \left(\frac{T}{\eta}\right)^{\beta-1} e^{-(1-t\eta)\frac{T}{\eta}}\, d\frac{T}{\eta}.
\end{aligned}
$$

Let $u = \frac{T}{\eta}$; then,

$$
M_G(t) = \int_0^\infty \frac{u^{\beta-1}}{(\beta-1)!} e^{-(1-t\eta)u}\, du.
$$

Integrating by parts repeatedly yields

$$M_G(t) = \frac{1}{(1 - t\eta)^\beta}.$$

By the property of the MGF, the MGF of the sum of independent random variables equals the product of the MGF's of those random variables, therefore the MGF of Z is given by

$$M_Z(t) = M_{T_1}(t) M_{T_2}(t) \cdots M_{T_N}(t).$$

But the T_i' s are iid exponential variables; thus,

$$M_Z(t) = \frac{1}{(1 - tm)^N}.$$

Comparing the MGF of Z with the MGF of the gamma distribution, it can be seen that Z is gamma distributed with parameters

$$\beta = N,$$

and

$$\eta = m.$$

Proof b: The *pdf* of the χ^2 distribution with γ degrees of freedom is

$$\chi^2(x, \gamma) = \frac{1}{2^{\frac{\gamma}{2}} \Gamma(\frac{\gamma}{2})} e^{-\frac{x}{2}} x^{\frac{\gamma}{2} - 1}, \quad x \geq 0;$$

therefore, its MGF is given by

$$M_{\chi^2(\gamma)}(t) = \int_0^\infty e^{xt} \frac{1}{2^{\frac{\gamma}{2}} \Gamma(\frac{\gamma}{2})} e^{-\frac{x}{2}} x^{\frac{\gamma}{2} - 1} \, dx,$$

$$= \int_0^\infty \frac{1}{\Gamma(\frac{\gamma}{2})} e^{-\frac{(1-2t)x}{2}} \left(\frac{x}{2}\right)^{\frac{\gamma}{2} - 1} d\frac{x}{2}.$$

Let $u = \frac{x}{2}$; then,

$$M_{\chi^2(\gamma)}(t) = \int_0^\infty \frac{1}{\Gamma(\frac{\gamma}{2})} e^{-(1-2t)u} u^{\frac{\gamma}{2} - 1} \, du,$$

$$= \frac{1}{(1 - 2t)^{\frac{\gamma}{2}}}.$$

Now consider the MGF of the $\frac{2Z}{m}$. Since

$$M_Z(t) = \frac{1}{(1 - tm)^N};$$

then, by the property of the MGF,

$$M_{\frac{2Z}{m}}(t) = M_Z(\frac{2}{m}t),$$

$$= \frac{1}{(1 - \frac{2}{m}tm)^N},$$

$$= \frac{1}{(1 - 2t)^N}.$$

Comparing the MGF of $\frac{2Z}{m}$ with the MGF of the χ^2 distribution, it can be seen that $\frac{2Z}{m}$ is χ^2 distributed with $2N$ degrees of freedom.

7C.2 THE MAXIMUM LIKELIHOOD ESTIMATOR (MLE) OF m

7C.2.1 FAILURE TERMINATED TEST WITHOUT REPLACEMENT

In this case, N units are being tested without replacing the failed units, and the test is terminated at the rth failure, where r is a preassigned number. If the ordered failure times are

$$0 \le T_1 \le T_2 \le \cdots \le T_r;$$

then, the likelihood function of the observations is given by [4]

$$L(T_1, T_2, \cdots, T_r \mid m) = \frac{N!}{(N-r)!m^r} exp\left[-\frac{\sum_{i=1}^{r} T_i + (N-r)T_r}{m}\right].$$

$$(7C.1)$$

Taking the natural logarithm of Eq. (7C.1) yields

$$\log L = \log \frac{N!}{(N-1)!} - r \log m - \frac{\sum_{i=1}^{r} T_i + (N-r)T_r}{m}. \qquad (7C.2)$$

Taking the derivative of Eq. (7C.2) with respect to m, and setting it equal to zero, yields

$$-\frac{r}{m} + \frac{\sum_{i=1}^{r} T_i + (N-r)T_r}{m^2} = 0. \qquad (7C.3)$$

Solving Eq. (7C.3) for m yields

$$\hat{m} = \frac{\sum\limits_{i=1}^{r} T_i + (N - r)T_r}{r},$$

which is the MLE of m, or

$$\hat{m} = \frac{T_a}{r}, \qquad (7C.4)$$

where

$$T_a = \sum_{i=1}^{r} T_i + (N - r)T_r$$

is the test time accumulated by all units in the test, nonfailed and failed.

7C.2.2 FAILURE TERMINATED TEST WITH REPLACEMENT

In this case N units are tested, whenever a failure occurs it is replaced by an identical one, and the test is terminated immediately after the rth failure occurs. If the ordered times to failure are

$$0 \le T_1 \le T_2 \le \cdots \le T_r;$$

then, the MLF of the observations is given by [4]

$$
\begin{aligned}
L\{T_1, T_2, \cdots, T_r \mid m\} &= \frac{n^r}{m^r} \left[exp\left(-\frac{T_1}{m} \right) \right] \cdot \left[exp\left(-\frac{T_2 - T_1}{m} \right) \right] \cdots \\
&\quad \cdot \left[exp\left(-\frac{T_r - T_{r-1}}{m} \right) \right] \\
&\quad \cdot \left[\int_{T_r}^{\infty} \frac{1}{m} exp\left(-\frac{T}{m} \right) dT \right]^{N-1}, \\
&= \left(\frac{N}{m} \right)^r exp\left(-\frac{NT_r}{m} \right).
\end{aligned}
$$

By the standard method for finding the MLE, the MLE of m can be found to be

$$\hat{m} = \frac{NT_r}{r},$$

or

$$\hat{m} = \frac{T_a}{r}. \qquad (7C.5)$$

7C.2.3 TIME TERMINATED TEST WITHOUT REPLACEMENT

In this case, N units are being tested without replacing the failed units, and the test is terminated at a preassigned test clock hours, t_d. If up to t_d , r failures are observed, and the ordered failure times are

$$0 \leq T_1 \leq T_2 \leq \cdots \leq T_r < t_d,$$

it can be seen that the failure number, say r^*, is a random variable.

Under the assumption of the exponential times-to-failure distribution, a unit's failure probability before t_d is given by

$$q = P(T \leq t_d) = 1 - exp(-\frac{t_d}{m}).$$

The probability that r failures occur before t_d, is given by

$$P(r^* = r) = \binom{N}{r} q^r p^{N-r}, \quad r = 0, 1, 2, \cdots, N,$$

where

$$p = 1 - q = exp(-\frac{t_d}{m}),$$

because $T_i = t_d$.

The Maximum Likelihood Function (MLF) of the observations $\{T_1, T_2, \cdots, T_r, r \mid m\}$ for $r = 0$ up to t_d, is given by

$$exp(-\frac{Nt_d}{m}).$$

For $m > 0$, the conditional *pdf* of the failure times, given that the unit has already failed before t_d, is given by

$$g(T \mid m) = \begin{cases} \frac{\frac{1}{m}exp(-T \ / \ m)}{1-exp(-t_d \ / \ m)}, & 0 < T \leq t_d; \\ 0, & \text{otherwise.} \end{cases}$$

Then the joint density of $T_1, T_2, \cdots, T_r$, is given by

$$g(T_1, T_2, \cdots, T_r, \ r \mid m) = \frac{\frac{r!}{m^r} \cdot exp\left(-\sum_{i=1}^{r} T_i \ / \ m\right)}{[1 - exp(-t_d \ / \ m)]^r}.$$

The MLF of the observations is then the joint density of $\{T_1, T_2, \cdots, T_r\}$ and r, or

$$L\{T_1, T_2, \cdots, T_r, r \mid m\} \ = \ g(T_1, T_2, \cdots, T_r) \cdot \binom{N}{r} q^r p^{N-r},$$

$$= \frac{r!}{m^r} \cdot \frac{exp\left(-\sum\limits_{i=1}^{r} T_i \,/\, m\right)}{[1 - exp(-t_d \,/\, m)]^r}$$

$$\cdot \frac{N!}{(N-r)!r!} \left[1 - exp\left(\frac{-t_d}{m}\right)\right]^r$$

$$\cdot \left[exp\left(-\frac{t_d}{m}\right)\right]^{(N-r)},$$

or

$$= \frac{N!}{(N-r)!} \cdot \frac{1}{m^r} \cdot exp\left[-\frac{\sum\limits_{i=1}^{r} T_i + (N-r)t_d}{m}\right].$$

Then, the MLE of m would be

$$\hat{m} = \frac{\sum\limits_{i=1}^{r} T_i + (N-r)t_d}{r}, \quad r > 0. \tag{7C.6}$$

For the case of $r = 0$, since

$$L\{T_1, T_2, \cdots, T_r, r \mid m\} = exp(-\frac{nt_d}{m})$$

is a strictly increasing function, the MLE of m would be

$$\hat{m} = \infty,$$

but this is not a good estimator. Bartholomew [5] recommends an estimator for m when $r = 0$ as

$$\hat{m} = Nt_d. \tag{7C.7}$$

Another, more realistic estimate of m, $\hat{m}$, is given by Eq. (7.29).

7C.2.4 TIME TERMINATED TEST WITH REPLACEMENT

In this case, N units are being tested, whenever a failure occurs that unit is replaced immediately by an identical unit, and the test is terminated at a preassigned test clock hours, t_d. If up to t_d , r failures are observed, and the ordered failure times are

$$0 \leq T_1 \leq T_2 \leq \cdots \leq T_r < t_d,$$

it was proved in Section 7C.1 that

$$Z_i = T_i - T_{i-1}, \quad i = 1, 2, \cdots, r,$$

are iid exponential variables with *pdf*

$$f(z_i) = \frac{N}{m} e^{-\frac{N}{m} z_i}, \quad i = 1, 2, \cdots, r.$$

Therefore, the failure number, r^*, in $[0, t_d]$ is Poisson distributed with parameter $r = N t_d / m$ [6], and the probability of r^* failures occurring in $[0, t_d]$ is given by

$$P(r^* = r \mid m) = \frac{(\frac{N t_d}{m})^r exp(-\frac{N t_d}{m})}{r!}, \quad r = 0, 1, 2, \cdots.$$

The MLE of r from a Poisson distribution is given by

$$r = \frac{N t_d}{\hat{m}};$$

therefore,

$$\hat{m} = \frac{N t_d}{r}. \tag{7C.8}$$

7C.3 REFERENCES

1. Epstein, B., and Sobel, M., "Life Testing", *Journal of the American Statistical Association*, Vol.48, pp. 480-502, 1953.

2. Mood, A.M., Graybill, F.A. and Boes,D.C., *Introduction to the Theory of Statistics*, Third Edition, McGraw-Hill Book Co., New York, 564 pp., 1974.

3. Sinha, S.K. and Kale, B.K., *Life Testing and Reliability Estimation*, John Wiley & Sons, Inc., New York, 196 pp., 1980.

4. Epstein, B., and Sobel, M., "Some Theorems Relevant to Life Testing from an Exponential Distribution", *Annals of Mathematical Statistics*, Vol.25, pp. 373-381, 1954.

5. Bartholomew, D.J., "A Problem in Life Tests", *Journal of the American Statistical Association*, Vol.65, pp. 350-355, 1957.

6. Epstein, B., "Truncated Life Tests in the Exponential Case", *Annals of Mathematical Statistics*, Vol. 25, pp. 555-564, 1954.

Chapter 8

TESTS OF COMPARISON OF THE MEAN LIFE FOR THE EXPONENTIAL CASE

8.1 INTRODUCTION

When the same product can be procured from two different manufacturers, it is often desired to know whether the *MTBF* of one manufacturer's product is significantly different than the other's, so that if there is a relatively small difference in price then the one with the significantly higher *MTBF* should be bought. To accomplish this, units from both manufacturers should be obtained on a trial basis and subjected to a comparative reliability test. The times-to-failure data should be recorded, the *MTBF* of each manufacturer's units should be determined, a level of significance chosen, and a statistical test of comparison of these two *MTBF's* conducted to determine if the *MTBF's* are indeed significantly different, or whether any difference in the *MTBF's* is due to chance causes alone and would have occurred anyway even if there were no difference in the reliability of the units obtained from both manufacturers.

Such tests of comparison can also be applied when a product's reliability is found not to meet the specified target and fixes are implemented through a design change, a component change, etc., to meet the *MTBF* goal. Now it is desired to establish whether the fixes have indeed significantly improved the product's reliability to enable it to meet its *MTBF* goal.

Also it may be desired to see if a specific change in the stress level of operation affects the *MTBF* of a product significantly. This test of

comparison enables the user to determine whether or not this is so.

The level of significance, α, gives the probability that in the long run one will be wrong only $\alpha\%$ of the time when a decision is reached to reject equality when in fact the two versions of such products are equal; i.e., they have essentially the same *MTBF*. In an inequality test, the probability that, in the long run, one will be wrong when a decision is reached that the two versions of such products have significantly different *MTBF's* when in fact they are equal, or the same, will not exceed α.

Two cases are discussed here [1]:

1. The failure terminated test case.

2. The time terminated test case.

8.2   TEST PROCEDURE

8.2.1   FAILURE TERMINATED TEST CASE

When both tests are failure terminated then use the following procedure *to determine that the MTBF's are equal*:

1. Calculate

$$F^* = \frac{\hat{m}_2}{\hat{m}_1} = \frac{T_{a2}/r_2}{T_{a1}/r_1}, \qquad (8.1)$$

   where

$$
\begin{aligned}
\hat{m}_2 &= \text{estimate of the } MTBF \text{ of Version 2,} \\
\hat{m}_1 &= \text{estimate of the } MTBF \text{ of Version 1,} \\
T_{a2} &= \text{accumulated unit-hours of operation of Version} \\
       & \quad \text{2 during the test,} \\
T_{a1} &= \text{accumulated unit-hours of operation of Version} \\
       & \quad \text{1 during the test,} \\
r_2 &= \text{number of failures observed on Version 2 during} \\
    & \quad \text{the test, which is terminated at the } r_2 th \text{ failure,} \\
r_1 &= \text{number of failures observed on Version 1 during} \\
    & \quad \text{the test, which is terminated at the } r_1 th \text{ failure.}
\end{aligned}
$$

2. Determine the lower, two-sided confidence limit of the critical value of the F distribution from the tables in Appendix F, or

$F_{\alpha/2;2r_2;2r_1}$.

Here F refers to the F, *Fisher* or ratio distribution. $F_{\alpha/2;2r_2;2r_1}$ is so determined that the *area to the left* of this F value, under the F *pdf* with $2r_2$ and $2r_1$ degrees of freedom, is equal to $\alpha/2$. As most F *pdf* area tables give areas to left of F and for areas greater than $\alpha = 0.90$, to find the $F_{\alpha/2}$ value use the identity

$$F_{\alpha/2;2r_2;2r_1} = \frac{1}{F_{1-\alpha/2;2r_1;2r_2}}. \tag{8.2}$$

Note that the degrees of freedom are interchanged at the right side of Eq. (8.2).

3. Determine the upper, two-sided confidence limit of the critical value of the F distribution, or

$$F_{1-\alpha/2;2r_2;2r_1},$$

such that, as in Step 2, the area to the left of this F value is equal to $(1 - \alpha/2)$.

4. Perform the following comparison:

$$F_{\alpha/2;2r_2;2r_1} < F^* < F_{1-\alpha/2;2r_2;2r_1}. \tag{8.3}$$

If this condition is satisfied then accept the equality of the *MTBF's* of the two versions, or that $m_2 = m_1$ at the 100 $\alpha\%$ significance level; i.e., we would have been wrong no more than 100 $\alpha\%$ of the time had we rejected the equality of their *MTBF's*, when in fact they were equal, or had essentially the same *MTBF*. This case is illustrated in Fig. 8.1.

To determine that the MTBF's it are not equal use the following procedure:

1. Make $F^* > 1$ by putting the larger of the two $m's$ in the numerator, then

$$F^* \geq 1.$$

2. Compare F^* with $F_{1-\alpha;2r_2;2r_1}$.

3. If

$$F^* > F_{1-\alpha;2r_2;2r_1};$$

then, accept that $m_2 > m_1$, otherwise reject that m_2 is greater than m_1 at the 100 $\alpha\%$ significance level. This is illustrated in Fig. 8.2.

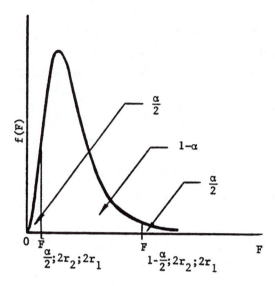

Fig. 8.1– Fisher *pdf* and and its two-sided confidence limits.

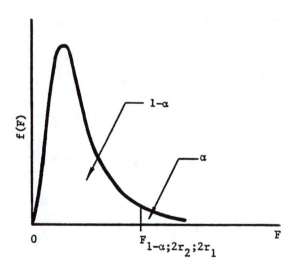

Fig. 8.2– Fisher *pdf* and its upper, one-sided confidence limit.

8.2.2 TIME TERMINATED TEST CASE

When both tests are terminated, not coincident with the occurrence of a failure and after a fixed unit-hours of operation are accumulated, use the following procedure *to determine that the MTBF's are equal:*

1. Find

$$F^* = \frac{T_{a2}}{2r_2 + 1} \bigg/ \frac{T_{a1}}{2r_1 + 1} \geq 1. \tag{8.4}$$

by putting the larger of the two $T_{ai}/(2r_i + 1)$ values in Eq. (8.4) in the numerator.

2. Determine

$$F_{\alpha/2;2r_2+1;2r_1+1}.$$

3. Determine

$$F_{1-\alpha/2;2r_2+1;2r_1+1}.$$

4. Perform the following comparison:

$$F_{\alpha/2;2r_2+1;2r_1+1} < F^* < F_{1-\alpha/2;2r_2+1;2r_1+1}. \tag{8.5}$$

If this condition is satisfied then accept the equality of the *MTBF's* of the two versions, or that $m_2 = m_1$ at the 100 $\alpha\%$ significance level.

To determine that the MTBF's are not equal, use the following procedure:

1. Make $F^* > 1$, by putting the larger of the two $T_{ai}/(2r_i+1)$ values in Eq. (8.4) in the numerator, then

$$F^* \geq 1.$$

2. Compare F^* with $F_{1-\alpha;2r_2+1;2r_1+1}.$

3. If

$$F^* > F_{1-\alpha;2r_2+1;2r_1+1},$$

then accept that $m_2 > m_1$, otherwise reject that m_2 is greater than m_1 at the 100 $\alpha\%$ significance level.

EXAMPLE 8–1

A product's reliability has been monitored and in 750 unit-hours of operation five failures have been observed. Three of these failures occurred in identical components. This component was redesigned and the product's reliability, with three of these redesigned components in it, was monitored again. In 2,250 unit-hours of operation five failures were observed.

1. If the product is exponential and the test was *failure terminated* did the redesign retain equality in the reliability of this product at the 10% significance level?

2. Is the *MTBF* of the redesigned product greater than before at the 10% significance level?

3. If the test were *time terminated*, repeat Case 1.

4. If the test were *time terminated*, repeat Case 2.

5. How many failures should have been observed in Case 3, after redesign, for the *MTBF* not to have been improved at the 5% significance level?

SOLUTIONS TO EXAMPLE 8–1

1. This case is a *failure terminated* test and given are $T_{a2} = 2,250$ unit hr, $r_2 = 5$ failures, $T_{a1} = 750$ unit hr, $r_1 = 5$ failures, and $\alpha = 0.10$.

   1.1 - From Eq. (8.1)

   $$F^* = \frac{2,250}{5} / \frac{750}{5} = \frac{450}{150},$$

   or

   $$F^* = 3.00.$$

   1.2 - From Eq. (8.2)

   $$F_{\alpha/2;2r_2;2r_1} = F_{0.05;2\times5;2\times5} = F_{0.05;10;10},$$

   $$F_{0.05;10;10} = \frac{1}{F_{0.95;10;10}} = \frac{1}{2.98},$$

   or

   $$F_{\alpha/2;2r_2;2r_1} = 0.336.$$

1.3 - Determine

$$F_{1-\alpha/2;2r_2;2r_1} = F_{0.95;10;10} = 2.98.$$

1.4 - The inequality of Eq. (8.3) becomes

$$0.336 < F^* = 3.00 \not< 2.98.$$

It may be seen that the inequality is *not* satisfied. Consequently, we reject the hypothesis that the two *MTBF's* are equal.

Observe that in this example F^* is very close to $F_{1-\alpha/2}$, and at $\alpha = 0.05$, instead of at $\alpha = 0.10$, the previous statement would not have been made, because $F_{0.975;10;10} = 3.72$. It is recommended that under these circumstances an inequality test be conducted.

2. In this example,

$$F_{1-\alpha;2r_2;2r_1} = F_{0.90;10;10} = 2.32,$$

and since

$$F^* = 3.00 > F_{1-\alpha;2r_2;2r_1} = 2.32,$$

accept the hypothesis that

$$m_2 > m_1,$$

and that the redesign has improved the *MTBF* of the product at the 10% significance level.

3. This case is a *time terminated* test.

3.1 - From Eq. (8.3)

$$F^* = \left(\frac{2,250}{2 \times 5 + 1}\right)\Big/\left(\frac{750}{2 \times 5 + 1}\right) = \left(\frac{2,250}{11}\right)\Big/\left(\frac{750}{11}\right),$$

or

$$F^* = 3.00.$$

3.2 - Find $F_{\alpha/2;2r_2+1;2r_1+1}$ using Eq. (8.2), then

$$F_{\alpha/2;2r_2+1;2r_1+1} = F_{0.05;2\times5+1;2\times5+1} = F_{0.05;11;11},$$

$$F_{0.05;11;11} = \frac{1}{F_{0.95;11;11}} = \frac{1}{2.82},$$

or

$$F_{\alpha/2;2r_2+1;2r_1+1} = 0.355.$$

3.3 - Determine

$$F_{1-\alpha/2;2r_2+1;2r_1+1} = F_{0.95;11;11} = 2.82.$$

3.4 - The inequality of Eq. (8.5) becomes

$$0.355 < F^* = 3.00 \not< 2.82.$$

It may be seen that the inequality is *not satisfied*; consequently, we reject the hypothesis that the two $MTBF's$ are equal.

4. In this example,

$$F_{1-\alpha;2r_2+1;2r_1+1} = F_{0.90;11;11} = 2.23.$$

Since

$$F^* = 3.00 > F_{1-\alpha;2r_2+1;2r_1+1} = 2.23,$$

the redesign did improve the product's $MTBF$ at the 10% significance level.

5. This is a *time terminated* test case and the inequality test should be used. Then, from Eq. (8.4),

$$F^* = \frac{2,250}{2r_2 + 1} \Big/ \frac{750}{2r_1 + 1}.$$

With $100\,\alpha\% = 5\%$ and $r_1 = 5$,

$$F_{1-\alpha;2r_2+1;2r_1+1} = F_{0.95;2r_2+1;11}.$$

Using the previous two relationships, find r_2 such that

$$F^* > F_{1-\alpha}.$$

Then $m_2 > m_1$ at the 5% significance level.

5.1 - Take $r_2 = 5$, then

$$F^* = \frac{2,250}{2 \times 5 + 1} \bigg/ \frac{750}{11} = 3.00,$$

$$F_{0.95;2\times5+1;1;11} = F_{0.95;11;11} = 2.82,$$

and

$$F^* = 3.00 > F_{0.95;11;11} = 2.82.$$

Consequently,

$$m_2 > m_1.$$

5.2 - Take $r_2 = 6$, then

$$F^* = \frac{2,250}{2 \times 6 + 1} \bigg/ \frac{750}{11} = 2.54,$$

$$F_{0.95;2\times6+1;11} = F_{0.95;13;11} = 2.77,$$

and

$$F^* = 2.54 \ngtr F_{0.95;13;11} = 2.77.$$

Therefore, if six or more failures had been observed there would not have been an improvement in the *MTBF* at the 5% significance level. Five or fewer failures should have been observed for the *MTBF* to have been improved through redesign at the 5% significance level.

8.3 CHOICE OF THE SIGNIFICANCE LEVEL

If it is desired to insure that one version or product, is better reliability-wise than another, or that a redesign effort has improved the reliability of the product, then lower levels of significance should be chosen which forces an equality decision on the two *MTBF's*, particularly when the differences in the two *MTBF's*, or the *MTBF* improvements, are not large enough. This choice would also reduce the risk of concluding that the *MTBF's* were different to only 100 α% of the time, when in fact they were the same.

   If the equality test says that the two are equal and the inequality test says that they are not, one procedure for resolving this problem

is to use a lower α value, or a higher confidence level. This results in higher values for $F_{1-\alpha/2}$ and $F_{1-\alpha}$, and thus forces the decision towards equality with less risk.

If the dichotomy in the decision still prevails then go to still lower values of α. Eventually this will lead to a decision of equality, which is conservative of the decisions, and to further redesign to come up with a better version or product, with a lower risk of making a wrong decision.

8.4   QUANTITATIVE BASES FOR SELECTING A MANUFACTURER

The following quantitative means and measures may also be used to select the better product and/or manufacturer:

1. Lower purchase price but with a product that also meets the reliability, maintainability, performance, safety and quality requirements.

2. Warranty availability with better coverage.

3. Reliability, mean life, maintainability, and mean-time-to-restore prediction, evaluation and demonstration capabilities.

4. Manufacturer's past performance and reputation.

5. Availability of manufacturer's scientifically written operating and maintenance manuals for its products.

6. Efficacy of manufacturer's service and repair organization.

7. Manufacturer's spare parts provisioning knowledge and organization.

8. Results of tests of comparison at the mean life with 90% or 95% CL, of the product with that of other manufacturers.

9. Results of comparison of one manufacturer's product's m_{L1}, at a 90% or 95% CL, with that of other manufacturers.

10. Comparison of one manufacturer's R_{L1}, on a parametric basis and at a 90% or 95% CL, with that of another manufacturer.

11. Use of other αs if comparisons are not conclusive enough.

12. Use of nonparametric tests of comparison techniques covered in [2, Chapter 5].

PROBLEMS

8–1. When conducting the exponential mean life comparison test, how do you verify that one version or product is reliabilitywise better than another?

8–2. If the equality test yields that two products are equal and the inequality test yields that they are not, and this situation remains unchanged even after you decrease the α values several times, what will you conclude about these two products?

8–3. Prove that the statistic $F^* = \frac{\hat{m}_2}{\hat{m}_1} = \frac{T_{a2}/r_2}{T_{a1}/r_1}$ is F-distributed with $2r_2$ and $2r_1$ degrees of freedom for the failure terminated test case.

8–4. A product's reliability was monitored, and in $2,000$ unit-hours of operation four failures were observed. The product was redesigned and its reliability was monitored again. In $2,000$ unit-hours of operation six failures were observed.

    (1) If the product is exponential and the test was failure terminated, did the redesign retain equality in the *MTBF* of this product at the 5% significance level?

    (2) Is the *MTBF* of the redesigned product greater than before at the 5% significance level?

    (3) If the test was time terminated, repeat Case 1.

    (4) If the test was time terminated, repeat Case 2.

    (5) How many failures should have been observed in Case 3, after redesign, for the *MTBF* to show an improvement at the 5% significance level?

8–5. A product's reliability was monitored, and in $1,000$ unit-hours of operation three failures were observed. The product was redesigned and its reliability was monitored again. In $2,000$ unit-hours of operation four failures were observed.

    (1) If the product is exponential and the test was failure terminated, did the redesign retain equality in the *MTBF* of this product at the 5% significance level?

    (2) Is the *MTBF* of the redesigned product greater than before at the 5% significance level?

    (3) If the test was time terminated, repeat Case 1.

    (4) If the test was time terminated, repeat Case 2.

(5) How many failures should have been observed in Case 3, after redesign, for the *MTBF* to show an improvement at the 5% significance level?

8-6. A product's reliability was monitored, and in 2,000 unit-hours of operation four failures were observed. The product was redesigned and its reliability was monitored again. In 2,000 unit-hours of operation six failures were observed.

(1) If the product is exponential and the test was failure terminated, did the redesign retain equality in the *MTBF* of this product at the 10% significance level?

(2) Is the *MTBF* of the redesigned product greater than before at the 5% significance level?

(3) If the test was time terminated, repeat Case 1.

(4) If the test was time terminated, repeat Case 2.

(5) How many failures should have been observed in Case 3, after redesign, for the *MTBF* to show an improvement at the 10% significance level?

8-7. Three manufacturers of the same equipment are being considered. One sample from each manufacturer is reliability tested with the results of Table 8.1.

If the equipment is exponential, the tests are failure terminated, all units are tested to failure, and the failed equipment is not replaced, determine the following:

(1) Which manufacturers' equipment have the same mean life at the 10% significance level?

(2) Which manufacturer's equipment would you recommend be bought on the basis of its mean life?

(3) If the equipment of Manufacturer C exhibited five failures, but after 1,550 unit-hours of operation, on what other *quantitative bases* would you base your decision as to which manufacturer's equipment should be bought?

(4) On what *qualitative bases* would you base your decision as to which manufacturer's equipment should be bought, in addition to the previous quantitative bases?

8-8. Three manufacturers of the same equipment are being considered. One sample from each manufacturer is reliability tested with the results of Table 8.2.

TABLE 8.1– Unit hours of test time and number of failures observed on equipment supplied by three different manufacturers.

Reliability test results	Manufacturer		
	A	B	C
Unit hours	1,800	2,600	1,100
Number of failures	5	5	5

TABLE 8.2– Unit hours of test time and number of failures observed on equipment supplied by three different manufacturers.

Reliability test results	Manufacturer		
	A	B	C
Unit hours	1,200	2,100	900
Number of failures	6	6	6

If the equipment are exponential, the tests are failure terminated, all units are tested to failure, and the failed equipment are not replaced, determine the following:

(1) Which manufacturers' equipment have the same mean life at the 20% significance level?

(2) Which manufacturer's equipment would you recommend be bought on the basis of its mean life?

(3) If the equipment of Manufacturer C exhibited six failures, but after 1,050 unit-hours of operation, on what other *quantitative bases* would you base your decision as to which manufacturer's equipment should be bought?

(4) On what *qualitative bases* would you base your decision as to which manufacturer's equipment should be bought, in addition to the previous quantitative bases?

8–9. Three manufacturers of the same equipment are being considered. One sample from each manufacturer is reliability tested

TABLE 8.3– Unit hours of test time and number of failures observed on equipment supplied by three different manufacturers.

Reliability test results	Manufacturer		
	A	B	C
Unit hours	1,000	1,800	700
Number of failures	4	4	4

with the results of Table 8.3.

If the equipment are exponential, the tests are failure terminated, all units are tested to failure, and the failed equipment are not replaced, determine the following:

(1) Which manufacturer's equipment have the same mean life at the 5% significance level?

(2) Which manufacturer's equipment would you recommend be bought on the basis of its mean life?

(3) If the equipment of Manufacturer C exhibited four failures, but after 900 unit hr of operation, on what other *quantitative bases* would you base your decision as to which manufacturer's equipment should be bought?

(4) On what *qualitative bases* would you base your decision as to which manufacturer's equipment should be bought, in addition to the previous quantitative bases?

REFERENCES

1. Epstein, B. and Tsao, C.K., "Some Tests on Ordered Observations from Two Exponential Populations," *Annals of Mathematical Statistics*, Vol. 24, pp. 458-466, 1953.

2. Kececioglu, Dimitri B., *Reliability and Life Testing Handbook*, prentice Hall, Inc., Englewood Cliffs, New Jersey, Vol.2, 950 pp., 1992.

Chapter 9

THE NORMAL DISTRIBUTION

9.1 NORMAL DISTRIBUTION CHARACTERISTICS

The normal (Gaussian) distribution is the most widely known distribution, and is given by

$$f(T) = \frac{1}{\sigma_T \sqrt{2\pi}} e^{-\frac{1}{2}(\frac{T-\overline{T}}{\sigma_T})^2}, \tag{9.1}$$

$$f(T) \geq 0, \quad -\infty < T < \infty, \quad -\infty < \overline{T} < \infty, \quad \sigma_T > 0,$$

where

$\overline{T}$ = mean of the normal times to failure, hr,

and

σ_T = standard deviation of the times to failure, hr.

It is a two-parameter distribution with parameters $\overline{T}$ and σ_T, which are the mean and the standard deviation of the times to failure, respectively. Figure 9.1 shows the normal distribution, with the effects on it of a change in the mean and in the standard deviation. Specific characteristics of the normal *pdf* are the following:

1. The mean life, or the MTBF, $\overline{T}$, is also the location parameter of the normal *pdf*, because it locates the *pdf* along the abscissa. It can assume the values of $-\infty < \overline{T} < \infty$. The larger the $\overline{T}$ the larger is the mean life of the components, or of the equipment, or of the

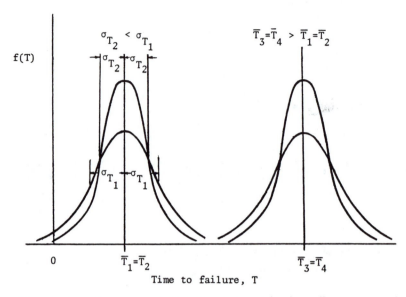

Fig. 9.1– The normal distribution with the effect on it of
changing the mean and the standard deviation.

systems. The better the design, or the longer the designed-in life, the
greater would be the value of $\overline{T}$. The normal *pdf* is bell shaped and
symmetrical about its mean.

2. The standard deviation, σ_T, is also the scale parameter of the
normal *pdf*. As σ_T decreases, the *pdf* gets pushed towards the mean of
the *pdf*, or it becomes narrower and taller, as indicated in Fig. 9.1. As
σ_T increases the *pdf* spreads out away from the mean, or it becomes
broader and shallower, as indicated in Fig. 9.1. The standard deviation
can assume values of $0 < \sigma_T < \infty$. The greater the variability, or the
spread, in the measurements or the data the larger the value of σ_T,
and vice versa. The standard deviation is also the distance between
the mean and the point of inflection of the *pdf*, as shown in Fig. 9.1, on
each side of the mean. The point of inflection is that point of the *pdf*
where the slope changes its value from a decreasing to an increasing
one, or where the second derivative of the *pdf* has a value of zero.

3. The normal *pdf* starts at $T = -\infty$ with an $f(T) = 0$. As T
increases $f(T)$ also increases, it goes through its point of inflection,
and $f(T)$ reaches its maximum value at $T = \overline{T}$. Thereafter $f(T)$
decreases, goes through its point of inflection, and assumes a value of
$f(T) = 0$ at $T = +\infty$.

4. The normal *pdf* has a mean, $\overline{T}$, which is equal to the median,
$\check{T}$, and also equal to the mode, $\tilde{T}$, or $\overline{T} = \check{T} = \tilde{T}$.

5. The normal *pdf* has no shape parameter, as may be seen from Eq. (9.1). This means that the normal *pdf* has only one and the same shape, the bell shape, and this shape does not change.

6. The coefficient of skewness of the normal *pdf* is zero, or $\alpha_3 = 0$. This says that the normal *pdf* is a perfectly symmetrical one.

7. The coefficient of kurtosis of the normal *pdf* is 3, or $\alpha_4 = 3$.

8. The normal function, which when plotted over the frequency histogram approximates it, or the *normal frequency distribution function per class interval width*, w, is given by

$$f(T)' = \frac{Nw}{\sigma_T \sqrt{2\pi}} e^{-\frac{1}{2}(\frac{T-\overline{T}}{\sigma_T})^2}, \tag{9.2}$$

where N is the sample size. As the ordinate scale for this distribution is N/w the total area under this function is N, or the total number of units in the sample, all tested to failure. Also [1, Vol.1, pp.126-127]

$$f(T)' = \frac{Nw}{\sigma_T} \phi(z). \tag{9.3}$$

9. *The normal probability density function,* which is the function per unit class interval width and per unit population, and whose total area is equal to unity, is given by

$$f(T) = \frac{1}{\sigma_T \sqrt{2\pi}} e^{-\frac{1}{2}(\frac{T-\overline{T}}{\sigma_T})^2}, \tag{9.4}$$

or

$$f(T) = \frac{f(T)'}{Nw}, \tag{9.4'}$$

Partial areas under this distribution give directly probabilities.

10. The *standardized normal probability density function,* which has a mean of zero and a standard deviation of unity, is given by

$$\phi(z) = \frac{1}{\sqrt{2\pi}} e^{-\frac{1}{2}z^2}. \tag{9.5}$$

This function is obtained from $f(T)$ using the two transformations

$$z = \frac{T - \overline{T}}{\sigma_T} \quad \text{and} \quad f(T)dT = \phi(z)dz.$$

The first transformation is to obtain a general normal distribution function independent of particular σ_T and $\overline{T}$ values. The second transformation is to maintain the equality of probabilities obtained from either $f(T)$ or $\phi(z)$, because for equality of probability

$$\int_{T_1}^{T_2} f(T)dT = \int_{z_1}^{z_2} \phi(z)dz.$$

Differentiating yields $f(T)dT = \phi(z)dz$, which is the second transformation. Hence

$$\phi(z) = \frac{f(T)dT}{dz}. \qquad (9.6)$$

From

$$z = \frac{T - \overline{T}}{\sigma_T}, \qquad (9.7)$$

$$dz = \frac{dT}{\sigma_T}. \qquad (9.8)$$

Substitution of Eq. (9.8) into Eq. (9.6) yields

$$\phi(z) = \frac{f(T)dT}{\frac{dT}{\sigma_T}} = \sigma_T f(T). \qquad (9.9)$$

Substitution of Eq. (9.7) into Eq. (9.4), and then into Eq. (9.9) yields

$$\phi(z) = \frac{\sigma_T}{\sigma_T \sqrt{2\pi}} e^{-\frac{1}{2}z^2} = \frac{1}{\sqrt{2\pi}} e^{-\frac{1}{2}z^2}.$$

11. The *normal distribution function per standard deviation,* which is used to get the ordinate values of any normal probability density function having a standard deviation, σ_T, from the ordinate values of the standardized normal probability density function, and which are tabulated in Appendix C, is given by

$$\phi(z)' = \frac{\phi(z)}{\sigma_T}. \qquad (9.10)$$

Equations (9.4) and (9.5) are called *probability* density functions because partial areas under these functions give directly the *probabilities* involved. These normal functions are illustrated by an example in Section 9.2.2.

9.2 COMPUTATIONAL METHODS FOR THE DETERMINATION OF THE PARAMETERS OF THE NORMAL DISTRIBUTION

9.2.1 WITH INDIVIDUAL DATA VALUES OR MEASUREMENTS

If few data are available from a small sample, with all units in the sample tested to failure, then their mean, $\overline{T}$, and standard deviation, σ_T, can be calculated from

$$\overline{T} = \frac{\sum\limits_{i=1}^{N} T_i}{N},$$ (9.11)

and

$$\sigma_T = \left[\frac{\sum\limits_{i=1}^{N} (T_i - \overline{T})^2}{N-1} \right]^{1/2} = \left[\frac{N \sum\limits_{i=1}^{N} T_i^2 - (\sum\limits_{i=1}^{N} T_i)^2}{N(N-1)} \right]^{1/2},$$ (9.12)

or

$$\sigma_T = \left[\frac{\sum\limits_{i=1}^{N} T_i^2}{N-1} - \frac{N}{N-1} (\overline{T})^2 \right]^{1/2},$$ (9.12')

respectively. Equations (9.11) and (9.12) give the mean and standard deviation of any set of data. However, only when it is known from past experience, or from goodness-of-fit tests, that the data came from a normally distributed population, these statistics also provide the parameters of a distribution, and in particular of a normal distribution.

EXAMPLE 9–1

Six units are tested to failure, with the following operating hours to failure: 1,260; 2,825; 125; 4,670; 2,080 and 3,550.

1. Find the mean of these data.

2. Find the standard deviation of these data.

3. If it is known that these data come from a normally distributed population, write the normal *pdf* representing these data.

SOLUTIONS TO EXAMPLE 9–1

1. The mean of these data is calculated using Eq. (9.11). Then,

$$\overline{T} = \frac{\sum\limits_{i=1}^{N} T_i}{N} = \frac{1,260 + 2,825 + 125 + 4,670 + 2,080 + 3,550}{6},$$

or

$$\overline{T} = \frac{14,510}{6} = 2,418.33 \text{ hr.}$$

2. The standard deviation of these data is calculated using Eq. (9.12′). Then,

$$\sigma_T = \left[\frac{1,260^2 + 2,825^2 + 125^2 + 4,670^2 + 2,080^2 + 3,550^2}{6-1} \right.$$

$$\left. - \frac{6}{6-1}(2,418.33)^2 \right]^{1/2},$$

$$\sigma_T = \left[\frac{48,321,650.00}{5} - \frac{6}{5}(5,848,319.99) \right]^{1/2},$$

$$\sigma_T = (9,664,330.00 - 7,017,983.99)^{1/2},$$

or

$$\sigma_T = 1,626.76 \text{ hr.}$$

3. Assuming that the mean and the standard deviation calculated in Cases 1 and 2, respectively, are representative of the parameters of a normally distributed population, the normal *pdf* is given by

$$f(T) = \frac{1}{1,626.76\sqrt{2\pi}} e^{-\frac{1}{2}\left(\frac{T-2,418.33}{1,626.76}\right)^2}.$$

TABLE 9.1 – Grouped data analysis for the data in Tables 3.1 and 9.2 to determine their mean and standard deviation.

1	2	3	4	5	6
Class number	Class mid-point, $\overline{T}_i$	Frequency of observations, f_i	u_i	$f_i u_i$	$f_i u_i^2$
1	427	14	-3	-42	126
2	430	28	-2	-56	112
3	433	37	-1	-37	37
4	436	52	0	0	0
5	439	39	1	39	39
6	442	26	2	52	104
7	445	19	3	57	171
8	448	6	4	24	96
9	451	2	5	10	50
10	454	2	6	12	72
		$\sum_{i=1}^{k} = N$ $= 225$		$\sum_{i=1}^{k} f_i u_i$ $= 59$	$\sum_{i=1}^{k} f_i u_i^2$ $= 807$

9.2.2   WITH GROUPED DATA

When there are a lot of data, say over 100 times to failure, or measurements, it is desirable to group the data using the methodology presented in [1, Chapter 4] using the data in Table 3.1. This methodology yields Table 9.1 and the information given in Columns 5 through 8 of Table 9.2. This information is called *grouped data*. These *grouped data* can be put into a histogram form, as was done in Fig. 9.2, to see the general behavior of the data and get an idea about an appropriate distribution to use. If the normal *pdf* looks like an appropriate one, then a recommended procedure for determining the mean and the standard deviation of the data, which are also the parameters of the normal *pdf* representing the population of the data, is illustrated in Table 9.1.

Fill in the first three columns of Table 9.1 with the Class number, Class midpoint, and Frequency of observations, which may be obtained from Table 9.2. Next, fill in the column headed by u_i, Column 4. To do this enter 0 in the row corresponding to the Class Midpoint of the class

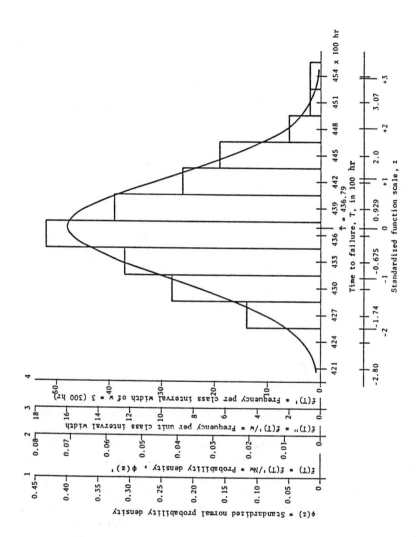

Fig. 9.2– The frequency histogram and various distribution functions and their curves for the data of Tables 3.1 and 9.2, assuming the data are normally distributed.

276

TABLE 9.2– Tally and frequency table for the data in Table 3.1.

Class number	Class starting value	Class end value	Tally	Frequency	Class lower bound	Class midpoint	Class upper bound
1	426	428		14	425.5	427	428.5
2	429	431		28	428.5	430	431.5
3	432	434		37	431.5	433	434.5
4	435	437		52	434.5	436	437.5
5	438	440		39	437.5	439	440.5
6	441	443		26	440.5	442	443.5
7	444	446		19	443.5	445	446.5
8	447	449		6	446.5	448	449.5
9	450	452		2	449.5	451	452.5
10	453	455		2	452.5	454	455.5

that has the greatest frequency of observations, or the modal class. In Table 9.2 this is Class 4 with 52 observations. For the next class above enter -1, next class higher -2, next class higher -3, etc., until all classes above the modal class are filled in.

For the class below the modal class enter $+1$, next class lower $+2$, next class lower $+3$, etc., until all classes below the modal class are filled in, or can use

$$u_i = \frac{\overline{T}_i - A}{w},$$
(9.13)

where A is chosen to be that value of the class midpoint of the modal class. Subsequently, fill in Column 5 by multiplying Column 3 by Column 4, and fill in Column 6 by multiplying Column 5 by Column 4. The sum of Column 5 is used to calculate the mean of the data from

$$\overline{T} = w \frac{\sum\limits_{i=1}^{k} f_i u_i}{N} + A,$$
(9.14)

where

$$w \;\; = \;\; \text{class interval width},$$

and

$$k \;\; = \;\; \text{total number of classes}.$$

The sum of Column 6 is used to calculate the standard deviation of the data from

$$\sigma_T = \frac{w}{[N(N-1)]^{1/2}} \left[N \sum_{i=1}^{k} (f_i u_i^2) - (\sum_{i=1}^{k} f_i u_i)^2 \right]^{1/2}.$$
(9.15)

If the data come from a normal population, then Eqs. (9.14) and (9.15) give the parameters of this distribution. These procedures are illustrated next by an example.

EXAMPLE 9–2

Using the data in Tables 9.1 and 9.2 determine the following:

1. The normal frequency distribution function per class interval width and plot it.

2. The normal probability density function and plot it.

3. The normal standardized probability density function and plot it.

4. The normal distribution function per standard deviation and plot it.

SOLUTIONS TO EXAMPLE 9–2

1. After the data in Table 3.1 are processed as in Table 9.2, Table 9.1 is prepared, as discussed previously. For these data we have the following:

$$w = 3, \quad \sum_{i=1}^{k=10} (f_i u_i) = 59, \quad N = 225, \quad A = 436 \quad \text{and} \quad k = 10.$$

Equation (9.14) with these values, yields the mean of

$$\overline{T} = 3\frac{59}{225} + 436,$$

or

$$\overline{T} = 436.79 \times 10^2 \text{ hr.}$$

Equation (9.15) with $\sum_{i=1}^{k=10} (f_i u_i^2) = 807$ and $\sum_{i=1}^{k=10} (f_i u_i) = 59$, from Table 9.1, yields the standard deviation of

$$\sigma_T = \frac{3}{[225(225-1)]^{1/2}}[(225)(807) - (59)^2]^{1/2},$$

or

$$\sigma_T = 5.64 \times 10^2 \text{ hr.}$$

Consequently, the normal frequency distribution function per class interval width of $w = 3$, as per Eq. (9.2), is

$$f(T)' = \frac{(225)(3)}{5.64\sqrt{2\pi}}e^{-\frac{1}{2}(\frac{T-436.79}{5.64})^2} = \frac{120.0}{\sqrt{2\pi}}e^{-\frac{1}{2}(\frac{T-436.79}{5.64})^2}.$$

This function has been plotted in Fig. 9.2 with scale $f(T)'$ as frequency per class interval width of $w = 3$, or $w = 300$ hr. As may be seen, it represents the histogram quite well.

Much calculation time can be saved in obtaining the necessary values of $f(T)'$ by using the computational form given by Eq. (9.3), or

$$f(T)' = \frac{Nw}{\sigma_T}\phi(z) = \frac{(225)(3)}{5.64}\phi(z) = 120.0\ \phi(z).$$

The $\phi(z)$ values are obtained directly from Appendix C. To get the $\phi(z)$ values, calculate z for each class midpoint, $\overline{T}_i$, from

$$z_i = \frac{\overline{T}_i - \overline{T}}{\sigma_T},$$

and enter with this z_i value Appendix C. For example, for

$$\overline{T}_1 = 427, \ z_1 = \frac{427 - 436.79}{5.64} = -1.74 \text{ and } \phi(z_1) = 0.0878.$$

Then,

$$f(T)' = 120.0 \times 0.0878 = 10.536 \text{ per class width of } w = 3.$$

All such values are given in Column 5 of Table 9.3. These values have been plotted in Fig. 9.2, and a smooth normal time-to-failure distribution curve has been drawn through them.

2. The normal probability density function, as per Eq. (9.4), is

$$f(T) = \frac{1}{5.64\sqrt{2\pi}} e^{-\frac{1}{2}(\frac{T-436.79}{5.64})^2}.$$

To obtain values of $f(T)$, to plot this function, use the form

$$f(T) = \frac{1}{5.64}\phi(z).$$

The values of $f(T)$ are given in Column 7 of Table 9.3.

It is easier, however, to use the curve for $f(T)'$ and add a new ordinate axis for $f(T)$, designated as follows: For $f(T) = 0.01$, $f(T)' = Nwf(T) = 225 \times 3 \times 0.01 = 6.75$. Hence, in line with $f(T)' = 6.75$ mark off 0.01 on the new $f(T)$ ordinate axis. Similarly, mark off 0.02 in line with $f(T)' = 13.5$, 0.03 with 20.25, and so on, until you mark off 9 in line with 60.75. The new ordinate axis with these values designated thereupon is shown in Fig. 9.2.

3. The normal standardized probability density function, per Eq. (9.5), is

$$\phi(z) = \frac{1}{\sqrt{2\pi}} e^{-\frac{1}{2}z^2}.$$

Values of $\phi(z)$ may be obtained either from $\sigma \cdot f(T)$, or from Appendix C, and plotted. These values are given in Column 4 of Table 9.3. It would be easier to use the normal curve in Fig. 9.2 and use a new ordinate axis and scale.

To locate the corresponding $\phi(z)$ values on this axis do the following: For $\phi(z) = 0.05$, $f(T)' = 120.0$ $\phi(z) = 120.0 \times 0.05 = 6$. Hence,

TABLE 9.3 – The ordinate values of various normal distributions for the data of Tables 3.1 and 9.2.

1	2	3	4	5	6	7	8
Class	Class midpoint, $\overline{T_i}$	z	$\phi(z)$	$f(T)'$	$f(T)''$	$f(T)$	$\phi(z)'$
–	421	-2.800	0.0079	0.948	0.316	0.0014	0.0014
–	424	-2.268	0.0305	3.648	1.216	0.0054	0.0054
1	427	-1.736	0.0884	10.582	3.527	0.0157	0.0157
2	430	-1.204	0.1933	23.130	7.710	0.0343	0.0343
3	433	-0.672	0.3183	38.098	12.699	0.0564	0.0564
4	436	-0.140	0.3951	47.285	15.762	0.0700	0.0701
5	439	0.392	0.3695	44.222	14.741	0.0655	0.0655
6	442	0.924	0.2604	31.163	10.388	0.0462	0.0462
7	445	1.456	0.1383	16.548	5.516	0.0245	0.0245
8	448	1.988	0.0553	6.621	2.207	0.0098	0.0100
9	451	2.520	0.0167	1.997	0.665	0.0030	0.0030
10	454	3.052	0.0038	0.454	0.151	0.0007	0.0007

in line with $f(T)' = 6$ mark off 0.05 on the new ordinate axis. Similarly, mark off 0.10 in line with 12, and so on, until you mark off 0.50 in line with 60. The new ordinate axes, with these values designated on it, are shown in Fig. 9.2.

4. The normal distribution function per standard deviation, per Eq. (9.10), is

$$\phi(z)' = \frac{\phi(z)}{\sigma_T}.$$

Numerically it is identical with $f(T)$. Hence, the $\phi(z)'$ curve is identical with the $f(T)$ curve. Thus we do not need a separate ordinate axis for $\phi(z)'$ and all we need to do is to designate the $f(T)$ ordinate axis also with $\phi(z)'$, as has been done in Fig. 9.2. The values of $\phi(z)'$ are given in Column 8 of Table 9.3.

Incidentally the plotting of any normal probability density function is best done for multiples of their standard deviation with the use of Eq. (9.10), the values of $\phi(z)$ given in Appendix C, and the value of σ_T of the normal distribution involved.

9.3 DETERMINATION OF THE PARAMETERS OF THE NORMAL DISTRIBUTION BY PROBABILITY PLOTTING

The probability plotting method of determining the parameters of the normal distribution representing a set of data involves the use of the normal probability plotting paper, a sample of which is given in Fig. 9.3, and plotting on it the MR_j versus the T_j values. The MR_j values are found in Appendix A. The normal distribution is symmetrical; therefore, the area from minus infinity up to the mean equals 0.50. Consequently the T value corresponding to $MR = \hat{Q}(T) = 0.50$ is the value of the mean, $\overline{T}$. the mean is also equal to the median and to the mode as well, or $\overline{T} = \check{T} = T$. The standard deviation is obtained by finding the T values corresponding to the dots along the ordinate. The difference between the T values at any two adjacent points is equal to $\sigma_T/2$ if the straight line truly represents the normal distribution. Also the interval from 15.0% under to 84.1% under covers two standard deviations, consequently the graphical standard deviation estimate is given by

$$\hat{\sigma}_T = \frac{T(84.1\% \text{ under }) - T(15.9\% \text{ under })}{2}. \tag{9.16}$$

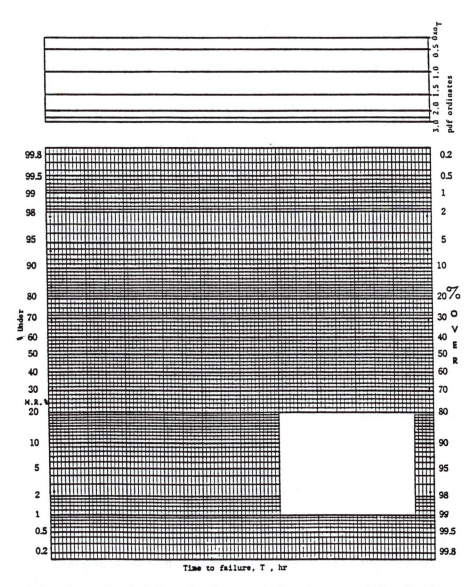

Fig. 9.3 – Probability plotting paper for the normal distribution.

Also the interval from 2.3% under to 97.7% under covers four standard deviations. Then,

$$\hat{\sigma}_T = \frac{T(97.7\% \text{ under }) - T(2.3\% \text{ under })}{4}, \tag{9.17}$$

and the interval from 0.135% under to 99.865% under covers six standard deviations. Then,

$$\hat{\sigma}_T = \frac{T(99.865\% \text{ under }) - T(0.135\% \text{ under })}{6}. \tag{9.18}$$

Any one of Eqs. (9.16), (9.17) or (9.18) may be used to obtain $\hat{\sigma}_T$.

The top section of the normal probability plotting paper is used for plotting the normal probability density function representing the data. The mean of the *pdf* is plotted at the uppermost line, the points $\sigma_T/2$ distance away on each side of the mean are plotted at the level of the second line starting at the top, the points σ_T distance away on each side of the mean are plotted at the level of the third line from the top, and so on, and finally the points $3\sigma_T$ distance away on each side of the mean are plotted at the level of the bottom line. The σ_T multiple these lines represent is indicated as "*pdf* ordinates" at the right side of this scale. The *pdf* plotting procedure is illustrated in the next example.

EXAMPLE 9-3

The normal time-to-failure data given in Table 9.4 were obtained from six units all tested to failure. Do the following:

1. Find the median rank (percent under) corresponding to each failure and tabulate it.

2. Find the corresponding reliability estimate.

3. Plot the MR_j versus T_j.

4. Draw the best possible straight line through these points, and find $\hat{\bar{T}}$ and $\hat{\sigma}$, using Eq. (9.16) with the corresponding dots.

5. Write down the $f(T)$.

6. Plot the $f(T)$ in the section provided for on top of the normal probability plotting paper.

SOLUTIONS TO EXAMPLE 9-3

1. The median ranks, which are also the unreliability estimates, $\hat{Q}(T_j)$, are tabulated in Column 3 of Table 9.5.

TABLE 9.4 – Ungrouped normal time-to-failure data of six units all tested to failure.

Failure order number, j	Time to failure, T_j, hr
1	10,125
2	11,260
3	12,080
4	12,825
5	13,550
6	14,670

2. The corresponding reliability estimates, $\hat{R}(T_j)$, are tabulated in Column 4 of Table 9.5.

3. The plot of the MR_j versus the T_j is given in Fig. 9.4.

4. The visually best fitting straight line is shown drawn in Fig. 9.4. The T value corresponding to $MR = 50\%$ is 12,430 hr. Consequently, this is the value of the estimate of the mean, or $\hat{\overline{T}} = 12,430$ hr.

Using the dots along the left and right sides of the probability plotting paper, the standard deviation is found as follows: From the dot at the 15.9% under level, draw a horizontal line until it intersects the fitted straight line. Drop vertically down from this intersection and read off the value at the intersection of this vertical line with the abscissa, or 10,560 hr in this case. Draw another horizontal line from the dot at the 84.1% under level until it intersects the fitted straight line. Drop vertically down from this intersection and read off the value at the intersection of this vertical line with the abscissa, or 14,260 hr in this case. The distance from 10,560 hr to 14,260 hr spans 2 σ_T. Therefore, from Eq. (9.16),

$$\sigma_T = \frac{14,260 - 10,560}{2},$$

or

$$\sigma_T = 1,850 \text{ hr.}$$

5. From Eq. (9.1),

$$f(T) = \frac{1}{1,850\sqrt{2\pi}} e^{-\frac{1}{2}\left(\frac{T-12,430}{1,850}\right)^2}.$$

6. To plot the $f(T)$ for these data proceed as follows: As illustrated in Fig. 9.4, draw a vertical line at the mean of the data, or at the

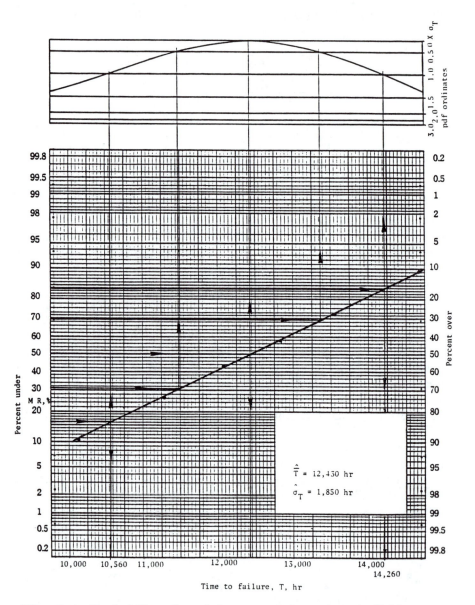

Fig. 9.4– Probability plot of the normal data given in Tables 9.4 and 9.5. $\hat{\bar{T}} = 12,430$ hr and $\hat{\sigma}_T = 1,850$ hr.

TABLE 9.5 – Ungrouped normal time-to-failure data, median ranks (percent under) and reliability estimates.

1	2	3	4
Failure order number, j	Time to failure, T_j, hr	$\hat{Q}(T_j) = MR$, median rank, %	$\hat{R}(T_j) = 1 - \hat{Q}(T_j)$, reliability estimate, %
1	10,125	10.91	89.09
2	11,270	26.45	73.56
3	12,080	42.14	57.86
4	12,825	57.86	42.14
5	13,550	73.56	26.45
6	14,670	89.09	10.91

intersection of the horizontal line drawn at the 50% level with the fitted straight line, until it intersects with the line at the top of the paper. This intersection locates the ordinate of the *pdf* at the mean; i.e., the modal point of the normal *pdf*.

Draw a horizontal line from the dot at the 30.9% level until it intersects with the fitted straight line. Draw a vertical line at this intersection until it intersects with the second line, counting from the top. This intersection provides the ordinate of the *pdf* $\sigma_T/2$ distance to the left of the mean.

Draw a horizontal line from the dot at the 15.9% level until it intersects with the fitted straight line. Draw a vertical line at this intersection until it intersects with third line, counting from the top. This intersection provides the ordinate of the *pdf* σ_T distance to the left of the mean. Continue this way to obtain the ordinate points at the left half of the normal *pdf*.

To obtain the ordinate at the right half of the *pdf*, draw a horizontal line from the dot at the 69.1% level until it intersects with the fitted straight line. Draw a vertical line at this intersection until it intersects with the second line, counting from the top. This intersection provides the ordinate of the *pdf* $\sigma_T/2$ distance to the right of the mean.

Draw a horizontal line from the dot at the 84.1% level until it intersects with the fitted straight line. Draw a vertical line at this intersection until it intersects with the third line, counting from the top. This intersection provides the ordinate of the *pdf* σ_T distance to the right of the mean.

This process is continued, using the dots on the probability plotting paper and the fitted straight line, until all possible points of the *pdf* are located at the upper portion thereof. Draw a smooth curve through all the available points, keeping in mind that the normal curve is bell shaped and has an inflection point one σ_T distance away from the mean on each side of the mean. This has been done in Fig. 9.4 and the portion of the normal *pdf* representing these data on this paper has been drawn in.

9.3.1 USES OF, AND POINTERS ON, NORMAL PROBABILITY PLOTS

The reader is referred to [1, Vol.1, pp. 351-355] for pointers on the use of normal probability plots.

9.4 NORMAL RELIABILITY CHARACTERISTICS

If the times to failure are well represented by the normal distribution, then the reliability of these units is given by [1, Vol.1, pp. 355-361]

$$R(T) = \int_T^\infty f(T)dT = \int_{z(T)}^\infty \phi(z)dz, \qquad (9.19)$$

where

$f(T)$ = normal *pdf* given by Eq. (9.1),

$$z(T) = \frac{T-\overline{T}}{\sigma_T}, \qquad (9.20)$$

T = mission duration when starting the mission at age zero,

and

$\phi(z)$ = standardized normal *pdf*.

Once $f(T)$ and T are given, $R(T)$ can be obtained by entering Appendix B with the $z(T)$ of Eq. (9.20). It must be noted that the normal $f(T)$ extends from $-\infty$ to $+\infty$, and only with this full range will it give a reliability of 1 for $T = -\infty$. This is impossible however for real-time events, as a new component, equipment or system enters service at $T = 0$ and not at $T = -\infty$. Consequently $R(T) + Q(T) \neq 1$ if only times $T \geq 0$ are considered with a normal $f(T)$. However, when $\overline{T} \geq 4.5\sigma_T$, then the area under the normal $f(T)$ for $T < 0$ is

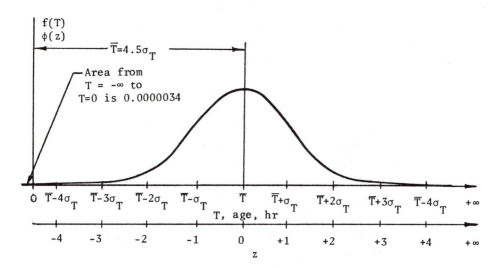

Fig. 9.5– The normal distribution and the effect of its negative tail on reliability.

insignificant by being equal to 0.00000340, as illustrated in Fig. 9.5. Again, however, it must be observed that if reliabilities greater than five nines are involved the use of the normal *pdf* even with $\overline{T} \geq 4.5\sigma_T$ would not be advisable, and the relative values of $\overline{T}$ and σ_T should be studied to insure that the area under $f(T)$ for $T < 0$ is small enough to give the desired accuracy in the predicted $R(T)$.

Another way of coping with this problem is to use the *truncated normal distribution*

$$f_T(T) = \frac{1}{k'\sigma_T\sqrt{2\pi}}e^{-\frac{1}{2}(\frac{T-\overline{T}}{\sigma_T})^2}, \quad 0 \leq T < \infty,$$

where

$\qquad k' \;=\;$ normalizing constant, which is determined from

$\qquad k' \;=\; \displaystyle\int_0^\infty f(T)dT,$

$\quad f(T) \;=\;$ regular normal *pdf*,

$\qquad \overline{T} \;=\;$ mean of the time-to-failure data as given by Eq. (9.11), using all the data,

and

$\qquad \sigma_T \;=\;$ standard deviation of the times-to-failure data as given

TABLE 9.6 – Wear-out failure rate and reliability from the standardized normal density function for the data in Tables 3.1 and 9.2.

1	2	3	4	5
				$\lambda(T) =$
		$R(z) =$	$\lambda(z) =$	$\lambda(z)/\sigma_T^*$,
z	$\phi(z)$	$\int_z^{+\infty} \phi(z)dz$	$\phi(z)/R(z)$	fr/hr
-3.5	0.0009	0.9998	0.0009	0.00000160
-3.0	0.0044	0.9987	0.0044	0.00000780
-2.5	0.0175	0.9938	0.0176	0.0000312
-2.0	0.0540	0.9773	0.0553	0.0000980
-1.5	0.1295	0.9332	0.1387	0.000246
-1.0	0.2420	0.8413	0.2877	0.000510
-0.5	0.3521	0.6915	0.5092	0.000903
0.0	0.3989	0.5000	0.7978	0.00142
0.5	0.3521	0.3085	1.1413	0.00202
1.0	0.2420	0.1587	1.5248	0.00270
1.5	0.1295	0.0668	1.9386	0.00344
2.0	0.0540	0.0227	2.3788	0.00422
2.5	0.0175	0.0062	2.8226	0.00500
3.0	0.0044	0.0013	3.3846	0.00600
3.5	0.0009	0.0002	4.5000	0.00801

* $\sigma_T = 564$ for the illustrative example.

by Eq. (9.12), using all the data.

For details about the truncated normal distribution the reader is referred to [1, Vol.1, pp. 356-357 and 363-390].

The reliability as given by Eq. (9.19) has been evaluated for various values of $z(T)$ in Table 9.6, Column 3, where z is given by Eq. (9.20), and the z values are in increments of 0.5. This table can be used for any normal *pdf* as long as the mission times are in multiples of $0.5\sigma_T$. Figure 9.6 is a plot of the normal *pdf* reliability for various values of z and for $T = \overline{T} \pm k\sigma_T$, where k is in increments of 1.

It may be seen in Fig. 9.6 that the reliability starts at the value of 1 at $T = 0$, or at a practically very small value of $\overline{T} - k\sigma_T$, and remains at essentially this level until wear-out type failures actually start to occur. As soon as wear-out failures set in, the reliability starts to drop and very sharply as the mean life is approached. The rate of drop decreases after $\overline{T}$, and $R(T) \rightarrow 0$ at $T \rightarrow +\infty$.

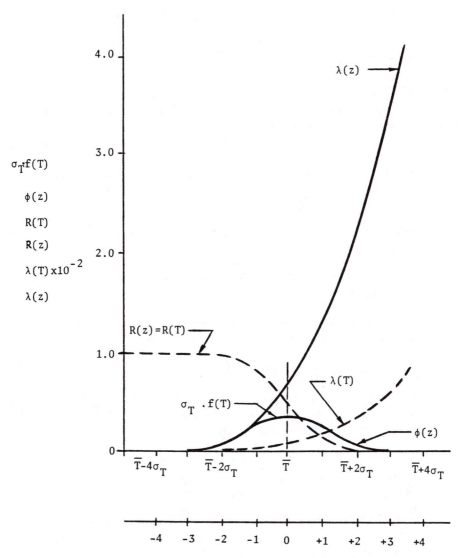

Fig. 9.6– Plots of the $\sigma_T \cdot f(T)$, $\phi(z)$, $R(T)$ and $\lambda(z)$ functions for the normal distribution.

The conditional reliability for a normal *pdf* is given by

$$R(T,t) = \frac{R(T+t)}{R(T)} = \frac{\int_{T+t}^{\infty} f(T)dT}{\int_{T}^{\infty} f(T)dT}, \tag{9.20}$$

where $f(T)$ is the normal *pdf*. If a normal *pdf* is substituted in Eq. (9.20) it may be seen that both the age, T, at the beginning of the mission, and the mission duration, t, remain.

EXAMPLE 9–4

Using the normal *pdf* found in Example 9–3, do the following:

1. Write the reliability function.

2. Find the reliability for a mission of 11,000 hr, starting the mission at age zero.

3. Find the reliability for a mission of 6,880 hr, starting the mission at age zero.

4. Find the reliability for an additional mission of 100 hr after completing 6,880 hr of operation from age zero.

SOLUTIONS TO EXAMPLE 9–4

1. In Example 9–3, the parameters of the normal *pdf* representing the times-to-failure data in Table 9.4 were found to be: $\overline{T} = 12,430$ hr and $\sigma_T = 1,850$ hr. Using Eq. (9.19), the reliability function is given by

$$R(T) = \int_{T}^{\infty} f(T)dT = \int_{T}^{\infty} \frac{1}{1,850\sqrt{2\pi}} e^{-\frac{1}{2}\left(\frac{T-12,430}{1,850}\right)^2} dT. \tag{9.19'}$$

2. Substitution into Eq. (9.19') of $T = 11,000$ hr yields

$$R(T = 11,000 \text{ hr}) = \int_{11,000}^{\infty} f(T)dT,$$

which is evaluated from

$$R(T = 11,000 \text{ hr}) = \int_{z(T=11,000 \text{ hr})}^{\infty} \phi(z)dz,$$

where

$$z(T = 11,000 \text{ hr}) = \frac{T - \overline{T}}{\sigma_T} = \frac{11,000 - 12,430}{1,850} = -0.773.$$

From Appendix B

$$R(T = 11,000 \text{ hr}) = \int_{-0.773}^{\infty} \phi(z)dz = 0.780.$$

This agrees well with the value obtained by probability plotting from Fig. 9.4, where for

$$T = 11,000 \text{ hr}, \quad MR = \hat{Q}(T = 11,000 \text{ hr}) = 22\%,$$

or

$$\hat{R}(T = 11,000 \text{ hr}) = 1 - \hat{Q}(T = 11,000 \text{ hr}) = 1 - 0.22 = 0.78.$$

3. The reliability for a mission of 6,880 hr, starting the mission at age zero, is given by

$$R(T = 6,880 \text{ hr}) = \int_{z(T=6,880 \text{ hr})}^{\infty} \phi(z)dz,$$

where

$$z(T = 6,880 \text{ hr}) = \frac{6,880 - 12,430}{1,850} = -3.00.$$

From Appendix B

$$R(T = 6,880 \text{ hr}) = 0.9987.$$

This value can also be obtained from Table 9.6, Column 3.

4. The reliability for an additional mission of $t = 100$ hr after completing $T = 6,880$ hr of operation, is given by Eq. (9.21), or

$$R(T = 6,880 \text{ hr}, \ t = 100 \text{ hr}) = \frac{\int_{6,880+100}^{\infty} f(T)dT}{\int_{T=6,880}^{\infty} f(T)dT},$$

where the denominator was found to be 0.9987 in the previous case. The numerator is evaluated from

$$\int_{6,880+100}^{\infty} f(T)dT = \int_{z(6,980 \text{ hr})}^{\infty} \phi(z)dz,$$

where

$$z(6,980 \text{ hr}) = \frac{6,980 - 12,430}{1,850} = -2.946.$$

Then, from Appendix B, the numerator is 0.9984. Consequently,

$$R(T = 6,880 \text{ hr}, t = 100 \text{ hr}) = \frac{0.9984}{0.9987} = 0.9997.$$

9.5   NORMAL FAILURE RATE CHARACTERISTICS

The instantaneous normal failure rate is given by

$$\lambda(T) = \frac{f(T)}{R(T)} = \frac{\frac{1}{\sigma_T\sqrt{2\pi}}e^{-\frac{1}{2}(\frac{T-\overline{T}}{\sigma_T})^2}}{\int_T^\infty \frac{1}{\sigma_T\sqrt{2\pi}}e^{-\frac{1}{2}(\frac{T-\overline{T}}{\sigma_T})^2}dT}. \tag{9.21}$$

The evaluation of the numerator of Eq. (9.21) involves finding the height of the normal *pdf* at the specified value of T. This can be obtained from Appendix C. The evaluation of the denominator involves the determination of the reliability for a mission of T duration starting the mission at age zero, as discussed in Section 9.4.

Another method of evaluating $\lambda(T)$ is to develop a standardized failure rate for the normal *pdf*, $\lambda(z)$, as follows:

From Eq. (9.9)

$$f(T) = \frac{\phi(z)}{\sigma_T}. \tag{9.22}$$

Substitution of Eq. (9.22) into Eq. (9.21) yields

$$\lambda(T) = \frac{\phi(z)}{\sigma_T R(T)}. \tag{9.23}$$

Also

$$R(T) = \int_T^\infty f(T)dT = \int_{z(T)}^\infty \phi(z)dz = R(z). \tag{9.24}$$

Therefore Eq. (9.23) may now be written as

$$\lambda(T) = \frac{\phi(z)}{\sigma_T R(z)}. \tag{9.25}$$

To make it independent of σ_T we may define

$$\lambda(z) = \sigma_T \cdot \lambda(T). \tag{9.26}$$

Then, Eq. (9.25) becomes

$$\lambda(z) = \frac{\phi(z)}{R(z)}. \tag{9.27}$$

From this we can calculate the *standardized failure rate* in *failures per standard deviation*, and then divide it by the σ_T of the particular

normal *pdf* in units of hours, to get the *actual failure rate* in units of *failures per hour.* This procedure enables one to obtain $\lambda(T)$ for any normal *pdf* from a table of standardized $\lambda(z)$ by just dividing these values by the σ_T of the *pdf* involved. Values of $\lambda(z)$ are given in Table 9.6, Column 4 for increments of 0.5 for z in the range of $z = -3.5$ to $z = +3.5$. These values of $\lambda(z)$ are plotted in Fig. 9.6, as well as the corresponding values of $\lambda(T)$ given in Column 5 of Table 9.6, for the *pdf* of the data in Tables 3.1 and 9.2. Table 9.6 can be used to obtain the $R(T)$ and $\lambda(T)$ for any normal *pdf* given its parameters and T.

A study of Fig. 9.6 shows that, if the times-to-failure data are well represented by a normal *pdf*, their failure rate increases with age, and the rate of increase becomes greater as their age increases. The failure rate starts at zero at $T = 0$ and becomes ∞ as $T \to \infty$.

EXAMPLE 9–5

For the normal *pdf* of Example 9–3 do the following:

1. Write down the associated failure rate function.

2. Find the failure rate at the age of 6,880 hr.

SOLUTIONS TO EXAMPLE 9–5

1. The failure rate function, with $\overline{T} = 12,430$ hr and $\sigma_T = 1,850$ hr, from Eq. (9.22), is

$$\lambda(T) = \frac{\frac{1}{1,850\sqrt{2\pi}}e^{-\frac{1}{2}(\frac{T-12,430}{1,850})^2}}{\int_T^\infty \frac{1}{1,850\sqrt{2\pi}}e^{-\frac{1}{2}(\frac{T-12,430}{1,850})^2}dT}. \tag{9.22'}$$

2. The failure rate at the age of $T = 6,880$ hr is obtained by substituting in Eq. (9.22') $T = 6,880$ hr. In this case an easier method would be to find

$$z(T = 6,880 \text{ hr}) = \frac{6,880 - 12,430}{1,850} = -3.00.$$

Then, from Appendix C and Table 9.6, $\phi(z = -3.0) = 0.0044$, and

$$R(T = 6,880 \text{ hr}) = R(z = -3.00) = 0.9987.$$

Consequently, from Eq. (9.27),

$$\lambda(z = -3.00) = \frac{0.0044}{0.9987} = 0.004406,$$

and, from Eq. (9.26),

$$\lambda(T = 6,880 \text{ hr}) = \frac{\lambda(z = -3.00)}{\sigma_T} = \frac{0.004406}{1,850},$$

or

$$\lambda(T = 6,880 \text{ hr} = 0.00000238 \text{ fr/hr}.$$

PROBLEMS

9-1. A sample of 50 identical components is tested to failure. The times to failure exhibit a normal distribution with a mean of 10,000 hr and a standard deviation of 1,000 hr.

  (1) There are 1,000 such components of age 9,000 hr. How many of these will survive a 1,000-hr mission after the age of 9,000 hr?

  (2) Determine the number of such components that should start out at $T = 0$, to end up with 1,000 such components of age 9,000 hr.

9-2. The field data of Table 9.7 have been obtained. Do the following:

  (1) Find the mean and standard deviation of the data assuming the times to failure are normally distributed.

  (2) Write down the frequency distribution function per class interval width, calculate it, tabulate it and plot it.

  (3) Write down the frequency distribution function per unit class interval width, calculate it, tabulate it and plot it.

  (4) Write down the probability density function, calculate it, tabulate it and plot it.

  (5) Write down the standardized probability density function, calculate it, tabulate it and plot it.

  (6) Find the reliability of such units for their first mission of 1,000 hr.

  (7) Find the reliability for a second mission of 1,000 hr.

9-3. The field data given in Table 9.8, Columns 1 and 2, have been obtained:

  (1) Fill in the table completely.

  (2) Plot the time-to-failure histogram and draw in the best fitting curve in your judgement.

TABLE 9.7 – Field data from a population exhibiting a normally distributed failure characteristic for Problem 9-2.

Life in 100 hr, starting at 500 hr	Number of failures, N_F
0 – 1	40
1 – 2	210
2 – 3	300
3 – 4	250
4 – 5	80
5 – 6	20
	$N_T = 900$

TABLE 9.8 – Field data from a population exhibiting a normally distributed failure characteristic for Problem 9-3.

Life, in 1,000 hr	Number of failures, N_F	Failure rate, fr/hr	Number of units surviving period, N_S	Reliability, $\hat{R} = N_S/N_T$
0 – 1	4			
1 – 2	21			
2 – 3	30			
3 – 4	25			
4 – 5	8			
5 – 6	2			
	$N_T = 90$			

TABLE 9.9 – Time-to-failure data, in hours, for Problem 9-4.

25	25	21	28	21	23	28	24	20	23
26	27	26	24	29	25	24	25	24	27
22	26	27	25	26	22	24	25	27	24
23	27	22	24	22	25	28	29	23	25
30	25	26	22	25	27	26	20	24	26

(3) Plot the failure rate histogram and draw in the best fitting curve in your judgement.

(4) Plot the reliability histogram and draw in the best fitting curve in your judgement.

(5) Calculate the mean and the standard deviation of the data.

(6) Write down the frequency distribution function per class interval width.

(7) Write down the frequency distribution function per unit class interval width.

(8) Write down the probability density function.

(9) Write down the standardized probability density function.

(10) Calculate, tabulate and plot the frequency distribution function per class interval width on the previous histogram. Label the axes.

(11) Calculate, tabulate and plot the frequency distribution function per unit class interval width. Label the axes.

(12) Calculate, tabulate and plot the probability density function. Label the axes.

(13) Calculate, tabulate and plot the standardized probability density function. Label the axes.

9-4. Using the data in Table 9.9 do the following:

(1) Fit a normal distribution to these data using probability plotting.

(2) Fit a Weibull distribution to these data using probability plotting.

(3) Find the reliability for a mission of 20 hr using:

(3.1) The fitted normal distribution.

(3.2) The fitted Weibull distribution.

(3.3) Comparatively discuss the previous two reliabilities.

(4) Find the reliability for a second mission of five hours, starting this mission at the age of 20 hr using:

(4.1) The fitted normal distribution.

(4.2) The fitted Weibull distribution.

(4.3) Comparatively discuss the previous two reliabilities.

(5) Determine the skewness and the kurtosis of:

(5.1) The raw data.

(5.2) The fitted normal distribution.

(5.3) The fitted Weibull distribution.

(5.4) Comparatively discuss the results.

(6) Determine the coefficient of skewness and kurtosis of:

(6.1) The fitted normal distribution.

(6.2) The fitted Weibull distribution.

(6.3) Comparatively discuss the results.

9-5. Given are the following time-to-failure data obtained from six units put to a reliability test:

Failure order number	Time to failure, hr
1	2,650
2	2,900
3	3,300
4	3,400
5	3,650
6	3,950

Using the *normal probability plotting paper,* do the following:

(1) Find the numerical values of the mean and of the standard deviation (a) graphically and (b) by calculating them, using the data and the equations for $\bar{T}$ and σ_T.

(2) Write down the probability density function for the times to failure of these units.

(3) Determine graphically the reliability of these units for a mission of 2,000 hr, starting the mission at age zero.

TABLE 9.10 – Grouped data of normally distributed car lives for Problem 9-6.

Class number	Endpoints of class, miles $\times 10^3$	Frequency of observations, N_F
1	36	1
2	40	3
3	44	10
4	48	23
5	52	39
6	56	48
7	60	39
8	64	23
9	68	10
10	72	3
11	76	1

(4) Determine the reliability for an additional mission of 1,500 hr duration.

(5) Plot the probability density function in the section provided for at the top of the probability plotting paper.

9-6. Given the data in Table 9.10, do the following:

(1) Using probability plotting find the parameters of the normal *pdf* representing these data.

(2) Determine the reliability of these cars for 50,000 miles of operation, if they are not maintained.

(3) Determine the reliability of these cars for 50,000 miles of operation, if they are checked out every 5,000 miles.

9-7. The grouped data of Table 9.11 are given. Do the following:

(1) By calculation, determine the parameters of the normal distribution that represents these data.

(2) By probability plotting determine the parameters of the normal distribution that represents these data.

(3) Discuss the results comparatively.

TABLE 9.11 – **The normally distributed data for Problem 9-7.**

Class number	Class intervals	Frequency of observations
1	4.00 – 4.04	2
2	4.05 – 4.09	4
3	4.10 – 4.14	5
4	4.15 – 4.19	9
5	4.20 – 4.24	12
6	4.25 – 4.29	8
7	4.30 – 4.34	6
8	4.35 – 4.39	3
9	4.40 – 4.44	1

TABLE 9.12 – **Ungrouped normal times-to-failure data for Problem 9-9.**

Failure order number, j	Time to failure, T_j, hr
1	125
2	1,260
3	2,080
4	2,825
5	3,550
6	4,670

9-8. A part has a mean wear-out life of 1,000 hr and a standard deviation of 200 hr. What is its reliability for a mission of 200 hr starting at an age of 600 hr?

9-9. The normal times-to-failure data given in Table 9.12 were obtained from six units tested to failure. Do the following:

(1) Find the median rank (percent under) corresponding to each failure and tabulate it.

(2) Find the corresponding reliability estimate.

(3) Plot the MR_j versus T_j.

(4) Draw the best possible straight line through these points

and find $\hat{\bar{T}}$ and $\hat{\sigma}_T$, using Eqs. (9.11) and (9.12), respectively, and also the dots on the probability plotting paper.

(5) Write down the $f(T)$.

(6) Write down the $R(T)$.

(7) Write down the $\lambda(T)$.

(8) Find the reliability for a mission of $1,000$ hr, starting the mission at age zero.

(9) Find the reliability for an additional mission of 100 hr duration.

(10) Plot the $f(T)$ in the section provided for at the top of the normal probability plotting paper.

9-10. Seven units are reliability tested and the following times to failure are observed (in hours): 950; 1,450; 1,700; 2,000; 2,300; 2,560. Using the *normal probability plotting paper* do the following:

(1) Find the numerical values of the mean and of the standard deviation graphically.

(2) Write down the probability density function for the times to failure of these units.

(3) Determine graphically the reliability of these units for a mission of 500 hr, starting the mission at age zero.

(4) Determine the reliability for an additional mission of 500 hr duration.

(5) Plot the probability density function in the section provided for at the top of the probability plotting paper.

REFERENCE

1. Kececioglu, Dimitri B., *Reliability Engineering Handbook*, Prentice Hall, Inc., Englewood Cliffs, New Jersey, Vol. 1, 720 pp. and Vol. 2, 568 pp., 1991.

Chapter 10

CONFIDENCE INTERVAL ON THE MEAN LIFE AND ON THE RELIABILITY OF NORMALLY DISTRIBUTED DATA

10.1 TWO-SIDED CONFIDENCE LIMITS ON THE MEAN LIFE WHEN σ_T IS KNOWN, OR $N \geq 25$

In reliability, "σ_T known" is taken to mean that we have a large enough sample size representative of the population, usually $N \geq 25$, from the times-to-failure data of which we calculate $\hat{\sigma}_T$. With such a sample size whose T_i are normally distributed, we obtain $\widehat{\bar{T}}$. If we test other such samples from the same population, we will obtain $\widehat{\bar{T}}_i$ values which vary from each other, therefore the $\widehat{\bar{T}}_i$ are distributed. The *distribution of the $\widehat{\bar{T}}_i$* can be proven to be also *normal* with parameters

$$\overline{T} = \widehat{\bar{T}} \text{ and } \sigma_{\bar{T}} = \hat{\sigma}_T/\sqrt{N}, \tag{10.1}$$

where $\widehat{\bar{T}}$ and $\hat{\sigma}_T$ are obtained from a single sample with size $N \geq 25$. These parameters are illustrated in Fig. 10.1. Now we can find the confidence intervals which contain the true mean life, $\overline{T}$, at a specified confidence level. From the properties of the normal distribution of the mean

303

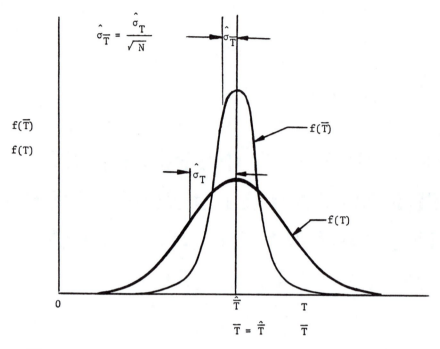

Fig. 10.1 – The parameters of the $f(T)$ and $f(\overline{T})$ distributions.

$$P(\overline{T}_{L2} \leq \overline{T} \leq \overline{T}_{U2}) = CL = 1 - \alpha,$$

where

$$\overline{T}_{L2} = \hat{\overline{T}} - K_{\alpha/2}\frac{\hat{\sigma}_T}{\sqrt{N}} \text{ and } \overline{T}_{U2} = \hat{\overline{T}} + K_{\alpha/2}\frac{\hat{\sigma}_T}{\sqrt{N}}, \qquad (10.2)$$

and $K_{\alpha/2}$ is the number of standard deviations away from the mean such that the area to the right of it is $\alpha/2$. These confidence limits are obtained as follows:

For the standardized normal variate,

$$P(-z_{\alpha/2} \leq z \leq z_{\alpha/2}) = CL = 1 - \alpha. \qquad (10.3)$$

For any other normally distributed variate x, z can be defined as

$$z = \frac{x - \bar{x}}{\sigma_x}.$$

Here the normally distributed variate is $\overline{T}$, then

$$z = \frac{\overline{T} - \bar{\overline{T}}}{\sigma_{\overline{T}}}.$$

Substituting this into Eq. (10.3) yields

$$P(-z_{\alpha/2} \leq \frac{\overline{T} - \bar{\overline{T}}}{\sigma_{\overline{T}}} \leq z_{\alpha/2}) = CL, \qquad (10.4)$$

from which, knowing that

$$\bar{\overline{T}} = \hat{\overline{T}} \text{ and } \sigma_{\overline{T}} = \hat{\sigma}_T/\sqrt{N},$$

Eq. (10.4) becomes

$$P(\hat{\overline{T}} - z_{\alpha/2}\frac{\hat{\sigma}_T}{\sqrt{N}} \leq \overline{T} \leq \hat{\overline{T}} + z_{\alpha/2}\frac{\hat{\sigma}_T}{\sqrt{N}}) = CL = 1 - \alpha.$$

Then,

$$\overline{T}_{L2} = \hat{\overline{T}} - z_{\alpha/2}\frac{\hat{\sigma}_T}{\sqrt{N}}, \qquad (10.5)$$

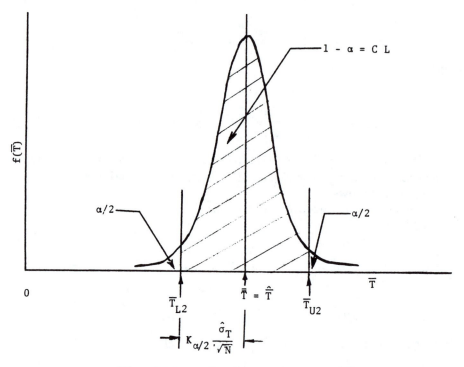

Fig. 10.2 – Confidence limits on $\overline{T}$.

and

$$\overline{T}_{U2} = \hat{\overline{T}} + z_{\alpha/2}\frac{\hat{\sigma}_T}{\sqrt{N}}. \tag{10.6}$$

These limits are illustrated in Fig. 10.2.

Most statistics books, for the confidence limit factors of the normal *pdf*, use $K_{\alpha/2}$ in place of $z_{\alpha/2}$, such that the area under the normal *pdf* to the right of $K_{\alpha/2}$ is $\alpha/2$, or

$$\alpha/2 = \int_{K_{\alpha/2}=z_{\alpha/2}}^{\infty} \phi(z)dz.$$

See Appendix B for the tables of z_δ.

EXAMPLE 10–1

A sample of 25 units is tested, all to failure, and from the T_i the following parameters are calculated:

$$\hat{\overline{T}} = \frac{1}{N}\sum_{i=1}^{N} T_i = 7,500 \text{ hr},$$

and

$$\hat{\sigma}_T = \Big[\frac{1}{N-1}\sum_{i=1}^{N}(T_i - \hat{\overline{T}})^2\Big]^{1/2} = 1,030 \text{ hr}.$$

What are the two-sided confidence limits on $\overline{T}$ at a 90% confidence level?

SOLUTION TO EXAMPLE 10-1

Here $\hat{\overline{T}} = 7,500$ hr, $\hat{\sigma}_T = 1,030$ hr, $N = 25$, $CL = 1 - \alpha = 0.90$, $\alpha = 0.10, \alpha/2 = 0.05$ and $z_{\alpha/2} = 1.645$.

The lower, two-sided confidence limit on the mean life, from Eq. (10.5), is

$$\overline{T}_{L2} = 7,500 - 1.645\frac{1,030}{\sqrt{25}} = 7,500 - 1.645 \times 206 = 7,500 - 339,$$

or

$$\overline{T}_{L2} = 7,161 \text{ hr}.$$

The upper, two-sided confidence limit on the mean life, from Eq. (10.6), is

$$\overline{T}_{U2} = 7,500 + 1.645\frac{1,030}{\sqrt{25}} = 7,500 + 339,$$

or

$$\overline{T}_{U2} = 7,839 \text{ hr.}$$

Therefore, at the 90% confidence level, the two-sided confidence interval on the mean life, for this sample, is

$$(7,161 \text{ hr} \; ; \; 7,839 \text{ hr}),$$

or, in 90% of such samples tested, such confidence intervals will contain the true mean life.

10.2   ONE-SIDED CONFIDENCE LIMITS ON THE MEAN LIFE WHEN σ_T IS KNOWN, OR $N \geq 25$

The one-sided confidence limits on $\overline{T}$ are given by

$$\overline{T}_{L1} = \hat{\overline{T}} - K_\alpha\frac{\hat{\sigma}_T}{\sqrt{N}}, \tag{10.7}$$

and

$$\overline{T}_{U1} = \hat{\overline{T}} + K_\alpha\frac{\hat{\sigma}_T}{\sqrt{N}}, \tag{10.8}$$

where K_α is the number of standard deviations away from the mean life such that the area under the normal *pdf* to the right of it is α, or

$$\alpha = \int_{K_\alpha = z_\alpha}^{\infty} \phi(z)dz.$$

These limits are illustrated in Figs. 10.3, and 10.4, respectively.

EXAMPLE 10-2

Find the one-sided confidence limits on $\overline{T}$ at a 90% confidence level for the data of Example 10-1.

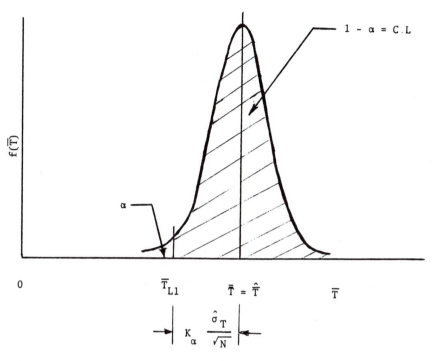

Fig. 10.3 – The one-sided, lower confidence limit on the mean life.

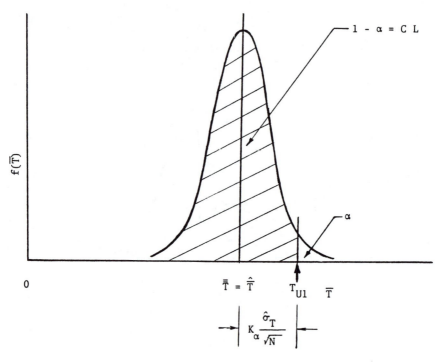

Fig. 10.4 – The one-sided, upper confidence limit on the mean life.

SOLUTION TO EXAMPLE 10-2

Here $\hat{\overline{T}} = 7,500$ hr, $\hat{\sigma}_T = 1,030$ hr, $N = 25$, $CL = 0.90 = 1 - \alpha$, $\alpha = 0.10$ and $z_\alpha = 1.282$.

The lower, one-sided confidence limit on the mean life, from Eq. (10.7), is

$$\overline{T}_{L1} = 7,500 - 1.282\frac{1,030}{\sqrt{25}} = 7,500 - 264,$$

or

$$\overline{T}_{L1} = 7,236 \text{ hr.}$$

Therefore, at the 90% confidence level, the lower, one-sided confidence interval on the mean life, for this sample, is

$$(7,236 \text{ hr} ; \infty).$$

The upper, one-sided confidence limit on the mean life, from Eq. (10.8), is

$$\overline{T}_{U1} = 7,500 + 1.282\frac{1,030}{\sqrt{25}} = 7,500 + 264,$$

or

$$\overline{T}_{U1} = 7,764 \text{ hr.}$$

Therefore, at the 90% confidence level, the upper, one-sided confidence interval on the mean life, for this sample, is

$$(0 ; 7,764 \text{ hr}).$$

10.3 CONFIDENCE LIMITS ON $\overline{T}$ WHEN σ_T IS UNKNOWN, OR $N < 25$

If $N < 25$ the estimate of σ_T, $\hat{\sigma}_T$, may be significantly different than that of σ_T. It can be shown that

$$\frac{\hat{\overline{T}} - \overline{T}}{\hat{\sigma}_T/\sqrt{N}} = t$$

is Student's t distributed, with $N - 1$ degrees of freedom, or

$$f(t) = \frac{1}{\sqrt{\pi\nu}}\frac{\Gamma(\frac{\nu+1}{2})}{\Gamma(\frac{\nu}{2})}\left(1 + \frac{t^2}{\nu}\right)^{-\frac{\nu+1}{2}}, \tag{10.9}$$

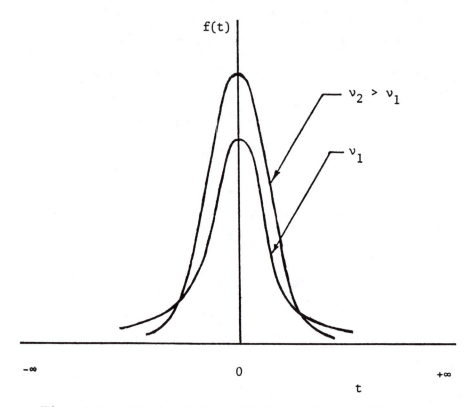

Fig. 10.5 – The Student's t distribution for two different degrees of freedom.

where $-\infty < t < \infty$ and $\nu =$ degrees of freedom. It is a symmetric distribution about $t = 0$, and has greater variability than a comparable normal *pdf*. It is illustrated in Fig. 10.5.

From the Student's t *pdf*

$$P(-t_{\alpha/2;\nu} \leq t \leq t_{\alpha/2;\nu}) = CL = 1 - \alpha,$$

where, for a normally distributed $\overline{T}$,

$$t = \frac{\widehat{\overline{T}} - \overline{T}}{\hat{\sigma}_T/\sqrt{N}}.$$

Therefore,

$$P(-t_{\alpha/2;\nu} \leq \frac{\widehat{\overline{T}} - \overline{T}}{\hat{\sigma}_T/\sqrt{N}} \leq t_{\alpha/2;\nu}) = CL,$$

or

$$P(\overline{T} - t_{\alpha/2;N-1}\frac{\hat{\sigma}_T}{\sqrt{N}} \leq \overline{T} \leq \widehat{\overline{T}} + t_{\alpha/2;N-1}\frac{\hat{\sigma}_T}{\sqrt{N}}) = CL.$$

The $t_{\alpha/2;N-1}$ values are illustrated in Fig. 10.6. See Appendix E for their values. Then, the two-sided confidence limits on the mean life are

$$\overline{T}_{L2} = \widehat{\overline{T}} - t_{\alpha/2;N-1}\frac{\hat{\sigma}_T}{\sqrt{N}}, \tag{10.10}$$

$$\overline{T}_{U2} = \widehat{\overline{T}} + t_{\alpha/2;N-1}\frac{\hat{\sigma}_T}{\sqrt{N}}, \tag{10.11}$$

and

$$P(\overline{T}_{L2} < \overline{T} < \overline{T}_{U2}) = CL = 1 - \alpha.$$

Similarly, the lower, one-sided confidence limit on the mean life is

$$\overline{T}_{L1} = \widehat{\overline{T}} - t_{\alpha;N-1}\frac{\hat{\sigma}_T}{\sqrt{N}}, \tag{10.12}$$

such that

$$P(\overline{T} > \bar{T}_{L1}) = CL = 1 - \alpha,$$

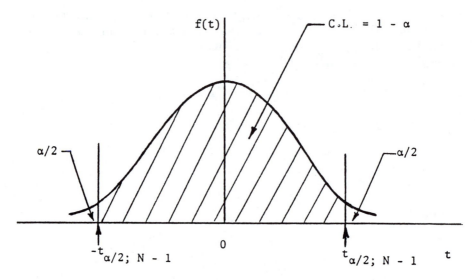

Fig. 10.6 – The two-sided confidence limit values on t for the Student's t distribution.

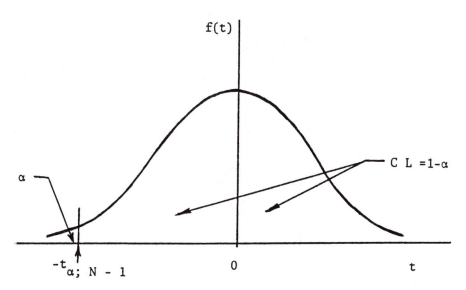

Fig. 10.7 – Lower, one-sided confidence limit value of t for the Student's t distribution.

and the upper, one-sided confidence limit on the mean life is

$$\overline{T}_{U1} = \widehat{\overline{T}} + t_{\alpha;N-1}\frac{\hat{\sigma}_T}{\sqrt{N}}, \tag{10.13}$$

such that

$$P(\overline{T} < \overline{T}_{L1}) = CL = 1 - \alpha.$$

The lower, one-sided confidence limit value of t is illustrated in Fig. 10.7.

EXAMPLE 10-3

Ten units are tested to failure. They come from a population having a normal times-to-failure *pdf*. The calculated parameters are $\widehat{\overline{T}} = 7,500$ hr and $\hat{\sigma}_T = 980$ hr. Do the following:

1. Find the two-sided confidence limits on $\overline{T}$ with a $CL = 90\%$.

2. Find the lower, one-sided confidence limit on $\overline{T}$ with a $CL = 90\%$.

3. Find the upper, one-sided confidence limit on $\overline{T}$ with a $CL = 90\%$.

4. Compare the results with those for $N = 25$.

SOLUTIONS TO EXAMPLE 10-3

1. In this case $N = 10$, $\nu = N - 1 = 10 - 1 = 9$, $CL = 1 - \alpha = 0.90$, $\alpha = 0.10$, $\alpha/2 = 0.05$, $t_{0.05;9} = 1.833$, $t_{0.10;9} = 1.383$, $\widehat{\overline{T}} = 7,500$ hr and $\hat{\sigma}_T = 980$ hr.

The lower, two-sided confidence limit on the mean life, from Eq. (10.10), is

$$\overline{T}_{L2} = 7,500 - 1.833\frac{980}{\sqrt{10}} = 7,500 - 568,$$

or

$$\overline{T}_{L2} = 6,932 \text{ hr.}$$

The upper, two-sided confidence limit on the mean life, from Eq. (10.11), is

$$\overline{T}_{U2} = 7,500 + 1.833\frac{980}{\sqrt{10}} = 7,500 + 568,$$

or

$$\overline{T}_{U2} = 8,068 \text{ hr.}$$

Therefore, at the 90% confidence level, the two-sided confidence interval on the mean life, for this sample, is

$$(6,932 \text{ hr ; } 8,068 \text{ hr}).$$

2. The lower, one-sided confidence limit on the mean life, from Eq. (10.12), is

$$\overline{T}_{L1} = 7,500 - 1.383\frac{980}{\sqrt{10}} = 7,500 - 429,$$

or

$$\overline{T}_{L1} = 7,071 \text{ hr.}$$

Therefore, at the 90% confidence level, the lower, one-sided confidence interval on the mean life, for this sample, is

$$(7,071 \text{ hr ; } \infty).$$

3. The upper, one-sided confidence limit on the mean life, from Eq. (10.13), is

$$\overline{T}_{U1} = 7,500 + 429,$$

or

$$\overline{T}_{U1} = 7,929 \text{ hr.}$$

Therefore, at the 90% confidence level, the upper, one-sided confidence interval on the mean life, for this sample, is

$$(0 \; ; \; 7,929 \text{ hr}).$$

4. The spread in the two-sided confidence limits on $\overline{T}$ is greater, in this case, with a smaller N, because

$$t_{\alpha/2;N-1} > K_{\alpha/2},$$

as

$$1.833 > 1.645,$$

and

$$\hat{\sigma}_{\overline{T}} \text{ is larger as } 310 > 206.$$

The lower, one-sided confidence limit for $N = 10$ is smaller than for $N = 25$, as

$$t_{\alpha;N-1} > K_{\alpha},$$

or

$$1.383 > 1.282,$$

and $\sigma_{\hat{T}}$ is larger as $310 > 206$; i.e., $\overline{T}_{L1} = 7,071$ hr for $N = 10$ versus $\overline{T}_{L1} = 7,236$ hr for $N = 25$.

The upper, one-sided confidence limit for $N = 10$ is larger than for $N = 25$ for the same reasons as before; i.e., $\overline{T}_{U1} = 7,929$ hr for $N = 10$ versus $\overline{T}_{U1} = 7,764$ hr for $N = 25$.

10.4   RANGE OF LIFE ABOUT THE MEAN LIFE WITH A GIVEN PROBABILITY WHEN σ_T IS KNOWN, OR $N \geq 25$

From the normal *pdf* and its parameters $(\overline{T}, \sigma_T)$ we can find the $T's$ about the mean life such that a specific proportion of the lives lies within this range, or

$$P(T_{L2} < T < T_{U2}) = CL = 1 - \alpha,$$

where

$$T_{L2} = \hat{\overline{T}} - z_{\alpha/2} \cdot \hat{\sigma}_T, \tag{10.14}$$

and

$$T_{U2} = \hat{\overline{T}} + z_{\alpha/2} \cdot \hat{\sigma}_T. \tag{10.15}$$

This $P = 1 - \alpha$ is also the a priori probability of failing during the age range of

$$T_{L2} \longrightarrow T_{U2},$$

as may be seen in Fig. 10.8.

Similarly we can find T_{L1} such that the probability of having a life greater than T_{L1} is equal to $CL = 1 - \alpha$, or

$$P(T \geq T_{L1}) = CL = 1 - \alpha, \tag{10.16}$$

where

$$T_{L1} = \hat{\overline{T}} - z_{\alpha}\hat{\sigma}_T. \tag{10.17}$$

Equation (10.16) also gives *the reliability for a mission of T_{L1} duration, starting the mission at age zero.* Therefore,

$$R(T_{L1}) = \int_{T_{L1}}^{\infty} f(T)dt = \int_{z_\delta}^{\infty} \phi(z)dz = CL = 1 - \alpha.$$

The z_δ values may be found in Appendix B.

EXAMPLE 10–4

Twenty-five units are reliability tested to failure. From their times-to-failure data the following normal distribution parameters are calculated: $\hat{\overline{T}} = 7,500$ hr and $\hat{\sigma}_T = 1,030$ hr. Do the following:

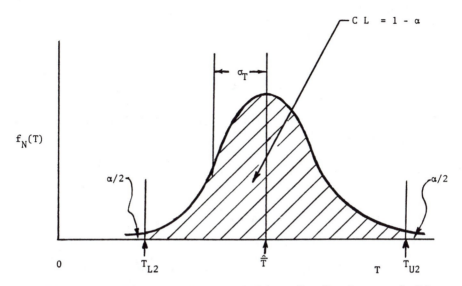

Fig.10.8 - Normal times-to-failure distribution and life range for a given confidence level.

1. Find the life range which contains 99% of the distribution about the mean life.

2. Find the life to the right of which 99% of the distribution lies, or $\alpha = 0.01$.

SOLUTIONS TO EXAMPLE 10–4

1. The lower limit of the life range which contains 99% of the distribution about the mean life for $1-\alpha = 0.99, \alpha = 0.01, \alpha/2 = 0.005$ and $z_{0.005} = 2.576$, from Eq. (10.14), is

$$T_{L2} = 7,500 - 2.576 \times 1,030 = 7,500 - 2,653,$$

or

$$T_{L2} = 4,847 \text{ hr.}$$

The upper limit of the life range which contains 99% of the distribution about the mean life, from Eq. (10.15), is

$$T_{U2} = 7,500 + 2.576 \times 1,030 = 7,500 + 2,653,$$

or

$$T_{U2} = 10,153 \text{ hr.}$$

Then,

$$P(T_{L2} = 4,847 \text{ hr} \le T \le T_{U2} = 10,153 \text{ hr}) = 0.99 = 1 - \alpha.$$

2. The life to the right of which 99% of the distribution lies, from Eq. (10.17), is

$$T_{L1} = 7,500 - 2.326 \times 1,030 = 7,500 - 2,396,$$

or

$$T_{L1} = 5,104 \text{ hr.}$$

Then,

$$P(T \ge T_{L1} = 5,104 \text{ hr}) = 0.99 = 1 - \alpha = R(T_{L1}).$$

10.5 RANGE OF LIFE ABOUT THE MEAN LIFE FOR A GIVEN PROBABILITY WHEN σ_T IS UNKNOWN, OR $N < 25$

If $N < 25$ then the $T's$ are Student's t distributed, and the range of the $T's$ about the mean that includes $(1-\alpha)$ proportion of the distribution, or of the lives, is obtained from

$$T_{L2} = \widehat{\overline{T}} - t_{\alpha/2;N-1} \cdot \hat{\sigma}_T, \tag{10.18}$$

and

$$T_{U2} = \widehat{\overline{T}} + t_{\alpha/2;N-1} \cdot \hat{\sigma}_T. \tag{10.19}$$

Similarly, we can find T_{L1} such that the probability of having a life greater than T_{L1} is equal to $CL = 1 - \alpha$, from

$$T_{L1} = \widehat{\overline{T}} - t_{\alpha;N-1} \cdot \hat{\sigma}_T. \tag{10.20}$$

The $t_{\delta;\nu}$ values are obtained from Appendix E.

EXAMPLE 10–5

Ten units are reliability tested to failure. From their times-to-failure data, the normal distribution parameters are found to be $\widehat{\overline{T}} = 7,500$ hr and $\hat{\sigma}_T = 980$ hr. Find the following:

1. The life range which contains 99% of the distribution about the mean.

2. The life to the right of which 99% of the distribution lies, or $\alpha = 0.01$.

SOLUTIONS TO EXAMPLE 10–5

1. Here $N = 10$, $1 - \alpha = 0.99$, $\alpha = 0.01$, $\alpha/2 = 0.005$, and $t_{0.005;9} = 3.250$. Then, from Eq. (10.18),

$$T_{L2} = 7,500 - 3.250 \times 980 = 7,500 - 3,185,$$

or

$$T_{L2} = 4,315 \text{ hr.}$$

Also, from Eq. (10.19),

$$T_{U2} = 7,500 + 3.250 \times 980 = 7,500 + 3,185,$$

or

$$T_{U2} = 10,685 \text{ hr.}$$

Then, the life range which contains 99% of the distribution about the mean life is

$$(4.315 \text{ hr} \ ; \ 10,685 \text{ hr}).$$

These results may also be stated as

$$P(T_{L2} = 4,315 \text{ hr} \leq T \leq T_{U2} = 10,685 \text{ hr}) = 0.99 = 1 - \alpha.$$

2. The life to the right of which 99% of the distribution lies, from Eq. (10.20), is

$$T_{L1} = 7,500 - 2.821 \times 980 = 7,500 - 2,765,$$

or

$$T_{L1} = 4,735 \text{ hr.}$$

Then, the life range to the right of which 99% of the distribution lies is

$$(4,735 \text{ hr} \ ; \ \infty).$$

This result may also be stated as

$$P(T \geq T_{L1} = 4,735 \text{ hr}) = 0.99 = 1 - \alpha.$$

10.6   TOLERANCE LIMITS ON LIFE

We can also find such limits on life that they contain $100(1 - \alpha)\%$ of the lives *with a prescribed probability of 100 γ %*, or

$$P(T_{TL2} \leq T \leq T_{TU2}) = 1 - \alpha; \gamma,$$

where

$$
\begin{aligned}
T_{TL2} &= \text{ lower, two-sided tolerance limit,} \\
T_{TU2} &= \text{ upper, two-sided tolerance limit,} \\
1 - \alpha &= \text{ proportion of the lives contained within the range} \\
&\quad\ (T_{TL2}\ ;\ T_{TU2}),
\end{aligned}
$$

and

$$
\begin{aligned}
\gamma \quad &= \text{ probability with which the } (1 - \alpha) \text{ proportion of the} \\
&\quad\ \text{lives is contained within the range } (T_{TL2}\ ;\ T_{TU2}).
\end{aligned}
$$

This is so because each finite sample yields its own $\widehat{\overline{T}}$ and $\hat{\sigma}_T$, and we don't know the exact $\overline{T}$ and σ_T. Therefore, from the $\widehat{\overline{T}}$ and $\hat{\sigma}_T$ of each sample we will get a different set of life limits for the same proportion $(1 - \alpha)$ of the lives, for each *pdf* with its own $\widehat{\overline{T}}$ and $\hat{\sigma}_T$.

The tolerance limits are so determined that they contain at least $(1 - \alpha)$ proportion of the lives (area under the normal *pdf*) with a desired probability γ, or

$$P\left(\int_{T_{TL2}}^{T_{TU2}} f(T)\, dT \geq 1 - \alpha \right) = \gamma.$$

Consider the one-sided, lower tolerance limit first; i.e., we are seeking a statistic $T_{TL1}(T_1, T_2, \cdots, T_N)$ such that

$$P\left\{ \int_{T_{TL1}(T-1, T_2, \cdots, T_N)}^{\infty} \frac{1}{\sqrt{2\pi}} e^{-\frac{1}{2}\left(\frac{T - \overline{T}}{\sigma_T}\right)^2} dT \geq 1 - \alpha \right\} = \gamma,$$

where $\overline{T}$ and σ_T are unknown and the $(T_1, T_2, \cdots, T_N)$ are the life values of the sample from $f(T)$. It is known that

$$P\left(\frac{T - \overline{T}}{\sigma_T} \geq z_\alpha \right) = 1 - \alpha,$$

or

$$P(T \geq \overline{T} + \sigma_T \cdot z_\alpha) = 1 - \alpha.$$

It may be seen that if $T_{TL1}(T_1, T_2, \cdots, T_N)$ is so chosen that

$$P\{T_{TL1}(T_1, T_2, \cdots, T_N) \leq \overline{T} + \sigma_T \cdot z_\alpha\} = \gamma;$$

then, by the definition of the tolerance limit, $T_{TL1}(T_1, T_2, \cdots, T_N)$ would be the one-sided, lower tolerance limit.

Let

$$T_{TL1}(T_1, T_2, \cdots, T_N) = \hat{\overline{T}} - K'_{N;\alpha;\gamma;1} \hat{\sigma}_T,$$

then

$$T_{TL1}(T_1, T_2, \cdots, T_N) \leq \overline{T} + \sigma_T \cdot z_\alpha.$$

Equating the right sides of the previous two equations yields

$$\frac{\hat{\overline{T}} - (\overline{T} + \sigma_T \cdot z_\alpha)}{\hat{\sigma}_T} \leq K'_{N;\alpha;\gamma;1}.$$

Therefore, $K'_{N;\alpha;\gamma;1}$ satisfies

$$P\{\frac{\sqrt{N}[\hat{\overline{T}} - (\overline{T} + \sigma_T \cdot z_\alpha)]}{\hat{\sigma}_T} \leq \sqrt{N} K'_{N;\alpha;\gamma;1})\} = \gamma.$$

Furthermore,

$$\frac{\hat{\overline{T}} - \overline{T} - \sigma_T \cdot z_\alpha}{\hat{\sigma}_T} = \frac{\frac{\hat{\overline{T}} - \overline{T}}{\sigma_T/\sqrt{N}} - \sqrt{N} \cdot z_\alpha}{\hat{\sigma}_T/\sigma_T},$$

$$\frac{(N-1)\hat{\sigma}_T^2}{\sigma_T^2} \sim \chi^2(N-1),$$

and

$$Z = \frac{\hat{\overline{T}} - \overline{T}}{\sigma/\sqrt{N}} \sim N(0\ ;\ 1).$$

Let

$$\delta = -\sqrt{N} \cdot z_\alpha;$$

Then,

$$\frac{\hat{\overline{T}} - \overline{T} - \sigma_T \cdot z_\alpha}{\hat{\sigma}_T} = \frac{Z + \delta}{\sqrt{\frac{\chi^2(N-1)}{N-1}}} \sim t(N-1\ ;\ \delta),$$

where $t(N-1\,;\,\delta)$ designates a noncentral t distribution with $(N-1)$ degrees of freedom and parameter of noncentrality, δ.

Let $t_{\gamma;N-1;\delta}$ be the γth percentile of $t(N-1\,;\,\delta)$; i.e.,

$$P[t(N-1\,;\,\delta) \le t_{\gamma;N-1;\delta}] = \gamma,$$

then,

$$K'_{N;\alpha;\gamma;1} = \frac{1}{\sqrt{N}}t_{\gamma;N-1;\delta},$$

and the one-sided, lower tolerance limit is given by

$$T_{TL1} = \widehat{\overline{T}} - K'_{N;\alpha;\gamma;1} \cdot \hat{\sigma}_T, \tag{10.21}$$

where

$K'_{N;\alpha;\gamma;1}$ = one-sided, tolerance factor for proportion $(1-\alpha)$ of inclusion with probability γ, given in Appendix I.

Similarly, the one-sided, upper tolerance limit is given by

$$T_{TU1} = \widehat{\overline{T}} + K'_{N;\alpha;\gamma;1} \cdot \hat{\sigma}_T. \tag{10.22}$$

The two-sided tolerance limits on the life are given by

$$T_{TL2} = \widehat{\overline{T}} - K'_{N;\alpha;\gamma;2} \cdot \hat{\sigma}_T, \tag{10.23}$$

and

$$T_{TU2} = \widehat{\overline{T}} + K'_{N;\alpha;\gamma;2} \cdot \hat{\sigma}_T, \tag{10.24}$$

where

$K'_{N;\alpha;\gamma;2}$ = two-sided tolerance factor for proportion $(1-\alpha)$ of inclusion with probability γ, given in Appendix H, and it is given by

$$K'_{N;\alpha;\gamma;2} = \frac{1}{\sqrt{N}}t_{\gamma/2;N-1;\delta}.$$

EXAMPLE 10-6

Ten units are reliability tested to failure. From their times-to-failure data the normal distribution parameters are determined to be $\widehat{\overline{T}} = 7,500$ hr, and $\hat{\sigma}_T = 980$ hr. Find the following:

1. The two-sided tolerance limits on the life of these units with $1 - \alpha = 0.99$ and $\gamma = 0.90$.

2. The lower, one-sided tolerance limit on the life with $1 - \alpha = 0.99$ and $\gamma = 0.90$.

SOLUTIONS TO EXAMPLE 10–6

1. Here $N = 10, \widehat{\overline{T}} = 7,500$ hr, $\hat{\sigma}_T = 980$ hr, $1-\alpha = 0.99$, $\alpha = 0.01$, $\gamma = 0.90$ and from Appendix H

$$K'_{N;\alpha;\gamma;2} = K'_{10;0.01;0.90;2} = 3.959.$$

Then, from Eq. (10.23),

$$T_{TL2} = 7,500 - 3.959 \times 980 = 7,500 - 3,880,$$

or

$$T_{TL2} = 3,620 \text{ hr.}$$

From Eq. (10.24)

$$T_{TU2} = 7,500 + 3.969 \times 980 = 7,500 + 3,880,$$

or

$$T_{TU2} = 11,380 \text{ hr,}$$

and

$$P(T_{TL2} = 3,620 \text{ hr} \leq T \leq T_{TU2} = 11,380 \text{ hr}) = 0.99; 0.90.$$

This compares with the life range statement of

$$P(T_{L2} = 4,315 \text{ hr} \leq T \leq T_{U2} = 10,685 \text{ hr}) = 0.99,$$

found in Example 10–5. Consequently, the tolerance limits are broader than the life range for the same probability of inclusion in this range of 0.99, when we wish the inclusion of 0.99 to be with 90% probability.

2. The lower, one-sided tolerance limit on life, using Appendix I where $K_{10;0.01;0.90;1} = 3.532$ and Eq. (10.21), is

$$T_{TL1} = 7,500 - 3.532 \times 980 = 7,500 - 3,461,$$

or

$$T_{TL1} = 4,039 \text{ hr},$$

and

$$P(T \geq T_{TL1} = 4,039 \text{ hr}) = 0.99; 0.90.$$

This compares with

$$P(T \leq T_{L1} = 4,735 \text{ hr}) = 0.99,$$

found in Example 10–5. Consequently, the tolerance range is greater than the life range.

10.7   LOWER, ONE-SIDED CONFIDENCE LIMIT ON THE RELIABILITY FOR THE NORMAL *pdf* CASE

Generally,

$$R(T_1) = \int_{T_1}^{\infty} f(T)dT,$$

or

$$R(T_1) = P(T \geq T_1) ;$$

i.e., reliability is the proportion, $(1 - \alpha)$, of the *pdf* to the right of T_1. If we want the reliability at a specified confidence level, probability γ; then, we need to find the mission duration, T_{TL1}, starting the mission at age 0, such that

$$P(T \geq T_{TL1}) = 1 - \alpha; \gamma = CL,$$

and

$$R(T_{L1}) = 1 - \alpha = R_{goal},$$

as illustrated in Fig. 10.9. We can also find R_{L1}, given T and CL.

Appendix J gives the one-sided tolerance factors at a confidence level of $\gamma = 50\%$.

EXAMPLE 10–7

Normal units have the distribution parameters of $\widehat{\overline{T}} = 7,500$ hr and $\hat{\sigma}_T = 980$ hr, based on 10 units tested to failure. Do the following:

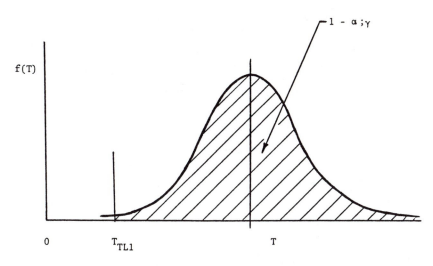

Fig. 10.9 – Reliability at a specified confidence level, or
$$R(T_{TL1}) = P(T < T_{TL1}) = 1 - \alpha \text{ and } CL = \gamma.$$

1. Find the mission duration for a mission reliability of 0.99 at a 90% confidence level.

2. Find R_{L1} for a mission of 3,000 hr duration, starting the mission at age 0, with a $CL = 90\%$.

3. Find R_{L1} for a mission of 3,000 hr duration, starting the mission at age 0, with a $CL = 50\%$.

SOLUTIONS TO EXAMPLE 10–7

1. Here $N = 10, R = 0.99 = 1 - \alpha, \alpha = 0.01, CL = 0.90 = \gamma$, and $K'_{10;0.01;0.90;1} = 3.532$; therefore, from Eq. (10.21),

$$T_{TL1} = 7,500 - 3.532 \times 980 = 7,500 - 3,461,$$

or

$$T_{TL1} = 4,039 \text{ hr.}$$

Consequently, the mission duration should be 4,039 hr from age 0. Such a mission duration would yield an average reliability of $\overline{R} = 0.99979!$

2. Here $T = T_{TL1} = 3,000$ hr and $\gamma = 0.90$; then, from Eq. (10.21),

$$3,000 = 7,500 - K'_{10;\alpha;0.90;1} \cdot 980.$$

Therefore,

$$K'_{10;\alpha;0.90;1} = \frac{7,500 - 3,000}{980} = \frac{4,500}{980},$$

or

$$K'_{10;\alpha;0.90;1} = 4.5918.$$

From the one-sided, tolerance factor tables of Appendix I, by interpolation,

$$\alpha \cong 0.0013,$$

or

$$R_{L1}(T = 3,000 \text{ hr}) = 1 - \alpha = 1 - 0.0013.$$

Therefore,

$$R_{L1}(T = 3,000 \text{ hr}) = 0.9987,$$

with $CL = 90\%$.

3. In this case $T = 3,000$ hr and $\gamma = 0.50$; then, from Eq. (10.21),

$$3,000 = \widehat{\overline{T}} - K_{10;\alpha;050;1} \cdot \hat{\sigma}_T = 7,500 - K_{10;\alpha;0.50;1} \cdot 980,$$

and

$$K_{10;\alpha;0.50;1} = 4.5918.$$

Then, from Appendix J, by interpolation,

$$R_{L1}(T = 3,000 \text{ hr}) = 0.9_5 5072$$

with $CL = 50\%$.

PROBLEMS

10-1. Mechanical components are exhibiting a normally distributed wear-out times-to-failure characteristic. Sixteen such identical components are tested to failure. The mean time to failure is estimated to be 15,000 hr and the standard deviation to be 1,500 hr. At a 90% confidence level, do the following:

(1) Calculate the two-sided confidence limits on the mean time to failure.

(2) Calculate the lower, one-sided confidence limit on the mean time to failure.

(3) Repeat Case 1 for a sample size of 30.

(4) Repeat Case 2 for a sample size of 30.

10-2. Mechanical components are exhibiting a normally distributed wear-out times-to-failure characteristic. Sixteen such identical components are tested to failure. The mean time to failure is estimated to be 10,000 hr and the standard deviation to be 1,500 hr. At a 90% confidence level, do the following:

(1) Calculate the two-sided confidence limits on the mean time to failure.

(2) Calculate the lower, one-sided confidence limit on the mean time to failure.

(3) Repeat Case 1 for a sample size of 37.

(4) Repeat Case 2 for a sample size of 37.

10-3. Identical mechanical components are exhibiting a wear-out times-to-failure characteristic. Sixty-four such identical components are tested to failure. The mean time to failure is estimated to be 15,000 hr and the standard deviation to be 2,000 hr. At a 90% confidence level, do the following:

(1) Calculate the two-sided confidence limits on the mean life.

(2) Calculate the lower, one-sided confidence limit on the mean life.

(3) Work out Case 1 for a sample size of 18.

(4) Work out Case 2 for a sample size of 18.

10-4. Nineteen identical units are reliability tested. Their times-to-failure distribution is determined to be normal with the following parameters:

$$\widehat{\overline{T}} = 2,722.29 \text{ hr and } \hat{\sigma}_T = 21.92 \text{ hr.}$$

Find the following:

(1) The minimum reliability that could be assured for this device for a mission of 2,655 hr with 90% confidence.

(2) The operating, or mission, time to assure a minimum reliability of 90% at the 95% confidence level.

(3) The confidence level that can be assured, for future production items of this device, to exhibit a minimum reliability of 90% for a required operating time of 2,655 hr.

10-5. Ten units are reliability tested to failure and the following normal distribution parameters of their times-to-failure distribution are calculated:

$$\widehat{\overline{T}} = 15,000 \text{ hr and } \hat{\sigma}_T = 1,410 \text{ hr.}$$

Determine the following:

(1) The reliability for a mission of 10,500 hr at the 50% confidence level.

(2) The minimum reliability for a mission of 10,500 hr at the 90% confidence level.

(3) The mission duration for a minimum mission reliability of 99% and a confidence level of 95%.

10-6. Ten units are reliability tested to failure and the following normal distribution parameters of their times-to-failure distribution are calculated:

$$\widehat{\overline{T}} = 15,000 \text{ hr and } \hat{\sigma}_T = 1,500 \text{ hr.}$$

Determine the following:

(1) The reliability for a mission of 10,500 hr at the 50% confidence level.

(2) The minimum reliability for a mission of 10,500 hr at the 90% confidence level.

(3) The mission duration for a minimum mission reliability of 99% and a confidence level of 95%.

10-7. A minimum reliability of 0.995 is required of a component exhibiting normally distributed wear-out times to failure. Twenty such components are tested to failure and the following estimates of the mean life and of the standard deviation are calculated from the times to failure:

$$\widehat{\overline{T}} = 10,000 \text{ hr and } \hat{\sigma}_T = 2,000 \text{ hr.}$$

(1) Determine the replacement time for this component which will, with at least 95% assurance, give the required reliability.

(2) Repeat Case 1 if the required minimum reliability is 0.975.

10-8. Electrical components are exhibiting a normally distributed wearout times-to-failure characteristic. Fourteen such identical components are tested to failure. The mean time to failure is estimated to be 12,500 hr and the standard deviation to be 1,100 hr. At a 95% confidence level, do the following:

(1) Calculate the two-sided confidence limits on the mean time to failure.

(2) Calculate the lower, one-sided confidence limit on the mean time to failure.

(3) Repeat Case 1 for a sample size of 35.

(4) Repeat Case 2 for a sample size of 35.

10-9. A minimum reliability of 0.95 is required of a component exhibiting normally distributed wear-out times to failure. Fifteen such components are tested to failure and the following estimates of the mean life and of the standard deviation are calculated from the times to failure:

$$\widehat{\overline{T}} = 7,000 \text{ hr and } \hat{\sigma}_T = 1,000 \text{ hr.}$$

(1) Determine the replacement time for this component which will, with at least 95% assurance, give the required reliability.

(2) Repeat Case 1 if the required minimum reliability is 0.975.

10-10. Eight units are reliability tested to failure and the following normal distribution parameters of their times-to-failure distribution are calculated:

$$\widehat{\overline{T}} = 12,500 \text{ hr and } \hat{\sigma}_T = 750 \text{ hr.}$$

Determine the following:

(1) The reliability for a mission of 10,500 hr at the 50% confidence level.

(2) The minimum reliability for a mission of 10,500 hr at the 90% confidence level.

(3) The mission duration for a minimum mission reliability of 99% and a confidence level of 95%.

10-11. Mechanical components are exhibiting a normally distributed wear-out times-to-failure characteristic. Thirteen such identical components are tested to failure. The mean time to failure is estimated to be 17,500 hr and the standard deviation to be 2,500 hr. At a 90% confidence level, do the following:

(1) Calculate the two-sided confidence limits on the mean time to failure.

(2) Calculate the lower, one-sided confidence limit on the mean time to failure.

(3) Repeat Case 1 for a sample size of 30.

(4) Repeat Case 2 for a sample size of 30.

10-12. A minimum reliability of 0.90 is required of a component exhibiting normally distributed wear-out times to failure. Ten such components are tested to failure and the following estimates of the mean life and of the standard deviation are calculated from the times to failure:

$$\hat{\overline{T}} = 9,000 \text{ hr and } \hat{\sigma}_T = 1,250 \text{ hr.}$$

Do the following:

(1) Determine the replacement time for this component which will, with at least 95% assurance, give the required reliability.

(2) Repeat Case 1 if the required minimum reliability is 0.975.

Chapter 11

THE LOGNORMAL DISTRIBUTION

11.1 LOGNORMAL DISTRIBUTION CHARACTERISTICS

The two-parameter lognormal distribution, illustrated in Figs. 11.1(a) and (b), is given by [1]

$$f(T) = \frac{1}{T\sigma_{T'}\sqrt{2\pi}}e^{-\frac{1}{2}(\frac{T'-\overline{T}'}{\sigma_{T'}})^2}, \tag{11.1}$$

$$f(T) \geq 0, T \geq 0, -\infty < \overline{T}' < \infty, \sigma_{T'} > 0,$$

$$T' = \log_e T, \tag{11.2}$$

where

$\overline{T}'$ = mean of the Naperian, or natural, logarithms of the times to failure, $\log_e$ hr,

and

$\sigma_{T'}$ = standard deviation of the Naperian, or natural, logarithms of the times to failure, $\log_e$ hr.

A random variable is lognormally distributed if the logarithm of the random variable is normally distributed. It is for this reason that

333

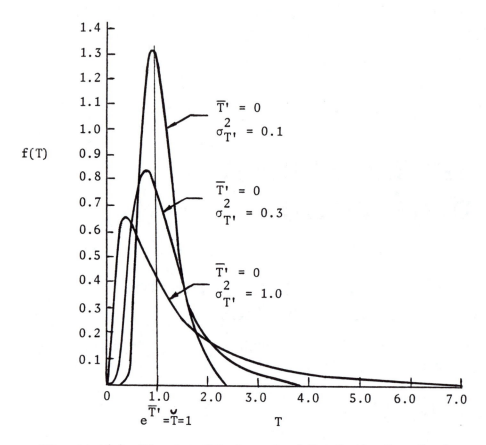

Fig. 11.1(a)– The plot of the lognormal distribution for a fixed value of $\overline{T}'$ and various values of $\sigma_{T'}^2$, [2, p. 98].

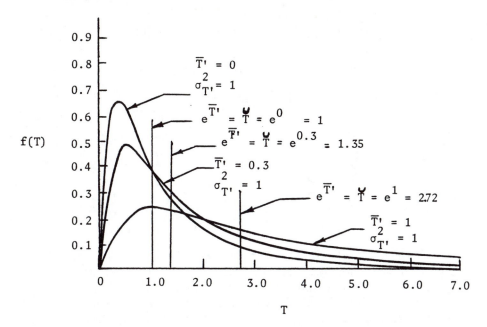

Fig. 11.1(b)– The plot of the lognormal distribution for various values of $\overline{T}'$ and fixed values of $\sigma^2_{T'}$, [2, p. 98].

Eq. (11.1) has its two parameters, $\overline{T}'$ and $\sigma_{T'}$, in terms of the logarithms of the data values. These parameters are determined as follows:

$$\overline{T}' = \frac{1}{N} \sum_{i=1}^{n} \log_e T_i = \frac{1}{N} \sum_{i=1}^{N} T_i', \tag{11.3}$$

and

$$\sigma_{T'} = \left[\frac{\sum\limits_{i=1}^{N} (T_i')^2 - N(\overline{T}')^2}{N-1} \right]^{1/2}, \tag{11.4}$$

where N is the total number of measurements made, or of the time-to-failure data obtained. Furthermore, to find reliabilities, areas under the *pdf* have to be determined by integration. But the lognormal *pdf* is not exactly and directly integrable, whereas the areas under the standardized normal *pdf* have been tabulated. As a lognormal *pdf* can be transformed to a normal *pdf* through the transform $\log_e T = T'$, and the parameters of this transformed, and now normal *pdf*, are $\overline{T}'$ and $\sigma_{T'}$, it is computationally more convenient to write the lognormal *pdf* in terms of $\overline{T}'$ and $\sigma_{T'}$, which have to be found anyway to calculate reliabilities with lognormal times-to-failure distributions.

Since the logarithms of a lognormally distributed random variable are normally distributed, the distribution of $\log_e T$, or of T', is

$$f(T') = \frac{1}{\sigma_{T'}\sqrt{2\pi}} e^{-\frac{1}{2}\left(\frac{T'-\overline{T}'}{\sigma_{T'}}\right)^2}, \tag{11.5}$$

where $\overline{T}'$ and $\sigma_{T'}$ are calculated using Eqs. (11.3) and (11.4), respectively, knowing that $T' = \log_e T$, where the T' are the straight times to failure in the unit of the measurements. The lognormal *pdf* can be obtained from Eq. (11.1) realizing that for equal probabilities under the normal and lognormal *pdf's*, incremental areas should also be equal, or

$$f(T)dT = f(T')dT'. \tag{11.6}$$

Taking the derivative of Eq. (11.2) yields

More

*You want it.
Citibank delivers.*

BUSINESS REPLY MAIL

FIRST CLASS MAIL   PERMIT NO. 348 LANSING, MI

POSTAGE WILL BE PAID BY ADDRESSEE

AMERICAN COLLEGIATE MARKETING
P.O. BOX 85003
LANSING MI 48908-9929

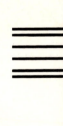

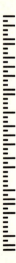

$$dT' = \frac{dT}{T}. \tag{11.7}$$

Substitution of Eq. (11.7) into Eq. (11.6) gives

$$f(T) = \frac{f(T')}{T}, \tag{11.8}$$

where $f(T')$, the numerator, is given by Eq. (11.5). Consequently, dividing the normal *pdf* in terms of the T' by T yields the lognormal *pdf* of Eq. (11.1)

If the logarithm is taken to the base 10 instead of e, then

$$T'' = \log_{10} T, \tag{11.9}$$

and

$$dT'' = \log_{10} e \, \frac{dT}{T} = 0.43429 \, \frac{dT}{T} = M \frac{dT}{T}, \tag{11.10}$$

where

$$M = 0.43429. \tag{11.11}$$

Also, as before,

$$f(T)dT = f(T'')dT''. \tag{11.12}$$

Substitution of Eq. (11.10) into Eq. (11.12) yields

$$f(T)dT = f(T'')M\frac{dT}{T},$$

or

$$f(T) = \frac{M}{T}f(T''). \tag{11.13}$$

Consequently,

$$f(T) = \frac{0.43429}{T\sigma_{T''}\sqrt{2\pi}}e^{-\frac{1}{2}(\frac{T''-\overline{T}''}{\sigma_{T''}})^2}. \tag{11.14}$$

Equation (11.14) gives the lognormal *pdf* of the data when their mean, $\overline{T}''$, and standard deviation, $\sigma_{T''}$, have been determined using Eqs. (11.3) and (11.4) where the T' are substituted by T'' as given by Eq. (11.9).

Specific characteristics of the lognormal *pdf* are the following:

1. As may be seen in Fig. 11.1, the lognormal distribution is a distribution skewed to the right. The *pdf* starts at zero, increases to its mode, and decreases thereafter. The degree of skewness *increases* as $\sigma_{T'}$ increases, for a given $\overline{T}'$, as may be seen in Fig. 11.1(a). For the same $\sigma_{T'}$ the *pdf*'s skewness increases as $\overline{T}'$ increases, as may be seen in Fig. 11.1(b). For $\sigma_{T'}$ values significantly greater than 1, the *pdf* rises very sharply in the beginning; i.e., for very small values of T near zero, and essentially follows the ordinate axis, peaks out early at the small mode, $\check{T}$, and then decreases sharply like an exponential *pdf*, or a Weibull *pdf* with $0 < \beta < 1$. For this reason with large $\sigma_{T'}$ the lognormal *pdf* could represent early failures if failure times slightly greater than zero and thereafter are considered only.

   The lognormal *pdf* can be plotted conveniently by using the values of $\phi(z)$ given in Appendix C and the relationships among $\phi(z)$, $f(T')$ and $f(T)$, which are

$$f(T') = \frac{\phi(z)}{\sigma_{T'}},\qquad\qquad(11.15)$$

   and

$$f(T) = \frac{f(T')}{T} = \frac{\phi(z)}{T\sigma_{T'}}.\qquad\qquad(11.16)$$

   Equation (11.15) yields the ordinate values of the normal *pdf*, or of the distribution of the $\log_e T$. Equation (11.16) yields the lognormal *pdf*'s ordinate values, or of the distribution of the T. These two equations are illustrated in Example 11–1.

2. The parameter, $\overline{T}'$, the mean life, or the *MTBF* in terms of the logarithm of the T is also the *scale parameter*, and not the location parameter as in the case of the normal *pdf*. $\overline{T}'$ can assume values of $-\infty < \overline{T}' < \infty$, if positive as well as negative variates are considered.

3. The parameter $\sigma_{T'}$, the standard deviation of the T in terms of their logarithm, or of their T', is also the *shape parameter* and not the scale parameter as in the case of normal *pdf*, and assumes only positive values.

4. The mean of the lognormal distribution, $\overline{T}$, is given by [3, pp. 19–20]

$$\overline{T} = e^{\overline{T}' + \frac{1}{2}\sigma_{T'}^2}. \tag{11.17}$$

5. The standard deviation of the lognormal distribution, σ_T, is given by [3, p. 20]

$$\sigma_T = [(e^{2\overline{T}' + \sigma_{T'}^2})(e^{\sigma_{T'}^2} - 1)]^{1/2}. \tag{11.18}$$

6. The median of the lognormal distribution, $\check{T}$, is given by [3, p. 19]

$$\check{T} = e^{\overline{T}'}. \tag{11.19}$$

7. The mode of the lognormal distribution, $\tilde{T}$, is given by [1]

$$\tilde{T} = e^{\overline{T}' - \sigma_{T'}^2}. \tag{11.20}$$

8. The coefficient of skewness of the lognormal *pdf* is given by [2, p. 128; 5, p. 247].

$$\alpha_3 = \frac{e^{3\sigma_{T'}^2} - 3e^{\sigma_{T'}^2} + 2}{(e^{\sigma_{T'}^2} - 1)^{3/2}} = (e^{\sigma_{T'}^2} - 1)^{1/2}(e^{\sigma_{T'}^2} + 2) > 0. \tag{11.21}$$

9. The coefficient of kurtosis of the lognormal *pdf* is given by [5, p. 248; 2, p. 128].

$$\alpha_4 = \frac{e^{6\sigma_{T'}^2} - 4e^{3\sigma_{T'}^2} + 6e^{\sigma_{T'}^2} - 3}{(e^{\sigma_{T'}^2} - 1)^2} > 0,$$

or

$$\alpha_4 = 3 + (e^{\sigma_{T'}^2} - 1)(e^{3\sigma_{T'}^2} + 3e^{2\sigma_{T'}^2} + 6e^{\sigma_{T'}^2} + 6) > 0. \tag{11.22}$$

10.   The *kth* moment of the lognormal *pdf* about the origin is given
       by [5, p. 245]

$$\mu'_k = e^{k\overline{T}' + \frac{1}{2}k^2 \sigma^2_{T'}}. \tag{11.23}$$

11.   The *kth* moment of the lognormal *pdf* about the mean is given
       by [4, p. 246]

$$\mu_k = e^{k\overline{T}'} \sum_{j=0}^{k} (-1)^j \binom{k}{j} e^{\frac{1}{2}\sigma^2_{T'}[(k-j)^2 + j]}. \tag{11.24}$$

12.   It must be noted that, given $\overline{T}$ and σ_T for lognormally distributed
       data, specific $\overline{T} \pm$ multiples of σ_T cannot be set up to obtain
       specific areas under the lognormal *pdf*. The lognormal *pdf* could
       have a large σ_T with respect to $\overline{T}$; then, the value of $(\overline{T} - k\sigma_T)$
       may become a negative value for which the lognormal *pdf* does
       not exist! For small values of $\sigma_{T'}$, say $\sigma_{T'} < 0.2$, the lognormal
       pdf is close to the normal *pdf*, and either *pdf* may then be used
       to represent the life characteristics of components and products
       [6].

       The *three-parameter lognormal distribution* exists also, and is [2,
       p. 99]

$$f(T) = \frac{1}{(T-\gamma)\sigma_{T'}\sqrt{2\pi}} e^{-\frac{1}{2}[\frac{\log_e(T-\gamma) - \overline{T}'}{\sigma_{T'}}]^2}, \tag{11.25}$$

$$T \geq \gamma, -\infty < \overline{T}' < \infty, \sigma_{T'} > 0, -\infty < \gamma < \infty, T' = \log_e T,$$

where

$$\gamma \;=\; \text{location parameter,}$$
$$\overline{T}' \;=\; \text{scale parameter,}$$

and

$$\sigma_{T'} \;=\; \text{shape parameter.}$$

Methods for determining these three parameters are given in [2, pp. 203–213 and pp. 282–284]. For the derivation of Eqs. (11.17) through (11.23) see [4, pp. 434-439].

See [4, pp. 439-441] for the formulas to calculate the mean, standard deviation, median, mode and the *kth* moment about the origin of the lognormal distribution when the logarithmic base used to calculate the mean and the standard deviation is 10.

EXAMPLE 11–1

Nine identical devices are tested continuously to failure, and the following times to failure, in hours, are obtained: 30.4, 36.7, 53.3, 58.5, 74.0, 99.3, 114.3, 140.1, 257.9. Do the following:

1. Find the parameters of the lognormal distribution which might represent the times-to-failure distribution of these devices.

2. Write down the *pdf* of the $\log_e T$ for these devices.

3. Write down the lognormal *pdf* for these devices to the base e.

4. Plot the $f(T')$ for these devices.

5. Plot the $f(T)$ for these devices.

6. Determine the mean of this lognormal *pdf*.

7. Determine the standard deviation for this lognormal *pdf*.

8. Determine the median of this lognormal *pdf*

9. Determine the mode of this lognormal *pdf*.

10. Determine the coefficient of skewness for this lognormal *pdf*.

11. Determine the coefficient of kurtosis for this lognormal *pdf*.

SOLUTIONS TO EXAMPLE 11–1

1. To determine the parameters of the lognormal distribution which represents the given data, prepare Table 11.1. Then, from Eq. (11.3),

$$\overline{T}' = \frac{1}{9}(39.198087) = 4.355343,$$

and, from Eq. (11.4),

$$\sigma_{T'} = [\frac{174.385255 - 9(4.355343)^2}{9 - 1}]^{\frac{1}{2}},$$

$$= (0.45801759)^{\frac{1}{2}} = 0.676770.$$

2. Retaining four decimal places in the values of $\overline{T}'$ and $\sigma_{T'}$ the *pdf* of the $\log_e T$ of these devices is

$$f(T') = \frac{1}{(0.6768)\sqrt{2\pi}} e^{-\frac{1}{2}(\frac{T'-4.3553}{0.6768})^2}, \tag{11.26}$$

where $T' = \log_e T$. This *pdf* is that of a normal distribution.

3. The lognormal *pdf* for these devices, to the base e, is

$$f(T) = \frac{1}{T(0.6768)\sqrt{2\pi}} e^{-\frac{1}{2}(\frac{T'-4.3553}{0.6768})^2}. \tag{11.27}$$

4. To plot the $f(T')$ prepare Table 11.2, and then plot the values in Column 5 against the corresponding values in Column 3 as shown in Fig. 11.2. A sample calculation for Point 11 is as follows:
From

$$z = \frac{T' - \overline{T}'}{\sigma_{T'}},$$

$$T' = z \cdot \sigma_{T'} + \overline{T}'. \tag{11.28}$$

For $z = 1.0$, Eq. (11.28) yields

$$T' = 1.0(0.6768) + 4.3553 = 5.0321.$$

The corresponding $f(T')$ value is

$$f(T') = \frac{\phi(z)}{\sigma_{T'}} = \frac{0.2420}{0.6768} = 0.35746,$$

the corresponding T value is

$$T = e^{T'} = e^{5.0321} = 153.2545,$$

TABLE 11.1– Analysis of the data given in Example 11–1 for the determination of the parameters of their lognormal *pdf*.

j	T_j	$\log_e T_j = T'_j$	$(\log_e T_j)^2 = (T'_j)^2$
1	30.4	3.414427	11.658312
2	36.7	3.602778	12.980009
3	53.3	3.975934	15.808051
4	58.5	4.069027	16.556981
5	74.0	4.304052	18.524864
6	99.3	4.598137	21.142864
7	114.3	4.738825	22.456462
8	140.1	4.942350	24.426823
9	257.9	5.552557	30.830889
$\sum$		39.198087	174.385255

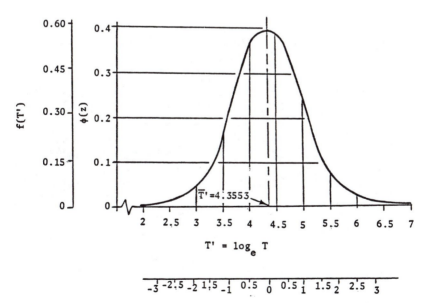

Fig. 11.2– The plot of $f(T')$ in Example 11–1.

TABLE 11.2 – Determination of the ordinate values of $f(T')$ and $f(T)$ for Example 11–1, where $\sigma_{T'} = 0.6768$.

1	2	3	4	5	6	7
Point number	z	T'	$\phi(z)$	$f(T') = \frac{\phi(z)}{\sigma_{T'}}$	$T,$ hr	$f(T) = \frac{f(T')}{T}$
1	∞	$-\infty$	0	0	0	0
2	-3.5	1.9865	0.0009	0.00132	7.3	0.0001824
3	-3.0	2.3249	0.0044	0.00650	10.2	0.0006356
4	-2.5	2.6633	0.0175	0.02585	14.3	0.0018022
5	-2.0	3.0017	0.0534	0.07888	20.1	0.0039204
6	-1.5	3.3401	0.1295	0.19129	28.2	0.0067779
7	-1.0	3.6785	0.2420	0.35746	39.6	0.0090297
8	-0.5	4.0169	0.3521	0.52009	55.5	0.0093661
9	0.0	4.3553	0.3989	0.58922	77.6	0.0075647
10	0.5	4.6937	0.3521	0.52009	109.3	0.0047602
11	1.0	5.0321	0.2420	0.35746	153.3	0.0023325
12	1.5	5.3705	0.1295	0.19129	215.0	0.0008898
13	2.0	5.7089	0.0534	0.07888	301.5	0.0002616
14	2.5	6.0473	0.0175	0.02585	423.0	0.0000611
15	3.0	6.3857	0.0044	0.00650	593.3	0.0000110
16	3.5	6.7241	0.0009	0.00132	832.2	0.0000016

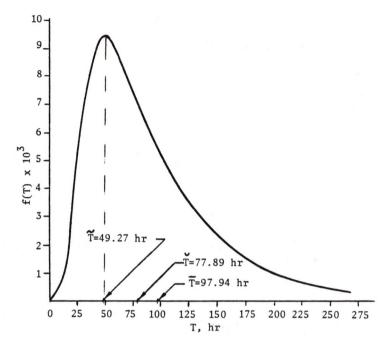

Fig. 11.3– The plot of $f(T)$ in Example 11–1.

and finally

$$f(T) = \frac{f(T')}{T} = \frac{0.35746}{153.2545} = 0.0023325.$$

Figure 11.2 illustrates the normality of $f(T')$, or the symmetrical bell-shaped nature thereof.

5. The plot of $f(T)$, the lognormal *pdf*, is given in Fig. 11.3, which is the result of plotting the values in Column 7 of Table 11.2 against those in Column 6. This figure illustrates that the lognormal distribution is skewed to the right, while starting at zero.

6. The mean of this lognormal distribution, using Eq. (11.17), is

$$\overline{T} = e^{\overline{T}' + \frac{1}{2}\sigma^2_{T'}} = e^{4.3553 + \frac{1}{2}(0.6768)^2} = 97.94 \text{ hr.}$$

7. The standard deviation of the lognormal distribution, using Eq. (11.18), is

$$\sigma_T = [(e^{2\overline{T}' + \sigma^2_{T'}})(e^{\sigma^2_{T'}} - 1)]^{\frac{1}{2}},$$

$$= \{[e^{2(4.3553)+(0.6768)^2}][e^{(0.6768)^2} - 1]\}^{\frac{1}{2}},$$

$$= [(9591.75)(0.5810)]^{\frac{1}{2}} = (5572.81)^{\frac{1}{2}},$$

or

$$\sigma_T = 74.65 \text{ hr.}$$

8. The median of the lognormal distribution, using Eq. (11.19), is

$$\check{T} = e^{\overline{T}'} = e^{4.3553} = 77.89 \text{ hr.}$$

The value of the median is smaller than the value of the mean, because this distribution is skewed to the right.

9. The mode of the lognormal distribution, using Eq. (11.20), is

$$\tilde{T} = e^{\overline{T}' - \sigma_{T'}^2} = e^{4.3553 - (0.6768)^2} = e^{3.8972},$$

or

$$\tilde{T} = 49.27 \text{ hr.}$$

The value of the mode is smaller than the value of the median because of the positive skewness of this distribution.

10. The coefficient of skewness of this lognormal distribution, using Eq. (11.21), is

$$\alpha_3 = (e^{\sigma_{T'}^2} - 1)^{\frac{1}{2}}(e^{\sigma_{T'}^2} + 2),$$

$$= [e^{(0.6768)^2} - 1]^{\frac{1}{2}}[e^{(0.6768)^2} + 2],$$

or

$$\alpha_3 = 2.08.$$

This positive α_3 value indicates that this lognormal distribution is significantly skewed to the right.

11. The coefficient of kurtosis of this lognormal distribution, using Eq. (11.22), is

$$\alpha_4 = 3 + (e^{\sigma_{T'}^2} - 1)(e^{3\sigma_{T'}^2} + 3e^{2\sigma_{T'}^2} + 6e^{\sigma_{T'}^2} + 6),$$

$$\alpha_4 = 3 + [e^{(0.6768)^2} - 1][e^{3(0.6768)^2} + 3e^{2(0.6768)^2} + 6e^{(0.6768)^2} + 6],$$

$$\alpha_4 = 3 + (0.5810)(26.9365),$$

or

$$\alpha_4 = 18.65.$$

The large value of α_4 indicates that this lognormal distribution is significantly more peaked than the normal distribution.

11.2 PROBABILITY PLOTTING OF THE LOGNORMAL DISTRIBUTION

Probability plotting enables the determination of the parameters of the distribution whose probability plotting paper is used, of the life characteristics of the units which yielded the data used for plotting, of the reliability of these units, and of the reliable life of these units. The probability plotting procedure for the lognormal distribution is identical to that used for the normal distribution, the difference being that the $\log_e T$ values are plotted along the abscissa, or a logarithmic scale to the base e is used along the abscissa, while the ordinate scale is the same probability scale as that used for the normal distribution. This is so because the $\log_e T$ values are normally distributed if the T values are lognormally distributed. Consequently, one may plot along the abscissa the $\log_e T = T'$ values, or the T values on a $\log_e$ scale. It must then be remembered that if a $\log_e$ scale is used, the values read off the abscissa should be converted to their $\log_e$ values to represent values belonging to $f(T')$. The procedure is illustrated by the next example.

EXAMPLE 11–2

Fifteen units are tested to failure and the times-to-failure data, given in the first two columns of Table 11.3, are obtained. Do the following:

1. Through probability plotting on the lognormal probability plotting paper determine the parameters of the lognormal distribution representing these data.

2. Write down the $f(T')$.

TABLE 11.3 – Times-to-failure data for Example 11–2 and their median ranks for a sample size of $N = 15$.

Failure order number, j	Time to failure, T_j, hr	Median ranks, MR_j, %
1	62.5	4.52
2	91.9	10.94
3	100.3	17.43
4	117.4	23.94
5	141.1	30.45
6	146.8	36.97
7	172.7	43.48
8	192.5	50.00
9	201.6	56.52
10	235.8	63.03
11	249.2	69.55
12	297.5	76.06
13	318.3	82.57
14	410.6	89.06
15	464.2	95.48

3. Write down the $f(T)$.

4. Determine graphically the median of this lognormal distribution.

SOLUTIONS TO EXAMPLE 11–2

1. Enter the median ranks in Column 3 of Table 11.3, for sample size $N = 15$, from Appendix A, and plot these median ranks versus their respective times to failure on the lognormal probability plotting paper given in Fig. 11.4, as shown in Fig. 11.5.

It appears that the points fall well on a straight line; consequently, the lognormal is an acceptable distribution for these data.

The mean of $f(T')$ is obtained by entering Fig. 11.5 at the $MR = 50\%$ level going to the fitted straight line, dropping vertically at the intersection, and reading off along the abscissa the value of 186 hr. Then

$$\overline{T}' = \log_e 186 = 5.22575.$$

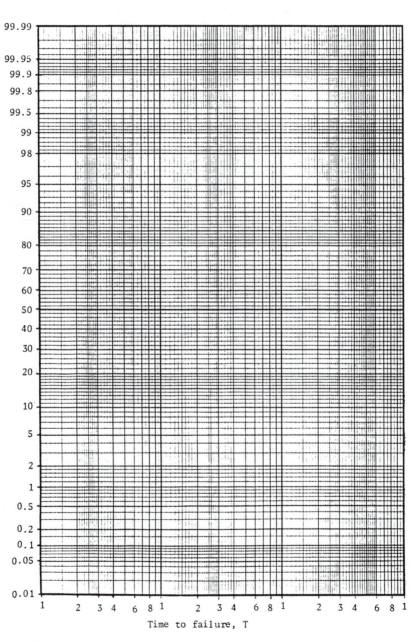

Fig. 11.4– Lognormal probability plotting paper.

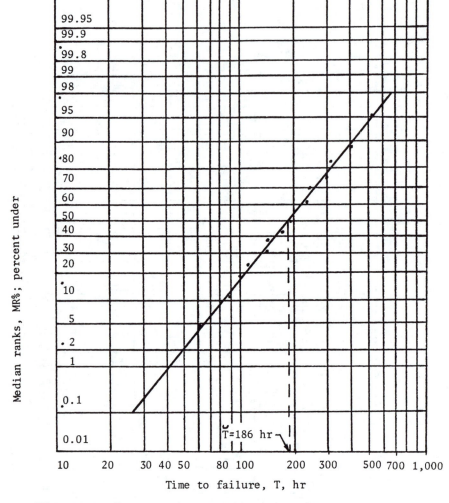

Fig. 11.5– Lognormal probability plot of the data in Table
11.3 for Example 11–2.

The standard deviation of $f(T')$ is obtained by either one of the following relationships:

$$\sigma_{T'} = \frac{T'(84.1\% \text{ under}) - T'(15.9\% \text{ under})}{2}, \tag{11.29}$$

$$\sigma_{T'} = \frac{T'(97.7\% \text{ under}) - T'(2.3\% \text{ under})}{4}, \tag{11.30}$$

or

$$\sigma_{T'} = \frac{T'(99.865\% \text{ under}) - T'(0.135\% \text{ under})}{6}. \tag{11.31}$$

Using Eq. (11.30) and entering the probability plot of Fig. 11.5 at the level of $MR = 2.3\%$ yields $T = 56$ hr or $T' = 4.02535$, and at the level of $MR = 97.7\%$ yields $T = 670$ hr or $T' = 6.50728$. Substitution of these values into Eq. (11.30) yields

$$\sigma_{T'} = \frac{6.50728 - 4.02535}{4} = \frac{2.48193}{4} = 0.62048.$$

2. Then, the equation of $f(T')$ is

$$f(T') = \frac{1}{0.62048\sqrt{2\pi}} e^{-\frac{1}{2}\left(\frac{T'-5.22575}{0.62048}\right)^2}. \tag{11.32}$$

3. The equation of $f(T)$; i.e., the lognormal distribution representing the data of Table 11.3, is

$$f(T) = \frac{1}{T(0.62048)\sqrt{2\pi}} e^{-\frac{1}{2}\left(\frac{T'-5.22575}{0.62048}\right)^2}. \tag{11.33}$$

4. The median of this lognormal distribution, from Eq. (11.19), is

$$\check{T} = e^{\overline{T}'} = e^{5.22575} = 186 \text{ hr},$$

or it is the value read off the abscissa of Fig. 11.5 when entering with $MR = 50\%$. This is so also because the mean of $f(T')$, which is normal, is that for $MR = 50\%$, but for the equality of areas under the $f(T')$ and $f(T)$ distributions the 50% area value of T for $f(T)$ is its median.

11.3   LOGNORMAL RELIABILITY CHARACTERISTICS

If the times to failure of units are well represented by the lognormal distribution, their reliability, starting the mission at the age of zero, is given by

$$R(T) = \int_T^\infty f(T)\,dT = \int_{T'}^\infty f(T')\,dT' = \int_{z(T')}^\infty \phi(z)\,dz, \quad (11.34)$$

where $f(T)$ is given by Eq. (11.1), $f(T')$ is given by Eq. (11.5), T is the mission duration starting the mission at age zero, and

$$z(T') = \frac{T' - \overline{T}'}{\sigma_{T'}}. \tag{11.35}$$

The conditional reliability is given by

$$R(T, t) = \frac{R(T + t)}{R(T)} = \frac{\displaystyle\int_{z[(T+t)']}^\infty \phi(z)\,dz}{\displaystyle\int_{z(T')}^\infty \phi(z)\,dz}. \tag{11.36}$$

The lognormal reliability function, given by Eq. (11.34) starts at the level of 1 at $T = 0$, essentially maintains this level thereafter for $\sigma_{T'} < 0.35$, approximately, until wear-out starts to set in. Subsequently it drops sharply, and $R(T) \to 0$ as $T \to \infty$, as shown in Fig. 11.6. However, for $\sigma_{T'} > 0.35$, approximately, the lognormal reliability function starts at the level of 1 at $T = 0$ but drops relatively sharply thereafter, rather than lingering at around the $R(T) = 1$ level, as also shown in Fig. 11.6, and $R(T) \to 0$ as $T \to \infty$. These behaviors again make the lognormal distribution a flexible one and it can represent early, exponential (useful life), as well as, wear-out life characteristics.

The reliable life for a specified reliability, R_s, is obtained by finding the $z(T')$ value for that specified reliability from Eq. (11.34), solving Eq. (11.35) for T', and taking its antilog, or

$$T(R_s) = e^{[\overline{T}' + z(R_s)\sigma_{T'}]}. \tag{11.37}$$

EXAMPLE 11–3

Given are the data of Table 11.1. Do the following:

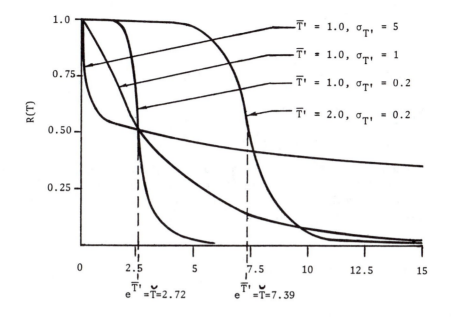

Time to failure, T

Fig. 11.6– Plot of the lognormal reliability function.

1. Find the reliability for a mission of 30 hr, or the probability such devices will function continuously for 30 hr without failure, starting this period of operation at the age of zero.

2. Find the reliability for an additional 60 hr mission starting this mission at the age of 30 hr.

3. Find the reliable life for a reliability of 95%.

SOLUTIONS TO EXAMPLE 11–3

1. The reliability for a mission of 30 hr, using Eqs. (11.34) and (11.35), is given by

$$R(T = 30 \text{ hr}) = \int_{z(T'=\log_e 30=3.4012)}^{\infty} \phi(z)\, dz,$$

where

$$z(T' = 3.4012) = \frac{T' - \overline{T}'}{\sigma_{T'}} = \frac{3.4012 - 4.3553}{0.6768},$$

or

$$z(T' = 3.4012) = \frac{-0.9541}{0.6768} = -1.4097.$$

From Appendix B,

$$R(T = 30 \text{ hr}) = 0.9207.$$

Consequently, the reliability, or the probability that such devices will complete 30 hr of continuous operation without failure is 92.07%.

2. The reliability for an additional mission of 60 hr, starting this mission at the age of 30 hr, is obtained from Eq. (11.36), or

$$R(T = 30 \text{ hr}; t = 60 \text{ hr}) = \frac{R(30 + 60)}{R(30)} = \frac{R(90)}{R(30)},$$

or

$$R(T = 30 \text{ hr}; t = 60 \text{ hr}) = \frac{\int_{z[\log_e(30+60)]}^{\infty} \phi(z)\, dz}{\int_{z(\log_e 30)}^{\infty} \phi(z)\, dz},$$

where

$$
\begin{aligned}
z[\log_e(30 + 60)] &= z(\log_e 90 = 4.4998), \\
&= \frac{\log_e 90 - \overline{T}'}{\sigma_{T'}}, \\
&= \frac{4.4998 - 4.3553}{0.6768}, \\
&= \frac{0.1445}{0.6768},
\end{aligned}
$$

or

$$z[\log_e(30 + 60)] = 0.2135.$$

From Appendix B, this z value yields

$$R(T + t = 90 \text{ hr}) = 0.4155.$$

This is the numerator in the solution. The denominator was determined in the previous case as 0.9207. Consequently,

$$R(T = 30 \text{ hr}; t = 60 \text{ hr}) = \frac{0.4155}{0.9207} = 0.4513,$$

or the reliability for an additional mission of 60 hr is only 45.13%, which is a very low reliability. This indicates the strong setting-in of wear-out in these devices during the second mission.

3. The reliable life for a reliability of 95%, from Eq. (11.37), is

$$T(R_s = 0.95) = e^{\{4.3553 + [z(R_s = 0.95)](0.6768)\}},$$

where

$$z(R_s = 0.95) = -1.645.$$

Therefore,

$$T(R_s = 0.95) = e^{[4.3553 + (-1.645)(0.6768)]},$$
$$T(R_s = 0.95) = e^{3.2420},$$

or

$$T(R_s = 0.95) = 25.58 \text{ hr}.$$

EXAMPLE 11–4

Given the data of Table 11.3, and the lognormal probability plot of Fig. 11.5, do the following:

1. Find the reliability for a mission of 50 hr, starting the mission at age zero, from the probability plot.

2. Find the reliability for an additional mission of 50 hr.

3. Find the reliable life for a reliability of 95%.

SOLUTIONS TO EXAMPLE 11–4

1. Entering Fig. 11.5 with $T = 50$ hr yields $MR = Q(T = 50 \text{ hr}) = 2.2\%$; consequently,

$$R(T = 50 \text{ hr}) = 100\% - Q(T = 50 \text{ hr})\% = 100\% - 2.2\% = 97.8\%,$$

or the reliability for a 50-hr mission, starting the mission at age zero, is 97.8%.

2. The reliability for an additional 50-hr mission is given by

$$R(T = 50 \text{ hr}; t = 50 \text{ hr}) = \frac{R(50 + 50)}{R(50)} = \frac{R(100)}{R(50)}.$$

From Fig. 11.6, $R(100)$ is found to be 84.0%. $R(50)$ was found to be 97.8% in the previous case; therefore,

$$R(T = 50 \text{ hr}; t = 50 \text{ hr}) = \frac{0.840}{0.978} = 0.859, \text{ or } 85.9\%.$$

3. The reliable life for a reliability of 95% is obtained by entering Fig. 11.4 with 5% along the ordinate and reading along the abscissa $T = 66.5$ hr, or the reliability for a 66.5 hr mission would be 95%.

11.4   LOGNORMAL FAILURE RATE CHARACTERISTICS

The instantaneous failure rate for the lognormal case is given by

$$\lambda(T) = \frac{f(T)}{R(T)} = \frac{\frac{1}{T\sigma_{T'}\sqrt{2\pi}} e^{-\frac{1}{2}(\frac{T'-\overline{T}'}{\sigma_{T'}})^2}}{\int_{z(T')}^{\infty} \phi(z)\, dz}. \tag{11.38}$$

Evaluation of the numerator of Eq. (11.38) involves finding the height of the lognormal *pdf* at the specified value of T. This may be obtained using the procedure illustrated in Table 11.2, or from

$$f(T) = \frac{\phi(z)}{T\sigma_{T'}}. \tag{11.39}$$

Evaluation of the denominator of Eq. (11.38) involves the use of Appendix B. Figure 11.7 shows the behavior of the lognormal failure rate function, where it may be seen that, for small values of $\sigma_{T'}$, $\lambda(T)$ increases with age, and sharper for smaller $\overline{T}'$ for the same $\sigma_{T'}$. One must be careful however because $\lambda(T)$ does increase with increasing T but it reaches a peak value and decreases thereafter with increasing T, as shown in Fig. 11.8. Actually, $\lambda(T)$ starts at zero, reaches a maximum and then decreases forever regardless of the value of its skewness [7].

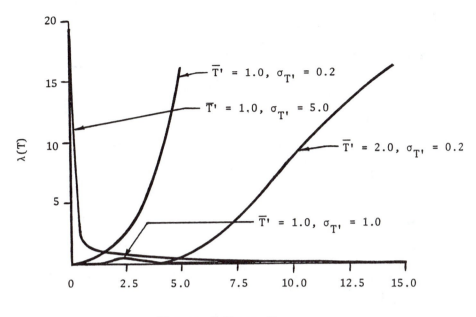

Time to failure, T

Fig. 11.7– The plot of the lognormal failure rate function for four combinations of $\overline{T}'$ and $\sigma_{T'}$ values [6, p. 21].

For a very skewed distribution, with $\sigma_{T'} > \sqrt{2/\pi} = 0.798$, $\lambda(T)$ peaks between the median and the mode; then it decreases continually. This happens even before half the population dies. If the skewness is much less, with $\sigma_{T'} < \sqrt{2/\pi} = 0.798$, there is still a maximum but it occurs beyond the median life, and even beyond the mean life. For very small values of skewness, with $\sigma_{T'} \leq 0.2$, most of the population is dead before the peak is reached. Nevertheless, if one waits long enough, the lognormal $\lambda(T)$ always has a decreasing value.

It may be seen that the lognormal *pdf* may represent the times to failure over most of the range of life. For a $\sigma_{T'}$ value of 0.5, the $\lambda(T)$ is essentially constant over much of the life range, like the exponential, thus it represents the relatively constant failure rate behavior of units during their useful life. For $\sigma_{T'} \leq 0.2$, $\lambda(T)$ increases over most of the life range and is like a normal distribution. For $\sigma_{T'} \geq 0.8$, $\lambda(T)$ increases sharply around $T = 0$, but after a very short operating period of such units, $\lambda(T)$ decreases over essentially most of their life range, like the Weibull with $0 < \beta < 1$, thus representing early failure characteristics. This flexibility of the lognormal $\lambda(T)$ and its *pdf* make them suitable for many components and products.

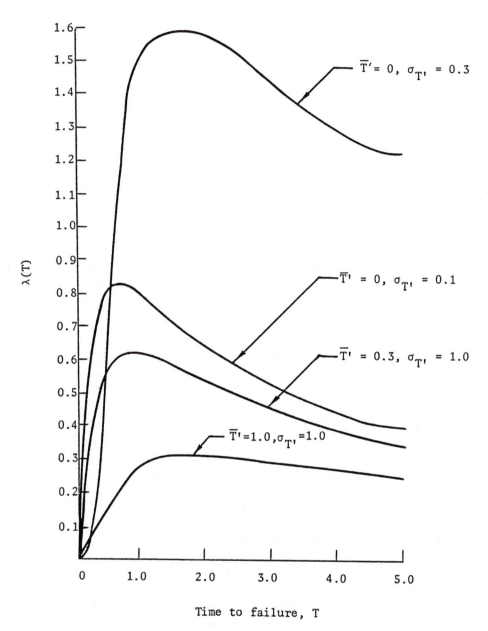

Fig. 11.8– The plot of the lognormal failure rate function for four combinations of $\overline{T}'$ and $\sigma_{T'}$ values different than those in Fig. 11.7 [2, p. 120].

EXAMPLE 11–5

A lognormal *pdf* has the following parameters:

$$\overline{T}' = 1.0, \quad \text{and } \sigma_{T'} = 1.0.$$

Do the following:

1. Write down the associated failure rate function.

2. Find the failure rate at the age of 2.5 hr.

SOLUTIONS TO EXAMPLE 11–5

1. The failure rate function with $\overline{T}' = 1.0$ and $\sigma_{T'} = 1.0$ is

$$\lambda(T) = \frac{\frac{1}{T\sqrt{2\pi}} e^{-\frac{1}{2}(T'-1.0)^2}}{\int_{z(T')}^{\infty} \phi(z)\,dz},$$

where the numerator can also be evaluated from

$$f(T) = \frac{\phi(z)}{T\sigma_{T'}} = \frac{\phi(z)}{T} \quad \text{with } \sigma_{T'} = 1.0.$$

2. $\lambda(T)$ for $T = 2.5$ hr is

$$\lambda(T = 2.5 \text{ hr}) = \frac{\frac{\phi(z)}{2.5}}{\int_{z(T'=\log_e 2.5)}^{\infty} \phi(z)\,dz},$$

$$z(T' = \log_e 2.5) = \frac{\log_e T - \overline{T}'}{\sigma_{T'}} = \frac{\log_e 2.5 - 1.0}{1.0},$$

or

$$z(T' = 0.9163) = \frac{0.9163 - 1.0}{1.0} = -0.0837.$$

From Appendix C,

$$\phi(z = -0.0837) = 0.39755.$$

Consequently,

$$f(T) = \frac{0.39755}{2.5} = 0.15902.$$

Also,

$$R(T) = \int_{z(T'=0.9163)=-0.0837}^{\infty} \phi(z)\, dz = 0.53335,$$

from Appendix B. Therefore,

$$\lambda(T = 2.5\ \text{hr}) = \frac{0.15902}{0.53335} = 0.29815\ \text{fr/ hr}\ .$$

11.5   APPLICATIONS OF THE LOGNORMAL DISTRIBUTION

For additional applications of the lognormal *pdf* see [4, p. 425].

11.6   PHENOMENOLOGICAL CONSIDERATIONS FOR USING THE LOGNORMAL DISTRIBUTION

Phenomenological considerations for using the lognormal *pdf* are given in [4, pp. 425-426].

PROBLEMS

11-1. Given the data in Table 11.1, do the following:

(1) Using the parameters of the lognormal *pdf* determined by calculation in Example 11-1, write down the failure rate function.

(2) Calculate at least 10 failure rate values and plot this $\lambda(T)$ function, paying attention to the very early life and very aged life portions of this function so that a representative $\lambda(T)$ plot is obtained.

(3) Discuss the behavior of this function with age.

11-2. Given the data of Example 11-1, do the following:

(1) Plot the data on lognormal probability plotting paper.

(2) State if the lognormal *pdf* represents these data well enough.

(3) Determine the parameters of the lognormal *pdf* representing these data using the plot of Case 1.

(4) Write down the $f(T')$ *pdf* for these data.

(5) Write down the $f(T)$ *pdf* for these data.

(6) Determine the hours of operation at which 5% of these devices will fail.

(7) Comparatively discuss the $\overline{T}'$ values determined in Example 11-1 and in Case 3.

(8) Comparatively discuss the $\sigma_{T'}$ values determined in Example 11-1 and in Case 3.

(9) Calculate the mission time during which 5% of these devices will fail using the parameters determined in Example 11-1 and compare this value with that determined in Case 6.

(10) Determine the reliability for a mission of 24 hr, starting the mission at age zero.

(11) Determine the reliability for an additional mission of 24 hr, starting this mission at the age of 24 hr.

(12) Plot the $f(T')$ determined in Case 4.

(13) Plot the $f(T)$ determined in Case 5.

11-3. Given the data in Example 11-2, do the following:

(1) Determine the $\overline{T}'$ and $\sigma_{T'}$ for these data by calculation.

(2) Write down the $f(T')$.

TABLE 11.4 – Time-to-failure data for Problem 11-4 and a sample size of $N = 15$.

Failure order number, j	Time to failure, T_j, hr	Median ranks, MR_j, %
1	75.0	
2	110.3	
3	120.4	
4	140.9	
5	169.3	
6	176.2	
7	207.2	
8	231.0	
9	241.9	
10	283.0	
11	299.0	
12	357.0	
13	382.0	
14	492.7	
15	660.6	

(3) Write down the $f(T)$.

(4) Calculate the median of $f(T)$.

(5) Find the reliability for a mission of 30 hr duration using the parameters determined in Case 1, starting the mission at age zero.

(6) Find the reliability for an additional mission of 60 hr duration, starting the mission at the age of 30 hr.

(7) Find the reliable life for a reliability of 95%.

(8) Comparatively discuss the results of Cases 1, 4, 5, 6, and 7 with those obtained in Example 11–2.

11-4. Given the data in Table 11.4, do the following:

(1) Determine the $\overline{T}'$ and $\sigma_{T'}$ for these data by calculation.

(2) Write down the $f(T')$.

(3) Write down the $f(T)$.

TABLE 11.5 — **Cycles-to-failure data for SAE 4340 steel R_c 35/40 hardness, subjected to fatigue tests under loads of combined reversed bending and steady torque, for Problem 11-5.**

Specimen number	Cycles to failure	Specimen number	Cycles to failure
1	84,544	19	87,152
2	126,031	20	80,010
3	78,262	21	99,853
4	68,394	22	76,602
5	100,961	23	77,817
6	91,923	24	89,789
7	102,620	25	72,305
8	71,268	26	71,713
9	84,218	27	90,767
10	85,788	28	71,624
11	127,690	29	61,667
12	88,337	30	62,882
13	89,759	31	72,039
14	110,799	32	69,372
15	71,179	33	69,342
16	98,679	34	112,577
17	96,131	35	85,433
18	90,797		

(4) Calculate the median of $f(T)$.

(5) Find the reliability for a mission of 50 hr duration using the parameters determined in Case 1, starting the mission at age zero.

(6) Find the reliability for an additional mission of 50 hr duration, starting the mission at the age of 50 hr.

(7) Find the reliable life for a reliability of 95%.

11-5. Given the cycles-to-failure data in Table 11.5 for 35 grooved specimens of AISI 4340 steel, R_c 35/40 hardness, subjected to fatigue tests under loads of combined reversed bending and steady torque, do the following:

(1) Calculate the mean of these data, assuming they are lognormally distributed.

(2) Calculate their standard deviation.

(3) Calculate their coefficient of skewness.

(4) Calculate their coefficient of kurtosis.

(5) Prepare a tally table for these data and draw a histogram.

(6) Draw on this histogram the lognormal distribution determined by the parameters calculated in Cases 1 and 2.

(7) Write down the reliability function and plot it.

(8) Write down the failure rate function and plot it.

11-6. The life of an electronic control for locomotives, in thousands of miles, has a lognormal *pdf* with the following parameters calculated to the base 10:

$$\overline{T}'' = 2.236 \text{ and } \sigma_{T''} = 0.320.$$

Do the following:

(1) Calculate the reliability for their first 80,000 miles of operation.

(2) Calculate the life by which 1.0% of these controls will fail.

(3) Calculate the life by which 50% of these controls will fail.

(4) Calculate the mean life of these controls in miles.

(5) Calculate the modal life of these controls in miles.

(6) Calculate the standard deviation of the lives of these controls in miles.

(7) Write down their failure rate function and discuss the shape this function has over its life.

11-7. Identical components have a lognormal times-to-failure *pdf* with parameters

$$\overline{T}' = 5 \text{ and } \sigma_{T'} = 1.$$

Do the following:

(1) Plot carefully $f(T')$, using at least 10 values, and discuss its behavior.

(2) Plot carefully $R(T)$, using the same points that were used in Case 1, and discuss its behavior.

(3) Plot carefully $\lambda(T)$, using the same points that were used in Case 1, and discuss its behavior.

(4) Find their reliability for a life of 25 time units.

(5) Find their failure rate at the life of 25 time units.

11-8. The lognormal times to failure of a particular electronic device have the following parameters:

$$\overline{T}' = 1.0 \text{ and } \sigma_{T'} = 1.0, \text{ in } \log_e \text{ hr.}$$

Do the following:

(1) Calculate, tabulate, and plot the $f(T)$.

(2) Calculate, tabulate, and plot the $f(T')$.

(3) Calculate, tabulate, and plot the $R(T)$.

(4) Calculate, tabulate, and plot the $\lambda(T)$.

(5) Determine the reliability of such units for a mission of 30 minutes, starting this mission at age 0.

(6) Determine the reliability of such units for a new mission of 30 minutes, starting this mission at age of 30 minutes.

(7) Find the failure rate of these devices for the age of 30 minutes.

(8) Find the failure rate of these devices for the age of 60 minutes.

(9) Find the failure rate of these devices for the age of 50 hr.

(10) Find $\lambda(T = \infty)$.

11-9. Given the lognormal probability plot of Fig. 11.9, for the cycles to failure of spindles used in a gear drive, do the following:

(1) Find the parameters of the lognormal *pdf* representing the lives of these spindles.

(2) Find the percent of these spindles that will fail up to 10^6 cycles of operation, by calculation, using the associated *pdf*.

(3) Find the percent of these spindles that will fail up to 10^6 cycles of operation, using the probability plot.

(4) If no more than 2% of these spindles can be afforded to fail, determine their replacement time in cycles.

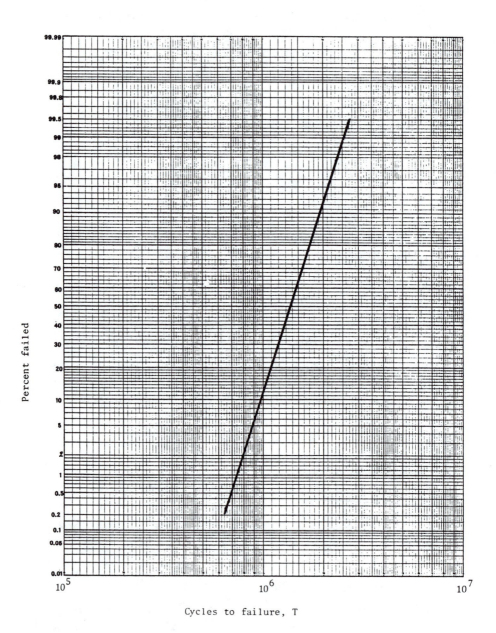

Cycles to failure, T

Fig. 11.9 – The lognormal probability plot of the cycles to
failure of Problem 11–9.

11-10. Identical components have a lognormal times-to-failure *pdf* with parameters

$$\overline{T}' = 5 \text{ and } \sigma_{T'} = 0.5, \text{ in } \log_e \text{ time units.}$$

Do the following:

(1) Plot carefully $f(T')$, using at least 10 values, and discuss its behavior.
(2) Plot carefully $R(T)$, using the same points that were used in Case 1, and discuss its behavior.
(3) Plot carefully $\lambda(T)$, using the same points that were used in Case 1, and discuss its behavior.
(4) Find their reliability for a life of 25 time units.
(5) Find their failure rate at the life of 25 time units.

11-11. In a life test of 10 identical electromechanical components, subjected to identical stresses, the following times to failure, in hours, were observed:

$$100, \quad 247, \quad 404, \quad 659, \quad 686,$$
$$714, \quad 736, \quad 789, \quad 898, \quad 1{,}200.$$

Assume that the population from which the 10 units were drawn has a lognormal *pdf* of the times to failure. Find the following:

(1) The estimates of the mode, the median and the mean, in hours, of the lognormal *pdf* of this population, using probability plotting.
(2) A second estimate of the mean and of the standard deviation of the lognormal distribution using the ordinary sample mean and standard deviation method.

11-12. Using the data in Table 11.6, do the following:

(1) Assuming the data are lognormally distributed, calculate the mean, median, mode and standard deviation of these data using the mean and standard deviation of the logarithms of these data to the base e.

TABLE 11.6 – Lognormally distributed times-to-failure data for Problem 11-12.

Data number	T, cycles
1	39
2	59
3	90
4	136
5	205
6	313
7	475
8	710
9	1,090
10	1,630
11	2,500
12	3,780
13	5,700
14	8,700
15	13,200
16	20,000
17	30,500
18	46,000
19	69,500
20	106,000
21	162,000
22	246,000
23	360,000
24	560,000
25	840,000
26	1,270,000
27	1,900,000
28	2,900,000
29	4,400,000
30	6,800,000
31	10,000,000

(2) Calculate the mean and standard deviation of the logarithms of these data to the base 10.

(3) Calculate the mean, median, mode and standard deviation of these data using the $\overline{T}''$ and $\sigma_{T''}$ values found in Case 2.

(4) Comparatively discuss the results of Cases 1 and 3.

(5) Write down the lognormal *pdf*, $f(T)$, to the base 10, and plot it, giving a table of all T'' and $f(T'')$ values used for the plot.

(6) Write down the normal *pdf*, $f(T)$, to the base 10, and plot it, giving a table of all T'' and $f(T'')$ values used for the plot.

(7) Find the reliability of these units for a mission of 10^5 cycles, starting the mission at the age of zero cycles.

(8) Find the reliability of these units for a mission of 10^4 cycles, starting the mission at the age of 10^5 cycles.

11-13. Given are the times-to-failure data of Table 11.7 for a specific part in a system. Do the following:

(1) Determine the *pdf* of the times to failure of this part in its system, assuming it is lognormal.

(2) Determine the failure rate function.

(3) Determine the reliability function.

(4) What is the reliability of this part for a five-hour mission?

(5) Same as Case 4, but for a 20-hr mission.

(6) What is the mean time to failure of this part?

(7) What is the median of the times to failure of this part?

(8) What is the time by which onehalf of such failures will occur?

(9) What is the most frequently occurring time to failure?

(10) What is the standard deviation of the times to failure?

TABLE 11.7 – Times-to-failure data for Problem 11-13.

Times to failure, t_r, hr	Frequency of observation, N
0.4	1
0.6	1
1.0	2
1.2	3
1.4	4
1.6	5
2.0	6
2.2	7
2.6	7
3.0	6
4.0	5
4.4	4
5.0	4
5.4	3
6.0	2
6.6	2
8.0	1
9.0	1
9.4	1
10.0	1
11.0	1
13.0	1
19.0	1
25.0	1
44.0	1

REFERENCES

1. Aitchison, Jr. and Brown, J.A.C., *The Lognormal Distribution*, Cambridge University Press, Cambridge, 176 pp., 1957.

2. Hahn, Gerald J. and Shapiro, Samuel S., *Statistical Models in Engineering*, John Wiley & Sons, Inc., New York, 355 pp., 1967.

3. Kapur, K.C. and Lamberson, L.R., *Reliability in Engineering Design*, John Wiley & Sons, Inc., New York, 586 pp., 1977.

4. Kececioglu, Dimitri, *Reliability Engineering Handbook*, Prentice Hall, Englewood Cliffs, N. J., Vol.1, 720 pp., 1991.

5. Green, A.E. and Bourne, A.J., *Reliability Technology*, Wiley-cience, 636 pp., 1972.

6. Nelson, W.B., *Basic Concepts and Distributions for Product Life*, General Electric Company, Corporate Research and Development, P.O. Box 43, Bldg. 5, Schenectady, New York 12301, 28 pp., December 1974.

7. Evans, Ralph A., " The Lognormal Distribution is Not A Wear-out Distribution", *Reliability Group Newsletter*, IEEE, Inc., 345 East 47th St., New York, 10017, Vol. XV, Issue 1, p. 9, January 1970.

Chapter 12

THE WEIBULL DISTRIBUTION

12.1 WEIBULL DISTRIBUTION CHARACTERISTICS

The Weibull distribution is one of the most commonly used distributions in reliability engineering because of the many shapes it attains for various values of β, thus it can model a great variety of data and life characteristics. The Weibull *pdf* is

$$f(T) = \frac{\beta}{\eta}(\frac{T-\gamma}{\eta})^{\beta-1}e^{-(\frac{T-\gamma}{\eta})^\beta}, \tag{12.1}$$

where

$f(T) \geq 0, T \geq \gamma, \beta > 0, \eta > 0, -\infty < \gamma < \infty,$

$\beta =$ shape parameter,

$\eta =$ scale parameter,

and

$\gamma =$ location parameter.

Figure 12.1 shows how the shape of the Weibull *pdf* changes as β changes from $\beta=1/5$ to $\beta=1/2$, $\beta=1$, $\beta=1\text{-}1/2$, $\beta=3$ and $\beta=5$. Some of the specific characteristics of the Weibull *pdf* are the following:

1. For $0 < \beta < 1$ as $T \rightarrow \gamma$ then $f(T) \rightarrow \infty$, as $T \rightarrow \infty$ then $f(T) \rightarrow 0$. $f(T)$ decreases monotonically and is convex as T increases beyond the value of γ, as may be seen in Fig. 12.1. The mode is nonexistent [1, p. 33].

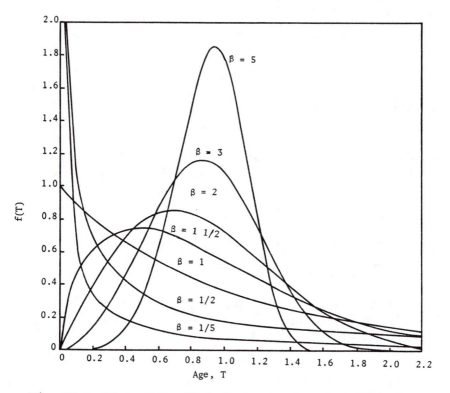

Fig. 12.1– Plot of the Weibull density function $f(T)$, for various values of β, $\eta = 1$ and $\gamma = 0$.

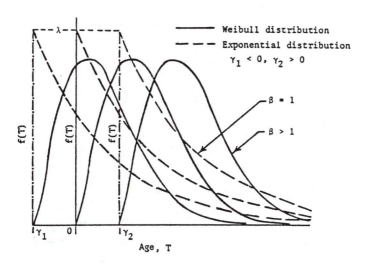

Fig. 12.2– The effect of the location parameter, γ, on the Weibull and the exponential distributions.

2. For $\beta=1$ it becomes the two-parameter exponential distribution, as a special case, or

$$f(T) = \frac{1}{\eta}e^{-\frac{T-\gamma}{\eta}}; \quad \gamma \geq 0, \quad \eta > 0, \quad T \geq \gamma, \tag{12.2}$$

which is illustrated in Fig. 12.2. Equation (12.2) may also be written as

$$f(T) = \lambda e^{-\lambda(T-\gamma)}, \tag{12.3}$$

where

$$\frac{1}{\eta} = \lambda = \text{chance, or useful life, failure rate,}$$

$$\eta = \frac{1}{\lambda} = \overline{T} - \gamma = m - \gamma,$$

and

$$\overline{T} = m = \gamma + \eta.$$

$f(T) = \frac{1}{\eta}$ at $T = \gamma$. As T increases beyond the value of γ, $f(T)$ decreases monotonically and is convex. $f(T) \to 0$ as $T \to \infty$, as may be seen in Fig. 12.2. The mode is at $T = \gamma$, or $\tilde{T} = \gamma$.

3. For $\beta > 1$, $f(T)$ assumes wear-out type shapes whereby $f(T)$ is zero at $T = \gamma$, increases as $T \to \tilde{T}$ and decrease thereafter or for

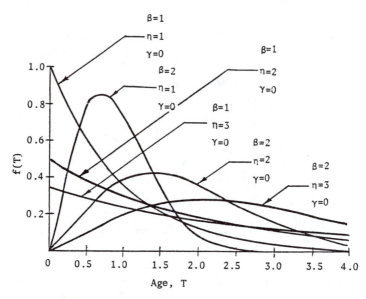

Fig. 12.3– The effect of the scale parameter, η, on the Weibull and exponential distributions.

$T > \tilde{T}$, as may be seen in Figs. 12.1 and 12.2. $f(T)=0$ at $T = \gamma$ and $f(T) \to 0$ as $T \to \infty$. For $\beta=2$ it becomes the Rayleigh distribution as a special case [3, p. 109]. For $\beta < 2.6$ the Weibull *pdf* is positively skewed, for $2.6 < \beta < 3.7$ its coefficient of skewness approaches zero; consequently, it may approximate the normal *pdf*, and for $\beta > 3.7$ it is negatively skewed [1, p. 33; 4, p. 54; 31].

4. A change in the scale parameter η has the same effect on the distribution as a change of the abscissa scale. If η is increased, while β and γ are kept the same, the distribution gets stretched out to the right and its height decreases, while maintaining its shape and location. If η is decreased, while β and γ are kept the same, the distribution gets pushed in towards the left, towards its beginning; i.e., toward 0 or γ, and its height increases. These effects of η on the $f(T)$ are illustrated in Fig. 12.3.

5. The location parameter γ, as the name implies, locates the distribution along the abscissa. When $\gamma=0$ the distribution starts at $T=0$, or at the origin. If γ is positive the distribution starts at the location γ to the right of the origin. If γ is negative the distribution starts at the location γ to the left of the origin. These effects of γ on the $f(T)$ are illustrated in Fig. 12.2. γ provides an estimate of the earliest time to failure of such units. The life period 0 to γ is also the only failure free operating period of such units.

6. The mean, $\overline{T}$, of the Weibull *pdf* is given by

$$\overline{T} = \gamma + \eta\Gamma(\frac{1}{\beta} + 1), \tag{12.4}$$

where $\Gamma(\frac{1}{\beta}+1)$ is the gamma function evaluated at the value of $(\frac{1}{\beta}+1)$. Table 12.1 gives the values of the Γ function, where $n = (\frac{1}{\beta} + 1)$. Figure 12.4 gives a plot of the Γ function. From Table 12.1, $\Gamma_{min}(n) \simeq 0.88560$, which is for $n \simeq 1.46$. Consequently, since $n = \frac{1}{\beta} + 1$, Γ_{min} is obtained when $1.46 = \frac{1}{\beta} + 1$, or when $\beta \simeq 2.17$.

7. The median, $\check{T}$, of the Weibull *pdf* is given by

$$\check{T} = \gamma + \eta(\log_e 2)^{\frac{1}{\beta}}. \tag{12.5}$$

8. The mode, $\tilde{T}$, of the Weibull *pdf* is given by

$$\tilde{T} = \gamma + \eta(1 - \frac{1}{\beta})^{\frac{1}{\beta}}. \tag{12.6}$$

9. The standard deviation, σ_T, of the Weibull *pdf* is given by

$$\sigma_T = \eta\{\Gamma(\frac{2}{\beta} + 1) - [\Gamma(\frac{1}{\beta} + 1)]^2\}^{\frac{1}{2}}. \tag{12.7}$$

10. The coefficient of skewness, α_3, of the Weibull *pdf* is given by

$$\alpha_3 = \frac{\Gamma(1 + \frac{3}{\beta}) - 3\Gamma(1 + \frac{2}{\beta})\Gamma(1 + \frac{1}{\beta}) + 2[\Gamma(1 + \frac{1}{\beta})]^3}{\{\Gamma(1 + \frac{2}{\beta}) - [\Gamma(1 + \frac{1}{\beta})]^2\}^{\frac{3}{2}}}. \tag{12.8}$$

If α_3 is negative then the *pdf* is skewed, or stretched out, to the left, if $\alpha_3=0$ then the *pdf* is symmetrical as is the case for the normal *pdf*, and if α_3 is positive then the *pdf* is skewed, or stretched out, to the right.

11. The skewness, α'_3, of the Weibull *pdf* is given by

$$\alpha'_3 = \eta^3\{\Gamma(1 + \frac{3}{\beta}) - 3\Gamma(1 + \frac{2}{\beta})\Gamma(1 + \frac{1}{\beta}) + 2[\Gamma(1 + \frac{1}{\beta})]^3\}. \tag{12.8'}$$

12. The coefficient of kurtosis, α_4, of the Weibull *pdf* is given by

$$\alpha_4 = \{\Gamma(1 + \frac{4}{\beta}) - 4\Gamma(1 + \frac{3}{\beta})[\Gamma(1 + \frac{1}{\beta})] + 6\Gamma(1 + \frac{2}{\beta})[\Gamma(1 + \frac{1}{\beta})]^2$$
$$- 3[\Gamma(1 + \frac{1}{\beta})]^4\}/\{\Gamma(1 + \frac{2}{\beta}) - [\Gamma(1 + \frac{1}{\beta})]^2\}^2. \tag{12.9}$$

TABLE 12.1– Tabulation of values of $\Gamma(n)$ for the gamma function versus n.

n	$\Gamma(n)$	n	$\Gamma(n)$	n	$\Gamma(n)$	n	$\Gamma(n)$
1.00	1.00000	1.25	0.90640	1.50	0.88623	1.75	0.91906
1.01	0.99433	1.26	0.90440	1.51	0.88659	1.76	0.92137
1.02	0.98884	1.27	0.90250	1.52	0.88704	1.77	0.92376
1.03	0.98355	1.28	0.90072	1.53	0.88757	1.78	0.92623
1.04	0.97844	1.29	0.89904	1.54	0.88818	1.79	0.92877
1.05	0.97350	1.30	0.89747	1.55	0.88887	1.80	0.93138
1.06	0.96874	1.31	0.89600	1.56	0.88964	1.81	0.93408
1.07	0.96415	1.32	0.89464	1.57	0.89049	1.82	0.93685
1.08	0.95973	1.33	0.89338	1.58	0.89142	1.83	0.93969
1.09	0.95546	1.34	0.89222	1.59	0.89243	1.84	0.94261
1.10	0.95135	1.35	0.89115	1.60	0.89352	1.85	0.94561
1.11	0.94739	1.36	0.89018	1.61	0.89468	1.86	0.94869
1.12	0.94359	1.37	0.88931	1.62	0.89592	1.87	0.95184
1.13	0.93993	1.38	0.88854	1.63	0.89724	1.88	0.95507
1.14	0.93642	1.39	0.88785	1.64	0.89864	1.89	0.95838
1.15	0.93304	1.40	0.88726	1.65	0.90012	1.90	0.96177
1.16	0.92980	1.41	0.88676	1.66	0.90167	1.91	0.96523
1.17	0.92670	1.42	0.88636	1.67	0.90330	1.92	0.96878
1.18	0.92373	1.43	0.88604	1.68	0.90500	1.93	0.97240
1.19	0.92088	1.44	0.88580	1.69	0.90678	1.94	0.97610
1.20	0.91817	1.45	0.88565	1.70	0.90864	1.95	0.97988
1.21	0.91558	1.46	0.88560	1.71	0.91057	1.96	0.98374
1.22	0.91311	1.47	0.88563	1.72	0.91258	1.97	0.98768
1.23	0.91075	1.48	0.88575	1.73	0.91466	1.98	0.99171
1.24	0.90852	1.49	0.88595	1.74	0.91683	1.99	0.99581
						2.00	1.00000

$\Gamma(n) = \int_0^\infty e^{-x} x^{n-1} dx.$

$\Gamma(n+1) = n\Gamma(n).$

$\Gamma(1) = 1.$

$\Gamma(\frac{1}{2}) = \sqrt{\pi}.$

$\Gamma(\frac{n}{2}) = (\frac{n}{2} - 1)! = (\frac{n}{2} - 1)(\frac{n}{2} - 2)...(3)(2)(1)$ for n even and $n > 2.$

$\Gamma(\frac{n}{2}) = (\frac{n}{2} - 1)! = (\frac{n}{2} - 1)(\frac{n}{2} - 2)...(\frac{3}{2})(\frac{1}{2})\sqrt{\pi}$ for n odd and $n > 2.$

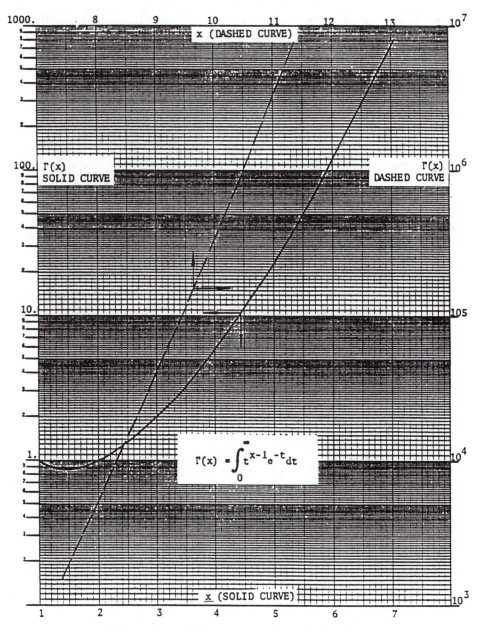

Fig. 12.4 –The gamma function's plot.

If α_4 is less than 3 in value then the *pdf* is flatter than the normal *pdf*, if α_4 is equal to 3 in value then it has the peakedness of a normal *pdf*, and if α_4 is greater than 3 in value then the *pdf* is more peaked than a normal *pdf*.

13. The kurtosis, α_4', of the Weibull *pdf* is given by

$$\alpha_4' = \eta^4 \{\Gamma(1+\frac{4}{\beta}) - 4\Gamma(1+\frac{3}{\beta})[\Gamma(1+\frac{1}{\beta})]$$
$$+ \quad 6\,\Gamma(1+\frac{2}{\beta})[\Gamma(1+\frac{1}{\beta})]^2 - 3[\Gamma(1+\frac{1}{\beta})]^4\}. \qquad (12.9')$$

14. The *kth* moment of the Weibull *pdf* about the origin, if $\gamma=0$, is given by

$$\mu_k' = \eta^k \Gamma(\frac{k}{\beta}+1). \qquad (12.4')$$

If $\gamma \neq 0$, to find the *kth* moment use

$$E(T-\gamma)^k = \eta^k \Gamma(\frac{k}{\beta}+1). \qquad (12.4'')$$

For example, if $k=2$ then

$$E(T-\gamma)^2 = E(T^2 - 2T\gamma + \gamma^2) = \eta^2 \Gamma(\frac{2}{\beta}+1).$$

Consequently,

$$E(T^2) = \eta^2 \Gamma(\frac{2}{\beta}+1) + 2\gamma E(T) - \gamma^2,$$
$$= \eta^2 \Gamma(\frac{2}{\beta}+1) + 2\gamma[\gamma + \eta\Gamma(\frac{1}{\beta}+1)] - \gamma^2,$$
$$= \eta^2 \Gamma(\frac{2}{\beta}+1) + 2\gamma^2 + 2\gamma\eta\Gamma(\frac{1}{\beta}+1) - \gamma^2,$$

or

$$E(T^2) = \eta^2 \Gamma(\frac{2}{\beta}+1) + 2\gamma\eta\Gamma(\frac{1}{\beta}+1) + \gamma^2.$$

The variance is then given by

$$\sigma_T^2 = E(T^2) - [E(T)]^2.$$

Consequently,

$$\sigma_T^2 = \eta^2 \Gamma(\frac{2}{\beta}+1) + 2\gamma\eta\Gamma(\frac{1}{\beta}+1) + \gamma^2 - [\gamma + \eta\Gamma(\frac{1}{\beta}+1)]^2,$$
$$= \eta^2 \Gamma(\frac{2}{\beta}+1) - \eta^2[\Gamma(\frac{1}{\beta}+1)]^2,$$

or

$$\sigma_T^2 = \eta^2\{\Gamma(\frac{2}{\beta}+1) - [\Gamma(\frac{1}{\beta}+1)]^2\},$$

where $(\sigma_T^2)^{1/2} = \sigma_T$, is the same result as in Eq. (12.7).

15. The *kth* moment of the Weibull *pdf about the mean* is given by

$$\mu_k = \eta^k \sum_{j=0}^{k}(-1)^j \binom{k}{j}\Gamma(\frac{k-j}{\beta}+1)[\Gamma(\frac{1}{\beta}+1)]^j. \qquad (12.4'')$$

16. The parameter β is a pure number. The parameters η and γ have the same units as T, such as hours, miles, cycles, actuations, etc. to failure. The parameter γ may assume all values and provides an estimate of the earliest time a failure may be observed. A negative γ may indicate that failures have occurred prior to the beginning of the test, namely during production, in storage, in transit, during checkout prior to the start of a mission, or prior to actual use.

12.2 WEIBULL RELIABILITY CHARACTERISTICS

The Weibull reliability function is

$$R(T) = e^{-(\frac{T-\gamma}{\eta})^\beta}, \qquad (12.10)$$

and is shown in Fig. 12.5. Some of the specific characteristics of this function are the following:

1. The Weibull reliability function, as may be seen in Fig. 12.5, starts at the value of 1 at $T = \gamma$, it decreases thereafter for $T > \gamma$. As $T \to \infty$ $R(T) \to 0$. $R(T)$ decreases sharply and monotonically for $0 < \beta < 1$, it is convex, and decreases less sharply for the same β but with an η larger than before.

2. For $\beta=1$ and the same η, $R(T)$ decreases monotonically but less sharply than for $0 < \beta < 1$, and is convex.

3. For $\beta > 1$, $R(T)$ decreases as T increases but less sharply than before, and as wear-out sets in it decreases sharply and goes through an inflection point.

4. The reliability for a mission of $(\gamma + \eta)$ duration, starting the mission at the age of zero, is always equal to 0.368. Consequently, for all Weibull *pdf's* 36.8% of the units survive for $T = \gamma + \eta$, and for any β and the same $(\gamma + \eta)$ value all $R(T)$ plots cross at $R(T) = 0.368$ and $T = \gamma + \eta$. This may be proven by substituting $T = \gamma + \eta$ in the $R(T)$ function of Eq. (12.10).

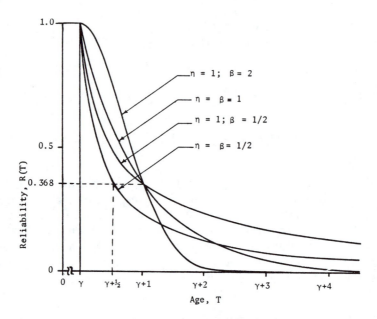

Fig. 12.5– The effect of the scale parameter, η, and of the shape parameter, β, on the reliability, with $\gamma=0$.

5. The Weibull conditional reliability function is given by

$$R(T,t) = \frac{R(T+t)}{R(T)} = \frac{e^{-(\frac{T+t-\gamma}{\eta})^{\beta}}}{e^{-(\frac{T-\gamma}{\eta})^{\beta}}}, \tag{12.11'}$$

or

$$R(T,t) = e^{-[(\frac{T+t-\gamma}{\eta})^{\beta}-(\frac{T-\gamma}{\eta})^{\beta}]}. \tag{12.11}$$

Equation (12.11') gives the reliability for a new mission of t duration, having already accumulated T hours of operation up to the start of this new mission, and the units are checked out to assure that they start the next mission successfully. For additional characteristics consult [1 through 8].

   6. The reliable life, T_R, of a unit for a specified reliability, starting the mission at age zero, is obtained as follows:

$$R(T) = e^{-(\frac{T-\gamma}{\eta})^{\beta}},$$

$$\log_e[R(T)] = -(\frac{T-\gamma}{\eta})^{\beta},$$

$$\{-\log_e[R(T)]\}^{\frac{1}{\beta}} = \frac{T-\gamma}{\eta}. \tag{12.12'}$$

Solving Eq. (12.12) for T, yields the reliable life

$$T_R = \gamma + \eta\{-\log_e[R(T_R)]\}^{\frac{1}{\beta}}. \tag{12.12}$$

This is the life for which the unit will be functioning successfully with a probability of $R(T_R)$. If $R(T_R)=0.50$ then $T_R = \tilde{T}$, the median life, or the life by which half of the units will survive.

12.3   WEIBULL FAILURE RATE CHARACTERISTICS

The Weibull failure rate function, $\lambda(T)$, is given by

$$\lambda(T) = \frac{\beta}{\eta}(\frac{T-\gamma}{\eta})^{\beta-1}, \tag{12.13}$$

and is shown plotted in Fig. 12.6. Some of the characteristics of this function are the following:

1. The Weibull failure rate for $0 < \beta < 1$, as may be seen in Fig. 12.6, starts at a value of ∞ at $T = \gamma$. $\lambda(T)$ decreases thereafter monotonically and is convex, approaching the value of zero as $T \to \infty$ or $\lambda(\infty) = 0$. This behavior makes it suitable for representing the failure rate of units exhibiting early-type failures for which the failure rate decreases with age.

2. For $\beta=1$, $\lambda(T)$ yields a constant value of $\frac{1}{\eta}$, or

$$\lambda(T) = \lambda = \frac{1}{\eta}. \tag{12.14}$$

This makes it suitable for representing the failure rate of chance-type failures and the useful life period failure rate of units.

3. For $\beta > 1$, $\lambda(T)$ increases as T increases and becomes suitable for representing the failure rate of units exhibiting wear-out type failures. For $1 < \beta < 2$ the $\lambda(T)$ curve is concave, consequently the failure rate increases at a decreasing rate as T increases, as shown in Fig. 12.6.

For $\beta=2$, or for the Rayleigh distribution case, the failure rate function is

$$\lambda(T) = \frac{2}{\eta}(\frac{T-\gamma}{\eta}), \tag{12.15}$$

hence there emerges a straight line relationship between $\lambda(T)$ and T, starting at a value of $\lambda(T) = 0$ at $T = \gamma$, and increasing thereafter with

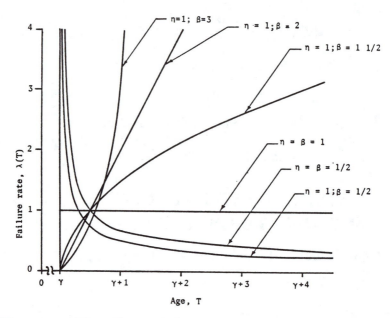

Fig. 12.6– The effects of the scale parameter, η, and of the shape parameter, β, on the failure rate, with $\gamma = 0$.

a slope of $\frac{2}{\eta^2}$. Consequently the failure rate increases at a constant rate as T increases. Furthermore, if also $\eta=1$ the slope becomes equal to 2, and when $\gamma=0$, $\lambda(T)$ becomes a straight line which passes through the origin with a slope of 2.

When $\beta > 2$ the $\lambda(T)$ curve is convex, with its slope increasing as T increases. Consequently, the failure rate increases at an increasing rate as T increases.

12.4 ESTIMATION OF THE PARAMETERS OF THE WEIBULL DISTRIBUTION BY PROBABILITY PLOTTING

12.4.1 WHEN THE DATA FALL ON A STRAIGHT LINE

Let's assume ten identical units ($N=10$) are being reliability tested at the same application and operation stress levels for a duration of 200 clock hours. Six ($j_{max} = 6$) of these fail during this test after operating the following number of hours, T_j: 93, 34, 16, 120, 53 and 75. The recommended steps for determining the parameters of the Weibull *pdf* representing these data are the following:

TABLE 12.2– Times-to-failure data and the corresponding median ranks for the determination of the Weibull *pdf* parameters for a sample of ten units ($N = 10$).

1	2	3
j	T_j	$\hat{Q}(T_j) \times 100$
Failure number	Time to failure,hr	Median rank, $MR, \%$
1	16	6.70
2	34	16.23
3	53	25.86
4	75	35.51
5	93	45.17
6	120	54.83

1. Prepare Table 12.2.

2. In Column 1 titled j, enter the failure number, for the earliest failure being number one, or $j=1$, for the next $j=2$, etc., the last number being equal to the total number of failures observed in the test, or $j=6$ in this case. The failed units are intended to be removed from the test, not necessarily physically though; i.e., they are turned off.

3. Arrange the data, the T_j, in increasing value of hours, cycles, revolutions, actuations, miles, etc., to failure (any time measure to failure will do) and enter this information in that order in Column 2.

4. As the graphical method requires the plotting of probability of failure by time T_j, or $Q(T_j)$, versus T_j, an estimate of $Q(T_j)$, or $\hat{Q}(T_j)$, must be found. One such estimate is the *Median Rank, MR* [9]. The median rank for the jth failure in N units tested is the MR value such that the probability the jth failure in N occurs before time T_j, for a large sample, is 50%. It is the value the true probability of failure, $Q(T_j)$, should have, with 50% confidence, at the time of the jth failure in N units. The general expression for the median rank [9; 10, p. 51] is

$$1 - (1 - MR)^N - (N)(MR)(1 - MR)^{N-1}$$
$$- \frac{N(N - 1)}{2!}(MR)^2(1 - MR)^{N-2} - \ldots$$

$$-\frac{N(N-1)\ldots(N-j+2)}{(j-1)!}(MR)^{j-1}(1-MR)^{N-j+1} = 0.50.$$

$$(12.16)$$

Equation (12.16) may be rearranged as

$$\sum_{k=j}^{N}\binom{N}{k}MR)^{k}(1-MR)^{N-k} = 0.50. \qquad (12.17)$$

Equation (12.17) is the binomial cumulative distribution function, and it may be solved for MR by iterative computer methods, given N and j. From Appendix A, under the column headed by the number 50.0, the MR values may be obtained for various combinations of N and j values. These tables go up to $N=25$.

The $MR's$ may also be obtained from [9]

$$Median\ \ Rank(\%) = MR(\%),$$

$$= \hat{Q}(T_j) \times 100 \simeq \frac{j-0.3}{N+0.4} \times 100, \quad (12.18)$$

when quick estimates are needed and also when the N and j values are beyond the range of Appendix A.

An easier and more exact way of obtaining any MR value is to apply two transformations to Eq. (12.17), first to the beta pdf and second to the F distribution equivalent of the beta pdf, and obtain

$$MR = \frac{1}{1+\frac{N-j+1}{j}F_{0.50;m;n}}, \qquad (12.18')$$

where

$$m = 2(N-j+1),$$

and

$$n = 2j.$$

The accuracy of Eq. (12.18') depends on the accuracy of the value of the F distribution's $50th$ percentile value. Consequently, either good $F_{50;m;n}$ value tables should be found or these values should be computer

generated [27]. These values should have at least four decimal place accuracy.

EXAMPLE 12–1

Given a sample of size 10, find the median rank for the sixth failure.

SOLUTION TO EXAMPLE 12–1

In this example $N=10$, $j=6$, $m = 2(10 - 6 + 1) = 10$, and $n = 2 \times 6 = 12$. Then, Eq. (12.18') becomes

$$MR = \frac{1}{1 + (\frac{10-6+1}{6})F_{0.50;10;12}},$$

where, from Appendix F, $F_{0.50;10;12} = 0.9886$. Consequently,

$$MR = \frac{1}{1 + \frac{5}{6} \times 0.9886} = 0.5483,$$

or

$$MR(\%) = 54.83\%.$$

This value compares very well with $MR(\%) = 54.8312\%$ given in Appendix A.

Other estimates of $\hat{Q}(T_j)$ are the following [5, pp. 2-10; 11]:

$$\text{Sample } cdf = \frac{j}{N}, \tag{12.19}$$

where cdf is the cumulative distribution function,

$$\text{Symmetrical sample } cdf = (j - \frac{1}{2})/N, \tag{12.20}$$

$$\text{Mean of } cdf, \text{ or mean rank} = j/(N + 1), \tag{12.21}$$

$$\text{Mode of } cdf = (j - 1)/(N - 1). \tag{12.22}$$

The median ranks given in Appendix A are used as the estimate of $Q(T_j)$, because they provide the 50% confidence line, or the median line, of the unreliability. When the confidence limits on the unreliability, and in turn on the reliability, need to be determined the corresponding confidence bands need to be plotted on a chart containing the *MR*, or the 50% confidence level, line. This provides consistency as all plots on such a chart now represent specific confidence levels.

The median rank values for the example are given in Column 3 of Table 12.2. They are obtained by finding in Appendix A the section titled SAMPLE SIZE 10, and recording the values under the column headed by 50.0 for rank orders $j=1$ through 6, for each one of the six (up to $j=6$) failures.

5. Obtain a copy of Weibull probability plotting paper of the right number of cycles along the abscissa, and of the right probability range along the ordinate. A variety of Weibull probability papers are available. Figures 12.7 and 12.8 are Ford Motor Company probability papers, and Fig. 12.9 is a TEAM probability paper which is available commercially [12]. The Ford Weibull paper of Fig. 12.8 is a two-cycle paper along the abscissa and has a PERCENT (failed), or unreliability, range of 1% to 99%, or a corresponding reliability range of 99% to 1% along the ordinate. If a reliability higher than 99% is sought then a Weibull probability paper starting with a percent range lower than 1% should be used. For example, if a reliability of up to 99.9% is sought then a probability paper starting with 0.1% value should be used, and so on. It must be noted that on the Ford probability paper the ordinate is labeled as PERCENT and it means percent failed, or the probability of failure, or the unreliability in percent, or $\hat{Q}(T_j)$ in percent, a good estimate of which is the median rank. On the TEAM probability papers the ordinate is labeled as "Percent Failure."

Label the abscissa with what has been measured; e.g., age, or time to failure, T, hours. Put the applicable number of zeros next to each number 1 appearing along the abscissa. In Fig. 12.10 a zero was added to the first 1, two zeros were added to the second 1, and three zeros to the third 1, such that the data range was adequately covered, including desired extrapolations thereof.

6. Plot each value of MR, or $\hat{Q}(T_j)$, on the ordinate against the corresponding value of T_j on the abscissa, as shown in Fig. 12.10.

7. Draw a straight line, through these points, that best represents the data, or in such a way that the total added up distances between each point and the fitted straight line is minimized. A good way of achieving this is to first draw with blue pencil, lightly, a good-eyeball-fit straight line using a transparent straight-edge. Then place a pencil point near the smallest plotted value and move the straight-edge until the points above and below the straight-edge are divided into two equal parts and draw a second blue line lightly. Now place the pencil point near the largest plotted value and move the straight-edge until the points above and below the straight-edge are divided into two equal parts and draw a third blue line lightly. Now draw a heavy blue line averaging the prior three lines dividing the points falling above and below this final line into two equal parts. This line is a median regression line which can be considered as a "practical best fit" for

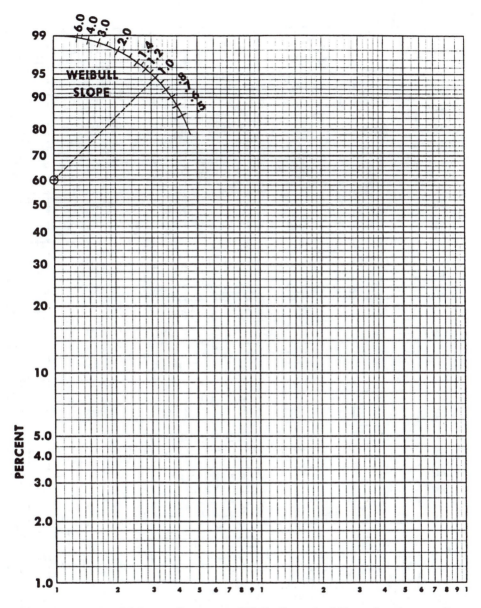

Fig. 12.7 –Ford Motor Company Weibull probability plotting paper.

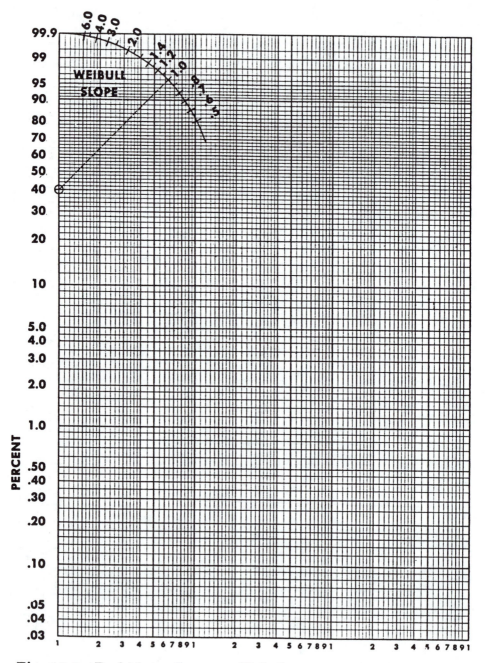

Fig. 12.8 –Ford Motor Company Weibull probability plotting paper.

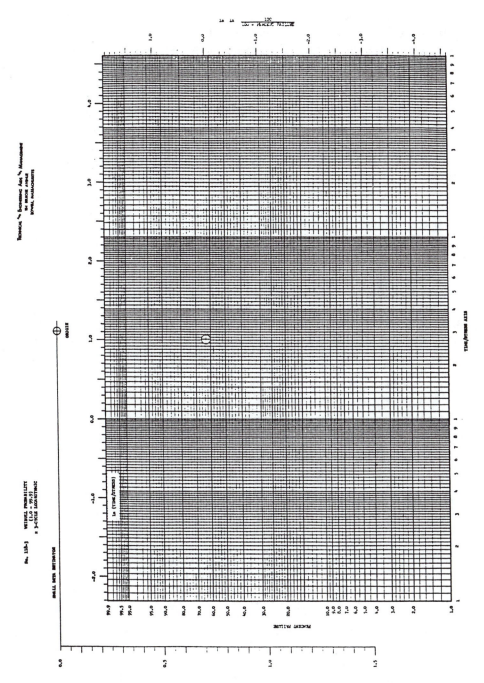

Fig. 12.9 –TEAM Weibull probability plotting paper.

most kinds of data without requiring any calculations.

It should be pointed out that if data points exist which are substantially out of line with the rest of the data, called outliers, they may be suspected as bad data. A reason for their cause should be sought however, as a study of their cause and implementation of the study results may improve the product or the data collection process. If no plausible reasons could be found, the outliers may be the very scarce points in the tails of the parent population's distribution which we seek, to justify the possible use of another distribution more skewed than the one chosen. Only, of course, the very pronounced peculiarities should be interpreted as bad data or a bad choice of a distribution. Inexperienced data analysts may tend to overinterpret plots and may expect them to be more orderly than they usually are.

If the straight line does indeed represent the points well then the location parameter is zero, or $\gamma=0$, and this line represents the $\hat{Q}(T_j)$ function.

8. Draw a line parallel to the straight line just drawn, through the center of a small circle to be found on the probability paper used. On the Ford paper of Fig. 12.7 it is located on the ordinate scale at the 40% level and of Fig. 12.8 at the 60% level. On the TEAM paper of Fig. 12.9 it is identified as "ORIGIN," and is located near the middle at the top of the paper. Extend this parallel line until it crosses a scale identified as "WEIBULL SLOPE" on the Ford paper, or as "SMALL BETA ESTIMATOR" on the TEAM paper. Read off on this scale the value crossed by the parallel line. This is the value of the shape parameter β. From Fig. 12.10 $\beta=1.2$ for the data in Table 12.2.

9. At the 63.2% level on the ordinate draw a horizontal line until it intersects the fitted straight line, or the $\hat{Q}(T_j)$ versus T_j line. Through this intersection draw a vertical line until it intersects the abscissa, and read off the abscissa value at this intersection, This is the value of the scale parameter η. From Fig. 12.10 $\eta=144$ hr.

10. If desired, write the Weibull *pdf* representing the data given in Table 12.2, then

$$f(T) = \frac{1.2}{144}(\frac{T-0}{144})^{1.2-1}e^{-(\frac{T-0}{144})^{1.2}},$$

or

$$f(T) = \frac{1.2}{144}(\frac{T}{144})^{0.2}e^{-(\frac{T}{144})^{1.2}}. \tag{12.23}$$

Actually it is more expeditious to use Fig. 12.10 directly, as illustrated in the examples that follow, to obtain unreliabilities, reliabilities, conditional reliabilities, and the duration of a mission to attain a specified

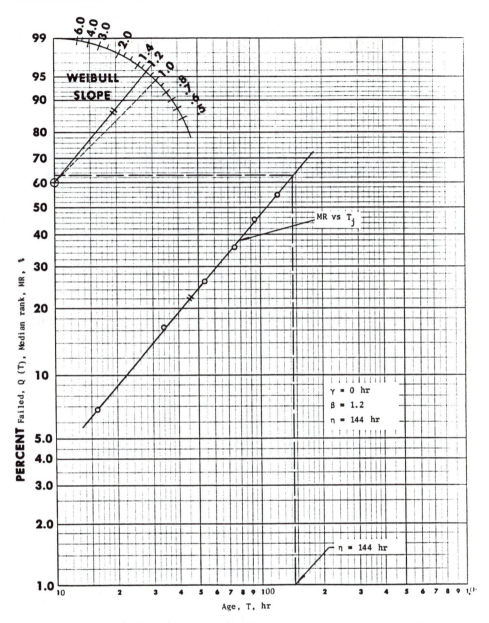

Fig. 12.10– Plot of data in Table 12.2 to obtain the reliability
characteristics and the parameters of the Weibull
pdf representing the data.

reliability, rather than spending time to calculate these values using Eqs. (12.10), (12.11) and (12.12).

EXAMPLE 12–2

What is the unreliability of these units for a mission of 30 hr duration starting the mission at age zero?

SOLUTION TO EXAMPLE 12–2

Enter Fig. 12.10 with $T=30$ hr go up vertically to the straight line fitted to the data, then go horizontally to the ordinate, and read off $MR = \hat{Q}(T = 30 \text{ hr}) = 14\%$. Then, a good estimate of the sought unreliability is 14%.

EXAMPLE 12–3

What is the corresponding reliability?

SOLUTION TO EXAMPLE 12–3

$$\hat{R}(T = 30 \text{ hr}) = 100 - \hat{Q}(T = 30 \text{ hr}) = 100 - 14 = 86, \text{ or } 86\%.$$

Then, a good estimate of the sought reliability is 86%.

EXAMPLE 12–4

What is the reliability for a new mission of $t=30$ hr duration starting the new mission at the age of $T=30$ hr?

SOLUTION TO EXAMPLE 12–4

$$R(T, t) = \frac{R(T + t)}{R(T)},$$

or

$$\hat{R}(30 \text{ hr}, 30 \text{ hr}) = \frac{\hat{R}(30 \text{ hr} + 30 \text{ hr})}{\hat{R}(30 \text{ hr})} = \frac{\hat{R}(60 \text{ hr})}{\hat{R}(30 \text{ hr})}.$$

From Fig. 12.10, $\hat{Q}(60 \text{ hr}) = 29\%$, $\hat{R}(60 \text{ hr}) = 1 - \hat{Q}(60 \text{ hr}) = 1 - 0.29 = 0.71$, and $\hat{R}(30 \text{ hr}) = 0.86$, or $\hat{R}(30 \text{ hr}, 30 \text{ hr}) = 0.71/0.86 = 0.83$. Consequently, the estimate of the reliability for the new mission of 30 hr is 83%.

EXAMPLE 12–5

What is the longest mission that this equipment should undertake

for a reliability of 95%, starting the mission at the age of 30 hr?

SOLUTION TO EXAMPLE 12–5

The conditional reliability in this example is given by

$$\hat{R}(30 \text{ hr}, t) = \frac{\hat{R}(30 \text{ hr} + t)}{\hat{R}(30 \text{ hr})} = \frac{\hat{R}(30 \text{ hr} + t)}{0.86} = 0.95,$$

where $\hat{R}(30 \text{ hr})$ was determined in Example 12–4. Then,

$$\hat{R}(30 \text{ hr} + t) = 0.95 \times 0.86 = 0.82,$$

where t is the answer. Consequently,

$$\hat{Q}(30 \text{ hr} + t) = 1 - \hat{R}(30 \text{ hr} + t) = 1 - 0.82 = 0.18 = MR.$$

Entering Fig. 12.10 with $MR=18\%$, going horizontally to the MR versus T_j line, dropping vertically down from the point of intersection with this line, and reading off the corresponding abscissa value yields $(30 \text{ hr} + t) = 38 \text{ hr}$; therefore, $\hat{t}=8$ hr. Consequently, the duration of the new mission should not exceed 8 hr to achieve a reliability of 95% with a confidence of 50%.

12.4.2 WHEN THE DATA DO NOT FALL ON A STRAIGHT LINE

12.4.2.1 – METHOD 1

When the MR versus T_j points plotted on the Weibull probability paper do not fall on a satisfactory straight line and it is detected that the points fall on a curve, then a location parameter, γ, might exist which may straighten out these points. The procedure given below may be pursued to determine the existence of a γ and to find its value, as well as the values of β and η.

Let's assume ten identical units $(N=10)$ are being reliability tested at the same application and operation stress levels for a duration of 200 clock hours. Six of these units fail during this test after operating the following number of hours, T_j: 150, 105, 83, 123, 64 and 46. Do the following:

1. Prepare Table 12.3 which is similar to Table 12.2 except for the addition of Column 4 for $(T_j - T_1)$.

2. Repeat Steps 2 through 6 of the previous case and plot the points MR versus T_j on Weibull probability paper, as shown in Fig. 12.11.

3. If the points MR versus T_j do not fall on a straight line, then the location parameter, γ, has a nonzero value and its actual value should

TABLE 12.3– Times-to-failure data, corresponding median ranks, and time-to-failure minus the first time to failure for the determination of the Weibull *pdf*'s parameters ($N = 10$).

1	2	3	4
j, Failure number	T_j, Time to failure, hr	$\check{Q}(T_j) \cdot 100$, Median rank, percent	$T_j - T_1$, T_j minus time to the earliest failure, or 46 hr
1	46	6.70	0
2	64	16.23	18
3	83	25.86	37
4	105	35.51	59
5	123	45.17	77
6	150	54.83	104

be determined first before β and η are determined. Draw a smooth curve through these points, as shown in Fig. 12.11.

4. Enter in Table 12.3, Column 4, the values $(T_j - T_1)$, where T_1 is the smallest observed time-to-failure, or $T_1 = 46$ hr in this case.

5. Plot the points *MR* versus $(T_j - T_1)$, on the same probability paper. Draw a smooth curve through these new points, as shown in Fig. 12.11. Use the following rules to obtain an estimate of γ, or $\hat{\gamma}$:

Case 1 – If the curve for *MR* versus T_j is concave and for *MR* versus $(T_j - T_1)$ is convex, as shown in Fig. 12.12(a), then there exists a γ such that $0 < \gamma < T_1$, or γ has a value between zero and the smallest time to failure, which will straighten out the points for a good straight line fit to the *MR* versus $(T_j - \gamma)$ points.

Case 2 – If the curves for *MR* versus T_j and *MR* versus $(T_j - T_1)$, are both convex, as shown in Fig. 12.12(b), then there exists a negative γ which will straighten out the curve of *MR* versus T_j.

Case 3 – If neither one of the previous two cases prevails, then reject the Weibull *pdf* as one capable of representing the data, because no other value of γ will straighten out the *MR* versus T_j points for a straight line fit to the data points, which is the requirement for accepting the Weibull as the *pdf* that represents the data best. This requirement follows from the basis on which the Weibull probability paper, as well as most all other probability plotting papers, are constructed. Returning to the example, the plotting of points *MR* versus T_j and *MR* versus $(T_j - T_1)$, as shown in Fig. 12.11, yields two curves

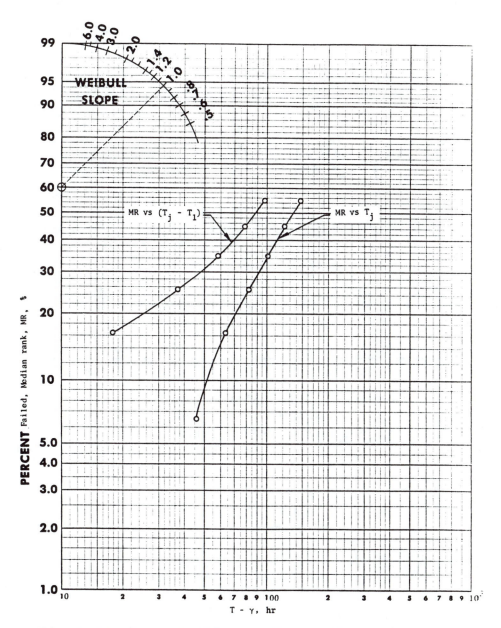

Fig. 12.11– Plot of the MR versus T_j, and MR versus $(T_j - T_1)$, using the data in Table 12.3 to obtain the reliability characteristics and the parameters of the Weibull pdf representing the data.

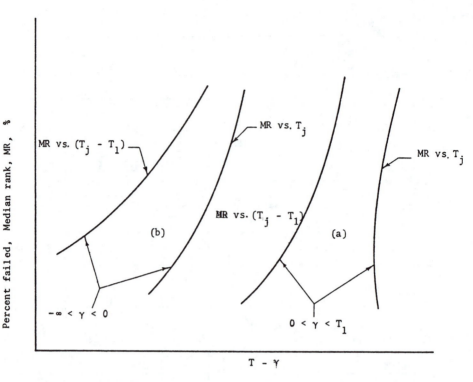

Fig. 12.12– Location parameter values for the Weibull *pdf*
with different curvatures for the *MR* versus T_j
and *MR* versus $(T_j - T_1)$ curves.

TABLE 12.4– Adjustment of the data given in Table 12.3 to find γ.

1	2	3	4
Failure number, j	Time-to-failure minus 15 hr, $T_j - 15$, hr	Time-to-failure minus 30 hr, $T_j - 30$, hr	Median rank, MR, %
1	31	16	6.70
2	49	34	16.23
3	68	53	25.86
4	90	75	35.51
5	108	93	45.17
6	135	120	54.83

conforming to Case 1; consequently, there exists a γ with a value between zero and $T_1 = 46$ hr, or $0 < \hat{\gamma} < 46$ hr.

6. Prepare Table 12.4 where values smaller than T_1 are subtracted from the T_j values; i.e., 15 hr and 30 hr are subtracted from each failure time T_j, in Columns 2 and 3, respectively. The median ranks remain the same because the failure order for each failure does not change.

7. Plot, as shown in Fig. 12.13, the MR versus T_j, MR versus $(T_j - T_1)$, MR versus $(T_j - 15)$, and MR versus $(T_j - 30)$ values from Tables 12.3 and 12.4. Determine which set of values plots the best straight line. Then $\hat{\gamma}$ is that value subtracted from T_j which results in the best straight line. In Fig. 12.13 the plot of MR versus $(T_j - 30)$ gives the best straight line; therefore, $\hat{\gamma} = 30$ hr.

If the values chosen for subtraction from T_j do not give a good straight line, then different values with smaller increments should be subtracted. The subtracted value giving the best straight line would then be the best estimate of γ.

8. If needed, find β and η using the same procedure as that used when the MR versus T_j plot gave a good straight line, using the straightened-out line, as shown in Fig. 12.13. Therefore, for the data of Table 12.3, $\hat{\gamma} = 30$ hr, $\hat{\beta} = 1.2$ and $\hat{\eta} = 144$ hr.

9. If desired, write the Weibull *pdf* representing the data given in Table 12.3, or

$$f(T) = \frac{1.2}{144} \left(\frac{T - 30}{144}\right)^{0.2} e^{-\left(\frac{T-30}{144}\right)^{1.2}}.$$ (12.24)

It is now more expeditious to use Fig. 12.13 directly to obtain unreliabilities, reliabilities, conditional reliabilities, and the duration of a

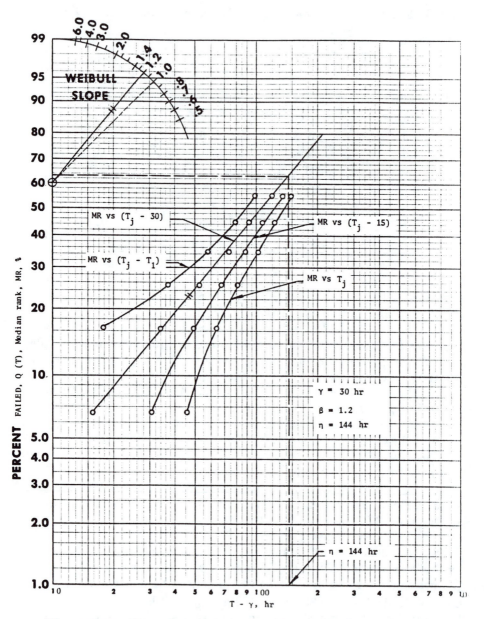

Fig. 12.13– Plot of the *MR* versus T_j, *MR* versus $(T_j - T_1)$, *MR* versus $(T_j - 30)$, and *MR* versus $(T_j - 15)$ data, in Tables 12.3 and 12.4, to obtain the reliability characteristics and the parameters of the Weibull *pdf* representing the data.

mission to attain a specified reliability, rather than calculating them from Eq. (12.24), as illustrated previously in Examples 12–2 through 12–5.

EXAMPLE 12–6

What is the reliability of these units for a mission of 30 hr duration, starting the mission at age zero?

SOLUTION TO EXAMPLE 12–6

Here $T=30$ hr; therefore, enter Fig. 12.13 with $T-\gamma = 30-30 = 0$. This value is not on Fig. 12.13; the implication is that as $(T-\gamma) \to 0$, $\hat{Q}(T) \to 0$ or $\hat{R}(T) \to 1$. This may be proven knowing that

$$R(T) = e^{-(\frac{T-\gamma}{\eta})^\beta} = e^{-(\frac{T-30}{144})^{1.2}},$$

and

$$R(T = 30 \text{ hr}) = e^{-(\frac{30-30}{144})^{1.2}} = e^{-(0)^{1.2}} = e^{-0} = 1.$$

Also, by the definition of γ, no failures can occur before γ hours of operation are accumulated during a mission starting at age zero; consequently, the reliability is 1 up to $T = \gamma$.

EXAMPLE 12–7

What is the reliability for a new mission of $t=30$ hr, starting the new mission at the age of $T=30$ hr?

SOLUTION TO EXAMPLE 12–7

The conditional reliability is being sought which is given by

$$R(T,t) = \frac{R(T+t)}{R(T)} = \frac{R(30 \text{ hr} + 30 \text{ hr})}{R(30 \text{ hr})} = \frac{R(60 \text{ hr})}{R(30 \text{ hr})}.$$

To be able to use Fig. 12.13, where the unreliabilities are given, the numerator is expressed in terms of Q, or

$$\hat{R}(30 \text{ hr}, 30 \text{ hr}) = \frac{1 - \hat{Q}(60 \text{ hr})}{R(30 \text{ hr})}.$$

To find $\hat{Q}(60 \text{ hr})$, Fig. 12.13 should be entered with $T-\gamma = 60-30 = 30$ hr, to read off $\hat{Q}(T = 60 \text{ hr}) = 14\%$. The denominator was already found in Example 12–5 to be $R(30 \text{ hr}) = 1$; therefore,

$$R(T,t) = R(30 \text{ hr}, 30 \text{ hr}) = \frac{1 - 0.14}{1} = \frac{0.86}{1} = 0.86, \text{ or } 86\%.$$

EXAMPLE 12–8

What is the reliability for another new mission of $t=30$ hr, starting the new mission at the age of $T=60$ hr?

SOLUTION TO EXAMPLE 12–8

As in Example 12–7

$$R(T,t) = \frac{R(T+t)}{R(T)},$$

but $T=60$ hr and $t=30$ hr, or

$$R(T,t) = \frac{R(60 \text{ hr} + 30 \text{ hr})}{R(60 \text{ hr})} = \frac{R(90 \text{ hr})}{R(60 \text{ hr})} = \frac{1 - \hat{Q}(90 \text{ hr})}{R(60 \text{ hr})}.$$

$\hat{Q}(90 \text{ hr})$ is found by entering Fig. 12.13 with $T - \gamma = 90 - 30 = 60$ hr, or $\hat{Q} = 29\%$. From Example 12–7 $R(60 \text{ hr})=0.86$. Substitution of these quantities yields

$$R(T,t) = R(60 \text{ hr}, 30 \text{ hr}) = \frac{1 - 0.29}{0.86} = \frac{0.71}{0.86} = 0.83, \text{ or } 83\%.$$

EXAMPLE 12–9

What is the longest mission these units should undertake for a reliability of 95% starting the mission at the age of 60 hr?

SOLUTION TO EXAMPLE 12–9

It is required that

$$R(T,t) = R(60 \text{ hr}, t) = \frac{R(60 \text{ hr} + t)}{R(60 \text{ hr})} = 0.95;$$

therefore,

$$R(60 \text{ hr} + t) = 0.95 \times R(60 \text{ hr}),$$

where R (60 hr) was already found in Example 12–7 to be 0.86, and the only remaining unknown quantity is t which is the longest allowable mission duration for $R(T,t)=0.95$, or $R(60 \text{ hr}+t) = 0.95 \times 0.86 = 0.82$, and $Q(60 \text{ hr} + t)=0.18$ or 18%.

If Fig. 12.13 is entered with $Q(60 \text{ hr} + t) = Q(T') = 18\%$, $T' - \gamma = 38$ hr is obtained where $\gamma = 30$ hr, or $T' - 30 = 38$, and $T' = 68$ hr. Therefore, $60 + t = 68$, and finally $t = 8$ hr. This means that the duration of the new mission should not exceed 8 hr, if already 60 hr of operation were accumulated, to achieve a reliability of 95% with a confidence of 50%.

12.4.2.2 – METHOD 2

An estimate of the location parameter may be obtained by plotting the T_j versus MR_j. If the points plot curvilinear with a curvature that is concave looking upward, then extend the smooth curve drawn through the points towards the abscissa until it intersects it almost vertically and read off the T value at this intersection along the abscissa. This value gives an estimate of γ, because this is the T value that gives a $\hat{Q}(T)$ value closest to $\hat{Q}(T)=0$ at which T value the probability of failure is zero, as is the case at $T = \gamma$. Of course the more extended the ordinate scale is at the lower end, the closest $\hat{Q}(T)$ will be to zero and the closer $\hat{\gamma}$ will be to the actual location parameter, γ.

If this estimate does not yield an acceptable straight line, but changes the curvature from concave upward to concave downward it means that the assumed $\hat{\gamma}$ value was too large and should be *reduced* until the plot of MR versus $(T_j - \gamma)$ is as straight as possible. If the curvature does not reverse itself with the use of this $\hat{\gamma}$ then its value should be increased until the plot of MR versus $(T_j - \hat{\gamma})$ is as straight as possible.

EXAMPLE 12–10

Find the location parameter of the Weibull *pdf* for the data in Table 12.3 using Method 2.

SOLUTION TO EXAMPLE 12–10

Plot the T_j versus the corresponding MR_j values in Table 12.3 on an appropriate Weibull probability plotting paper as shown in Fig. 12.14. Extend the smooth curve fitted to the points, which is concave looking upward as shown, and find its intersection with the abscissa in Fig. 12.14. This intersection is at $T = 31$ hr, consequently $\hat{\gamma} = 31$ hr, which compares favorably with the value of $\hat{\gamma} = 30$ hr found using Method 1.

12.4.2.3 – METHOD 3

Another estimate of the location parameter, $\hat{\gamma}$, is obtained [33, pp. 75-77] by drawing three horizontal parallel lines equally spaced vertically, finding the corresponding $(T - \gamma)$ values where these three lines intersect the curve fitted to the $(T_j - \gamma)$ versus MR_j plot, as illustrated in Fig. 12.14, reading off the values of T_1, T_2, and T_3, corresponding to these intersections, and calculating $\hat{\gamma}$ from

$$\hat{\gamma} = T_2 - \frac{(T_3 - T_2)(T_2 - T_1)}{(T_3 - T_2) - (T_2 - T_1)}. \tag{12.25}$$

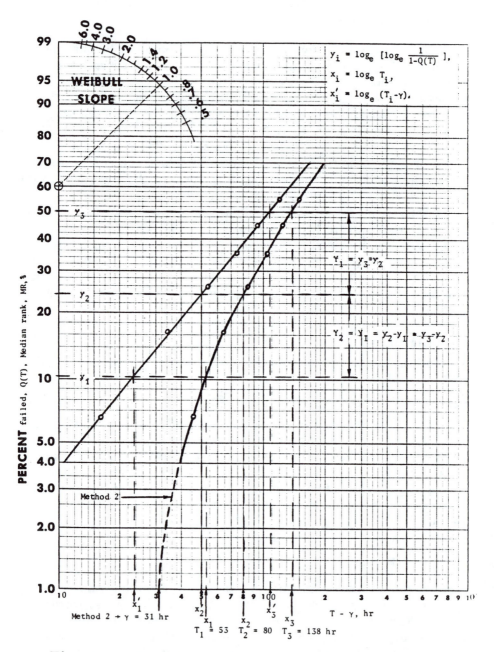

Fig. 12.14– Methods 2 and 3 for determining the location
parameter, γ, of the Weibull *pdf* using the data
in Table 12.3.

For the derivation of Eq. (12.25) see [33, Vol. 1, pp. 301-302].

Do keep in mind that the T_i's are obtained from the original times-to-failure plot, or from the T_j versus MR_j curve. It is also recommended that the three points chosen for the T_i values be located in the curved portion of the T_j versus MR_j plot.

EXAMPLE 12–11

Find the location parameter of the Weibull *pdf* for the data in Table 12.3 using Method 3.

SOLUTION TO EXAMPLE 12–11

Plot the T_j versus the corresponding MR_j values in Table 12.3 on an appropriate Weibull probability plotting paper, as shown in Fig. 12.14. The points fall on a curved line. Choose the first point (x_3, y_3) on this curve at the $Q(T_3) = 50\%$ level. Choose the second point (x_2, y_2) at the $Q(T_2) = 24\%$ level. The vertical geometric distance between the $Q(T_3) = 50\%$ level and the $Q(T_2) = 24\%$ level, or $Y_1 = y_3 - y_2$, is 1-3/8 inches (3.4925 cm). Find point (x_1, y_1) such that the vertical geometric distance between the $Q(T_2)$ level and the $Q(T_1)$ level, or $Y_2 = y_2 - y_1$, is equal to $Y_1 = y_3 - y_2$, or 1-3/8 inches (3.4925 cm) also. This establishes the level of $Q(T_1)$ at 10.3%. Drop down vertically from the intersection of the $Q(T_1)$, $Q(T_2)$ and $Q(T_3)$ levels with the T_j versus MR_j curve and read off $T_1=53$ hr, $T_2=80$ hr, and $T_3=138$ hr, as illustrated in Fig. 12.14.

Substitution of these values into Eq. (12.25) yields

$$\hat{\gamma} = 80 - \frac{(138 - 80)(80 - 53)}{(138 - 80) - (80 - 53)} = 80 - 50.52 = 29.48,$$

or

$$\hat{\gamma} = 29.5 \text{ hr.}$$

This value compares with $\hat{\gamma} = 30$ hr using Method 1, and with $\hat{\gamma} = 31$ hr using Method 2. The three values are very close to each other in this example. Averaging the $\hat{\gamma}$ values obtained by these three methods may yield a better $\hat{\gamma}$. In this case $\hat{\gamma} = 30$ hr is a good estimate to the nearest hour.

12.4.3   THE DETERMINATION OF A NEGATIVE γ

As discussed earlier and illustrated in Fig. 12.12, if the T_j versus MR_j values plot convex looking upward, and the $(T_j - T_1)$ versus MR_j values also plot convex looking upward and with a greater curvature, then there exists a $\gamma < 0$ which will straighten the points for a Weibull *pdf* fit to the data. Because subtracting T_1 from the T_j increases the

curvature of the curve passing through the points, instead of decreasing it, the only way to straighten the data are to push them to the right, so that they plot with a smaller curvature and eventually as a straight line: our goal. This requires adding a T value to the original T_j values rather than subtracting, so that the points get shifted to the right to straighten them. In the Weibull *pdf* to increase the value of $(T - \gamma)$, γ must be negative. Consequently, a variety of T_i values are *added* to all of the T_j's and the $(T_j + T_i)$ values are plotted, as illustrated in Table 12.6 and in Fig. 12.15. The T_i value that gives the best straight line plot is $\hat{\gamma}$.

EXAMPLE 12–12

Given the times-to-failure data of Table 12.5, find the parameters of the Weibull *pdf* that represents these data.

SOLUTION TO EXAMPLE 12–12

Prepare Columns 1, 2, and 3 in Table 12.6, and plot the T_j versus the corresponding MR_j values, as illustrated in Fig. 12.15. The points fall on a curve that is convex looking upward. Prepare Column 4 by subtracting T_1 from the T_j values and plot the $(T_j - T_1)$ values versus the corresponding MR_j values. As illustrated in Fig. 12.15, these points fall on a curve with an increased curvature; consequently, there may be a negative γ that will straighten out the points. Prepare Columns 5, 6 and 7 by adding to the T_j's the values of 200, 300, and 400. Plot these $(T_j + T_i)$ values versus the corresponding MR_j values, as illustrated in Fig. 12.15. The T_i value added to the T_j's that gives the best straight line is 300 hr. This value then is the estimate of the location parameter; therefore,

$$\hat{\gamma} = -300 \text{ hr.}$$

Using techniques discussed previously, $\hat{\beta} = 3.0$ and $\hat{\eta} = 1,220$ hr, thus completing the determination of the parameters of the Weibull *pdf* that best represents the data of Table 12.5.

The age or real positive time, T, by which 63.2% of such components will fail is given by

$$T - \gamma = \eta,$$

$$T = \eta + \gamma = 1,220 + (-300),$$

or

$$T = 920 \text{ hr.}$$

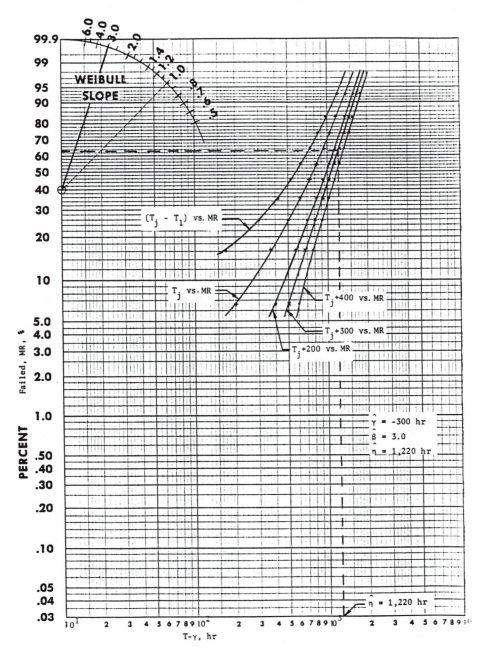

Fig. 12.15— Plot of the data in Table 12.6 to determine the
parameters of the Weibull distribution representing the data.

TABLE 12.5– Times-to-failure data, in hours, for ten identical units, all tested to failure.

j	T_j, hr
1	200
2	370
3	500
4	620
5	730
6	840
7	950
8	1,050
9	1,160
10	1,400

12.4.4 GROUPED DATA WEIBULL ANALYSIS

If the available data are in grouped form, as in Table 12.7, to estimate the Weibull *pdf's* parameters by probability plotting, plot the class end point T_j versus MR_j, where the MR_j are determined from the total number of failures observed until the class end, or from $r_j = \sum_{i=1}^{j} f_i$ which is the total number of failures observed by time T_j.

Using this procedure proceed as with ungrouped data and find a set of points $(T_j - \gamma)$ versus MR_j which give a straight line plot on Weibull probability paper, as illustrated in Fig. 12.16 and by the next example.

EXAMPLE 12–13

Twenty-five units ($N=25$) are put to a 1,000 hr reliability test. Every 200 hr the number of units which fail is counted. After 200 hr of testing, no failures are observed; by 400 hr two failures are observed; by 600 hr four more failures are observed; by 800 hr five more failures are observed; and by 1,000 hr five more units fail, as tabulated in Column 2 of Table 12.7. Find the parameters of the Weibull *pdf* representing these data.

SOLUTION TO EXAMPLE 12–13

Complete Columns 1, 2, 3, and 4 in Table 12.7. Note that $T_1 = 200$ hr, $T_2 = 400$ hr, etc., and $T_{jmax} = 1,000$ hr.

The corresponding median ranks are determined as follows and entered in Column 9: MR_1 is zero because no failures were observed

TABLE 12.6– Time-to-failure data in hours for ten identical units, all tested to failure.

1	2	3	4	5	6	7
j	T_j, hr	MR, %	$T_j - T_1$, hr	$T_j + 200$, hr	$T_j + 300$, hr	$T_j + 400$, hr
1	200	6.70	0	400	500	600
2	370	16.23	170	570	670	770
3	500	25.86	300	700	800	900
4	620	35.51	420	820	920	1,020
5	730	45.17	530	930	1,030	1,130
6	840	54.83	640	1,040	1,140	1,240
7	950	64.49	750	1,150	1,250	1,350
8	1,050	74.14	850	1,250	1,350	1,450
9	1,160	83.77	960	1,360	1,460	1,560
10	1,400	93.30	1,200	1,600	1,700	1,800

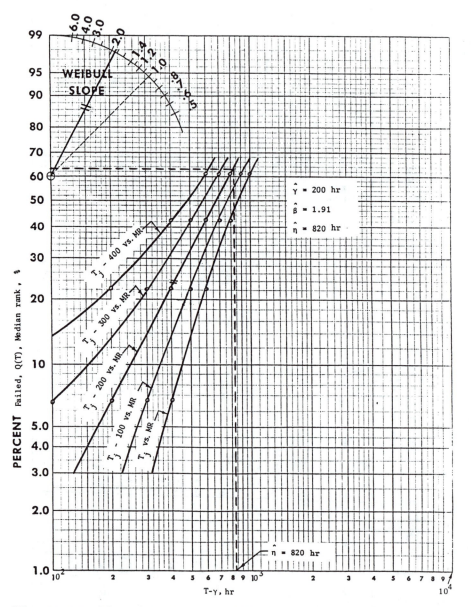

Fig. 12.16— Plot of the T_j versus MR, $(T_j - 400)$ versus MR, $(T_j - 100)$ versus MR, $(T_j - 200)$ versus MR, and $(T_j - 300)$ versus MR data in Table 12.7, to determine the parameters of the Weibull *pdf* representing the grouped data.

TABLE 12.7– Grouped data Weibull analysis to determine the *pdf*'s parameters.

1	2	3	4	
Class or group, j,	Number of failures in each class, f_i,	Total failures up to the end of each class, $r_j = \sum_{i=1}^{j} f_i$,	Time-to-failure at class end point, T_j, hr	
1	0	0	200	
2	2	2	$400 \leftarrow T_2$	
3	4	6	600	
4	5	11	800	
5	5	16	1,000	

5	6	7	8	9
Class end point minus 400 hr, $T_j - 400$, hr	Class end point minus 100 hr, $T_j - 100$, hr	Class end point minus 200 hr, $T_j - 200$, hr	Class end point minus 300 hr, $T_j - 300$, hr	Median rank, MR,%
–	100	–	–	0.00
0	300	200	100	6.62
200	500	400	300	22.38
400	700	600	500	42.11
600	900	800	700	61.84

during the first 200 hr of the test, as shown in Column 3. MR_2 is obtained by entering the tables in Appendix A with $N = 25$ and Order Number $r_2 = 2$, because two failures were observed up to $T_j = 400$ hr, as shown in Column 3, and reading under the column headed by 50.0, $MR_2 = 6.62\%$. MR_3 is obtained by entering the same table and column with Order Number $r_3 = 6$, as shown in Column 3, and reading off $MR_3 = 22.38\%$, etc., and $MR_5 = 61.84\%$.

Now the pairs of values T_j and MR_j are plotted, as shown in Fig. 12.16. These points fall on a curve which may be straightened by finding an appropriate γ.

Next fill in Column 5 with $(T_j - T_2)$, because $T_2 = 400$ hr is the earliest endpoint time-to-failure for these data. Plot the pairs of values $(T_j - T_2)$ versus the same MR_j as before, because the failure order number did not change. These points fall on a curve with an opposite curvature to the previous curve. These two curves comply with Case (a) in Fig. 12.12; consequently, there does exist a γ in the range of $0 < \gamma < 400$.

Next fill in Columns 6, 7, and 8, and plot these times versus the corresponding MR_j values. The plot $(T_j - 200)$ versus MR yields a good straight line, consequently $\hat{\gamma} = 200$ hr. The procedure discussed previously, and illustrated in Fig. 12.16, yields $\hat{\beta} = 1.91$ and $\hat{\eta} = 820$ hr using the straight line with $\hat{\gamma}=200$ hr.

12.4.5   WHEN IS THE WEIBULL NOT AN APPROPRIATE DISTRIBUTION FOR THE TIMES-TO-FAILURE DATA?

If the plot of the $(T_j - \hat{\gamma})$ values versus the corresponding MR_j values cannot be straightened, because the data points are too distant from the fitted curve, or the plot is S-shaped, or the lower end or the upper end of the data deviates substantially from the best straight line possible, then the Weibull *pdf* should be rejected as one that can represent the data adequately. Other distributions should then be fitted to the data, such as the lognormal, the gamma, the extreme value, the χ^2, F, etc. In that case the probability plotting paper of the chosen *pdf* should be used to plot the data and find its parameters. Of course other methods, such as the least squares, matching moments, and maximum likelihood estimators, may also be used to find the parameters of the chosen distribution. One major advantage of the probability plotting method is that the complete behavior of the data may be seen, whereas the other methods do not provide this possibility right away. Nevertheless, the value of the correlation coefficient of the least squares method, goodness-of-fit tests, and computer graphics, provide means

to determine which distribution is the best.

12.5 CONSTRUCTION OF THE WEIBULL PROBABILITY PAPER

The Weibull probability plotting paper is constructed as follows: Since

$$Q(T) = 1 - e^{-(\frac{T-\gamma}{\eta})^\beta}, \tag{12.26}$$

taking the natural logarithm of both sides twice yields **A PROGRAM FOR ESTIMATING THE PARAMETERS OF THE TWO-PARAMETER WEIBULL DISTRIBUTION USING MLE METHOD.**

$$\log_e\{\log_e[\frac{1}{1-Q(T)}]\} = \beta\log_e(T-\gamma) - \beta\log_e\eta, \tag{12.27}$$

which is of the form

$$y = \beta x + \text{ constant.} \tag{12.28}$$

Hence, if $\log_e\{\log_e[\frac{1}{1-Q(T)}]\}$ is plotted as a function of $\log_e(T-\gamma)$ on ordinary cartesian graph paper, a straight line will be obtained, according to Eqs. (12.27) and (12.28).

The Weibull probability plotting paper, by using a $\log_e$ scale along the abscissa and a $\log_e\log_e$ scale along the ordinate, makes it possible to plot $\hat{Q}(T)$, or MR, directly as a "linear" function of $(T-\gamma)$ without first having to compute $\log_e(T-\gamma)$ and $\log_e\{\log_e[\frac{1}{1-Q(T)}]\}$. On the ordinate scale, instead of putting the value of $\log_e\{\log_e[\frac{1}{1-Q(T)}]\}$ corresponding to the value of $\hat{Q}(T) = MR$ used, the value of $\hat{Q}(T)$ from which the value of $\log_e\{\log_e[\frac{1}{1-Q(T)}]\}$ is calculated is put, thus minimizing the labor of plotting the data to obtain the Weibull *pdf*'s parameters.

From Eq. (12.27) it may be seen that β, the shape parameter, is the slope of the Weibull line $Q(T)$ versus $(T-\gamma)$. It is for this reason that the *shape parameter* is also called the Weibull *slope*.

If by any chance there is no β scale on the Weibull paper on hand, once a straight-line fit to the data points is found, then an estimate of β may be calculated from

$$\hat{\beta} = \frac{\log_e\{\log_e[\frac{1}{1-Q(T)}]\}}{\log_e(T_j - \hat{\gamma}) - \log_e\hat{\eta}}. \tag{12.29}$$

Equation (12.29) is obtained by solving Eq. (12.27) for β. In Eq. (12.29) the value of a particular chosen time-to-failure, T_j, the value of the corresponding $\hat{Q}(T_j)$, and the values of $\hat{\gamma}$ and $\hat{\eta}$ which were found from the fitted Weibull straight line are substituted and $\hat{\beta}$ is calculated. To obtain a better estimate several sets of T_j and $\hat{Q}(T_j)$ values may be substituted into Eq. (12.29), and all of the calculated $\hat{\beta}$ values averaged.

EXAMPLE 12–14

Determine $\hat{\beta}$ for the data of Table 12.3 using Fig. 12.13 and Eq. (12.29).

SOLUTION TO EXAMPLE 12–14

Take $T_1 = 50$ hr and $T_2 = 130$ hr. From Fig. 12.13, $Q(T_1 = 50$ hr$) = 0.087$, $Q(T_2 = 130$ hr$) = 0.475$, $\hat{\gamma} = 30$ hr, and $\hat{\eta} = 144$ hr. Substitution of these values into Eq. (12.29) yields

$$\hat{\beta}_1 = \frac{\log_e[\log_e(\frac{1}{1-0.087})]}{\log_e(50-30) - \log_e 144} = \frac{-2.396683}{-1.974081},$$

or

$$\hat{\beta}_1 = 1.214,$$

and

$$\hat{\beta}_2 = \frac{\log_e[\log_e(\frac{1}{1-0.475})]}{\log_e(130-30) - \log_e 144} = \frac{-0.4395023}{-0.3646431},$$

or

$$\hat{\beta}_2 = 1.205.$$

Averaging the $\hat{\beta}_1$ and $\hat{\beta}_2$ values yields a better estimate of $\hat{\beta} = 1.2097$. This compares very favorably with the value of $\beta = 1.21$ obtained by the graphical method of Fig. 12.13.

The "Weibull slope" scale on the Weibull probability paper is so scaled as to measure the slope of a line drawn through the "origin," located at some specific point along the ordinate, parallel to the fitted $(T - \gamma)$ versus MR line.

The reason $\hat{\eta}$ is found by entering the ordinate at the 63.2% level is the following: For $T - \gamma = \eta$,

$$Q(T) = 1 - e^{-(\frac{\eta}{\eta})^\beta} = 1 - e^{-1} = 0.632, \quad \text{or} \quad 63.2\%.$$

Hence the value of $(T - \gamma)$ at which $Q(T) = MR = 63.2\%$ gives the estimate of η.

Estimates of β, γ, and η may also be obtained using the least squares, matching moments or the maximum likelihood estimators methods, given here in the order of increasing sophistication.

12.6 PROBABILITY OF PASSING A RELIABILITY TEST

One reliability test used is to put N identical units on test for T hours and allow only $r < N$ units to fail. The probability, P, of passing such a test is the probability failing, or

$$P(k \le r) = \sum_{k=0}^{r} \binom{N}{k} [Q(T)]^k \, [R(T)]^{N-k}, \tag{12.30}$$

where $R(T) = 1 - Q(T)$, and $Q(T)$ may be obtained from the Weibull plot, from the plot of any distribution that represents the data or the times-to-failure of the units involved, or by calculation using Eq. (12.10). Equation (12.30) is the cumulative binomial distribution. P may be calculated from Eq. (12.30) or obtained from tables of the cumulative binomial distribution by entering them with N, $Q(T)$, and r.

EXAMPLE 12–15

Ten identical units, whose *pdf* is given by Eq. (12.24) are put to a reliability test of $T = 50$ hr. To pass this test no more than one failure is allowed. What is the probability that these units will pass this test?

SOLUTION TO EXAMPLE 12–15

Here $N = 10$ and $r = 1$. Figure 12.13, entering with $(T - \gamma) = 50 - 30 = 20$ hr, yields $Q(T = 50 \text{ hr}) = 8.7\%$, or 0.087. Hence,

$$P(k \le 1) = \sum_{k=0}^{1} \binom{10}{k} (0.087)^k (1 - 0.087)^{10-k},$$

$$= \sum_{k=0}^{1} \frac{10!}{k!(10-k)!} (0.087)^k (1 - 0.087)^{10-k},$$

$$= \frac{10!}{0!(10-0)!} (0.087)^0 (0.913)^{10-0} + \frac{10!}{1!(10-1)!} (0.087)^1 (0.913)^{10-1},$$

$$= (0.913)^{10} + 10(0.087)(0.913)^9 = 0.4024 + 0.3835,$$

or

$$P(k \le 1) = 0.786, \text{ or } 78.6\%.$$

Consequently, the probability that these units will pass this test is 78.6%.

12.7   APPLICATIONS OF THE WEIBULL DISTRIBUTION

For applications of the Weibull *pdf* please see [3, p. 134; 6; 14 through 18; 19, p. 185; 20, p. 36; 25; 26; 29; 34, Vol. 1, p. 313].

12.8   PHENOMENOLOGICAL CONSIDERATIONS FOR USING THE WEIBULL DISTRIBUTION

For phenomenological considerations for using the Weibull *pdf* please see [5, pp. 26-29; 3, p. 116; 14; 21; 22; 23; 30; 34, Vol. 1, pp. 313-315].

12.9   ANALYSIS OF PROBABILITY PLOTS AND CHOOSING THE RIGHT PROBABILITY PLOTTING PAPER

For a detailed analysis of probability plots and choosing the right probability plotting paper please see [28; 34, Vol. 1, pp. 315-323].

12.10   THE LEAST SQUARES, MATCHING MOMENTS AND MAXIMUM LIKELIHOOD ESTIMATORS METHODS OF DETERMINING THE PARAMETERS OF THE TWO-PARAMETER WEIBULL DISTRIBUTION

12.10.1   INTRODUCTION

In addition to direct parameters calculation, whenever this is possible, and probability plotting, the following methods may be used to determine the parameters of the distribution representing the data: (1) Least Squares. (2) Matching Moments. (3) Maximum Likelihood Estimators. These three methods of determining the parameters of the *two-parameter Weibull distribution* are the subject of this section.

12.10.2 PARAMETER ESTIMATION BY THE METHOD OF LEAST SQUARES FOR THE TWO-PARAMETER WEIBULL DISTRIBUTION

The *cdf* of the Weibull distribution is given by

$$F(T) = Q(T) = 1 - e^{-(\frac{T}{\eta})^{\beta}}. \tag{12.31}$$

Taking the natural logarithm of both sides twice yields

$$\log_e \left\{ \log_e \left[\frac{1}{1 - Q(T)} \right] \right\} = \beta \log_e T - \beta \log_e \eta, \tag{12.32}$$

which is of the form

$$y = a + bx. \tag{12.33}$$

The parameters of Eq. (12.33) are given by [35, p. 250].

$$b = \frac{N \sum\limits_{i=1}^{N} x_i y_i - \sum\limits_{i=1}^{N} x_i \sum\limits_{i=1}^{N} y_i}{N \sum\limits_{i=1}^{N} x_i^2 - (\sum\limits_{i=1}^{N} x_i)^2}, \tag{12.34}$$

and

$$a = \bar{y} - b\bar{x}, \tag{12.35}$$

where

$$\bar{y} = \frac{1}{N} \sum_{i=1}^{N} y_i,$$

and

$$\bar{x} = \frac{1}{N} \sum_{n=1}^{N} x_i.$$

For Eq. (12.33) the variables and parameters are as follows:

$$y_i = \log_e \left\{ \log_e \left[\frac{1}{1 - Q(T_i)} \right] \right\}, \tag{12.36}$$

where the median ranks are used for the $Q(T_i)$,

$$x_i = \log_e T_i,\qquad\qquad\qquad(12.37)$$

where

$$T_i = \text{\textit{ith} time to failure,}$$

$$b = \beta,\qquad\qquad\qquad(12.38)$$

and

$$a = -\beta \,\log_e \eta,$$

or

$$\eta = e^{-\frac{a}{b}}.\qquad\qquad\qquad(12.39)$$

EXAMPLE 12–16

Determine the estimates of the parameters of the *two-parameter Weibull distribution* using the least squares method for the data in Table 12.8.

SOLUTION TO EXAMPLE 12–16

Using the data in Table 12.8 the required values of y and x are given by

$$y_1 = \log_e \left(\log_e \left[\frac{1}{1 - 0.06697}\right]\right) = \log_e[\log_e(1.0717769)],$$
$$= \log_e(0.06932),$$

or

$$y_1 = -2.6690518,$$

and

$$x_1 = \log_e 16 = 2.7725887.$$

Table 12.9 contains the data needed to determine parameters a and b. Parameter b, from Eq. (12.34) and using the nonrounded 7-decimal place values, is

$$b = \frac{(10)(-14.4302385) - (45.4763683)(-5.2408075)}{(10)(214.663446) - (45.4763683)^2},$$

TABLE 12.8– Times-to-failure data for Example 12–16 from a sample of $N = 10$ units all tested to failure.

Failure number, j	Time to failure, T_j, hr	Median rank, %
	1	2	3
1	16	6.697
2	34	16.226
3	53	25.857
4	75	35.510
5	93	45.169
6	120	54.831
7	150	64.490
8	191	74.143
9	240	83.774
10	339	93.303

or

$$\beta = b = \frac{94.0305}{78.5344} = 1.20.$$

Since

$$\bar{x} = \frac{45.4763683}{10} = 4.54763683,$$

and

$$\bar{y} = \frac{-5.2408075}{10} = -0.52408075;$$

parameter a, from Eq. (12.35), is

$$a = -0.52408075 - 1.20 \times 4.54763683,$$

or

$$a = -5.97.$$

The estimated Weibull parameters, from Eqs. (12.38) and (12.39), are

$$\hat{\beta} = b = 1.20,$$

TABLE 12.9– Data calculations required for the determination of the least squares parameters for Example 13-16.

j	x	y	xy	x^2	y^2
1	2.7726	-2.6690	-7.4002	7.6872	7.1238
2	3.5264	-1.7313	-6.1051	12.4352	2.9973
3	3.9703	-1.2067	-4.7911	15.7632	1.4562
4	4.3175	-0.8241	-3.5579	18.6407	0.6791
5	4.5326	-0.5093	-2.3083	20.5444	0.2594
6	4.7875	-0.2297	-1.0999	22.9201	0.0528
7	5.0106	0.0348	0.1742	25.1065	0.0012
8	5.2523	0.3020	1.5863	27.5864	0.0912
9	5.4806	0.5980	3.2775	30.0374	0.3576
10	5.8260	0.9946	5.7942	33.9423	0.9891
	$\sum = 45.4764$	$\sum = -5.2408$	$\sum = -14.4302$	$\sum = 214.0077$	$\sum = 14.0077$

and

$$\hat{\eta} = e^{-\frac{a}{b}} = e^{\frac{5.97}{1.20}} = e^{4.985},$$

or

$$\hat{\eta} = 146.2 \text{ hr.}$$

To determine whether the Weibull *pdf* represents the data accept-ably well, when the parameters are determined using the least squares method, it is desirable to establish the goodness of fit by studying the relative value of the sample correlation coefficient, ρ. Values of $\rho \geq 0.75$ are desirable, however values of $\rho \geq 0.90$ are more desirable. The sample correlation coefficient, $\hat{\rho}$, is given by

$$\hat{\rho} = \frac{\sum\limits_{i=1}^{10} x_i y_i - \left(\sum\limits_{i=1}^{10} x_i \sum\limits_{i=1}^{10} y_i \right)/10}{\sqrt{\left[\sum\limits_{i=1}^{10} x_i^2 - (\sum\limits_{i=1}^{10} x_i)^2/10 \right] \left[\sum\limits_{i=1}^{10} y_i^2 - (\sum\limits_{i=1}^{10} y_i)^2/10 \right]}},$$

$$\hat{\rho} = \frac{-14.4302385 - (45.4763683)(-5.2408075)/10}{\sqrt{\left[214.663446 - \frac{(45.4763683)^2}{10} \right] \left[14.007955 - \frac{(-5.2408075)^2}{10} \right]}}$$

or

$$\hat{\rho} = 0.9998703.$$

Since $\hat{\rho}$ is close to 1, it can be concluded that the Weibull *pdf* represents the data in Table 12.8 very well.

EXAMPLE 12–17

Given the times-to-failure data in Table 12.10, estimate the param-eters of the *two-parameter Weibull distribution* that represents these data using the least squares method. Also determine the sample cor-relation coefficient, $\hat{\rho}$, and comment on the goodness of fit.

SOLUTION TO EXAMPLE 12–17

The computer program given in Table 12.11 was used to estimate the parameters. It utilizes the previously given equations. The results are

$$\hat{\beta} = 1.4204411, \text{ or } \hat{\beta} = 1.420,$$

TABLE 12.10– Times-to-failure data for Example 12–17 from a sample of $N = 100$ units all tested to failure.

Failure number, j	Time to failure, T_j, hr	Failure number, j	Time to failure, T_j, hr	Failure number, j	Time to failure, T_j, hr
1	360	34	2,980	67	5,130
2	380	35	3,040	68	5,520
3	420	36	3,140	69	5,710
4	490	37	3,160	70	5,750
5	570	38	3,180	71	5,850
6	620	39	3,190	72	5,860
7	670	40	3,380	73	6,160
8	780	41	3,470	74	6,250
9	880	42	3,490	75	6,290
10	1,030	43	3,640	76	6,360
11	1,200	44	3,890	77	6,550
12	1,210	45	3,910	78	7,100
13	1,380	46	3,960	79	7,390
14	1,480	47	3,980	80	7,550
15	1,560	48	4,000	81	7,890
16	1,590	49	4,150	82	8,380
17	1,620	50	4,290	83	8,410
18	1,700	51	4,300	84	8,460
19	1,760	52	4,430	85	8,540
20	1,770	53	4,450	86	8,880
21	1,820	54	4,550	87	9,250
22	1,920	55	4,580	88	9,630
23	2,140	56	4,610	89	9,680
24	2,250	57	4,670	90	10,440
25	2,290	58	4,670	91	10,870
26	2,410	59	4,730	92	11,840
27	2,560	60	4,740	93	12,230
28	2,670	61	4,750	94	12,340
29	2,830	62	4,830	95	12,420
30	2,840	63	4,880	96	12,890
31	2,850	64	5,020	97	14,650
32	2,890	65	5,090	98	14,850
33	2,950	66	5,120	99	15,120
				100	16,070

TABLE 12.11– A program for estimating the parameters
of the two-parameter Weibull distribution
using the least squares method.

```
          REAL Y(100),X(100),T(100)
          READ(1,100) T
100       FORMAT(5F7.1)
          SUM1=0.0
          SUM2=0.0
          SUM3=0.0
          SUM4=0.0
          DO 110 I=1,100
          Y(I)=ALOG(ALOG(1/1-((I-.3)/(100.0+.4))))
          X(I)=ALOG(T(I))
          SUM1=SUM1+Y(I)
          SUM2=SUM2+X(I)
          SUM3=SUM3+X(I)**2
          SUM4=SUM4+X(I)*Y(I)
110       CONTINUE
          B=(100*SUM4-SUM1*SUM2)/(100*SUM3-SUM2**2)
          A=(SUM1/100)-B*(SUM2/100)
          BATE=B
          ATE=EXP(-(A/B))
          WRITE(1,200) BATE,ATE
200       FORMAT(5X,'BATE=',F13.7/
         +        5X,'ATE =',F13.7)
          STOP
          END
```

and

$$\hat{\eta} = 5,482.8013 \text{ hr, or } \hat{\eta} = 5,482.8 \text{ hr.}$$

The calculation of $\hat{\rho}$ requires the following values:

$$\sum_{i=1}^{100} x_i \quad = \quad 820.9173504,$$

$$\sum_{i=1}^{100} y_i \quad = \quad -56.8457579,$$

$$\sum_{i=1}^{100} x_i y_i \quad = \quad -359.6178096,$$

$$\sum_{i=1}^{100} x_i^2 \quad = \quad 6814.1668909,$$

and

$$\sum_{i=1}^{100} y_i^2 \quad = \quad 186.1396540.$$

Then,

$$\hat{\rho} = \frac{-359.6178096 - (820.9173504)(-56.8457579)/100}{\sqrt{\left[6814.1668909 - \frac{(820.9173504)^2}{100}\right]\left[186.1396540 - \frac{(-56.8457579)^2}{100}\right]}}$$

or

$$\hat{\rho} = 0.9925894.$$

Since the value of $\hat{\rho}$ is close to 1, the Weibull distribution represents the data of Table 12.10 very well.

For additional coverage of this subject see Chapter 18.

12.10.3   PARAMETER ESTIMATION BY THE METHOD OF MATCHING MOMENTS FOR THE WEIBULL DISTRIBUTION

The method of matching moments, or of moments, equates the sample moments and the population moments, yielding as many equations as there are parameters to be found. Solving these equations simultaneously yields the sought parameters.

For the Weibull *pdf* the mean and the variance are given by

$$\overline{T} = \eta \, \Gamma(\frac{1}{\beta} + 1), \tag{12.40}$$

and

$$\sigma_T^2 = \eta^2 \Big[\Gamma(\frac{2}{\beta} + 1) - \Gamma^2(\frac{1}{\beta} + 1) \Big], \tag{12.41}$$

respectively.

Using the raw data, estimates of $\overline{T}$ and σ_T may be calculated from

$$\widehat{\overline{T}} = \Big(\sum_{i=1}^{N} T_i \Big)/N, \tag{12.42}$$

and

$$\hat{\sigma}_T^2 = \frac{\displaystyle\sum_{i=1}^{N} T_i^2 - N(\widehat{\overline{T}})^2}{N - 1}, \tag{12.43}$$

respectively.

Matching the moments; i.e., equating Eqs. (12.42) and (12.40), and Eqs. (12.43) and (12.41) yields

$$\widehat{\overline{T}} = \eta \, \Gamma(\frac{1}{\beta} + 1), \tag{12.44}$$

and

$$\hat{\sigma}_T^2 = \eta^2 [\Gamma(\frac{2}{\beta} + 1) - \Gamma^2(\frac{1}{\beta} + 1)]. \tag{12.45}$$

Equations (12.44) and (12.45) may now be solved simultaneously to determine β and η. However, an easier procedure may be developed using the concept of the coefficient of variation. The coefficient of variation, COV, is defined as the ratio of the standard deviation to the mean, or

$$\text{COV} = \frac{\sigma_T}{\overline{T}}. \tag{12.46}$$

Using Eqs. (12.40) and (12.41) the coefficient of variation becomes

$$\text{COV} = \frac{\eta[\Gamma(\frac{2}{\beta}+1) - \Gamma^2(\frac{1}{\beta}+1)]^{1/2}}{\eta\Gamma(\frac{1}{\beta}+1)},$$

$$= \left[\frac{\Gamma(\frac{2}{\beta}+1) - \Gamma^2(\frac{1}{\beta}+1)]}{\Gamma^2(\frac{1}{\beta}+1)}\right]^{1/2},$$

$$= \left[\frac{\Gamma(\frac{2}{\beta}+1)}{\Gamma^2(\frac{1}{\beta}+1)} - \frac{\Gamma^2(\frac{1}{\beta}+1)}{\Gamma^2(\frac{1}{\beta}+1)}\right]^{1/2},$$

or

$$\text{COV} = \left[\frac{\Gamma(\frac{2}{\beta}+1)}{\Gamma^2(\frac{1}{\beta}+1)} - 1\right]^{1/2}. \tag{12.47}$$

In Eq. (12.47) the left side can be calculated from Eq. (12.46) in conjunction with Eqs. (12.42) and (12.43), and then β can be solved for by an iterative process.

A more convenient method of finding β would be to use Table 12.12 [36, p. 45], where values of COV for typical values of β are given. Thus entering Table 12.12 with the calculated value of COV yields the value of β directly, or by interpolation.

Another way of finding β is to plot Eq. (12.47) as shown in Fig. 12.17, and through linear regression of the linear portion of Fig. 12.17 to find the best equation representing it. Such an equation is [31, p. 22-26]

$$\hat{\beta} = \text{COV}^{-1.08}, \tag{12.48}$$

which, as may be seen in Fig. 12.17, is valid in the range of

$$0.02 \le \text{COV} \le 1.5.$$

After $\hat{\beta}$ is determined from Eq. (12.48), Eq. (12.40) may be used to determine $\hat{\eta}$ from

$$\hat{\eta} = \frac{\hat{\bar{T}}}{\Gamma(\frac{1}{\beta}+1)}. \tag{12.49}$$

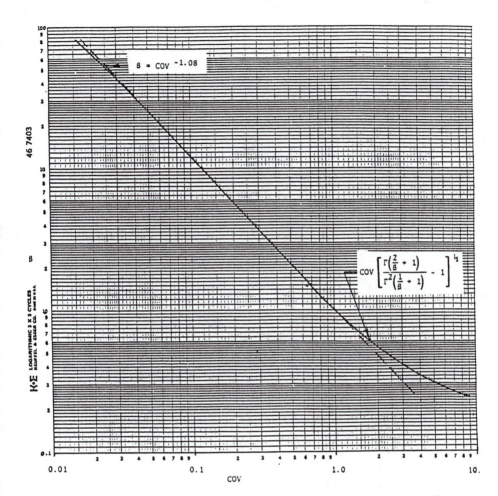

Fig. 12.17– Plot of β versus the coefficient of variation for the two-parameter Weibull distribution.

TABLE 12.12– The two-parameter Weibull *pdf's* shape parameter, β, corresponding to the coefficient of variation, COV, of the life test data.

COV	β	COV	β	COV	β
429.8314	0.10	0.7238	1.40	0.3994	2.70
47.0366	0.15	0.7006	1.45	0.3929	2.75
15.8430	0.20	0.6790	1.50	0.3866	2.80
8.3066	0.25	0.6588	1.55	0.3805	2.85
5.4077	0.30	0.6399	1.60	0.3747	2.90
3.9721	0.35	0.6222	1.65	0.3690	2.95
3.1409	0.40	0.6055	1.70	0.3634	3.00
2.6064	0.45	0.5897	1.75	0.3581	3.05
2.2361	0.50	0.5749	1.80	0.3529	3.10
1.9650	0.55	0.5608	1.85	0.3479	3.15
1.7581	0.60	0.5474	1.90	0.3430	3.20
1.5948	0.65	0.5348	1.95	0.3383	3.25
1.4624	0.70	0.5227	2.00	0.3336	3.30
1.3529	0.75	0.5112	2.05	0.3292	3.35
1.2605	0.80	0.5003	2.10	0.3248	3.40
1.1815	0.85	0.4898	2.15	0.3206	3.45
1.1130	0.90	0.4798	2.20	0.3165	3.50
1.0530	0.95	0.4703	2.25	0.3124	3.55
1.0000	1.00	0.4611	2.30	0.3085	3.60
0.9527	1.05	0.4523	2.35	0.3047	3.65
0.9102	1.10	0.4438	2.40	0.3010	3.70
0.8718	1.15	0.4341	2.45	0.2974	3.75
0.8369	1.20	0.4279	2.50	0.2938	3.80
0.8050	1.25	0.4204	2.55	0.2904	3.85
0.7757	1.30	0.4131	2.60	0.2870	3.90
0.7487	1.35	0.4062	2.65	0.2838	3.95

EXAMPLE 12–18

Rework Example 12–17 using the method of matching moments for the data in Table 12.10.

SOLUTION TO EXAMPLE 12–18

The sample mean for the data in Table 12.10, from Eq. (12.42) with $\sum_{i=1}^{n} T_i = 500,380$ hr and $N = 100$, is

$$\widehat{\overline{T}} = \frac{500,380}{100},$$

or

$$\widehat{\overline{T}} = 5,003.80 \text{ hr.}$$

The sample variance is given by Eq. (12.43), where

$$\sum_{i=1}^{n} T_i^2 = 3,849.5364 \times 10^6 \text{ hr}^2,$$

and

$$\hat{\sigma}_T^2 = \frac{(3,849.5364 \times 10^6) - 100(5,003.80)^2}{100 - 1},$$

$$\hat{\sigma}_T^2 = \frac{134,573.5}{99} = 13.593282 \times 10^6 \text{ hr}^2,$$

or

$$\hat{\sigma}_T = 3,686.9069 \text{ hr.}$$

Then, from Eq. (12.46),

$$\text{COV} = \frac{3,686.9069}{5,003.80} = 0.7368.$$

Entering Table 12.12 with this value yields the estimate of the shape parameter, or

$$\hat{\beta} = 1.374.$$

The scale parameter is estimated from Eq. (12.49), or

$$\hat{\eta} = \frac{5,003.80}{\Gamma(\frac{1}{1.374} + 1)} = \frac{5,003.80}{\Gamma(1.73)},$$

where

$$\Gamma(1.73) = 0.91467.$$

Therefore,

$$\hat{\eta} = \frac{5,003.80}{0.91467}$$

or

$$\hat{\eta} = 5,470.6 \text{ hr.}$$

EXAMPLE 12–19

Work out Example 12–18 using Eq. (12.48).

SOLUTION TO EXAMPLE 12–19

From the solution of Example 12–18

$$\text{COV} = 0.7368.$$

Therefore, from Eq. (12.48),

$$\hat{\beta} = 0.7368^{-1.08},$$

or

$$\hat{\beta} = 1.391.$$

Since

$$\Gamma\left(\frac{1}{1.391} + 1\right) = \Gamma(1.72) = 0.91258;$$

then, from Eq. 12.49,

$$\hat{\eta} = \frac{5,003.80}{0.91258} = 5,483 \text{ hr.}$$

These parameter estimates compare with the following values obtained in Examples 12-17 and 12-18 using the least squares and the matching

moments methods, respectively: $\hat{\beta} = 1.420$ and $\hat{\eta} = 5,482.8$ hr; and $\hat{\beta} = 1.374$ and $\hat{\eta} = 5,470.6$ hr.

EXAMPLE 12–20

Rework the times-to-failure data in Table 12.8 and estimate the parameters of the two-parameter Weibull *pdf* using the method of matching moments, Table 12.12 and Eq. (12.48).

SOLUTION TO EXAMPLE 12–20

The sample mean, from Eq. (12.42), is

$$\overline{T} = \frac{1,311}{10},$$

or

$$\overline{T} = 131.1 \text{ hr.}$$

The sample variance, from Eq. (12.43), is

$$\hat{\sigma}_T^2 = 10,280.5444 \text{ hr}^2;$$

therefore, the standard deviation is

$$\hat{\sigma}_T = 101.393 \text{ hr.}$$

Then,

$$\text{COV} = \frac{101.393}{131.1} = 0.7734.$$

Entering Table 12.12 with $\text{COV} = 0.7734$ yields

$$\hat{\beta} = 1.31,$$

and the estimate of β from Eq. (12.48) is

$$\hat{\beta} = \text{COV}^{-1.08} = (0.7734)^{-1.08},$$

or

$$\hat{\beta} = 1.32$$

The estimate of the scale parameter, from Eq. (12.49), is

$$\hat{\eta} = \frac{131.1}{\Gamma(\frac{1}{1.32} + 1)},$$
$$= \frac{131.1}{\Gamma(1.76)},$$

where

$$\Gamma(1.76) = 0.92137.$$

Consequently,

$$\hat{\eta} = \frac{131.1}{0.92137} = 142.3 \text{ hr.}$$

These parameter estimates compare with the following values obtained in Example 12–16 using the least squares method:

$$\hat{\beta} = 1.20 \text{ and } \hat{\eta} = 146.2 \text{ hr.}$$

For additional coverage of this subject see Chapter 18.

12.10.4   PARAMETER ESTIMATION BY THE MAXIMUM LIKELIHOOD METHOD FOR THE TWO-PARAMETER WEIBULL DISTRIBUTION

The Weibull probability density function is

$$f(T) = \left(\frac{\beta}{\eta}\right) \left(\frac{T}{\eta}\right)^{\beta-1} \exp\left[-\left(\frac{T}{\eta}\right)^{\beta}\right]. \tag{12.50}$$

In a failure terminated test of N units, terminated at the rth failure, if the test data are

$$T_1 \leq T_2 \leq T_3 \leq \cdots \leq T_r;$$

then, the likelihood function is [19, pp. 185-191; 37, pp. 356-386]

$$L(\beta, \eta) = f(T_1, T_2, \cdots T_r),$$

$$= \frac{N!}{(N-r)!} \left[\prod_{i=1}^{r} f(T_i) \right] \left[1 - F(T_r) \right]^{N-r},$$

or

$$L(\beta, \eta) = \frac{N!}{(N-r)!} \left\{ \prod_{i=1}^{r} \left\{ \frac{\beta}{\eta} \left(\frac{T_i}{\eta} \right)^{\beta-1} \exp\left[-\left(\frac{T_i}{\eta} \right)^{\beta} \right] \right\} \right\}$$

$$\cdot \left\{ \exp\left[-\left(\frac{T_r}{\eta} \right)^{\beta} \right] \right\}^{N-r}.$$

Rearranging this equation yields

$$L(\beta, \eta) = \frac{N!}{(N-r)!} \frac{\beta^r}{\eta^{\beta r}} \left[\prod_{i=1}^{r} (T_i)^{\beta-1} \right] \exp\left[-\frac{1}{\eta^\beta} \right.$$

$$\left. \cdot \left(\sum_{i=1}^{r} T_i^\beta + (N-r) T_r^\beta \right) \right]. \tag{12.51}$$

Taking the natural logarithm of both sides of Eq. (12.51) yields

$$\log_e L(\beta, \eta) = \log_e \left[\frac{N!}{(N-r)!} \right] + r \log_e \beta - r\beta \log_e \eta$$

$$+ (\beta - 1) \sum_{i=1}^{r} \log_e (T_i)$$

$$- \frac{1}{\eta^\beta} \left[\sum_{i=1}^{r} T_i^\beta + (N-r) T_r^\beta \right]. \tag{12.52}$$

Taking the partial derivatives of Eq. (12.52) with respect to η and β and setting them equal to zero, respectively, yields

$$\frac{\partial \log_e L}{\partial \eta} = -\frac{r\beta}{\eta} + \frac{\beta}{\eta^{\beta+1}} \left[\sum_{i=1}^{r} T_i^\beta + (N-r) T_r^\beta \right] = 0, \tag{12.53}$$

$$\frac{\partial \log_e L}{\partial \beta} = \frac{r}{\beta} - r \log_e \eta + \sum_{i=1}^{r} \log_e T_i$$

$$- \frac{1}{\eta^\beta} \Big[\sum_{i=1}^{r} T_i^\beta \, \log_e T_i + (N - r) T_r^\beta \, \log_e T_r,$$

$$- \log_e \eta \Big(\sum_{i=1}^{r} T_i^\beta + (N - r) T_r^\beta \Big) \Big] = 0. \tag{12.54}$$

Rearranging Eq. (13.53) yields

$$\eta^\beta = \frac{1}{r} \Big[\sum_{i=1}^{r} T_i^\beta + (N - r) T_r^\beta \Big]. \tag{12.55}$$

Rearranging Eq. (12.54), and substituting Eq. (12.55) into it, yields

$$\frac{\sum_{i=1}^{r} T_i^\beta \, \log_e T_i + (N - r) T_r^\beta \, \log_e T_r}{\sum_{i=1}^{r} T_i^\beta + (N - r) T_r^\beta} - \frac{1}{\beta} - \frac{1}{r} \sum_{i=1}^{r} \log_e T_i = 0. \tag{12.56}$$

Then, the Maximum Likelihood Estimators (MLE) of β and η for the failure terminated test are the solutions of Eqs. (12.55) and (12.56).

To solve these two equations, first solve Eq. (12.56) for $\hat{\beta}$, using the Newton-Raphson iteration method, or other numerical methods, then substitute $\hat{\beta}$ into Eq. (12.55) and solve for $\hat{\eta}$. The use of a computer program is recommended.

In the case of the time terminated test, the MLE equations are identical to Eqs. (12.55) and (12.56) except that T_r should be replaced by T_o which is the time at which the test is terminated.

In the case of the complete life test of N units, replacing r in Eqs. (12.55) and (12.56) by N yields

$$\eta^\beta = \frac{1}{N} \sum_{i=1}^{N} T_i^\beta, \tag{12.57}$$

and

$$\frac{\sum_{i=1}^{N} T_i^\beta \, \log_e T_i}{\sum_{i=1}^{N} T_i^\beta} - \frac{1}{\beta} - \frac{1}{N} \sum_{i=1}^{N} \log_e T_i = 0. \tag{12.58}$$

TABLE 12.13– Unbiasing factors, $B(N)$, for the MLE of β as a function of sample size, N.

N	$B(N)$	N	$B(N)$	N	$B(N)$	N	$B(N)$
5	0.669	18	0.923	42	0.968	66	0.980
6	0.752	20	0.931	44	0.970	68	0.981
7	0.792	22	0.938	46	0.971	70	0.981
8	0.820	24	0.943	48	0.972	72	0.982
9	0.842	26	0.947	50	0.973	74	0.982
10	0.859	28	0.951	52	0.974	76	0.983
11	0.872	30	0.955	54	0.975	78	0.983
12	0.883	32	0.958	56	0.976	80	0.984
13	0.893	34	0.960	58	0.977	85	0.985
14	0.901	36	0.962	60	0.978	90	0.986
15	0.908	38	0.964	62	0.979	100	0.987
16	0.914	40	0.966	64	0.980	120	0.990

The MLE of β, obtained by solving Eq. (12.56), is quite biased for small sample sizes, N. But it has been proven [38, pp. 445-460] that $\hat{\beta}/\beta$ is independent of β; therefore,

$$E\left[\frac{\hat{\beta}}{E(\hat{\beta}/\beta)}\right] = \beta.$$

Thus $\left[B(N)\right]\hat{\beta}$ is the unbiased estimate of β, where

$$B(N) = \frac{1}{E(\hat{\beta}/\beta)}.$$

The values of $B(N)$ can be found in Table 12.13 for a given sample of size N.

EXAMPLE 12–21

Rework Example 12–16 using the MLE method.

SOLUTION TO EXAMPLE 12–21

A series of iterations is required to find $\hat{\beta}$ and $\hat{\eta}$. As an initial guess, let

$$\hat{\beta}_1 = 1.20,$$

and choose an iteration error of $\epsilon = 0.01$. It means that if

$$\epsilon = |\hat{\beta}_{n-1} - \hat{\beta}_n| \leq 0.01,$$

the iteration is stopped at the nth iteration, then $\hat{\beta} = \beta_n$. From Eq. (12.56), with $r = N$,

$$\hat{\beta} = \cfrac{1}{\cfrac{\sum\limits_{i=1}^{N} T_i^{\beta} \, \log_e T_i}{\sum\limits_{i=1}^{N} T_i^{\beta}} - \cfrac{1}{N} \sum\limits_{i=1}^{N} \log_e T_i}, \qquad (12.59)$$

Substituting $\beta = 1.20$ into the right side of Eq. (12.59) yields

$$\hat{\beta}_2 = \cfrac{1}{\cfrac{19{,}297.93}{3{,}695.18} - \cfrac{45.48}{10}},$$

or

$$\beta_2 = 1.49.$$

To facilitate the calculations using Eq. (12.59), Table 12.14 may be constructed. Since $\epsilon = |\hat{\beta}_2 - \hat{\beta}_1| = |1.49 - 1.20| = 0.28 > 0.01$, a second iteration should be performed. For this example eight iterations were performed. The results are given in Table 12.15, yielding

$$\hat{\beta} = 1.37.$$

Substituting the value of $\hat{\beta} = 1.37$ into Eq. (12.57) yields

$$\hat{\eta} = \left(\frac{1}{N} \sum_{i=1}^{N} T_i^{\hat{\beta}} \right)^{1/\hat{\beta}},$$

or

$$\hat{\eta} = 143.52 \text{ hr.}$$

TABLE 12.14– Calculation table for the MLE of the data in Example 12–16 for Example 12–21.

Order number of the data, i	T_i	T_i^β	$\log_e T_i$	$T_i^\beta \log_e T_i$
1	16	27.86	2.77	77.17
2	34	68.83	3.53	242.97
3	53	117.26	3.97	465.52
4	75	177.86	4.32	768.36
5	93	230.24	4.53	1,042.99
6	120	312.62	4.79	1,497.45
7	150	408.61	5.01	2,063.48
8	191	546.06	5.25	2,866.82
9	240	718.21	5.48	3,935.79
10	329	1,087.03	5.83	6,337.38
$\sum$		3,695.18	45.48	19,297.93

TABLE 12.15– Results of the iterations to find β, using the MLE method, for Example 12–21.

Number of iteration, n	β
1	1.48
2	1.31
3	1.41
4	1.34
5	1.38
6	1.36
7	1.38
8	1.37

From Table 12.13, for $N = 10$, $B(10) = 0.859$. Then, the unbiased estimate of β, $\hat{\beta}'$, is

$$\hat{\beta}' = 0.859 \times 1.37,$$

or

$$\hat{\beta}' = 1.18.$$

Substituting this value into Eq. (12.57) yields

$$\hat{\eta}' = 137.25 \text{ hr.}$$

These compare with the least squares method values of

$$\hat{\beta} = 1.20 \text{ and } \hat{\eta} = 146.2 \text{ hr,}$$

and with the matching moments method values of

$$\hat{\beta} = 1.32 \text{ and } \hat{\eta} = 142.3 \text{ hr.}$$

EXAMPLE 12–22

Rework Example 12–17 using the MLE method.

SOLUTION TO EXAMPLE 12–22

For a large data base, it is difficult and time consuming to perform the iterations; consequently, a computer program, like the one given

TABLE 12.16– A program for estimating the of the two-parameter Weibull distribution using the MLE method.

```
          REAL T1(100)
          READ(1,150) T1
150       FORMAT(5F7.1)
          BATE0=1.2
50        SUM1=0.0
          SUM2=0.0
          SUM3=0.0
          DO 100 I=1,100
          TB=T1(I)**BATE0
          PLNT=ALOG(T1(I))
          TL=TB*KLNT
          SUM1=SUM1+TB
          SUM2=SUM2+TL
          SUM3=SUM3+RLNT
100       CONTINUE
          BATE=1.0/(SUM2/SUM1-SUM3/100.0)
          WRITE(1,200) BATE
200       FORMAT(5X,F13.7)
          IF(ABS(BATE-BATE0).LE.1.0E-5) GOTO 300
          BATE0=BATE
          GOTO 50
300       SUM4=0.0
          DO 400 J=1,100
          TAT=T1(J)**BATE
          SUM4=SUM4+TAT
400       CONTINUE
          ATE=(SUM4/100.0)**(1.0/BATE)
          WRITE(1,450) ATE
450       FORMAT(5X,F13.7)
          STOP
          END
```

in Table 12.16 should be used. For this example 27 iterations were performed to fulfill the requirements that $\epsilon \leq 10^{-5}$.

Table 12.17 gives the iteration results of β. Thus, the estimate of β is

$$\hat{\beta} = 1.3764, \text{ or } \hat{\beta} = 1.38.$$

Substituting the value of $\hat{\beta} = 1.3764$ into Eq. (12.57), yields

$$\hat{\eta} = \left(\frac{1}{N} \sum_{i=1}^{N} T_i^{\hat{\beta}} \right)^{1/\hat{\beta}},$$

or

$$\hat{\eta} = \left[\frac{1}{100}(13,966,431.2) \right]^{\frac{1}{1.3764}} = 5,471.1 \text{ hr.}$$

From Table 12.13, for $N = 100$, $\beta(100) = 0.987$. Then, the unbiased estimate of β, $\hat{\beta}'$ is

$$\hat{\beta}' = 0.987 \times 1.3764,$$

or

$$\hat{\beta}' = 1.3585.$$

Substituting this value into Eq. (12.57) yields

$$\hat{\eta}' = \left[\frac{1}{100}(11,907,421.5) \right]^{\frac{1}{1.3585}},$$

or

$$\hat{\eta}' = 5,449.1 \text{ hr.}$$

12.10.5 SUMMARY OF RESULTS

Tables 12.18 and 12.19 summarize the results for the data given in Example 12–16 where $N = 10$ and in Example 12–17 where $N = 100$. It may be seen from these two tables that the MLE method yields the lowest parameter values for both sample sizes. Comparing the Least Squares with the Matching Moments method, it may be seen that the former yields the lower β estimate but the higher η estimate for $N = 10$, but both the opposite for $N = 100$.

The β and η estimates for $N = 100$ are both the higher for the Least Squares method, and the second higher values for both β and η are the Matching Moments method.

For more detailed coverage of methods of parameter estimation see Chapter 18.

TABLE 12.17– Iteration results of β using the program of Table 12.16.

k	β_k
1	1.5113964
2	1.2943841
3	1.4346024
4	1.3391437
5	1.4019591
6	1.3596610
7	1.3877141
8	1.3689178
9	1.3814257
10	1.3730640
11	1.3786370
12	1.3749157
13	1.3773974
14	1.3757407
15	1.3768456
16	1.3761084
17	1.3766000
18	1.3762723
19	1.3764904
20	1.3763448
21	1.3764421
22	1.3763771
23	1.3764206
24	1.3763917
25	1.3764105
26	1.3763985
27	1.3764060

TABLE 12.18– Summary of the results for the parameters of the two-parameter Weibull distribution for the data given in Example 12–22 for sample size 10.

Method	Estimated parameters	
	β	η, hr
Least Squares	1.20	146.2
Matching Moments	1.32	142.3
MLE	1.18	137.3

TABLE 12.19– Summary of the results for the parameters of the two-parameter Weibull distribution for the data given in Example 12–22 for sample size 100.

Method	Estimated parameters	
	β	η, hr
Least Squares	1.420	5,482.8
Matching Moments	1.374	5,470.6
MLE	1.359	5,449.1

12.11 THE LINEAR ESTIMATION OF THE PARAMETERS OF THE WEIBULL DISTRIBUTION – RELATIONSHIP BETWEEN THE WEIBULL AND THE EXTREME VALUE DISTRIBUTION

In previous sections methods of estimating the parameters of the Weibull distribution such as maximum likelihood, matching moments, and least squares were presented. In this section, linear estimates of the parameters of the two-parameter Weibull distribution are discussed.

Let T be the random variable of the life time of a product, Weibull distributed with *cdf*

$$F(T) = 1 - e^{-(\frac{T}{\eta})^{\beta}}.$$

Consider a new random variable, X, which is a function of T given by

$$X = \log_e T.$$

Then

$$
\begin{aligned}
F(x) &= P(X \leq x) = P(\log_e T \leq x), \\
&= P(T \leq e^x), \\
&= 1 - e^{-(\frac{e^x}{\eta})^{\beta}},
\end{aligned}
$$

or

$$F(x) = 1 - \exp\left[-\exp\left(x - \mu/\sigma\right)\right], \tag{12.60}$$

where

$$\mu = \log_e \eta, \tag{12.61}$$

and

$$\sigma = \frac{1}{\beta}; \tag{12.62}$$

therefore, it may be seen from Eq. (12.60) that $X = \log_e T$ is extreme-value distributed; and μ and σ are the location and scale parameters, respectively.

In the following sections the linear estimates of μ and σ are given. Then, the estimates of η and β can be obtained through Eqs. (12.61) and (12.62).

12.12   THE GENERAL LINEAR MODEL OF THE ORDER STATISTICS

Consider a random variable, X, which has a two-parameter distribution with location parameter μ, scale parameter σ, and a *pdf*

$$f(x;\mu,\sigma) = \frac{1}{\sigma}\, g\left(\frac{x-\mu}{\sigma}\right),$$

where g is a function independent of μ and σ, and $Z = \frac{x-\mu}{\sigma}$ has the density of $f(z;0,1)$.

Assume a random sample of size n was drawn from the population of X, and the life test of these n items was determined after the *rth* failure occurred. Let

$$y_1 \le y_2 \le \cdots \le y_r$$

be the ordered observations; i.e.,

$$Y = \begin{pmatrix} y_1 \\ y_2 \\ \vdots \\ y_r \end{pmatrix} = \begin{pmatrix} x_{1:n} \\ x_{2:n} \\ \vdots \\ x_{r:n} \end{pmatrix};$$

then,

$$z_i = \frac{y_i - \mu}{\sigma}; i = 1, 2, \cdots, r,$$

are the ordered observations from Z.

Furthermore assume that the expectation and the variance of the z_i's exist, and are given, respectively, by

$$E(Z) = \begin{pmatrix} E(z_1) \\ E(z_2) \\ \vdots \\ E(z_r) \end{pmatrix} = \begin{pmatrix} a_{1,n} \\ a_{2,n} \\ \vdots \\ z_{r,n} \end{pmatrix} = A,$$

and

$$V(Z) = [\text{COV } (z_i, z_j)] = V_{ij,n} = V; \; i, j = 1, 2, \cdots, r.$$

Then, since

$$y_i = \mu + \sigma z_i; \; i = 1, 2, \cdots, r,$$

$$\begin{aligned} E(y_i) &= \mu + \sigma E(z_i), \\ E(y_i) &= \mu + \sigma \, a_{i,n}; \; i = 1, 2, \cdots, r, \end{aligned}$$

and

$$\begin{aligned} \text{COV } (y_i, y_j) &= \sigma^2 \text{ COV } (z_i, z_j), \\ \text{COV } (y_i, y_j) &= \sigma^2 \, V_{ij,n}; \; i, j = 1, 2, \cdots, r, \end{aligned}$$

or in matrix form

$$\left\{ \begin{aligned} E(Y) &= \begin{pmatrix} E(y_1) \\ E(y_2) \\ \vdots \\ E(y_r) \end{pmatrix} = \begin{pmatrix} 1 & a_{1,n} \\ 1 & a_{2,n} \\ & \vdots \\ 1 & a_{r,n} \end{pmatrix} \begin{pmatrix} \mu \\ \sigma \end{pmatrix}, \\ V(Y) &= \sigma^2 V, \end{aligned} \right\}$$

which is the linear model of the order statistics.

12.13   THE BEST LINEAR UNBIASED ESTIMATE (BLUE) OF μ AND σ

The preceding regression model is a general linear model with correlated observations. Thus, by the Gauss-Markov theorem, the Best Linear Unbiased Estimates for μ and σ are given by [39; 40],

$$\begin{pmatrix} \hat{\mu} \\ \hat{\sigma} \end{pmatrix} = \left[(J, A)' \, V^{-1} \, (J, A) \right]^{-1} (J, A)' \, V^{-1} \, Y, \tag{12.63}$$

where

$$J = \begin{pmatrix} 1 \\ 1 \\ \vdots \\ 1 \end{pmatrix}_{r \times 1}, (J, A) = \begin{pmatrix} 1 & a_{1,n} \\ 1 & a_{2,n} \\ & \vdots \\ 1 & a_{r,n} \end{pmatrix}.$$

Let

$$\Gamma = V^{-1}(JA' - AJ') \, V^{-1}/\Delta,$$

and

$$\Delta = |(J, A)' \, V^{-1}(J, A)|;$$

then,

$$\hat{\mu} = -A' \, \Gamma \, Y, \qquad (12.64)$$

and

$$\hat{\sigma} = J' \, \Gamma \, Y, \qquad (12.65)$$

It can be seen that $\hat{\mu}$ and $\hat{\sigma}$ are linear functions of $(y_1, y_2, \cdots, y_r)$.
The variance and covariance of $\hat{\mu}$ and $\hat{\sigma}$ are given, respectively, by

$$\mathrm{Var}\ \hat{\mu} = \frac{A' \, V^{-1} \, A}{\Delta} \sigma^2,$$

$$\mathrm{Var}\ \hat{\sigma} = \frac{J' \, V^{-1} \, J}{\Delta} \sigma^2,$$

and

$$\mathrm{COV}\ (\hat{\mu}, \hat{\sigma}) = \frac{-J' \, V^{-1} \, A}{\Delta} \sigma^2.$$

Knowing the distribution of X, the matrices A and V can be calculated [41] from the order statistics of Z which is independent of the parameters μ and σ. When X is extreme value distributed, the tables for the values of A and V can be found from White [42] for n from 1 to 20.

12.14 ESTIMATING THE WEIBULL PARAMETERS' BLUE'S

Assume that T is Weibull distributed, and $T_1, T_2, \cdots, T_r$ are the ordered observations from a failure terminated life test. Let

$$Y_i = \log T_i, \quad i = 1, 2, \cdots, r,$$

then the Y_i's are the ordered observations from the extreme-value distribution. The estimates of μ and σ are then given by

$$\hat{\mu} = b(1; n, r)\, y_1 + \cdots + b(r; n, r)\, y_r \qquad (12.66)$$

and

$$\hat{\sigma} = c(1; n, r)\, y_1 + \cdots + c(r; n, r)\, y_r, \qquad (12.67)$$

respectively, where $b(i; n, r)$ and $c(i; n, r)$ can be found in Appendix K. Thus, from Eqs. (12.61) and (12.62),

$$\hat{\eta} = e^{\hat{\mu}}, \qquad (12.68)$$

and

$$\hat{\beta} = \frac{1}{\hat{\sigma}}. \qquad (12.69)$$

EXAMPLE 12–23

Ten items were put in a life test and the test was terminated after the sixth failure occurred. The failure times are 16, 34, 53, 75, 93, and 120 hr. Assume the *pdf* of the life times is the two-parameter Weibull. Estimate its parameters.

SOLUTION TO EXAMPLE 12–23

From the data

$$Y_1 = \log_e 16 = 2.7726,$$
$$Y_2 = \log_e 34 = 3.5264,$$
$$Y_3 = \log_e 53 = 3.9703,$$
$$Y_4 = \log_e 75 = 4.3175,$$
$$Y_5 = \log_e 93 = 4.5326,$$

and

$$Y_6 = \log_e 120 = 4.7875.$$

$n = 10, r = 6$, from Appendix K, and Eq. (12.66),

$$
\begin{aligned}
\hat{\mu} &= (-0.0690324) \times 2.7726 + (-0.0506448) \times 3.5284 \\
&+ (-0.0225702) \times 3.9703 + (0.0140717) \times 4.3175 \\
&+ (0.0601887) \times 4.5376 + (1.0679869) \times 4.7875 \\
\hat{\mu} &= 4.9873.
\end{aligned}
$$

And from Eq. (12.67),

$$
\begin{aligned}
\hat{\sigma} &= (-0.1748478) \times 2.7726 + (-0.1753912) \times 3.5264 \\
&+ (-0.1596414) \times 3.9703 + (-0.1308280) \times 4.3175 \\
&+ (-0.0882729) \times 4.5376 + (0.7289813) \times 4.7875, \\
\hat{\sigma} &= 0.7875.
\end{aligned}
$$

Then, from Eqs. (12.68) and (12.69),

$$\hat{\eta} = e^{4.9873} = 146.54,$$

and

$$\hat{\beta} = \frac{1}{0.7875} = 1.2699.$$

12.15   THE BEST LINEAR INVARIANT ESTIMATES (BLIE) OF μ AND σ

The Mean Squared Error (MSE) of an estimator is defined as

$$\text{MSE}(\hat{\theta}) = E(\hat{\theta} - \theta)^2. \tag{12.70}$$

Adding to and subtracting from the right side of Eq. (12.70) the term $E(\hat{\theta})$ yields

$$
\begin{aligned}
\text{MSE}(\hat{\theta}) &= E[\hat{\theta} + E(\hat{\theta}) - E(\hat{\theta}) - \theta]^2, \\
&= E\{[\hat{\theta} - E(\hat{\theta})]^2 + 2[\hat{\theta} - E(\hat{\theta})] - \theta \\
&+ [E(\hat{\theta} - \theta]^2\}.
\end{aligned}
$$

Since,

$$
\begin{aligned}
E\{[\hat{\theta} - E(\hat{\theta})][E(\hat{\theta}) - \theta]\} &= [E(\hat{\theta}) - \theta] \, E[\hat{\theta} - E(\hat{\theta})], \\
&= [E(\hat{\theta}) - \theta][E(\hat{\theta}) - E(\hat{\theta})], \\
&= 0;
\end{aligned}
$$

then,

$$\text{MSE}(\hat{\theta}) = E[\hat{\theta} - E(\hat{\theta})]^2 + [E(\hat{\theta}) - \theta]^2,$$

or

$$\text{MSE}(\hat{\theta}) = \text{var}\,(\hat{\theta}) + [\text{bias}(\hat{\theta})]^2. \tag{12.71}$$

It can be seen from Eq. (12.71) that the MSE reflects both the bias and the variance of the estimator, and if $\hat{\theta}$ is unbiased, then the MSE is just the variance of the estimator. An estimator is called the Best Linear Invariant Estimator if (1) the estimator is a linear function of the ordered observations, (2) its MSE does not depend on the location parameter μ, and (3) it minimizes the MSE among all linear estimators.

Most best linear invariant estimators are biased, and can be obtained through the BLUE of μ and σ.
Let

$$
\begin{aligned}
\hat{\mu} &= \text{BLUE of } \mu, \\
\hat{\sigma} &= \text{BLUE of } \sigma, \\
\text{Var } \hat{\mu} &= A\sigma^2, \\
\text{Var } \hat{\sigma} &= B\sigma^2,
\end{aligned}
$$

and

$$\text{COV}\,(\hat{\mu}, \hat{\sigma}) = C\sigma^2,$$

where A, B and C depend on n and r but not on μ or σ. Then the BLUE of μ and σ are given by

$$\hat{\mu}^* = \hat{\mu} - \hat{\sigma}[C/(1 + B)], \tag{12.72}$$

and

$$\hat{\sigma}^* = \hat{\sigma}/(1 + B). \tag{12.73}$$

In the case of the extreme-value distribution, Mann, Schafer, and Singpurwalla [43] tabulate the coefficients of the BLIE's for μ and σ, so that $\hat{\mu}^*$ and $\hat{\sigma}^*$ can be calculated from the ordered observation directly using the following equations:

$$\hat{\mu}^* = A(n,r,1)\, y_1 + A(n,r,2)\, y_2 + \cdots + A(n,r,r)\, y_r, \quad (12.74)$$

and

$$\hat{\sigma}^* = C(n,r,1)\, y_1 + C(n,r,2)\, y_2 + \cdots + C(n,r,r)\, y_r, \quad (12.75)$$

where the values of $A(n,r,i)$ and $C(n,r,i)$ are given in Appendix L, and the y_i's are the ordered observations from the extreme-value distribution.

EXAMPLE 12–24

Rework on Example 12–23 to find the BLIE of the parameters β and η of the two-parameter Weibull distribution.

SOLUTION TO EXAMPLE 12–24

From Eqs. (12.74) and (12.75) and Appendix L,

$$
\begin{aligned}
\hat{\mu}^* &= (-0.058017) \times 2.7726 + (-0.039595) \times 3.5264 \\
&+ (-0.012513) \times 3.9703 + (0.022314) \times 4.3175 \\
&+ (0.065750) \times 4.5326 + (1.022062) \times 4.7875 \\
\hat{\mu}^* &= 4.9373,
\end{aligned}
$$

and

$$
\begin{aligned}
\hat{\sigma}^* &= (-0.149985) \times 2.7726 + (0.150451) \times 3.5264 \\
&+ (-0.136941) \times 3.9703 + (-0.112224) \times 4.3175 \\
&+ (-0.075721) \times 4.5326 + (0.625321) \times 4.7875 \\
\hat{\sigma}^* &= 0.6759.
\end{aligned}
$$

Then, from Eqs. (12.68) and (12.69)

$$\hat{\eta} = e^{4.9373} = 139.19 \text{ hr},$$

and

$$\hat{\beta} = \frac{1}{0.6759} = 1.4795.$$

Comparing these results with those of Example 12–23, where $\hat{\eta} = 146.54$ and $\hat{\beta} = 1.2699$, it may be seen that for the BLIE's $\hat{\eta}$ is smaller and $\hat{\beta}$ is larger than those obtained using the BLUE's.

PROBLEMS

12-1. Fifteen items are put to a 1,600 hr life test and 10 failures are observed at 500; 670; 800; 920; 1,030; 1,140; 1,250; 1,350; 1,460 and 1,560 hr. Do the following:

   (1) Find the parameters of the Weibull distribution which fits the above data best using probability plotting, and assuming $\gamma \neq 0$.

   (2) Find the mission duration for a reliability of 95%, starting the mission at age zero.

   2.1 From the Weibull plot.

   2.2 By calculation using the three parameters found.

   (3) Find the mission duration for a reliability of 90%, starting the mission at the age of 250 hr.

   (4) Find the unreliability for a mission of 1,000 hr duration, starting the mission at age zero.

   (4.1) From the Weibull plot.

   (4.2) By calculation using the three parameters found.

   (5) Find the reliability for a new mission of 250 hr duration, starting this new mission at the age of 250 hr.

   (6) Find the mean life of these items.

   (7) Find the 99% reliable life.

12-2. Work out Problem 12-1 assuming $\gamma = 0$, and compare the results obtained for each case.

12-3. Ten items are put to a 2,000 hr life test and seven failures are observed at 250; 460; 670; 870; 1,100; 1,320 and 1,600 hr.

   (1) Determine graphically the three parameters of the best Weibull distribution which fits these data. Do you believe the MR versus T_i values are well represented by a straight line?

   (2) What should the mission duration be for a reliability of 95% starting the mission at age zero?

   (3) What is the unreliability for a mission of 50 hr starting the mission at age zero?

   (4) What is the reliability for a new 10-hr mission starting the new mission at the age of 50 hr?

12-4. Ten items are put to a 200 hr life test and six failures are observed at 35, 46, 85, 110 and 145 hr.

(1) Determine the three parameters of the best weibull distribution which fits these data. (Assume $\gamma \neq 0$).

(2) What should the mission duration be for a reliability of 99%?

(3) What is the unreliability for a mission of 50 hr?

12-5. Twenty units are put to a 1,500 hr life test and 12 failures are observed at 540; 700; 800; 900; 980; 1,060; 1,120; 1,180; 1,250; 1,340; 1,400 and 1,450 hr.

(1) Find the parameters of the Weibull distribution that best represents these data.

(2) Find the reliability for a mission of 375 hr, starting the mission at age zero.

(3) Find the reliability for an additional mission of 375 hr.

(4) Find the maximum mission duration for a reliability of 99.5%, starting the mission at age zero.

12-6. Specific identical components have a times-to-failure distribution which is Weibullian, with the following parameters:

$$\beta = 0.80; \quad \eta = 500 \text{ hr}; \quad \gamma = -125 \text{ hr}.$$

(1) Write down the probability density function for these components.

(2) Write down the reliability function for these components.

(3) Write down the conditional reliability function for these components.

(4) Write down the failure rate function for these components.

(5) Write down the frequency distribution function per 50-hr class interval width (w) and a sample size of 200 such components (N).

(6) Plot the probability density function.

(7) Plot the failure rate function.

(8) Plot the reliability function.

(9) Find the reliability for a 125-hr mission, starting the mission at age zero.

(10) Find the reliability for a second 125-hr mission starting this new mission at the age of 125 hr.

12-7. Specific identical components have a times-to-failure distribution which is Weibullian, with the following parameters:

$$\beta = 0.30; \quad \eta = 100 \text{ hr}; \quad \gamma = -100 \text{ hr}.$$

(1) Write down the probability density function for these components.

(2) Write down the reliability function for these components.

(3) Write down the conditional reliability function for these components.

(4) Write down the failure rate function for these components.

(5) Write down the frequency distribution function per 50-hr class interval width (w) and a sample size of 200 such components (N).

(6) Plot the probability density function.

(7) Plot the failure rate function.

(8) Plot the reliability function.

(9) Find the reliability for a mission of 125-hr duration starting the mission at age zero.

(10) Find the reliability for a second 125-hr mission starting this new mission at the age of 125 hr.

12-8. A bearing comes from a population having the *pdf*

$$f(T) = \frac{\beta}{\eta}\left(\frac{T-\gamma}{\eta}\right)^{\beta-1} e^{-\left(\frac{T-\gamma}{\eta}\right)^{\beta}},$$

where $\beta = 1.15$, $\eta = 5.0 \times 10^6$ revolutions, and $\gamma = 0$.

(1) What is the reliability of this bearing for a 10-hr mission, starting at an age of 10^4 revolutions? The bearing speed is 1,000 revolutions per minute.

(2) If 1,000 such bearings were being used, how many spares would be required?

12-9. A bearing comes from a population having the *pdf*

$$f(T) = \frac{\beta}{\eta}\left(\frac{T-\gamma}{\eta}\right)^{\beta-1} e^{-\left(\frac{T-\gamma}{\eta}\right)^{\beta}},$$

where $\beta = 1.50$, $\eta = 2.5 \times 10^6$ revolutions, and $\gamma = 0$.

(1) What is the reliability of this bearing for a 10-hr mission, starting at an age of 10^4 revolutions? The bearing speed is 1,000 revolutions per minute.

(2) If 1,000 such bearings were being used, how many spares would be required?

TABLE 12.20 — Field data for the units of Problem 12-10 exhibiting a wear-out characteristic.

Life in 100 hr, starting at 7,500 hr	Number of failures, N_F
0 – 1	40
1 – 2	210
2 – 3	300
3 – 4	250
4 – 5	80
5 – 6	20
	$N_T = 900$

12-10. Using the data in Table 12.20, do the following:

(1) Fit a Weibull distribution to the grouped data and write
down its probability density function.

(2) Calculate, tabulate and plot the probability density function.

(3) Write down the failure frequency distribution function per
class interval width.

(4) Calculate and tabulate the failure frequency distribution
function per class interval width and superimpose it on the
number failing versus the time-to-failure histogram. Comment on the goodness of fit.

(5) Calculate, tabulate and plot the reliability function.

(6) Calculate, tabulate and plot the failure rate function.

12-11. The field data given in Table 12.21, Columns 1 and 2, have been
obtained:

(1) Fill in the table completely.

(2) Plot the time-to-failure versus the number-of-failures histogram, and draw in the best fitting curve in your judgement.

(3) Plot the failure rate versus the time-to-failure histogram,
and draw in the best fitting curve in your judgement.

TABLE 12.21 - Field data for the population of Problem 12-11 exhibiting an early failure characteristic.

Life, hr	Number of failures, N_F	Failure rate, $\hat{\lambda}$, fr/10^6 hr	Number of units surviving period, N_S	Reliability, $\hat{R} = \frac{N_S}{N_T}$
0 – 1	80			
1 – 2	43			
2 – 3	25			
3 – 4	15			
4 – 5	10			
5 – 6	7			
6 – 7	5			
7 – 8	4			
8 – 9	3			
9 – 10	2			
Over 10	6			
	$N_T = 200$			

(4) Plot the reliability versus the time-to-failure (mission time, starting the mission at age zero) histogram and draw in the best fitting curve in your judgement.

(5) Find the parameters of the Weibull distribution representing the data.

(6) Write down the frequency distribution function per class interval width.

(7) Write down the frequency distribution function per unit class interval width.

(8) Write down the probability density function.

(9) Write down the standardized probability density function.

(10) Calculate, tabulate and plot the frequency distribution function per class interval width on the previous histogram. Label the axes.

(11) Calculate, tabulate and plot the frequency distribution per unit class interval width on the previous histogram. Label the axes.

(12) Calculate, tabulate and plot the probability density function on the previous histogram. Label the axes.

(13) Calculate, tabulate and plot the standardized probability density function on the previous histogram. Label the axes.

(14) Find the reliability of such units for their first mission of 50 hr duration.

(15) Find the reliability of such units for their second mission of 50 hr duration, starting their second mission at the age of 50 hr.

12-12. Using the data in Table 12.22, do the following:

(1) Fit a Weibull distribution to the grouped data and write down its probability density function.

(2) Calculate, tabulate and plot the probability density function.

(3) Write down the failure frequency distribution function per class interval width.

(4) Calculate and tabulate the failure frequency distribution function per class interval width and superimpose it on the number failing versus the time-to-failure histogram. Comment on the goodness of fit.

(5) Calculate, tabulate and plot the reliability function.

TABLE 12.22 – Grouped field failure data for the identical units of Problem 12-12.

Life in 100 hr	N_F
0 – 1	30
1 – 2	20
2 – 3	14
3 – 4	9
4 – 5	6
5 – 6	4
Over 6	7
	$N_T = 90$

(6) Calculate, tabulate and plot the failure rate function.

12-13. Through a joint effort between the design and reliability engineers, it was determined that the optimum reliability to be designed, manufactured, tested and delivered to the customers for their new product is 97.0%. They come up with a design that in their good judgement will meet this goal. Fifteen such products are manufactured and reliability tested with the following results: One fails after 390 hr, another after 1,800 hr, a third after 4,000 hr, a fourth after 7,300 hr, a fifth after 12,200 hr and a sixth fails after 17,500 hr at which time the test is terminated.

(1) How would you analyze these test results?

(2) What key points are missing in the reliability specifications for this product?

(3) Does this product meet its reliability goal if the representative, cumulative function period without any checkout is 500 hr? How do you determine that the product meets its reliability goal?

(4) What life period are these products in? How do you know this?

(5) What should the function period be to meet the reliability goal?

(6) If the function period cannot be shortened what should the design and reliability engineers do?

(7) What are the parameters of the Weibull distribution that represents these data?

(8)　What is the confidence level of the reliabilities obtained from your Weibull plot?

12-14. Twentyfour transistors are tested to failure. The observed data in hours to failure per transistor, are presented in Table 12.23. Determine if the Weibull *pdf* fits the observed data well and then determine its parameters.

12-15. Specific identical components have a times-to-failure distribution which is Weibullian, with the following parameters:

$$\beta = 0.50; \qquad \eta = 400 \text{ hr}; \qquad \gamma = -250 \text{ hr}.$$

(1)　Write down the probability density function for these components.

(2)　Write down the reliability function for these components.

(3)　Write down the conditional reliability function for these components.

(4)　Write down the failure rate function for these components.

(5)　Write down the frequency distribution function per 50-hr class interval width (w) and a sample size of 200 such components (N).

(6)　Calculate the reliability of these components for a mission of 100 hr when used as received.

(7)　Calculate the reliability of these components for a mission of 100 hr when used after being checked out.

12-16. Seven items were put in a life test and the test was terminated after the seventh failure occurred. The failure times are 1,260; 2,040; 2,695; 3,300; 4,015; 4,850 and 6,140 hr. Assume the *pdf* of the life times is the two-parameter Weibull. Find the Best Linear Unbiased Estimates (BLUE) of β and η.

12-17. Nine items were put in a life test and the test was terminated after the seventh failure occurred. The failure times are 4,210; 5,360; 6,155; 6,840; 7,515; 8,260 and 9,290 hr. Assume the *pdf* of the life times is the two-parameter Weibull. Find the Best Linear Unbiased Estimates (BLUE) of β and η.

12-18. Seven items were put in a life test and the test was terminated after the seventh failure occurred. The failure times are 1,260; 2,040; 2,695; 3,300; 4,015; 4,850 and 6,140 hr. Assume the *pdf* of the life times is the two-parameter Weibull. Find the Best Linear Invariant Estimates (BLIE) of β and η and compare the results with those of Problem 12-16.

TABLE 12.23 – Transistor times to failure for Problem 12-14.

j	T_j, hr
1	260
2	350
3	420
4	440
5	480
6	480
7	530
8	580
9	680
10	710
11	740
12	780
13	820
14	840
15	920
16	930
17	1,050
18	1,060
19	1,070
20	1,270
21	1,340
22	1,370
23	1,880
24	2,130

12-19. Nine items were put in a life test and the test was terminated after the seventh failure occurred. The failure times are 4,210; 5,360; 6,155; 6,840; 7,515; 8,260 and 9,290 hr. Assume the *pdf* of the life times is the two-parameter Weibull. Find the Best Linear Invariant Estimates (BLIE) of β and η and compare the results with those of Problem 12-17.

12-20. Twelve items were put in a life test and the test was terminated after the eighth failure occurred. The failure times are 2,225; 2,740; 3,025; 3,850; 4,015; 4,900; 5,540 and 5,950 hr. Assume the *pdf* of the life times is the two-parameter Weibull.

(1) Find the Best Linear Unbiased Estimates (BLUE) of β and η.

(2) Find the Best Linear Invariant Estimates (BLIE) of β and η.

(3) Compare the results of Cases 1 and 2.

REFERENCES

1. Lehman, Jr., Eugene H., "Shapes, Moments, and Estimators of the Weibull Distribution," *IEEE Transactions on Reliability*, pp. 32-38, Sept. 1963.

2. Mischke, Charles R., "Some Tentative Weibullian Descriptions of the Properties of Steels, Aluminums, and Titaniums," Vibrations Conference and the International Design Automation Conference, Design Engineering Division, ASME, Toronto, Canada, 71-Vibr-64, 20 pp., September 8-10, 1971.

3. Hahn, Gerald J. and Shapiro, Samuel S., *Statistical Models in Engineering*, John Wiley & Sons, Inc., New York, 355 pp., 1967.

4. Ravenis II, Joseph V. J., "Estimating Weibull Distribution Parameters," Electro-Technology, pp. 46-54, March 1964.

5. Ireson, Grant (Editor), *Reliability Handbook*, McGraw-Hill Book Co., Inc., New York, 720 pp., 1966.

6. Weibull, Waloddi, "A Statistical Distribution Function of Wide Applicability," *Journal of Applied Mechanics*, Vol. 18, pp. 293-297, 1951.

7. Kao, J.H.K., "A Graphical Estimation of Mixed Weibull Parameters in Life Testing of Electron Tubes," *Technometrics*, Vol. 1, No. 4, pp. 389-407, November 1959.

8. Kao, J.H.K., "A Summary of Some New Techniques of Failure Analysis," *Proceedings 6th National Symposium on Reliability and Quality Control*, pp. 190-201, 1960.

9. Johnson, Leonard G., "The Median Ranks of Sample Values in Their Population with an Application to Certain Fatigue Studies," *Industrial Mathematics*, Vol. 2, pp. 1-9, 1951.

10. Johnson, Leonard G., "The Statistical Treatment of Fatigue Experiments," Elsevier Publishing Co., New York, 114 pp., 1964.

11. Hald, A., *Statistical Theory with Engineering Applications*, John Wiley & Sons, Inc., New York, 783 pp., 1952.

12. TEAM (Technical and Engineering Aids for Management), Box 25, Tamworth, N.H. 03886. Telephone: (603) 323-8843.

13. Kao, John H. K., "Computer Methods for Estimating Weibull Parameters in Reliability Studies," *IRE Transactions on Reliability and Quality Control*, PGRQC 13, pp. 15-22, July 1958.

14. Lieblein, J. and Zelen, M., " Statistical Investigation of the Fatigue Life of Deep-Groove Ball Bearings," *Journal of Research*, National Bureau of Standards, Vol. 57, p. 273, 1956.

15. Kao, J.H.K., "A New Life Quality Measure for Electron Tubes," *IRE Transactions on Reliability and Quality Control*, Vol. 7, p. 1, 1956.

16. Perry, J. N., "Semiconductor Burn-in and Weibull Statistics," *Semiconductor Reliability*, Vol. 2, Engineering Publishers, Elizabeth, N.J., pp. 8-90, 1962.

17. Procassini, A. and Romano, A., "Semiconductor Burn-in and Weibull Statistics," *Semiconductor Reliability*, Vol. 2, Engineering Publishers, Elizabeth, N.J., pp. 29-34, 1962.

18. Procassini, A. and Romano, A., "Transistor Reliability Estimates Improved with Weibull Distribution Function," Motorola Military Products Division, Engineering Bulletin, Vol. 9, No. 2, pp. 16-18, 1961.

19. Mann, Nancy R., Schafer, Ray E. and Singpurwalla, Nozer D., *Methods for Statistical Analysis of Reliability and Life Data*, John Wiley & Sons, New York, 564 pp., 1974.

20. Lipson, Charles and Sheth, Narendra, J., *Statistical Design and Analysis of Engineering Experiments*, McGraw-Hill, Inc., New York, 518 pp., 1973.

21. Fisher, R. A. and Tippett, L.H.C., "Limiting Forms of the Frequency Distribution of the Largest or Smallest Member of a Sample," *Proc. Cambridge Phil. Soc.*, Vol. 24, No. 2, p. 180, 1928. Reprinted in

Fisher, R. A., *Contributions to Mathematical Statistics,* John Wiley & Sons, New York, 1950.

22. Freudenthal, A. M., and Gumbel, E. J., "On the Statistical Interpretation of Fatigue Tests," *Proc. Royal Society,* Great Britain, Vol. 216-A, pp. 309-322, 1953.

23. Downton, F., "Linear Estimates of Parameters in the Extreme Value Distribution," *Technometrics,* Vol. 8, p. 3, 1966.

24. Johnson, L. G., *Ball Bearing Engineer's Statistical Guide Book,* New Departure Division of General Motors Corporation, April 1957.

25. Weibull, W., "A Statistical Representation of Fatigue Failure in Solids," *Trans. Royal Institute of Technology,* No. 27, Stockholm, 1949.

26. Freudenthal, A. M. and Gumbel, E. J., "Minimum Life in Fatigue," *Journal of the American Statistical Association,* Vol. 49, pp. 575-597, September 1954.

27. Parsons, Frederick, G., "Calculation of the Order Statistics of Failures with the F-Distribution Approximation to the Cumulative Binomial," Advanced Reliability Engineering Course Term Paper at The University of Arizona, submitted to Dr. Dimitri Kececioglu, 22 pp., May 1978.

28. King, James R., "Graphical Data Analysis with Probability Papers," *TEAM* (see [12]), 20 pp., 1966.

29. Le Mense, R. A., "Use of the Weibull Distribution in Analyzing Life Test Data from Vehicle Structural Components," *Proc. Aerospace Reliability and Maintainability Conference,* pp. 628-638, 1964.

30. King, James, R., "TEAM Easy Analysis Methods," *TEAM,* Vol. 3, No. 4, 1976.

31. Marks, David L., *Computer Applications for the Weibull Distribution,* The University of Arizona Master's Research Report, submitted to Dr. Dimitri B. Kececioglu, 200 pp., 1980.

32. Bompas-Smith, J. H., *Mechanical Survival: The Use of Reliability Data,* McGraw-Hill Book Co., (UK) Limited, Maidenhead, Berkshire, England, 199 pp., 1973.

33. O'Connor, Patrick D. T., *Practical Reliability Engineering,* 2nd ed., John Wiley & Sons, New York, 398 pp., 1985.

34. Kececioglu, Dimitri B., *Reliability Engineering Handbook,* Prentice Hall, Inc., Englewood Cliffs, N. J., Vol. 1, 720 pp. and Vol. 2, 568 pp., 1991.

35. Spiegel, M. R., *Probability and Statistics,* McGraw-Hill, New York, 372 pp., 1975.

36. Sinha, S. K. and Kale, B. K., *Life Testing and Reliability Estimation,* John Wiley & Sons, Inc., New York, 196 pp., 1980.

37. Nelson, W., *Applied Life Data Analysis,* John Wiley & Sons, Inc., New York, 654 pp., 1982.

38. Thoman, D. R., Bain, L. J. and Antle, C. E., "Inferences on the Parameters of the Weibull Distribution," *Technometrics,* Vol. 11, No. 3, pp. 445-460, 1969.

39. Lloyd, E. H., "Least-Squares Estimation of Location and Scale Parameters Using Order Statistics," *Biometrika,* Vol. 39, pp. 89-95, 1952.

40. Plackett, R. L., "Linear Estimation from Censored Data," *Ann. Math. Statist.,* Vol. 29, pp. 131-142, 1958.

41. Lieblein, J., "On the Exact Evaluation of the Variances and Covariances of Order Statistics in Samples from the Extreme-Value Distribution," *Ann. Math. Statist.,* Vol. 24, pp. 282-287, 1953.

42. White, J. S., "Least-Squares Unbiased Censored Linear Estimation for The Log Weibull (Extreme-Value) Distribution, " *Industrial Mathematics,* Vol. 14, pp. 21-60, 1964.

43. Mann, N. R., Schafer, R. E., and Singpurwalla, N. D., "Methods for Statistical Analysis of Reliability and Life Data," John Wiley & Sons, Inc., New York, 564 pp., 1974.

Chapter 13

CONFIDENCE LIMITS ON THE RELIABILITY WITH WEIBULL DISTRIBUTED TIMES TO FAILURE, AS WELL AS ON THE MEAN LIFE, MISSION DURATION, β AND η

13.1 PROCEDURE

To obtain the confidence limit on the Weibullian reliability for a specified mission, the following procedure is recommended:

1. Prepare Table 13.1 where in Column 2 the times-to-failure data are listed in ascending order, and in Column 3 the corresponding median ranks are entered using the procedure presented previously.

2. Plot the median rank values versus the corresponding times-to-failure values and draw the best fitting straight line through these points, as illustrated in Fig. 13.1. If the points do not fall on a good straight line, straighten them out using the procedure presented in the previous chapter and determine the values of γ, β and η.

465

TABLE 13.1 -- The times-to-failure, median ranks, 5% rank and 95% rank values to obtain the straight-line fit to the data, and the 5% and the 95% confidence band values to determine the confidence limits on the unreliability and on the reliability of the test units involved. The sample size, N, is 10.

j	T_j, hr	MR,%	5% Rank	95% Rank
1	10	6.70	0.51	25.89
2	26	16.23	3.68	39.42
3	42	25.86	8.73	50.69
4	65	35.51	15.00	60.66
5	90	45.17	22.24	69.65
6	120	54.83	30.35	77.76

3. After a good straight line is obtained, get the ranks for the desired confidence level from the table in Appendix A. This has been done for the 5% and the 95% ranks in Columns 4 and 5 of Table 13.1.

   If 80% two-sided confidence limits are desired, enter the 10% and 90% rank values. If 90% two-sided confidence limits are desired, enter the 5% and 95% rank values. If 95% two-sided confidence limits are desired, enter the 2.5% and 97.5% rank values.

4. Plot these rank values above their respective times to failure, as shown in Fig. 13.1. Join these points for the same rank by a smooth curve to obtain the corresponding confidence band on the *unreliability* of the units tested.

From the information in Table 13.1 the two-sided 90% confidence limits on the unreliability are obtained. For example, from Fig. 13.1, the two-sided confidence limits on the unreliability, for a mission of 45 hr are obtained as follows:

Enter the abscissa with $T = 45$ hr, go vertically until the 5% and 95% confidence bands are intersected. The $Q_{L2}(T)$ value for $T = 45$ hr is read off at the intersection with the 5% band, or

$$Q_{L2}(T = 45 \text{ hr}) = 9.5\%,$$

and the $Q_{U2}(T)$ value is read off the 95% band intersection, or

$$Q_{U2}(T = 45 \text{ hr}) = 52\%.$$

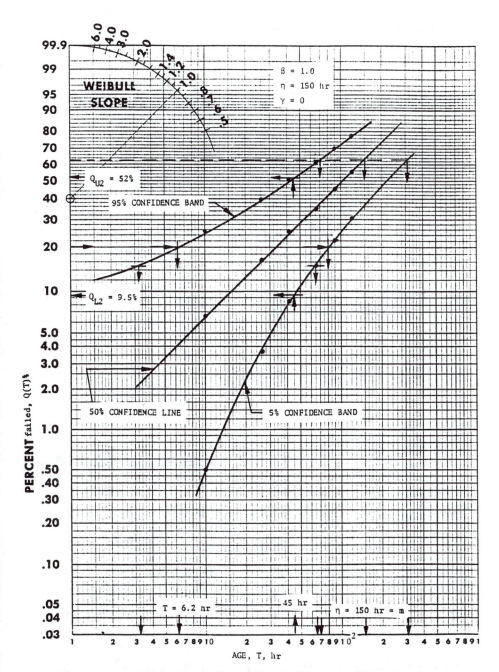

Fig. 13.1– The plot of the median, 5% and 95% rank values
for the data in Table 13.1.

The corresponding reliabilities are then

$$R_{L2}(T) = 1 - Q_{U2}(T),$$

and

$$R_{U2}(T) = 1 - Q_{L2}(T).$$

In this example

$$R_{L2}(T = 45 \text{ hr}) = 1 - 0.52 = 0.48,$$

and

$$R_{U2}(T = 45 \text{ hr}) = 1 - 0.095 = 0.905.$$

Consequently, the two-sided 90% confidence interval on the reliability of these units for a mission of 45 hr is

$$(0.48 \, ; \, 0.905),$$

while the 50% confidence level value is

$$R(T = 45 \text{ hr}) = 0.74,$$

from the straight line fitted to the data.

From Fig. 13.1 we can also obtain the one-sided confidence limits on the reliability but at a *95% confidence level*, as follows:

To find $R_{L1}(T = 45 \text{ hr})$ enter the abscissa with $T = 45$ hr, go to the 95% confidence band and read off $Q_{U1}(T = 45 \text{ hr}) = 52\%$. Therefore,

$$R_{L1}(T = 45 \text{ hr}) = 1 - Q_{U1}(T = 45 \text{ hr}) = 1 - 0.52,$$

or

$$R_{L1}(T = 45 \text{ hr}) = 0.48.$$

Consequently, the one-sided, 95% confidence interval on the reliability of these units, for a mission of 45 hr, is

$$(0.48 \, ; \, 1.0).$$

13.2 MISSION DURATION FOR GIVEN RELIABILITY AND CONFIDENCE LEVEL

We can also find what the *mission duration* should be for a reliability of at least, say, 80% at a confidence level of 95%. In this case

$$R_{L1}(T) = 80\% = 1 - Q_{U1}(T),$$

or

$$Q_{U1}(T) = 20\%.$$

Entering the Weibull plot at this level, going to the 95% band and dropping down vertically at the intersection yields

$$T_{U1} = 6.2 \text{ hr.}$$

Therefore, the mission duration should not exceed 6.2 hr for a reliability at least 80% at a $CL = 95\%$.

13.3 MISSION RANGE FOR A GIVEN CONFIDENCE LEVEL AND RELIABILITY

From the confidence bands we can also obtain the mission duration's range at a desired confidence level that will give us a desired reliability. For example let's find the mission range at a $CL = 90\%$ for a reliability of 85%.

From Fig. 13.1, entering at the level

$$Q(T)\% = 100\% - R(T)\% = 100\% - 85\% = 15\%,$$

and reading off the T values associated with the intersection with the 95% and 5% bands, we get

$$T_{L2} = 3.2 \text{ hr and } T_{U2} = 65 \text{ hr.}$$

Consequently, the two-sided, 90% confidence level interval on the mission duration, for a reliability of 85%, is

$$(3.2 \text{ hr} ; 65 \text{ hr}).$$

13.4   THE TRUE WEIBULL SLOPE, β

Given the number of failures used to plot the Weibull line and using Figs. 13.2 and 13.3 titled "Weibull Slope Error" for CL = 90% and 50%, respectively, the confidence interval on the true β can be determined.

For example, in Fig. 13.1 six failures occurred and $\hat{\beta} = 1.0$. What is the error in β at a $CL = 90\%$? From Fig. 13.2 with six failures, and $CL = 90\%$, the Weibull slope error is $\pm47\%$, or

$$P(\hat{\beta} - 0.47\hat{\beta} \le \beta \le \hat{\beta} + 0.47\hat{\beta}) = 90\%.$$

For this case, $\hat{\beta} = 1.0$; consequently, the two-sided, 90% confidence interval on the true β is

(0.53 ; 1.47).

In other words, if ten exponential units are tested simultaneously until six of them fail, their Weibull shape parameter may be as low as 0.53 and as high as 1.47. So one has to be very careful how such β values should be interpreted or acted upon. For example, such test data may yield a $\beta = 0.65$, which lies inside the confidence interval just found. However, the conclusion may be drawn that early failures are occurring and more burn-in is required, because a $0 < \beta < 1$ is obtained only when the units are in their early life. But, it is known that these units are in their useful life, because they were already adequately burned-in, debugged and stress screened. The original conclusion could have initiated unnecessary, costly and time consuming actions had it not been known that small test sample sizes could yield such β values which are different than one, even though the units are in their useful life. As a matter of fact, if many more such units were tested, we would have obtained a $\beta \cong 1$.

As another example, consider the case when such test data yield a $\beta = 1.35$, which also lies inside the confidence interval just found. The conclusion may be drawn that these production units are exhibiting wear-out life characteristics, whereas we know that they are in their useful life. Such a conclusion may have led to redesign for longer life, better materials that do not wear out so fast, more derating of the offending components, etc., which of course are wrong, very costly and time consuming. The best way to resolve such matters is to conduct exhaustive failure analyses to identify the actual cause of the failures. These examples and the topics covered in this handbook and in [6], bring out the great value of these reliability engineering and testing techniques in designing highly reliable products which are easy to maintain, safe to operate, of the highest quality and are sold at competitive prices.

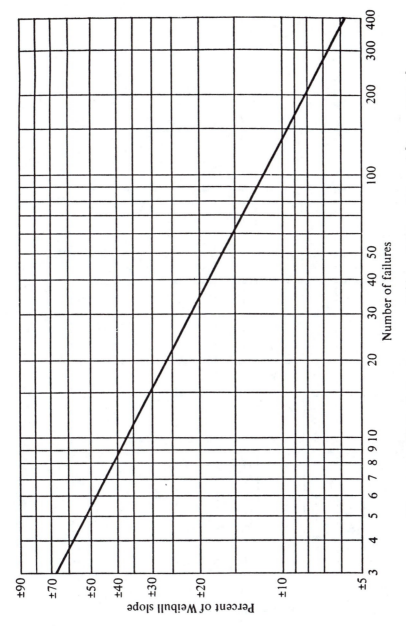

Fig. 13.2–Weibull slope error – 90% confidence interval [1, p. 469; 2].

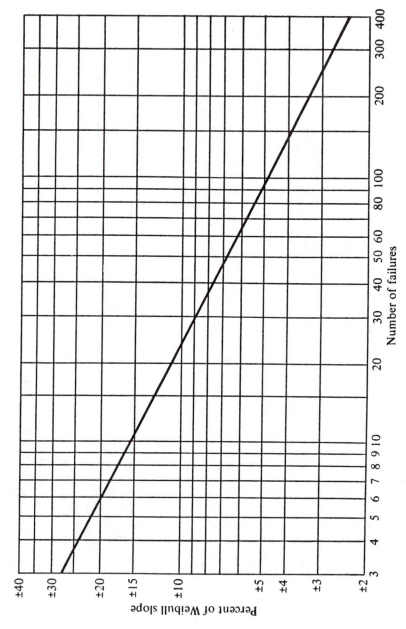

Fig. 13.3 – Weibull slope error – 50% confidence interval [1, p. 469; 2].

472

13.5 THE WEIBULL POPULATION MEAN AND ITS CONFIDENCE LIMITS

Given a Weibull plot, like that of Fig.13.1, and the curve in Fig. 13.4, we can find the mean life of the population. The equation for the curve in Fig 13.4 is given in Appendix 13A. For example with $\hat{\beta} = 1$ enter Fig. 13.4 with this value and find the percent failed at the mean life. This is 63.2%, which means that 63.2% of the units will fail by the time each one has accumulated operating time equal to its mean life. Enter with this $Q(\overline{T}) = 63.2\%$ value the ordinate of the Weibull plot of Fig. 13.1 and find the corresponding T value, as read off the fitted straight line. This is the mean life at a $CL = 50\%$, or

$$\overline{T} = 150 \text{ hr.}$$

This value is also equal to η because $\beta = 1$.

The confidence interval on this mean life can also be found from the Weibull confidence bands. In this case the T values can be read off the 95% and 5% confidence bands at the 63.2% level to be: $\overline{T}_{L2} = 70$ hr and $\overline{T}_{U2} = 310$ hr. Therefore, the confidence interval on the mean life at a $CL = 90\%$ is

$$(70 \text{ hr} \ ; \ 310 \text{ hr}).$$

The mean life can also be found from

$$\overline{T} = \gamma + \eta \, \Gamma(\frac{1}{\beta} + 1).$$

Here $\gamma = 0$, $\eta = 150$ and $\beta = 1$. Therefore,

$$\overline{T} = 0 + (150)\Gamma(2) = 150(1),$$

or

$$\overline{T} = 150 \text{ hr.}$$

Table 13.2 may be used to find the values of the gamma function, or of $\Gamma(\frac{1}{\beta} + 1)$, where $n = \frac{1}{\beta} + 1$.

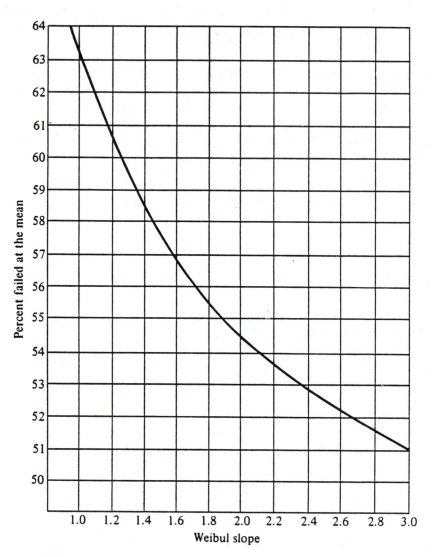

Fig. 13.4–Position of the Weibull mean [1, p. 468].

TABLE 13.2–Values of $\Gamma(n)$, the gamma function, versus n.

n	$\Gamma(n)$	n	$\Gamma(n)$	n	$\Gamma(n)$	n	$\Gamma(n)$
1.00	1.00000	1.25	0.90640	1.50	0.88623	1.75	0.91906
1.01	0.99433	1.26	0.90440	1.51	0.88659	1.76	0.92138
1.02	0.98884	1.27	0.90250	1.52	0.88704	1.77	0.92376
1.03	0.98355	1.28	0.90072	1.53	0.88757	1.78	0.92623
1.04	0.97844	1.29	0.89904	1.54	0.88818	1.79	0.92877
1.05	0.97350	1.30	0.89747	1.55	0.88887	1.80	0.93138
1.06	0.96874	1.31	0.89600	1.56	0.88964	1.81	0.93408
1.07	0.96415	1.32	0.89464	1.57	0.89049	1.82	0.93685
1.08	0.95973	1.33	0.89338	1.58	0.89142	1.83	0.93969
1.09	0.95546	1.34	0.89222	1.59	0.89243	1.84	0.94261
1.10	0.95135	1.35	0.89115	1.60	0.89352	1.85	0.94561
1.11	0.94740	1.36	0.89018	1.61	0.89468	1.86	0.94869
1.12	0.94359	1.37	0.88931	1.62	0.89592	1.87	0.95184
1.13	0.93993	1.38	0.88854	1.63	0.89724	1.88	0.95507
1.14	0.93642	1.39	0.88785	1.64	0.89864	1.89	0.95838
1.15	0.93304	1.40	0.88726	1.65	0.90012	1.90	0.96177
1.16	0.92980	1.41	0.88676	1.66	0.90167	1.91	0.96523
1.17	0.92670	1.42	0.88636	1.67	0.90330	1.92	0.96877
1.18	0.92373	1.43	0.88604	1.68	0.90500	1.93	0.97240
1.19	0.92089	1.44	0.88581	1.69	0.90678	1.94	0.97610
1.20	0.91817	1.45	0.88566	1.70	0.90864	1.95	0.97988
1.21	0.91558	1.46	0.88560	1.71	0.91057	1.96	0.98374
1.22	0.91311	1.47	0.88563	1.72	0.91258	1.97	0.98769
1.23	0.91075	1.48	0.88575	1.73	0.91467	1.98	0.99171
1.24	0.90852	1.49	0.88595	1.74	0.91683	1.99	0.99581

$\Gamma(n) = \int_0^\infty e^{-x} x^{n-1}\, dx.$

$\Gamma(n+1) = n\Gamma(n).$

$\Gamma(\frac{1}{2}) = \pi.$

$\Gamma(\frac{n}{2}) = (\frac{n}{2} - 1)! = \begin{cases} (\frac{n}{2} - 1)(\frac{n}{2} - 2)\cdots(3)(2)(1), & \text{for } n \text{ even and } n > 2. \\ (\frac{n}{2} - 1)(\frac{n}{2} - 2)\cdots(\frac{3}{2})(\frac{1}{2})\sqrt{\pi}, & \text{for } n \text{ odd and } n > 2. \end{cases}$

13.6 CONFIDENCE INTERVALS ON THE PARAMETERS OF THE WEIBULL DISTRIBUTION

13.6.1 INTRODUCTION

More sophisticated methods for determining the confidence interval on β and on η are given in this section. For the two-parameter Weibull distribution, the Maximum Likelihood Estimates (MLE) of β and η, for the failure terminated (censored) test, are the solutions of

$$\frac{\sum_{i=1}^{r} T_i^{\hat{\beta}} \log_e T_i + (N - r)T_r^{\hat{\beta}} \log_e T_r}{\sum_{i=1}^{r} T_i^{\hat{\beta}} + (N - r)T_r^{\hat{\beta}}}$$

$$-\frac{1}{\hat{\beta}} - \frac{1}{r}\sum_{i=1}^{r} \log_e T_i = 0, \tag{13.1}$$

and

$$\hat{\eta}^{\hat{\beta}} = \frac{\sum_{i=1}^{r} T_i^{\hat{\beta}} + (N - r)T_r^{\hat{\beta}}}{r}, \tag{13.2}$$

where $T_1 \leq T_2 \leq \dots \leq T_r$ are the ordered values of the failure times of the sample, N is the sample size and r is the failure number. See Appendix 13A, Eqs. (13A.25) and (13A.26).

Billman, Antele and Bain [3, pp. 831-840] developed the confidence intervals for β and η for the failure terminated test case; however, the tables they provide to obtain the confidence intervals are for some given censored ratios (ratio of the number of units tested to failure to the total sample size), the smallest sample size being 40. They give three censored ratios which are 0.50, 0.75 and 1.00. Here more complete tables are given, namely Tables 13.3 and 13.4, to obtain the confidence intervals for β and η, respectively. The sample size range is 5 to 30, and the smallest value of the failure number is 5.

13.6.2 THE DISTRIBUTION OF THE MLE OF THE WEIBULL DISTRIBUTION'S PARAMETERS

$\hat{\beta}_{11}$ and $\hat{\eta}_{11}$ are used to denote the MLEs of β and η when $\beta = \eta = 1$. The distribution is called the standardized exponential in the case of $\beta = \eta = 1$. It may be shown that $\hat{\beta}/\beta$ is independent of β and η,

TABLE 13.3– Confidence limits on $\hat{\beta}/\beta$ for various N, r and $\alpha \times 100\%$, or $P(\hat{\beta} < l_\alpha) = \alpha$.

N	r	$\alpha \times 100$ (%)															
		2	2.5	5	10	12.5	25	40	50	60	75	80	87.5	90	95	97.5	98
5	5	0.642	0.652	0.710	0.792	0.826	0.972	1.122	1.241	1.398	1.681	1.843	2.131	2.288	2.792	3.464	3.562
6	5	0.556	0.570	0.674	0.769	0.803	0.958	1.117	1.255	1.417	1.730	1.890	2.179	2.338	2.819	3.458	3.717
6	6	0.619	0.637	0.709	0.797	0.831	0.945	1.101	1.203	1.310	1.570	1.695	1.946	2.078	2.410	2.770	2.823
7	5	0.589	0.606	0.682	0.759	0.799	0.965	1.121	1.228	1.393	1.780	1.944	2.312	2.452	3.103	3.707	3.939
7	6	0.613	0.622	0.712	0.790	0.821	0.962	1.101	1.203	1.330	1.561	1.691	1.987	2.109	2.572	3.152	3.409
7	7	0.634	0.649	0.728	0.806	0.831	0.943	1.074	1.178	1.275	1.440	1.544	1.763	1.848	2.199	2.625	2.728
8	5	0.583	0.600	0.661	0.766	0.799	0.957	1.138	1.267	1.433	1.783	1.954	2.303	2.549	3.228	3.890	4.238
8	6	0.617	0.630	0.690	0.769	0.802	0.952	1.099	1.204	1.333	1.608	1.763	2.003	2.093	2.444	3.070	3.265
8	7	0.643	0.656	0.703	0.792	0.819	0.955	1.095	1.182	1.290	1.482	1.614	1.803	1.894	2.214	2.577	2.668
8	8	0.683	0.689	0.735	0.809	0.838	0.934	1.054	1.137	1.232	1.392	1.468	1.636	1.683	1.994	2.325	2.430
9	5	0.563	0.586	0.664	0.751	0.799	0.958	1.129	1.283	1.461	1.811	1.965	2.349	2.584	3.298	4.152	4.382
9	6	0.593	0.606	0.679	0.767	0.797	0.949	1.085	1.196	1.344	1.626	1.729	2.988	2.120	2.608	3.063	3.214
9	7	0.567	0.598	0.659	0.750	0.794	0.942	1.072	1.173	1.297	1.512	1.658	1.858	1.940	2.309	2.738	2.843
9	8	0.651	0.663	0.724	0.790	0.820	0.922	1.049	1.134	1.233	1.415	1.489	1.715	1.809	2.090	2.442	2.533
9	9	0.687	0.703	0.736	0.799	0.821	0.930	1.042	1.119	1.202	1.368	1.442	1.597	1.680	1.878	2.168	2.255
10	5	0.546	0.576	0.681	0.758	0.793	0.967	1.168	1.301	1.485	1.874	2.064	2.453	2.615	3.426	4.204	4.868
10	6	0.579	0.610	0.676	0.747	0.780	0.933	1.085	1.204	1.331	1.590	1.729	2.020	2.195	2.646	3.210	3.378

TABLE 13.3– Continued.

N	r							$\alpha \times 100$ (%)									
		2	2.5	5	10	12.5	25	40	50	60	75	80	87.5	90	95	97.5	98
10	7	0.595	0.609	0.691	0.757	0.797	0.917	1.069	1.197	1.306	1.547	1.646	1.878	1.979	2.359	2.730	2.888
	8	0.627	0.633	0.684	0.773	0.805	0.929	1.057	1.138	1.238	1.437	1.550	1.741	1.819	2.142	2.398	2.451
	9	0.668	0.675	0.715	0.791	0.817	0.929	1.037	1.125	1.212	1.391	1.472	1.638	1.702	1.870	2.110	2.167
	10	0.678	0.690	0.764	0.812	0.832	0.926	1.035	1.091	1.172	1.305	1.387	1.531	1.577	1.802	2.011	2.099
11	5	0.593	0.606	0.665	0.761	0.803	0.990	1.161	1.323	1.502	1.802	1.982	2.392	2.590	3.146	3.538	3.723
	6	0.580	0.612	0.664	0.724	0.773	0.937	1.110	1.242	1.397	1.657	1.787	2.100	2.232	2.721	3.205	3.292
	7	0.597	0.621	0.684	0.759	0.788	0.926	1.055	1.155	1.270	1.534	1.654	1.904	2.059	2.312	2.754	2.831
	8	0.621	0.631	0.683	0.793	0.330	0.949	1.076	1.169	1.272	1.455	1.563	1.753	1.884	2.298	2.669	2.835
	9	0.642	0.670	0.726	0.798	0.832	0.944	1.059	1.127	1.221	1.405	1.473	1.620	1.713	1.911	2.165	2.300
	10	0.672	0.679	0.736	0.812	0.837	0.939	1.057	1.137	1.216	1.382	1.452	1.596	1.685	1.916	2.129	2.170
	11	0.687	0.704	0.753	0.801	0.820	0.904	1.003	1.061	1.149	1.291	1.350	1.483	1.547	1.719	1.870	1.958
12	5	0.539	0.570	0.671	0.754	0.791	0.961	1.174	1.319	1.507	1.890	2.048	2.443	2.686	3.543	4.716	5.042
	6	0.584	0.599	0.654	0.741	0.776	0.927	1.105	1.220	1.371	1.975	1.792	2.113	2.299	2.758	3.114	3.180
	7	0.593	0.615	0.677	0.749	0.794	0.915	1.079	1.176	1.276	1.546	1.669	1.891	1.998	2.342	2.708	2.770
	8	0.630	0.639	0.695	0.769	0.809	0.916	1.054	1.162	1.282	1.466	1.559	1.715	1.805	2.125	2.377	2.541
	9	0.666	0.674	0.717	0.793	0.818	0.928	1.050	1.130	1.222	1.396	1.480	1.679	1.763	2.115	2.512	2.824
	10	0.674	0.683	0.735	0.802	0.831	0.937	1.054	1.127	1.215	1.382	1.467	1.596	1.686	1.905	2.071	2.093

TABLE 13.3— Continued.

N	r	2	2.5	5	10	12.5	25	40	50	60	75	80	87.5	90	95	97.5	98
									α x 100 (%)								
	11	0.672	0.684	0.729	0.791	0.821	0.923	1.025	1.096	1.178	1.327	1.399	1.547	1.631	1.853	2.078	2.124
	12	0.677	0.683	0.752	0.823	0.840	0.928	1.030	1.089	1.159	1.292	1.352	1.494	1.557	1.711	1.893	1.914
	5	0.554	0.570	0.640	0.745	0.796	0.978	1.189	1.330	1.513	1.908	2.151	2.589	2.887	3.671	4.480	4.804
	6	0.614	0.640	0.703	0.802	0.842	1.004	1.168	1.274	1.408	1.708	1.839	2.064	2.200	2.686	3.084	3.213
	7	0.598	0.612	0.667	0.745	0.776	0.923	1.077	1.178	1.290	1.529	1.658	1.903	2.030	2.371	2.792	2.961
	8	0.626	0.633	0.692	0.771	0.795	0.918	1.048	1.144	1.241	1.477	1.583	1.756	1.839	2.172	2.490	2.549
13	9	0.636	0.654	0.708	0.778	0.809	0.935	1.055	1.145	1.249	1.454	1.557	1.716	1.809	2.152	2.565	2.694
	10	0.646	0.657	0.696	0.771	0.798	0.923	1.052	1.129	1.211	1.384	1.448	1.616	1.721	2.016	2.219	2.267
	11	0.674	0.681	0.722	0.787	0.812	0.929	1.036	1.105	1.190	1.333	1.405	1.535	1.592	1.795	1.900	2.051
	12	0.678	0.684	0.741	0.801	0.825	0.905	1.001	1.071	1.146	1.281	1.346	1.500	1.557	1.701	1.902	1.902
	13	0.713	0.730	0.774	0.846	0.866	0.951	1.038	1.102	1.159	1.280	1.332	1.450	1.498	1.651	1.791	1.870
	5	0.558	0.573	0.641	0.752	0.778	0.960	1.169	1.334	1.529	1.875	2.112	2.487	2.643	3.254	4.050	4.208
	6	0.584	0.595	0.681	0.766	0.808	0.963	1.147	1.264	1.392	1.705	1.870	2.212	2.351	2.850	3.570	3.848
	7	0.592	0.619	0.681	0.774	0.806	0.958	1.126	1.263	1.389	1.639	1.752	1.995	2.085	2.410	2.744	2.807
14	8	0.639	0.656	0.706	0.770	0.804	0.926	1.053	1.148	1.260	1.474	1.558	1.752	1.855	2.174	2.509	2.578
	9	0.636	0.646	0.706	0.766	0.791	0.905	1.027	1.113	1.215	1.411	1.480	1.654	1.731	1.980	2.182	2.273
	10	0.639	0.648	0.701	0.767	0.810	0.925	1.042	1.122	1.213	1.367	1.454	1.623	1.706	1.935	2.193	2.258

TABLE 13.3– Continued.

N	r	2	2.5	5	10	12.5	25	40	50	60	75	80	87.5	90	95	97.5	98
	11	0.671	0.678	0.738	0.793	0.821	0.914	1.023	1.093	1.173	1.354	1.426	1.572	1.639	1.907	2.197	2.279
	12	0.681	0.692	0.740	0.795	0.829	0.920	1.023	1.092	1.168	1.315	1.376	1.511	1.595	1.776	1.956	2.073
	13	0.704	0.719	0.756	0.816	0.839	0.930	1.015	1.082	1.153	1.297	1.358	1.479	1.543	1.734	1.884	1.974
	14	0.705	0.714	0.756	0.813	0.836	0.924	1.006	1.079	1.144	1.267	1.323	1.418	1.466	1.659	1.767	1.777
15	5	0.575	0.596	0.663	0.761	0.792	0.943	1.137	1.286	1.490	1.851	2.026	2.417	2.626	3.382	4.519	4.783
	6	0.572	0.603	0.661	0.768	0.801	0.961	1.149	1.279	1.412	1.761	1.913	2.249	2.476	3.077	3.711	3.841
	7	0.585	0.598	0.662	0.738	0.775	0.920	1.095	1.205	1.336	1.579	1.720	1.929	2.025	2.373	2.814	2.956
	8	0.615	0.626	0.676	0.751	0.782	0.914	1.052	1.151	1.256	1.460	1.560	1.747	1.853	2.126	2.557	2.739
	9	0.639	0.655	0.700	0.764	0.795	0.919	1.042	1.134	1.233	1.433	1.506	1.687	1.802	2.079	2.458	2.513
	10	0.648	0.663	0.719	0.796	0.824	0.925	1.056	1.138	1.231	1.404	1.496	1.667	1.743	2.027	2.329	2.398
	11	0.665	0.677	0.718	0.791	0.816	0.920	1.025	1.116	1.185	1.340	1.425	1.555	1.633	1.807	2.052	2.159
	12	0.685	0.704	0.741	0.804	0.834	0.927	1.045	1.110	1.191	1.325	1.400	1.560	1.602	1.766	2.037	2.044
	13	0.707	0.717	0.755	0.810	0.834	0.934	1.024	1.078	1.143	1.280	1.347	1.480	1.542	1.720	1.943	1.975
	14	0.706	0.728	0.756	0.805	0.822	0.912	1.024	1.079	1.139	1.271	1.328	1.444	1.487	1.617	1.754	1.857
	15	0.726	0.732	0.782	0.829	0.852	0.939	1.009	1.071	1.135	1.242	1.290	1.383	1.419	1.598	1.747	1.770
8	8	0.630	0.638	0.699	0.764	0.793	0.925	1.067	1.173	1.280	1.529	1.636	1.859	1.952	2.315	2.542	2.677
9	9	0.626	0.642	0.706	0.765	0.797	0.904	1.032	1.125	1.211	1.420	1.498	1.701	1.801	2.092	2.310	2.444

TABLE 13.3– Continued.

N	r	2	2.5	5	10	12.5	25	40	50	60	75	80	87.5	90	95	97.5	98
								α × 100 (%)									
16	10	0.658	0.667	0.716	0.775	0.798	0.922	1.036	1.132	1.221	1.378	1.446	1.617	1.693	1.864	2.095	2.151
	11	0.664	0.676	0.735	0.812	0.832	0.929	1.037	1.116	1.192	1.358	1.430	1.561	1.646	1.826	2.004	2.061
	12	0.682	0.700	0.741	0.794	0.821	0.924	1.016	1.097	1.174	1.336	1.399	1.552	1.623	1.765	1.918	1.963
	13	0.679	0.686	0.728	0.806	0.834	0.923	1.016	1.081	1.159	1.290	1.347	1.500	1.550	1.742	1.946	1.998
	14	0.709	0.718	0.762	0.820	0.839	0.933	1.019	1.070	1.142	1.251	1.304	1.415	1.466	1.576	1.725	1.774
	15	0.718	0.724	0.763	0.831	0.856	0.936	1.013	1.074	1.142	1.255	1.305	1.440	1.494	1.610	1.763	1.823
	16	0.711	0.718	0.757	0.810	0.840	0.919	0.993	1.047	1.108	1.200	1.246	1.335	1.392	1.506	1.637	1.697
17	8	0.620	0.647	0.688	0.758	0.792	0.911	1.057	1.182	1.285	1.518	1.609	1.864	1.999	2.294	2.677	2.857
	9	0.626	0.645	0.712	0.773	0.801	0.915	1.043	1.139	1.241	1.427	1.533	1.715	1.808	2.019	2.293	2.343
	10	0.646	0.656	0.711	0.783	0.808	0.918	1.039	1.113	1.200	1.377	1.445	1.590	1.677	1.939	2.211	2.258
	11	0.653	0.674	0.732	0.787	0.814	3.924	1.032	1.114	1.208	1.382	1.456	1.584	1.647	1.843	2.044	2.168
	12	0.676	0.685	0.733	0.797	0.818	0.915	1.023	1.091	1.160	1.320	1.382	1.528	1.577	1.749	1.979	2.047
	13	0.671	0.682	0.719	0.787	0.809	0.919	1.016	1.083	1.165	1.284	1.340	1.443	1.508	1.653	1.847	1.875
	14	0.731	0.740	0.788	0.838	0.858	0.943	1.023	1.084	1.155	1.254	1.313	1.415	1.444	1.596	1.720	1.778
	15	0.688	0.704	0.758	0.813	0.832	0.918	1.008	1.073	1.138	1.249	1.299	1.411	1.479	1.615	1.736	1.773
	16	0.722	0.731	0.772	0.820	0.845	0.924	0.999	1.052	1.109	1.208	1.256	1.379	1.430	1.555	1.730	1.772
	17	0.704	0.350	0.784	0.832	0.849	0.930	1.003	1.052	1.103	1.212	1.273	1.351	1.389	1.553	1.654	1.690

TABLE 13.3– Continued.

N	r	$\alpha \times 100$ (%)															
		2	2.5	5	10	12.5	25	40	50	60	75	80	87.5	90	95	97.5	98
18	9	0.605	0.631	0.692	0.776	0.815	0.946	1.080	1.173	1.283	1.474	1.549	1.741	1.815	2.138	2.458	2.593
	10	0.654	0.666	0.698	0.769	0.791	0.912	1.032	1.105	1.212	1.392	1.472	1.650	1.731	1.947	2.304	2.474
	11	0.642	0.653	0.712	0.780	0.804	0.918	1.033	1.099	1.190	1.361	1.437	1.570	1.662	1.930	2.083	2.177
	12	0.683	0.690	0.746	0.807	0.826	0.934	1.035	1.107	1.191	1.346	1.406	1.553	1.604	1.832	2.032	2.124
	13	0.667	0.676	0.720	0.774	0.816	0.911	1.015	1.094	1.176	1.296	1.363	1.476	1.529	1.740	1.871	1.927
	14	0.686	0.701	0.757	0.813	0.835	0.928	1.029	1.086	1.150	1.290	1.366	1.510	1.550	1.710	1.880	1.910
	15	0.704	0.713	0.760	0.815	0.834	0.927	1.010	1.065	1.133	1.249	1.308	1.413	1.468	1.643	1.804	1.826
	16	0.683	0.695	0.747	0.788	0.814	0.911	0.995	1.040	1.097	1.209	1.258	1.346	1.385	1.538	1.693	1.722
	17	0.728	0.741	0.771	0.834	0.849	0.928	1.006	1.062	1.125	1.228	1.281	1.365	1.404	1.526	1.632	1.714
	18	0.731	0.737	0.784	0.836	0.856	0.936	1.010	1.062	1.115	1.198	1.237	1.310	1.347	1.455	1.620	1.643
19	9	0.635	0.648	0.695	0.750	0.790	0.915	1.049	1.150	1.244	1.479	1.569	1.805	1.919	2.233	2.626	2.704
	10	0.626	0.638	0.690	0.768	0.799	0.942	1.063	1.152	1.246	1.440	1.518	1.667	1.733	1.985	2.303	2.358
	11	0.677	0.683	0.716	0.781	0.813	0.918	1.023	1.106	1.190	1.360	1.429	1.549	1.629	1.842	2.130	2.284
	12	0.668	0.683	0.727	0.788	0.813	0.927	1.034	1.107	1.176	1.308	1.398	1.560	1.614	1.828	2.032	2.075
	13	0.674	0.694	0.742	0.807	0.828	0.915	1.011	1.095	1.178	1.351	1.425	1.571	1.644	1.794	1.989	2.038
	14	0.674	0.693	0.745	0.797	0.818	0.915	0.998	1.063	1.140	1.269	1.326	1.433	1.472	1.613	1.720	1.752
	15	0.670	0.694	0.740	0.795	0.818	0.911	0.996	1.049	1.132	1.254	1.312	1.443	1.480	1.630	1.775	1.809

TABLE 13.3– Continued.

N	r	2	2.5	5	10	12.5	25	40	50	60	75	80	87.5	90	95	97.5	98
	16	0.708	0.719	0.763	0.814	0.839	0.930	1.032	1.085	1.148	1.252	1.313	1.402	1.449	1.579	1.712	1.831
	17	0.720	0.726	0.774	0.819	0.846	0.940	1.014	1.067	1.127	1.229	1.279	1.383	1.411	1.540	1.639	1.656
	18	0.700	0.713	0.752	0.815	0.844	0.922	1.008	1.055	1.115	1.210	1.259	1.328	1.383	1.525	1.692	1.762
	19	0.728	0.748	0.784	0.838	0.856	0.925	0.999	1.051	1.116	1.212	1.251	1.339	1.367	1.485	1.589	1.632
	10	0.633	0.649	0.704	0.778	0.809	0.928	1.061	1.146	1.243	1.413	1.494	1.683	1.762	2.010	2.299	2.431
	11	0.635	0.647	0.714	0.774	0.806	0.925	1.044	1.126	1.213	1.375	1.476	1.625	1.695	1.934	2.266	2.356
	12	0.649	0.662	0.710	0.785	0.823	0.927	1.046	1.121	1.198	1.348	1.418	1.552	1.606	1.818	2.060	2.112
	13	0.672	0.686	0.742	0.795	0.817	0.916	1.016	1.077	1.165	1.293	1.354	1.517	1.569	1.736	1.931	1.951
	14	0.684	0.693	0.736	0.791	0.819	0.928	1.015	1.077	1.145	1.272	1.338	1.438	1.506	1.613	1.751	1.791
	15	0.672	0.679	0.739	0.805	0.829	0.924	1.013	1.075	1.149	1.279	1.334	1.449	1.500	1.675	1.820	1.878
20	16	0.694	0.700	0.750	0.811	0.837	0.917	1.016	1.074	1.130	1.247	1.297	1.396	1.452	1.612	1.774	1.828
	17	0.743	0.750	0.786	0.835	0.850	0.938	1.015	1.064	1.115	1.224	1.269	1.364	1.416	1.528	1.629	1.649
	18	0.723	0.729	0.777	0.818	0.841	0.917	1.008	1.063	1.121	1.220	1.266	1.349	1.412	1.528	1.589	1.644
	19	0.729	0.740	0.774	0.838	0.856	0.932	1.009	1.067	1.125	1.234	1.266	1.356	1.387	1.478	1.567	1.613
	20	0.725	0.751	0.782	0.841	0.854	0.921	.989	1.034	1.083	1.165	1.212	1.289	1.327	1.422	1.487	1.534
	11	0.635	0.660	0.709	0.781	0.811	0.923	1.038	1.123	1.209	1.377	1.461	1.589	1.636	1.822	2.102	2.130
	12	0.655	0.657	0.722	0.783	0.816	0.910	1.041	1.117	1.201	1.362	1.449	1.619	1.652	1.828	2.018	2.104

483

TABLE 13.3- Continued.

N	r	2	2.5	5	10	12.5	25	40	50	60	75	80	87.5	90	95	97.5	98
	13	0.689	0.700	0.746	0.801	0.828	0.918	1.013	1.084	1.161	1.308	1.380	1.491	1.531	1.699	1.895	1.920
	14	0.660	0.682	0.725	0.784	0.810	0.928	1.029	1.091	1.173	1.301	1.369	1.477	1.534	1.734	1.877	1.898
	15	0.688	0.698	0.733	0.786	0.815	0.915	1.012	1.078	1.152	1.280	1.329	1.446	1.504	1.631	1.839	1.882
	16	0.698	0.705	0.743	0.802	0.826	0.923	1.015	1.077	1.138	1.253	1.314	1.406	1.458	1.662	1.778	1.839
22	17	0.698	0.705	0.744	0.798	0.823	0.907	1.007	1.068	1.130	1.228	1.280	1.386	1.427	1.536	1.662	1.682
	18	0.720	0.730	0.777	0.825	0.848	0.930	1.015	1.079	1.132	1.233	1.274	1.394	1.430	1.589	1.684	1.728
	19	0.724	0.733	0.773	0.819	0.846	0.920	1.003	1.050	1.106	1.209	1.254	1.333	1.379	1.482	1.601	1.639
	20	0.719	0.739	0.776	0.831	0.852	0.938	1.015	1.054	1.118	1.219	1.254	1.347	1.377	1.465	1.550	1.578
	21	0.746	0.757	0.790	0.833	0.850	0.927	1.001	1.041	1.089	1.199	1.232	1.303	1.326	1.415	1.512	1.543
	22	0.760	0.764	0.800	0.845	0.867	0.928	.994	1.038	1.086	1.163	1.212	1.282	1.326	1.443	1.539	1.558
	12	0.637	0.653	0.710	0.771	0.801	0.910	1.027	1.099	1.193	1.359	1.415	1.556	1.646	1.838	2.024	2.092
	13	0.655	0.670	0.717	0.787	0.815	0.913	1.024	1.096	1.174	1.328	1.403	1.543	1.618	1.789	1.924	1.982
	14	0.683	0.704	0.742	0.801	0.825	0.930	1.023	1.079	1.152	1.291	1.351	1.471	1.533	1.654	1.847	1.909
	15	0.688	0.696	0.742	0.801	0.824	0.909	1.003	1.065	1.149	1.280	1.333	1.467	1.521	1.695	1.897	1.966
	16	0.707	0.737	0.765	0.826	0.849	0.929	1.017	1.068	1.130	1.259	1.312	1.416	1.459	1.621	1.705	1.759
24	17	0.713	0.724	0.763	0.826	0.851	0.926	1.021	1.070	1.122	1.243	1.295	1.384	1.443	1.562	1.671	1.706
	18	0.700	0.723	0.769	0.817	0.848	0.930	1.016	1.071	1.126	1.245	1.285	1.386	1.449	1.575	1.682	1.724

$\alpha \times 100$ (%)

TABLE 13.3– Continued.

N	r	α × 100 (%)															
		2	2.5	5	10	12.5	25	40	50	60	75	80	87.5	90	95	97.5	98
	19	0.713	0.726	0.767	0.823	0.835	0.927	0.999	1.046	1.106	1.221	1.264	1.349	1.382	1.494	1.593	1.608
	20	0.739	0.747	0.782	0.837	0.857	0.936	1.019	1.071	1.123	1.231	1.264	1.343	1.377	1.508	1.616	1.653
	21	0.727	0.747	0.778	0.828	0.841	0.924	1.000	1.049	1.097	1.197	1.231	1.312	1.339	1.449	1.569	1.617
	22	0.742	0.757	0.792	0.841	0.862	0.930	1.002	1.046	1.103	1.194	1.232	1.302	1.335	1.429	1.530	1.603
	23	0.749	0.758	0.801	0.848	0.866	0.924	0.994	1.036	1.089	1.176	1.216	1.279	1.310	1.398	1.472	1.508
	24	0.757	0.763	0.797	0.846	0.867	0.932	0.992	1.026	1.067	1.146	1.183	1.251	1.284	1.378	1.460	1.489
	13	0.673	0.693	0.734	0.801	0.824	0.928	1.033	1.093	1.164	1.311	1.373	1.482	1.556	1.721	1.923	1.992
	14	0.679	0.695	0.736	0.794	0.823	0.920	1.014	1.078	1.156	1.280	1.349	1.460	1.522	1.653	1.833	1.877
	15	0.685	0.698	0.743	0.801	0.828	0.928	1.006	1.069	1.146	1.271	1.324	1.456	1.506	1.690	1.835	1.865
	16	0.672	0.689	0.743	0.801	0.827	0.925	1.007	1.067	1.135	1.266	1.318	1.418	1.468	1.631	1.793	1.817
	17	0.716	0.724	0.772	0.834	0.857	0.932	1.022	1.074	1.135	1.262	1.305	1.401	1.460	1.605	1.725	1.791
26	18	0.698	0.715	0.763	0.816	0.840	0.931	1.000	1.061	1.124	1.226	1.271	1.359	1.401	1.520	1.667	1.721
	19	0.713	0.719	0.759	0.810	0.836	0.915	1.004	1.064	1.123	1.234	1.274	1.366	1.418	1.571	1.636	1.666
	20	0.724	0.731	0.777	0.822	0.844	0.924	1.017	1.062	1.115	1.208	1.255	1.342	1.380	1.503	1.628	1.647
	21	0.742	0.752	0.778	0.830	0.849	0.935	1.003	1.040	1.095	1.193	1.235	1.308	1.342	1.451	1.523	1.556
	22	0.751	0.758	0.794	0.843	0.867	0.941	1.016	1.058	1.108	1.212	1.256	1.333	1.362	1.463	1.553	1.586
	23	0.726	0.733	0.768	0.830	0.846	0.909	0.981	1.027	1.076	1.148	1.183	1.259	1.305	1.399	1.531	1.540

TABLE 13.3– Continued.

N	r	α x 100 (%)															
		2	2.5	5	10	12.5	25	40	50	60	75	80	87.5	90	95	97.5	98
24	24	0.744	0.758	0.799	0.857	0.871	0.927	0.991	1.034	1.086	1.172	1.209	1.282	1.312	1.428	1.521	1.536
	25	0.760	0.766	0.799	0.852	0.867	0.933	0.998	1.045	1.087	1.177	1.213	1.285	1.314	1.391	1.475	1.506
	26	0.758	0.765	0.809	0.860	0.877	0.941	1.000	1.034	1.076	1.155	1.184	1.235	1.253	1.350	1.401	1.420
28	14	0.670	0.680	0.725	0.790	0.820	0.917	1.010	1.082	1.147	1.297	1.352	1.477	1.528	1.704	1.869	1.938
	15	0.679	0.699	0.759	0.813	0.839	0.928	1.019	1.077	1.150	1.280	1.339	1.463	1.512	1.653	1.773	1.841
	16	0.673	0.683	0.746	0.814	0.833	0.926	1.021	1.076	1.148	1.272	1.316	1.403	1.455	1.600	1.732	1.774
	17	0.701	0.719	0.761	0.822	0.836	0.925	1.009	1.069	1.137	1.252	1.307	1.394	1.436	1.566	1.678	1.716
	18	0.702	0.715	0.758	0.820	0.844	0.930	1.018	1.071	1.127	1.230	1.276	1.389	1.426	1.541	1.678	1.742
	19	0.726	0.738	0.786	0.833	0.848	0.924	1.007	1.064	1.117	1.221	1.276	1.358	1.395	1.506	1.624	1.666
	20	0.727	0.730	0.769	0.819	0.842	0.921	1.002	1.062	1.112	1.211	1.262	1.353	1.385	1.488	1.584	1.617
	21	0.733	0.744	0.779	0.830	0.851	0.937	1.017	1.064	1.113	1.204	1.244	1.339	1.373	1.463	1.609	1.657
	22	0.720	0.734	0.771	0.827	0.852	0.928	1.008	1.051	1.108	1.196	1.236	1.311	1.341	1.457	1.549	1.557
	23	0.731	0.740	0.788	0.836	0.850	0.922	0.990	1.033	1.083	1.168	1.218	1.304	1.351	1.450	1.563	1.609
	24	0.732	0.742	0.792	0.833	0.850	0.919	0.986	1.033	1.086	1.168	1.205	1.306	1.342	1.452	1.533	1.570
	25	0.760	0.769	0.793	0.839	0.858	0.935	1.007	1.047	1.099	1.182	1.220	1.286	1.315	1.402	1.470	1.508
	26	0.767	0.778	0.806	0.850	0.864	0.932	0.989	1.028	1.063	1.149	1.182	1.262	1.297	1.413	1.514	1.536
	27	0.756	0.767	0.815	0.866	0.882	0.942	1.007	1.043	1.090	1.159	1.194	1.257	1.279	1.367	1.463	1.495

TABLE 13.3– Continued.

N	r								α x 100 (%)								
		2	2.5	5	10	12.5	25	40	50	60	75	80	87.5	90	95	97.5	98
	28	0.777	0.787	0.820	0.859	0.877	0.933	0.995	1.038	1.077	1.157	1.185	1.245	1.272	1.342	1.437	1.444
30	15	0.680	0.693	0.751	0.798	0.822	0.911	1.008	1.068	1.149	1.287	1.334	1.466	1.518	1.694	1.858	1.895
	16	0.694	0.708	0.759	0.811	0.834	0.916	1.007	1.081	1.142	1.254	1.316	1.428	1.479	1.599	1.731	1.829
	17	0.696	0.701	0.745	0.805	0.834	0.916	1.015	1.069	1.132	1.258	1.312	1.422	1.465	1.629	1.763	1.804
	18	0.715	0.719	0.754	0.821	0.851	0.945	1.021	1.081	1.136	1.249	1.305	1.405	1.431	1.543	1.699	1.722
	19	0.710	0.716	0.765	0.814	0.833	0.913	0.997	1.049	1.107	1.208	1.254	1.360	1.413	1.523	1.601	1.628
	20	0.729	0.736	0.765	0.816	0.843	0.927	1.008	1.057	1.110	1.218	1.266	1.369	1.424	1.546	1.649	1.704
	21	0.705	0.718	0.760	0.821	0.849	0.935	1.010	1.059	1.121	1.201	1.239	1.324	1.365	1.483	1.593	1.630
	22	0.731	0.741	0.783	0.821	0.850	0.921	1.001	1.053	1.101	1.198	1.236	1.306	1.346	1.463	1.571	1.610
	23	0.735	0.740	0.786	0.836	0.852	0.920	0.992	1.044	1.094	1.189	1.241	1.331	1.362	1.503	1.602	1.621
	24	0.719	0.734	0.774	0.817	0.840	0.912	0.986	1.031	1.078	1.161	1.194	1.269	1.315	1.420	1.531	1.562
	25	0.748	0.756	0.793	0.846	0.870	0.939	1.000	1.043	1.086	1.172	1.213	1.286	1.319	1.401	1.495	1.513
	26	0.758	0.766	0.801	0.849	0.877	0.936	0.998	1.036	1.076	1.163	1.198	1.259	1.285	1.351	1.448	1.515
	27	0.760	0.772	0.810	0.855	0.868	0.936	0.997	1.037	1.083	1.160	1.193	1.258	1.287	1.361	1.435	1.455
	28	0.768	0.774	0.812	0.846	0.857	0.933	0.993	1.037	1.080	1.159	1.187	1.252	1.276	1.360	1.454	1.479
	29	0.766	0.777	0.815	0.861	0.879	0.948	0.999	1.041	1.081	1.154	1.181	1.241	1.272	1.349	1.425	1.463
	30	0.789	0.796	0.824	0.861	0.877	0.929	0.988	1.029	1.064	1.145	1.175	1.226	1.257	1.319	1.418	1.429

TABLE 13.4— Confidence limits on $\hat{\beta} log_e(\hat{\eta}/\eta)$ for various N, r and $\alpha \times 100\%$, or $P(\hat{\beta} log_e(\hat{\eta}/\eta) < \kappa_\alpha) = \alpha$.

N	r	2	2.5	5	10	12.5	25	40	50	60	75	80	87.5	90	95	97.5	98
									$\alpha \times 100$ (%)								
5	5	-1.727	-1.510	-1.105	-0.809	-0.685	-0.417	-0.194	-0.059	0.097	0.326	0.432	0.633	0.746	1.194	1.551	1.728
6	5	-1.958	-1.738	-1.405	-1.010	-0.862	-0.475	-0.242	-0.091	0.058	0.260	0.379	0.568	0.669	0.948	1.201	1.242
	6	-1.420	-1.308	-0.977	-0.716	-0.630	-0.392	-0.194	-0.031	0.072	0.291	0.398	0.606	0.692	0.962	1.188	1.265
7	5	-2.402	-2.301	-1.756	-1.212	-1.040	-0.549	-0.288	-0.113	0.018	0.239	0.319	0.495	0.562	0.789	0.973	1.034
	6	-1.766	-1.638	-1.248	-0.864	-0.705	-0.414	-0.181	-0.078	0.037	0.253	0.353	0.527	0.576	0.807	0.987	1.058
	7	-1.413	-1.249	-0.857	-0.655	-0.577	-0.340	-0.167	-0.039	0.072	0.302	0.380	0.543	0.596	0.825	0.994	1.054
8	5	-3.036	-2.715	-2.051	-1.464	-1.220	-0.678	-0.316	-0.167	0.001	0.203	0.267	0.393	0.464	0.666	0.897	0.936
	6	-2.067	-1.761	-1.326	-0.932	-0.797	-0.471	-0.234	-0.096	0.033	0.212	0.282	0.445	0.502	0.706	0.928	0.958
	7	-1.427	-1.276	-1.009	-0.688	-0.631	-0.364	-0.163	-0.043	0.063	0.230	0.306	0.440	0.505	0.714	0.936	0.984
	8	-1.032	-0.942	-0.770	-0.602	-0.528	-0.306	-0.136	-0.020	0.082	0.251	0.338	0.455	0.515	0.732	0.955	0.992
9	5	-3.794	-3.320	-2.358	-1.557	-1.350	-0.748	-0.378	-0.193	-0.039	0.163	0.245	0.384	0.429	0.577	0.786	0.854
	6	-1.833	-1.682	-1.325	-0.883	-0.776	-0.485	-0.252	-0.111	0.014	0.173	0.245	0.378	0.441	0.590	0.747	0.798
	7	-1.764	-1.476	-1.129	-0.803	-0.704	-0.416	-0.207	-0.081	0.031	0.211	0.282	0.440	0.515	0.668	0.793	0.831
	8	-1.083	-1.067	-0.863	-0.644	-0.567	-0.343	-0.143	-0.052	0.049	0.239	0.299	0.441	0.520	0.709	0.884	0.936
	9	-0.945	-0.857	-0.720	-0.542	-0.496	-0.300	-0.120	-0.031	0.077	0.240	0.294	0.427	0.497	0.657	0.843	0.911
5	5	-4.891	-4.315	-2.921	-1.905	-1.599	-0.849	-0.432	-0.224	-0.054	0.168	0.246	0.367	0.406	0.560	0.693	0.739
6	6	-2.139	-2.084	-1.519	-1.101	-0.945	-0.556	-0.289	-0.150	-0.029	0.181	0.248	0.386	0.434	0.578	0.735	0.768

TABLE 13.4– Continued.

N	r	2	2.5	5	10	12.5	25	40	50	60	75	80	87.5	90	95	97.5	98
									α × 100 (%)								
10	7	-1.659	-1.614	-1.263	-0.910	-0.810	-0.455	-0.223	-0.111	0.022	0.200	0.257	0.388	0.439	0.605	0.773	0.815
	8	-1.287	-1.245	-0.904	-0.720	-0.631	-0.374	-0.168	-0.054	0.051	0.209	0.305	0.430	0.477	0.614	0.736	0.771
	9	-1.046	-0.973	-0.762	-0.600	-0.552	-0.323	-0.141	-0.049	0.051	0.211	0.265	0.402	0.458	0.614	0.771	0.818
	10	-0.978	-0.863	-0.661	-0.485	-0.430	-0.252	-0.125	-0.032	0.051	0.224	0.274	0.404	0.471	0.589	0.794	0.859
11	5	-3.695	-3.455	-2.521	-1.786	-1.610	-0.855	-0.459	-0.258	-0.087	0.161	0.247	0.373	0.427	0.571	0.669	0.697
	6	-2.695	-2.380	-1.819	-1.297	-1.123	-0.641	-0.293	-0.152	-0.034	0.209	0.285	0.401	0.464	0.592	0.705	0.756
	7	-1.887	-1.741	-1.289	-0.920	-0.787	-0.443	-0.224	-0.105	-0.001	0.189	0.259	0.376	0.420	0.605	0.729	0.766
	8	-1.273	-1.171	-0.937	-0.723	-0.646	-0.375	-0.191	-0.065	0.033	0.194	0.253	0.393	0.452	0.595	0.766	0.810
	9	-1.040	-0.983	-0.814	-0.638	-0.555	-0.348	-0.147	-0.056	0.039	0.181	0.243	0.356	0.408	0.537	0.657	0.712
	10	-0.996	-0.900	-0.661	-0.507	-0.458	-0.282	-0.131	-0.052	0.050	0.206	0.276	0.403	0.454	0.595	0.757	0.784
	11	-0.803	-0.725	-0.618	-0.457	-0.396	-0.242	-0.105	-0.024	0.068	0.223	0.293	0.394	0.447	0.587	0.770	0.790
12	5	-5.084	-4.633	-3.131	-2.039	-1.782	-0.989	-0.517	-0.295	-0.124	0.116	0.211	0.367	0.419	0.548	0.659	0.678
	6	-2.692	-2.489	-1.835	-1.356	-1.196	-0.692	-0.385	-0.210	-0.061	0.148	0.223	0.355	0.391	0.538	0.664	0.690
	7	-2.209	-1.971	-1.356	-0.994	-0.843	-0.506	-0.250	-0.116	0.011	0.179	0.247	0.373	0.426	0.558	0.706	0.717
	8	-1.576	-1.501	-1.109	-0.820	-0.703	-0.417	-0.215	-0.112	0.009	0.159	0.230	0.333	0.394	0.550	0.695	0.711
	9	-1.274	-1.212	-0.921	-0.684	-0.594	-0.356	-0.162	-0.053	0.026	0.179	0.247	0.339	0.389	0.545	0.673	0.740
	10	-0.955	-0.858	-0.722	-0.571	-0.507	-0.268	-0.122	-0.017	0.071	0.220	0.280	0.380	0.424	0.582	0.693	0.728

TABLE 13.4– Continued.

N	r	2	2.5	5	10	12.5	25	40	50	60	75	80	87.5	90	95	97.5	98
	11	-0.813	-0.783	-0.619	-0.498	-0.457	-0.264	-0.122	-0.041	0.031	0.172	0.243	0.378	0.428	0.594	0.749	0.805
	12	-0.795	-0.754	-0.592	-0.448	-0.398	-0.213	-0.085	-0.012	0.066	0.203	0.276	0.367	0.423	0.573	0.706	0.745
13	5	-5.095	-4.400	-3.414	-2.268	-1.965	-1.116	-0.601	-0.398	-0.147	0.121	0.225	0.352	0.413	0.588	0.657	0.672
	6	-2.926	-2.609	-2.112	-1.580	-1.385	-0.815	-0.431	-0.268	-0.103	0.108	0.181	0.311	0.368	0.491	0.642	0.670
	7	-2.089	-2.034	-1.492	-0.983	-0.858	-0.488	-0.259	-0.139	-0.011	0.168	0.226	0.351	0.401	0.535	0.634	0.685
	8	-1.476	-1.396	-1.128	-0.814	-0.715	-0.427	-0.213	-0.099	0.010	0.180	0.230	0.334	0.379	0.513	0.604	0.634
	9	-1.350	-1.304	-0.965	-0.682	-0.620	-0.389	-0.173	-0.085	0.02	0.149	0.208	0.323	0.365	0.487	0.611	0.687
	10	-1.036	-0.968	-0.813	-0.582	-0.517	-0.301	-0.137	-0.045	0.042	0.180	0.249	0.345	0.394	0.505	0.657	0.745
	11	-0.895	-0.827	-0.669	-0.503	-0.442	-0.263	-0.117	-0.034	0.044	0.187	0.243	0.336	0.381	0.515	0.665	0.702
	12	-0.788	-0.738	-0.586	-0.425	-0.383	-0.227	-0.097	-0.019	0.056	0.201	0.266	0.382	0.420	0.572	0.738	0.795
	13	-0.682	-0.658	-0.553	-0.426	-0.365	-0.232	-0.090	-0.013	0.068	0.191	0.247	0.357	0.407	0.529	0.639	0.650
14	5	-4.779	-4.302	-3.214	-2.267	-2.010	-1.188	-0.618	-0.398	-0.169	0.112	0.206	0.361	0.429	0.574	0.676	0.726
	6	-2.766	-2.657	-2.124	-1.573	-1.408	-0.862	-0.455	-0.285	-0.122	0.111	0.180	0.326	0.387	0.539	0.633	0.673
	7	-2.217	-2.031	-1.600	-1.161	-1.042	-0.624	-0.319	-0.161	-0.016	0.143	0.208	0.314	0.362	0.477	0.622	0.651
	8	-1.680	-1.512	-1.090	-0.819	-0.731	-0.437	-0.219	-0.099	0.005	0.192	0.265	0.366	0.417	0.530	0.626	0.658
	9	-1.268	-1.183	-0.957	-0.705	-0.623	-0.374	-0.178	-0.076	0.020	0.170	0.232	0.338	0.376	0.538	0.663	0.690
	10	-1.117	-1.082	-0.828	-0.608	-0.548	-0.326	-0.166	-0.076	0.021	0.174	0.226	0.332	0.391	0.507	0.612	0.685

TABLE 13.4– Continued.

N	r	2	2.5	5	10	12.5	25	40	50	60	75	80	87.5	90	95	97.5	98
									α x 100 (%)								
	11	-1.002	-0.969	-0.696	-0.531	-0.460	-0.290	-0.144	-0.051	0.037	0.164	0.228	0.318	0.356	0.476	0.616	0.666
	12	-0.876	-0.823	-0.625	-0.481	-0.423	-0.238	-0.102	-0.023	0.052	0.194	0.254	0.362	0.403	0.569	0.712	0.747
	13	-0.864	-0.810	-0.597	-0.452	-0.409	-0.256	-0.089	-0.007	0.085	0.205	0.251	0.352	0.404	0.529	0.629	0.642
	14	-0.708	-0.687	-0.545	-0.396	-0.354	-0.212	-0.096	-0.013	0.057	0.191	0.256	0.365	0.409	0.524	0.660	0.745
15	5	-5.292	-5.012	-3.241	-2.312	-2.005	-1.222	-0.632	-0.393	-0.179	0.071	0.161	0.331	0.397	0.553	0.633	0.679
	6	-3.985	-3.423	-2.650	-1.742	-1.550	-0.912	-0.455	-0.271	-0.069	0.149	0.247	0.378	0.433	0.554	0.664	0.685
	7	-2.624	-2.254	-1.761	-1.225	-1.085	-0.608	-0.318	-0.164	-0.046	0.146	0.207	0.342	0.392	0.496	0.599	0.640
	8	-1.642	-1.596	-1.261	-0.944	-0.850	-0.480	-0.259	-0.134	-0.027	0.142	0.207	0.328	0.363	0.480	0.564	0.577
	9	-1.376	-1.340	-1.051	-0.771	-0.681	-0.401	-0.217	-0.108	0.003	0.143	0.198	0.304	0.367	0.467	0.547	0.578
	10	-1.387	-1.262	-0.863	-0.672	-0.619	-0.359	-0.191	-0.096	0.009	0.152	0.211	0.313	0.362	0.508	0.628	0.644
	11	-1.101	-1.019	-0.787	-0.574	-0.493	-0.273	-0.133	-0.043	0.032	0.160	0.208	0.297	0.337	0.478	0.576	0.619
	12	-0.941	-0.910	-0.701	-0.499	-0.422	-0.242	-0.112	-0.031	0.057	0.181	0.237	0.325	0.362	0.519	0.634	0.661
	13	-0.802	-0.747	-0.588	-0.459	-0.404	-0.247	-0.111	-0.046	0.030	0.148	0.214	0.312	0.360	0.489	0.625	0.659
	14	-0.738	-0.679	-0.560	-0.420	-0.363	-0.207	-0.091	-0.018	0.070	0.209	0.247	0.352	0.384	0.479	0.567	0.597
	15	-0.688	-0.638	-0.522	-0.395	-0.347	-0.202	-0.082	-0.013	0.051	0.167	0.207	0.300	0.331	0.459	0.559	0.597
	8	-1.977	-1.817	-1.439	-1.091	-0.918	-0.536	-0.255	-0.113	0.001	0.126	0.187	0.299	0.357	0.473	0.586	0.614
	9	-1.375	-1.290	-1.063	-0.800	-0.671	-0.404	-0.210	-0.109	-0.011	0.155	0.208	0.325	0.365	0.504	0.588	0.615

TABLE 13.4– Continued.

N	r	2	2.5	5	10	12.5	25	40	50	60	75	80	87.5	90	95	97.5	98
	10	-1.299	-1.232	-0.974	-0.675	-0.591	-0.358	-0.195	-0.113	-0.22	0.152	0.210	0.301	0.351	0.432	0.522	0.540
	11	-1.119	-1.042	-0.816	-0.563	-0.502	-0.301	-0.128	-0.043	0.047	0.157	0.198	0.293	0.336	0.486	0.573	0.603
	12	-0.991	-0.911	-0.711	-0.528	-0.464	-0.267	-0.116	-0.023	0.047	0.179	0.216	0.320	0.356	0.475	0.595	0.613
16	13	-0.789	-0.760	-0.597	-0.450	-0.404	-0.227	-0.101	-0.033	0.045	0.162	0.231	0.317	0.361	0.485	0.561	0.599
	14	-0.696	-0.641	-0.543	-0.419	-0.364	-0.213	-0.096	-0.028	0.055	0.176	0.217	0.312	0.341	0.461	0.570	0.580
	15	-0.651	-0.612	-0.524	-0.411	-0.358	-0.221	-0.091	-0.025	0.032	0.165	0.216	0.311	0.351	0.472	0.554	0.562
	16	-0.611	-0.557	-0.505	-0.369	-0.332	-0.182	-0.075	-0.015	0.058	0.180	0.226	0.324	0.361	0.464	0.544	0.586
	8	-2.071	-2.005	-1.583	-1.055	-0.966	-0.550	-0.278	-0.140	-0.017	0.139	0.203	0.302	0.357	0.469	0.571	0.593
	9	-1.535	-1.445	-1.216	-0.858	-0.776	-0.444	-0.227	-0.121	-0.027	0.145	0.209	0.316	0.343	0.464	0.569	0.603
	10	-1.363	-1.284	-0.954	-0.674	-0.594	-0.342	-0.187	-0.105	-0.006	0.142	0.195	0.284	0.324	0.435	0.513	0.549
	11	-1.080	-1.056	-0.833	-0.595	-0.529	-0.320	-0.170	-0.069	0.014	0.146	0.200	0.287	0.326	0.452	0.556	0.574
17	12	-0.902	-0.835	-0.687	-0.516	-0.466	-0.286	-0.126	-0.051	0.049	0.167	0.205	0.307	0.360	0.475	0.583	0.598
	13	-0.754	-0.698	-0.581	-0.430	-0.378	-0.233	-0.101	-0.031	0.058	0.169	0.222	0.324	0.357	0.465	0.598	0.608
	14	-0.719	-0.681	-0.575	-0.452	-0.416	-0.239	-0.121	-0.039	0.036	0.144	0.198	0.272	0.308	0.405	0.513	0.532
	15	-0.677	-0.641	-0.536	-0.395	-0.360	-0.203	-0.085	-0.021	0.043	0.169	0.205	0.287	0.344	0.462	0.574	0.592
	16	-0.635	-0.606	-0.482	-0.350	-0.322	-0.193	-0.095	-0.024	0.048	0.166	0.202	0.303	0.350	0.441	0.529	0.571
	17	-0.552	-0.507	-0.438	-0.348	-0.318	-0.174	-0.067	-0.006	0.048	0.164	0.210	0.270	0.305	0.421	0.582	0.600

$\alpha \times 100$ (%)

TABLE 13.4– Continued.

N	r	2	2.5	5	10	12.5	25	40	50	60	75	80	87.5	90	95	97.5	98
									α × 100 (%)								
	9	-1.740	-1.646	-1.274	-0.960	-0.822	-0.481	-0.244	-0.123	-0.022	0.139	0.202	0.285	0.327	0.444	0.529	0.545
	10	-1.274	-1.239	-1.076	-0.768	-0.675	-0.360	-0.172	-0.077	0.016	0.143	0.201	0.291	0.350	0.457	0.570	0.609
	11	-1.234	-1.171	-0.903	-0.612	-0.546	-0.306	-0.143	-0.066	0.006	0.140	0.198	0.285	0.337	0.453	0.556	0.577
	12	-0.947	-0.805	-0.705	-0.527	-0.469	-0.274	-0.119	-0.045	0.050	0.153	0.196	0.273	0.311	0.386	0.507	0.537
18	13	-0.890	-0.814	-0.632	-0.479	-0.421	-0.244	-0.103	-0.025	0.048	0.155	0.205	0.291	0.333	0.433	0.528	0.550
	14	-0.824	-0.799	-0.596	-0.457	-0.417	-0.266	-0.110	-0.049	0.028	0.140	0.193	0.266	0.303	0.412	0.501	0.530
	15	-0.723	-0.632	-0.528	-0.379	-0.339	-0.195	-0.076	-0.007	0.071	0.187	0.227	0.308	0.358	0.481	0.564	0.607
	16	-0.652	-0.592	-0.481	-0.372	-0.332	-0.205	-0.098	-0.026	0.054	0.162	0.211	0.294	0.321	0.434	0.568	0.601
	17	-0.649	-0.595	-0.488	-0.365	-0.325	-0.206	-0.090	-0.024	0.041	0.158	0.206	0.303	0.340	0.465	0.569	0.589
	18	-0.524	-0.507	-0.426	-0.320	-0.289	-0.175	-0.077	-0.002	0.065	0.167	0.212	0.297	0.341	0.470	0.532	0.571
	9	-1.797	-1.698	-1.250	-0.945	-0.854	-0.483	-0.230	-0.114	-0.010	0.137	0.185	0.296	0.335	0.437	0.516	0.553
	10	-1.463	-1.410	-1.187	-0.834	-0.760	-0.384	-0.181	-0.062	0.038	0.162	0.202	0.297	0.341	0.442	0.530	0.550
	11	-1.179	-1.113	-0.891	-0.685	-0.571	-0.344	-0.177	-0.073	0.008	0.149	0.207	0.303	0.337	0.422	0.510	0.549
19	12	-1.127	-1.069	-0.852	-0.608	-0.533	-0.317	-0.148	-0.050	0.029	0.152	0.199	0.276	0.308	0.398	0.481	0.515
	13	-0.941	-0.902	-0.712	-0.524	-0.452	-0.255	-0.112	-0.016	0.052	0.160	0.203	0.286	0.334	0.450	0.533	0.548
	14	-0.739	-0.710	-0.594	-0.433	-0.397	-0.245	-0.098	-0.023	0.060	0.157	0.201	0.292	0.324	0.434	0.559	0.582
	15	-0.707	-0.641	-0.537	-0.433	-0.392	-0.230	-0.106	-0.036	0.037	0.144	0.187	0.274	0.301	0.451	0.546	0.566

TABLE 13.4– Continued.

N	r	2	2.5	5	10	12.5	25	40	50	60	75	80	87.5	90	95	97.5	98
									α × 100 (%)								
	16	-0.671	-0.634	-0.530	-0.413	-0.354	-0.219	-0.077	-0.006	0.059	0.166	0.214	0.311	0.358	0.448	0.532	0.569
	17	-0.698	-0.629	-0.521	-0.399	-0.359	-0.214	-0.091	-0.015	0.065	0.180	0.223	0.307	0.337	0.440	0.528	0.552
	18	-0.587	-0.549	-0.454	-0.356	-0.318	-0.181	-0.071	-0.010	0.056	0.156	0.193	0.276	0.330	0.456	0.524	0.539
	19	-0.539	-0.523	-0.450	-0.352	-0.323	-0.191	-0.096	-0.025	0.041	0.156	0.196	0.285	0.321	0.410	0.508	0.537
	10	-1.655	-1.564	-1.184	-0.832	-0.733	-0.433	-0.229	-0.116	-0.015	0.144	0.206	0.315	0.358	0.470	0.573	0.603
	11	-1.260	-1.170	-0.975	-0.718	-0.610	-0.364	-0.191	-0.093	0.005	0.129	0.168	0.254	0.289	0.414	0.497	0.532
	12	-1.112	-1.074	-0.850	-0.651	-0.556	-0.334	-0.147	-0.048	-0.024	0.150	0.191	0.275	0.317	0.408	0.530	0.580
	13	-0.967	-0.945	-0.771	-0.555	-0.497	-0.281	-0.138	-0.057	0.020	0.133	0.171	0.265	0.296	0.389	0.498	0.522
	14	-0.716	-0.686	-0.593	-0.470	-0.417	-0.268	-0.131	-0.053	0.021	0.146	0.197	0.283	0.316	0.424	0.492	0.529
20	15	-0.903	-0.814	-0.631	-0.476	-0.421	-0.248	-0.112	-0.036	0.031	0.146	0.194	0.280	0.316	0.392	0.517	0.570
	16	-0.788	-0.729	-0.562	-0.404	-0.363	-0.209	-0.088	-0.016	0.051	0.147	0.189	0.258	0.294	0.425	0.502	0.519
	17	-0.679	-0.635	-0.513	-0.382	-0.343	-0.198	-0.092	-0.015	0.040	0.141	0.196	0.272	0.300	0.400	0.503	0.534
	18	-0.612	-0.584	-0.462	-0.351	-0.308	-0.191	-0.083	-0.012	0.072	0.182	0.225	0.294	0.330	0.423	0.494	0.505
	19	-0.533	-0.511	-0.434	-0.333	-0.304	-0.186	-0.074	-0.012	0.051	0.151	0.204	0.277	0.323	0.413	0.494	0.525
	20	-0.542	-0.499	-0.434	-0.328	-0.293	-0.184	-0.083	-0.023	0.031	0.137	0.176	0.256	0.291	0.384	0.479	0.501
	11	-1.230	-1.165	-0.967	-0.784	-0.700	-0.414	-0.199	-0.087	0.009	0.152	0.198	0.289	0.330	0.425	0.525	0.557
	12	-1.159	-1.103	-0.882	-0.671	-0.592	-0.363	-0.175	-0.087	-0.003	0.121	0.169	0.281	0.310	0.392	0.490	0.546

TABLE 13.4– Continued.

N	r	α × 100 (%)															
		2	2.5	5	10	12.5	25	40	50	60	75	80	87.5	90	95	97.5	98
22	13	-0.905	-0.871	-0.700	-0.509	-0.477	-0.277	-0.125	-0.040	0.037	0.146	0.194	0.281	0.308	0.387	0.469	0.484
	14	-0.949	-0.906	-0.750	-0.534	-0.469	-0.283	-0.129	-0.048	0.016	0.126	0.166	0.244	0.292	0.388	0.466	0.481
	15	-0.753	-0.707	-0.601	-0.472	-0.425	-0.257	-0.100	-0.032	0.038	0.158	0.198	0.275	0.308	0.395	0.485	0.501
	16	-0.732	-0.678	-0.591	-0.457	-0.406	-0.244	-0.101	-0.024	0.046	0.146	0.186	0.262	0.297	0.381	0.489	0.514
	17	-0.757	-0.688	-0.518	-0.389	-0.352	-0.213	-0.095	-0.034	0.029	0.143	0.182	0.260	0.298	0.394	0.479	0.519
	18	-0.621	-0.586	-0.497	-0.363	-0.331	-0.207	-0.101	-0.030	0.035	0.159	0.200	0.272	0.309	0.388	0.461	0.483
	19	-0.613	-0.572	-0.465	-0.359	-0.317	-0.190	-0.088	-0.022	0.034	0.149	0.194	0.263	0.301	0.397	0.491	0.508
	20	-0.544	-0.531	-0.426	-0.343	-0.307	-0.186	-0.084	-0.016	0.040	0.128	0.167	0.245	0.299	0.410	0.492	0.519
	21	-0.502	-0.460	-0.390	-0.319	-0.277	-0.151	-0.061	-0.004	0.046	0.140	0.176	0.263	0.299	0.381	0.450	0.469
	22	-0.510	-0.489	-0.378	-0.290	-0.259	-0.149	-0.054	0.009	0.063	0.169	0.209	0.281	0.324	0.421	0.492	0.513
24	12	-1.359	-1.273	-0.988	-0.749	-0.656	-0.376	-0.178	-0.064	0.017	0.138	0.185	0.277	0.309	0.414	0.512	0.532
	13	-1.119	-1.019	-0.828	-0.598	-0.544	-0.319	-0.149	-0.078	0.003	0.129	0.173	0.265	0.294	0.378	0.445	0.489
	14	-0.862	-0.819	-0.684	-0.524	-0.482	-0.291	-0.119	-0.044	0.024	0.146	0.192	0.265	0.286	0.388	0.465	0.500
	15	-0.953	-0.823	-0.686	-0.510	-0.446	-0.238	-0.112	-0.041	0.032	0.128	0.172	0.250	0.286	0.374	0.436	0.448
	16	-0.721	-0.675	-0.539	-0.426	-0.377	-0.222	-0.094	-0.029	0.032	0.140	0.179	0.254	0.280	0.369	0.421	0.445
	17	-0.712	-0.653	-0.544	-0.428	-0.365	-0.224	-0.109	-0.044	0.024	0.127	0.169	0.271	0.301	0.405	0.489	0.513
	18	-0.678	-0.650	-0.507	-0.378	-0.320	-0.195	-0.083	-0.018	0.059	0.166	0.204	0.282	0.304	0.410	0.484	0.506

TABLE 13.4– Continued.

N	r	align							$\alpha \times 100$ (%)								
		2	2.5	5	10	12.5	25	40	50	60	75	80	87.5	90	95	97.5	98
	19	-0.608	-0.599	-0.477	-0.365	-0.326	-0.189	-0.089	-0.021	0.038	0.132	0.173	0.250	0.283	0.353	0.444	0.472
	20	-0.582	-0.555	-0.463	-0.355	-0.310	-0.175	-0.073	-0.016	0.046	0.155	0.186	0.259	0.299	0.389	0.480	0.505
	21	-0.506	-0.480	-0.413	-0.316	-0.281	-0.179	-0.084	-0.022	0.029	0.132	0.168	0.244	0.277	0.374	0.448	0.454
	22	-0.518	-0.490	-0.427	-0.328	-0.287	-0.177	-0.080	-0.011	0.043	0.125	0.166	0.243	0.280	0.359	0.465	0.498
	23	-0.531	-0.520	-0.411	-0.306	-0.264	-0.155	-0.067	-0.060	0.048	0.142	0.174	0.245	0.282	0.357	0.429	0.446
	24	-0.433	-0.420	-0.356	-0.287	-0.259	-0.161	-0.063	0.000	0.059	0.160	0.201	0.267	0.295	0.374	0.460	0.482
26	13	-1.168	-1.074	-0.858	-0.651	-0.578	-0.333	-0.174	-0.088	-0.001	0.123	0.160	0.274	0.304	0.390	0.466	0.485
	14	-1.057	-0.991	-0.758	-0.574	-0.522	-0.319	-0.168	-0.077	0.011	0.140	0.177	0.252	0.288	0.375	0.458	0.473
	15	-0.899	-0.847	-0.713	-0.513	-0.479	-0.279	-0.126	-0.047	0.028	0.145	0.181	0.251	0.295	0.372	0.447	0.470
	16	-0.837	-0.723	-0.586	-0.440	-0.388	-0.251	-0.114	-0.042	0.032	0.152	0.181	0.273	0.306	0.382	0.443	0.476
	17	-0.755	-0.716	-0.578	-0.433	-0.397	-0.229	-0.112	-0.050	0.015	0.119	0.159	0.234	0.252	0.340	0.410	0.432
	18	-0.658	-0.631	-0.524	-0.408	-0.360	-0.211	-0.089	-0.032	0.027	0.135	0.171	0.238	0.267	0.351	0.433	0.457
	19	-0.665	-0.649	-0.501	-0.365	-0.320	-0.184	-0.094	-0.036	0.028	0.127	0.161	0.238	0.260	0.340	0.409	0.437
	20	-0.543	-0.503	-0.414	-0.319	-0.286	-0.173	-0.072	-0.011	0.045	0.136	0.177	0.247	0.284	0.370	0.452	0.462
	21	-0.563	-0.525	-0.430	-0.317	-0.291	-0.162	-0.071	-0.015	0.046	0.150	0.180	0.237	0.259	0.329	0.401	0.420
	22	-0.583	-0.525	-0.434	-0.321	-0.290	-0.174	-0.081	-0.016	0.040	0.124	0.166	0.232	0.272	0.329	0.428	0.447
	23	-0.529	-0.479	-0.383	-0.295	-0.255	-0.142	-0.054	-0.002	0.051	0.141	0.189	0.241	0.267	0.360	0.475	0.495

TABLE 13.4– Continued.

								α × 100 (%)									
N	r	2	2.5	5	10	12.5	25	40	50	60	75	80	87.5	90	95	97.5	98
24	24	-0.496	-0.473	-0.398	-0.316	-0.278	-0.164	-0.074	-0.010	0.041	0.125	0.162	0.223	0.254	0.343	0.415	0.436
	25	-0.473	-0.450	-0.385	-0.298	-0.262	-0.148	-0.056	0.000	0.051	0.146	0.185	0.273	0.306	0.370	0.452	0.484
	26	-0.436	-0.423	-0.361	-0.295	-0.256	-0.151	-0.064	-0.006	0.045	0.132	0.180	0.250	0.280	0.360	0.433	0.459
28	14	-1.157	-1.061	-0.803	-0.561	-0.490	-0.305	-0.161	-0.072	0.008	0.129	0.170	0.254	0.282	0.362	0.441	0.473
	15	-1.043	-0.976	-0.750	-0.563	-0.502	-0.303	-0.164	-0.074	0.006	0.124	0.161	0.243	0.280	0.364	0.432	0.454
	16	-0.835	-0.798	-0.664	-0.502	-0.448	-0.244	-0.115	-0.042	0.037	0.136	0.170	0.249	0.287	0.383	0.449	0.478
	17	-0.773	-0.719	-0.580	-0.434	-0.388	-0.227	-0.098	-0.033	0.038	0.140	0.188	0.246	0.271	0.348	0.415	0.429
	18	-0.723	-0.698	-0.588	-0.421	-0.382	-0.222	-0.103	-0.043	0.016	0.127	0.164	0.231	0.260	0.336	0.424	0.452
	19	-0.635	-0.604	-0.503	-0.400	-0.354	-0.209	-0.091	-0.035	0.025	0.127	0.155	0.217	0.243	0.324	0.392	0.424
	20	-0.565	-0.551	-0.473	-0.355	-0.318	-0.190	-0.086	-0.026	0.036	0.131	0.172	0.242	0.262	0.354	0.417	0.452
	21	-0.520	-0.498	-0.400	-0.313	-0.280	-0.172	-0.079	-0.013	0.047	0.122	0.158	0.236	0.279	0.355	0.415	0.424
	22	-0.580	-0.558	-0.432	-0.335	-0.305	-0.183	-0.080	-0.024	0.036	0.124	0.154	0.220	0.252	0.327	0.384	0.418
	23	-0.477	-0.466	-0.384	-0.304	-0.273	-0.160	-0.068	-0.005	0.049	0.138	0.173	0.230	0.277	0.354	0.428	0.456
	24	-0.534	-0.489	-0.395	-0.298	-0.269	-0.162	-0.070	-0.026	0.030	0.130	0.170	0.236	0.268	0.365	0.465	0.496
	25	-0.481	-0.438	-0.383	-0.291	-0.252	-0.142	-0.058	-0.002	0.045	0.127	0.167	0.254	0.298	0.387	0.455	0.469
	26	-0.459	-0.443	-0.366	-0.277	-0.249	-0.153	-0.066	-0.019	0.035	0.121	0.165	0.224	0.251	0.322	0.373	0.408
	27	-0.492	-0.477	-0.367	-0.276	-0.247	-0.149	-0.062	-0.006	0.049	0.137	0.168	0.232	0.255	0.345	0.408	0.421

TABLE 13.4– Continued.

N	r	α × 100 (%)															
		2	2.5	5	10	12.5	25	40	50	60	75	80	87.5	90	95	97.5	98
	28	-0.408	-0.393	-0.324	-0.252	-0.229	-0.127	-0.053	-0.007	0.055	0.135	0.167	0.230	0.260	0.356	0.404	0.425
	15	-0.950	-0.910	-0.741	-0.553	-0.509	-0.317	-0.148	-0.064	0.008	0.122	0.166	0.251	0.288	0.365	0.447	0.467
	16	-0.940	-0.905	-0.730	-0.511	-0.467	-0.277	-0.144	-0.071	0.004	0.113	0.146	0.239	0.270	0.348	0.417	0.443
	17	-0.800	-0.730	-0.616	-0.450	-0.407	-0.244	-0.121	-0.044	0.019	0.136	0.169	0.247	0.281	0.370	0.438	0.451
	18	-0.731	-0.699	-0.580	-0.454	-0.409	-0.242	-0.121	-0.042	0.021	0.127	0.164	0.226	0.255	0.327	0.386	0.396
	19	-0.613	-0.602	-0.491	-0.373	-0.332	-0.202	-0.088	-0.022	0.028	0.123	0.163	0.227	0.249	0.323	0.386	0.399
	20	-0.644	-0.630	-0.494	-0.386	-0.334	-0.205	-0.096	-0.025	0.026	0.116	0.150	0.223	0.247	0.309	0.375	0.390
	21	-0.552	-0.534	-0.419	-0.321	-0.290	-0.165	-0.071	-0.015	0.041	0.134	0.176	0.233	0.263	0.369	0.425	0.444
30	22	-0.598	-0.578	-0.446	-0.335	-0.294	-0.178	-0.080	-0.020	0.043	0.127	0.158	0.231	0.257	0.331	0.399	0.419
	23	-0.510	-0.497	-0.413	-0.306	-0.275	-0.161	-0.064	-0.007	0.051	0.132	0.162	0.229	0.257	0.327	0.381	0.402
	24	-0.512	-0.497	-0.413	-0.326	-0.291	-0.162	-0.075	-0.017	0.036	0.126	0.154	0.221	0.248	0.354	0.431	0.450
	25	-0.502	-0.478	-0.387	-0.313	-0.265	-0.157	-0.068	-0.009	0.043	0.137	0.174	0.233	0.280	0.358	0.420	0.433
	26	-0.455	-0.434	-0.362	-0.286	-0.264	-0.152	-0.062	-0.014	0.039	0.121	0.151	0.206	0.228	0.307	0.379	0.388
	27	-0.451	-0.420	-0.347	-0.275	-0.250	-0.150	-0.060	-0.014	0.037	0.110	0.144	0.211	0.244	0.324	0.391	0.408
	28	-0.413	-0.395	-0.343	-0.264	-0.238	-0.137	-0.055	-0.006	0.043	0.134	0.172	0.233	0.266	0.358	0.420	0.435
	29	-0.433	-0.390	-0.316	-0.245	-0.217	-0.130	-0.046	-0.005	0.050	0.130	0.165	0.222	0.250	0.321	0.379	0.413
	30	-0.380	-0.372	-0.325	-0.262	-0.229	-0.141	-0.064	-0.009	0.038	0.128	0.160	0.224	0.249	0.329	0.406	0.424

498

and has the same distribution as $\hat{\beta}_{11}$. Let $y_i, i = 1, 2, ..., r$, be a failure terminated test sample of size N from the standardized exponential distribution, then

$$Q(y) = 1 - e^{-y}. \tag{13.3}$$

Using the transformation

$$y_i = \left(\frac{T_i}{\eta}\right)^\beta \tag{13.4}$$

in Eq. (13.3) yields

$$Q(T_i) = 1 - e^{-\left(\frac{T_i}{\eta}\right)^\beta}.$$

Solving Eq. (13.4) for T yields

$$T_i = \eta \, (y_i)^{1/\beta}, i = 1, 2, ..., r, \tag{13.5}$$

where the T_i are values from a random, failure terminated test sample of size N from a Weibull distribution. $\hat{\beta}$ is the MLE of β, based on T_i, and satisfies Eq. (13.1).

Substitution of Eq. (13.5) into Eq. (13.1) yields

$$\frac{\sum\limits_{i=1}^{r} \eta^{\hat{\beta}} y_i^{\hat{\beta}/\beta} \log_e(\eta y_i^{1/\beta}) + (N-r)\eta^{\hat{\beta}} y_r^{\hat{\beta}/\beta} \log_e(\eta y_r^{1/\beta})}{\sum\limits_{i=1}^{r} \eta^{\hat{\beta}} y_i^{\hat{\beta}/\beta} + (N-r)\eta^{\hat{\beta}} y_r^{\hat{\beta}/\beta}}$$

$$-\frac{1}{\hat{\beta}} - \frac{1}{r}\sum\limits_{i=1}^{r}\log_e(\eta \, y_i^{1/\beta}) = 0. \tag{13.6}$$

Simplification of Eq. (13.6) yields

$$\frac{\sum\limits_{i=1}^{r} y_i^{\hat{\beta}/\beta} \log_e y_i + (N-r) \, y_r^{\hat{\beta}/\beta}\log_e y_r}{\sum\limits_{i=1}^{r} y_i^{\hat{\beta}/\beta} + (N-r) \, y_r^{\hat{\beta}/\beta}\log_e y_r}$$

$$-\frac{1}{\hat{\beta}/\beta} - \frac{1}{r}\sum\limits_{i=1}^{r}\log_e y_i = 0. \tag{13.7}$$

In the case of $\beta = \eta = 1$, Eq. (13.1) becomes

$$\frac{\sum\limits_{i=1}^{r} y_i^{\hat{\beta}_{11}} \log_e y_i + (N-r) y_r^{\hat{\beta}_{11}} \log_e y_r}{\sum\limits_{i=1}^{r} y_i^{\hat{\beta}_{11}} + (N-r) y_r^{\hat{\beta}_{11}}}$$

$$-\frac{1}{\hat{\beta}_{11}} - \frac{1}{r}\sum_{i=1}^{r}\log_e y_i = 0. \tag{13.8}$$

A comparison of Eq. (13.7) with Eq. (13.8) shows that they have the same solution; therefore,

$$\hat{\beta}/\beta = \hat{\beta}_{11}. \tag{13.9}$$

Furthermore, $\hat{\beta}/\beta$ is independent of β and η, because in Eq. (13.8) β and η do not appear, and has the same distribution as $\hat{\beta}_{11}$.

It may also be shown that $\hat{\beta} \log_e(\hat{\eta}/\eta)$ is independent of β and η, and has the same distribution as $\hat{\beta}_{11} \log_e(\hat{\eta}_{11})$. From Eq. (13.2)

$$\hat{\beta} \, \log_e(\hat{\eta}) = \log_e\left[\frac{\sum\limits_{i=1}^{r} T_i^{\hat{\beta}} + (N-r)T_r^{\hat{\beta}}}{r}\right], \tag{13.10}$$

where $\hat{\beta}$ satisfies Eq. (13.1). Substitution of Eq. (13.5) into Eq. (13.10) yields

$$\hat{\beta} \, \log_e(\hat{\eta}) = \log_e\left[\frac{\sum\limits_{i=1}^{r}[\eta\,(y_i)^{1/\beta}]^{\hat{\beta}} + (N-r)[\eta\,(y_r)^{1/\beta}]^{\hat{\beta}}}{r}\right], \tag{13.11}$$

where $y_i, i = 1, 2, ..., r$, is a failure terminated test sample from the standard exponential distribution. From Eq. (13.11)

$$\hat{\beta} \, \log_e(\hat{\eta}) = \hat{\beta} \, \log_e \eta + \log_e\left[\frac{\sum\limits_{i=1}^{r} y_i^{\hat{\beta}/\beta} + (N-r)\, y_r^{\hat{\beta}/\beta}}{r}\right]. \tag{13.12}$$

But if $y_i, i = 1, 2, ..., r$, is a real sample from the standard exponential distribution; then,

$$\hat{\beta}_{11} \, \log_e(\hat{\eta}_{11}) = \log_e\left[\frac{\sum\limits_{i=1}^{r} y_i^{\hat{\beta}_{11}} + (N-r)\, y_r^{\hat{\beta}_{11}}}{r}\right]. \tag{13.13}$$

From Eq. (13.9)

$$\hat{\beta}/\beta = \hat{\beta}_{11}; \tag{13.14}$$

therefore, a comparison of Eqs. (13.12) and (13.13) yields

$$\hat{\beta} \, \log_e(\hat{\eta}/\eta) = \hat{\beta}_{11} \, \log_e(\hat{\eta}_{11}). \tag{13.15}$$

13.6.3 CONFIDENCE INTERVAL ON β

From Eq. (13.9), $\hat{\beta}/\beta = \hat{\beta}_{11}$, and $\hat{\beta}/\beta$ is independent of β and η and has the same distribution as $\hat{\beta}_{11}$. But the distribution of $\hat{\beta}_{11}$ can be obtained by Monte-Carlo simulation. Given the sample size N and the truncation failure number, r, the Monte-Carlo simulation is carried out as follows:

1. Generate a random sample of size N from the standard exponential distribution, and truncate the sample at the rth failure after the sample is ordered.

2. Estimate $\hat{\beta}_{11}$ from the truncated sample by solving Eqs. (13.1) and (13.2) using the Newton-Raphson iterative method [4, pp. 575-578]. The initial values of $\hat{\beta}_{11}$ are obtained using the equation [5, pp. 188-189]

$$\hat{\beta}_{11} = \frac{2.99}{\log_e T_{([0.9737N]+1)} - \log_e T_{([0.1673N]+1)}},$$

where the quantities in the brackets in the denominator are taken to be the largest integer which is less than the calculated value. Then find the $([0.9737N]+1)$ ordered value of T, and then the natural logarithm of this T value.

3. Repeat Steps 1 and 2 for 1,000 or more times, then order the 1,000 or more values of $\hat{\beta}_{11}$ and find out the percentiles of the $\hat{\beta}_{11}$ distribution.

Table 13.3 gives the percentile values of the distribution of $\hat{\beta}_{11}$ for different sample sizes, N, and failure numbers, r; and it can be used to construct the confidence intervals on β, when η is unknown.

The two-sided confidence interval on β at the $(1 - \alpha)$ confidence level is given by

$$(\beta_{L2} \, ; \, \beta_{U2}) = (\hat{\beta}/l_2 \, ; \, \hat{\beta}/l_1), \tag{13.16}$$

where the values of l_1 and l_2 are obtained from Table 13.3, and should satisfy the probabilistic equation

$$P(l_1 \leq \hat{\beta}/\beta \leq l_2) = 1 - \alpha. \tag{13.17}$$

13.6.4   CONFIDENCE INTERVAL ON η

From Eq. (13.14)

$$\hat{\beta} \, \log_e(\hat{\eta}/\eta) = \hat{\beta}_{11} \, \log_e(\hat{\eta}_{11}).$$

The distribution of $\hat{\beta}_{11} \, \log_e(\hat{\eta}_{11})$ can also be obtained by Monte-Carlo simulation. Table 13.4 gives the percentile values of the distribution of $\hat{\beta}_{11} \, \log_e(\hat{\eta}_{11})$ for different sample sizes and failure numbers. It can be used to construct the confidence interval on η, when β is unknown. The two-sided confidence interval on η, at the $(1 - \alpha)$ confidence level, is given by

$$(\eta_{L2} \; ; \; \eta_{U2}) = \left(\hat{\eta} \, e^{-k_2/\hat{\beta}} \; ; \; \hat{\eta} \, e^{-k_1/\hat{\beta}} \right), \tag{13.18}$$

where the values of k_1 and k_2 are obtained from Table 13.4 and should satisfy the probabilistic equation

$$P(k_1 \leq \hat{\beta} \, \log_e(\hat{\eta}/\eta) \leq k_2) = 1 - \alpha. \tag{13.19}$$

EXAMPLE 13–1

The lives of specific units follow the two-parameter Weibull distribution. Eight of them are tested, the test is terminated when the seventh failure occurs, and the following ordered times to failure data are obtained:

$$0.09, 1.18, 1.25, 1.28, 1.79, 2.27, \text{ and } 3.34.$$

Find the 90% confidence intervals on β and η.

SOLUTION TO EXAMPLE 13–1

The MLE's of β and η are obtained using the Newton-Raphson iterative method, with Eqs. (13.1) and (13.2), yielding

$$\hat{\beta} = 1.33,$$

and

$$\hat{\eta} = 2.13.$$

Since $\alpha = 5\%$ and 95%, $N = 8$ and $r = 7$, entering Tables 13.3 and 13.4, yields

$$l_1 = 0.703,$$

$$l_2 = 2.214,$$

and

$$k_1 = -1.009,$$

$$k_2 = 0.714.$$

Substitution of the values of l_1 and l_2 into Eq. (13.16) yields

$$(\beta_{L2} ; \beta_{U2}) = \left(\frac{1.33}{2.214} ; \frac{1.33}{0.703}\right),$$

or the confidence interval on β, at the 90% confidence level, is

$$(0.60 ; 1.89).$$

Substitution of the values of k_1 and k_2 into Eq. (13.18) yields

$$(\eta_{L2} ; \eta_{U2}) = (2.13\, e^{-0.714/1.33} ;\quad 2.13\, e^{1.009/1.33}),$$

or the confidence interval on η, at the 90% confidence level, is

$$(1.25 ; 4.55).$$

PROBLEMS

13-1. Ten units are tested and the results of Table 13.5 are obtained:

  (1) Plot on Weibull probability paper the population line using the median ranks. Draw the best-fit line. It should have a slope close to one.

  (2) Plot the 5% and 95% confidence bands.

  (3) Find the two-sided confidence limits on the reliability for a mission of 30 hr starting the mission at age zero, at a 90% confidence level.

  (4) Find the lower, one-sided confidence limit on the reliability for a mission of 30 hr, starting the mission at age zero, at a 95% confidence level.

  (5) Repeat Case 3, using the χ^2 approach for the exponential case, and assuming this is a failure terminated test.

  (6) Compare the results of Cases 3 and 5.

  (7) Repeat Case 4, using the χ^2 approach for the exponential case, and assuming this is a failure terminated test.

  (8) Compare the results of Cases 4 and 7.

TABLE 13.5 – Times-to-failure data for a sample of size $N = 10$ for Problem 13-1.

Failure order number	Time to failure, hr
1	10
2	26
3	42
4	65
5	90
6	120

13-2. Ten units are put to a reliability test with the following results:

Failure order number	Time to failure, hr
1	800
2	1,300
3	1,400
4	1,850
5	1,900
6	2,400
7	2,600
8	2,650

Do the following:

(1) Find the parameters of the Weibull distribution representing these data.

(2) Determine the 2.5% and the 97.5% ranks for these data and plot them directly on the Weibull probability plot of the these data.

(3) Determine the reliability, at a 50% confidence level, for a mission of 500 hr, starting the mission at age zero.

(4) Determine the two-sided confidence limits on the reliability, at a 95% confidence level, for a mission of 2,000 hr, starting the mission at age zero, using the results in Case 2, and write the associated probabilistic statement.

(5) Repeat Case 3, but for an additional mission of 100 hr.

(6) Determine the lower, one-sided confidence limit on the reliability, at a 97.5% confidence level, from the results obtained in Case 2 for a mission of 2,000 hr.

13-3. Ten units are tested and the results of Table 13.6 are obtained.

(1) Plot on Weibull probability paper the population line using the median ranks. Draw in the best-fit line. It should have a slope close to one.

(2) Plot the 5% and 95% confidence bands.

(3) Find the two-sided confidence limits on the reliability for a mission of 300 hr, starting the mission at age zero, at a 90% confidence level.

(4) Find the lower, one-side confidence limit on the reliability for a mission of 300 hr, starting the mission at age zero, at a 95% confidence level.

(5) Repeat Case 3, using the χ^2 approach for the exponential case, and assuming this is a failure terminated test.

(6) Compare the results of Case 3 and Case 5.

(7) Repeat Case 4, using the χ^2 approach for the exponential case, and assuming this is a failure terminated test.

(8) Compare the results of Cases 4 and 7.

TABLE 13.6 – Times-to-failure data for a sample size of ten for Problems 13-3 and 13-4.

Failure order number j	Time to failure, hr, T_j
1	105
2	268
3	427
4	652
5	901
6	1,200

13-4. For the data in Table 13.6, do the following:

(1) Determine the parameters of the Weibull *pdf* representing these data.

(2) Plot the 95% confidence level bands about the median rank line.

(3) Determine the two-sided, 95% confidence level interval on the reliability of these units for a mission of 500 hr, starting the mission at age zero.

(4) Determine the lower, one-sided confidence interval on the reliability of these units for a mission of 500 hr, starting the mission at age zero, at a 97.5% confidence level.

(5) Determine the two-sided mission duration's interval, at a 95% confidence level, for a reliability of 80%.

(6) Determine the one-sided mission duration's interval at a 97.5% confidence level, for a reliability of 80%.

(7) Find the confidence interval on the Weibull slope at a confidence level of 50%, using the procedure of Section 13.4.

(8) Find the population mean, and its confidence interval, at a 95% confidence level.

(9) Find the 95% confidence interval on β, using the procedure of Section 13.6.3.

(10) Find the 95% confidence interval on η, using the procedure of Section 13.6.4.

13-5. Given are the following endurance life data obtained from fatigue testing five units:

Failure order number	Endurance life, cycles
1	14,000
2	23,000
3	29,900
4	39,000
5	53,300

Do the following:

(1) Determine the parameters of the Weibull *pdf* representing these data.

(2) Plot the 90% confidence level bands about the median rank line.

(3) Determine the two-sided, 90% confidence level interval on the reliability of these units for a mission of 25,000 cycles, starting the mission at age zero.

(4) Determine the lower, one-sided confidence interval on the reliability of these units for a mission of 25,000 cycles, starting the mission at age zero, at a 95% confidence level.

(5) Determine the two-sided mission duration's interval, at a 90% confidence level, for a reliability of 80%.

(6) Determine the one-sided mission duration's interval, at a 95% confidence level, for a reliability of 80%.

(7) Find the confidence interval on the Weibull slope at a confidence level of 90%, using the procedure of Section 13.4.

(8) Find the population mean, and its confidence interval, at a 90% confidence level.

(9) Find the 90% confidence interval on β, using the procedure of Section 13.6.3.

(10) Find the 90% confidence interval on η, using the procedure of Section 13.6.4.

(11) Comparatively discuss the results in Cases 7 and 9.

13-6. Twenty-five units were tested simultaneously. The test was terminated after the first ten failures, with the following results:

Failure order number	Time to failure, hr
1	920
2	1,840
3	2,600
4	3,300
5	4,700
6	6,400
7	8,700
8	10,450
9	13,260
10	17,250

Do the following:

(1) Determine the parameters of the Weibull *pdf* representing these data.

(2) Plot the 90% confidence level bands about the median rank line.

(3) Determine the two-sided, 90% confidence level interval on the reliability of these units for a mission of 5,000 hr, starting the mission at age zero.

(4) Determine the lower, one-sided confidence interval on the reliability of these units for a mission of 5,000 hr, starting the mission at age zero, at a 95% confidence level.

(5) Determine the two-sided mission duration's interval, at a 90% confidence level, for a reliability of 80%.

(6) Determine the one-sided mission duration's interval, at a 95% confidence level, for a reliability of 80%.

(7) Find the confidence interval on the Weibull slope at a confidence level of 90%, using the procedure of Section 13.4.

(8) Find the population mean, and its confidence interval, at a 90% confidence level.

(9) Find the 90% confidence interval on β, using the procedure of Section 13.6.3.

(10) Find the 90% confidence interval on η, using the procedure of Section 13.6.4.

13-7. The following times to failure were obtained by testing eight units to failure.

Time to failure, hr
130
190
80
185
265
240
260
140

Do the following:

(1) Determine the parameters of the Weibull *pdf* representing these data.

(2) Plot the 90% confidence level bands about the median rank line.

(3) Determine the two-sided, 90% confidence level interval on the reliability of these units for a mission of 150 hr, starting the mission at age zero.

(4) Determine the lower, one-sided confidence interval on the reliability of these units for a mission of 150 hr, starting the mission at age zero, at a 95% confidence level.

(5) Determine the two-sided mission duration's interval, at a 90% confidence level, for a reliability of 80%.

(6) Determine the one-sided mission duration's interval, at a 95% confidence level, for a reliability of 80%.

(7) Find the confidence interval on the Weibull slope at a confidence level of 90%, using the procedure of Section 13.4.

(8) Find the population mean and its confidence interval at a 90% confidence level.

(9) Find the 90% confidence interval on β, using the procedure of Section 13.6.3.

(10) Find the 90% confidence interval on η, using the procedure of Section 13.6.4.

13-8. Twenty units were tested simultaneously. The test was terminated after the 15th failure, with the following results:

Failure order number	Time to failure, hr
1	435
2	442
3	458
4	465
5	472
6	480
7	488
8	510
9	525
10	525
11	540
12	585
13	593
14	615
15	640

Do the following:

(1) Determine the parameters of the Weibull *pdf* representing these data.

(2) Plot the 90% confidence level bands about the median rank line.

(3) Determine the two-sided, 90% confidence level interval on the reliability of these units for a mission of 475 hr, starting the mission at age zero.

(4) Determine the lower, one-sided confidence interval on the reliability of these units for a mission of 475 hr, starting the mission at age zero, at a 95% confidence level.

(5) Determine the two-sided mission duration's interval, at a 90% confidence level, for a reliability of 80%.

(6) Determine the one-sided mission duration's interval, at a 95% confidence level, for a reliability of 80%.

(7) Find the confidence interval on the Weibull slope at a confidence level of 90%, using the procedure of Section 13.4.

(8) Find the population mean, and its confidence interval, at a 90% confidence level.

(9) Find the 90% confidence interval on β, using the procedure of Section 13.6.3.

(10) Find the 90% confidence interval on η, using the procedure of Section 13.6.4.

13-9. Given are the data in Table 13.7 obtained from testing $N = 50$ solder joints simultaneously by subjecting them to cyclic thermal stress. Do the following:

(1) Determine the parameters of the Weibull *pdf* representing these data.

(2) Plot the 90% confidence level bands about the median rank line.

(3) Determine the two-sided, 90% confidence level interval on the reliability of these units for a mission of 5,000 cycles, starting the mission at age zero.

(4) Determine the lower, one-sided confidence interval on the reliability of these units for a mission of 5,000 cycles , starting the mission at age zero, at a 95% confidence level.

(5) Determine the two-sided mission duration's interval, at a 90% confidence level, for a reliability of 85%.

(6) Determine the one-sided mission duration's interval, at a 95% confidence level, for a reliability of 85%.

(7) Find the confidence interval on the Weibull slope at a confidence level of 90%, using the procedure of Section 13.4.

(8) Find the population mean, and its confidence interval, at a 90% confidence level.

(9) Find the 90% confidence interval on β, using the procedure of Section 13.6.3.

(10) Find the 90% confidence interval on η, using the procedure of Section 13.6.4.

13-10. From data in Table 13.7, do the following:

(1) Determine the parameters of the Weibull *pdf* representing these data.

(2) Plot the 95% confidence level bands about the median rank line.

(3) Determine the two-sided, 95% confidence level interval on the reliability of these units for a mission of 7,000 cycles, starting the mission at age zero.

(4) Determine the lower, one-sided confidence interval on the reliability of these units for a mission of 7,000 cycles, starting the mission at age zero, at a 97.5% confidence level.

(5) Determine the two-sided mission duration's interval, at a 95% confidence level, for a reliability of 80%.

(6) Determine the one-sided mission duration's interval, at a 97.5% confidence level, for a reliability of 80%.

(7) Find the confidence interval on the Weibull slope at a confidence level of 50%, using the procedure of Section 13.4.

(8) Find the population mean, and its confidence interval, at a 95% confidence level.

(9) Find the 95% confidence interval on β, using the procedure of Section 13.6.3.

(10) Find the 95% confidence interval on η, using the procedure of Section 13.6.4.

TABLE 13.7 – Cycles-to-failure data for solder joints for Problems 13-9 and 13-10, $N=50$.

Cycles to failure	Failures
8,375	2
9,420	2
10,748	3
12,033	5
13,566	3
15,238	4
16,072	4
16,665	1
17,406	3
17,918	1
18,484	1
18,893	1
19,744	5
20,534	5
21,573	8
22,656	5

REFERENCES

1. Lipson, Charles and Sheth, Narendra, J., *Statistical Design and Analysis of Experiments*, McGraw-Hill Book Company, New York, 518 pp., 1973.

2. Johnson, Leonard G., *The Statistical Treatment of Fatigue Experiments*, Elsevier Publishing Company, Amsterdam, 116 pp., 1964.

3. Billman, B. R., Antle, C. E. and Bain, L. J., "Statistical Inference from Censored Weibull Samples," *Technometrics*, Vol. 14, pp. 831-840, 1972.

4. Hildebrand, F. B., *Introduction to Numerical Analysis*, McGraw-Hill Book Company, New York, 669 pp., 1974.

5. Mann, N. R., Schafer, R. E., and Signpurwalla, N. D., *Methods for Statistical Analysis of Reliability and Life Data*, John Wiley & Sons, Inc., New York, 654 pp., 1974.

6. Kececioglu, Dimitri, Reliability Engineering Handbook, Prentice Hall, Inc.,Englewood Cliffs, N. J., Vol. 1, 720 pp. and Vol. 2, 568 pp., 1991.

APPENDIX 13A

PERCENT FAILED AT THE MEAN LIFE
FOR THE WEIBULL DISTRIBUTION

The percent failed at the mean life, or $100\, Q(\bar{T})$, may be found from

$$Q(T) = 1 - e^{-(\frac{T-\gamma}{\eta})^{\beta}}, \qquad (13A.1)$$

by substituting in it

$$\bar{T} = \gamma + \eta\left[\Gamma(\frac{1}{\beta} + 1)\right], \qquad (13A.2)$$

or

$$100Q(\bar{T}) = 100\left[1 - e^{-(\frac{\bar{T}-\gamma}{\eta})^{\beta}}\right],$$

$$= 100\left[1 - e^{-\left\{\frac{\gamma+\eta[\Gamma(\frac{1}{\beta}+1)]-\gamma}{\eta}\right\}^{\beta}}\right],$$

or

$$100Q(\overline{T}) = 100\left\{1 - e^{-\left[\Gamma\left(\frac{1}{\beta}+1\right)\right]^{\beta}}\right\}. \qquad (13A.3)$$

Figure 13.4 is the plot of Eq. (13A.3), or of $100\,Q(\overline{T})$, which is the percent failed at the mean life, versus β the Weibull slope. It maybe seen that the percent failed at the mean life is independent of γ and η.

Chapter 14

RANKS NOT AVAILABLE IN TABLES

14.1 COMPUTATION OF ANY RANK

When the ranks and the sample size needed are not in available tables, the general formula for computing the rank Z for the jth failure, or the mean order number, of a sample of size N with a desired probability P is [1; 2, p. 51]

$$1 - (1 - Z)^N - NZ(1 - Z)^{N-1} - \frac{N(N - 1)}{2!} \times \quad (14.1)$$

$$Z^2(1 - Z)^{N-2} - \cdots - \frac{N(N - 1)...(N - j + 2)}{(j - 1)!} Z^{j-1} \times$$

$$(1 - Z)^{N-j+1} = P,$$

where Z is a value between zero and unity, and $j = 1, 2, ..., N$.

Equation (14.1) can be rearranged as

$$\sum_{k=j}^{N} \binom{N}{k} Z^k(1 - Z)^{N-k} = P. \quad (14.2)$$

It is apparent that Eq. (14.2) is the cumulative binomial distribution function, and it may be solved for Z by iterative computer methods given N, j and P. This is a tedious process, however. An easier method is to apply two transformations to Eq. (14.2), first to the beta distribution equivalent of the binomial and second to the F

517

distribution equivalent of the beta distribution, and obtain [3, p.498; 4, p. 398]

$$LR_\delta = \frac{1}{1 + \frac{N-j+1}{j}F_{1-\delta;2(N-j+1);2j}}, \qquad (14.3)$$

and

$$UR_{1-\delta} = \frac{1}{1 + \frac{N-j+1}{j}F_{\delta;2(N-j+1);2j}}, \qquad (14.4)$$

where

LR_δ = lower rank value in decimals for the rank of δ in decimals; e.g., for the 5% rank, LR_δ is the $LR_{0.05}$ value for the rank of $\delta = 0.05$,

N = sample size tested,

j = mean rank order number or failure order number,

$F_{1-\delta;2(N-j+1);2j}$ = F distribution value such that the area under the F distribution with $m = 2(N - j + 1)$ and $n = 2j$ degrees of freedom to the *left* of this F value is equal to $1 - \delta$,

$UR_{1-\delta}$ = upper rank value in decimals for the rank of $(1 - \delta)$ in decimals; e.g., for the 95% rank, $UR_{1-\delta}$ is the $UR_{0.95}$ value for the rank of $(1 - \delta) = 0.95$,

and

$F_{\delta;2(N-j+1);2j}$ = F distribution value such that the area under the F distribution with $m = 2(n - j + 1)$ and $n = 2j$ degrees of freedom, which are the same as those in Eq. (14.3), to the *left* of this F value is equal to δ.

EXAMPLE 14–1

Seventy units have been reliability tested. For the fortieth failure find the following:

1. The 2.5% rank.

2. The 97.5% rank.

SOLUTIONS TO EXAMPLE 14–1

1. The 2.5% rank for $\delta = 0.025$, $j = 40$ and $N = 70$, using Eq. (14.3), is given by

$$LR_{0.025} = \frac{1}{1 + \frac{70-40+1}{40}F_{0.975;2(70-40+1);2\times40}},$$

where

$$F_{0.975;62;80} = 1.6182.$$

Then,

$$LR_{0.025} = \frac{1}{1 + \frac{31}{40} \times 1.6182} = 0.4436,$$

or

$$LR_{0.025} = 44.36\%.$$

2. The 97.5% rank for $(1 - \delta) = 0.975$, $j = 40$ and $N = 70$, using Eq. (14.4), is given by

$$UR_{0.975} = \frac{1}{1 + \frac{70-40+1}{40}F_{0.025;62;80}},$$

where

$$F_{0.025;62;80} = \frac{1}{F_{0.975;80;62}} = \frac{1}{1.6352} = 0.61155.$$

Consequently,

$$UR_{0.975} = \frac{1}{1 + \frac{31}{40} \times 0.61155} = 0.6785,$$

or

$$UR_{0.975} = 67.85\%.$$

14.2 ALTERNATE METHODS FOR THE COMPUTATION OF THE MEDIAN RANK

Johnson [2, pp. 30-31] gives the following two equations for the computation of the approximate median rank:

$$MR = \frac{j - 0.30685 - 0.3863(\frac{j-1}{N-1})}{N} \tag{14.5}$$

for $N > 20$, and

$$MR = 1 - 2^{-\frac{1}{N}} + (\frac{j-1}{N-1})[2^{(1-\frac{1}{N})} - 1] \qquad (14.6)$$

for $N \leq 20$.

EXAMPLE 14–2

1. Find the median ranks for the following cases:

   1.1– $j = 1$ and $N = 6$.
   1.2– $j = 1$ and $N = 25$.
   1.3– $j = 1$ and $N = 70$.
   1.4– $j = 5$ and $N = 6$.
   1.5– $j = 5$ and $N = 25$.
   1.6– $j = 5$ and $N = 70$.

2. Compare these results with those obtained from the more exact tables in Appendix A.

SOLUTIONS TO EXAMPLE 14–2

1. Use Eqs. (14.5) and (14.6) to find the median ranks.

   1.1 From Eq. (14.6),

   $$MR_{1/6} = 1 - 2^{-1/6} = 0.10910.$$

   1.2 From Eq. (14.5),

   $$MR_{1/25} = \frac{1 - 0.30685}{25} = 0.02773.$$

   1.3 From Eq. (14.5),

   $$MR_{1/70} = \frac{1 - 0.30685}{70} = 0.00990.$$

   1.4 From Eq. (14.6),

   $$MR_{5/6} = 1 - 2^{-1/6} + \left(\frac{5-1}{6-1}\right)[2^{(1-1/6)} - 1],$$

   or

   $$MR_{5/6} = 0.73454.$$

TABLE 14.1– Comparison of the median rank values obtained from Eqs. (14.5) and (14.6) with Appendix A values, and values given by Eq. (14.3) as asterisked.

Case	$MR_{j/N}$	MR from Case 1, %	MR from Case 2, %
1.1	$MR_{1/6}$	10.910	10.910
1.2	$MR_{1/25}$	2.773	2.735
1.3	$MR_{1/70}$	0.990	0.985*
1.4	$MR_{5/6}$	73.454	73.555
1.5	$MR_{5/25}$	18.515	18.435
1.6	$MR_{5/70}$	6.673	6.638*

1.5 From Eq. (14.5),

$$MR_{5/25} = \frac{5 - 0.30685 - 0.3863(\frac{5-1}{25-1})}{25},$$

or

$$MR_{5/25} = 0.18515.$$

1.6 From Eq. (14.5),

$$MR_{5/70} = \frac{5 - 0.30685 - 0.3863(\frac{5-1}{70-1})}{70},$$

or

$$MR_{5/70} = 0.06673.$$

2. The median ranks of Cases 1.1, 1.2, 1.4 and 1.5 can be found in Appendix A. They are listed in Table 14.1. The median ranks of Cases 1.3 and 1.6 can be calculated from Eq. (14.3) or Eq. (14.4). For Case 1.3,

$$MR_{1/70} = \frac{1}{1 + \frac{70-1+1}{1}F_{0.50;2(70-1+1);2}},$$

where

$$F_{0.50;140;2} = 1.43575.$$

Therefore,

$$MR_{1/70} = \frac{1}{1 + 70 \times 1.43575} = 0.00985.$$

For Case 1.6,

$$MR_{5/70} = \frac{1}{1 + \frac{70-5+1}{5} F_{0.50;2(70-5+1);2\times5}},$$

where
$$F_{0.50;132;10} = 1.06553.$$

Therefore,

$$MR_{5/70} = \frac{1}{1 + \frac{66}{5} \times 1.06553} = 0.06638.$$

It may be seen that Eqs. (14.5) and (14.6) yield accurate enough values for median ranks.

PROBLEMS

14-1. One hundred units have been reliability tested. For the sixtieth failure find the following:

(1) The 5% rank.

(2) The 95% rank.

14-2. Find the median rank of the following cases:

(1) $j = 1$ and $N = 10$.

(2) $j = 1$ and $N = 22$.

(3) $j = 7$ and $N = 12$.

(4) $j = 8$ and $N = 27$.

14-3. If $N = 50$ and $j = 8$, find the following:

(1) The 2.5% rank.

(2) The 97.5% rank.

(3) The median rank.

14-4. Find the 2.5% lower rank, the 97.5% upper rank and the median rank for the following:

(1) $j = 5$ and $N = 15$.
(2) $j = 5$ and $N = 35$.
(3) $j = 27$ and $N = 35$.

14-5. If the 2.5% lower rank is 0.314 and $j = 6$, find N.

14-6. If the 97.5% upper rank is 0.245 and $N = 69$, find j.

14-7. If $N = 15$ and $MR = 0.435$, find j.

14-8. If $N = 30$ and $MR = 0.253$, find j.

14-9. If the median rank is 44.26% and $j = 8$, find N.

14-10. If the median rank is 27.59% and $j = 12$, find N.

REFERENCES

1. Johnson, Leonard G., "The Median Ranks of Sample Values in Their Population With an Application to Certain Fatigue Studies," *Industrial Mathematics*, Vol. 2, pp. 1-9, 1951.

2. Johnson, Leonard G., *The Statistical Treatment of Fatigue Experiments*, Elsevier Publishing Company, New York, 114 pp., 1964.

3. Hald, A., *Statistical Theory with Engineering Applications*, John Wiley & Sons, Inc., New York, 783 pp., 1952.

4. Hald, A., *Statistical Tables and Formulas*, John Wiley & Sons, Inc., New York, 97 pp., 1952.

Chapter 15

TESTS OF COMPARISON FOR THE WEIBULL DISTRIBUTION

15.1   APPLICABILITY

It is often desired to determine whether two different products which perform the same functions differ in reliability, whether a redesign has indeed improved the reliability of a product, or whether two different stress levels change the reliability of the same product. Such determinations may be made using the Weibull distribution and comparing the resulting lives of the two different products, or of the former and the redesigned product, or of the same product at two different stress levels. This chapter provides tools to enable such comparisons to be made statistically, by determining the confidence level at which the two versions differ at their mean lives and at their 10% failure lives.

15.2   METHODOLOGY AT THE MEAN LIFE

The test of comparison at the mean life, when the data are Weibull distributed, can be conducted for two cases [1, pp. 111-113]:

1. When the β's for the two samples are equal.

2. When the β's are not equal.

These cases are presented next.

15.2.1   WHEN THE β's ARE EQUAL

1. Test the two different products, the two different designs of

525

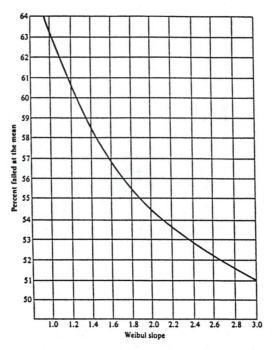

Fig. 15.1 – Position of the Weibull mean [1, p. 468].

the same product, or the same product at two different stress levels, and obtain two sets of times to failure data, one for each product, design, or stress level.

2. Plot the data on Weibull probability paper, or analyze the data by other methods to determine the Weibull *pdf's* parameters.

3. Determine the Weibull slope of each set of data, β_1 and β_2.

4. Determine the degrees of freedom, *d.f.*, of each set of data from $(n_1 - 1)$ and $(n_2 - 1)$, respectively, where n_1 is the number of failures observed in Set 1, and n_2 is the number of failures observed in Set 2.

5. Obtain the *combined degrees of freedom, c.d.f.*, from $(n_1 - 1)(n_2 - 1)$.

6. Enter Fig. 15.1 with the Weibull slope determined in Step 3, if the two sets of data yield the same Weibull slope, and find the percent failed at the mean life. For the derivation of the equation for the plot in Fig. 15.1 see Appendix 13A.

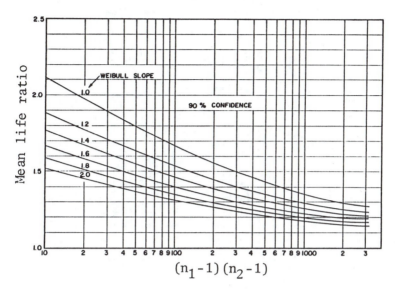

Fig. 15.2 — Test for significant difference in the mean lives of Weibull distributions at a 90% confidence level [1, p. 470; 2, pp. 16-33].

7. Find the mean life of Set 1 by entering their Weibull plot at the percent failed at the mean level found in Step 6. Repeat for Set 2.

8. Find the *mean life ratio* (MLR) by dividing the larger mean life found in Step 7 by the smaller mean life found in Step 7.

9. To determine whether this ratio is significant for the populations from which the two sets of reliability test data were obtained, enter Figs. 15.2, 15.3 and 15.4 for 90%, 95% and 99% confidence levels, respectively, with the combined degrees of freedom found in Step 5 and the Weibull slope found in Step 3. See at what confidence level the MLR is approached. Then conclude that with at least the approached confidence level the product with the higher mean life has a significantly higher reliability or a significantly improved design.

EXAMPLE 15-1

The main shaft of a motor has been exhibiting the cycles-to-failure data given in the first two columns of Table 15.1. This behavior was

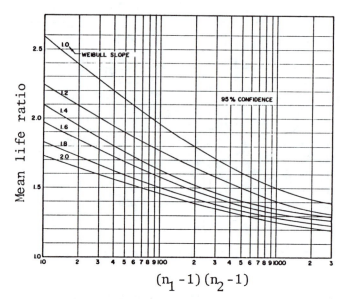

Fig. 15.3 − Test for significant difference in the mean lives of
Weibull distributions at a 95% confidence level
[1, p. 470; 2, pp. 16-33].

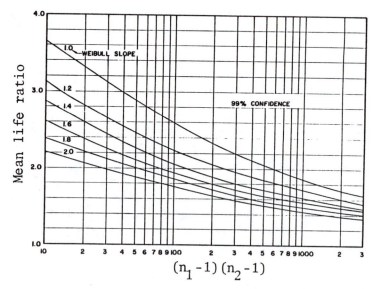

Fig. 15.4 − Test for significant difference in the mean lives of
Weibull distributions at a 99% confidence level
[1, p. 471; 2, pp. 16-33].

TABLE 15.1 – Cycles-to-failure data of seven motor shafts before and after redesign, and their median ranks for Example 15–1 with $N=7$.

Motor shaft before redesign		Motor shaft after redesign		Median rank
Failure number	Cycles to failure, 10^5 cycles	Failure number	Cycles to failure, 10^5 cycles	%
1	0.83	1	1.95	9.43
2	1.29	2	2.95	22.95
3	1.77	3	4.05	36.48
4	2.30	4	5.05	50.00
5	2.70	5	5.85	63.52
6	3.15	6	7.35	77.05
7	3.80	7	9.10	90.57

not satisfactory. The shaft was redesigned and its reliability tested with the results given in the next two columns of Table 15.1.

Find the confidence level at which the redesigned shaft is better than the former one, from its life and reliability point of view.

SOLUTION TO EXAMPLE 15-1

1. The reliability test data obtained from fatigue testing seven shafts of the original design and seven redesigned shafts are given in Table 15.1.

2. The data are shown plotted on Weibull probability paper in Fig. 15.5.

3. Both sets of data have a Weibull slope of 2.0, hence $\beta_1 = 2.0$ and $\beta_2 = 2.0$.

4. The degrees of freedom for the original shaft data are $d.f._{.1} = (7 - 1) = 6$ and for the redesigned shaft data are $d.f._{.2} = (7 - 1) = 6$.

5. The *combined degrees of freedom* are $c.d.f. = (n_1 - 1)(n_2 - 1) = 6 \times 6 = 36$.

6. The percent failed at the mean life from Fig. 15.1 with $\beta_1 = \beta_2 = 2.0$ is 54.5%.

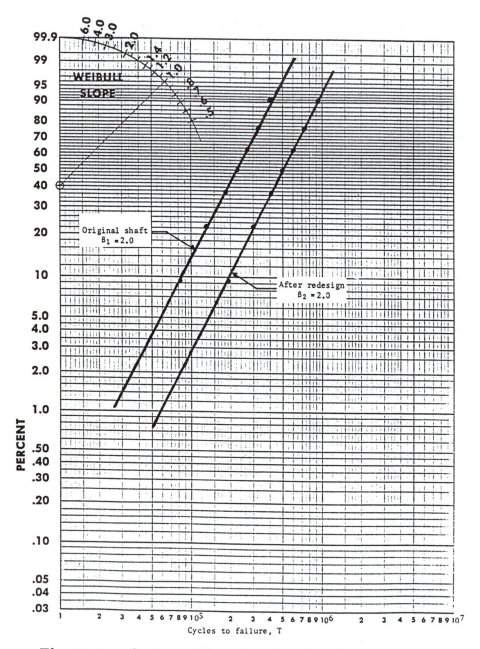

Fig. 15.5 – Cycles to failure data for original and redesigned shafts for Example 15–1.

7. The mean life of Set 1 is found by entering the ordinate of Fig. 15.5 at the 54.5% level, which gives a mean life of $\overline{T}_1 = 2.40 \times 10^5$ cycles. The mean life of Set 2 is found by entering the ordinate of Fig. 15.5 at the 54.5% level also, which gives a mean life of $\overline{T}_2 = 5.45. \times 10^5$ cycles.

8. The *mean life ratio, MLR*, is obtained by dividing the larger of the two mean lives with the smaller one, or

$$MLR = \frac{\overline{T}_2}{\overline{T}_1} = \frac{5.45 \times 10^5}{2.40 \times 10^5} = 2.27.$$

9. Entering Fig. 15.2 with the *combined degrees of freedom* of *c.d.f.* = 36 and $\beta = 2.0$, we find that the $MLR = 1.40$ at a 90% confidence level, which is lower than 2.27. Similarly entering Fig. 15.3 with *c.d.f.* = 36 and $\beta = 2.0$, we find that the $MLR = 1.57$ at a 95% confidence level, which is still lower than 2.27. Finally entering Fig. 15.4 we find that the $MLR = 1.94$ at a 99% confidence level, which is still lower than 2.27; consequently, we are more than 99% confident that the redesign has improved the mean life and thereby the reliability of the motor shaft.

15.3 TEST OF COMPARISON AT THE MEAN LIFE USING A NOMOGRAPH

The nomograph in Fig. 15.6 may be used to determine the confidence level at which the mean lives of the product supplied by two different suppliers, or of two different designs of the same product, are significantly different, or two different stress levels on the same product significantly affect the mean life of the product.

The procedure for using the nomograph in Fig. 15.6 is as follows:

1. Join by a straight line the combined degrees of freedom and the Weibull slope found in Steps 3 and 5, in the previous section, and mark its intersection with the auxiliary scale, as shown in Fig. 15.7.

2. Join the mean life ratio value found in Step 8 and the point of intersection on the auxiliary scale found in the previous step by another straight line until it intersects the confidence level scale, and read off the confidence level thus found.

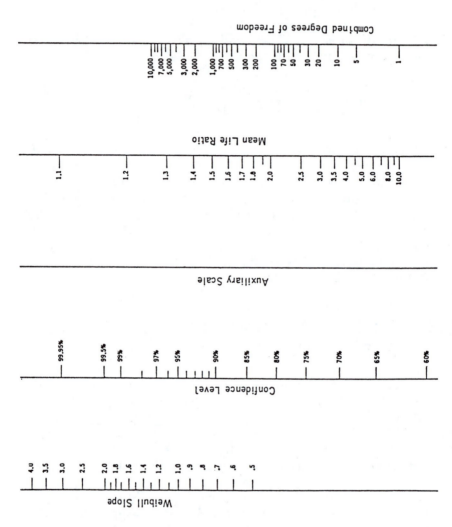

Fig. 15.6a– Confidence level at the mean life nomograph.

532

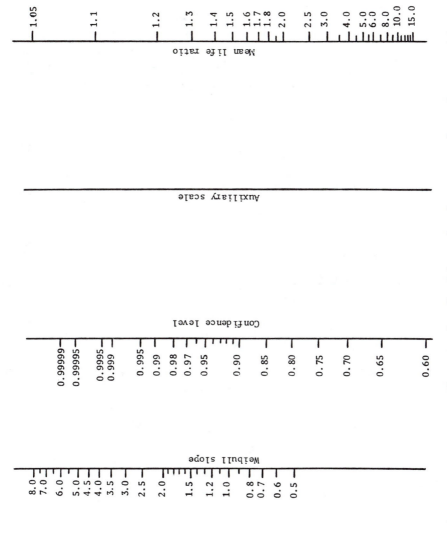

Fig. 15.6b– Confidence level nomograph at the mean life extended.

533

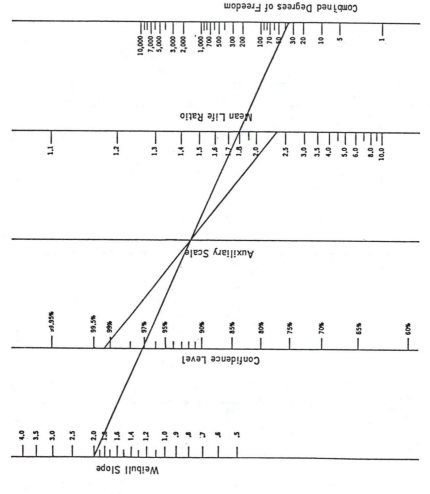

Fig. 15.7– Application of the test of comparison for the Weibull *pdf* at the mean life through a nomograph, for Example 15–2.

534

EXAMPLE 15–2

In Example 15-1 find the confidence level at which the mean lives are different, using the nomograph approach.

SOLUTION TO EXAMPLE 15–2

Figure 15.7 gives the nomograph as applied to this example. It may be seen that the confidence level is 99.2%. Consequently, we are 99.2% sure that the mean life of the redesigned shaft is greater than before redesign.

In Example 15-1 we could state that we are more than 99% confident that the redesign has improved the mean life. Using the nomograph, the confidence level can be stated more precisely as being 99.2%. Thus the advantage of the nomograph.

15.4 TEST OF COMPARISON AT THE MEAN LIFE WHEN THE β's ARE UNEQUAL

When the slopes of the two Weibull lines are different, or $\beta_1 \neq \beta_2$, the procedure for comparing the mean lives is as follows:

1. Assume that both sets of data have the same slope but equal to that of the first set, and determine the confidence level at which the MLR of the two designs is significant using the procedure discussed previously and illustrated with two examples.

2. Assume that both sets of data have the same slope but equal to that of the second set, and determine the confidence level at which the MLR of the two designs is significant using the same procedure as before.

3. Take the average of the two confidence levels thus determined and use it as the estimate of the true confidence level.

EXAMPLE 15–3

Two alternate designs of an equipment are to be compared. Eight units are built of Design A and six of Design B. To compare the mean lives of these two designs through life data, they are subjected to the same accelerated test. Care is exercised to assure that the resulting failures are representative of failures that will occur under normal operating conditions. The data given in Table 15.2 are obtained. Determine which design is better at the mean life.

TABLE 15.2 – Accelerated test times to failure data obtained on two alternate designs of the same equipment in Example 15–3 with $N_1 = 8$ and $N_2 = 6$.

Times to failure in hours	
Design A	Design B
185	640
220	184
130	470
90	330
280	260
245	92
180	–
150	–

SOLUTION TO EXAMPLE 15–3

Arranging the data in Table 15.2 as in Table 15.3, obtaining their median ranks, and plotting them on Weibull probability plotting paper yields Fig. 15.8.

We could have, of course, averaged the times to failure of each design and compared them to see if their mean lives are different. In this case the mean life for Design A is 185 hr and for Design B is 330 hr. On this basis Design B would have been chosen. However, such differences may occur in two samples drawn from the same design and production. Furthermore, we would not have known how sure we were that Design B was better.

Figure 15.8 tells us that the two slopes are significantly different; hence, the failure rates at a given life are quite different. It may also be seen that Design B will experience higher failure rates at lives of less than about 100 hr.

Let's determine how confident we are that there is a statistical difference in the mean lives of Designs A and B.

1. For Design A, $\beta_A = 3.0$ and for Design B, $\beta_B = 1.5$. From Fig. 15.1 the percent failed at the mean life for Design A is 51.0% and for Design B is 57.5%. From Fig. 15.8 the mean life for Design A is 185 hr and for Design B the mean life is 350 hr. Consequently, the MLR is $350/185 = 1.9$, and the *combined degrees of freedom* is $(8-1)(6-1) = 35$. Using the nomograph of Fig. 15.9, place

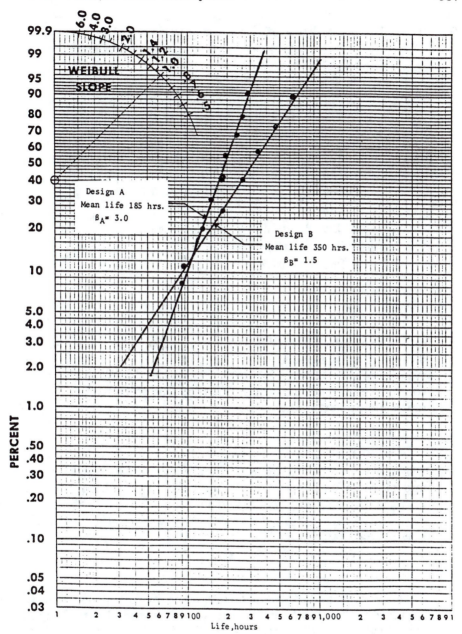

Fig. 15.8– Weibull plot of data in Table 15.3 of Example 15–
3 to determine the confidence level of difference
in their mean lives.

TABLE 15.3 – Ranking of the data given in Table 15.2 for Example 15–3, and their median ranks.

Design A		Design B	
T_j, hr	MR_j, %	T_j, hr	MR_j, %
90	8.3	92	10.9
130	20.2	184	26.6
150	32.1	260	42.2
180	44.0	330	57.8
185	56.0	470	73.4
220	67.9	640	89.1
245	79.8	–	–
280	91.7	–	–

on it the total *combined degrees of freedom* of 35. Assume both designs have a slope equal to that of Design A and place this value of $\beta_A = 3.0$ on Fig. 15.9. Using the *MLR* of 1.9, find a confidence level estimate of 99.5% that Design B, with the higher mean life, is better than Design A.

2. Assume both designs have a slope equal to that of Design B and place this value of $\beta_B = 1.5$ on Fig. 15.9. Using the same *combined degrees of freedom* and mean life ratio as before, from Fig. 15.9 find a confidence level estimate of 95.0% that Design B is better than Design A.

3. Take the average of these two confidence levels and find that the estimate of the confidence level at which the mean life of Design B is better than that of Design A is $(99.5\% + 95.0\%)/2 = 97.2\%$. In other words, we are highly confident that the mean life of Design B is different than that of Design A, and better. It must be pointed out that at lives of less than 10% failed, $T_{10\%}$, Design A appears to be better than Design B. It must be determined which life is of concern, based on reliability and warranty requirements, before a final conclusion is made as to which design to choose.

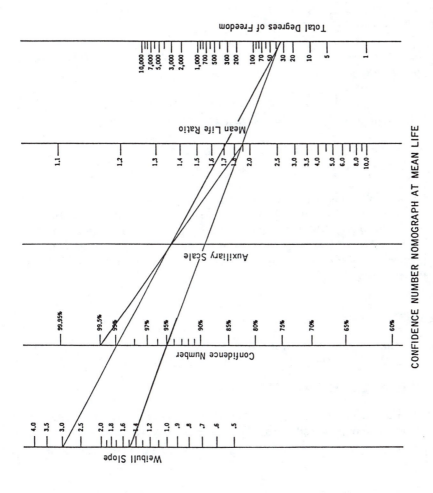

CONFIDENCE NUMBER NOMOGRAPH AT MEAN LIFE

Fig. 15.9— Nomographic solution of Example 15–3 to determine the confidence level of difference in their mean lives.

15.5   TEST OF COMPARISON AT THE 10% FAILED LIFE

To determine if there is a significant difference in the products of two different suppliers, or after redesigning, or at two different stress levels at the 10% failed life, in addition to that at the mean life, or in place of, use the following procedure:

1. Perform Steps 1 through 5 prescribed for the *mean life test of comparison* case.

2. Enter the Weibull plots of each set of data at the $Q(T_{10\%}) = 10\%$ level and find $T_{1-10\%}$ and $T_{2-10\%}$ lives for each set.

3. Find the $T_{10\%}$ life ratio by dividing the larger of the $T_{10\%}$ lives with the smaller one found in Step 2.

4. Enter Figs. 15.10, 15.11 and 15.12 for 90%, 95% and 99% confidence levels, respectively, and determine at which confidence level the $T_{10\%}$ life ratio is approached. We can then conclude that, with at least this confidence level, the product with the higher $T_{10\%}$ life has a higher $T_{10\%}$ life and an improved design for this life.

EXAMPLE 15–4

Using the data in Example 15–1 and Fig. 15-2 find the confidence level at which the redesigned shafts have a higher $T_{10\%}$ failed life.

SOLUTION TO EXAMPLE 15–4

From Fig. 15.5, entering at the $Q(T) = 10\%$ level, find that, for the motor shaft before redesign, $T_{1-10\%} = 0.86 \times 10^5$ cycles, and after redesign, $T_{2-10\%} = 1.95 \times 10^5$ cycles. The $T_{10\%}$ failed life ratio is

$$T_{10\%}\text{failed life ratio} = \frac{1.95 \times 10^5}{0.85 \times 10^5} = 2.27.$$

Entering Fig. 15.10 with the *combined degrees of freedom* of 36 and $\beta = 2.0$, we find that the $T_{10\%}$ failed life ratio is 2.7 at the 90% confidence level; and entering Fig. 15.11 we find that the $T_{10\%}$ failed life ratio is 3.5 at the 95% confidence level. As the actual $T_{10\%}$ failed life ratio is 2.27, we are less than 90% confident that the redesign has an improved 10% failed life.

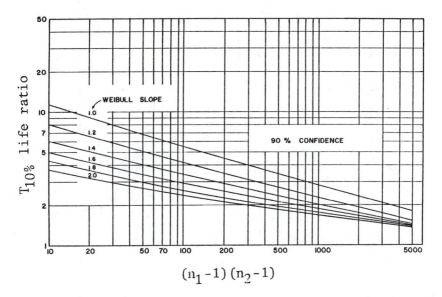

Fig. 15.10 – Test for significant difference at the 90% confidence level in the $T_{10\%}$ failed lives for the Weibull distribution [1, p. 471].

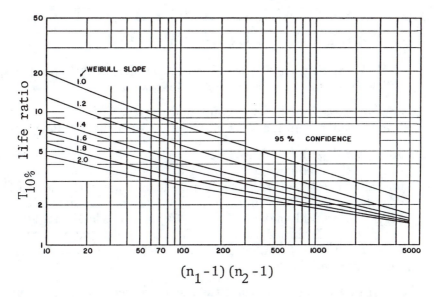

Fig. 15.11 – Test for significant difference at the 95% confidence level in the $T_{10\%}$ failed lives for the Weibull distribution [1, p. 471].

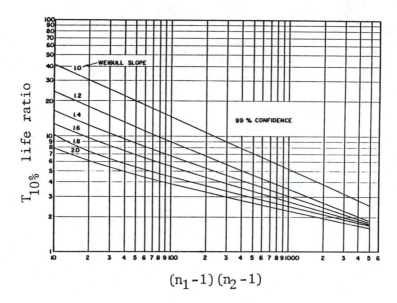

Fig. 15.12 – Test for significant difference at the 99% confidence level in the $T_{10\%}$ failed lives for the Weibull distribution [1, p. 471].

15.6   TEST OF COMPARISON AT THE 10% FAILED LIFE USING A NOMOGRAPH

Figure 15.13 may be used to determine the confidence level approached at the $T_{10\%}$ failed life more precisely, as follows:

1. Join by a straight line the combined degrees of freedom and the Weibull slope, and mark its intersection with the auxiliary scale.

2. Join the $T_{10\%}$ failed life ratio value and the point of intersection on the auxiliary scale found in Step 1 by another straight line until it intersects the confidence level scale, and read off the confidence level thus found. In this example, as illustrated in Fig.15.14 we find that the $CL = 86\%$, which is less than 90% as indicated in Example 15-4. We then conclude that we are 86% sure that the $T_{10\%}$ failed life of the redesigned shaft is significantly greater than before redesign.

   Comparing lives at lower than the mean life becomes important in the following situations:

1. The product's manufacturer wants to know what warranty cost improvements can be attained by redesign. Usually warranties

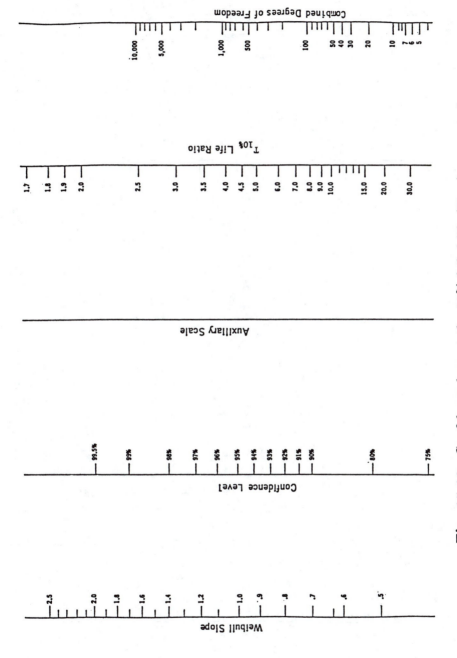

Fig. 15.13– Confidence leve at the 10% failed life.($T_{10\%}$) nomograph.

543

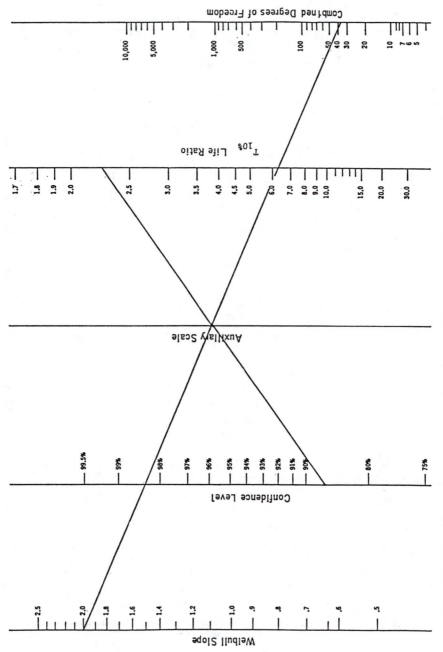

Fig. 15.14– Use of confidence leve at the 10% failed life ($T_{10\%}$) nomograph illustrated.

544

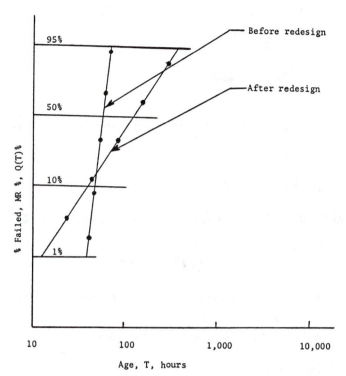

Fig. 15.15 – Two designs with crossing Weibull lines compared.

are for shorter periods of operation than the mean life.

2. The redesigned product's mean life may be significantly better, but the $T_{10\%}$ life may not be significantly better particularly if the two Weibull lines cross each other, as in Fig. 15.15.

15.7 CONSTRUCTION OF THE NOMOGRAPH GIVEN IN FIG.15.6a

The nomograph of Fig. 15.6a is based on the empirical formula [3, p. 34] for the confidence level of the *Mean Life Ratio* (*MLR*) of

$$CL = 1 - \frac{1}{2}(MLR)^{-\beta\rho^{1/4}}, \tag{15.1}$$

where

$$CL = \text{confidence level,}$$

$$\beta \quad = \text{Weibull slope,}$$

$$\rho \quad = \text{combined degrees of freedom} = (n_1 - 1)(n_2 - 1),$$

and

$$MLR = \text{mean life ratio.}$$

Rearranging Eq. (15.1), and taking the natural logarithm of both sides, yields

$$\log_e(2 - 2CL) = -\beta\rho^{1/4}\log_e(MLR). \tag{15.2}$$

Taking the natural logarithm of both sides of Eq. (15.2) yields

$$\log_e[-\log_e(2 - 2CL)] = \log_e\beta + \frac{1}{4}\log_e\rho$$

$$+\log_e\log_e(MLR), \tag{15.3}$$

or

$$\log_e[-\log_e(2 - 2CL)] - \log_e\log_e(MLR)$$

$$= \log_e\beta + \frac{1}{4}\log_e\rho. \tag{15.4}$$

Let

$$W = \log_e[-\log_e(2 - 2CL)] - \log_e\log_e(MLR), \tag{15.5}$$

and

$$W' = \log_e\beta + \frac{1}{4}\log_e\rho, \tag{15.6}$$

where W and W' are auxiliary variables.

Then the nomograph for Eq. (15.4) is equal to the combined nomograph of Eqs. (15.5) and (15.6) which have the general form of

$$f_1(U) + f_2(V) = f_3(W) \text{ or } f_3(W'). \tag{15.7}$$

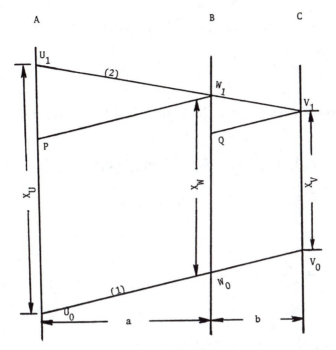

Fig. 15.16 – Designation of the scales of a nomograph.

Now consider the construction of the nomograph for Eq. (15.7). Choose three parallel scales, as shown in Fig. 15.16, A, B and C. So graduate them that Lines 1 and 2 cut the scales in values which satisfy Eq. (15.7).

Let m_U, m_V, and m_W be the scale moduli for A, B and C, respectively, then

$$X_U = m_U[f_1(U_1) - f_1(U_0)], \tag{15.8}$$
$$X_V = m_V[f_2(V_1) - f_2(V_0)], \tag{15.9}$$

and

$$X_W = m_W[f_3(W_1) - f_3(W_0)]. \tag{15.10}$$

Assume further that the spacing of the scales is in the ratio of a/b. In Fig. 15.16 draw lines through Points W_1 and V_1 parallel to line $U_0 V_0$. Then triangles $U_1 P W_1$ and $W_1 Q V_1$ are similar; therefore,

$$\frac{X_U - X_W}{X_W - X_V} = \frac{a}{b},$$

or

$$X_U b + X_V a = X_W(a + b). \tag{15.11}$$

Dividing both sides of Eq. (15.11) by ab yields

$$\frac{X_U}{a} + \frac{X_V}{b} = \frac{X_W}{\frac{ab}{a+b}}. \tag{15.12}$$

Substitution of Eqs. (15.8), (15.9) and (15.10) into Eq. (15.12) yields

$$\frac{1}{a}m_U[f_1(U_1) - f_1(U_0)] + \frac{1}{b}m_V[f_2(V_1) - f_2)V_0)]$$
$$= \frac{a+b}{ab}m_W[f_3(W_1) - f_3(W_0)]. \tag{15.13}$$

Then, if we set

$$\frac{a}{b} = \frac{m_U}{m_V}, \tag{15.14}$$

and

$$m_W = \frac{ab}{a+b} = \frac{m_U m_V}{m_U + m_V}, \tag{15.15}$$

Eqs. (15.14) and (15.15) give the construction requirements for the nomograph of Eq. (15.7). Thus, to construct the nomograph for an equation of the form given by Eq. (15.7) do the following:

1. Place the parallel lines for the scales of U and V a convenient distance apart.

2. Graduate them in accordance with their scale equation, or

$$X_U = m_U f_1(U),$$

   and

$$X_V = m_V f_2(V).$$

3. So locate the line for the scale of W that its distance from the U scale is to its distance from the V scale as $m_U/m_V = a/b$.

4. Graduate the W scale from its scale equation

$$X_W = \frac{m_U m_V}{m_U + m_V} f_3(W).$$

It can be seen that if we want to so place the W scale that the distance from the W scale to the U scale equals the distance from the W scale to the V scale; that is, if

$$\frac{a}{b} = 1,$$

then we should choose

$$\frac{m_U}{m_V} = 1.$$

Now, getting back to our problem and considering Eq. (15.5) first; that is,

$$W = \log_e[-\log_e(2 - 2CL)] - \log_e\log_e(MLR),$$

to make the nomograph look nice, choose

$$\frac{a}{b} = 1,$$

that means that we have to choose

$$m_{CL} = m_{MLR}.$$

The three scales of CL, MLR and W are placed as shown in Fig. 15.17. If we graduate the W scale increasing upward then the CL scale should also be graduated increasing upward; however, the MLR scale should be increasing downwards since it has a minus sign in the function, or $-\log_e\log_e(MLR)$.

5. Construct the CL scale.

   (a) Graduate the CL scale increasing upwards, as shown in Fig. 15.6b.

   (b) CL values range from 60% to 99.999%.

   (c) Choose the length of the CL scale, L_{CL}, to be 19.55835382 cm.

   (d) Then,

$$m_{CL} = \frac{L_{CL}}{\log_e[-\log_e(2 - 2 \times 0.99999)] - \log_e(2 - 2 \times 0.60)]},$$

or

$$m_{CL} = \frac{19.55835382}{2.381375782 + 1.499939987} = 5.039104.$$

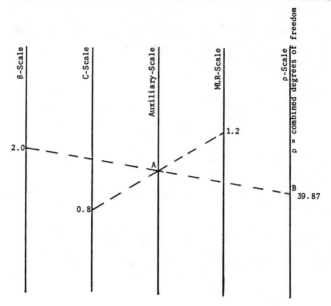

Fig. 15.17 — Determining the value of ρ at Point B.

6. Construct the MLR scale.

   (a) Graduate the MLR scale increasing downwards, as shown in Fig. 15.6b, and place it to the right of the CL scale.

   (b) MLR values range from 1.05 to 15.0.

   (c) Choose the length of the MLR scale to be $L_{MLR} = 20.24$ cm, and $m_{MLR} = m_{CL} = 5.039104$; then,

$$m_{MLR} = \frac{L_{MLR}}{\log_e\log_e MLR_{max} - \log_e\log_e MLR_{min}},$$

or

$$m_{MLR} = \frac{L_{MLR}}{\log_e\log_e(15.0) - \log_e\log_e(1.05)} = 5.0398104.$$

Therefore,

$$L_{MLR} = 5.039104[\log_e\log_e(15.0) - \log_e\log_e(1.05)],$$

or

$$L_{MLR} \cong 20.24.$$

7. Construct the W scale.

    (a) The W scale is an auxiliary scale, graduated upwards, and placed in the middle of the CL and the MLR scales.

    (b) Then,

$$m_W = \frac{m_{CL} \cdot m_{MLR}}{m_{CL} + m_{MLR}} 2.519552.$$

Now consider the second chart for Eq. (15.6), or

$$W' = \log_e \beta + \frac{1}{4}\log_e \rho.$$

Since the same W scale will be used for this chart, then we must have

$$\frac{m_\beta m_\rho}{m_\beta + m_\rho} = 2.519552. \tag{15.16}$$

Also we want to place the W scale in the middle of the β and ρ scales. Therefore, the same scale modulus should be used for the β and ρ scales as for the CL and MLR scales. Thus, we have

$$m_\beta = m_\rho = 5.039104.$$

8. Construct the β scale.

    (a) Graduate the β scale increasing upwards and place it at the very left side.

    (b) The β scale values range from 0.5 to 8.0.

    (c) $m_\beta = 5.039104$.

    (d) $L_\beta = 5.039104(\log_e 8.0 - \log_e 0.5)$,

or

$$L_\beta \cong 14.0 \text{ cm.}$$

9. Construct the ρ scale, or the *combined degrees of freedom* scale.

    (a) Graduate the ρ scale increasing upwards and place it at the very right side.

    (b) The ρ scale values range from 1.0 to 100,000.

    (c) $m_\rho = 5.039104$.

(d) $L_\rho = 5.039104[\log_e(100,000) - \log_e(1.0)]/4.0$,

or

$$L_\rho \cong 14.5 \text{ cm.}$$

10. Write the equations for graduating each scale.

Now we obtain the equations for graduating each scale from

$$X_{CL} = 5.039104\{\log_e[-\log_e(2 - 2CL)]$$

$$-\log_e(2 - 2 \times 0.6]\}, \tag{15.17}$$

$$X_{MLR} = 5.039104[\log_e\log_e(MLR)$$

$$-\log_e\log_e(1.05)], \tag{15.18}$$

$$X_\beta = 5.039104(\log_e\beta - \log_e 0.5), \tag{15.19}$$

and

$$X_\rho = 5.039104(\log_e\rho)/4.0. \tag{15.20}$$

Equations (15.17) through (15.20) were used to calculate the values given in Table 15.4.

Before making the nomograph, the position of the origins for each scale needs to be determined. Actually, the graduation of any three of the four scales can be placed in a convenient way, and the fourth one placed according to the position of its paired scale; e.g., for β its paired scale is the ρ scale.

After the CL, MLR and β scales are placed on the nomograph, as shown in Fig. 15.6b, to decide the position of the fourth scale, first draw a line through any two points on the CL and the MLR scales; e.g., $CL = 0.80$ and $MLR = 1.2$, and denote the intersection of this line with the auxiliary scale by A, as shown in Fig. 15.17. Then, draw another line through any point on the β scale; e.g., $\beta = 2.0$ and point A, denote the intersection of this line with the ρ scale by B. Then, the ρ value at point B can be calculated as follows:

Substitution of $CL = 0.80$ and $MLR = 1.2$ into Eq. (15.5) yields

TABLE 15.4– Values for graduating the confidence level of the mean life ratio nomograph scales.

β-scale		C-scale		MLR-scale		ρ-scale	
β value	Graduation position	C value	Graduation position	MLR value	Graduation position	ρ value	Graduation position
0.5	0.0000000	0.6	0.0000000	1.05	0.0000000	1.0	0.0000000
0.6	0.9187373	0.65	2.3633879	1.10	3.3742239	5.0	2.0275312
0.7	1.6955186	0.70	4.1734514	1.20	6.6427650	10.0	2.9007414
0.8	2.3683972	0.75	5.7114569	1.30	8.4768068	20.0	3.7739516
0.9	2.9619182	0.80	7.1178272	1.40	9.7304442	30.0	4.2847468
1.0	3.4928408	0.85	8.4937461	1.50	10.6703340	40.0	4.6471618
1.1	3.9731186	0.90	9.9563876	1.60	11.4146370	50.0	4.9282727
1.2	4.4115780	0.91	10.2759199	1.70	12.0259880	60.0	5.1579570
1.3	4.8149215	0.92	10.6106679	1.80	12.5415000	70.0	5.3521523
1.4	5.1883592	0.93	10.9650823	1.90	12.9849250	80.0	5.5203720
1.5	5.5360215	0.94	11.3454438	2.00	13.3723390	90.0	5.6687523
1.6	5.8612378	0.95	11.7611298	2.50	14.7787100	100.0	5.8014829
1.7	6.1667316	0.96	12.2272326	3.00	15.6931530	200.0	6.6746930
1.8	6.4547588	0.97	12.770790	3.50	16.3548050	300.0	7.1854883
1.9	6.7272091	0.98	13.449228	4.00	16.8651800	400.0	7.5479032
2.0	6.9856813	0.99	14.4319668	4.50	17.2760960	500.0	7.8290141
2.5	8.1101249	0.995	15.2539706	5.00	17.6172700	600.0	8.0586984

553

TABLE 15.4— Continued.

β-scale

β value	Graduation position
3.0	9.0288622
3.5	9.8056436
4.0	10.4785220
4.5	11.0720430
5.0	11.6029660
5.5	12.0832440
6.0	12.5217030
6.5	12.9250470
7.0	13.2984840
7.5	13.6461470
8.0	13.9713630

C-scale

C value	Graduation position
0.999	16.7643061
0.9995	17.2971514
0.99995	18.74681128
0.99999	19.55835382

MLR-scale

MLR value	Graduation position
5.50	17.9071820
6.00	18.1580320
7.00	18.5739180
8.00	18.9083610
9.00	19.1859930
10.00	19.4220120
11.00	19.6263930
12.00	19.8060050
13.00	19.9657630
14.00	20.1092920
15.00	20.2393370

ρ-scale

ρ value	Graduation position
700.0	8.2528938
800.0	8.4211134
900.0	8.5694937
1,000.0	8.7022244
2,000.0	9.5754346
3,000.0	10.0862300
4,000.0	10.4486450
5,000.0	10.7297560
6,000.0	10.9594400
7,000.0	11.1536350
8,000.0	11.3218550
9,000.0	11.4702350
10,000.0	11.6029660
20,000.0	12.4761760
50,000.0	13.6304970
70,000.0	14.0543770
100,000.0	14.5037070

$$W = \log_e[-\log_e(2 - 2 \times 0.8)] - \log_e\log_e(1.2),$$

or

$$W = 1.614561783.$$

Then, from Eq. (15.6),

$$1.614561783 = \log_e(2.0) + \frac{1}{4}\log_e\rho. \tag{15.21}$$

Solving Eq. (15.21) for ρ yields

$$\rho = e^{4(1.614561783 - \log_e 2.0)}, \tag{15.22}$$

or

$$\rho = 39.87136551.$$

Thus, after the ρ value of point B on the ρ scale is obtained, the graduation position on the ρ scale can be uniquely determined.

15.8   EQUATION FOR CONSTRUCTING THE NOMOGRAPH GIVEN IN FIG. 15.13

The nomograph given in Fig. 15.13 is based on an empirical formula for the confidence level at the $T_{10\%}$ life ratio. The formula is [3, p. 40]

$$CL = 1 - \frac{1}{2}(T_{10\%}LR)^{-0.86169\beta\rho^{1/14}}, \tag{15.23}$$

where

CL = confidence level,

β = Weibull slope,

ρ = *combined degrees of freedom* = $(n_1 - 1)(n_2 - 1)$,

and

$T_{10\%}LR = T_{10\%}$ failed life ratio.

Rearranging Eq. (15.23), and then taking the natural logarithm of both sides, yields

$$\log_e(2 - 2CL) = -0.86169\beta\rho^{1/14}\log_e(T_{10\%}LR).\qquad(15.24)$$

Taking the natural logarithm again of both sides of Eq. (15.24) yields the basic equation for constructing Fig. 15.13, or

$$\log_e[-\log_e(2 - 2CL)] = \log_e 0.86169 + \log_e\beta$$

$$+\frac{1}{14}\log_e\rho + \log_e\log_e(T_{10\%}LR).\qquad(15.25)$$

EXAMPLE 15–5

Using the data in Example 15–3 and Fig. 15.8 find the confidence level at which the $T_{10\%}$ failed life of Design A is the better one.

SOLUTION TO EXAMPLE 15–5

From Fig. 15.8, and entering at the $Q(T) = 10\%$ level, find that for Design A $T_{A-10\%} = 96$ hr and for Design B $T_{B-10\%} = 88$ hr. The $T_{10\%}$ failed life ratio is

$$T_{10\%}\text{failed life ratio} = \frac{96}{88} = 1.09.$$

This ratio is beyond the corresponding scale in Fig. 15.13; consequently, we should use Eq. (15.23). Then for design A, with a $T_{10\%}$ failed life ratio of 1.09 and $\beta = 3.0$,

$$CL_A = 1 - \frac{1}{2}(1.09)^{-0.86169(3.0)35^{1/14}}$$

or

$$CL_A = 0.648125, \text{say } 62.5\%,$$

and for Design B, with $T_{10\%}$ failed life ratio of 1.09 and $\beta = 1.5$,

$$CL_B = 1 - \frac{1}{2}(1.09)^{-0.86169(1.5)35^{1/14}}$$

or

$$CL_B = 0.566879, \text{say } 56.7\%.$$

Averaging these two values yields a confidence level of 59.6%. In other words, we are only 59.6% confident that the 10% failed life of Design A is better than that of Design B.

PROBLEMS

15-1. The design providing the results in Problem 13-2 is altered. Six of the redesigned units are tested with the results given in Table 15.5. Determine if the redesign has improved the mean life of these units and the associated level of confidence, using the Weibull *pdf*.

Table 15.5 – Times-to-failure data from the six redesigned units, for Problem 15.1.

Failure order number	Times to failure, hr
1	920
2	1,840
3	2,600
4	3,300
5	4,700
6	6,400

15-2. We wish to compare two alternative designs, A and B. Samples of each are built and reliability tested to failure, and the data given in Table 15.6 are obtained.

(1) Determine the confidence level at which Design B is better than Design A at the mean life, using the Weibull *pdf*.

(2) Determine the confidence level at which Design A is better than Design B at the 10% failed life, using the Weibull *pdf*.

15-3. The same product is purchased from two different manufacturers. Fourteen of these products are purchased from each, and all are tested to failure under identical application and operation stress levels. The times-to-failure data obtained from these tests are given in Table 15.7. Determine the confidence level at which the mean lives are different. Which manufacturer would you recommend and why?

Table 15.6 – Times-to-failure data for $N=8$ from Manufacturer A and $N=6$ from Manufacturer B, for Problem 15.2.

Design A Times to failure, hr	Design B Times to failure, hr
130	130
190	300
80	660
185	425
265	60
240	200
260	
140	

Table 15.7 – Times-to-failure data for 14 units each from Manufacturers A and B, for Problem 15.3.

Failure order number	Manufacturer A Times to failure, hr	Manufacturer B Times to failure, hr
1	290	295
2	295	295
3	305	305
4	310	320
5	315	325
6	320	345
7	325	355
8	340	365
9	350	375
10	350	375
11	360	381
12	390	395
13	395	404
14	410	440

15-4. We wish to compare two alternative designs, A and B. Samples of each are built and reliability tested to failure, and the data given in Table 15.8 are obtained.

(1) Determine the confidence level at which Design B is better than Design A at the mean life, using the Weibull *pdf*. Assume $\gamma = 0$.

(2) Determine the confidence level at which Design A is better than Design B at the 10% failed life, using the Weibull *pdf*.

Table 15.8 – Times-to-failure data for eight units each from Manufacturers A and B, for Problem 15.4.

Design A Times to failure, hr	Design B Times to failure, hr
1,900	2,000
2,600	4,250
1,400	1,300
1,300	600
800	3,000
2,650	6,600
1,850	9,000
2,400	12,000

15-5. We wish to compare two alternative designs, A and B. Samples of each are built and reliability tested to failure, and the data given in Table 15.9 are obtained.

(1) Determine the confidence level at which Design B is better than Design A at the mean life, using the Weibull *pdf*. Assume $\gamma = 0$.

(2) Determine the confidence level at which Design A is better than Design B at the 10% failed life, using the Weibull *pdf*.

15-6. The same product is purchased from two different manufacturers. Fourteen of these products are purchased from each, and all are tested under identical application and operation stress levels. The test is terminated when the first six units fail in each sample. The times to failure data obtained from these tests are given in

Table 15.10. Determine the confidence level at which the mean lives are different. Which manufacturer would you recommend and why?

Table 15.9 – Times-to-failure data for $N=9$ from Manufacturer A and $N=7$ from Manufacturer B, for Problem 15.5.

Design A Times to failure, hr	Design B Times to failure, hr
1,305	1,300
1,901	3,002
802	6,605
1,300	4,259
2,659	601
2,650	6,600
2,400	2,000
2,605	
1,400	

Table 15.10 – Times-to-failure data for the six units that failed in each sample from Manufacturers A and B, for Problem 15.6. $N=14$.

Failure order number	Manufacturer A Times to failure, hr	Manufacturer B Times to failure, hr
1	2,902	2,950
2	2,955	2,955
3	3,050	3,050
4	3,108	3,203
5	3,157	3,253
6	3,203	3,451

15-7. The same product is manufactured by Manufacturers A and B. To decide which manufacturer to be chosen, two samples of size

eight each are purchased from Manufacturers A and B, and all are tested to failure under identical application and operation stress levels. The times-to-failure data are given in Table 15.11. Determine the confidence level at which the mean lives are different using the Weibull *pdf*. Which manufacturer would you recommend and why?

Table 15.11 – Times-to-failure data for eight units each from Manufacturers A and B, for Problem 15.7.

Failure order number	Manufacturer A Times to failure, hr	Manufacturer B Times to failure, hr
1	460	610
2	850	920
3	1,250	1,160
4	1,600	1,350
5	1,900	1,600
6	2,400	1,800
7	2,800	2,050
8	4,100	2,500

15-8. We wish to compare two alternative designs, A and B. Samples of each are built and reliability tested to failure, and the data given in Table 15.12 are obtained.

(1) Determine the confidence level at which Design B is better than Design A at the mean life, using the Weibull *pdf*. Assume $\gamma = 0$.

(2) Determine the confidence level at which Design A is better than Design B at the 10% failed life, using the Weibull *pdf*.

15-9. The same product is manufactured by Manufacturers A and B. To decide which manufacturer to be chosen, two samples of size ten each are purchased from Manufacturers A and B, and are tested to failure under identical application and operation stress levels. The test is terminated when the first five units fail in each sample. The times to failure data are given in Table 15.13.

Determine the confidence level at which the mean lives are different, using the Weibull *pdf*. Which manufacturer would you recommend and why?

Table 15.12 – Times-to-failure data for eight units each from Manufacturers A and B, for Problem 15.8.

Design A Times to failure, hr	Design B Times to failure, hr
195	200
267	425
145	125
129	60
80	300
266	660
187	905
235	1,195

Table 15.13 – Times-to-failure data for ten units for a failure terminated test, after five failures occur in Manufacturers' A and B samples, for Problem 15.9.

Failure order number	Manufacturer A Times to failure, hr	Manufacturer B Times to failure, hr
1	610	460
2	920	850
3	1,160	1,250
4	1,350	1,600
5	1,600	1,900

15-10. The same product is manufactured by Manufacturers A and B. To decide which manufacturer to be chosen, two samples of size 15 each are purchased from Manufacturers A and B, and are

tested to failure under identical application and operation stress levels. The test is terminated when the first five units fail in each sample. The times-to-failure data are given in Table 15.14. Determine the confidence level at which the mean lives are different using the Weibull *pdf*. Which manufacturer would you recommend and why?

Table 15.14 – Times-to-failure data for 15 units for a failure terminated test, after five failures occur in Manufacturers' A and B samples, for Problem 15.10.

Failure order number	Manufacturer A Times to failure, hr	Manufacturer B Times to failure, hr
1	290	295
2	305	305
3	315	325
4	325	355
5	350	375

REFERENCES

1. Johnson, Leonard G., *The Statistical Treatment of Fatigue Experiments,* Elsevier Publishing Company, Amsterdam, 116 pp., 1964.

2. Johnson, Leonard G., *Theory and Technique of Variation Research,* Elsevier Publishing Company, New York, 105 pp., 1964.

3. Levens, A. S., *Graphical Methods in Research,* John Wiley & Sons, Inc., New York, 217 pp., 1965.

4. Lipson, Charles and Sheth, Narendra J., *Statistical Design and Analysis of Experiments,* McGraw-Hill Book Company, New York, 518 pp., 1973.

Chapter 16

THE GAMMA DISTRIBUTION

16.1 GAMMA DISTRIBUTION CHARACTERISTICS

The two-parameter gamma distribution, illustrated in Figs. 16.1 and 16.2, is [1, p. 85]

$$f(T) = \frac{1}{\eta\Gamma(\beta)}(\frac{T}{\eta})^{\beta-1}e^{-\frac{T}{\eta}}, \tag{16.1}$$

$$f(T) \geq 0, \; T \geq 0, \; \beta > 0, \; \eta > 0,$$

where

β =shape parameter,

η =scale parameter,

and

$\Gamma(\beta)$ =gamma function evaluated at the value of β. See Table 12.1

for this function's equation and its values.

Specific characteristics of the gamma *pdf* are the following:

1. As may be seen from Figs. 16.1 and 16.2, the gamma distribution is skewed to the right. For $\beta < 1$ as $T \to 0$ $f(T) \to \infty$. For $\beta = 1$, $f(T = 0) = \frac{1}{\eta}$ and $f(T)$ decreases monotonically when $T > 0$ until at $T \to \infty$, $f(T) \to 0$. For $\beta > 1$ the distribution starts at zero for $T = 0$; increases, as T increases, to its mode; and decreases thereafter to zero as $T \to \infty$. This behavior is similar to that of the Weibull *pdf* for various values of its β. This flexibility makes it suitable for describing product life.

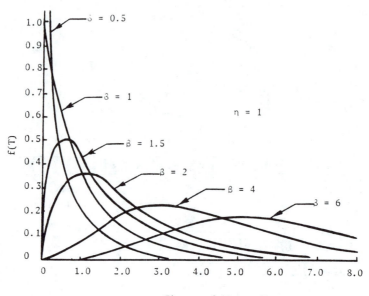

Fig.16.1 – The gamma distribution with various values of β
and $\eta = 1$.

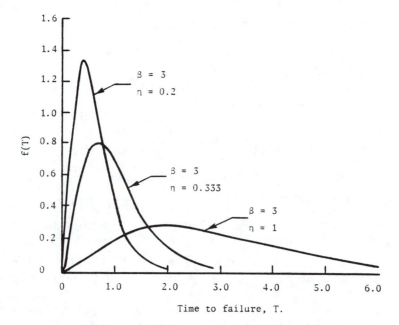

Fig.16.2 – The gamma distribution with various values of η
and $\beta = 3$.

2. The effect of η is identical to that of η in the Weibull *pdf*. As η is decreased the distribution gets pushed toward the origin, and as η is increased the distribution gets stretched out away from the origin towards the right. The effect is that of changing the scale of the abscissa.

3. The mean, $\bar{T}$, of the gamma *pdf* is

$$\bar{T} = \eta\,\beta. \tag{16.2}$$

4. The median, $\check{T}$, of the gamma *pdf* is obtained from

$$0.5 = \int_0^{\check{T}} f(T)\,dT, \tag{16.3}$$

where $f(T)$ is the gamma *pdf* of Eq. (16.1). Given β and η, Eq. (16.3) may be evaluated to find $\check{T}$ by using the tables of the incomplete gamma function [2; 3; 4] by trial and error. An approximation of $\check{T}$ may be obtained from

$$\check{T} \cong \eta\,(\beta - 0.3). \tag{16.4}$$

5. The mode, $\tilde{T}$, of the gamma *pdf* is

$$\tilde{T} = \begin{cases} \eta\,(\beta - 1), & \text{for } \beta > 1; \\ 0, & \text{otherwise.} \end{cases} \tag{16.5}$$

6. The standard deviation, σ_T, of the gamma *pdf* is

$$\sigma_T = \eta\,(\beta)^{1/2}. \tag{16.6}$$

7. The kth moment about the origin, μ'_k, of the gamma *pdf* is

$$\mu'_k = \eta^k\,[\beta]^k, \tag{16.7}$$

where

$$[\beta]^k = \beta\,(\beta + 1)...(\beta + k - 1). \tag{16.7'}$$

8. The kth moment about the mean, μ_k, of the gamma *pdf* is

$$\mu_k = \eta^k\,\{[\beta] - \beta\}^k. \tag{16.8}$$

When quantifying Eq. (16.8) first expand binomially the term $\{[\beta] - \beta\}^k$, then use Eq. (16.7') to quantify the $[\beta]^k$ terms in the expansion.

9. The coefficient of skewness, α_3, of the gamma *pdf* is

$$\alpha_3 = \frac{2}{(\beta)^{1/2}}. \tag{16.9}$$

10. The coefficient of kurtosis, α_4, of the gamma *pdf* is

$$\alpha_4 = 3 + \frac{6}{\beta}. \tag{16.10}$$

11. For the special case of $\beta = 1$, the gamma *pdf* becomes the exponential *pdf*.

12. For the special case of $\eta = 2$ and $\beta = \frac{\nu}{2}$, where ν is an integer, the gamma *pdf* becomes the chi-square *pdf* with a single parameter ν, the degrees of freedom.

13. When β is restricted to integers, the gamma *pdf* is referred to as the Erlang *pdf* used in queueing theory and specifically in event traffic analysis. The Erlang *pdf* then is

$$f(T) = \frac{1}{\eta(\beta - 1)!}(\frac{T}{\eta})^{\beta-1}e^{-\frac{T}{\eta}}. \tag{16.11}$$

14. The gamma *pdf* approaches the normal *pdf* for large β, with the normal *pdf* parameters [1, p. 86] of

$$\bar{T} = \eta\,\beta,$$

and

$$\sigma_T = \eta\,(\beta)^{1/2}.$$

15. The three-parameter gamma *pdf* exists also, and is

$$f(T) = \frac{1}{\eta\Gamma(\beta)}(\frac{T - \gamma}{\eta})^{\beta-1}e^{-(\frac{T-\gamma}{\eta})}, \tag{16.12}$$

$$f(T) \geq 0,\ T \geq \gamma,\ -\infty < \gamma < \infty,\ \beta > 0,\ \eta > 0,$$

where

$$\gamma = \text{location parameter.}$$

The effect of γ on the behavior of the gamma *pdf* is identical to that of γ in the three-parameter Weibull *pdf*.

The parameters of the gamma *pdf* can be determined most conveniently using the matching moments method, which yields

$$\beta = (\frac{\bar{T}}{\sigma_T})^2, \tag{16.13}$$

and

$$\eta = \frac{\bar{T}}{\beta}, \tag{16.14}$$

where $\bar{T}$ is obtained from Eq. (9.11), and σ_T from Eq. (9.12).

EXAMPLE 16–1

Times-to-failure data are obtained from identical units from which the following are calculated:

$$\bar{T} = 7,502 \text{ hr and } \sigma_T = 3,013 \text{ hr.}$$

Find the parameters of the gamma *pdf* representing these data.

SOLUTION TO EXAMPLE 16–1

From Eq. (16.13) the shape parameter of the gamma *pdf* representing these data is

$$\beta = \left(\frac{7,502}{3,013}\right)^2 = 6.1995, \text{ or } 6.2,$$

and from Eq. (16.14) the scale parameter is

$$\eta = \frac{7,502}{6.1995} = 1,210.0998, \text{ or } 1,210 \text{ hr.}$$

EXAMPLE 16–2

The times-to-failure of a unit have a gamma distribution with $\beta = 6.2$ and $\eta = 1,210$ hr. Do the following:

1. Write the gamma *pdf*.

2. Determine the mean of this *pdf*.

3. Determine the median of this *pdf*.

4. Determine the mode of this *pdf*.

5. Determine the standard deviation of this *pdf*.

6. Determine the coefficient of skewness of this *pdf*.

7. Determine the coefficient of kurtosis of this *pdf*.

8. Determine the probability density of this *pdf* for the age of 1,000 hr.

SOLUTIONS TO EXAMPLE 16–2

1. The gamma *pdf* with $\beta = 6.2$ and $\eta = 1,210$ hr, using Eq. (16.1), is

$$f(T) = \frac{1}{1,210 \; \Gamma(6.2)} \left(\frac{T}{1,210}\right)^{5.2} e^{-\frac{T}{1,210}} \tag{16.15}$$

2. The mean of this *pdf*, from Eq. (16.2) is

$$\bar{T} = \eta \; \beta = 1,210(6.2) = 7,502 \text{ hr.}$$

3. The median of this *pdf*, from Eq. (16.4), is

$$\check{T} \cong \eta \; (\beta - 0.3) = 1,210(6.2 - 0.3) = 7,139 \text{ hr.}$$

4. The mode of this *pdf*, from Eq. (16.5), is

$$\tilde{T} = \eta \; (\beta - 1) = 1,210(6.2 - 1) = 6,292 \text{ hr.}$$

5. The standard deviation of this *pdf*, from Eq. (16.6), is

$$\sigma_T = \eta \; (\beta)^{1/2} = 1,210(6.2)^{1/2} = 3,012.88 \text{ hr.}$$

6. The coefficient of skewness for this *pdf*, from Eq. (16.9), is

$$\alpha_3 = \frac{2}{(\beta)^{1/2}} = \frac{2}{(6.2)^{1/2}} = 0.8032.$$

This indicates that this distribution is positively skewed, or skewed to the right.

7. The coefficient of kurtosis for this *pdf*, from Eq. (16.10), is

$$\alpha_4 = 3 + \frac{6}{\beta} = 3 + \frac{6}{6.2} = 3.9677.$$

This indicates that this distribution is more peaked than a normal distribution fitted to this population.

8. The probability density for the age of 1,000 hr, from Eq. (16.15), is

$$f(T = 1,000 \text{ hr}) = \frac{1}{1,210\Gamma(6.2)}\left(\frac{1,000}{1,210}\right)^{5.2} e^{-\left(\frac{1,000}{1,210}\right)},$$

or

$$f(T = 1,000 \text{ hr}) = \frac{1}{1,210\Gamma(6.2)}(0.37112)(0.43760),$$

where from Table 12.1

$$\Gamma(6.2) = (5.2)(4.2)(3.2)(2.2)(1.2)\Gamma(1.2),$$

and

$$\Gamma(1.2) = 0.91817.$$

Therefore,

$$\Gamma(6.2) = 169.40633.$$

Consequently,

$$f(T = 1,000 \text{ hr}) = \frac{1}{1,210(169.40633)}(0.162403),$$

or

$$f(T = 1,000 \text{ hr}) = 0.000000792.$$

16.2   ESTIMATION OF THE PARAMETERS OF THE GAMMA DISTRIBUTION BY PROBABILITY PLOTTING

There is no general probability plotting paper for the gamma *pdf* on which all sample data from a gamma distributed population plot as straight lines. Consequently, gamma probability plots can be prepared for only known values of β. Special probability papers are available for $\beta = 0.5, 1.0, 1.5, 2.0, 3.0, 4.0$ and 5.0. Also, since the chi-square *pdf* is a special case of the gamma *pdf* when $\eta = 2$ and $\beta = \frac{\nu}{2}$, where ν is an integer and is known as the degrees of freedom, the chi-square probability plotting paper can also be used as a gamma probability plotting paper, such that the chi-square paper of $\nu = 1$ would be used for $\beta = 1/2$, of $\nu = 2$ for $\beta = 1$, and so on, with $\eta = 2$.

The plotting procedure is identical to that for the Weibull, and involves ordering the times to failure, obtaining their median ranks, and plotting these median ranks against their corresponding times to

failure. These points would plot approximately on a straight line if the gamma *pdf* and the preselected β value do describe the data adequately. If they do not, then either the data do not come from a gamma population or the actual value of β differs from the prechosen β value used to select the probability plotting paper.

When a satisfactory straight line is obtained, with β known only η needs to be determined. η is given by two times the reciprocal of the slope of this line on the chi-square probability plot, or

$$\hat{\eta} = 2\frac{T_2 - T_1}{z_2 - z_1},\tag{16.16}$$

where the $T_i's$ are the measurements, like time to failure, and the $z's$ are the corresponding values obtained from the special scale at the top of the paper. If the gamma variate is also a chi-square variate, this slope would be approximately 2.

EXAMPLE 16–3

The following times to failure were obtained from 16 identical equipment in hr: 305, 171, <100, 503, 968, <100, 120, 402, <100, 128, 231, 250, 689, <100, 139, 363.

The equipment in the past exhibited exponential behavior. Do the following:

1. Through probability plotting on gamma probability paper, determine the parameters of the gamma distribution representing these data.

2. Write down the probability density function.

3. Determine the mean of these data from the probability plot.

4. Determine the median of these data from the probability plot.

SOLUTIONS TO EXAMPLE 16–3

1. Order the data as in Table 16.1, Columns 1 and 2. It may be seen that the data are censored at the left because the precise times to failure for the first four ordered entries are not known; consequently, these points cannot be plotted, but their median ranks could be determined if needed. They were not entered in the table because they cannot be used. Enter the remaining median ranks for a sample of size $N = 16$, using Appendix A.

   It is stated that in the past such equipment exhibited exponential behavior, or $\beta = 1$; consequently, a chi-square probability plotting paper for $\nu = 2\beta = 2(1) = 2$ degrees of freedom is chosen,

TABLE 16.1– Times-to-failure data for Example 16–3 and their median ranks for probability plotting.

1	2	3
Failure order number, j	Time to failure, T_j, hr	Median ranks, MR_j, %
1	< 100	–
2	< 100	–
3	< 100	–
4	< 100	–
5	120	28.59
6	128	34.71
7	139	40.82
8	171	46.94
9	231	53.06
10	250	59.18
11	305	65.30
12	363	71.41
13	402	77.53
14	503	83.64
15	689	89.73
16	968	95.76

which is shown in Fig. 16.3. Plotting the pairs of values given in Columns 2 and 3 in Fig. 16.3 yields a satisfactory straight line fit; consequently, these equipment still are exhibiting an exponential behavior, and the plotting paper chosen is correct and $\beta = 1$. The scale parameter η is estimated from the slope of the fitted straight line using two points on it. These are for the values on the special scale along the top of the paper of 1.00 and 6.00, to which correspond the time-to-failure values of 145 hr and 885 hr, respectively. Therefore, from Eq. (16.16),

$$\hat{\eta} = 2\frac{885-145}{6.00-1.00} = 296.00 \text{ hr.}$$

The slope is not equal to 2; consequently, the gamma variate is not a chi-square variate in this case.

2. The *pdf* representing these data is

$$f(T) = \frac{1}{296\Gamma(1)}(\frac{T}{296})^{1-1}e^{-\frac{T}{296}},$$

or

$$f(T) = \frac{1}{296}e^{-\frac{T}{296}},$$

which is indeed the exponential *pdf*.

3. Having determined that the data are exponentially distributed, it is known that the mean life is given by the life corresponding to the probability of failure of 63.2%. Entering with this value, the probability plot yields 296 hr. This confirms the value of $\hat{\eta} = 296$ hr, which for the exponential case is indeed equal to the mean life!

4. The median of these data are obtained by entering the plot at the 50% failure probability level. This yields a median life of 200 hr.

16.3   GAMMA RELIABILITY CHARACTERISTICS

If the times to failure of units are well represented by the gamma distribution, the reliability, starting the mission at age zero, is

$$R(T) = \int_T^\infty f(T)dT = \int_T^\infty \frac{1}{\eta\Gamma(\beta)}(\frac{T}{\eta})^{\beta-1}e^{-\frac{T}{\eta}}dT. \qquad (16.17)$$

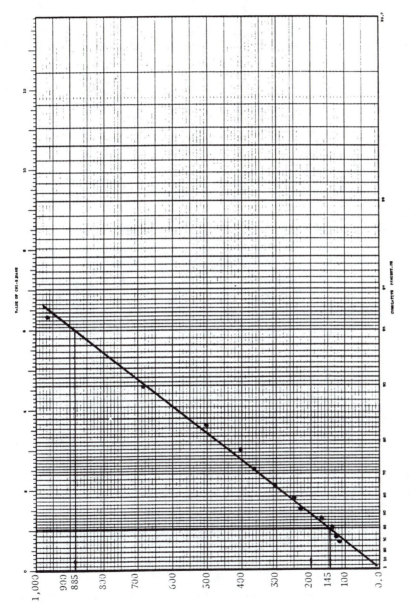

Fig. 16.3— Probability plot of the data in Table 16.1 for Example 16-3 on chi-square probability plotting paper with $\nu = 2$.

Equation (16.17) is the complement of the *incomplete gamma function* which can be found tabulated in [2; 3; 4]. Sample pages of such a table are given in Table 16.2. Consequently, Eq. (16.17) may be evaluated using these tables, keeping in mind that the tables give $Q(T)$ and subtracting the table value from 1 yields the desired $R(T)$. To use these tables determine

$$u = \frac{T}{\eta\,(\beta)^{1/2}} = \frac{T}{\sigma_T},$$

$$p = \beta - 1,$$

and then find $I(u, p)$ from the tables which is $Q(T)$. Then $R(T) = 1 - Q(T)$.

Theoretically, the gamma distribution approaches a normal distribution as β increases, with the normal *pdf's* parameters being given by Eqs. (16.2) and (16.6). But the speed of convergence is slow. The maximum error in the reliability, which occurs at $\bar{T}$, is given in Table 16.3. The behavior of the reliability is interesting, as shown in Figs. 16.4, 16.5 and 16.6.

If β is an integer, successive integration by parts of Eq. (16.17) yields

$$R(T) = \sum_{k=0}^{\beta-1} \frac{1}{k!} \left(\frac{T}{\eta}\right)^k e^{-\frac{T}{\eta}}. \tag{16.18}$$

The right side of Eq. (16.18) is the cumulative Poisson distribution. It can be evaluated by entering the tables of the cumulative Poisson distribution with the values of $\left(\frac{T}{\eta}\right)$ and $(\beta - 1)$, and coming out with the value of $R(T)$. For tables of the cumulative Poisson distribution see [5; 6]. Table 16.4 gives a sample page from [6].

If β is a multiple of $\frac{1}{2}$ and $\eta = 2$ the gamma *pdf* becomes a chi-square *pdf* with the single parameter $\nu = 2\beta$. In this case, tables of the cumulative chi-square distribution can be used to evaluate $R(T)$.

Using the preceding methodologies the reliability function can be determined and plotted as shown in Fig. 16.7. It may be seen that the reliability function starts at 1 for T = 0, decreases sharply for small β values and less sharply for large β values; and as $T \to \infty$, $R(T) \to 0$. Increasing the value of η results in a reliability plot which does not decrease as sharply with time; i.e., it is stretched out to the right more.

TABLE 16.2 – A sample page of the tables of the incomplete gamma function [3, pp. 9-10].

THE INCOMPLETE GAMMA-FUNCTION RATIO, $I(u, p)$

u	$I(u, 4.5)$	$I(u+5, 4.5)$	$I(u+10, 4.5)$	$I(u, 5)$	$I(u+5, 5)$	$I(u+10, 5)$
0.0	.000000000	.984748276	.999997764	.000000000	.982593662	.999997900
0.1	.000000979	.986930115	.999998154	.000000243	.985104911	.999998276
0.2	.000036386	.988815080	.999998477	.000012640	.987273118	.999998586
0.3	.000278188	.990440803	.999998744	.000116995.	.989141614	.999998840
0.4	.001113968	.991840650	.999998964	.000534703	.990748909	.999999049
0.5	.003131349	.993044108	.999999146	.001660930	.992129102	.999999221
0.6	.007040162	.994077163	.999999297	.004042978	.993312287	.999999362
0.7	.013572628	.994962647	.999999421	.008320464	.994324938	.999999478
0.8	.023390120	.995720571	.999999523	.015149095	.995190278	.999999572
0.9	.037012019	.996368428	.999999608	.025126455	.995928618	.999999650
1.0	.054772077	.996921474	.999999677	.038731819	.996557676	.999999714
1.1	.076801297	.997392984	.999999734	.056285817	.997092871	.999999766
1.2	.103032775	.997794485	.999999782	.077930936	.997547586	.999999809
1.3	.133222636	.998135965	.999999821	.103630686	.997933413	.999999844
1.4	.166981125	.998426061	.999999853	.133183572	.998260369	.999999873
1.5	.203808671	.998672232	.999999879	.166247392	.998537093	.999999896
1.6	.243132786	.998880901	.999999901	.202369556	.998771022	.999999915
1.7	.284342809	.999057597	.999999918	.241019709	.998968542	.999999931
1.8	.326820537	.999207066	.999999933	.281621731	.999135133	.999999944
1.9	.369965670	.999333379	.999999945	.323583003	.999275483	.999999954
2.0	.413215693	.999440020	.999999955	.366319631	.999393599	.999999962
2.1	.456060291	.999529970	.999999963	.409276912	.999492901	.999999969
2.2	.498050759	.999605773	.999999970	.451944894	.999576302	.999999975
2.3	.538805041	.999669596	.999999975	.493869209	.999646279	.999999980
2.4	.578009147	.999723288	.999999980	.534657638	.999704937	.999999984
2.5	.615415687	.999768418	.999999983	.573982996	.999754062	.999999987
2.6	.650840250	.999806322	.999999986	.611582988	.999795166	.999999989
2.7	.684156269	.999838132	.999999989	.647257711	.999829528	.999999991
2.8	.715288922	.999864806	.999999991	.680865407	.999858229	.999999993
2.9	.744208540	.999887157	.999999993	.712317036	.999882183	.999999994
3.0	.770923880	.999905872	.999999994	.741570140	.999902158	.999999995
3.1	.795475556	.999921531	.999999995	.768622399	.999918802	.999999996
3.2	.817929824	.999934625	.999999996	.793505180	.999932659	.999999997
3.3	.838372857	.999945565	.999999997	.816277338	.999944188	.999999997
3.4	.856905612	.999954701	.999999997	.837019402	.999953773	.999999998
3.5	.873639302	.999962325	.999999998	.855828300	.999961736	.999999998
3.6	.888691509	.999968683	.999999999	.872812651	.999968846	.999999999
3.7	.902182898	.999973981	.999999999	.888088671	.999973830	.999999999
3.8	.914234507	.999978395	.999999999	.901776681	.999978377	.999999999
3.9	.924965566	.999982069	.999999999	.913998199	.999982144	.999999999
4.0	.934491787	.999985126	.999999999	.924873566	.999985263	.999999999
4.1	.942924070	.999987668	.999999999	.934520071	.999987844	1.000000000
4.2	.950367574	.999989780	.999999999	.943050507	.999989978	
4.3	.956921084	.999991535	1.000000000	.950572116	.999991742	
4.4	.962676648	.999992992		.957185858	.999993199	
4.5	.967719408	.999994200		.962985957	.999994402	
4.6	.972127605	.999995203		.968059675	.999995394	
4.7	.975972718	.999996033		.972487268	.999996212	
4.8	.979319695	.999996722		.976342087	.999996887	
4.9	.982227262	.999997292		.979690796	.999997442	

TABLE 16.2 – Continued

THE INCOMPLETE GAMMA-FUNCTION RATIO, $I(u, p)$

u	$I(u, 5.5)$	$I(u+5, 5.5)$	$I(u+10, 5.5)$	$I(u, 6)$	$I(u+5, 6)$	$I(u+10, 6)$
0.0	.000000000	.980143735	.999997986	.000000000	.977373484	.999998033
0.1	.000000059	.983025419	.999998355	.000000014	.980669376	.999998402
0.2	.000004319	.985512973	.999998658	.000001453	.983515006	.999998702
0.3	.000048422	.987655851	.999998905	.000019741	.985966331	.999998946
0.4	.000252709	.989498130	.999999107	.000117703	.988073393	.999999145
0.5	.000867892	.991078934	.999999272	.000447136	.989880746	.999999307
0.6	.002288444	.992432865	.999999407	.001277734	.991427891	.999999438
0.7	.005030104	.993590413	.999999517	.003001019	.992749711	.999999545
0.8	.009680818	.994578357	.999999607	.006108033	.993876900	.999999631
0.9	.016838979	.995420146	.999999680	.011147198	.994836374	.999999702
1.0	.027051851	.996136249	.999999740	.018672085	.995651655	.999999759
1.1	.040763667	.996744487	.999999789	.029188730	.996343239	.999999805
1.2	.058278786	.997260330	.999999828	.043110116	.996928933	.999999842
1.3	.079741730	.997697178	.999999861	.060722638	.997424163	.999999873
1.4	.105133210	.998066607	.999999887	.082166625	.997842259	.999999897
1.5	.134279581	.998378595	.999999908	.107430750	.998194710	.999999917
1.6	.166872315	.998641727	.999999925	.136358600	.998491394	.999999933
1.7	.202493933	.998863367	.999999940	.168664745	.998740785	.999999946
1.8	.240647092	.999049827	.999999951	.203957358	.998950139	.999999956
1.9	.280784085	.999206502	.999999960	.241764458	.999125651	.999999965
2.0	.322334635	.999337994	.999999968	.281561230	.999272603	.999999972
2.1	.364730516	.999448226	.999999974	.322796387	.999395491	.999999977
2.2	.407426149	.999540533	.999999979	.364916030	.999498130	.999999982
2.3	.449914773	.999617746	.999999983	.407384067	.999583757	.999999985
2.4	.491740214	.999682267	.999999986	.449698649	.999655110	.999999988
2.5	.532504537	.999736126	.999999989	.491404505	.999714502	.999999990
2.6	.571872034	.999781041	.999999991	.532101328	.999763885	.999999992
2.7	.609570130	.999818461	.999999993	.571448574	.999804903	.999999994
2.8	.645387791	.999849608	.999999994	.609167157	.999838937	.999999995
2.9	.679172038	.999875509	.999999995	.645038584	.999867148	.999999996
3.0	.710823109	.999897028	.999999996	.678902085	.999890510	.999999997
3.1	.740288746	.999914892	.999999997	.710650269	.999909838	.999999997
3.2	.767558028	.999929709	.999999997	.740223787	.999925814	.999999998
3.3	.792655072	.999941988	.999999998	.767605416	.999939007	.999999998
3.4	.815632869	.999952155	.999999998	.792813914	.999949893	.999999999
3.5	.836567450	.999960568	.999999999	.815897924	.999958866	.999999999
3.6	.855552508	.999967523	.999999999	.836930132	.999966258	.999999999
3.7	.872694563	.999973269	.999999999	.856001834	.999972341	.999999999
3.8	.888108717	.999978013	.999999999	.873218018	.999977343	.999999999
3.9	.901914999	.999981926	.999999999	.888692999	.999981454	1.000000000
4.0	.914235298	.999985152	1.000000000	.902546651	.999984829	
4.1	.925190837	.999987810		.914901211	.999987598	
4.2	.934900161	.999989998		.925878639	.999989868	
4.3	.943477578	.999991798		.935598499	.999991728	
4.4	.951031990	.999993278		.944176297	.999993251	
4.5	.957666085	.999994494		.951722245	.999994496	
4.6	.963475813	.999995492		.958340371	.999995515	
4.7	.968550103	.999996311		.964127949	.999996347	
4.8	.972970786	.999996983		.969175176	.999997027	
4.9	.976812671	.999997534		.973565061	.999997581	

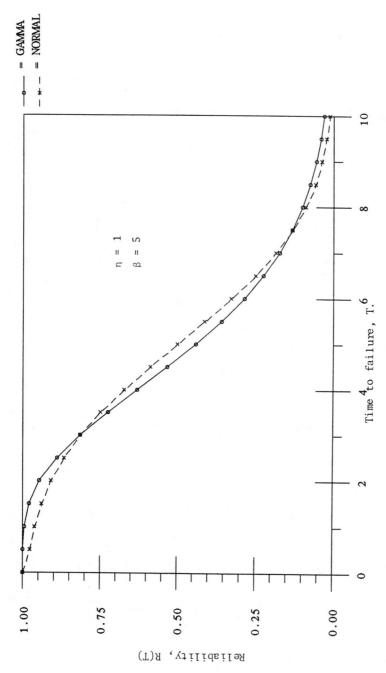

Fig. 16.4– Plot of the reliability function of the gamma distribution with $\eta = 1$ and $\beta = 5$, and its normal approximation.

579

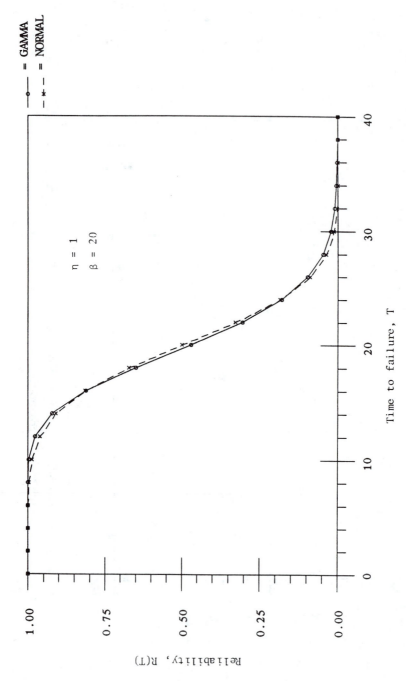

Fig. 16.5– Plot of the reliability function of the gamma distribution with $\eta = 1$ and $\beta = 20$, and its normal approximation.

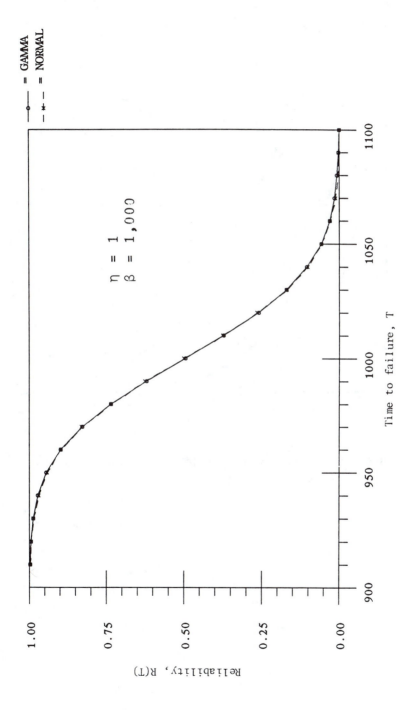

Fig 16.6– Plot of the reliability function of the gamma distribution with $\eta = 1$ and $\beta = 1,000$, and its normal approximation.

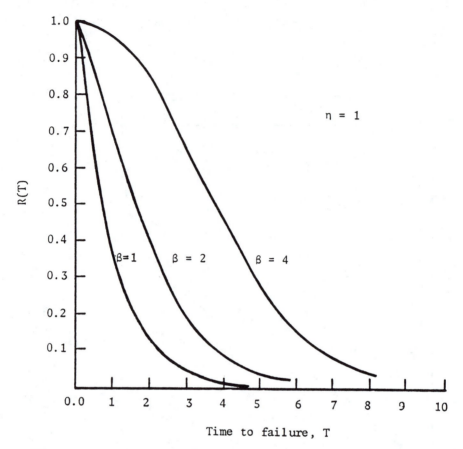

Fig. 16.7 - Plot of the gamma reliability function for various values of β [7, p. 26].

TABLE 16.3– The maximum error in the reliability using the normal approximation to the gamma *pdf* with $\eta = 1$.

β	Maximum error in the reliability
1	0.13212
5	0.05951
20	0.02974
50	0,01881
1,000	0.00421
10,000	0.00133

EXAMPLE 16–4

Using the gamma *pdf* of Example 16–2, find the reliability for a mission of 1,000 hr (1) from tables, and (2) from the normal approximation of this gamma *pdf*.

SOLUTIONS TO EXAMPLE 16–4

1. The parameters of the gamma *pdf* are $\beta = 6.2$ and $\eta = 1,210$ hr. From Eq. (16.18)

$$u = \frac{1,000}{1,210\,(6.2)^{1/2}} = \frac{1,000}{3,013} = 0.3319,$$

and from Eq. (16.19)
$$p = 6.2 - 1 = 5.2.$$
Entering Table 16.2 with these values, and performing multiple interpolations, yields

$$Q(T = 1,000 \text{ hr}) = I(u = 0.3319, p = 5.2) = 0.0001956.$$

Consequently,

$$R(T = 1,000 \text{ hr}) = 1 - Q(T) = 1 - 0.0001956,$$

or

TABLE 16.4 – A sample page of the tables of the cumulative Poisson distribution [6, p.72].

TABLE OF THE POISSON DISTRIBUTION
INDIVIDUAL AND CUMMULATIVE TERMS

U = 0.4000000 | | | U = 0.4050000

X	P(X)	C(X)	D(X)	P(X)	C(X)	D(X)
0	.67032005	.67032005	1.00000000	.66697682	.66697682	1.00000000
1	.26812802	.93844806	.32967995	.27012561	.93710242	.33302319
2	.05362560	.99207366	.06155194	.05470044	.99180286	.06289758
3	.00715008	.99922375	.00792633	.00738456	.99918741	.00819714
4	.00071501	.99993876	.00077625	.00074769	.99993510	.00081259
5	.00005720	.99999595	.00006124	.00006056	.99999566	.00006490
6	.00000381	.99999977	.00000404	.00000409	.99999975	.00000434
7	.00000022	.99999999	.00000023	.00000024	.99999999	.00000025

U = 0.4100000 | | | U = 0.4150000

X	P(X)	C(X)	D(X)	P(X)	C(X)	D(X)
0	.66365025	.66365025	1.00000000	.66034028	.66034028	1.00000000
1	.27209660	.93574686	.33634975	.27404121	.93438149	.33965972
2	.05577980	.99152666	.06425314	.05686355	.99124505	.06561850
3	.00762324	.99914990	.00847334	.00786612	.99911117	.00875495
4	.00078138	.99993128	.00085010	.00081611	.99992728	.00088882
5	.00006407	.99999535	.00006872	.00006774	.99999502	.00007271
6	.00000438	.99999973	.00000464	.00000469	.99999970	.00000498
7	.00000026	.99999999	.00000027	.00000028	.99999999	.00000029

U = 0.4200000 | | | U = 0.4250000

X	P(X)	C(X)	D(X)	P(X)	C(X)	D(X)
0	.65704682	.65704682	1.00000000	.65376979	.65376979	1.00000000
1	.27595966	.93300649	.34295318	.27785216	.93162195	.34623021
2	.05795153	.99095801	.06699351	.05904358	.99066553	.06837805
3	.00811321	.99907123	.00904198	.00836451	.99903003	.00933447
4	.00085189	.99992311	.00092877	.00088873	.99991877	.00096996
5	.00007156	.99999467	.00007688	.00007554	.99999431	.00008123
6	.00000501	.99999968	.00000532	.00000535	.99999966	.00000569
7	.00000030	.99999999	.00000032	.00000032	.99999998	.00000034
8				.00000002	.99999999	.00000002

U = 0.4300000 | | | U = 0.4350000

X	P(X)	C(X)	D(X)	P(X)	C(X)	D(X)
0	.65050909	.65050909	1.00000000	.64726467	.64726467	1.00000000
1	.27971891	.93022799	.34949091	.28156013	.92882480	.35273533
2	.06013956	.99036755	.06977201	.06123933	.99006412	.07117520
3	.00862000	.99898756	.00963244	.00887970	.99894382	.00993588
4	.00092665	.99991421	.00101244	.00096567	.99990949	.00105618
5	.00007969	.99999391	.00008578	.00008401	.99999350	.00009051
6	.00000571	.99999961	.00000609	.00000609	.99999959	.00000649
7	.00000035	.99999996	.00000038	.00000038	.99999997	.00000040
8	.00000002	.99999999	.00000003	.00000002	.99999999	.00000002

U = 0.4400000 | | | U = 0.4450000

X	P(X)	C(X)	D(X)	P(X)	C(X)	D(X)
0	.64403642	.64403642	1.00000000	.64082428	.64082428	1.00000000
1	.28337602	.92741244	.35596358	.28516680	.92599108	.35917572
2	.06234273	.98975517	.07258755	.06344961	.98944069	.07400892
3	.00914360	.99889877	.01024483	.00941169	.99885239	.01055930

$$R(T = 1,000 \text{ hr}) = 0.9998044.$$

2. From Example 16–2, using Eqs. (16.2) and (16.6), $\bar{T} = 7,502$ hr and $\sigma_T = 3,013$ hr. For $T = 1,000$ hr the standardized normal variate is

$$z(T = 1,000 \text{ hr}) = \frac{1,000 - 7,502}{3,013} = -2.1580.$$

Since $\beta = 6.2$, and therefore not relatively large, the normal approximate will not yield acceptable reliability results.

For this z value Appendix B yields

$$R(T = 1,000 \text{ hr}) = 0.9845.$$

This value compares with the value of 0.9998 obtained in Case 1 using the incomplete gamma function tables. It may be seen that the discrepancy is substantial when the normal approximation to the gamma *pdf* is used to calculate reliabilities with relatively small β values.

EXAMPLE 16–5

The times to failure of a unit are found to be gamma distributed with parameters $\beta = 4$ and $\eta = 250$ hr. Find the reliability of such units for a mission of 100 hr duration.

SOLUTION TO EXAMPLE 16–5

In this example $\beta = 4$ is an integer; consequently, Eq. (16.20) can be used to find the sought reliability, then

$$R(T = 100 \text{ hr}) = \sum_{k=0}^{3} \frac{1}{k!}(\frac{100}{250})^k e^{-\frac{100}{250}},$$

or

$$R(T = 100 \text{ hr}) = 0.999224.$$

The values needed to find $R(T = 100 \text{ hr})$ from Table 16.4 are $\frac{T}{\eta} = \frac{100}{250} = 0.40$ and $\beta - 1 = 4 - 1 = 3$. Table 16.4 yields $R(T = 100 \text{ hr}) = 0.999224$. The two results are of course identical, and when the cumulative Poisson distribution tables are handy, the latter procedure is faster.

16.4 GAMMA DISTRIBUTION FAILURE RATE CHARACTERISTICS

The gamma distribution's instantaneous failure rate is given by

$$\lambda(T) = \frac{f(T)}{R(T)} = \frac{\frac{1}{\eta\Gamma(\beta)}\left(\frac{T}{\eta}\right)^{\beta-1}e^{-\frac{T}{\eta}}}{\int_T^\infty \frac{1}{\eta\Gamma(\beta)}\left(\frac{T}{\eta}\right)^{\beta-1}e^{-\frac{T}{\eta}}\,dT}. \tag{16.19}$$

Evaluation of the numerator of Eq. (16.19) involves finding the height of the gamma *pdf* for a given value of T. The denominator can be evaluated by the methods given in the previous section for the same given value of T.

Figure 16.8 shows the behavior of $\lambda(T)$ for various values of β and for $\eta = 1$. It may be seen that for $\beta < 1$, $\lambda(T)$ starts at ∞ for $T = 0$, and drops sharply thereafter until at $T \to \infty$, $\lambda(T) \to 1$.

For $\beta = 1$ the gamma failure rate is constant with age at the value of $\frac{1}{\eta}$. For $\beta > 1$ the gamma failure rate starts at zero for $T = 0$, increases sharply thereafter, and asymptotically approaches $\frac{1}{\eta}$ as $T \to \infty$.

EXAMPLE 16–6

For the gamma distribution of Example 16–1, with $\beta = 6.2$ and $\eta = 1,210$ hr, do the following:

1. Find $f(T = 1,000$ hr$)$.

2. Find $R(T = 1,000$ hr$)$.

3. Find $\lambda(T = 1,000$ hr$)$.

SOLUTIONS TO EXAMPLE 16–6

1. The gamma *pdf* for this case was found in Example 16–2, Case 8, to be

   $$f(T = 1,000 \text{ hr}) = 0.0_6792.$$

2. The reliability for this case was found in Example 16–4, Case 1, to be

   $$R(T = 1,000 \text{ hr}) = 0.9998044.$$

3. From Eq. (16.19)

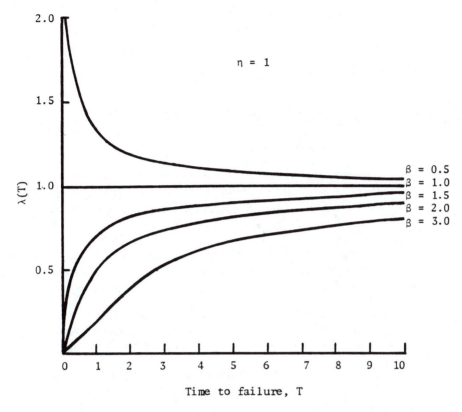

Fig. 16.8 — Plot of the gamma failure rate function for various values of β with $\eta = 1$ [1, p. 119].

$$\lambda(T = 1,000 \text{ hr}) = \frac{f(T=1,000 \text{ hr})}{R(T=1,000 \text{ hr})} = \frac{0.0_6 792}{0.9998044},$$

or

$$\lambda(T = 1,000 \text{ hr}) = 0.0_6 7922 \text{ fr/hr, or } 0.7922 \text{ fr}/10^6 \text{ hr.}$$

16.5 APPLICATIONS OF THE GAMMA DISTRIBUTION

1. If a system fails when β failures occur, such as in a standby system, and the failing subsystems have a constant failure rate of $\frac{1}{\eta}$, the system's times to failure are gamma distributed [1, p. 83].

2. If a system is preventively maintained after β uses, or missions, the time between consecutive maintenance actions is gamma distributed under the appropriate conditions [1, p. 83].

3. Phenomena that phenomenologically cannot be represented by the gamma *pdf,* can empirically be approximated well by the gamma *pdf;* for example, the times to failure of capacitors.

4. The gamma *pdf* is used in Bayesian testing as a prior model describing initial uncertainty about the failure rate of the system [8; 1, p. 85].

16.6 PHENOMENOLOGICAL CONSIDERATIONS FOR USING THE GAMMA DISTRIBUTION

1. The gamma distribution is the model for the time required for a total of exactly β (integer) independent events to take place, if events occur at a constant rate of $\frac{1}{\eta}$. For example, a system's time to failure is gamma distributed if the system fails as soon as exactly β failures have taken place and if subfailures occur independently at a constant rate $\frac{1}{\eta}$. Also, the time between consecutive maintenance actions, for a system that is scheduled for preventive maintenance or inspection, after exactly β uses is gamma distributed under appropriate conditions [1, p. 83; 9 ;10 ;11].

2. If a system has a constant failure rate, the time to the *nth* failure of the system is gamma distributed [11, p. 25]. For example, if the

times to failure, X_i, are exponentially distributed with a mean time to a failure of $m = \frac{1}{\lambda}$, then $T_n = x_1 + x_2 + ... + x_n$ is gamma distributed with parameters $\beta = n$ and $\eta = \lambda = \frac{1}{m}$, where the x_i are the times between failures.

PROBLEMS

16-1. Ten electromechanical components failed at 100, 247, 404, 659, 686, 714, 736, 789, 898 and 1,200 hr. Assuming the gamma distribution applies to these components, calculate the following:

    (1) The scale parameter.

    (2) The shape parameter.

    (3) The mean life.

    (4) The standard deviation.

16-2. Prove Eq. (16.7).

16-3. Prove Eq. (16.8).

16-4. Prove Eq. (16.9).

16-5. Prove Eq. (16.10).

16-6. The following times-to-failure data were obtained from 12 identical components in hr: 300, 412, 416, 350, 365, 312, 380, 325, 305, 101, 214 and 307.

    (1) Determine the parameters of the gamma *pdf* representing these data.

    (2) Write down the gamma *pdf*.

    (3) Plot $f(T)$.

    (4) Plot $R(T)$.

    (5) Plot $\lambda(T)$.

    (6) Find $R(T = 300 \text{ hr})$.

    (7) Find $F(T = 300 \text{ hr})$.

    (8) Find $\lambda(T = 300 \text{ hr})$.

16-7. The following times-to-failure data were obtained from 17 identical components in hr: 1,240; 2,320; 2,400; 2,407; 2,522; 1,862; 1,620; 2,120; 2,100; 1,900; 1,831; 2,822; 800; 2,405; 2,172; 2,150 and 1,972.

(1) Determine the parameters of the gamma *pdf* representing these data.

(2) Write down the gamma *pdf*.

(3) Plot $f(T)$.

(4) Plot $R(T)$.

(5) Plot $\lambda(T)$.

(6) Find $R(T = 1,800 \text{ hr})$.

(7) Find $F(T = 1,800 \text{ hr})$.

(8) Find $\lambda(T = 1,800 \text{ hr})$.

16-8. The times-to-failure of a unit have a gamma distribution with a mean of 5,300 hr and a standard deviation of 1,280 hr.

(1) Write down the gamma *pdf*.

(2) Determine the median of this *pdf*.

(3) Determine the mode of this *pdf*.

(4) Determine the coefficient of skewness of this *pdf*.

(5) Determine the coefficient of kurtosis of this *pdf*.

(6) Determine the probability density of this *pdf* at the age of 1,500 hr.

16-9. A unit exhibiting a gamma distribution for its times to failure has a mean wear-out life of 10,000 hr and a standard deviation of 1,120 hr. Determine the following:

(1) The reliability for a mission of 6,000 hr.

(2) The reliability for a second mission of 5,000 hr starting the mission at the age of 6,000 hr.

16-10. A system consists of three units which function reliabilitywise in series. All three units exhibit a gamma times to failure distribution characteristic, with the following parameters:
$Unit\ 1 - \beta = 3.2, \quad \eta = 1,450 \text{ hr}.$
$Unit\ 2 - \beta = 4.2, \quad \eta = 1,570 \text{ hr}.$
$Unit\ 3 - \beta = 4.7, \quad \eta = 1,680 \text{ hr}.$
Determine the reliability of the system for a mission of 1,100 hr.

16-11. The times to failure of a unit have a gamma distribution with a mean of 500 hr and a standard deviation of 95 hr.

(1) Write down the gamma *pdf*.

(2) Determine the median of this *pdf*.

(3) Determine the mode of this *pdf.*

(4) Determine the coefficient of skewness of this *pdf.*

(5) Determine the coefficient of kurtosis of this *pdf.*

(6) Determine the probability density of this *pdf* at the age of 250 hr.

16-12. The times to failure of a unit have a gamma distribution with a mean of 1.3×10^6 cycles and a standard deviation of 2.5×10^4 cycles.

(1) Write down the gamma *pdf.*

(2) Determine the median of this *pdf.*

(3) Determine the mode of this *pdf.*

(4) Determine the coefficient of skewness of this *pdf.*

(5) Determine the coefficient of kurtosis of this *pdf.*

(6) Determine the probability density of this *pdf* at the age of 5×10^5 cycles.

REFERENCES

1. Hahn, Gerald, Jr. and Shapiro, Samuel S., *Statistical Models in Engineering,* John Wiley & Sons, Inc., New York, 355 pp., 1967.

2. Abramowitz, M. and Stegun, I.A., Editors, *Handbook of Mathematical Functions and Formulas, Graphs, and Mathematical Tables,* Dover Publications Inc., New York 10014, 1,046 pp., 1970.

3. Harter, H.L., *New Tables of the Incomplete Gamma Function Ratio and of Percentage Points of the Chi-Square and Beta Distributions,* Aerospace Research Laboratories, U.S. Air Force and available from Superintendent of Documents, U.S. Government Printing Office, Washington, D.C., 1964.

4. Pearson, Karl, *Tables of the Incomplete Gamma-Function,* Biometrika Office, University College, London, 164 pp., 1951.

5. General Electric Company, Defense Systems Department, *Tables of the Individual and Cumulative Terms of Poisson Distribution,* D. Van Nostrand Company, Inc., 24 West 40th St., New York 10018, 202 pp., 1962.

6. Molina, E.C., *Poisson's Exponential Binomial Limit Tables I & II,* Robert E. Krieger Publishing Co., P.O. Box 542, Huntington, New York 11743, 47 pp., 1949.

7. Kapur, K.C. and Lamberson, L.R., *Reliability in Engineering Design,* John Wiley & Sons, New York, 586 pp., 1977.

8. Schlaifer, R., *Probability and Statistics for Business Decisions,* McGraw-Hill Book Co., New York, 732 pp., 1959.

9. Cox, D.R. and Smith, W.L., *Queues,* John Wiley & Sons, New York, 180 pp., 1961.

10. Saaty, T.L., *Elements of Queueing Theory, with Applications,* McGraw Hill Book Co., New York, 423 pp., 1961.

11. Morse, P.M., *Queues, Inventories and Maintenance,* John Wiley & Sons, New York, 202 pp., 1958.

Chapter 17

THE BETA DISTRIBUTION

17.1 BETA DISTRIBUTION'S CHARACTERISTICS

The beta distribution allows the representation of a wide diversity of distributional shapes over the values of the variable between zero and unity. The beta probability density function is [1, p. 91]

$$f(x) = \frac{\Gamma(\alpha + \beta + 2)}{\Gamma(\alpha + 1)\Gamma(\beta + 1)} x^{\alpha}(1 - x)^{\beta}, \qquad (17.1)$$

where

$$0 < x < 1, \ \alpha > -1, \ \beta > -1,$$

$$\alpha = \text{shape parameter},$$

$$\beta = \text{shape parameter},$$

and

$$\Gamma(n) = \text{gamma function}.$$

Figure 17.1 shows the various shapes of the beta distribution for a variety of α and β values.

Some of the specific characteristics of the beta distribution are the following [1, p. 91]:

593

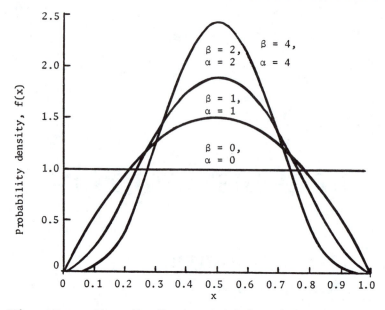

Fig. 17.1a– Beta distributions with four different pairs of parameter values when $\alpha = \beta$ [1, p. 93].

1. When both parameters α and β have the value of zero, it becomes the uniform distribution with $f(x) = 1$, for $0 \leq x \leq 1$, and $f(x) = 0$ elsewhere, as shown in Fig. 17.1a.

2. When one parameter is zero and the other is one, it becomes a straight line with a slope of $\tan\theta = +2$ for $\alpha = 1$ and $\beta = 0$, and a slope of $\tan\theta = -2$ for $\beta = 1$ and $\alpha = 0$, as shown in Fig. 17.1b.

3. When $\alpha > 0$ and $\beta > 0$, the distribution is single peaked with a modal value of

$$\tilde{x} = \frac{\alpha}{\alpha + \beta},$$ (17.2)

as shown in Fig. 17.1c.

4. When $\alpha < 0$ and $\beta < 0$ the distribution is U-shaped, as shown in Fig. 17.1d.

5. When $\alpha < 0$ and $\beta \geq 0$ the distribution is reverse J-shaped, as shown in Fig. 17.1b.

6. When $\beta < 0$ and $\alpha \geq 0$ the distribution is J-shaped.

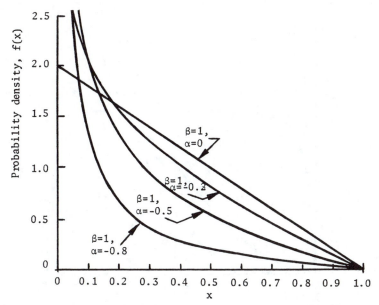

Fig. 17.1b– Beta distributions with four different pairs of parameter values when $\alpha \neq \beta$, $\beta = 1$ and $\alpha = 0, -0.2, -0.5$ and -0.8 [1, p. 93].

7. When $\alpha = \beta$ the distribution is symmetrical, as shown in Fig. 17.1a.

8. The mean, $\bar{x}$, of the beta distribution is given by

$$\bar{x} = \frac{\alpha + 1}{\beta + \alpha + 2}. \tag{17.3}$$

9. The median, $\check{x}$, of the beta distribution is determined from

$$0.5 = B(x; \alpha, \beta), \tag{17.4}$$

where $B(x; \alpha, \beta)$ is the incomplete beta function [2; 3; 4].

10. The mode, $\tilde{x}$, of the beta distribution is given by

$$\tilde{x} = \frac{\alpha}{\alpha + \beta}; \alpha > 0, \beta > 0. \tag{17.5}$$

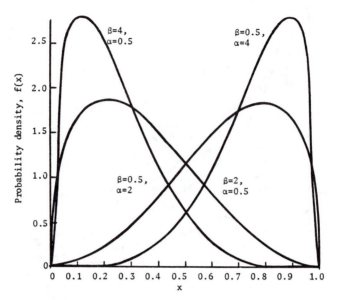

Fig. 17.1c– Beta distributions with different pairs of positive parameter values illustrating their single peakedness [1, p. 92].

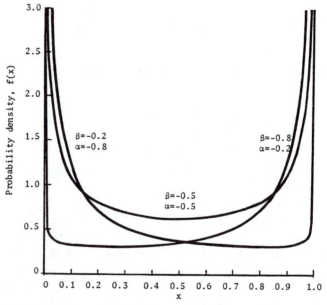

Fig. 17.1d– Beta distributions with three different pairs of parameter values and both negative, yielding *U*-shaped distributions [1, p. 92].

11. The standard deviation, σ_x, of the beta distribution is given by

$$\sigma_x = [\frac{(\alpha+1)(\beta+1)}{(\alpha+\beta+2)^2(\alpha+\beta+3)}]^{1/2}. \qquad (17.6)$$

12. The coefficient of skewness, α_3, of the beta distribution is given by

$$\alpha_3 = \frac{2(\beta-\alpha)(\alpha+\beta+3)^{1/2}}{(\alpha+\beta+4)[(\alpha+1)(\beta+1)]^{1/2}}. \qquad (17.7)$$

If α_3 is negative, which occurs when $\alpha > \beta$, the *pdf* peaks to the right and stretches out to the left. As the value of β increases relative to a fixed value of α, the *pdf* peaks more to the left and the stretching to the right is reduced.

When α_3 is positive, which occurs when $\beta > \alpha$, the *pdf* peaks to the left and stretches out to the right. As the value of α increases relative to a fixed value of β, the *pdf* peaks more to the left and the stretching to the right is reduced.

If $\alpha = \beta$, then $\alpha_3 = 0$ and the *pdf* is symmetrical.

13. The coefficient of kurtosis, α_4, of the beta distribution is given by

$$\alpha_4 = \frac{3(\alpha+\beta+3)^2 + [2(\alpha+\beta+2)^2(\alpha+1)(\beta+1)(\alpha+\beta-4)]}{(\alpha+1)(\beta+1)(\alpha+\beta+4)(\alpha+\beta+5)}. \qquad (17.8)$$

When $\alpha = \beta$, the beta distribution becomes symmetrical and its peakedness increases with increasing values of $\alpha = \beta$.

14. The *kth* moment of the beta *pdf* about the origin is

$$\mu_k' = \frac{(\alpha+k)(\alpha+k-1)....(\alpha+1)}{(\alpha+\beta+k+1)(\alpha+\beta+k)....(\alpha+\beta+2)}. \qquad (17.9)$$

15. The beta distribution can be generalized to cover the interval (a, b). This leads to the probability density function

$$f(T) = \frac{\Gamma(\alpha + \beta + 2)}{(b-a)\Gamma(\alpha+1)\Gamma(\beta+1)}\left(\frac{T-a}{b-a}\right)^\alpha (1 - \frac{T-a}{b-a})^\beta, \quad (17.10)$$

where

$$a \leq T \leq b, \ \alpha > -1, \ \beta > -1.$$

This *pdf* is very useful in times to failure, reliability and failure rate analysis, and is discussed further in the next two sections.

17.2 BETA DISTRIBUTION'S RELIABILITY CHARACTERISTICS

The generalized beta distribution reliability function, $R(T)$, will exist when the random variable is time to failure, T. Then

$$R(T) = 1 - \int_0^T f(T)dT,$$

or

$$R(T) = 1 - \frac{\Gamma(\alpha + \beta + 2)}{(b-a)\Gamma(\alpha+1)\Gamma(\beta+1)} \int_a^T (\frac{T-a}{b-a})^\alpha (1 - \frac{T-a}{b-a})^\beta dT.$$
$$(17.11)$$

Let $t = \frac{T-a}{b-a}$, then Eq. (17.11) becomes

$$R(T) = 1 - \frac{\Gamma(\alpha + \beta + 2)}{\Gamma(\alpha+1)\Gamma(\beta+1)} \int_0^{\frac{T-a}{b-a}} t^\alpha (1-t)^\beta dt, \quad (17.12)$$

or

$$R(T) = 1 - B(\frac{T-a}{b-a}; \alpha, \beta), \quad (17.13)$$

and

$$B(x; \alpha, \beta) = \frac{\Gamma(\alpha + \beta + 2)}{\Gamma(\alpha+1)\Gamma(\beta+1)} \int_0^x t^\alpha (1-t)^\beta dt \quad (17.14)$$

is the incomplete beta function. Its value can be found in tables or charts given in [2; 3; 4], or by using appropriate computer software packages, available at the Computer Center of The University of Arizona. They can also be obtained by inserting Eq. (17.14) in commercially available mathematical software or advanced scientific calculators.

Reliability function plots of the beta distribution with various values of α and β are given in Fig. 17.2.

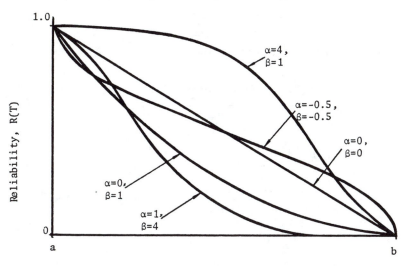

Fig. 17.2– Plot of the beta reliability function for various values of α and β [5; pp. 4-40 – 4-42].

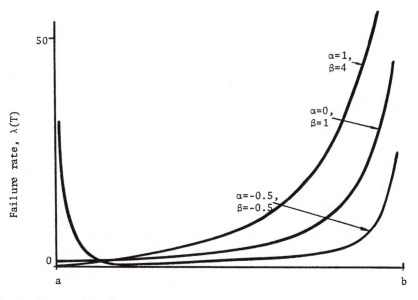

Fig. 17.3– Plots of the beta distribution failure rate function with various values of α and β.

17.3 BETA DISTRIBUTION'S FAILURE RATE CHARACTERISTICS

The generalized beta failure rate function, $\lambda(T)$, is given by

$$\lambda(T) = \frac{\left(\frac{T-a}{b-a}\right)^{\alpha}\left(1 - \frac{T-a}{b-a}\right)^{\beta}}{\int_T^b \left(\frac{T-a}{b-a}\right)^{\alpha}\left(1 - \frac{T-a}{b-a}\right)^{\beta} dt}, \tag{17.15}$$

for $a \leq T \leq b, a < b$, or

$$\lambda(x) = \frac{\Gamma(\alpha + \beta + 2)x^{\alpha}(1 - x)^{\beta}}{\Gamma(\alpha + 1)\Gamma(\beta + 2)(b - a)[1 - B(x; \alpha, \beta)]}, \quad \text{for } 0 \leq x \leq 1, \tag{17.15'}$$

where

$a =$ minimum life,

$b =$ maximum life,

$x = \frac{T-a}{b-a}$, a normalized time to failure,

and

$B(x; \alpha, \beta) =$ incomplete beta function, as defined by Eq. (17.14).

Plots of the beta failure rate function with various values of α and β are given in Fig. 17.3. Specific characteristics of the beta distribution's failure rate function are the following:

1. When $\alpha > 0$ the failure rate starts at zero at $T = a$, then monotonically increases and approaches infinity as T approaches b.

2. When $\alpha = 0$ the failure rate starts at the value of $1/(b - a)$ at $T = a$, then increases monotonically and finally approaches infinity as T approaches b.

3. When $\alpha < 0$ the failure rate approaches infinity as T approaches a, then decreases sharply to the lowest values of the failure rate, increases slowly thereafter, finally it increases sharply and approaches infinity as T approaches b. In this case, the curve of the failure rate is the bath-tub curve.

17.4 ESTIMATION OF THE PARAMETERS OF THE BETA DISTRIBUTION

The maximum likelihood estimates of the beta distribution's parameters are difficult to obtain. The easier matching moments method provides satisfactory estimates. The loss in precision is small for large sample sizes. Using this method

$$\hat{\bar{x}} = \frac{\alpha + 1}{\alpha + \beta + 2},$$

and

$$S^2 = \frac{(\alpha + 1)(\beta + 1)}{(\beta + \alpha + 2)^2 (\beta + \alpha + 3)}.$$

Solving these equations for the parameters yields

$$\hat{\beta} = \frac{\hat{\bar{x}}(1 - \hat{\bar{x}})^2 + S^2(\hat{\bar{x}} - 2)}{S^2}, \qquad (17.16)$$

and

$$\hat{\alpha} = \frac{\hat{\bar{x}}(\beta + 2) - 1}{1 - \hat{\bar{x}}}, \qquad (17.17)$$

where $\hat{\bar{x}}$ and S^2 are the sample mean and variance of the normalized data. For the general times-to-failure data, this method is illustrated next.

EXAMPLE 17–1

Ten identical devices are tested continuously to failure and the following times to failure are obtained: 23.5, 50.1, 65.3, 68.9, 70.4, 77.3, 81.6, 85.7, 89.9, 95.3 hr. From past experience, the minimum time to failure may be taken to be zero, whereas the maximum time to failure never exceeds 100 hr. Do the following:

1. Find the parameters of the generalized beta distribution which might represent the times-to-failure distribution of these devices.

2. Write down and plot the *pdf*.

3. Write down and plot the reliability function.

4. Write down and plot the failure rate function.

5. Find the reliability for a mission of 20 hr starting at the age of 0.

6. Find the reliability for a mission of 20 hr starting at the age of 20 hr.

SOLUTIONS TO EXAMPLE 17–1

1. From the problem statement, the minimum and the maximum times to failure are $a = 0$ hr and $b = 100$ hr, respectively. The normalized times-to-failure data are determined from

$$x_i = \frac{T_i - a}{b - a} = \frac{T_i}{100}.$$

Then the sample mean and variance of the normalized data are

$$\hat{\bar{x}} = \frac{1}{10} \sum_{i=1}^{10} \frac{T_i}{100} = 0.708,$$

and

$$S^2 = \frac{1}{10 - 1} \sum_{i=1}^{10} (\frac{T_i}{100} - \hat{\bar{x}})^2 = 0.044953.$$

From Eqs. (17.16) and (17.17)

$$\hat{\beta} = \frac{(0.708)(1 - 0.708)^2 + (0.044953)(0.708 - 2)}{0.044953},$$

or

$$\hat{\beta} = 0.05089,$$

and

$$\hat{\alpha} = \frac{0.708(0.05089 + 2) - 1}{1 - 0.708},$$

or

$$\hat{\alpha} = 1.5480.$$

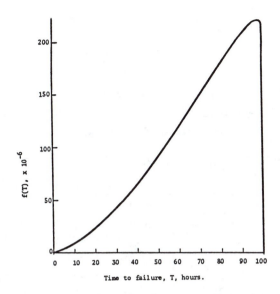

Fig. 17.4– The plot of the generalized beta *pdf* for Example 17–1.

2. The probability density function, from Eq. (17.10), is

$$f(T) = \frac{\Gamma(1.5480 + 0.05089 + 2)}{(100 - 0)\Gamma(1.5480 + 1)\Gamma(0.05089 + 1)}(\frac{T}{100})^{1.5480}(1 - \frac{T}{100})^{0.05089},$$

or

$$f(T) = (0.02773)(\frac{T}{100})^{1.5480}(1 - \frac{T}{100})^{0.05089}.$$

The values for plotting this $f(T)$ are given in Table 17.1 and the plot is given in Fig. 17.4.

3. The reliability function is given by Eq. (17.11), or

$$R(T) = 1 - B(x; \alpha, \beta),$$

where $x = \frac{T-a}{b-a}$, $a = 0$ hr, $b = 100$ hr, $\alpha = 1.5480$ and $\beta = 0.05089$.
Therefore,

$$R(T) = 1 - B(\frac{T}{100}; 1.5480, 0.05089),$$

where $B(x; \alpha, \beta)$ is the incomplete beta function, defined by Eq. (17.14). The values for plotting the reliability function are given in Table 17.1 and the plot is given in Fig. 17.5.

Table 17.1– Values for the *pdf*, reliability and failure rate plots of Example 17–1.

Time to failure, T, hr	$f(T)$ $\times 10^4$	$R(T)$	$\lambda(T)$, fr/10^3 hr
0.0	0.00	1.000	0.000
5.0	2.68*	0.999	0.268
10.0	7.81	0.997	0.783
15.0	14.59	0.991	1.472
20.0	22.70	0.982	2.312
25.0	31.96	0.969	3.298
30.0	42.23	0.950	4.445
35.0	53.41	0.926	5.768
40.0	65.41	0.896	7.300
45.0	78.15	0.861	9.077
50.0	91.55	0.818	11.192
55.0	105.5	0.769	13.719
60.0	120.0	0.713	16.830
65.0	134.9	0.649	20.786
70.0	150.2	0.578	25.986
75.0	165.5	0.499	33.166
80.0	180.9	0.412	43.908
85.0	195.8	0.318	61.572
90.0	209.5	0.216	96.991
92.0	214.3	0.174	123.161
94.0	218.4	0.131	166.718
96.0	221.0	0.087	254.023
98.0	220.3	0.043	512.326

*The actual value is 2.68×10^{-4}

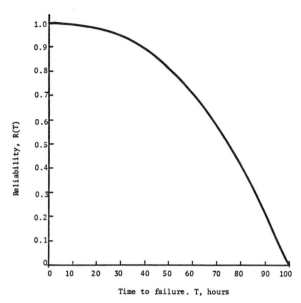

Fig. 17.5– The plot of the generalized beta reliability function for Example 17–1.

4. The failure rate function from Eq. (17.15') is

$$\lambda(T) = \frac{\Gamma(\alpha + \beta + 2)(\frac{T}{100})^{\alpha}(1 - \frac{T}{100})^{\beta}}{\Gamma(\alpha + 1)\Gamma(\beta + 1)(100)[1 - B(\frac{T}{100}; \alpha, \beta)]},$$

$$\lambda(T) = \frac{\Gamma(1.5480 + 0.05089 + 2)(\frac{T}{100})^{1.5480}(1 - \frac{T}{100})^{0.05089}}{\Gamma(1.5480 + 1)\Gamma(0.05089 + 1)(100)[1 - B(\frac{T}{100}; 1.5480, 0.05089)]},$$

or

$$\lambda(T) = \frac{(0.02773)(\frac{T}{100})^{1.5480}(1 - \frac{T}{100})^{0.05089}}{[1 - B(\frac{T}{100}; 1.5480, 0.05089)]}.$$

The values for plotting this $\lambda(T)$ are given in Table 17.1, and the plot is given in Fig. 17.6.

5. From Eq. (17.13)

$$R(20) = 1 - B(\frac{20}{100}; 1.5480, 0.05089),$$

and from Table 17.1

$$R(20) = 0.982.$$

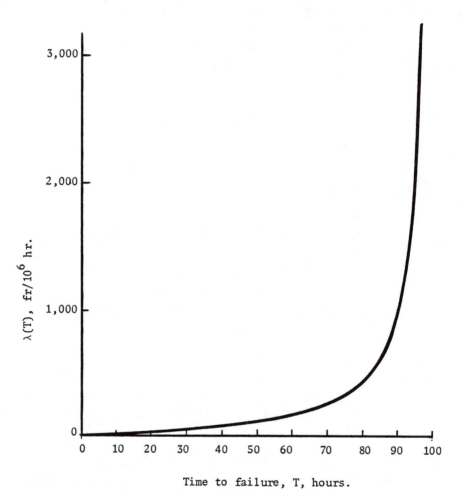

Fig. 17.6– The plot of the generalized beta failure rate function for Example 17–1.

6. The reliability for a mission of 20 hr, starting at the age of 0 hr, is

$$R(20, 20) = \frac{R(40)}{R(20)},$$

$$R(40) = 1 - B(\frac{40}{100}; 1.5480, 0.05089).$$

From Table 17.1

$$R(40) = 0.896;$$

therefore,

$$R(20, 20) = \frac{0.896}{0.982} = 0.912.$$

17.5   APPLICATIONS OF THE BETA DISTRIBUTION

Because of its many shapes, the beta distribution may represent a wide variety of times-to-failure data whose values are restricted to an identifiable interval, such as the data given in Example 17-1. And because of its relationships with the binomial, normal, Poisson, gamma, chi-square and F-distributions [6], the beta distribution is used widely in reliability engineering. Two typical applications of the beta distribution are illustrated next.

1. Take n independent random observations of variable z which has an arbitrary probability density function. Rank these observations in increasing order of magnitude, and let z_r and z_{n-s+1} be the rth smallest and the sth largest observations, respectively. It can be shown [7, p. 400] that the proportion, p, of the original population lying between z_r and z_{n-s+1} follows a beta distribution with parameters

$$\alpha = n - r - s, \tag{17.18}$$

and

$$\beta = r + s - 1. \tag{17.19}$$

Substitution of these values of α and β into Eq. (17.1), for integer values of α and β, yields

$$f(p; n-r-s, r+s-1) = \frac{(n)!}{(n-r-s)!(r+s-1)!}p^{n-r-s}(1-p)^{r+s-1},$$

$$0 < p < 1. \tag{17.20}$$

2. Another application of the beta distribution is its use in the development of Bayesian reliability demonstration test plans [8]. Here the beta distribution is used as a measure of the reliability of a system, or component, in terms of the proportion of the number of successes to the number of tests, or trials, undertaken. The Bayesian technique involves determining a posterior distribution knowing the distribution from which the sample was taken assuming a prior distribution before the data were taken. Bayesian testing is a very useful tool in reliability testing and is discussed in Volume 2, Chapter 11.

EXAMPLE 17–2

A missile system's target detection device has been designed to receive signals of varying power levels. Fifty independent signals are sampled at random from each new signal source. The target detection device is then adjusted to receive the future signals whose power levels fall between the lowest and the highest sample values. What is the probability that the target detection device, after adjustment, will receive at least 95% of a large number of signals from a given signal source?

SOLUTION TO EXAMPLE 17–2

The proportion, p, of the population between the lowest and highest observations in a random sample of size $n = 50$ is a beta variate. With $r = 1$, $s = 1$, $\alpha = 50-1-1 = 48$, and $\beta = 1+1-1 = 1$. The probability that the accepted proportion of signals is at least 0.95 is determined as follows: The cumulative beta distribution, or the incomplete beta function, is

$$
\begin{aligned}
F(x; \alpha, \beta) &= \frac{\Gamma(\alpha + \beta)}{\Gamma(\alpha)\Gamma(\beta)} \int_0^x t^{\alpha-1}(1 - t)^{\beta-1} dt, 0 \le x \le 1, \\
&= 0, x < 0, \\
&= 1, x > 1,
\end{aligned}
\tag{17.20}
$$

and is tabulated in [2; 3; 4]. Using these tables when $\alpha \geq \beta$ we get $p = \alpha$ and $q = \beta$, and find $B(x; p, q) = F(x; \alpha; \beta)$. Therefore, using the values required for the problem we have:

$$P(p \geq 0.95) = 1 - F(0.95; 48; 1),$$

and

$$F(0.95; 48, 1) = B(0.95; 48, 1) = 0.279.$$

The probability that at least 95% of the signals will be received is

$$P(p \geq 0.95) = 1 - 0.279 = 0.721.$$

PROBLEMS

17–1. Ten identical devices are tested continuously to failure and the following times to failure, in hours, are obtained: 27.5, 56.1, 63.3, 68.9, 80.4, 81.3, 83.6, 88.7, 89.9, 92.3. From past experience, the minimum time to failure may be taken to be zero, whereas the maximum time to failure never exceeds 100 hr. Do the following:

   (1) Find the parameters of the generalized beta distribution which might represent the times-to-failure distribution of these devices.

   (2) Write down and plot the *pdf*.

   (3) Write down and plot the reliability function.

   (4) Write down and plot the failure rate function.

   (5) Find the reliability for a mission of 20 hr, starting the mission at the age of 0 hr.

   (6) Find the reliability for a mission of 20 hr, starting the mission at the age of 20 hr.

17–2. Twelve identical devices are tested continuously to failure and the following times to failure, in hours, are obtained: 272.5, 564.1, 668.3, 681.7, 800.4, 801.3, 823.6, 858.7, 859.9, 862.0, 882.3, 962.3. From past experience, the minimum time to failure may be taken to be zero, whereas the maximum time to failure never exceeds 1,000 hr. Do the following:

   (1) Find the parameters of the generalized beta distribution which might represent the times-to-failure distribution of these devices.

(2) Write down and plot the *pdf*.

(3) Write down and plot the reliability function.

(4) Write down and plot the failure rate function.

(5) Find the reliability for a mission of 300 hr, starting the mission at the age of 0 hr.

(6) Find the reliability for a mission of 300 hr, starting the mission at the age of 200 hr.

(7) Find the reliability for a mission of 400 hr, starting the mission at the age of 400 hr.

17–3. Ten identical capacitors are tested continuously to failure and the following times to failure, in hours, are obtained: 57.5, 58.1, 60.3, 61.9, 70.4, 81.3, 83.2, 84.7, 85.9, 86.3. From past tests, the minimum time to failure was found to be 50 hr, whereas the maximum time to failure never exceeds 90 hr. Do the following:

(1) Find the parameters of the generalized beta distribution which might represent the times-to-failure distribution of these devices.

(2) Write down and plot the *pdf*.

(3) Write down and plot the reliability function.

(4) Write down and plot the failure rate function.

(5) Find the reliability for a mission of 60 hr, starting the mission at the age of 0 hr.

(6) Find the reliability for a mission of 70 hr, starting the mission at the age of 50 hr.

17–4. A missile system's target detection device has been designed to receive signals of varying power levels. Fifty independent signals are sampled at random from each new signal source. The target detection device is then adjusted to receive the future signals whose power levels fall between the lowest and the highest sample values. What is the probability that the target detection device, after adjustment, will receive at least 90% of a large number of signals from a given signal source?

17–5. A newly designed automated quality control system, for a major aerospace company, is being calibrated. During calibration the system examines 100 random independent samples. The system is then adjusted to test only the samples whose quality falls between the lowest and highest sample values. Determine the probability that the automated quality control system, after adjustment will test at least 95% of a large number of products from a given production line?

17-6. Thirteen identical automobile tires are tested continuously to failure and the following miles to failure are obtained: 21,234; 24,356; 27,657; 31,678; 40,010; 45,890; 46,870; 46,970; 47,200; 47,707; 48,409; 49,907; 50,102. From past tests, the minimum miles to failure was found to be 18,000 miles, whereas the maximum miles to failure never exceeds 51,500 miles. Do the following:

(1) Find the parameters of the generalized beta distribution which might represent the times-to-failure distribution of these devices.

(2) Write down and plot the *pdf*.

(3) Write down and plot the reliability function.

(4) Write down and plot the failure rate function.

(5) Find the reliability for a mission of 27,000 miles, starting the mission at the age of 0 miles.

(6) Find the reliability for a mission of 40,000 miles, starting the mission at the age of 10,000 miles.

17-7. During the calibration of a voltage regulator 200 voltage fluctuations are sampled at random, at different power periods. The voltage regulator is then adjusted to accept and correct only the voltage fluctuations that fall between the highest and the lowest sampled values. What is the probability that at most 5% of the voltage fluctuations fall outside the range of the device after adjustment?

17-8. Develop a computer program in either FORTRAN, PASCAL or BASIC that will generate the values found in Table 17.1. To test your program solve Example 17-1 and generate Table 17.1 again. Using your program, generate similar tables for the data found in Problems 17-1, 17-2, 17-3 and 17-6. Turn in all pages of output as well the computer program.

17-9. Twelve identical items are tested continuously to failure and the following times to failure, in hours, are obtained: 234, 256, 277, 398, 410, 458, 468, 497, 510, 537, 549, 557. From past tests, the minimum time to failure was found to be 100 hr, whereas the maximum time to failure never exceeds 575 hr. Do the following:

(1) Find the parameters of the generalized beta distribution which might represent the times-to-failure distribution of these devices.

(2) Write down and plot the *pdf*.

(3) Write down and plot the reliability function.

(4) Write down and plot the failure rate function.

(5) Find the reliability for a mission of 300 hr, starting the mission at the age of 100 hr.

(6) Find the reliability for a mission of 400 hr, starting the mission at the age of 400 hr.

REFERENCES

1. Hahn, Gerald J. and Shapiro, Samuel S., *Statistical Models in Engineering,* John Wiley & Sons Inc., New York, 1967.

2. Thompson, C.M., "Tables of Percentage Points of the Incomplete Beta Function," *Biometrika,* Vol. 32, pp. 151-171, 1941.

3. Hartley, H.O. and Fitch, E.R., "A Chart for the Incomplete Beta Function and the Cumulative Binomial Distribution," *Biometrika,* Vol. 38, pp. 423-426, 1951.

4. Pearson, Karl, *Tables of the Incomplete Beta Function,* Cambridge University Press, London, 494 pp., 1956.

5. Ireson, Grant (ed), *Reliability Handbook,* McGraw Hill Book Co., New York, 720 pp., 1966.

6. Kao, J.H.K., "The Beta Distribution in Reliability and Quality Control," *Proceedings of the Seventh National Symposium on Reliability and Quality Control,* pp. 496-511, 1961.

7. Wilks, S. S., "Statistical Prediction with Special Reference to the Problem of Tolerance Limits," *Ann. Math. Statist.,* Vol. 13, pp. 400-409, 1942.

8. Drnas, Tom, "Bayesian Reliability Demonstration," Book III, Section II, Lecture Notes for the Fourth Annual Reliability Testing Institute, The University of Arizona, Tucson, Arizona, April 3-7, 1978.

Chapter 18

METHODS OF PARAMETER ESTIMATION

18.1 INTRODUCTION

Parameter estimation plays a very important role in the quantitative evaluation of system reliability and maintainability. Fig. 18.1 is a typical flow chart for data analysis in reliability and maintainability engineering. From field operation or from lab tests, data such as times to (or between) failure(s) or times to repair can be collected. Through failure analysis or using statistical techniques, the outlier or unusual data in the sample can be found. With the purified data, we then evaluate the system's reliability, or maintainability, parametrically. To do this, the most important task is to find the best, or the most appropriate distribution to describe or represent these data. The optimum distribution can be obtained using the following process:

1. Choose one of the distributions summarized in Fig. 18.1. Write down its *pdf*.

2. Estimate the parameter(s) of this chosen distribution from the sample data using any one or more of the following methods: least squares, matching moments, modified moments, maximum likelihood, probability plotting, or others.

3. Conduct goodness-of-fit test(s) to ascertain that the selected distribution fits the data. Such tests are the chi-squared, Kolmogorov-Smirnov, modified Kolmogorov-Smirnov, correlation coefficient, likelihood ratio, Anderson-Darling, Cramer-von Mises tests, or others.

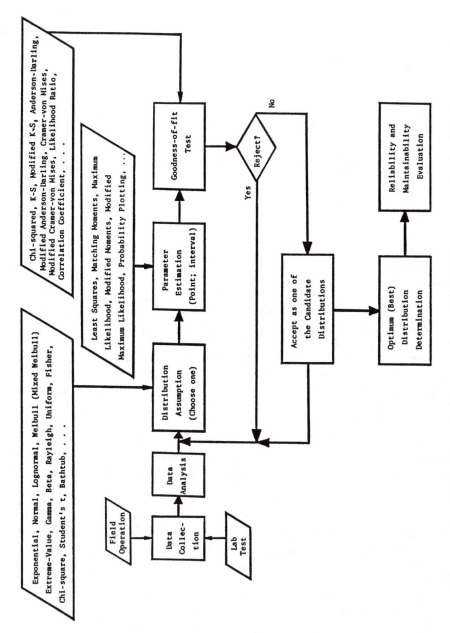

Fig. 18.1 – Flow chart for data analysis in reliability and maintainability engineering.

614

4. If the selected distribution is rejected, then go to Step 1 in Fig. 18.1, choose another distribution, and repeat Steps 2 and 3. If the selected distribution is not rejected, then take it as a possible candidate and repeat Steps 1, 2 and 3 by choosing another distribution.

5. After exhausting all of the possible distributions by repeating Steps 1, 2, 3 and 4, we may get several (could be only one) candidate distributions which were not rejected in the goodness-of-fit tests. To decide which one is the best distribution, compare the sample values of the goodness-of-fit statistic of all candidates; such as, χ^2, D_{max}, A_n^2, W_n^2, or others. The best distribution will be the one with the *minimum difference test statistic value* or the *maximum likelihood test statistic value*.

6. Once the best distribution is determined, we can use it to evaluate the reliability or maintainability characteristics of the system, such as $MTBF$, $MTTR$, $R(t)$, $M(t)$, $\lambda(t)$, and others.

The preceding process is fully illustrated in Fig. 18.1.

The whole process just discussed is actually a hypothesis-and-test process from the mathematical statistics viewpoint. It may be seen that without parameter estimation it is impossible to determine the best distribution quantitatively and to evaluate system reliability and maintainability parametrically. The parameter estimates can be classified into *point estimators* with 50% confidence level and *interval estimators* with any desired confidence levels. The estimation methods can be classified into two categories: (1) the *graphical approach*, such as the probability plotting method, and (2) the *analytical approach*, such as the least squares, matching moments and maximum likelihood methods.

Meanwhile since the sample completeness also has an impact on the selection of the methods and the mathematical models for parameter estimation, we have to be very careful about the behavior of the given sample data. The sample can be classified into a *complete sample* or an *incomplete (truncated* or *censored) sample*. In this chapter only analytical point estimation methods for complete samples are presented. For interval estimates and graphical methods, readers should refer to other chapters in this book where specific distributions are covered. For incomplete sample data, the readers are referred to [1].

Different estimation statistics often yield different properties. We are frequently asked to find some parameter estimates possessing one or more particular characteristics. The goodness of estimators can be measured by consistency, biasedness, efficiency, variance and so on.

Correspondingly, there are *consistent estimate, unbiased estimate, efficient estimate, least variance estimate (LVE), best linear unbiased estimate (BLUE), minimum variance unbiased estimate (MVUE)*, and others. Some of these are covered for specific distributions in other chapters of this book.

18.2 IDENTIFICATION OF DATA OUTLIERS

18.2.1 INTRODUCTION

In the times-to-failure data obtained from a sample of identical units, all tested to failure, there might be values which are either very low or very high; or in a probability plot there might be points which are substantially below the straight line fitted to the data points at the low-value end or there might be points which are substantially above the straight line at the high-value end. This forces us to decide which observations should be retained, or rejected as not belonging to the population the data come from. Methods for rejecting data that do not belong to the population are presented in this section.

18.2.2 A SIMPLE BUT APPROXIMATE OUTLIER TEST

To reject an observation, calculate the sample mean, $\hat{\bar{T}}$, and sample standard deviation, $\hat{\sigma}_T$, while using all test data. If the suspected observation is larger in value than $\hat{\bar{T}} + k\hat{\sigma}_T$ for the high-value end of the data, or smaller in value than $\hat{\bar{T}} - k\hat{\sigma}_T$ for the low-value end of the data, then there is a chance the observation is an outlier. If we want to establish this chance at 99%, then the value of k would be 2.3264, because the probability that an observation is greater than $\hat{\bar{T}} + 2.326\hat{\sigma}_T$ is 1%, or the chance that it is an outlier is 99%. The value of k would be 0.1282 for a 90% chance, and 1.645 for a 95% chance. It may be seen that this method would be relatively exact if the data are normally, or approximately normally, distributed, but it would be quite approximate for other cases. It may also be applied if the data are Weibull distributed with the shape parameter in the following range: $2.6 \leq \beta \leq 3.7$. In this range the Weibull distribution approximates the normal distribution. To apply this outlier test [2, pp. 90-91], do the following:

1. Calculate the sample mean, $\hat{\bar{T}}$, and the sample standard deviation, $\hat{\sigma}_T$, using all observations, including the suspected outlier.

2. Calculate

$$z_\alpha = |\frac{T_S - \hat{\bar{T}}}{\hat{\sigma}_T}|,$$ (18.1)

where T_S is the suspected outlier observation's value, which might be T_1 and/or T_N if all the observations are arranged in an increasing order, or $T_1 < T_2 < ... < T_N$.

3. From Appendix B, find α corresponding to z_α obtained in Step 2, such that

$$\alpha = \begin{cases} P(T > T_S) = 1 - \Phi(z_S), & \text{if } T_S > \hat{\bar{T}}, \\ P(T < T_S) = \Phi(-z_S), & \text{if } T_S < \hat{\bar{T}}. \end{cases}$$

4. If

$$\alpha \leq \frac{1}{N},$$ (18.2)

reject T_S, because it is very likely an outlier. In Eq. (18.2) N is the total number of observations, including the suspected outlier. The criterion $1/N$ is used, because the probability that one observation out of N may be an outlier is equal to $1/N$, in a given sample of size N. Therefore "$\alpha < 1/N$" means that T_1 and/or T_N are/is away from the position(s) [value(s)] they/it should be.

EXAMPLE 18–1

Nineteen units are all tested to failure, and the following times-to-failure data, in hours, are obtained:

2,511 2,691 2,697 2,702 2,706 2,709 2,712
2,716 2,720 2,722 2,726 2,728 2,730 2,736
2,739 2,744 2,747 2,754 2,908.

A study of these data indicates that the first and the last observations might not belong to the population of such units. Determine the following:

1. If the first time to failure belongs to the population from which the tested sample was obtained.

2. If the last time to failure belongs to the population from which the tested sample was obtained.

SOLUTIONS TO EXAMPLE 18-1

1. From the data the calculated mean is $\hat{\overline{T}} = 2720.95$ hr and the standard deviation is $\hat{\sigma}_T = 68.47$ hr. Then, with $T_S = 2,511$ hr,

$$z_\alpha = |\frac{2,511 - 2,720.95}{68.47}| = 3.0663,$$

and from Appendix B , $\alpha = 0.0011$. The sample size is $N = 19$. Therefore,

$$\frac{1}{N} = \frac{1}{19} = 0.05263,$$

and

$$\alpha = 0.0011 < \frac{1}{N} = 0.05263.$$

Consequently, the first observation, $T_1 = 2,511$ hr, is an outlier. The one percentile value of the times to failure is

$$\hat{T}_{1\%} = \hat{\overline{T}} - 2.3264\hat{\sigma}_T,$$
$$= 2,720.95 - (2.3264)(68.47),$$

or

$$\hat{T}_{1\%} = 2,561.66 \text{ hr.}$$

It is seen that

$$T_1 = 2,511 \text{ hr} < T_{1\%} = 2,561.66 \text{ hr.}$$

Consequently, the chance, or the confidence, that T_1 is an outlier is slightly over 99%; or, the risk in the decision that T_1 is an outlier is slightly less than 1%.

2. For $T_N = T_{19} = 2,908$ hr,

$$z_\alpha = |\frac{2,908 - 2,720.95}{68.47}| = 2.7319,$$

and from Appendix B, $\alpha = 0.003166$. Since

$$\alpha = 0.003166 < \frac{1}{N} = \frac{1}{19} = 0.05263,$$

the 19th observation, $T_{19} = 2,908$ hr, is an outlier.

The 99th percentile value of the times to failure is

$$
\begin{aligned}
T_{99\%} &= \hat{\bar{T}} + 2.3264 \hat{\sigma}_T, \\
&= 2,720.95 + (2.3264)(68.47),
\end{aligned}
$$

or

$$T_{99\%} = 2,880.23 \text{ hr.}$$

It is seen that

$$T_{19} = 2,908 \text{ hr} > T_{99\%} = 2,880.23.$$

Consequently, the chance, or the confidence that T_{19} is an outlier is slightly over 99%; or the risk in the decision that T_{19} is an outlier is slightly less than 1%.

18.2.3 THE NATRELLA-DIXON OUTLIER TEST

Natrella [3] and Dixon [4] introduced another method for identifying outliers, which may be used when the mean and the standard deviation of the population are not known. Then arrange the data in increasing value, or

$$T_1 < T_2 < T_3 < ... < T_{N-1} < T_N,$$

where N is the sample size, all tested to failure. If

$$
\begin{aligned}
3 &\le N \le 7, &\text{compute } r_{10}, \\
8 &\le N \le 10, &\text{compute } r_{11}, \\
11 &\le N \le 13, &\text{compute } r_{21},
\end{aligned}
$$

and if

$$14 \le N \le 25, \quad \text{compute } r_{22},$$

where the r_{ij} are computed as follows:

r_{ij}	If T_N is suspect	If T_1 is suspect
r_{10}	$(T_N - T_{N-1})/(T_N - T_1)$	$(T_2 - T_1)/(T_N - T_1)$
r_{11}	$(T_N - T_{N-1})/(T_N - T_2)$	$(T_2 - T_1)/(T_{N-1} - T_1)$
r_{21}	$(T_N - T_{N-2})/(T_N - T_2)$	$(T_3 - T_1)/(T_{N-1} - T_1)$
r_{22}	$(T_N - T_{N-2})/(T_N - T_3)$	$(T_3 - T_1)/(T_{N-2} - T_1)$

Look up $r_{\alpha/2}$ for the given sample size N from Table 18.1, where $(1-\alpha)$ is the confidence level desired. If $r_{ij} > r_{\alpha/2}$, reject the suspect observation with $100(1-\alpha)\%$ confidence.

The preceding applies when the suspect observations are present on both high and low sides of the sample. In cases where only one of the two situations arises, the value of $r_{\alpha/2}$ corresponds to the confidence level of $(1-\alpha/2)$ instead of $(1-\alpha)$.

EXAMPLE 18–2

Repeat the analysis of the test data in Example 18–1, using the Natrella-Dixon outlier test. Furthermore, test the case when the suspect observations are present at the both high and low sides of the sample.

SOLUTIONS TO EXAMPLE 18–2

1. From the data given in Example 18–1, $T_1 = 2,511$ hr is the suspect observation, $N = 19$, $T_3 = 2,697$ hr and $T_{19-2} = T_{17} = 2,747$ hr. Here, with $N = 19$, we compute r_{22}. Then,

$$r_{22} = \frac{T_3 - T_1}{T_{N-2} - T_1} = \frac{T_3 - T_1}{T_{17} - T_1} = \frac{2,697 - 2,511}{2,747 - 2,511} = 0.788.$$

The outlier test is for the low-side observation. We want a risk of at most 1%, or a confidence level, CL, of at least 99%, then

$$CL = 1 - \frac{\alpha}{2} = 0.99,$$

or

$$\frac{\alpha}{2} = 0.01.$$

Table 18.1 yields $r_{0.01} = 0.547$ for $N = 19$. Since

$$r_{22} = 0.788 > r_{\alpha/2} = r_{0.01} = 0.547,$$

we conclude with 99% confidence that $T_1 = 2,511$ hr is an outlier, and should be excluded from the analysis of these data.

2. From the data given in Example 18–1, $T_N = T_{19} = 2,908$ hr is the suspect observation, $N = 19$, $T_{N-2} = T_{17} = 2,747$ hr, and $T_3 = 2,697$ hr. We compute r_{22} because $N = 19$. Then

$$r_{22} = \frac{T_N - T_{N-2}}{T_N - T_3} = \frac{T_{19} - T_{17}}{T_{19} - T_3} = \frac{2,908 - 2,747}{2,908 - 2,697},$$

Table 18.1– Criteria for rejecting suspect observations using the Natrella-Dixon outlier test.

Statistic	Sample size, N	Critical values at the probability levels of $\alpha/2$						
		0.30	0.20	0.10	0.05	0.02	0.01	0.005
	3	0.684	0.781	0.886	0.941	0.976	0.988	0.994
	4	0.471	0.560	0.679	0.765	0.846	0.889	0.926
r_{10}	5	0.373	0.451	0.557	0.642	0.729	0.780	0.821
	6	0.318	0.386	0.482	0.560	0.644	0.698	0.740
	7	0.281	0.344	0.434	0.507	0.586	0.637	0.680
	8	0.318	0.385	0.479	0.554	0.631	0.683	0.725
r_{11}	9	0.288	0.352	0.441	0.512	0.587	0.635	0.677
	10	0.265	0.325	0.409	0.477	0.551	0.597	0.639
	11	0.391	0.442	0.517	0.576	0.638	0.679	0.713
r_{21}	12	0.370	0.419	0.490	0.546	0.605	0.642	0.675
	13	0.351	0.399	0.467	0.521	0.578	0.615	0.649
	14	0.370	0.421	0.492	0.547	0.602	0.641	0.674
	15	0.353	0.402	0.472	0.525	0.579	0.616	0.647
	16	0.338	0.386	0.454	0.507	0.559	0.595	0.624
	17	0.325	0.373	0.438	0.490	0.542	0.577	0.605
	18	0.314	0.361	0.424	0.475	0.527	0.561	0.589
	19	0.304	0.350	0.412	0.462	0.514	0.547	0.575
r_{22}	20	0.295	0.340	0.401	0.450	0.502	0.535	0.562
	21	0.287	0.331	0.391	0.440	0.491	0.524	0.551
	22	0.280	0.323	0.382	0.430	0.481	0.514	0.541
	23	0.274	0.316	0.374	0.421	0.472	0.505	0.532
	24	0.268	0.310	0.367	0.413	0.464	0.497	0.524
	25	0.262	0.304	0.360	0.406	0.457	0.489	0.516

or

$$r_{22} = \frac{161}{211} = 0.763.$$

The outlier test is for the high-side observation; consequently, for a confidence level of 99% or $\frac{\alpha}{2} = 0.01$, Table 18.1 yields $r_{0.01} = 0.547$. Since

$$r_{22} = 0.763 > r_{\frac{\alpha}{2}} = r_{0.01} = 0.547,$$

we conclude, with at least 99% confidence, that $T_{19} = 2,908$ hr is an outlier and should be excluded from the analysis of these data. The risk in making this decision is no more than 1%.

3. If we suspect that both the first and the last observations may be outliers, for a confidence level of 99%; then,

$$\begin{aligned} 1 - \alpha &= 0.99, \\ \alpha &= 0.01, \end{aligned}$$

and

$$\frac{\alpha}{2} = 0.005.$$

Consequently, from Table 18.1, $r_{0.005} = 0.575$. Since, for the first observation

$$r_{22} = 0.788 > r_{\frac{\alpha}{2}} = r_{0.005} = 0.575,$$

and for the last observation

$$r_{22} = 0.763 > r_{\frac{\alpha}{2}} = r_{0.005} = 0.575,$$

we conclude, with 99% confidence, that $T_1 = 2,511$ hr and $T_{19} = 2,908$ hr are both outliers.

18.2.4 THE GRUBBS OUTLIER TEST

For the Grubbs [5] outlier test, arrange the data in an increasing order, or $T_1 < T_2 < T_3 < ... < T_{N-1} < T_N$, where N is the test sample size.

Let

$$\nu^2 = \sum_{i=1}^{N} (T_i - \hat{\bar{T}})^2, \tag{18.3}$$

$$\nu_1^2 = \sum_{i=2}^{N} (T_i - \hat{\bar{T}}_1)^2, \tag{18.4}$$

$$\nu_N^2 = \sum_{i=1}^{N-1} (T_i - \hat{\bar{T}}_N)^2, \tag{18.5}$$

where

$$\hat{\bar{T}} = \frac{\sum\limits_{i=1}^{N} T_i}{N}, \tag{18.6}$$

$$\hat{\bar{T}}_1 = \frac{\sum\limits_{i=2}^{N} T_i}{N-1}, \tag{18.7}$$

and

$$\hat{\bar{T}}_N = \frac{\sum\limits_{i=1}^{N-1} T_i}{N-1}. \tag{18.8}$$

If T_N is the suspect value, calculate ν_N^2/ν^2. If T_1 is the suspect value, calculate ν_1^2/ν^2. If these values are less than those in Table 18.2, reject the suspect observations at the appropriate confidence level.

EXAMPLE 18–3

Repeat the analysis of the test data in Example 18–1, using the Grubbs outlier test.

SOLUTIONS TO EXAMPLE 18–3

1. The data are already arranged in increasing order. Then,

$$\hat{\bar{T}} = 2,720.95 \text{ hr},$$

$$\hat{\bar{T}}_1 = \frac{\sum\limits_{i=2}^{19} T_i}{19-1} = 2,732.61 \text{ hr},$$

Table 18.2– Criteria for rejecting suspect observations using the Grubbs outlier test.

Sample size, N	Critical values at the confidence levels of $(1 - \alpha)$			
	99%	97.5%	95%	90%
3	0.0001	0.0007	0.0027	0.0109
4	0.0100	0.0248	0.0494	0.0975
5	0.0442	0.0808	0.1270	0.1984
6	0.0928	0.1453	0.2032	0.2826
7	0.1447	0.2066	0.2696	0.3503
8	0.1948	0.2616	0.3261	0.4050
9	0.2411	0.3101	0.3742	0.4502
10	0.2831	0.3526	0.4154	0.4881
11	0.3221	0.3901	0.4511	0.5204
12	0.3554	0.4232	0.4822	0.5483
13	0.3864	0.4528	0.5097	0.5727
14	0.4145	0.4792	0.5340	0.5942
15	0.4401	0.5030	0.5559	0.6134
16	0.4634	0.5241	0.5755	0.6306
17	0.4848	0.5442	0.5933	0.6461
18	0.5044	0.5611	0.6095	0.6601
19	0.5225	0.5785	0.6243	0.6730
20	0.5393	0.5937	0.6379	0.6848
21	0.5548	0.6076	0.6504	0.6958
22	0.5692	0.6206	0.6621	0.7058
23	0.5827	0.6327	0.6728	0.7151
24	0.5953	0.6439	0.6829	0.7238
25	0.6071	0.6544	0.6923	0.7319

$$\hat{T}_N = \frac{\sum\limits_{i=1}^{18} T_i}{19 - 1} = 2,710.56 \text{ hr},$$

$$\nu^2 = \sum_{i=1}^{19} \left(T_i - \hat{T}\right)^2 = 84,380.95,$$

$$\nu_1^2 = \sum_{i=2}^{19} \left(T_1 - \hat{T}_1\right)^2 = 37,854.28,$$

and

$$\nu_N^2 = \sum_{i=1}^{18} \left(T_1 - \hat{T}_N\right)^2 = 47,448.45.$$

T_1 is the suspect observation; consequently, we calculate

$$\frac{\nu_1^2}{\nu^2} = \frac{37,854.28}{84,380.95},$$

or

$$\frac{\nu_1^2}{\nu^2} = 0.4486.$$

This value is less than the value of $r_{99\%} = 0.5225$ from Table 18.2, for $N = 19$ and confidence level 99%. Consequently, with at least 99% confidence, we conclude that the suspect observation, $T_1 = 2,511$ hr, is an outlier and should be excluded from the analysis of these data.

2. In this case, we calculate

$$\frac{\nu_N^2}{\nu^2} = \frac{47,448.45}{84,380.95},$$

or

$$\frac{\nu_N^2}{\nu^2} = 0.5623.$$

This value is greater than the value of $r_{99\%} = 0.5225$, from Table 18.2, for $N = 19$ and a confidence level of 99%, but less than the value of $r_{97.5\%} = 0.5785$ from Table 18.2, for level 97.5%. By interpolation it is found that the confidence level corresponding to $\nu_N^2/\nu^2 = 0.5623$ is 98.43%. Consequently, with at least 98.43% confidence, we conclude that the suspect observation, $T_{19} = 2,908$ hr, is an outlier and should be excluded from the analysis of these data.

18.2.5   RANK LIMITS METHOD

The rank limits method, proposed by Dimitri Kececioglu and Feng-Bin Sun in 1992, is easy to implement. More importantly, it can be used to identify the multiple outliers in any given sample data not only at the lower and upper ends but the outliers between them as well. On the other hand, this method applies to any population distribution not just to the normal distribution as is the case for the previous methods. To apply this test do the following:

1. **To check the first and the last observations, T_1 and T_N, using the one-sided rank limits.**

    (a) Arrange the data in an increasing order, or $T_1 < T_2 < T_3 < ... < T_{N-1} < T_N$, where N is the test sample size.

    (b) Estimate the distribution's parameters from the sample for either the known or the assumed population distribution; e.g., $\hat{\lambda}$ for exponential population; $\hat{\bar{T}}$ and $\hat{\sigma}_T$ for the normal population; $\hat{\bar{T}}'$ and $\hat{\sigma}_{T'}$ for the lognormal population; $\hat{\gamma}, \hat{\beta}$ and $\hat{\eta}$ for the Weibull population; etc.

    (c) Find the cumulative probabilities for the first and the last observations, T_1 and T_N, of the fitted *pdf*, $\hat{F}(T_1)$ and $\hat{F}(T_N)$.

    (d) For a given significance level, α, sample size N, find in Appendix A, the lower one-sided rank limit for the first observation T_1, or $r_{1;N;\alpha}$, and the upper one-sided rank limit for the last observation T_N, or $r_{N;N;1-\alpha}$. These ranks are distribution free.

    (e) Compare $\hat{F}(T_1)$ with $r_{1;N;\alpha}$, and compare $\hat{F}(T_N)$ with $r_{N;N;1-\alpha}$, and conclude that

    $$\begin{cases} T_1 \text{ is an outlier,} & \text{if } \hat{F}(T_1) \leq r_{1;N;\alpha}, \\ T_N \text{ is an outlier,} & \text{if } \hat{F}(T_N) \geq r_{N;N;1-\alpha}, \end{cases}$$

    at the confidence level of $(1 - \alpha)\%$.

2. **To check the other observations between T_1 and T_N using the two-sided rank limits.**

    Case 1– If both T_1 and T_N are not outliers, then check for outliers between them by doing the following:

    (a) Find the cumulative probability at each ordered observation T_i of the fitted *pdf*, $\hat{F}(T_i)$, $i = 1, 2, ..., N$.

(b) For the same significance level, α, sample size, N, find in Appendix A, the two-sided rank limits for each ordered observation T_i, $r_{i;N;\alpha/2}$ and $r_{i;N;1-\alpha/2}$.

(c) Compare $\hat{F}(T_i)$ with $r_{i;N;\alpha/2}$ and $r_{i;N;1-\alpha/2}$ to see whether or not the interval $[r_{i;N;\alpha/2} \quad ; \quad r_{i;N;1-\alpha/2}]$ contains $\hat{F}(T_i)$, and conclude that

$$T_i \text{ is an} \begin{cases} \text{inlier,} & \text{if } \hat{F}(T_i) \in [r_{i;N;\alpha/2} \quad ; \quad r_{i;N;1-\alpha/2}], \\ \text{outlier,} & \text{otherwise,} \end{cases}$$

at the confidence level of $(1-\alpha)\%$ for each T_i, $i = 1, 2, ..., N$.

Case 2– If one or both of T_1 and T_N are outliers, then remove the outlier(s) (T_1 and/or T_N) from the sample and let $N' = (N-1)$ for one outlier or let $N' = (N-2)$ for two outliers, and check for outliers between T_1 and T_N by doing the following:

(a) Reestimate the distribution parameters based on the remaining sample data of size N'.

(b) Find the cumulative probability at each newly-ordered observation $T_i^\dagger$ of the fitted *pdf*, $\hat{F}(T_i^\dagger)$, $i = 1, 2, ..., N'$.

(c) For the same significance level, α, but new sample size, N', find in Appendix A, the two-sided rank limits for each newly-ordered observation $T_i^\dagger$, $r_{i;N';\alpha/2}$ and $r_{i;N';1-\alpha/2}$.

(d) Compare $\hat{F}(T_i^\dagger)$ with $r_{i;N';\alpha/2}$ and $r_{i;N';1-\alpha/2}$ to see whether or not the interval $[r_{i;N';\alpha/2} \quad ; \quad r_{i;N';1-\alpha/2}]$ contains $\hat{F}(T_i^\dagger)$, and conclude that

$$T_i^\dagger \text{ is an} \begin{cases} \text{inlier,} & \text{if } \hat{F}(T_i^\dagger) \in [r_{i;N';\alpha/2} \quad ; \quad r_{i;N';1-\alpha/2}], \\ \text{outlier,} & \text{otherwise,} \end{cases}$$

at the confidence level of $(1-\alpha)\%$ for each $T_i^\dagger$, $i = 1, 2, ..., N'$.

EXAMPLE 18–4

Repeat the analysis of the test data in Example 18–1, using the rank limits method at the risk level of 10%.

SOLUTION TO EXAMPLE 18–4

1. The data are already arranged in increasing order.

2. Assume these data are from a normal population. The parameters have been estimated in Example 18–1 as

$$\hat{\bar{T}} = 2,720.95 \text{ hr},$$

and

$$\hat{\sigma}_T = 68.47 \text{ hr}.$$

3. The cumulative probabilities at the first and the last observations T_1 and T_{19} of the fitted normal *pdf*, $\hat{F}(T_1)$ and $\hat{F}(T_{19})$, are calculated as follows:

$$z_1 = \frac{2,511 - 2,720.95}{68.47} = -3.07,$$

$$\hat{F}(T_1) = \hat{F}(2,511 \text{ hr}) = \Phi(z_1) = 0.107\%,$$

and

$$z_{19} = \frac{2,908 - 2,720.95}{68.47} = 2.73,$$

$$\hat{F}(T_{19}) = \hat{F}(2,908 \text{ hr}) = \Phi(z_{19}) = 99.683\%.$$

4. For the given significance level, $\alpha = 10\%$, and sample size $N = 19$, the corresponding lower, one-sided rank limit for the first observation T_1, $r_{1;19;0.10}$, and the upper, one-sided rank limit for the last observation T_{19}, $r_{19;19;0.90}$, are obtained from Appendix A as follows:

$$r_{1;19;0.10} = 0.553\%,$$

and

$$r_{19;19;0.90} = 99.447\%.$$

5. Comparing $\hat{F}(T_1)$ with $r_{1;19;0.10}$ and comparing $\hat{F}(T_{19})$ with $r_{19;19;0.90}$ yields

$$\begin{cases} \bullet \ T_1 \text{ is an outlier,} \quad \text{because} \\ \hat{F}(T_1) = 0.107\% < r_{1;19;0.10} = 0.553\%, \\ \bullet \ T_{19} \text{ is an outlier,} \quad \text{because} \\ \hat{F}(T_{19}) = 99.683\% > r_{19;19;0.90} = 99.447\%, \end{cases}$$

at the confidence level of 90%.

Table 18.3– Data outlier identification by the rank limits method for Example 18–4.

i	$T_i^\dagger$, hr	z_i ††	$\hat{F}(T_i^\dagger)$, %	$r_{i;17;0.005}$, %	$r_{i;17;0.995}$, %	I/O‡
1	2,691	-1.72	4.272	0.301	16.157	I
2	2,697	-1.39	8.226	3.132	25.013	I
3	2,702	-1.12	13.136	4.990	32.619	I
4	2,706	-0.90	18.406	8.465	39.564	I
5	2,709	-0.73	23.270	12.377	46.055	I
6	2,712	-0.57	28.434	16.636	52.192	I
7	2,716	-0.35	36.317	21.191	58.030	I
8	2,720	-0.13	44.828	26.011	63.599	I
9	2,722	-0.02	49.202	31.083	68.917	I
10	2,726	0.20	57.926	36.401	73.989	I
11	2,728	0.31	62.172	41.970	78.809	I
12	2,730	0.42	66.276	47.808	83.364	I
13	2,736	0.75	77.337	53.945	87.623	I
14	2,739	0.92	82.121	60.436	91.535	I
15	2,744	1.19	88.298	67.381	95.010	I
16	2,747	1.36	91.308	74.987	97.868	I
17	2,754	1.74	95.907	83.843	99.699	I

†† $z_i = \dfrac{T_i^\dagger - 2{,}722.29}{18.17}$.

‡ I/O = Inlier/Outlier.

6. Remove the outliers T_1 and T_{19} from the sample. Then $N' = 19 - 2 = 17$ for the remaining data.

7. Reestimate the distribution parameters based on the remaining sample data of size $N' = 17$. This yields

$$\hat{\bar{T}} = 2{,}722.29 \text{ hr,}$$

and

$$\hat{\sigma}_T = 18.17 \text{ hr.}$$

8. Find the cumulative probability at each newly-ordered observation $T_i^\dagger$ which is entered in Column 2 of Table 18.3, of the fitted *pdf*, $\hat{F}(T_i^\dagger)$, $i = 1, 2, ..., 17$, as shown in Column 4 of Table 18.3.

9. For the same significance level, $\alpha = 10\%$, but the new sample size, $N' = 17$, find in Appendix A the two-sided rank limits for

each newly-ordered observation $T_i^\dagger$, $r_{i;17;0.05}$ and $r_{i;17;0.95}$, which are entered in Columns 5 and 6 of Table 18.3.

10. Comparing $\hat{F}(T_i^\dagger)$ with $r_{i;17;0.05}$ and $r_{i;17;0.95}$ yields no more outliers at the confidence level of 90%.

It may be seen that this method yields the same conclusion as the three preceding methods. Therefore T_1 and T_{19} should be excluded from the analysis of these data.

EXAMPLE 18–5

Eight identical electronic equipment are all tested to failure with the following times-to-failure data in hours: 100; 100; 100; 100; 500; 1,000; 2,000 and 5,000. Assume these data are from the one-parameter exponential distribution population. Using the rank limits method, check to see if there are any outliers in the given sample at the significance level of 5%.

SOLUTION TO EXAMPLE 18–5

1. The data are already arranged in increasing order.

2. The estimate of the failure rate for this sample with $r = N$, because all equipment are tested to failure, is

$$\hat{\lambda} = N / \sum_{i=1}^{N} T_i,$$

$$= 8 / \sum_{i=1}^{8} T_i,$$

or

$$\hat{\lambda} = 0.0008988 \text{ fr/hr.}$$

3. The cumulative probabilities for the first and the last observations, T_1 and T_8, of the fitted normal *pdf*, $\hat{F}(T_1)$ and $\hat{F}(T_8)$, are calculated as follows:

$$\hat{F}(T_1) = \hat{F}(100 \text{ hr}) = 1 - e^{-0.0008988 \times 100} = 8.570\%,$$

and

$$\hat{F}(T_8) = \hat{F}(5,000 \text{ hr}) = 1 - e^{-0.0008988 \times 5,000} = 98.883\%.$$

4. For the given significance level, $\alpha = 5\%$ and sample size $N = 8$, the corresponding lower, one-sided rank limit for the first observation T_1, $r_{1;8;0.05}$, and the upper, one-sided rank limit for the last observation T_8, $r_{8;8;0.95}$, are obtained from Appendix A as follows:

$$r_{1;8;0.05} = 0.639\%,$$

and

$$r_{8;8;0.95} = 99.361\%.$$

5. Comparing $\hat{F}(T_1)$ with $r_{1;8;0.05}$, and comparing $\hat{F}(T_8)$ with $r_{8;8;0.95}$, yields

$$\begin{cases} \bullet\ T_1 \text{ is not an outlier, \quad because} \\ \hat{F}(T_1) = 8.570\% > r_{1;8;0.05} = 0.316\%, \\ \bullet\ T_8 \text{ is not an outlier, \quad because} \\ \hat{F}(T_8) = 98.883\% < r_{8;8;0.95} = 99.684\%, \end{cases}$$

at the confidence level of 95%.

6. Keep the original sample with $N' = N = 8$.

7. The distribution parameter, λ, is the same as for the original sample, or

$$\hat{\lambda} = 0.0008988 \text{ fr/hr.}$$

8. Find the cumulative probability of the fitted *pdf*, $\hat{F}(T_i)$, $i = 1, 2, ..., 8$, for each ordered observation T_i in Column 2 of Table 18.4, and enter them in Column 3 of Table 18.4.

9. For the same significance level, $\alpha = 5\%$, and the same sample size, $N = 8$, find in Appendix A, the two-sided rank limits for each ordered observation T_i, $r_{i;8;0.025}$ and $r_{i;8;0.975}$, and enter them in Columns 4 and 5 of Table 18.4.

10. Comparing $\hat{F}(T_i)$ with $r_{i;8;0.025}$ and $r_{i;8;0.975}$ yields that T_4 is an outlier at the confidence level of 90%, because

$$F(T_4) = 8.570\% < r_{4;8;0.025} = 15.701\%.$$

Therefore, T_4 should be excluded from the analysis of these data.

Table 18.4– Data outlier identification by the rank limits method for Example 18–5.

i	T_i, hr	$\hat{F}(T_i)^{\dagger\dagger}$, %	$r_{i;8;0.025}$, %	$r_{i;8;0.975}$, %	I/O‡
1	100	8.570	0.316	36.942	I
2	100	8.570	3.185	52.651	I
3	100	8.570	8.523	65.086	I
4	100	8.570	15.701	75.513	O
5	500	36.201	24.487	84.299	I
6	1,000	59.297	34.914	91.477	I
7	2,000	83.433	47.349	96.815	I
8	5,000	98.883	63.058	99.684	I

†† $\hat{F}(T_i) = 1 - e^{-0.0008988\, T_i}$.
‡ I/O = Inlier/Outlier.

18.2.6   OTHER DATA OUTLIER IDENTIFICATION METHODS

In addition to the four methods presented in the preceding sections, there are some other approaches attacking the outlier identification problem.

Nelson [6; 7, p. 258, p. 275, p. 292] and Lawless [8] gave both the approximate and the accurate prediction limits for the $N th$ observation for data from the exponential, normal, lognormal, Weibull and extreme value distribution populations based on order statistics. By comparing the actual $N th$ observation with the upper prediction limit, we can decide whether or not to reject it as an outlier. If the $N th$ observation is above a $100(1 - \alpha)\%$, it is a statistically significant outlier at the $100(1 - \alpha)\%$ confidence level. Otherwise accept it as an inlier at the $100(1-\alpha)\%$ confidence level. For more methods the readers are referred to Hawkins [9] who did a comprehensive study on the identification methods of the data outliers in his book.

18.3 LEAST SQUARES METHOD OF PARAMETER ESTIMATION

18.3.1 PARAMETER ESTIMATION

The method of least squares requires that a straight line be fitted to a set of data points such that the sum of the squares of the vertical deviations from the points to the line is minimized.

If the random variable y is a linear function of a measurable independent variable x, or if

$$y = a + bx, \qquad (18.9)$$

then

$$y_i = a + bx_i + \epsilon_i, \qquad (18.10)$$

$$E(y_i) = a + b\, E(x_i),$$

$$E(\epsilon_i) = 0,$$

and

$$COV(\epsilon_i, \epsilon_j) = 0, \quad i = 1, 2, ..., N, \quad j = 1, 2, ..., N, \quad i \neq j,$$

where a and b are unknown constant parameters which need to be estimated, and ϵ_i and ϵ_j are the model errors.

Assume that a set of data pairs $(x_1, y_1), (x_2, y_2),..., (x_N, y_N)$, was obtained and plotted, as in Fig. 18.2. It may be seen that estimating a and b is equivalent to finding the straight line in Fig. 18.2 which best fits the data. According to the *least squares principle,* which minimizes the vertical distance between the data points and the straight line fitted to the data, the best fitting straight line to these data is the straight line $y = \hat{a} + \hat{b}x$ such that

$$\sum_{i=1}^{N}(\hat{a} + \hat{b}x_i - y_i)^2 = min(a, b) \sum_{i=1}^{N}(a + bx_i - y_i)^2,$$

and $\hat{a}$ and $\hat{b}$ are the *least squares estimates* of a and b [10, pp. 35-39; 11, pp. 3-12].

To obtain $\hat{a}$ and $\hat{b}$, let

$$S = \sum_{i=1}^{N}(a + bx_i - y_i)^2. \qquad (18.11)$$

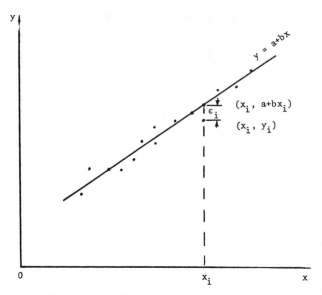

Fig. 18.2 – Data and fitted line for the least squares method
of parameter estimation.

Differentiating S with respect to a and b, respectively, yields

$$\frac{\partial S}{\partial a} = 2\sum_{i=1}^{N}(a + bx_i - y_i), \tag{18.12}$$

and

$$\frac{\partial S}{\partial b} = 2\sum_{i=1}^{N}(a + bx_i - y_i)x_i. \tag{18.13}$$

Setting Eqs. (18.12) and (18.13) equal to zero yields

$$\sum_{i=1}^{N}(a + bx_i - y_i) = \sum_{i=1}^{N}(\hat{y}_i - y_i) = -\sum_{i=1}^{N}(y_i - \hat{y}_i) = 0, \tag{18.14}$$

and

$$\sum_{i=1}^{N}(a + bx_i - y_i)x_i = \sum_{i=1}^{N}(\hat{y}_i - y_i)x_i = -\sum_{i=1}^{N}(y_i - \hat{y}_i)x_i = 0. \tag{18.15}$$

Solving Eqs. (18.14) and (18.15) simultaneously yields

$$\hat{a} = \bar{y} - \hat{b}\bar{x}, \tag{18.16'}$$

and

$$\hat{b} = \frac{\sum\limits_{i=1}^{N} x_i y_i - (\sum\limits_{i=1}^{N} x_i \sum\limits_{i=1}^{N} y_i)/N}{\sum\limits_{i=1}^{N} x_i^2 - (\sum\limits_{i=1}^{N} x_i)^2/N}, \tag{18.16''}$$

or

$$\hat{b} = \frac{L_{xy}}{L_{xx}},$$

where

$$\bar{x} = \frac{1}{N} \sum_{i=1}^{N} x_i,$$

$$\bar{y} = \frac{1}{N} \sum_{i=1}^{N} y_i,$$

$$L_{xy} = \sum_{i=1}^{N} (x_i - \bar{x})(y_i - \bar{y}),$$

and

$$L_{xx} = \sum_{i=1}^{N} (x_i - \bar{x})^2.$$

18.3.2   MEASURES OF GOODNESS-OF-FIT

1. COEFFICIENT OF DETERMINATION

After the least squares line and its parameters are determined, one frequently wishes to know how well this line fits the data. Two descriptive measures of association are frequently used in practice to describe the degree of relation between x and y.

The first measure, r^2, is called the *coefficient of determination* [10, pp. 89-90], and is defined by

$$r^2 = \frac{\sum\limits_{i=1}^{N} (\hat{y}_i - \bar{y})^2}{\sum\limits_{i=1}^{N} (y_i - \bar{y})^2} = \frac{SSR}{SSTO} = \frac{SSTO - SSE}{SSTO}, \tag{18.17}$$

where $SSTO$, the *total sum of squares*, is given by

$$SSTO = \sum_{i=1}^{N}(y_i - \bar{y})^2 = \sum_{i=1}^{N} y_i^2 - N\bar{y}^2, \tag{18.18}$$

and it is the measure of the total variation,

$$SSE = \sum_{i=1}^{N}(y_i - \hat{y}_i)^2, \tag{18.19}$$

and SSR, the *regression sum of squares*, is given by

$$SSR = \sum_{i=1}^{N}(\hat{y}_i - \bar{y})^2 = \sum_{i=1}^{N} \hat{y}_i^2 - N\bar{y}^2. \tag{18.20}$$

The relationship of $SSTO = SSR + SSE$ can be proven as follows: Consider the separation of the *total sum of squares*, where

$$SSTO = \sum_{i=1}^{N}(y_i - \hat{y}_i + \hat{y}_i - \bar{y})^2,$$

or

$$SSTO = \sum_{i=1}^{N}(y_i - \hat{y}_i)^2 + \sum_{i=1}^{N}(\hat{y}_i - \bar{y})^2 + 2\sum_{i=1}^{N}(y_i - \hat{y}_i)(\hat{y}_i - \bar{y}), \tag{18.21}$$

where

$$\sum_{i=1}^{N}(y_i - \hat{y}_i)(\hat{y}_i - \bar{y}) = \sum_{i=1}^{N}\hat{y}_i(y_i - \hat{y}_i) - \bar{y}\sum_{i=1}^{N}(y_i - \hat{y}_i),$$

and

$$\sum_{i=1}^{N}(y_i - \hat{y}_i) = \sum_{i=1}^{N}(y_i - \hat{a} - \hat{b}x_i) = 0$$

by Eq. (18.14). Also

$$\sum_{i=1}^{N}\hat{y}_i(y_i - \hat{y}_i) = \hat{a}\sum_{i=1}^{N}(y_i - \hat{y}_i) + \hat{b}\sum_{i=1}^{N}x_i(y_i - \hat{y}_i) = 0$$

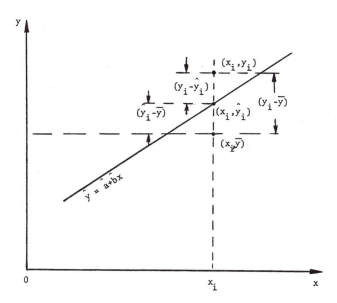

Fig.18.3 – An illustration for the relations among the variations using the least squares method of parameter
estimation.

by Eqs. (18.14) and (18.15). Then Eq. (18.21) becomes

$$SSTO = \sum_{i=1}^{N}(y_i - \hat{y}_i)^2 + \sum_{i=1}^{N}(\hat{y}_i - \bar{y})^2. \tag{18.22}$$

It can be seen from Eq. (18.22) that the total variation, $SSTO$,
can be separated into two parts, the first part is SSE, and the second
part is the *regression sum of squares, SSR*, or

$$SSTO = SSR + SSE.$$

SSR is the sum of the deviations, $\hat{y}_i - \bar{y}$, squared. Each deviation is
simply the difference between the fitted value on the regression line
and the mean of the fitted value, $\bar{y}$, as shown in Fig. 18.3.

Since $0 \le SSE \le SSTO$, it follows that

$$0 \le r^2 \le 1. \tag{18.23}$$

r^2 may be interpreted as the proportionate reduction of the total variation associated with the use of the independent variable x. Thus, the
larger r^2 is, the more the total variation of y is reduced by introducing
the independent variable x.

The statistic r^2 should be used with caution [12, p. 380], since it is always possible to make r^2 larger by simply adding enough terms to the model. It should be noted that r^2 does not measure the magnitude of the slope of the regression line. A large value of r^2 does not imply a steep slope. Furthermore, r^2 does not measure the appropriateness of the model, since it can be artificially inflated by adding higher-order polynomial terms.

2. CORRELATION COEFFICIENT

The second measure is the *correlation coefficient* [13, pp. 372-377], ρ, when x and y are both random variables, and it is defined by

$$\rho = \frac{\sigma_{xy}}{\sigma_x \sigma_y}, \tag{18.24}$$

where

σ_{xy} = covariance of x and y,

σ_x = standard deviation of x,

and

σ_y = standard deviation of y.

The estimator of ρ is the sample *correlation coefficient*, $\hat{\rho}$, given by

$$\hat{\rho} = \frac{\sum\limits_{i=1}^{N} x_i y_i - (\sum\limits_{i=1}^{N} x_i \sum\limits_{i=1}^{N} y_i)/N}{\left\{\left[\sum\limits_{i=1}^{N} x_i^2 - (\sum\limits_{i=1}^{N} x_i)^2/N\right]\left[\sum\limits_{i=1}^{N} y_i^2 - (\sum\limits_{i=1}^{N} y_i)^2/N\right]\right\}^{1/2}}, \tag{18.25}$$

or

$$\hat{\rho} = \frac{L_{xy}}{\sqrt{L_{xx} L_{yy}}},$$

where

$$L_{xy} = \sum_{i=1}^{N}(x_i - \bar{x})(y_i - \bar{y}),$$

$$L_{xx} = \sum_{i=1}^{N}(x_i - \bar{x})^2,$$

and

$$L_{yy} = \sum_{i=1}^{N}(y_i - \bar{y})^2.$$

The range of $\hat{\rho}$ is $-1 \leq \hat{\rho} \leq 1$, depending on the degree of associa-tion. This may be arrived at as follows:

If

$$y = a + bx$$

is true, then

$$\sigma_y^2 = Var(y) = Var(a + bx) = b^2 \sigma_x^2.$$

Therefore,

$$\sigma_y = |b|\sigma_x,$$

and the covariance

$$
\begin{aligned}
\sigma_{xy} &= E[(x - \bar{x})(y - \bar{y})], \\
&= E[(x - \bar{x})(a + bx - a - b\bar{x})], \\
&= bE[(x - \bar{x})^2],
\end{aligned}
$$

or

$$\sigma_{xy} = b\sigma_x^2.$$

Substituting σ_{xy} and σ_y into Eq. (18.24) yields

$$
\begin{aligned}
\rho &= \frac{\sigma_{xy}}{\sigma_x \sigma_y}, \\
&= \frac{b\sigma_x^2}{\sigma_x |b| \sigma_x}, \\
&= \frac{b}{|b|},
\end{aligned}
$$

or

$$\rho = \begin{cases} 1, & \text{if } b > 0 \\ -1, & \text{if } b < 0 \end{cases}$$

Figure 18.4 shows the different $\hat{\rho}$ values for the different correlations between x and y. Clearly, the closer to 1 is the absolute value of $\hat{\rho}$, the better the linear relationship between x and y. From the practical point of view, the minimum value of $\hat{\rho}$ should be 0.90 to provide a better linear relationship between x and y.

In general, the random variable y should be a linear function of k measurable independent variables $(x_1, x_2, ..., x_k)$ given by

$$\mathbf{y} = \mathbf{xb} + \boldsymbol{\epsilon}, \tag{18.26}$$

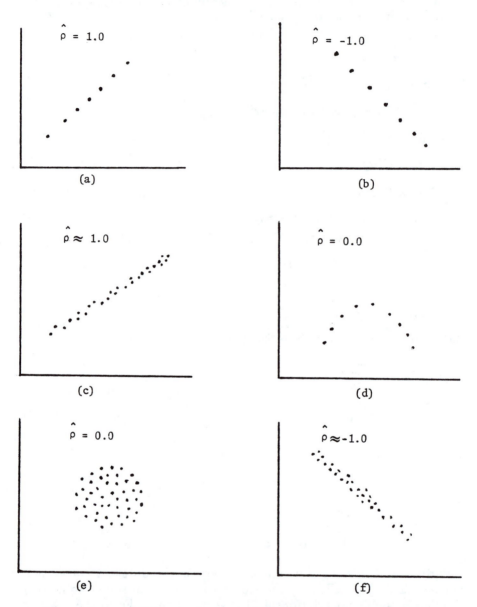

Fig. 18.4 – Different $\hat{\rho}$ values for different relationships between x and y for the least squares method of parameter estimation.

where

$$y = \begin{pmatrix} y_1 \\ y_2 \\ \vdots \\ y_N \end{pmatrix}, \quad b = \begin{pmatrix} b_0 \\ b_1 \\ \vdots \\ b_k \end{pmatrix}, \quad \epsilon = \begin{pmatrix} \epsilon_1 \\ \epsilon_2 \\ \vdots \\ \epsilon_N \end{pmatrix},$$

$$x = \begin{pmatrix} 1 & x_{11} & x_{12} & \cdots & x_{1k} \\ 1 & x_{21} & x_{22} & \cdots & x_{2k} \\ \vdots & \vdots & \vdots & \cdots & \vdots \\ 1 & x_{N1} & x_{N2} & \cdots & x_{Nk} \end{pmatrix},$$

and

$$\begin{cases} E(y) = xb, \\ Var(\epsilon) = \sigma^2 I. \end{cases} \tag{18.27}$$

where I is a $k \times k$ unit matrix.

Then, the least squares estimate of b, $\hat{b}$, is given by

$$\hat{b} = (x'x)^{-1}x'y, \tag{18.28}$$

where x' is the transpose of x.

In the more general case, in Eq. (18.28) replace

$$Var(\epsilon) = \sigma^2 I$$

with

$$Var(\epsilon) = \sigma^2 V, \tag{18.29}$$

where V is a $k \times k$ positive definite matrix. The least squares estimate of b, is then given by

$$\hat{b} = (x'V^{-1}x)^{-1}x'V^{-1}y, \tag{18.30}$$

where V^{-1} is the inverse of V.

EXAMPLE 18–6

Fit a least squares straight line to the following data:

x	1	3	4	6	8	9	11	14
y	1	2	4	4	5	7	8	9

TABLE 18.5 – Data analysis for the least-squares solution of Example 18–6.

i	x	y	x^2	xy	y^2
1	1	1	1	1	1
2	3	2	9	6	4
3	4	4	16	16	16
4	6	4	36	24	16
5	8	5	64	40	25
6	9	7	81	63	49
7	11	8	121	88	64
8	14	9	196	126	81
$\sum$	56	40	524	364	256

SOLUTION TO EXAMPLE 18–6

Using the analyzed data in Table 18.5, Eq. (18.16) yields

$$\hat{b} = \frac{\sum_{i=1}^{8} x_i y_i - (\sum_{i=1}^{8} x_i \sum_{i=1}^{8} y_i)/8}{\sum_{i=1}^{8} x_i^2 - (\sum_{i=1}^{8} x_i)^2/8},$$

$$\hat{b} = \frac{364 - 56 \times 40/8}{524 - 56^2/8} = 0.636,$$

and

$$\hat{a} = \bar{y} - \hat{b}\bar{x},$$

$$\hat{a} = \frac{40}{8} - 0.636 \times \frac{56}{8} = 0.548.$$

Now the least squares line is given by

$$y = 0.548 + 0.636x.$$

EXAMPLE 18–7

The following life test data were obtained: 800; 860; 950; 1,020; 1,100; 1,200; 1,300; 1,400; 1,500; 1,800. Assuming they follow a two-parameter exponential cumulative distribution, or

$$F(T) = 1 - e^{-\lambda(T-\gamma)}, \tag{18.31}$$

estimate the parameters λ and γ, and determine ρ using the least squares method.

SOLUTION TO EXAMPLE 18–7

Taking the natural logarithm of both sides of Eq. (18.31) yields

$$\log_e[1 - F(T)] = -\lambda(T - \gamma),$$

or

$$\log_e[1 - F(T)] = \lambda\gamma - \lambda T. \tag{18.32}$$

Let

$$y = \log_e[1 - F(T)], \tag{18.33}$$

$$a = \lambda\gamma, \tag{18.34}$$

and

$$b = -\lambda. \tag{18.34'}$$

Substitution of Eqs. (18.33), (18.34) and (18.34') into Eq. (18.32) yields the linear equation

$$y = a + bT.$$

Estimating λ and γ is equivalent to estimating a and b, where the y_i are calculated from

$$y_i = \log_e[1 - F(T_i)],$$

and the $F(T_i)$ may be estimated by

$$\hat{F}(T_i) \cong \frac{i - 0.3}{N + 0.4},$$

if the sample size is larger than 25, where N is the sample size, and

$$\hat{F}(T_i) = MR = \text{Median Rank.}$$

TABLE 18.6 — Data analysis for the least squares solution of Example 18–7. The T_i are in 100 hr.

N	T_i	$\hat{F}(T_i)$	y_i	T_i^2	y_i^2	$T_i y_i$
1	8.0	0.0673	-0.0697	64.00	0.0049	-0.5576
2	8.6	0.1635	-0.1785	73.96	0.0319	-1.5351
3	9.5	0.2596	-0.3006	90.25	0.0904	-2.8557
4	10.2	0.3558	-0.4397	104.04	0.1933	-4.4849
5	11.0	0.4519	-0.6013	121.00	0.3616	-6.6143
6	12.0	0.5481	-0.7942	144.00	0.6308	-9.5304
7	13.0	0.6442	-1.0335	169.00	1.0681	-13.4355
8	14.0	0.7404	-1.3486	196.00	1.8186	-18.8804
9	15.0	0.8365	-1.8112	225.00	3.2804	-27.1680
10	18.0	0.9327	-2.6985	324.00	7.2818	-48.5730
$\sum$	119.3		-9.2758	1,511.25	14.7618	-133.6349

More accurate values of MR can be found in Appendix A.

For this example, using the values of y_i and T_i given in Table 18.6,

$$\hat{b} = \frac{\sum\limits_{i=1}^{10} T_i y_i - (\sum\limits_{i=1}^{10} T_i)(\sum\limits_{i=1}^{10} y_i)/10}{\sum\limits_{i=1}^{10} T_i^2 - (\sum\limits_{i=1}^{10} T_i)^2/10},$$

$$\hat{b} = \frac{-133.6349 - (119.3)(-9.2758)/10}{1,511.25 - (119.3)^2/10},$$

or

$$\hat{b} = -0.2611,$$

and

$$\hat{a} = \bar{y} - \hat{b}\bar{T} = \frac{\sum\limits_{i=1}^{N} y_i}{N} - \hat{b}\frac{\sum\limits_{i=1}^{N} T_i}{N},$$

or

$$\hat{a} = \frac{-9.2758}{10} - (-0.2611)\frac{119.3}{10} = 2.19.$$

Therefore, from Eq. (18.34')

$$\hat{\lambda} = -\hat{b} \times 10^{-2} = -(-0.2611) \times 10^{-2} = 0.002611 \text{ fr/hr},$$

and from Eq. (18.34)

$$\hat{\gamma} = \frac{\hat{a}}{\hat{\lambda}},$$
$$\hat{\gamma} = 2.19/(0.002611),$$

or

$$\hat{\gamma} = 839 \text{ hr.}$$

Then,

$$f(T) = (0.002611)e^{-0.002611(T-839)}.$$

The sample *correlation coefficient*, $\hat{\rho}$, from Eq. (18.25), is

$$\hat{\rho} = \frac{\sum\limits_{i=1}^{10} T_i y_i - (\sum\limits_{i=1}^{10} T_i \sum\limits_{i=1}^{10} y_i)/10}{\left\{\left[\sum\limits_{i=1}^{10} T_i^2 - (\sum\limits_{i=1}^{10} T_i)^2/10\right]\left[\sum\limits_{i=1}^{10} y_i^2 - (\sum\limits_{i=1}^{10} y_i)^2/10\right]\right\}^{1/2}},$$

$$\hat{\rho} = \frac{-133.6349 - (119.3)(-9.2758)/10}{\left\{[1,511.25 - (119.3)^2/10][14.7618 - (-9.2758)^2/10]\right\}^{1/2}},$$

or

$$\hat{\rho} = \frac{-22.975}{23.279} = -0.987.$$

It must be observed that, with such a high value of the *correlation coefficient*, the exponential distribution represents these life test data very well.

Table 18.7 provides the variables and the coefficients for various distributions for the least squares method. Table 18.8 provides the linearized forms of some nonlinear equations for use in other situations.

18.3.3 COMMENTS ON THE LEAST SQUARES METHOD OF PARAMETER ESTIMATION (LSE)

1. The least squares estimation method is quite good for functions which can be linearized.

TABLE 18.7 – Variables and coefficients for different distributions for the least squares method.

cdf	y	x	a	b
Exponential distribution				
$F(T) = 1 - e^{-\lambda(T-\gamma)}$	$\log_e[1 - F(T)]$	T	$\lambda\gamma$	$-\lambda$
Weibull distribution				
$F(T) = 1 - e^{-\left(\frac{T}{\eta}\right)^\beta}$	$\log_e\{-\log_e[1 - F(T)]\}$	$\log_e T$	$-\beta\log_e \eta$	β
Normal distribution				
$F(T) = \Phi(\frac{T-\mu}{\sigma})^*$	$\Phi^{-1}[F(T)]$	T	$-\frac{\mu}{\sigma}$	$\frac{1}{\sigma}$
Lognormal distribution				
$F(T) = \Phi(\frac{\log_e T - \mu_{T'}}{\sigma_{T'}})^{**}$	$\Phi^{-1}[F(T)]$	$\log_e T$	$-\frac{\mu_{T'}}{\sigma_{T'}}$	$\frac{1}{\sigma_{T'}}$

$^*\Phi(x) = \int_{-\infty}^{x} \frac{1}{\sqrt{2\pi}} e^{-T^2/2} dT.$

$^{**}T' = \log_e T.$

TABLE 18.8 – Linearized form of some nonlinear equations [13, p. 402].

Nonlinear equation	Linearized equation	Linearized variables		
		y	x_1	x_2
$y = a_0 + a_1\sqrt{x}$	$y = a_0 + a_1 x_1$	y	$\sqrt{x}$	
$y = ax^b$	$\log_e y = \log_e a + b\log_e x$	$\log_e y$	$\log_e x$	
$y = a + \frac{b}{x}$	$y = a + bx$	y	$\frac{1}{x}$	
$y = e^{a+bx}$	$\log_e y = a + bx$	$\log_e y$	x	
$e^y = a_0 x^{a_1}$	$y = \log_e a_0 + a_1 \log_e x$	y	$\log_e x$	
$y = a_0(x_1)^{a_1}(x_2)^{a_2}$	$\log_e y = \log_e a_0 + a_1 \log_e x_1$ $+a_2 \log_e x_2$	$\log_e y$	$\log_e x_1$	$\log_e x_2$
$y = a_0 + a_1 x_1$ $+ a_2\sqrt{x_2}$	$y = a_0 + a_1 x_1 + a_2 x_2$	y	x_1	$\sqrt{x_2}$
$y = a_0 e^{a_1 x_1 + a_2 x_2}$	$\log_e y = \log_e a_0 + a_1 x_1$ $+a_2 x_2$	$\log_e y$	x_1	x_2

2. Calculation is easy and straightforward.

3. The correlation coefficient provides a good measure of the goodness-of-fit of the chosen distribution.

4. For some complex distributions, it is difficult and sometimes impossible to implement.

18.4 MATCHING MOMENTS METHOD OF PARAMETER ESTIMATION

18.4.1 PARAMETER ESTIMATION

The matching moments method was first proposed by Karl Pearson around 1900. If x is a random variable with *pdf* $f(x; \theta_1, \theta_2, ..., \theta_k)$, where $(\theta_1, \theta_2, ..., \theta_k)$ are k unknown parameters which need to be estimated, and $x_1, x_2, ..., x_N$ is a random sample of size of N from x, the first k sample moments about the origin would be given by

$$m'_r = \frac{1}{N} \sum_{i=1}^{N} x_i^r, \quad r = 1, 2, ..., k. \tag{18.35}$$

The first k population moments about the origin are given by

$$\mu_r' = E(x^r) = \int_{-\infty}^{\infty} x^r f(x; \theta_1, \theta_2, ..., \theta_k) dx, \quad r = 1, 2, ..., k.$$

(18.36)

The population moments, μ_r', will, in general, be functions of k unknown parameters, θ_i.

Setting

$$\mu_r' = m_r', \quad r = 1, 2, ..., k,$$

(18.37)

and solving Eq. (18.37) simultaneously for $\theta_1, \theta_2, ..., \theta_k$ yields the estimates $\hat{\theta}_1, \hat{\theta}_2, ..., \hat{\theta}_k$ [12, pp. 239-240; 14, pp. 541-544] which are called moments estimators (ME).

EXAMPLE 18–8

Determine the estimators of μ and σ^2 of the normal distribution using the matching moments method.

SOLUTION TO EXAMPLE 18–8

For the normal distribution, the first moment about the origin is given by

$$\mu_1' = \int_{-\infty}^{\infty} \frac{x}{\sigma\sqrt{2\pi}} e^{-\frac{1}{2}(\frac{x-\mu}{\sigma})^2} dx.$$

If we let $z = (x - \mu)/\sigma$, then $x = \mu + \sigma z$, $dx = \sigma dz$, and the preceding equation becomes

$$\mu_1' = \int_{-\infty}^{\infty} \frac{1}{\sqrt{2\pi}} (\mu + \sigma z) e^{-\frac{z^2}{2}} dz,$$

or

$$\mu_1' = \mu \int_{-\infty}^{\infty} \frac{1}{\sqrt{2\pi}} e^{-\frac{z^2}{2}} dz + \sigma \int_{-\infty}^{\infty} \frac{z}{\sqrt{2\pi}} e^{-\frac{z^2}{2}} dz.$$

Since

$$\int_{-\infty}^{\infty} \frac{1}{\sqrt{2\pi}} e^{-\frac{z^2}{2}} dz = 1,$$

and

$$\int_{-\infty}^{\infty} \frac{z}{\sqrt{2\pi}} e^{-\frac{z^2}{2}} dz = \int_{-\infty}^{\infty} \frac{1}{\sqrt{2\pi}} e^{-\frac{z^2}{2}} d(\frac{z^2}{2}) = -\frac{1}{\sqrt{2\pi}} e^{-\frac{z^2}{2}} \Big|_{-\infty}^{\infty} = 0,$$

then $\mu'_1 = \mu$.

The second moment about the origin is given by

$$\mu'_2 = \int_{-\infty}^{\infty} \frac{x^2}{\sigma\sqrt{2\pi}} e^{-\frac{1}{2}(\frac{x-\mu}{\sigma})^2} dx.$$

Letting $z = (x - \mu)/\sigma$, the above equation becomes

$$\mu'_2 = \int_{-\infty}^{\infty} \frac{1}{\sqrt{2\pi}}(\mu + \sigma z)^2 e^{-\frac{z^2}{2}} dz,$$

$$\mu'_2 = \mu^2 \int_{-\infty}^{\infty} \frac{1}{\sqrt{2\pi}} e^{-\frac{z^2}{2}} dz + 2\sigma\mu \int_{-\infty}^{\infty} \frac{z}{\sqrt{2\pi}} e^{-\frac{z^2}{2}} dz$$
$$+ \sigma^2 \int_{-\infty}^{\infty} \frac{z^2}{\sqrt{2\pi}} e^{-\frac{z^2}{2}} dz.$$

Since

$$\int_{-\infty}^{\infty} \frac{1}{\sqrt{2\pi}} e^{-\frac{z^2}{2}} dz = 1,$$

$$\int_{-\infty}^{\infty} \frac{z}{\sqrt{2\pi}} e^{-\frac{z^2}{2}} dz = 0,$$

and

$$\int_{-\infty}^{\infty} \frac{z^2}{\sqrt{2\pi}} e^{-\frac{z^2}{2}} dz = -\frac{1}{\sqrt{2\pi}} \int_{-\infty}^{\infty} z \; d\left(e^{-\frac{z^2}{2}}\right),$$

which, integrating by parts, becomes

$$-\frac{1}{\sqrt{2\pi}} \int_{-\infty}^{\infty} z \; d\left(e^{-\frac{z^2}{2}}\right)$$

$$= -\frac{1}{\sqrt{2\pi}} z e^{-\frac{z^2}{2}} \Big|_{-\infty}^{\infty} + \frac{1}{\sqrt{2\pi}} \int_{-\infty}^{\infty} e^{-\frac{z^2}{2}} dz = 0 + 1 = 1.$$

Then,

$$\mu'_2 = \mu^2 + \sigma^2.$$

Setting μ'_1 and μ'_2 equal to m'_1 and m'_2, respectively, yields

$$\mu = \mu'_1 = \frac{1}{N} \sum_{i=1}^{N} x_i = \bar{x} = m'_1, \qquad (18.38)$$

and

$$\mu_2' = \mu^2 + \sigma^2 = \frac{1}{N}\sum_{i=1}^{N} x_i^2 = m_2',$$ (18.39)

which gives the solution

$$\hat{\mu} = \frac{1}{N}\sum_{i=1}^{N} x_i = \bar{x},$$ (18.40)

and

$$\hat{\sigma}^2 = \frac{1}{N}\sum_{i=1}^{N} x_i^2 - \hat{\mu}^2 = \frac{1}{N}\sum_{i=1}^{N} x_i^2 - \bar{x}^2,$$

$$\hat{\sigma}^2 = \frac{1}{N}\left(\sum_{i=1}^{N} x_i^2 - N\bar{x}^2\right),$$

or

$$\hat{\sigma}^2 = \frac{1}{N}\sum_{i=1}^{N}(x_i - \bar{x})^2.$$ (18.41)

EXAMPLE 18–9

Determine the estimators of η and β of the gamma distribution using the matching moments method.

SOLUTION TO EXAMPLE 18–9

The *pdf* of the gamma distribution is

$$f(x) = \frac{\eta^\beta}{\Gamma(\beta)} e^{-\eta x} x^{\beta-1}, \quad 0 \le x < \infty.$$

The first moment about the origin is given by

$$\mu_1' = \int_0^\infty x \frac{\eta^\beta}{\Gamma(\beta)} e^{-\eta x} x^{\beta-1} dx,$$

or

$$\mu_1' = -\frac{1}{\eta}\int_0^\infty \frac{\eta^\beta}{\Gamma(\beta)} x^\beta d(e^{-\eta x}).$$

Integrating the previous expression by parts yields

$$\mu_1' = -\frac{1}{\eta}\frac{\eta^\beta}{\Gamma(\beta)}x^\beta e^{-\eta x}\Big|_0^\infty + \frac{\beta}{\eta}\int_0^\infty \frac{\eta^\beta}{\Gamma(\beta)}e^{-\eta x}x^{\beta-1}dx.$$

Since the first term on the right is zero and in the second term

$$\int_0^\infty \frac{\eta^\beta}{\Gamma(\beta)}e^{-\eta x}x^{\beta-1}dx = 1,$$

then

$$\mu_1' = \frac{\beta}{\eta} = \bar{x}. \tag{18.42}$$

The second moment about the origin is given by

$$\mu_2' = \int_0^\infty x^2 \frac{\eta^\beta}{\Gamma(\beta)}e^{-\eta x}x^{\beta-1}dx,$$

or

$$\mu_2' = \int_0^\infty \frac{\eta^\beta}{\Gamma(\beta)}e^{-\eta x}x^{\beta+1}dx.$$

Integrating by parts yields

$$\mu_2' = -\frac{1}{\eta}\frac{\eta^\beta}{\Gamma(\beta)}e^{-\eta x}x^{\beta+1}\Big|_0^\infty + \frac{\beta+1}{\eta}\int_0^\infty \frac{\eta^\beta}{\Gamma(\beta)}e^{-\eta x}x^\beta dx.$$

Since

$$e^{-\eta x}x^{\beta+1}\Big|_0^\infty = 0,$$

and

$$\int_0^\infty \frac{\eta^\beta}{\Gamma(\beta)}e^{-\eta x}x^\beta dx = \frac{\beta}{\eta},$$

then

$$\mu_2' = \frac{\beta}{\eta^2}(\beta+1).$$

Setting

$$\mu_1' = \frac{\beta}{\eta} = \frac{1}{N}\sum_{i=1}^N x_i = \bar{x},$$

and

$$\mu_2' = \frac{\beta}{\eta^2}(\beta + 1) = \frac{1}{N}\sum_{i=1}^{N} x_i^2, \qquad (18.43)$$

and then substituting Eq. (18.42), or

$$\frac{\beta}{\eta} = \bar{x}$$

into Eq. (18.43) yields

$$\frac{\bar{x}}{\eta} + \bar{x}^2 = \frac{1}{N}\sum_{i=1}^{N} x_i^2,$$

or

$$\frac{\bar{x}}{\eta} = \frac{1}{N}\sum_{i=1}^{N} x_i^2 - \bar{x}^2 = \frac{1}{N}\sum_{i=1}^{N}(x_i - \bar{x})^2.$$

Therefore,

$$\hat{\eta} = \frac{N\bar{x}}{\displaystyle\sum_{i=1}^{N} x_i^2 - N\bar{x}^2} = \frac{N\bar{x}}{\displaystyle\sum_{i=1}^{N}(x_i - \bar{x})^2}, \qquad (18.44)$$

and from Eq. (18.42) we have

$$\hat{\beta} = \hat{\eta}\bar{x}. \qquad (18.45)$$

18.4.2   COMMENTS ON THE MATCHING MOMENTS METHOD OF PARAMETER ESTIMATION (ME)

1. The ME are usually easy to calculate, provided that samples are not censored or truncated.

2. The estimate variances, or the sampling errors, are often unacceptably large if higher order moments are used.

3. The ME only represent part of the whole picture of the sample; therefore, they are not as accurate for skewed distributions.

4. The ME are generally inefficient; consequently, they are not used except in the absence of better estimators.

5. The ME have little theoretical justification, with the result that its appeal is primarily intuitive!!!

18.5   MAXIMUM LIKELIHOOD METHOD OF PARAMETER ESTIMATION

If x is a continuous random variable with *pdf*

$$f(x; \theta_1, \theta_2, ..., \theta_k),$$

or if x is a discrete random variable with *pdf*

$$p(x; \theta_1, \theta_2, ..., \theta_k),$$

where $\theta_1, \theta_2, ..., \theta_k$ are k unknown constant parameters which need to be estimated, conduct an experiment and obtain N independent observations, $x_1, x_2, ..., x_N$, then the likelihood function is given by

$$L(x_1, x_2, ..., x_N | \theta_1, \theta_2, ..., \theta_k) = L = \prod_{i=1}^{N} f(x_i; \theta_1, \theta_2, ..., \theta_k),$$

$$i = 1, 2, ..., N, \tag{18.46}$$

for continuous random variables, and by

$$L(x_1, x_2, ..., x_N | \theta_1, \theta_2, ..., \theta_k) = L = \prod_{i=1}^{N} p(x_i; \theta_1, \theta_2, ..., \theta_k),$$

$$i = 1, 2, ..., N, \tag{18.46'}$$

for discrete random variables.

The logarithmic likelihood function is

$$\log_e L = \sum_{i=1}^{N} \log_e f(x_i; \theta_1, \theta_2, ..., \theta_k) \tag{18.47}$$

for continuous random variables, and

$$\log_e L = \sum_{i=1}^{N} \log_e p(x_i; \theta_1, \theta_2, ..., \theta_k) \tag{18.47'}$$

for discrete random variables. Maximizing Eqs. (18.46) and (18.46') is equivalent to maximizing Eqs. (18.47) and (18.47'), which is easier to work with computationally.

The maximum likelihood estimators (MLE) of $\theta_1, \theta_2, ..., \theta_k$ are the solutions of equations [12, pp. 236-239; 14, pp.544-546; 15, pp. 238-240]

$$\frac{\partial(\log_e L)}{\partial \theta_j} = 0, \quad j = 1, 2, ..., k, \tag{18.48}$$

which are the necessary conditions, and the

$$\frac{\partial^2(\log_e L)}{\partial^2 \theta_j} < 0, \quad j = 1, 2, ..., k \tag{18.49}$$

are the sufficient conditions. Usually Eq. (18.48) yields the desired MLE.

EXAMPLE 18–10

Find the MLE of the two parameters of the two-parameter exponential distribution.

SOLUTION TO EXAMPLE 18–10

If $x_1, x_2, ..., x_N$ are N observations, then the likelihood function for the two-parameter exponential *pdf* is given by

$$L(x_1, x_2, ..., x_N | \lambda, \gamma) = L = \prod_{i=1}^{N} [\lambda e^{-\lambda(x_i - \gamma)}],$$

or

$$L = \begin{cases} \lambda^N e^{-\lambda \sum\limits_{i=1}^{N}(x_i - \gamma)}, & x_i \geq \gamma, \quad i = 1, 2, ..., N, \\ 0, & \text{elsewhere.} \end{cases} \tag{18.50}$$

Taking the natural logarithm of both sides of Eq. (18.50) yields

$$\log_e(L) = N \log_e \lambda - \lambda \sum_{i=1}^{N}(x_i - \gamma). \tag{18.51}$$

Taking the first derivative of Eq. (18.51) yields

$$\frac{\partial(\log_e L)}{\partial \lambda} = \frac{N}{\lambda} - \sum_{i=1}^{N}(x_i - \gamma) = 0, \tag{18.52}$$

from which

$$\hat{\lambda} = \frac{N}{\sum\limits_{i=1}^{N}(x_i - \hat{\gamma})} \neq 0; \tag{18.53}$$

but it requires an estimate of γ, $\hat{\gamma}$.
    Also

$$\frac{\partial(\log_e L)}{\partial \gamma} = 0 - \lambda \sum_{i=1}^{N}(-1) = 0, \tag{18.54}$$

from which

$$\hat{\lambda} = 0!$$ (18.55)

It may be seen that Eqs. (18.53) and (18.55) are incompatible, because no $\hat{\gamma}$ may be found for $\hat{\gamma} \leq x_1$ which satisfies the constraint on γ in Eq. (18.50). This means that there is no global MLE for the two-parameter exponential *pdf*; however, it is possible to find a local MLE for the given sample, which is discussed next.

Consider λ is fixed and $\lambda = \lambda_0$. The maximization of $L(x_1, x_2, ..., x_N | \lambda, \gamma)$, is attained at the value of $\hat{\gamma}$ which minimizes $\sum_{i=1}^{N}(x_i - \gamma)$. Referring to Fig. 6.20, it may be seen that λ is less than any x_i. Hence, the maximum possible value of λ would be x_1. Consequently, when $\hat{\gamma} = x_1$, $\sum_{i=1}^{N}(x_i - \gamma)$ attains the minimum value. Therefore, $\hat{\gamma} = x_1$ is the MLE of γ.

Substituting $\hat{\gamma} = x_1$ into Eq. (18.51), taking the derivative of both sides with respect to λ and setting it equal to 0 yields

$$\frac{\partial(\log_e L)}{\partial \lambda} = \frac{N}{\lambda} - \sum_{i=1}^{N}(x_i - x_1) = 0.$$ (18.56)

Solving Eq. (18.56) for λ yields

$$\hat{\lambda} = \frac{N}{\sum_{i=1}^{N}(x_i - x_1)} = \frac{1}{\frac{1}{N}\sum_{i=1}^{N} x_i - x_1} = \frac{1}{\hat{\bar{T}} - \hat{\gamma}}.$$ (18.57)

If $\gamma = 0$, then

$$\hat{\lambda} = \frac{1}{\hat{\bar{T}}},$$

and

$$\hat{\bar{T}} = \frac{1}{N}\sum_{i=1}^{N} x_i.$$

For the data given in Example 18–7, for example, $\hat{\gamma} = 8.0 \times 100$ hr. With $N = 10$ and the data in Table 18.5, Eq. (18.57) yields

$$\hat{\lambda} = \frac{1}{\frac{119.3 \times 100}{10} - 800} = \frac{1}{1,193 - 800} = \frac{1}{393},$$

or

$$\hat{\lambda} = 0.002545 \text{ fr/hr.}$$

EXAMPLE 18–11

Find the MLE of $\bar{T}$ and σ_T for the normal distribution.

SOLUTION TO EXAMPLE 18–11

The *pdf* of the normal distribution is

$$f(T) = \frac{1}{\sigma_T\sqrt{2\pi}}e^{-\frac{1}{2}(\frac{T-\bar{T}}{\sigma_T})^2}.$$

If $T_1, T_2, ..., T_N$ is a sample from a normal population, then the likelihood function is given by

$$L(T_1, T_2, ..., T_N|\bar{T}, \sigma_T) = L = \prod_{i=1}^{N}\left[\frac{1}{\sigma_T\sqrt{2\pi}}e^{-\frac{1}{2}(\frac{T_i-\bar{T}}{\sigma_T})^2}\right],$$

$$L = \frac{1}{(\sigma_T\sqrt{2\pi})^N}e^{-\frac{1}{2}\sum_{i=1}^{N}(\frac{T_i-\bar{T}}{\sigma_T})^2}, \tag{18.58}$$

then

$$\log_e L = -\frac{N}{2}\log_e(2\pi) - N\log_e\sigma_T - \frac{1}{2}\sum_{i=1}^{N}\left(\frac{T_i - \bar{T}}{\sigma_T}\right)^2. \tag{18.59}$$

Taking the derivative of both sides of Eq. (18.59) with respect to $\bar{T}$ and σ_T, respectively, and setting them equal to zero yields

$$\frac{\partial(\log_e L)}{\partial\bar{T}} = \frac{1}{\sigma_T^2}\sum_{i=1}^{N}(T_i - \bar{T}) = 0, \tag{18.60}$$

and

$$\frac{\partial(\log_e L)}{\partial\sigma_T} = -\frac{N}{\sigma_T} + \frac{1}{\sigma_T^3}\sum_{i=1}^{N}(T_i - \bar{T})^2 = 0. \tag{18.61}$$

Solving Eqs. (18.60) and (18.61) simultaneously yields

$$\hat{\bar{T}} = \frac{1}{N}\sum_{i=1}^{N}T_i, \tag{18.62}$$

and

$$\hat{\sigma}_T^2 = \frac{1}{N} \sum_{i=1}^{N} (T_i - \bar{T})^2. \qquad (18.63)$$

EXAMPLE 18–12

Determine the MLE for parameter c of the Poisson distribution.

SOLUTION TO EXAMPLE 18–12

The Poisson *pdf* is

$$p(x) = \frac{c^x e^{-c}}{x!}, \quad x = 0, 1, 2, ...,$$

and the corresponding likelihood function is

$$L(x_1, x_2, ..., x_N | c) = \prod_{i=1}^{N} \frac{c^{x_i} e^{-c}}{x_i!},$$

or

$$L(x_1, x_2, ..., x_N | c) = L = \frac{c^{\sum_{i=1}^{N} x_i} e^{-Nc}}{\prod_{i=1}^{N} (x_i!)}. \qquad (18.64)$$

Taking the natural logarithm of both sides of Eq. (18.64) yields

$$\log_e L = (\log_e c) \sum_{i=1}^{N} x_i - Nc - \log_e [\prod_{i=1}^{N} (x_i!)]. \qquad (18.65)$$

Setting the derivative of $\log_e L$ with respect to c equal to zero yields

$$\frac{1}{c} \sum_{i=1}^{N} x_i - N = 0,$$

or

$$\hat{c} = \frac{1}{N} \sum_{i=1}^{N} x_i. \qquad (18.66)$$

18.5.1 COMMENTS ON THE MLE

1. The MLE are consistent and asymptotically efficient.

2. The probability distribution of the estimator is asymptotically normal.

3. The MLE of the Weibull and of the gamma *pdf's* parameters are badly behaved in the vicinity of the transition point; i.e., the point ($\alpha_3 = 2$ or $\beta = 1$) where the shape changes from bell-shaped ($\alpha_3 \leq 2$) to reverse J-shaped ($\alpha_3 \geq 2$).

4. In the three-parameter lognormal distribution, it is necessary to settle for local MLEs, because the global estimators lead to inadmissible estimates.

5. When the location parameter is known, the MLE in the resulting distributions are relatively free from computational problems. However, introduction of the threshold (or location) parameter creates complications, so, the MLE are sometimes difficult to calculate.

6. It is not considered advisable to employ the MLE for the three-parameter Weibull distribution unless there is reason to expect that $\beta \geq 2.2$, because of the following reasons:

   (a) Computational problems arise when β is near 1.
   (b) The usual asymptotic properties of MLE do not hold unless $\beta \geq 2$.

7. For some parameter combinations the MLE break down and fail to exist.

8. The MLE are good for one- and two-parameter distributions.

9. The MLE should be used with caution for three-or-more-parameter distributions.

18.6 MODIFIED MOMENTS METHOD OF PARAMETER ESTIMATION

18.6.1 INTRODUCTION

In an effort to alleviate problems encountered with the ME and MLE, especially for skewed distributions, various modifications of these estimates have from time to time been proposed. Kao (1959) [16], Dubey

(1966) [17], Wycoff, Bain, and Engelhardt (1980) [18], Zanakis (1977, 1979 a, b) [19; 20; 21], and Zanakis and Mann (1981) [22] proposed estimators based on percentiles. Cohen and Whitten (1980, 1982, 1984, 1985, 1986, 1990) [23; 24; 25; 26; 27; 28; 29] and Chan (1982) [30] proposed several modifications of MLE and ME which employ the first-order statistic. Although several of the various modified estimators were reasonably satisfactory, primary consideration here is limited to a modification of the moments estimators (MME) in which the sample first-order statistic is introduced. It has been proven and will be shown here that the modified moments estimators (MME) are unbiased with respect to both the mean and the variance; their performance is particularly good when the skewness is large, in contrast to the MLE, which are most unsatisfactory in this case; the first-order statistic as the key of MME contains more information about the threshold (or location) parameter than do any of the other sample observations and often more than all of the other observations combined.

18.6.2   THE FIRST-ORDER STATISTIC OF THE SAMPLE

Let $x_1 \leq x_2 \leq x_3 \leq ... \leq x_N$ be an ordered random sample of size N from a distribution with *pdf* $f(x; \theta_1, \theta_2, ..., \theta_N)$, where $(\theta_1, \theta_2, ..., \theta_N)$ are k unknown parameters which need to be estimated from the sample. Let X_1 be the first-order statistic (a random variable) of this random sample and $F(X_1)$ be the *cdf* of X_1. Then x_1 is the corresponding sample value of X_1 and $F(x_1)$ is the corresponding sample value of $F(X_1)$.

It may be proven that

$$E(X_1) = x_1, \tag{18.67}$$

and

$$E[F(X_1)] = F(x_1) = \frac{1}{1+N}, \tag{18.68}$$

where symbol $E(\cdot)$ designates expected value. These two equations are extremely important in the MME method.

For distributions which possess reproductive properties, such as the Weibull, exponential, Rayleigh and so on, the first-order statistic, X_1, has the same type of distribution as X, but with different parameters.

18.6.3   PARAMETER ESTIMATION

Since the MME was proposed mainly to overcome the disadvantages of ME and MLE for skewed distributions, the first-order statistic was introduced to give more information about the threshold parameters.

Therefore, the MME are used mainly for (highly) skewed (not normal) distributions with location parameters.

For two-parameter distributions (with location parameter), the first moment about the origin and the first-order statistic are employed to establish two equations for solving (estimating) the two unknown parameters. For three-parameter distributions (with location parameter), the first moment about the origin, the second moment about the mean and the first-order statistic are employed to establish three equations for solving (estimating) the three unknown parameters. The equations for estimating the parameters of the two-parameter exponential, three-parameter Weibull, three-parameter lognormal, and the three-parameter gamma distributions are summarized in the following:

For the **two-parameter exponential distribution** with *pdf*

$$f(x) = \begin{cases} \lambda\,e^{-\lambda(x-\gamma)}, & x > \gamma, \\ 0, & x \le \gamma, \end{cases} \tag{18.69}$$

the equations used for the MME are

$$\begin{cases} \mu_1' & = m_1' = \bar{x}, \\ E(X_1) & = x_1, \end{cases} \tag{18.70}$$

or

$$\begin{cases} \gamma + \frac{1}{\lambda} & = \frac{1}{N}\sum_{i=1}^{N} x_i = \bar{x}, \\ \gamma + \frac{1}{N\lambda} & = x_1. \end{cases} \tag{18.71}$$

Then,

$$\begin{cases} \hat{\lambda} & = \frac{N-1}{N(\bar{x}-x_1)}, \\ \hat{\gamma} & = \frac{Nx_1-\bar{x}}{N-1}. \end{cases} \tag{18.72}$$

It has been proven that these two estimators are both best linear unbiased (BLUE) and minimum variance unbiased (MVUE) [29] estimators.

For the **three-parameter Weibull distribution** with *pdf*

$$f(x) = \begin{cases} \frac{\beta}{\eta}(\frac{x-\gamma}{\eta})^{\beta-1}e^{-(\frac{x-\gamma}{\eta})^{\beta}}, & x > \gamma, \\ 0, & x \le \gamma, \end{cases} \tag{18.73}$$

the equations used for the MME are

$$\begin{cases} \mu_1' & = m_1' = \bar{x}, \\ \mu_2 & = m_2, \\ E(X_1) & = x_1, \end{cases} \tag{18.74}$$

or

$$\begin{cases} \gamma + \eta\Gamma(\frac{1}{\beta}+1) & = \frac{1}{N}\sum_{i=1}^{N} x_i = \bar{x}, \\ \eta^2[\Gamma(\frac{2}{\beta}+1) - \Gamma^2(\frac{1}{\beta}+1)] & = \frac{1}{N-1}\sum_{i=1}^{N}(x_i - \bar{x})^2 = m_2, \\ \gamma + \frac{\eta}{N^{1/\beta}}\Gamma(\frac{1}{\beta}+1) & = x_1. \end{cases} \quad (18.75)$$

Then

$$\begin{cases} W(\hat{\beta}) & = \frac{m_2}{(\bar{x}-x_1)^2}, \\ \hat{\eta} & = \sqrt{\frac{m_2}{\Gamma(\frac{2}{\beta}+1)-\Gamma^2(\frac{1}{\beta}+1)}}, \\ \hat{\gamma} & = \bar{x} - \hat{\eta}\Gamma(\frac{1}{\beta}+1), \end{cases} \quad (18.76)$$

where

$$W(\hat{\beta}) = \frac{\Gamma(\frac{2}{\beta}+1) - \Gamma^2(\frac{1}{\beta}+1)}{[(1 - N^{-\frac{1}{\beta}})\Gamma(\frac{1}{\beta}+1)]^2}, \quad (18.77)$$

which can be evaluated either by tables or charts [29] or by computer programs (numerical integration).

It should be pointed out that although the MME are applicable over the entire parameter space, they are most useful when the skewness is large. When the skewness is small, the MLE or the ME might be preferred. For very small values of α_3, it might be better to replace the Weibull distribution, as a model, with the normal distribution, in which case the MLE and the ME are identical.

For the **three-parameter lognormal distribution** with *pdf*

$$f(x) = \begin{cases} \frac{1}{\sqrt{2\pi}\sigma_{x'}(x-\gamma)}e^{-\frac{1}{2}(\frac{\log_e(x-\gamma)-\mu_{x'}}{\sigma_{x'}})^2}, & x > \gamma, \\ 0, & x \leq \gamma, \end{cases} \quad (18.78)$$

the equations used for the MME are

$$\begin{cases} \mu_1' & = m_1' = \bar{x}, \\ \mu_2 & = m_2, \\ E[\log_e(X_1 - \gamma)] & = \log_e(x_1 - \gamma), \end{cases} \quad (18.79)$$

or

$$\begin{cases} \gamma + \xi\omega^{\frac{1}{2}} & = \frac{1}{N}\sum_{i=1}^{N} x_i = \bar{x}, \\ \xi^2\omega(\omega - 1) & = \frac{1}{N-1}\sum_{i=1}^{N}(x_i - \bar{x})^2 = m_2, \\ \gamma + \xi\, e^{\sqrt{\log_e \omega}\cdot E(Z_{1,N})} & = x_1, \end{cases} \quad (18.80)$$

where

$$\xi = e^{\mu_{x'}}, \tag{18.81}$$

$$\omega = e^{\sigma^2_{x'}}, \tag{18.82}$$

and

$E(Z_{1,N})$ = expected value of the first-order statistic in a random sample of size N from a standardized normal distribution $N(0,1)$, which can be obtained either from associated tables [29] or by a computer program using Monte Carlo simulation.

Then

$$\begin{cases} J(\hat{\omega}, N) &= \frac{m_2}{(\bar{x}-x_1)^2}, \\ \hat{\xi} &= \sqrt{\frac{m_2}{\hat{\omega}(\hat{\omega}-1)}}, \\ \hat{\gamma} &= \bar{x} - \sqrt{\frac{m_2}{\hat{\omega}-1}}, \end{cases} \tag{18.83}$$

where

$$J(\hat{\omega}, N) = \frac{\hat{\omega}(\hat{\omega} - 1)}{[\sqrt{\hat{\omega}} - e^{\sqrt{\log_e \hat{\omega}} \cdot E(Z_{1,N})}]^2}, \tag{18.84}$$

which can be evaluated either by tables or charts [29], or by computer programs.

Therefore, $\hat{\mu}_{x'}$ and $\hat{\sigma}_{x'}$ can be obtained by substituting Eq. (18.83) into Eqs. (18.81) and (18.82) as follows:

$$\begin{cases} \hat{\mu}_{x'} &= \log_e \hat{\xi}, \\ \hat{\sigma}_{x'} &= \sqrt{\log_e \hat{\omega}}. \end{cases} \tag{18.85}$$

In many practical applications, the MME are the preferred estimators. These estimators are unbiased with respect to the population mean and variance. They are easy to calculate, and their variances are minimal or at least near minimal. They do not suffer from regularity problems.

For the **three-parameter gamma distribution** with *pdf*

$$f(x) = \begin{cases} \frac{1}{\eta \Gamma(\beta)} (\frac{x-\gamma}{\eta})^{\beta-1} e^{-(\frac{x-\gamma}{\eta})}, & x > \gamma, \\ 0, & x \le \gamma, \end{cases} \tag{18.86}$$

the equations used for the MME are

$$\begin{cases} \mu_1' & = & m_1' = \bar{x}, \\ \mu_2 & = & m_2, \\ E[F(X_1)] & = & F(x_1), \end{cases} \qquad (18.87)$$

or

$$\begin{cases} \gamma + \beta\eta & = & \frac{1}{N}\sum_{i=1}^{N} x_i = \bar{x}, \\ \beta\eta & = & \frac{1}{N-1}\sum_{i=1}^{N}(x_i - \bar{x})^2 = m_2, \\ \int_{-\frac{2}{\alpha_3}}^{z_1} g(t;0,1,\alpha_3)dt & = & \frac{1}{N+1}, \end{cases} \qquad (18.88)$$

where

$$g(t;0,1,\alpha_3) = \begin{cases} (\frac{2}{\alpha_3})^{\frac{4}{\alpha_3^2}}(t+\frac{2}{\alpha_3})^{\frac{4}{\alpha_3^2}-1}e^{-\frac{2}{\alpha_3}(t+\frac{2}{\alpha_3})}, & t > -\frac{2}{\alpha_3}, \\ 0, & t \le -\frac{2}{\alpha_3}, \end{cases}$$

$$(18.89)$$

and

$$z_1 = \frac{x_1 - \bar{x}}{\sqrt{m_2}}. \qquad (18.90)$$

Then

$$\begin{cases} \hat{\eta} & = & \frac{\sqrt{m_2}\,\hat{\alpha}_3}{2}, \\ \hat{\gamma} & = & \bar{x} - \frac{2\sqrt{m_2}}{\hat{\alpha}_3}, \\ \hat{\beta} & = & \frac{4}{\hat{\alpha}_3^2}. \end{cases} \qquad (18.91)$$

The MME given in Eq. (18.91) are unbiased with respect to the population mean and variance. They are applicable over the entire parameter space, and their estimate variances are minimal or at least nearly minimal in comparison with the corresponding variances of the ME and the MLE. Notice that to get $\hat{\eta}$, $\hat{\gamma}$ and $\hat{\beta}$ in Eq. (18.91), the third equation of Eq. (18.88) should be solved first for $\hat{\alpha}_3$ by inverse interpolation in tables of the standardized gamma *cdf*, or by referring to the associated tables or figures of Cohen [29].

EXAMPLE 18–13

Rework Example 18–7 using the modified moments method and compare the results with those obtained in Example 18–7 by the least

TABLE 18.9– **Parameter estimates for two-parameter exponential distribution using LSE, MLE and MME.**

Parameters	Methods		
	LSE	MLE	MME
$\hat{\lambda}$, fr/hr	0.002611	0.002545	0.002290
$\hat{\gamma}$, hr	839	800	756

squares method, and those in Example 18–10 by the maximum likelihood method.

SOLUTION TO EXAMPLE 18–13

From Example 18–7 we know that $T_1 = 800$ hr, $N = 10$, and

$$\bar{T} = \frac{1}{N} \sum_{i=1}^{N} T_i = \frac{1}{10} \sum_{i=1}^{10} T_i = 1,193 \text{ hr.}$$

Substituting T_1, N and $\bar{T}$ into Eq. (18.72) yields

$$\hat{\lambda} = \frac{N-1}{N(\bar{T} - T_1)} = \frac{10-1}{10(1,193 - 800)},$$

or

$$\hat{\lambda} = 0.002290 \text{ fr/hr,}$$

and

$$\hat{\gamma} = \frac{NT_1 - \bar{T}}{N-1} = \frac{10 \times 800 - 1,193}{10 - 1},$$

or

$$\hat{\gamma} = 756 \text{ hr.}$$

A comparison is made in Table 18.9 on the difference between the MME, LSE and MLE. It may be seen that the MME values for both $\hat{\lambda}$ and $\hat{\gamma}$ are lower than the LSE and MLE values. The MME are preferred not only because they are quite easy to calculate, but more importantly, because they are both the best linear unbiased (BLUE) and the minimum variance unbiased (MVUE) estimates.

EXAMPLE 18–14

The maximum flood levels, in millions of cubic feet per second, for the Susquehanna River at Harrisburg, Pennsylvania, over 20 four-year periods from 1890 to 1969 are 0.654, 0.613, 0.315, 0.449, 0.297, 0.402, 0.379, 0.423, 0.379, 0.3235, 0.269, 0.740, 0.418, 0.412, 0.494, 0.416, 0.338, 0.392, 0.484 and 0.265. Assuming these data are from a gamma distribution, find the corresponding distribution parameters using the modified moments method (MME). Compare the results with those obtained by the matching moments method (ME) and the maximum likelihood method (MLE).

SOLUTION TO EXAMPLE 18–14

For this example, $N = 20, x_1 = 0.265$,

$$\bar{x} = \frac{1}{N} \sum_{i=1}^{N} x_i = \frac{1}{20} \sum_{i=1}^{20} x_i = 0.423125,$$

$$m_2 = \frac{1}{N-1} \sum_{i=1}^{N} (x_i - \bar{x})^2 = \frac{1}{20-1} \sum_{i=1}^{20} (x_i - 0.423125)^2,$$

or

$$m_2 = 0.0156948,$$

and

$$z_1 = \frac{x_1 - \bar{x}}{\sqrt{m_2}} = \frac{0.265 - 0.423125}{\sqrt{0.0156948}} = -1.262.$$

From Fig. 6.2 or Table 6.3, of [29], with $N = 20$ and $z_1 = -1.262$, we read $\hat{\alpha}_3 = 11.174$. Substituting $\hat{\alpha}_3 = 1.174$ into Eqs. (18.91) yields

$$\hat{\eta} = 0.0735,$$
$$\hat{\gamma} = 0.210,$$

and

$$\hat{\beta} = 2.902.$$

Table 18.10 gives a comparison of the estimation results by ME, MLE and MME. The detailed computations for the ME and MLE are omitted here. The readers are encouraged to verify these on their own, as an exercise.

It may be seen that the MME are different from the ME and the MLE, and the estimate variance of the MME is minimal in comparison with (at least not larger than) those of the ME and the MLE. Therefore, the MME are preferable.

TABLE 18.10– Parameter estimates for the three-parameter gamma distribution using the ME, MLE and MME for Example 18–14.

Parameters	Methods		
	ME	MLE	MME
$\hat{\eta}$	0.0669	0.1343	0.0735
$\hat{\gamma}$	0.1882	0.2627	0.210
$\hat{\beta}$	3.5113	1.1940	2.902
$\hat{E}(X)$	0.4231	0.4231	0.4231
$\sqrt{\hat{V}(X)}$	0.1253	0.1468	0.1253
$\hat{\alpha}_3(X)$	1.0673	1.8303	1.174

18.6.4   COMMENTS ON THE MME

1. For skewed distributions, the first-order statistic contains more information concerning the threshold (or location or origin) parameter than any other sample observation, and often more than all of the other observations combined.

2. The MME retain most of the desirable, while eliminating most of the undesirable, properties of both the MLE and the ME.

3. The MME are unbiased with respect to distribution means and variances.

4. The MME apply over the entire parameter space, and with the aid of tables and charts, they are easy to calculate from the sample data.

5. The MME provide an *attractive* alternative to both the MLE and the ME.

PROBLEMS

18-1. Using the data in Table 18.11 do the following:

(1) Fit a three-parameter Weibull distribution to the grouped data using the (1) matching moments, (2) least squares, and (3) modified moments methods, and write down the corresponding probability density functions, respectively. Assume $\hat{\gamma} = 500$ hr.

(2) Calculate, tabulate and plot the probability density function using the results of the least squares method.

(3) Write down the failure frequency distribution function per class interval width, as in Case 2.

(4) Calculate and tabulate the failure frequency distribution function per class interval width and superimpose it on the number failing versus the time-to-failure histogram, as in Case 2. Comment on the goodness of fit.

(5) Calculate, tabulate and plot the reliability function, as in Case 2.

(6) Calculate, tabulate and plot the failure rate function, as in Case 2.

TABLE 18.11 – Field data for Problem 18-1.

Life in 100 hr, starting at 500 hr	Number of failures, N_F
0 – 1	40
1 – 2	210
2 – 3	300
3 – 4	250
4 – 5	80
5 – 6	20

Total = 900

18-2. Fifteen items were put to a 1,600-hr life test and 10 failures were observed at 500, 670, 800, 920, 1,030, 1,140, 1,250, 1,350, 1,460 and 1560 hr. Do the following:

(1) Fit a two-parameter Weibull distribution to the data using the (1) matching moments, (2) least squares, and (3) modified moments methods, and write down the corresponding probability density functions, respectively.

(2) Calculate, tabulate and plot the probability density function, using the results of the matching moments method.

(3) Calculate, tabulate and plot the reliability function, as in Case 2.

(4) Calculate, tabulate and plot the failure rate function, as in Case 2.

18-3. In a reliability test 10 identical units are tested for a duration of 200 hr. Six of the units fail after operating the following number of hours: 93, 34, 16, 120, 53 and 75. Using the method of MLE, estimate the Weibull parameters. Assume $\gamma = 0$.

18-4. Maximum flood levels, in millions of cubic feet per second, for the Susquehanna River at Harrisburg, Pennsylvania, over 20 four-year periods from 1890 to 1969 are 0.654, 0.613, 0.315, 0.449, 0.297, 0.402, 0.379, 0.423, 0.379, 0.324, 0.269, 0.740, 0.418, 0.412, 0.494, 0.416, 0.338, 0.392, 0.484 and 0.265. Using the least squares, matching moments and modified moments methods, find the parameters of the following *pdf*'s:

(1) Weibull with the *pdf* of

$$f(x) = \frac{\beta}{\eta}(\frac{x - \gamma}{\eta})^{\beta - 1} e^{-(\frac{x-\gamma}{\eta})^\beta}.$$

(2) Extreme value of the maxima with the *pdf* of

$$f_{max}(x) = \frac{1}{\eta_{max}} e^{-(\frac{x-\gamma_{max}}{\eta_{max}}) - e^{-(\frac{x-\gamma_{max}}{\eta_{max}})}}.$$

18-5. The fatigue lives, in hours, of 10 bearings of a particular type were 152.7, 172.0, 172.5, 173.3, 193.0, 204.7, 216.5, 234.9, 262.6 and 422.6. Assuming these data are from a Weibull population, with the *pdf* of

$$f(T) = \frac{\beta}{\eta}(\frac{T - \gamma}{\eta})^{\beta - 1} e^{-(\frac{T-\gamma}{\eta})^\beta},$$

find the distribution parameters using the least squares, matching moments and the modified moments methods.

18-6. Given in Table 18.12 are 110 miles-to-failure data of one brand of passenger car tires in 100 miles. From past experience they can be described by a normal distribution with the *pdf* of

$$f(T) = \frac{1}{\sigma_T \sqrt{2\pi}} e^{-\frac{1}{2}(\frac{T-\overline{T}}{\sigma_T})^2}.$$

(1) Find the parameters using the matching moments method.
(2) Find the parameters using the maximum likelihood method.

18-7. Given are the times-to-failure data of Table 18.13, for 10 identical units, all tested to failure. Assuming they are from a Weibull population with the *pdf* of

$$f(T) = \frac{\beta}{\eta}(\frac{T - \gamma}{\eta})^{\beta - 1} e^{-(\frac{T-\gamma}{\eta})^\beta},$$

TABLE 18.12 — Data of miles to failure, in 100 miles, for 110 of one brand of passenger car tires for Problem 18-6.

390	393	395	405	420	376	381	381	383	401
380	387	395	397	407	377	383	387	390	393
393	395	403	405	414	376	388	395	397	400
387	400	400	403	410	391	392	394	397	405
390	391	395	401	405	379	391	393	394	410
390	397	400	406	428	380	382	389	391	399
375	383	392	395	404	387	390	398	400	408
390	395	395	397	403	382	399	401	406	406
390	395	395	400	410	381	390	394	397	399
387	389	398	401	415	372	378	396	400	405
387	389	391	391	400	376	380	391	406	412

find the parameters using the following methods:

(1) Probability plotting.
(2) Matching moments.
(3) Least squares.
(4) Modified moments.

Discuss comparatively these four methods, and give your preference and the reasons for it.

18-8. Given are the data in Table 18.14 from a lognormal population with the *pdf* of

$$f(x) = \frac{1}{\sigma_{x'}(x - \gamma)\sqrt{2\pi}} \, e^{-\frac{1}{2}(\frac{\log_e(x-\gamma)-\mu_{x'}}{\sigma_{x'}})^2}.$$

Find the distribution parameters using the following methods:

(1) Probability plotting.
(2) Least squares.
(3) Matching moments.
(4) Modified moments.
(5) Maximum likelihood.

TABLE 18.13 – Times-to-failure data, in hours, for 10 identical units, all tested to failure for Problem 18–7.

j	T_j, hr
1	200
2	370
3	500
4	620
5	730
6	840
7	950
8	1,050
9	1,160
10	1,400

Discuss comparatively the results of Cases (1) through (5), give your preference and the reasons for it.

18-9. A set of reliability test data for a special equipment, was obtained as given in Table 18.15. Assume that these data are Rayleigh distributed with the *pdf* of

$$f(T) = \frac{T}{\eta^2} \, e^{-\frac{1}{2}(\frac{T}{\eta})^2}.$$

Find the parameter of this *pdf* using the following methods:

(1) Probability plotting.

(2) Least squares.

(3) Matching moments.

(4) Maximum likelihood.

Discuss comparatively the results obtained, give your preference and your reasons for it.

18-10. A set of times-to-failure data is given in Table 18.16. These data are from a uniform population with the *pdf* of

$$f(T) = \begin{cases} \frac{1}{\beta-\alpha}, & \alpha \le T \le \beta, \\ 0, & \text{elsewhere.} \end{cases}$$

Find the parameters using the following methods:

**TABLE 18.14 – Times-to-failure data for Problem
18-8 and sample size of $N = 15$.**

Failure order number, j	Time to failure, T_j, hr
1	75.0
2	110.3
3	120.4
4	140.9
5	169.3
6	176.2
7	207.2
8	231.0
9	241.9
10	283.0
11	299.0
12	357.0
13	382.0
14	492.7
15	660.6

TABLE 18.15 – **Times-to-failure data and corresponding median ranks and reliability for the determination of the Rayleigh parameter for Problem 18-9.**

Number of failures, j	Time to failure T_j, hr	Median rank, %	Reliability, R_j, %
1	8.3	5.19	94.81
2	13.0	12.58	87.42
3	16.9	20.05	79.96
4	20.1	27.53	72.47
5	23.8	35.02	64.98
6	26.5	42.51	57.49
7	29.8	50.00	50.00
8	33.2	57.49	42.51
9	36.5	64.98	35.02
10	41.0	72.47	27.53
11	45.1	79.96	20.05
12	51.7	87.42	12.58
13	61.3	94.81	5.19

TABLE 18.16 – Times-to-failure data and corresponding median ranks for the determination of the parameters of the uniform distribution for Problem 18–10.

Number of failures, j	Time to failure, T_j, hr	Median rank, %
1	90.0	3.78
2	92.5	9.15
3	115.0	14.58
4	119.0	20.02
5	125.5	25.47
6	134.9	30.92
7	161.0	36.37
8	167.5	41.82
9	170.0	47.27
10	182.0	52.73
11	204.0	58.18
12	208.5	63.63
13	217.5	69.08
14	235.0	74.53
15	240.5	79.98
16	254.0	85.42
17	272.0	90.85
18	275.0	96.22

(1) Least squares.
(2) Matching moments.
(3) Maximum likelihood.

Discuss comparatively the results obtained, give your preference and your reasons for it.

18-11. Twentyfive specimens were tensile tested. Their tensile strengths are listed in Table 18.17. Assuming the breaking strength, S_b, is represented by the EVD of the minima with the *pdf* of

$$f_{min}(S_b) = \frac{1}{\eta_{min}} \ e^{\frac{S_b-\gamma_{min}}{\eta_{min}} -e^{\frac{S_b-\gamma_{min}}{\eta_{min}}}},$$

find the parameters using the following methods:

(1) Least squares.

(2) Matching moments.

(3) Maximum likelihood.

Discuss comparatively these three methods and give your preference and the reasons for it.

18-12. Given the times-to-failure data, in hours, of 16, 34, 53, 75, 93 and 120, obtained by testing a sample of six units all tested to failure, find the parameters of the two-parameter Weibull *pdf* using the following methods:

(1) Least squares.

(2) Matching moments.

(3) MLE.

18-13. Given the times-to-failure data, in hours, of 10,125; 11,260; 12,080; 12,825; 13,550 and 14,670, obtained by testing a sample of six units, all tested to failure, find the parameters of the normal *pdf* using the following methods:

(1) Least squares.

(2) Matching moments.

(3) MLE.

18-14. Given the times-to-failure data, in hours, of 30.4, 36.7, 53.5, 58.5, 74.0, 99.3, 114.3, 140.1 and 257.9, obtained by testing a sample of nine units all tested to failure, find the parameters of the lognormal *pdf* using the following methods:

(1) Least squares.

(2) Matching moments.

(3) MLE.

18-15. Given the times-to-failure data, in hours, of 450; 760; 1,200; 1,590 and 2,210, obtained by testing five units to failure, find the parameters of the two-parameter exponential *pdf* using the following methods:

(1) Least squares.

(2) Matching moments.

(3) MLE.

TABLE 18.17 – The breaking strengths of 25 specimens for Problem 18–11.

Specimen number, j	Breaking strength, S_j, kg/cm^2
1	88.40
2	90.70
3	94.10
4	95.02
5	97.00
6	97.20
7	97.50
8	98.30
9	98.90
10	99.50
11	99.90
12	100.40
13	100.82
14	101.30
15	101.70
16	102.11
17	102.50
18	102.90
19	103.39
20	103.80
21	104.30
22	104.81
23	105.50
24	106.15
25	107.30

REFERENCES

1. Cohen, A. C., *Truncated and Censored Samples*, Marcel Dekker, Inc., New York, 312 pp., 1991.

2. Lipson, Charles and Seth, Narendra J., *Statistical Design and Analysis of Engineering Experiments*, McGraw-Hill Book Company, New York, 518 pp., 1973.

3. Natrella, M. G., *Experimental Statistics*, National Bureau of Standards Handbook, Vol. 91, 1963.

4. Dixon, W. J. and Massey F. J. Jr., *Introduction to Statistical Analysis*, 2nd edition, McGraw-Hill Book Company, New York, 1957.

5. Grubbs, F. E., "Sample Criteria for Testing Outlying Observations," *Annals of Mathematical Statistics*, Vol. 21, 1958.

6. Nelson, W. and Schmee J., "Prediction Limits for the Last Failure Time of a (Log)Normal Sample from Early Failures," *IEEE Transactions on Reliability*, Vol. R-30, No. 5, pp. 461-463, December 1981.

7. Nelson, W., *Applied Life Data Analysis*, John Wiley & Sons, Inc., New York, 634 pp., 1982.

8. Lawless, J. F., "A Prediction Problem Concerning Samples from the Exponential Distribution, with Application in Life Testing," *Technometrics*, Vol. 13, No. 4, November 1971.

9. Hawkins, D. M., *Identification of Outliers*, Chapman and Hall Ltd., 188 pp., 1980.

10. Neter, J. and Wasserman, W., *Applied Linear Statistical Models*, Richard D. Irwin, Inc., Homewood, 842 pp., 1977.

11. Guttman, I., *Linear Models: An Introduction*, John Wiley & Sons, Inc., New York, 358 pp. 1982.

12. Hines, W.W. and Montgomery, D.C., *Probability and Statistics in Engineering and Management Science*, (Second edition), John Wiley & Sons, Inc., New York, 634 pp., 1980.

13. Lipson, C. and Sheth, N.J., *Statistical Design and Analysis of Engineering Experiments*, McGraw-Hill Book Company, New York, 518 pp., 1973.

14. Goldberg, M.A., *An Introduction to Probability Theory with Statistical Applications*, Plenum Press, New York, 662 pp., 1984.

15. Neter, J., Wasserman, W. and Whitmore, G.A., *Applied Statistics*, (Second edition), Allyn and Bacon, Inc., Boston, 773 pp., 1982.

16. Kao, J.H.K., "A Graphical Estimation of Mixed Weibull Parameters in Life Testing of Electronic Tubes," *Technometrics*, Vol. 1, pp. 389-407, 1959.

17. Dubey, S. D., "Hyper-efficient Estimator of the Location Parameter of the Weibull Laws," *Naval Research Logistics Quarterly*, Vol. 13, pp. 253-263, 1966.

18. Wycoff, J., Bain, L. and Engelhardt, M., "Some Complete and Censored Sampling Results for the Three-parameter Weibull Distribution," *Journal of Statistics – Computation and Simulation*, Vol. 11, pp. 139-151, 1980.

19. Zanakis, S. H., "Computational Experience with Some Nonlinear Optimization Algorithms in Deriving MLE for the Three-parameter Weibull Distribution," In *Algorithmic methods in probability: TIMS studies in Management Science*, Vol. 7, North-Holland, Amsterdam, 1977.

20. Zanakis, S. H., "Monte Carlo Study of Some Simple Estimators of the Three-parameter Weibull Distribution," *Journal of Statistics – Computation and Simulation*, Vol. 9, pp. 101-116, 1979a.

21. Zanakis, S. H., "Extended Pattern Search with Transformations for the Three-parameter Weibull MLE Problem," *Management Science*, Vol. 25, pp. 1149-1161, 1979b.

22. Zanakis, S. H. and Mann, N. R., "A Good Simple Percentile Estimator of the Weibull Shape Parameter for Use When All Three Parameters are Unknown," Unpublished manuscript, 1981.

23. Cohn, A. C. and Whitten, B. J., "Estimation in the Three-parameter Lognormal Distribution," *Journal of the American Statistical Association*, Vol. 75, pp. 399-404, 1980.

24. Cohn, A. C. and Whitten, B. J., "Modified Moment and Maximum Likelihood Estimators for Parameters of the Three-parameter Gamma Distribution," *Communications in Statistics – Simulation and Computation*, Vol. 11, pp. 197-216, 1982.

25. Cohn, A. C., Whitten, B. J. and Ding, Y., "Modified Moment Estimation for the Three-parameter Weibull Distribution," *Journal of Quality Technology*, Vol. 16, pp. 159-167, 1984.

26. Cohn, A. C., Whitten, B. J. and Ding, Y., "Modified Moment Estimation for the Three-parameter Lognormal Distribution," *Journal of Quality Technology*, Vol. 17, pp. 92-99, 1985.

27. Cohn, A. C. and Whitten, B. J., "Modified Moment Estimation for the Three-parameter Inverse Gaussian Distribution," *Journal of Quality Technology*, Vol. 17, pp. 147-154, 1985.

28. Cohn, A. C. and Whitten, B. J., "Modified Moment Estimation for the Three-parameter Gamma Distribution," *Journal of Quality Technology*, Vol. 18, pp. 53-62, 1986.

29. Cohen, A. C., *Parameter Estimation*, Marcel Dekker, Inc., New York, 400 pp., 1990.

30. Chan, Micah Y., "Modified Moment and Maximum Likelihood Estimators for Parameters of the Three-parameter Inverse Gaussian Distribution," Ph.D. Dissertation, University of Georgia, Athens, 1982.

Chapter 19

CHI-SQUARED
GOODNESS–OF–FIT TEST

19.1 WHEN AND HOW TO APPLY

In testing programs, the experimental results are usually not identical to those obtained from theoretical distributions. Therefore, several theoretical distributions are fitted to the experimental data and the chi–squared goodness–of–fit test is used to determine which distribution fits the data best.

If the normal distribution is to be fitted to the data, then, since the population mean and standard deviation; i.e., the two parameters of the distribution are unknown, their estimates are calculated first. The estimate of the mean, $\hat{\overline{T}}$, is defined as the arithmetic mean value of the data, T_i, or

$$\hat{\overline{T}} = \frac{1}{N} \sum_{i=1}^{N} T_i.$$

The estimate of the standard deviation, $\hat{\sigma}_T$, is given by

$$\hat{\sigma}_T = \left[\frac{1}{N-1} \sum_{i=1}^{N} (T_i - \overline{T})^2 \right]^{1/2}.$$

If a histogram is to be developed, the data are grouped first into a suitable number of class intervals, and the observed frequency that falls in each class interval is counted. The optimal number of class intervals, k, may be estimated from Sturges' Rule, or

$$k = 1 + 3.322 \ \log_{10}(N), \tag{19.1}$$

where N is the sample size. The class width, w, is given by

$$w = r/k,$$

where $r = T_{max} - T_{min}$. w is rounded off to the nearest measurement accuracy figure. The class starting values are determined by beginning with the minimum observed value and adding to it successively the class width, w. The class end values are determined by adding to each class starting value the quantity $(w - \varepsilon)$, where ε is the measurement accuracy. The lower bound of a class interval is determined by subtracting the quantity $\varepsilon/2$ from the class starting value, and the upper bound by adding $\varepsilon/2$ to the class end value. The frequencies observed in each class are determined by counting the number of observations falling in a particular class. For additional details of this method see [1, Vol.1, pp. 109-122].

Having grouped the data and determined the observed frequency for each class, the next step is to determine the expected frequency in each class. This can be obtained by multiplying the total number of observations, N, by the area under the *pdf* of the assumed distribution, between the lower and upper bounds of the particular class. To assure that the total expected frequency agrees with the total observed frequency, the area under the *pdf* to the left of the lower bound of the first class is included in the area under the first class, and the area under the *pdf* to the right of the upper bound of the *kth* class is included in the area under the *kth* class. Then the chi–squared value, χ_o^2, is calculated from

$$\chi_o^2 = \sum_{i=1}^{k} \frac{(O_i - E_i)^2}{E_i}, \tag{19.2}$$

where

$$O_i = \text{observed frequency in the } ith \text{ class,}$$

and

$$E_i = \text{expected frequency in the } ith \text{ class.}$$

If the observed frequencies are identical to the expected frequencies; i.e., the experimental data exactly fit the theoretical distribution, then the chi–squared value, χ_o^2, will be zero. The χ_o^2 value is always positive and finite; therefore, the larger the value of χ_o^2, the worse the "fit" of the experimental data to the assumed distribution.

For the implementation of the chi–squared goodness–of–fit test, the χ_o^2 value calculated using Eq. (19.2) is compared with a critical χ_{cr}^2 value obtained from Appendix D. The χ_{cr}^2 value is chosen so that the

area under the χ^2 *pdf*, to the right of χ^2_{cr}, is equal to the risk level, α. This χ^2_{cr} value depends on the number of degrees of freedom, n, which is the parameter of the χ^2 *pdf* whose χ^2_{cr} value is sought. The larger the number of degrees of freedom, the larger will be the χ^2_{cr} value for a specified confidence level. The number of degrees of freedom, n, is given by

$$n = k - 1 - m,$$

where

$$k = \text{ number of classes,}$$

and

$$m = \text{ number of parameters estimated from the sample.}$$

This is so because m restrictions are imposed on the χ^2_{cr} value, and m equals zero if the mean and the standard deviation of the distribution are known. For the normal distribution $m = 2$, since the mean and the standard deviation must be estimated from the sample data. Also $(k - 1)$, because if $(k - 1)$ of the expected frequencies are known, the frequency in the remaining class is uniquely determined.

The chi-squared value calculated using Eq. (19.2), or χ^2_o, is compared with the critical value obtained from Appendix D, or with the χ^2_{cr} value corresponding to the desired confidence level, $1 - \alpha$, or the level of significance (risk level), α, and the degrees of freedom, n. If the calculated chi–squared value is less that the value obtained from the chi–square table, the *fit is considered to be good at the desired confidence level*, $1 - \alpha$; i.e., if

$$\chi^2_o < \chi^2_{cr},$$

accept the fitted distribution as representing the data acceptably well, or with confidence level, $1 - \alpha$. If

$$\chi^2_o \geq \chi^2_{cr},$$

reject the fitted distribution.

The probability that an acceptable distribution is rejected by this test cannot exceed the risk level, α.

EXAMPLE 19–1

Illustrate the procedure to prepare histograms that is a part of the chi–squared goodness–of–fit test data reduction process, by applying it to the data given in Table 19.1.

TABLE 19.1– Forty cycles-to-failure data obtained using a fatigue testing machine.

Specimen number	Cycles to failure	Specimen number	Cycles to failure
1	50,400	21	43,000
2	44,500	22	31,700
3	35,000	23	37,000
4	41,800	24	44,000
5	55,300	25	48,700
6	70,600	26	46,400
7	45,400	27	43,400
8	37,200	28	45,600
9	48,100	29	39,300
10	68,800	30	36,100
11	40,800	31	48,500
12	54,100	32	38,000
13	43,800	33	36,400
14	56,600	34	37,100
15	42,400	35	38,400
16	38,500	36	41,300
17	58,100	37	52,100
18	48,100	38	55,000
19	42,900	39	54,200
20	46,000	40	54,300

SOLUTION TO EXAMPLE 19–1

1. Determine the number of classes from

$$k = 1 + (3.322) \log_{10} 40,$$

which is Sturges' Rule for 40 data points, or

$$k = 1 + (3.322)(1.60) = 1 + 5.32 = 6.32 \cong 6.$$

2. Determine the range of the data, in 100's of cycles, from

$$r = T_{max} - T_{min},$$

or

$$r = 706 - 317 = 389.$$

3. Determine the class width from

$$w = \frac{r}{k} = \frac{389}{6} = 65.$$

4. Determine the class starting values, in 100's of cycles, from

$1st$ class $= T_{min} = 317,$
$2nd$ class and thereafter $=$ class starting value plus $w,$

or

$2nd$ class $= 317 + 65 = 382,$
$3rd$ class $= 382 + 65 = 447,$
$4th$ class $= 447 + 65 = 512,$
$5th$ class $= 512 + 65 = 577,$

and

$6th$ class $= 577 + 65 = 642.$

5. Determine the class end value in 100′s of cycles from the class starting value plus $(w - \varepsilon)$, where $\varepsilon =$ measurement accuracy $=1$ (100 cycles), or

$$1st \text{ class} = 317 + (65 - 1) = 317 + 64 = 381,$$
$$2nd \text{ class} = 382 + 64 = 446,$$
$$3rd \text{ class} = 447 + 64 = 511,$$
$$4th \text{ class} = 512 + 64 = 576,$$
$$5th \text{ class} = 576 + 64 = 641,$$

and

$$6th \text{ class} = 642 + 64 = 706.$$

6. Determine the class mid-value from

$$1st \text{ class} = x_1 + \frac{w - \varepsilon}{2} = 317 + \frac{(65 - 1)}{2},$$
$$= 317 + 32 = 349,$$
$$2nd \text{ class} = 382 + 32 = 414,$$
$$3rd \text{ class} = 447 + 32 = 479,$$
$$4th \text{ class} = 512 + 32 = 544,$$
$$5th \text{ class} = 577 + 32 = 609,$$

and

$$6th \text{ class} = 642 + 32 = 674.$$

7. Determine the class lower bound from the class starting value $-\varepsilon/2$, where $\varepsilon/2 = 0.5$, or

$$1st \text{ class} = 317 - 0.5 = 316.5,$$
$$2nd \text{ class} = 382 - 0.5 = 381.5,$$
$$3rd \text{ class} = 447 - 0.5 = 446.5,$$
$$4th \text{ class} = 512 - 0.5 = 511.5,$$
$$5th \text{ class} = 577 - 0.5 = 576.5,$$

and

$$6th \text{ class} = 642 - 0.5 = 641.5.$$

TABLE 19.2– Tally table for the data in Table 19.1.

Class number	Class starting value, $\times 10^2$	Class end value, $\times 10^2$	Tally	Fre-quency	Class lower bound, $\times 10^2$	Class mid-value, $\times 10^2$	Class upper bound, $\times 10^2$
1	317	381	////////	8	316.5	349	381.5
2	382	446	//////// /////	13	381.5	414	446.5
3	447	511	///////// /	9	446.5	479	511.5
4	512	576	///////	7	511.5	544	576.5
5	577	641	/	1	576.5	609	641.5
6	642	706	//	2	641.5	674	706.5

8. Determine the class upper bound from the class end value $+\varepsilon/2$, or

$$1st \text{ class} = 381 + 0.5 = 381.5,$$
$$2nd \text{ class} = 446 + 0.5 = 446.5,$$
$$3rd \text{ class} = 511 + 0.5 = 511.5,$$
$$4th \text{ class} = 576 + 0.5 = 576.5,$$
$$5th \text{ class} = 641 + 0.5 = 641.5,$$

and

$$6th \text{ class} = 706 + 0.5 = 706.5.$$

9. Prepare a tally table, as in Table 19.2.

Tally under each class those cycles-to-failure values which are equal to or greater than the class starting value or equal to or less than the class end value. Then count the tally to obtain the frequency, or the number of observations, falling in each class.

10. Prepare a histogram.

Now the frequency can be plotted versus the class mid-value, as shown in Fig. 19.1.

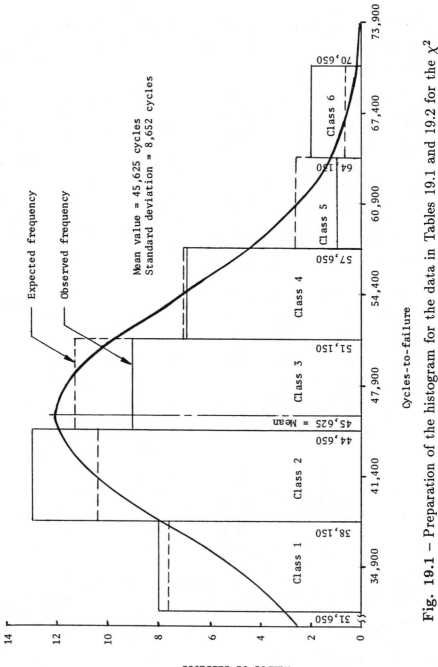

Fig. 19.1 – Preparation of the histogram for the data in Tables 19.1 and 19.2 for the χ^2 goodness-of-fit test.

TABLE 19.3– Calculation of the mean and of the standard deviation of the data in Table 19.1.

Class number	Class midvalue, $T_i \times 10^{-2}$	Frequency, f_i	u_i	$f_i u_i$	$f_i u_i^2$
1	349	8	-2	-16	32
2	414	13	-1	-13	13
3	479	9	0	0	0
4	544	7	1	7	7
5	609	1	2	2	4
6	674	2	3	6	18
		$\sum = 40$		-14	74

11. Fit a distribution to the data.

Choose a distribution whose shape you believe will fit the histogram. In this case the normal distribution is used. Hence, calculate the parameters of the normal *pdf* representing the data as follows:

$$\overline{T} = \frac{w}{N}\sum_{i=1}^{k}(f_i u_i) + A,$$

where, from Table 19.3 and [1, Vol.1, pp. 340-343],

$$w = 65(10^2), \quad \sum_{i=1}^{k}(f_i u_i) = -14,$$

$$N = 40, \quad A = 479(10^2), \quad \text{and } k = 6,$$

then

$$\overline{T} = \frac{65(10^2)}{40}(-14) + 479(10^2) = 45,625 \text{ cycles},$$

and

$$\sigma_T = \frac{w}{N}\left[\frac{N^2}{N-1}\sum_{i=1}^{k}(f_i u_i^2) - \frac{N}{N-1}(\sum_{i=1}^{k}f_i k u_i)^2\right]^{1/2},$$

$$\sigma_T = \frac{65(10^2)}{40}\left[\frac{40^2}{40-1}(74) - \frac{40}{40-1}(-14)^2\right]^{1/2},$$

TABLE 19.4– Calculation of the chi-squared value of the data in Table 19.1.

Class number	Expected frequency, E_i	Observed frequency, O_i	$\frac{(O_i - E_i)^2}{E_i}$
1	7.748	8.00	0.0082
2	10.452	13.00	0.6212
3	11.356	9.00	0.4888
$\left.\begin{array}{c}4\\5\\6\end{array}\right\}4$	$\left.\begin{array}{c}7.152\\2.664\\0.648\end{array}\right\}10.464$	$\left.\begin{array}{c}7.00\\1.00\\2.00\end{array}\right\}10.00$	0.0206
	39.960	40.00	$\chi_o^2 = 1.1388$

or
$$\sigma_T = 8,652 \text{ cycles.}$$

12. Calculate the chi-squared value and determine the goodness-of-fit.

The results of these calculations appear in Table 19.4, where the χ_o^2 value is found to be 1.1388, on the basis of a total of four classes, because to find the χ_o^2 value this goodness-of-fit test requires that each class have *at least five observations*, although this restriction may be relaxed to as few as three observations. Nevertheless, in this example we will follow the former restriction of five. Consequently, we start with the last class that has two observations, or fewer than five. The policy to use is to merge this number of observations with the adjacent class which has one observation. That makes a total of three observations, which is still fewer than five. So we merge the new class with the next adjacent class. This makes a total of 10 observations which is more than five; consequently, we have a new, merged Class 4 of 10 observations, as may be seen in Table 19.4. A similar procedure is used if Class 1 or other classes have fewer than five observations, which consists of merging adjacent classes until the new merged class has five or more observations.

Next determine the critical χ_{cr}^2 value to compare it with that found in the previous analysis, or with χ_o^2. For this, find the

degrees of freedom, n, from

$$n = k - 1 - m = 4 - 1 - 2 = 1.$$

Choose a confidence level of 95%, or a level of significance or risk level, α, for this test of 5%, a commonly used value. The critical χ^2_{cr} value, or the $\chi^2_{cr} = \chi^2_{\alpha;n} = \chi^2_{0.05;1}$ value, from Appendix D, is

$$\chi^2_{cr} = \chi^2_{0.05;1} = 3.841.$$

Now, since

$$\chi^2_o = 1.1388 < \chi^2_{cr} = \chi^2_{0.05;1} = 3.841,$$

we accept the hypothesis that the normal *pdf* represents the data with 95% confidence.

19.2 COMMENTS ON CHI-SQUARED GOODNESS–OF–FIT TEST

1. This test is well adapted for the discrete distributions.

2. This test can be used both for the case when the expected distribution was completely specified, and for the case when the expected distribution parameters are unknown but need to be estimated from the sample.

3. The power of this test is not very high. Generally speaking it is less powerful than the *Kolmogorov-Smirnov* and the other empirical distribution function (EDF) based tests, covered in Chapters 20 and 21.

4. This test can not be used effectively in the small sample size cases; e.g., for sample sizes smaller than 25, due to the constraint that each interval should contain at least five data to calculate the χ^2 statistic.

PROBLEMS

19-1. The data given in Table 19.5 are in 1,000 miles to the first major motor failure of 191 buses operated by a large city bus company. Failure was defined as either abrupt, in which case some part broke and the motor would not run; or the maximum power produced, as could be measured by a dynamometer, fell below a fixed acceptable percentage of the normal rated value. Do the following:

TABLE 19.5 – Field data for the first bus motor failures of Problem 19–1.

Life, in 1,000 miles	Number of failures
0 – 20	6
20 – 40	11
40 – 60	16
60 – 80	25
80 – 100	34
100 – 120	46
120 – 140	33
140 – 160	16
160 – 180	2
> 180	2

Total = 191

(1) Apply the chi-squared goodness-of-fit test to the normal distribution fitted to these data.

(2) Apply the chi-squared goodness-of-fit test to the Weibull distribution fitted to these data.

(3) Comparatively discuss the results of the previous two cases at the 5% and the 10% levels of significance.

19-2. Given are the data of Table 19.6 for the second field failure of the bus motors described in Problem 19-1. Do the following:

(1) Apply the chi-squared goodness-of-fit test to the normal distribution fitted to these data.

(2) Apply the chi-squared goodness-of-fit test to the Weibull distribution fitted to these data.

(3) Comparatively discuss the results of the previous two cases at the 5% and the 10% levels of significance.

19-3. The situation is the same as in Problem 19-1 but the times to the third bus motor failure are analyzed. The data are given in Table 19.7. Do the following:

(1) Apply the chi-squared goodness-of-fit test to the normal distribution fitted to these data.

TABLE 19.6 – Field data for the second bus motor failures of Problem 19-2.

Life, in 1,000 miles	Number of failures
0 – 20	13
20 – 40	19
40 – 60	13
60 – 80	15
80 – 100	15
100 – 120	12
120 – 140	4
> 140	4

Total = 95

(2) Apply the chi-squared goodness-of-fit test to the Weibull distribution fitted to these data.

(3) Comparatively discuss the results of the previous two cases at the 5% and the 10% levels of significance.

19-4. The situation is the same as in Problem 19-1 but the times to the fourth bus motor failure are analyzed. The data are given in Table 19.8. Do the following:

(1) Apply the chi-squared goodness-of-fit test to the two-parameter Weibull distribution fitted to these data.

(2) Apply the chi-squared goodness-of-fit test to the exponential distribution fitted to these data.

(3) Comparatively discuss the results of the previous two cases at the 5% and the 10% levels of significance.

19-5. The situation is the same as in Problem 19-1 but the times to the fifth bus motor failure are analyzed. The data are given in Table 19.9. Do the following:

(1) Apply the chi-squared goodness-of-fit test to the two-parameter Weibull distribution fitted to these data.

(2) Apply the chi-squared goodness-of-fit test to the exponential distribution fitted to these data.

(3) Comparatively discuss the results of the previous two cases at the 5% and the 10% levels of significance.

TABLE 19.7 – Field data for the third bus motor failures of Problem 19-3.

Life, in 1,000 miles	Number of failures
0 – 20	27
20 – 40	18
40 – 60	16
60 – 80	13
80 – 100	9
> 100	8

Total = 91

TABLE 19.8 – Field data for the fourth bus motor failures of Problem 19-4.

Life, in 1,000 miles	Number of failures
0 – 20	34
20 – 40	19
40 – 60	16
60 – 80	12
> 80	11

Total = 100

TABLE 19.9 – Field data for the fifth bus motor failures of Problem 19-5.

Life, in 1,000 miles	Number of failures
0 - 20	29
20 - 40	27
40 - 60	14
60 - 80	8
> 80	7

Total = 95

19-6. Apply the chi-squared goodness-of-fit test to the fitted distributions in Problem 18-1, where the parameters were estimated from the sample data using different methods. Comparatively discuss the goodness-of-fit of the fitted distributions at the 5% and the 10% levels of significance.

19-7. Apply the chi-squared goodness-of-fit test to the fitted distributions in Problem 18-2, where the parameters were estimated from the sample data using different methods. Comparatively discuss the goodness-of-fit of the fitted distributions at the 5% and the 10% levels of significance.

19-8. Apply the chi-squared goodness-of-fit test to the fitted distributions in Problem 18-3, where the parameters were estimated from the sample data using different methods. Comparatively discuss the goodness-of-fit of the fitted distributions at the 5% and the 10% levels of significance.

19-9. Apply the chi-squared goodness-of-fit test to the fitted distributions in Problem 18-4, where the parameters were estimated from the sample data using different methods. Comparatively discuss the goodness-of-fit of the fitted distributions at the 5% and the 10% levels of significance.

19-10. Apply the chi-squared goodness-of-fit test to the fitted distributions in Problem 18-5, where the parameters were estimated from the sample data using different methods. Comparatively discuss the goodness-of-fit of the fitted distributions at the 5% and the 10% levels of significance.

19-11. Apply the chi-squared goodness-of-fit test to the fitted distributions in Problem 18-6, where the parameters were estimated from the sample data using different methods. Comparatively discuss the goodness-of-fit of the fitted distributions at the 5% and the 10% levels of significance.

19-12. Apply the chi-squared goodness-of-fit test to the fitted distributions in Problem 18-7, where the parameters were estimated from the sample data using different methods. Comparatively discuss the goodness-of-fit of the fitted distributions at the 5% and the 10% levels of significance.

19-13. Apply the chi-squared goodness-of-fit test to the fitted distributions in Problem 18-8, where the parameters were estimated from the sample data using different methods. Comparatively discuss the goodness-of-fit of the fitted distributions at the 5% and the 10% levels of significance.

19-14. Apply the chi-squared goodness-of-fit test to the fitted distributions in Problem 18-9, where the parameters were estimated from the sample data using different methods. Comparatively discuss the goodness-of-fit of the fitted distributions at the 5% and the 10% levels of significance.

19-15. Apply the chi-squared goodness-of-fit test to the fitted distributions in Problem 18-10, where the parameters were estimated from the sample data using different methods. Comparatively discuss the goodness-of-fit of the fitted distributions at the 5% and the 10% levels of significance.

19-16. Apply the chi-squared goodness-of-fit test to the fitted distributions in Problem 18-11, where the parameters were estimated from the sample data using different methods. Comparatively discuss the goodness-of-fit of the fitted distributions at the 5% and the 10% levels of significance.

19-17. Apply the chi-squared goodness-of-fit test to the fitted distributions in Problem 18-12, where the parameters were estimated from the sample data using different methods. Comparatively discuss the goodness-of-fit of the fitted distributions at the 5% and the 10% levels of significance.

19-18. Apply the chi-squared goodness-of-fit test to the fitted distributions in Problem 18-13, where the parameters were estimated from the sample data using different methods. Comparatively discuss the goodness-of-fit of the fitted distributions at the 5% and the 10% levels of significance.

19-19. Apply the chi-squared goodness-of-fit test to the fitted distributions in Problem 18-14, where the parameters were estimated from the sample data using different methods. Comparatively discuss the goodness-of-fit of the fitted distributions at the 5% and the 10% levels of significance.

19-20. Apply the chi-squared goodness-of-fit test to the fitted distributions in Problem 18-15, where the parameters were estimated from the sample data using different methods. Comparatively discuss the goodness-of-fit of the fitted distributions at the 5% and the 10% levels of significance.

REFERENCES

1. Cochran, W. G., "The χ^2 Test of Goodness of Fit," *Annals of Mathematical Statistics*, Vol. 23, pp. 315-345, 1952.

2. Kececioglu, Dimitri B., *Reliability Engineering Handbook*, Prentice Hall, Inc., Englewood Cliffs, New Jersey, Vol. 1, 720 pp., and Vol. 2, 568 pp., 1991.

3. Williams, C. A., "On the Choice of the Number and Width of Classes for the Chi-square Test of Goodness of Fit," *Journal of the American Statistical Association*, Vol. 45, pp. 77-86, 1950.

4. Mann, H. B. and Wald, A., "On the Choice of the Number of Intervals in the Application of the χ^2 Test," *Annals of Mathematical Statistics*, Vol. 13, pp. 306-317, 1942.

Chapter 20

KOLMOGOROV-SMIRNOV GOODNESS-OF-FIT TEST

20.1 INTRODUCTION

When evaluating test, or operational use, data to determine the reliability of a device, it is necessary to examine the applicability of the selected times-to-failure distribution. After a distribution has been selected that apparently fits the data, and its parameters quantified, it remains to determine whether the generated data, in fact ,"fit" the selected distribution. This analysis of the fit of a selected distribution to the data may be accomplished by means of the *Kolmogorov-Smirnov goodness-of-fit test* [1; 2; 3], or simply the *K-S test*, which is easy to use. The step-by-step application of this test to the normal, lognormal, exponential and Weibull distributions are presented in this chapter and illustrated by examples.

It must be assumed that the data being evaluated were obtained under uniform test conditions, the devices tested were of identical design, the data were not compromised by the repair and re-use of the failed units, provided that the repair returned the units to essentially their original performance capability, and that the repair operation resulted in no significant alteration of the original design configuration and the times-to-failure distribution.

It should be pointed out that the *standard Kolmogorov-Smirnov goodness-of-fit test* is only valid for continuous distributions and *with known parameters*. It cannot be used for discrete distributions. In this test the maximum of the difference between the expected cumulative distribution at the time of a failure, and the cumulative distribution observed at the time of this failure is used.

Also a *refined Kolmogorov-Smirnov goodness-of-fit test* is presented in this chapter which uses the maximum difference between the expected cumulative distribution, and either the observed cumulative distribution at the failure being analysed or the observed cumulative distribution for the previous failure. This *refined K-S test* provides a more accurate measure of the goodness-of-fit.

If the (continuous) distribution parameters are not known but estimated from the given sample data, then the test critical values should be modified. This case is presented later as the *modified Kolmogorov-Smirnov goodness-of-fit test*.

The *K-S test* critical values for complete data are different than those for censored data. The test procedures for ungrouped data and grouped data are somewhat different also because the difference in the behavior of the empirical distributions in these two cases must be considered. With grouped data $Q_O(T_i^-)$ cannot be determined because we do not know when the previous to the last failure in each group occurred, although the $Q_O(T_i)$ can be determined.

In this chapter the *standard*, the *refined* and the *modified K-S tests* for complete and ungrouped data are covered. For incomplete and grouped data cases, the readers are referred to [13; 14].

20.2 STEP-BY-STEP METHOD FOR THE STANDARD $K\text{-}S$ GOODNESS-OF-FIT TEST WITH DISTRIBUTION PARAMETERS KNOWN

Step 1– Test the whole sample of size N to failure. Arrange all times to failure in increasing value, or rank them.

Step 2– Determine the total number of failures, which equals N.

Step 3– After each successive failure, determine the total number of failures already observed, f_i. Divide this number by the total number of failures, N. Each one of these calculations gives the observed probability of failure, or

$$Q_O(T_i) = \frac{f_i}{N}, \quad i = 1, 2, ..., N.$$

Step 4– Calculate in the same order, using the equation of the selected distribution, the theoretical or expected probability of failure, $Q_E(T_i)$, for each failure that has occurred.

Step 5– Determine the absolute difference, D_i, between $Q_E(T_i)$ and $Q_O(T_i)$, by calculating $D_i = |Q_E(T_i) - Q_O(T_i)|$ for each failure.

Step 6– Determine the *maximum absolute difference*, D_{max}; i.e., ignoring the plus or minus signs.

Step 7– In the *K-S test* critical value tables given in Appendix G-1, look up the allowable or critical difference value, D_{cr}, for the sample size N, at the desired or required level of significance, usually 0.05.

Step 8– If the calculated maximum absolute difference is less than the allowable critical value from the table, or if

$$D_{max} < D_{cr},$$

then the selected distribution is accepted as representing the actual test, or operational, data to the indicated level of significance. If the calculated difference is equal to or greater than the allowable critical value, or if

$$D_{max} \geq D_{cr},$$

then the selected distribution is not accepted, as representing the data, at the indicated level of significance.

EXAMPLE 20–1

The times-to-failure data presented in Column 2 of Table 20.1 may be described by a normal distribution having a mean of $\overline{T} = 2,722.26$ hr and a standard deviation of $\sigma_T = 21.92$ hr. Does this normal *pdf* fit the data at the significance level of 0.05?

SOLUTION TO EXAMPLE 20–1

Step 1– The data are arranged in the order of increasing times to failure in Column 2 of Table 20.1.

Step 2– The total number of failures is 19, or $N = 19$, from Column 1.

Step 3– The observed probability of failure for each successive failure, $Q_O(T_i)$, is calculated and is entered in Column 3. For example, for Failure No. 5 at $T_5 = 2,706$ hr,

$$Q_O(T_5 = 2,706 \text{ hr}) = \frac{f_5}{N} = \frac{5}{19} = 0.2632.$$

TABLE 20.1– Kolmogorov-Smirnov goodness-of-fit test for the normally distributed data of Example 20–1.

Failure number, i	Time to failure, T_i, hr	Observed unreliability, $Q_O(T_i)$	Expected unreliability, $Q_E(T_i)$	Absolute difference, D_i^*
1	2,681	0.0526	0.0301	0.0225
2	2,691	0.1053	0.0764	0.0289
3	2,697	0.1579	0.1251	0.0328
4	2,702	0.2105	0.1788	0.0317
5	2,706	0.2632	0.2296	0.0336
6	2,709	0.3158	0.2743	0.0415
7	2,712	0.3684	0.3192	0.0492**
8	2,716	0.4211	0.3859	0.0352
9	2,720	0.4737	0.4602	0.0135
10	2,722	0.5263	0.4960	0.0303
11	2,726	0.5789	0.5675	0.0114
12	2,728	0.6316	0.6026	0.0290
13	2,730	0.6842	0.6368	0.0474
14	2,736	0.7368	0.7357	0.0011
15	2,739	0.7895	0.7764	0.0131
16	2,744	0.8421	0.8389	0.0032
17	2,747	0.8947	0.8708	0.0239
18	2,754	0.9474	0.9265	0.0209
19	2,763	1.0000	0.9686	0.0314

* $D_i = |Q_E(T_i) - Q_O(T_i)|$.
** Maximum absolute difference, $D_{max} = 0.0492$.

Step 4– The *expected probability of failure* is calculated, for each time to failure, from

$$Q_E(T_i) = \Phi\left(\frac{T_i - \overline{T}}{\sigma_T}\right),$$

and entered in Column 4. For example, for Failure No. 5, $T_5 = 2,706$ hr, $\overline{T} = 2,722.26$ hr and $\sigma_T = 21.92$ hr. Then,

$$Q_E(T_5 = 2,706 \text{ hr}) = \Phi(\frac{2,706 - 2,722.26}{21.92}) = \Phi(-0.74),$$

or, from Appendix B,

$$Q_E(T_5 = 2,706 \text{ hr}) = 0.2296.$$

Step 5– The absolute difference, $D_i = |\, Q_E(T_i) - Q_O(T_i)\,|$, is calculated for each failure and entered in Column 5. For example, for Failure No. 5,

$$D_5 = |\, Q_E(T_5) - Q_O(T_5)\,| = |\, 0.2296 - 0.2632\,| = 0.0336.$$

Step 6– Failure No. 7 has the maximum difference, $D_7 = D_{max} = 0.0492$.

Step 7– From the Kolmogorov-Smirnov tables in Appendix G-1, it may be seen that the maximum allowable, or critical, difference for $N = 19$ failures, at the $\alpha = 0.05$ significance level, is $D_{cr} = 0.30143$.

Step 8– Since the maximum difference, $D_{max} = 0.0492$, is less than the allowable, $D_{cr} = 0.30143$, or

$$D_{max} = 0.0492 < D_{cr} = 0.30143,$$

the normal distribution is accepted as adequately describing the times-to-failure data at the 0.05 significance level.

EXAMPLE 20–2

The times-to-failure data in Column 2 of Table 20.2 may be described by a lognormal *pdf* having a mean of $\overline{T}' = 4.3553$ and a standard deviation of $\sigma_{T'} = 0.67670$. Does this lognormal *pdf* fit these data at the significance level of 0.05 ?

TABLE 20.2– Kolmogorov-Smirnov goodness-of-fit test for the lognormally distributed data of Example 20–2.

Failure number, i	Time to failure, T_i, hr	$\log_e T_i$	Observed unreliability, $Q_o(T_i)$	Expected unreliability, $Q_E(T_i)$	Absolute difference, D_i^*
1	30.4	3.41443	0.1111	0.0823	0.0288
2	36.7	3.60278	0.2222	0.1335	0.0887
3	53.3	3.97593	0.3333	0.2877	0.0456
4	58.5	4.06903	0.4444	0.3372	0.1072**
5	74.0	4.30405	0.5556	0.4681	0.0875
6	99.3	4.59814	0.6667	0.6406	0.0261
7	114.3	4.73883	0.7778	0.7157	0.0621
8	140.1	4.94235	0.8889	0.8078	0.0811
9	257.9	5.55256	1.0000	0.9616	0.0384

* $D_i = |Q_E(T_i) - Q_o(T_i)|$
** Maximum absolute difference, $D_{max} = 0.1072$.

SOLUTION TO EXAMPLE 20–2

Step 1– The data are arranged, in order of increasing times to failure, in Column 2 of Table 20.2, and the natural logarithms of these are entered in Column 3.

Step 2– The total number of failures is $N = 9$ from Column 1.

Step 3– The $Q_O(T_i)$ for each successive failure is given in Column 4. For example, for Failure No. 3 at $T_3 = 53.3$ hr,

$$Q_O(T_3 = 53.3 \text{ hr}) = \frac{f_3}{N} = \frac{3}{9} = 0.3333.$$

Step 4– The *expected probability of failure* is calculated, for each time to failure, from

$$Q_E(T_i) = \Phi\left(\frac{\log_e T_i - \overline{T}'}{\sigma_{T'}}\right),$$

and entered in Column 5. For example, for Failure No. 3,

$T_3 = 53.3$ hr,

$\log_e T_3 = 3.97593, \overline{T}' = 4.3553$ and $\sigma_{T'} = 0.67670$. Then,

$$Q_E(T_3 = 53.3 \text{ hr}) = \Phi\left(\frac{3.97593 - 4.3553}{0.67670}\right) = \Phi(-0.56),$$

or, from Appendix B,

$$Q_E(T_3 = 53.3 \text{ hr}) = 0.2877.$$

Step 5– The absolute difference, $D_i = |Q_E(T_i) - Q_O(T_i)|$ is calculated for each failure and entered in Column 6. For example, for Failure No. 3,

$$D_3 = |Q_E(T_3) - Q_O(T_3)| = |0.2877 - 0.3333| = 0.0456.$$

Step 6– Failure No. 4 has the maximum difference, $D_4 = D_{max} = 0.1072.$

Step 7– From the Kolmogorov-Smirnov tables in Appendix G-1, it may be seen that the maximum allowable, or critical, difference for $N = 9$ failures, at the $\alpha = 0.05$ significance level, is $D_{cr} = 0.43001.$

Step 8–   Since the maximum difference, $D_{max} = 0.1072$, is less than the allowable, $D_{cr} = 0.43001$, or

$$D_{max} = 0.1072 < D_{cr} = 0.43001,$$

the lognormal distribution is accepted as adequately describing the times-to-failure data at the 0.05 significance level.

EXAMPLE 20–3

The times-to-failure data presented in Column 2 of Table 20.3 may be described by an exponential distribution with a mean time between failures of $MTBF = m = 101.4$ hr. Does this exponential *pdf* fit these data at the significance level of 0.05 ?

SOLUTION TO EXAMPLE 20–3

Step 1–   The data are arranged in order of increasing times to failure in Column 2 of Table 20.3.

Step 2–   The total number of failures is 49, or $N = 49$, from Column 1.

Step 3–   The $Q_O(T_i)$ for each successive failure is given in Column 3. For example, for Failure No. 26 at $T_{26} = 95.1$ hr,

$$Q_O(T_{26} = 95.1 \text{ hr}) = \frac{f_{26}}{N} = \frac{26}{49} = 0.5306.$$

Step 4–   The *expected probability of failure* is calculated, for each time to failure, from

$$Q_E(T_i) = 1 - e^{-\frac{T_i}{m}},$$

and entered in Column 4. For example, for Failure No. 26,

$T_{26} = 95.1$ hr and $m = 101.4$ hr. Then,

$$Q_E(T_{26} = 95.1 \text{ hr}) = 1 - e^{-\frac{95.1}{101.4}},$$

or

$$Q_E(T_{26} = 95.1 \text{ hr}) = 0.6086.$$

Step 5–   The absolute difference, $D_i = |\,Q_E(T_i) - Q_O(T_i)\,|$, is calculated for each failure and entered in Column 6. For Failure No. 26,

$$D_{26} = |\,Q_E(T_{26}) - Q_O(T_{26})\,| = |\,0.6086 - 0.5306\,| = 0.0780.$$

TABLE 20.3– Kolmogorov-Smirnov goodness-of-fit test for the exponentially distributed data of Example 20–3.

Failure number, i	Time to failure, T_i, hr	Observed unreliability, $Q_O(T_i)$	Expected unreliability, $Q_E(T_i)$	Absolute difference, D_i^*
1	1.2	0.0204	0.0117	0.0087
2	1.5	0.0408	0.0147	0.0261
3	2.8	0.0612	0.0272	0.0340
4	4.9	0.0816	0.0472	0.0344
5	6.8	0.1020	0.0648	0.0372
6	7.0	0.1224	0.0667	0.0557
7	12.1	0.1429	0.1126	0.0303
8	13.7	0.1633	0.1264	0.0369
9	15.1	0.1837	0.1384	0.0453
10	15.2	0.2041	0.1392	0.0649
11	23.9	0.2245	0.2100	0.0145
12	24.3	0.2449	0.2131	0.0318
13	25.1	0.2653	0.2193	0.0460
14	35.8	0.2857	0.2974	0.0117
15	38.9	0.3061	0.3186	0.0125
16	47.9	0.3265	0.3764	0.0499
17	48.9	0.3469	0.3826	0.0357
18	49.3	0.3673	0.3850	0.0177
19	53.2	0.3878	0.4082	0.0204
20	55.6	0.4082	0.4420	0.0138
21	62.7	0.4286	0.4611	0.0325
22	72.4	0.4490	0.5103	0.0613
23	73.6	0.4694	0.5160	0.0466
24	76.8	0.4898	0.5310	0.0412
25	83.8	0.5102	0.5623	0.0521

* $D_i = |Q_E(T_i) - Q_O(T_i)|$.

TABLE 20.3– Continued.

Failure number, i	Time to failure, T_i, hr	Observed unreliability, $Q_O(T_i)$	Expected unreliability, $Q_E(T_i)$	Absolute difference, D_i^*
26	95.1	0.5306	0.6086	0.0780
27	97.9	0.5510	0.6191	0.0681
28	99.6	0.5714	0.6255	0.0541
29	102.8	0.5918	0.6371	0.0453
30	108.5	0.6122	0.6569	0.0447
31	128.7	0.6326	0.7189	0.0863*
32	133.6	0.6531	0.7322	0.0791
33	144.1	0.6735	0.7585	0.0850
34	147.6	0.6939	0.7667	0.0728
35	150.6	0.7143	0.7735	0.0592
36	151.6	0.7347	0.7757	0.0410
37	152.6	0.7551	0.7779	0.0228
38	164.2	0.7755	0.8019	0.0264
39	166.8	0.7959	0.8069	0.0110
40	178.6	0.8163	0.8281	0.0118
41	185.21	0.8367	0.8390	0.0023
42	187.1	0.8571	0.8420	0.0151
43	203.0	0.8775	0.8649	0.0126
44	204.3	0.8980	0.8666	0.0314
45	229.5	0.9184	0.8960	0.0224
46	233.1	0.9388	0.8996	0.0392
47	254.1	0.9592	0.9184	0.0408
48	291.7	0.9796	0.9437	0.0359
49	304.4	1.0000	0.9503	0.0497

* Maximum absolute difference, $D_{max} = 0.0863$.

Step 6– Failure No. 31 has the maximum difference, $D_{31} = D_{max} = 0.0863$.

Step 7– From the Kolmogorov-Smirnov tables in Appendix G-1, it may be seen that the maximum allowable, or critical, difference for $N = 49$ failures, at the $\alpha = 0.05$ significance level, is $D_{cr} = 0.19028$.

Step 8– Since the maximum difference, $D_{max} = 0.0863$, is less than the allowable, $D_{cr} = 0.19028$, or

$$D_{max} = 0.0863 < D_{cr} = 0.19028,$$

the exponential distribution is accepted as adequately describing these times-to-failure data at the 0.05 significance level.

EXAMPLE 20–4

The Weibull distribution was found to fit the transistor times-to-failure data presented in Column 2 of Table 20.4 with reasonable accuracy. According to past experience the parameters are known to be

$$\gamma = 200 \text{ hr}, \quad \beta = 1.39 \text{ and } \eta = 9,916 \text{ hr}.$$

Does this Weibull *pdf* fit these data at the significance level of 0.05?

SOLUTION TO EXAMPLE 20–4

Step 1– The data are arranged in order of increasing times to failure in Column 2 of Table 20.4.

Step 2– The total number of failures is 24, or $N = 24$, from Column 1.

Step 3– The $Q_O(T_i)$ for each successive failure is given in Column 3. For example, for Failure No. 11 at 740 hr,

$$Q_O(T_{11} = 740 \text{ hr}) = \frac{f_{11}}{N} = \frac{11}{24} = 0.4583.$$

Step 4– The *expected probability of failure* is calculated, for each time to failure, from

$$Q_E(T_i) = 1 - e^{-(\frac{T_i - \gamma}{\eta})^\beta},$$

and entered in Column 4. For example, for Failure No. 11, $T_{11} = 740$ hr, $\gamma = 200$ hr, $\beta = 1.39$ and $\eta = 9,916$ hr. Then

$$Q_E(T_{11} = 740 \text{ hr}) = 1 - e^{-(\frac{740-200}{9,916})^{1.39}} = 0.0174.$$

TABLE 20.4– Kolmogorov-Smirnov goodness-of-fit test for the Weibull distributed data of Example 20–4.

Failure number, i	Time to failure, T_i, hr	Observed unreliability, $Q_O(T_i)$	Expected unreliability, $Q_E(T_i)$	Absolute difference, D_i^*
1	260	0.0417	0.0008	0.0409
2	350	0.0833	0.0029	0.0804
3	420	0.1250	0.0050	0.1200
4	440	0.1667	0.0057	0.1610
5	480	0.2083	0.0070	0.2013
6	480	0.2500	0.0070	0.2430
7	530	0.2917	0.0088	0.2829
8	580	0.3333	0.0107	0.3226
9	680	0.3750	0.0148	0.3602
10	710	0.4167	0.0160	0.4007
11	740	0.4583	0.0174	0.4409
12	780	0.5000	0.0191	0.4809
13	820	0.5417	0.0210	0.5207
14	840	0.5833	0.0219	0.5614
15	920	0.6250	0.0258	0.5992
16	930	0.6667	0.0263	0.6404
17	1,050	0.7083	0.0323	0.6760
18	1,060	0.7500	0.0329	0.7171
19	1,070	0.7917	0.0334	0.7583
20	1,270	0.8333	0.0443	0.7890
21	1,340	0.8750	0.0482	0.8268
22	1,370	0.9167	0.0500	0.8667
23	1,880	0.9583	0.0813	0.8770
24	2,130	1.0000	0.0977	0.9023**

* $D_i = |Q_E(T_i) - Q_O(T_i)|$.
** Maximum absolute difference, $D_{max} = 0.9023$.

Step 5– The absolute difference, $D_i = |Q_E(T_i) - Q_O(T_i)|$ is calculated for each failure and entered in Column 5. For Failure No. 11,

$$D_{11} = |Q_E(T_{11}) - Q_O(T_{11})| = |0.0174 - 0.4583| = 0.4409.$$

Step 6– Failure No. 24 has the maximum difference, $D_{24} = D_{max} = 0.9023$.

Step 7– From the Kolmogorov-Smirnov tables in Appendix G-1, it may be seen that the maximum allowable, or critical, difference for $N = 24$ failures, at the $\alpha = 0.05$ significance level, is $D_{cr} = 0.26931$.

Step 8– Since the maximum difference, $D_{max} = 0.9023$, is greater than the allowable difference, $D_{cr} = 0.26931$, or

$$D_{max} = 0.9023 > D_{cr} = 0.26931,$$

the Weibull distribution cannot be accepted as adequately describing these times-to-failure data at the 0.05 significance level.

20.3   THE REFINED *K-S* GOODNESS-OF-FIT TEST FOR CONTINUOUS DISTRIBUTIONS WITH PARAMETERS KNOWN

Let $T_1 \leq T_2 \leq ... \leq T_N$ be an ordered random sample of size N from a population with unknown $Q(T)$. Let $Q_E(T)$ be a completely specified distribution function. Our goal is to test

$$H_0: \quad Q(T) \quad = \quad Q_E(T) \quad \text{for all } T,$$

versus

$$H_1: \quad Q(T) \quad \neq \quad Q_E(T). \tag{20.1}$$

Kolmogorov and Smirnov [5; 6; 7; 10] suggested computing the empirical, or observed, cumulative distribution function $Q_O(T)$ based on the data $T_1, T_2, ..., T_N$, and then considering the statistic

$$D_{max} = \sup_{-\infty < T < +\infty} |Q_E(T) - Q_O(T)| \tag{20.2}$$

as a measure of agreement between the empirical and the specified (assumed) distributions; i.e., $Q_O(T)$ and $Q_E(T)$. Here $Q_O(T)$ is defined by

$$Q_O(T) = \begin{cases} 0, & T < T_1, \\ f_i/N, & T_i \le T < T_{i+1}, \quad i = 1, 2, ..., N-1, \\ 1, & T \ge T_N, \end{cases} \quad (20.3)$$

where

$$f_i = \text{total number of observations up to } T_i.$$

Note that D_{max} is the maximum vertical distance between $Q_O(T)$ and $Q_E(T)$. Since $Q_E(T)$ is an increasing continuous function, while $Q_O(T)$ is a nondecreasing step function, the supremum of the difference occurs at, or just before, a jump point of $Q_O(T)$; i.e., at T_i, or T_i^+, because $Q_O(T)$ is a right continuous function; or T_i^-, where T_i is one of the given observations $T_1, T_2, ..., T_N$ at which $Q_O(T)$ jumps by $\frac{1}{N}$. This is shown in Fig. 20.1.

Therefore,

$$\begin{aligned} D_{max} &= \sup_{-\infty < T < +\infty} |Q_E(T) - Q_O(T)|, \\ &= max\{D_1, D_2, ..., D_N\}; \end{aligned} \quad (20.4)$$

where,

$$D_i = max\{D_i^+, D_i^-\},$$

or

$$D_i = max\{|Q_E(T_i) - \frac{f_i}{N}|, |Q_E(T_i) - \frac{f_i-1}{N}|\}. \quad (20.5)$$

Notice that for a continuous distribution, $Q_E(T)$,

$$Q_E(T_i^-) = Q_E(T_i) = Q_E(T_i^+),$$

but for the step function $Q_O(T)$, which is right continuous,

$$Q_O(T_i^+) = Q_O(T_i) = Q_O(T_i^-) + \frac{1}{N},$$

and

$$Q_O(T_i^-) = Q_O(T_{i-1}) = \frac{f_i-1}{N}. \quad (20.6)$$

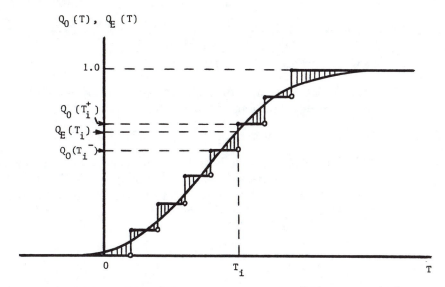

Fig. 20.1– The difference between $Q_E(T_i)$ and $Q_O(T_i)$.

Therefore,

$$D_{max} = max\{max\{D_1^+, D_1^-\}, max\{D_2^+, D_2^-\}, ..., \\ max\{D_N^+, D_N^-\}\},$$

or

$$D_{max} = max\{D_{max}^+, D_{max}^-\}, \tag{20.7}$$

where

$$D_{max}^+ = max\{D_1^+, D_2^+, ..., D_N^+\}$$

and

$$D_{max}^- = max\{D_1^-, D_2^-, ..., D_N^-\}.$$

Then for a given level of significance, α, the decision can be made on rejecting or not rejecting the hypothesis H_0 by applying the following rule:

$$\text{Reject, if } D_{max} \geq D_{cr},$$
$$\text{Do not reject, if } D_{max} < D_{cr},$$

where D_{cr} is the critical, or allowable, value for D_{max} corresponding to the specified significance level, α. The step-by-step procedure of this *K-S test*, and its application to different distributions, are illustrated and discussed next.

20.3.1 STEP-BY-STEP METHOD FOR THE REFINED *K-S* GOODNESS-OF-FIT TEST WITH PARAMETERS KNOWN

Step 1– Test the whole sample of size N to failure. Arrange all times to failure in increasing value, or rank them.

Step 2– Determine the total number of failures, which equals N.

Step 3– After each successive failure, determine the total number of failures already observed, f_i, at T_i^+, and f_i-1 at T_i^-. Divide these two numbers by the total number of failures, N. Each one of these calculations gives the observed probability of failure at T_i^+ and at T_i^-, or

$$Q_O(T_i^+) = \frac{f_i}{N},$$

and

$$Q_O(T_i^-) = \frac{f_i - 1}{N}, \quad i = 1, 2, ..., N.$$

Step 4– Calculate in the same order, using the equation of the selected distribution, the theoretical, or expected, probability of failure, $Q_E(T_i)$, for each failure that has occurred.

Step 5– Determine D_i^+ and D_i^-, which are the maximum values of the absolute difference between $Q_E(T_i)$ and $Q_O(T_i^+)$, and between $Q_E(T_i)$ and $Q_O(T_i^-)$, from

$$D_i^+ = |Q_E(T_i) - Q_O(T_i^+)|,$$

and

$$D_i^- = |Q_E(T_i) - Q_O(T_i^-)|,$$

for each failure.

Step 6– Determine the *maximum absolute differences*, D_{max}^+, D_{max}^- and finally D_{max} from

$$D_{max} = max\{D_{max}^+, D_{max}^-\}.$$

Step 7– In the *K-S* goodness-of-fit test critical value tables, given in Appendix G-1, look up the allowable, or critical, difference value, D_{cr}, for the sample size N, at the desired, or required, level of significance, usually 0.05.

Step 8– If the calculated maximum absolute difference is less than the allowable critical value from the table, or if

$$D_{max} < D_{cr},$$

then the selected distribution is accepted as representing the actual test, or operational, data at the indicated level of significance. If the calculated difference is equal to or greater than the allowable critical value, or if

$$D_{max} \geq D_{cr},$$

then the selected distribution is not accepted, as representing the data, at the indicated level of significance.

EXAMPLE 20–5

The times-to-failure data presented in Column 2 of Tables 20.1 and 20.5 may be described by a normal distribution having a mean of $\overline{T} = 2,722.26$ hr and a standard deviation of $\sigma_T = 21.92$ hr. Does this normal *pdf* fit the data at the significance level of 0.05?

SOLUTION TO EXAMPLE 20–5

Step 1– The data are arranged in the order of increasing times to failure in Column 2 of Table 20.5.

Step 2– The total number of failures is 19, or $N = 19$, from Column 1.

Step 3– The observed probabilities of failure for each successive failure, $Q_O(T_i^+)$ and $Q_O(T_i^-)$, are calculated and are given in Columns 3 and 4. For example, for Failure No. 5 at $T_5 = 2,706$ hr,

$$Q_O(T_5^+ = 2,706^+ \text{ hr}) = \frac{5}{19} = 0.2632,$$

and

$$Q_O(T_5^- = 2,706^- \text{ hr}) = \frac{5-1}{19} = 0.2105.$$

Step 4– The *expected probability of failure* is calculated, for each time to failure, from

$$Q_E(T_i) = \Phi\left(\frac{T_i - \overline{T}}{\sigma_T}\right),$$

TABLE 20.5– Kolmogorov-Smirnov goodness-of-fit test for the normally distributed data of Example 20–5.

i^*	T_i, hr	$Q_O(T_i^+)^{**}$	$Q_O(T_i^-)^{***}$	$Q_E(T_i)$	D_i^{+****}	D_i^{-*****}
1	2,681	0.0526	0.0000	0.0301	0.0225	0.0301
2	2,691	0.1053	0.0526	0.0764	0.0289	0.0238
3	2,697	0.1579	0.1053	0.1251	0.0328	0.0198
4	2,702	0.2105	0.1579	0.1788	0.0317	0.0209
5	2,706	0.2632	0.2105	0.2296	0.0336	0.0191
6	2,709	0.3158	0.2632	0.2743	0.0415	0.0111
7	2,712	0.3684	0.3158	0.3192	0.0492†	0.0034
8	2,716	0.4211	0.3684	0.3859	0.0352	0.0175
9	2,720	0.4737	0.4211	0.4602	0.0135	0.0391
10	2,722	0.5263	0.4737	0.4960	0.0303	0.0223
11	2,726	0.5789	0.5263	0.5675	0.0114	0.0412
12	2,728	0.6316	0.5789	0.6026	0.0290	0.0237
13	2,730	0.6842	0.6316	0.6368	0.0474	0.0052
14	2,736	0.7368	0.6842	0.7357	0.0011	0.0515‡
15	2,739	0.7895	0.7368	0.7764	0.0131	0.0396
16	2,744	0.8421	0.7895	0.8389	0.0032	0.0494
17	2,747	0.8947	0.8421	0.8708	0.0239	0.0287
18	2,754	0.9474	0.8947	0.9265	0.0209	0.0318
19	2,763	1.0000	0.9474	0.9686	0.0314	0.0212

* i = failure number.
** $Q_O(T_i^+) = \frac{i}{N}$.
*** $Q_O(T_i^-) = \frac{i-1}{N}$.
**** $D_i^+ = |Q_E(T_i) - Q_O(T_i^+)|$.
***** $D_i^- = |Q_E(T_i) - Q_O(T_i^-)|$.
† Maximum absolute difference, $D_{max}^+ = 0.0492$.
‡ Maximum absolute difference, $D_{max}^- = 0.0515$.

and entered in Column 5. For example, for Failure No. 5,

$T_5 = 2,706$ hr, $\overline{T} = 2,722.26$ hr and $\sigma_T = 21.92$ hr. Then,

$$Q_E(T_5 = 2,706 \text{ hr}) = \Phi(\frac{2,706 - 2,722.26}{21.92}) = \Phi(-0.74),$$

or, from Appendix B,

$$Q_E(T_5 = 2,706 \text{ hr}) = 0.2296.$$

Step 5– The absolute differences, D_i^+ and D_i^-, are calculated for each failure and entered in Columns 6 and 7. For example, for Failure No. 5,

$$\begin{aligned} D_5^+ &= |Q_E(T_5) - Q_0(T_5^+)|, \\ &= |0.2296 - 0.2632|, \\ &= 0.0336, \end{aligned}$$

and

$$\begin{aligned} D_5^- &= |Q_E(T_5) - Q_0(T_5^-)|, \\ &= |0.2296 - 0.2105|, \\ &= 0.0191. \end{aligned}$$

Step 6– Failure No. 7 has the maximum difference, $D_7^+ = D_{max}^+ = 0.0492$. Failure No. 14 has the maximum difference, $D_{14}^- = D_{max}^- = 0.0515$. Therefore $D_{max} = 0.0515$.

Step 7– From the Kolmogorov-Smirnov tables in Appendix G-1, it may be seen that the maximum allowable, or critical, difference for $N = 19$ failures, at the $\alpha = 0.05$ significance level, is $D_{cr} = 0.30143$.

Step 8– Since the maximum difference, $D_{max} = 0.0515$, is less than the allowable, $D_{cr} = 0.30143$, or

$$D_{max} = 0.0515 < D_{cr} = 0.30143,$$

the normal distribution is accepted as adequately describing the times-to-failure data at the 0.05 significance level.

EXAMPLE 20–6

The times-to-failure data in Column 2 of Tables 20.2 and 20.6 may be described by a lognormal *pdf* having a mean of $\overline{T}' = 4.3553$ and a standard deviation of $\sigma_{T'} = 0.67670$. Does this lognormal *pdf* fit these data at the significance level of 0.05?

TABLE 20.6– Kolmogorov-Smirnov goodness-of-fit test for the lognormally distributed data of Example 20–6.

i^*	T_i, hr	$\log_e T_i$	$Q_O(T_i^+)^{**}$	$Q_O(T_i^-)^{***}$	$Q_E(T_i)$	D_i^{+*****}	D_i^{-*****}
1	30.4	3.41443	0.1111	0.0000	0.0823	0.0288	0.0823
2	36.7	3.60278	0.2222	0.1111	0.1335	0.0887	0.0224
3	53.3	3.97593	0.3333	0.2222	0.2877	0.0456	0.0655
4	58.5	4.06903	0.4444	0.3333	0.3372	0.1072†	0.0039
5	74.0	4.30405	0.5556	0.4444	0.4681	0.0875	0.0237
6	99.3	4.59814	0.6667	0.5556	0.6406	0.0261	0.0850‡
7	114.3	4.73883	0.7778	0.6667	0.7157	0.0621	0.0490
8	140.1	4.94235	0.8889	0.7778	0.8078	0.0811	0.0300
9	257.9	5.55256	1.0000	0.8889	0.9616	0.0384	0.0727

$*$ i = failure number.

$**$ $Q_O(T_i^+) = \frac{i}{N}$.

$***$ $Q_O(T_i^-) = \frac{i-1}{N}$.

$****$ $D_i^+ = |Q_E(T_i) - Q_O(T_i^+)|$.

$*****$ $D_i^- = |Q_E(T_i) - Q_O(T_i^-)|$.

$\dagger$ Maximum absolute difference, $D_{max}^+ = 0.1072$.

$\ddagger$ Maximum absolute difference, $D_{max}^- = 0.0850$.

SOLUTION TO EXAMPLE 20–6

Step 1– The data are arranged, in order of increasing times to failure, in Column 2 of Table 20.6, and the natural logarithms of these are entered in Column 3.

Step 2– The total number of failures is $N = 9$ from Column 1.

Step 3– The observed probabilities of failure for each successive failure, $Q_O(T_i^+)$ and $Q_O(T_i^-)$, are calculated and are given in Columns 4 and 5. For example, for Failure No. 3 at $T_3 = 53.3$ hr,

$$Q_O(T_3^+ = 53.3^+ \text{ hr}) = \frac{3}{9} = 0.3333,$$

and

$$Q_O(T_3^- = 53.3^- \text{ hr}) = \frac{3-1}{9} = 0.2222.$$

Step 4– The *expected probability of failure* is calculated, for each time to failure, from

$$Q_E(T_i) = \Phi\left(\frac{\log_e T_i - \overline{T}'}{\sigma_{T'}}\right),$$

and entered in Column 6. For example, for Failure No. 3,

$T_3 = 53.3$ hr,

$\log_e T_3 = 3.97593, \overline{T}' = 4.3553$ and $\sigma_{T'} = 0.67670$. Then,

$$Q_E(T_3 = 53.3 \text{ hr}) = \Phi\left(\frac{3.97593 - 4.3553}{0.67670}\right) = \Phi(-0.56),$$

or, from Appendix B,

$Q_E(T_3 = 53.3 \text{ hr}) = 0.2877.$

Step 5– The absolute differences, D_i^+ and D_i^-, are calculated for each failure and entered in Columns 7 and 8. For example, for Failure No. 3,

$$\begin{aligned} D_3^+ &= |Q_E(T_3) - Q_O(T_3^+)|, \\ &= |0.2877 - 0.3333|, \\ &= 0.0456, \end{aligned}$$

and

$$\begin{aligned} D_3^- &= |Q_E(T_3) - Q_O(T_3^-)|, \\ &= |0.2877 - 0.2222|, \\ &= 0.0655. \end{aligned}$$

Step 6– Failure No. 4 has the maximum difference, $D_4^+ = D_{max}^+ = 0.1072$. Failure No. 6 has the maximum difference, $D_6^- = D_{max}^- = 0.0850$. Therefore $D_{max} = 0.1072$.

Step 7– From the Kolmogorov-Smirnov tables in Appendix G-1, it may be seen that the maximum allowable, or critical, difference for $N = 9$ failures, at the $\alpha = 0.05$ significance level, is $D_{cr} = 0.43001$.

Step 8– Since the maximum difference, $D_{max} = 0.1072$, is less than the allowable, $D_{cr} = 0.43001$, or

$$D_{max} = 0.1072 < D_{cr} = 0.43001,$$

the lognormal distribution is accepted as adequately describing the times-to-failure data at the 0.05 significance level.

EXAMPLE 20–7

The times-to-failure data presented in Column 2 of Tables 20.3 and 20.7 may be described by an exponential distribution with a mean time between failures of $MTBF = m = 101.4$ hr. Does this exponential *pdf* fit these data at the significance level of 0.05?

SOLUTION TO EXAMPLE 20–7

Step 1– The data are arranged in order of increasing times to failure in Column 2 of Table 20.7.

Step 2– The total number of failures is 49, or $N = 49$, from Column 1.

Step 3– The $Q_O(T_i^+)$ and $Q_O(T_i^-)$ for each successive failure are given in Columns 3 and 4. For example, for Failure No. 26 at $T_{26} = 95.1$ hr,

$$Q_O(T_{26}^+ = 95.1^+ \text{ hr}) = \frac{26}{49} = 0.5306,$$

and

$$Q_O(T_{26}^- = 95.1^- \text{ hr}) = \frac{26 - 1}{49} = 0.5102.$$

TABLE 20.7– Kolmogorov-Smirnov goodness-of-fit test for the exponentially distributed data of Example 20–7.

i^*	T_i, hr	$Q_O(T_i^+)^{**}$	$Q_O(T_i^-)^{***}$	$Q_E(T_i)$	D_i^{+****}	D_i^{-*****}
1	1.2	0.0204	0.0000	0.0117	0.0087	0.0117
2	1.5	0.0408	0.0204	0.0147	0.0261	0.0057
3	2.8	0.0612	0.0408	0.0272	0.0340	0.0136
4	4.9	0.0816	0.0612	0.0472	0.0344	0.0140
5	6.8	0.1020	0.0816	0.0648	0.0372	0.0168
6	7.0	0.1224	0.1020	0.0667	0.0557	0.0353
7	12.1	0.1429	0.1224	0.1126	0.0303	0.0098
8	13.7	0.1633	0.1429	0.1264	0.0369	0.0165
9	15.1	0.1837	0.1633	0.1384	0.0453	0.0249
10	15.2	0.2041	0.1837	0.1392	0.0649	0.0445
11	23.9	0.2245	0.2041	0.2100	0.0145	0.0059
12	24.3	0.2449	0.2245	0.2131	0.0318	0.0114
13	25.1	0.2653	0.2449	0.2193	0.0460	0.0256
14	35.8	0.2857	0.2653	0.2974	0.0117	0.0321
15	38.9	0.3061	0.2857	0.3186	0.0125	0.0329
16	47.9	0.3265	0.3061	0.3764	0.0499	0.0703
17	48.9	0.3469	0.3265	0.3826	0.0357	0.0561
18	49.3	0.3673	0.3469	0.3850	0.0177	0.0519
19	53.2	0.3878	0.3673	0.4082	0.0204	0.0409
20	55.6	0.4082	0.3878	0.4420	0.0138	0.0542
21	62.7	0.4286	0.4082	0.4611	0.0325	0.0529
22	72.4	0.4490	0.4286	0.5103	0.0613	0.0817
23	73.6	0.4694	0.4490	0.5160	0.0466	0.0670
24	76.8	0.4898	0.4694	0.5310	0.0412	0.0616
25	83.8	0.5102	0.4898	0.5623	0.0521	0.0725

TABLE 20.7– Continued.

i^*	T_i, hr	$Q_O(T_i^+)^{**}$	$Q_O(T_i^-)^{***}$	$Q_E(T_i)$	$D_i^{+}{}^{****}$	$D_i^{-}{}^{*****}$
26	95.1	0.5306	0.5102	0.6086	0.0780	0.0984
27	97.9	0.5510	0.5306	0.6191	0.0681	0.0885
28	99.6	0.5714	0.5510	0.6255	0.0541	0.0745
29	102.8	0.5918	0.5714	0.6371	0.0453	0.0657
30	108.5	0.6122	0.5918	0.6569	0.0447	0.0651
31	128.7	0.6327	0.6122	0.7189	0.0862†	0.1067‡
32	133.6	0.6531	0.6327	0.7322	0.0791	0.0995
33	144.1	0.6735	0.6531	0.7585	0.0850	0.1054
34	147.6	0.6939	0.6735	0.7667	0.0728	0.0932
35	150.6	0.7143	0.6939	0.7735	0.0592	0.0796
36	151.6	0.7347	0.7143	0.7757	0.0410	0.0614
37	152.6	0.7551	0.7347	0.7779	0.0228	0.0432
38	164.2	0.7755	0.7551	0.8019	0.0264	0.0468
39	166.8	0.7959	0.7755	0.8069	0.0110	0.0314
40	178.6	0.8163	0.7959	0.8281	0.0118	0.0322
41	185.2	0.8367	0.8163	0.8390	0.0023	0.0227
42	187.1	0.8571	0.8367	0.8420	0.0151	0.0053
43	203.0	0.8776	0.8571	0.8649	0.0127	0.0078
44	204.3	0.8980	0.8776	0.8666	0.0314	0.0110
45	229.5	0.9184	0.8980	0.8960	0.0224	0.0020
46	233.1	0.9388	0.9184	0.8996	0.0392	0.0188
47	254.1	0.9592	0.9388	0.9184	0.0408	0.0204
48	291.7	0.9796	0.9552	0.9437	0.0359	0.0155
49	304.4	1.0000	0.9796	0.9503	0.0497	0.0293

* i = failure number.

** $Q_O(T_i^+) = \frac{i}{N}$.

*** $Q_O(T_i^-) = \frac{i-1}{N}$.

**** $D_i^+ = |Q_E(T_i) - Q_O(T_i^+)|$.

***** $D_i^- = |Q_E(T_i) - Q_O(T_i^-)|$.

† Maximum absolute difference, $D_{max}^+ = 0.0862$.

‡ Maximum absolute difference, $D_{max}^- = 0.1067$.

Step 4– The *expected probability of failure* is calculated, for each time to failure, from

$$Q_E(T_i) = 1 - e^{-\frac{T_i}{m}},$$

and entered in Column 5. For example, for Failure No. 26, $T_{26} = 95.1$ hr and $m = 101.4$ hr. Then,

$$Q_E(T_{26} = 95.1 \text{ hr}) = 1 - e^{-\frac{95.1}{101.4}},$$

or

$$Q_E(T_{26} = 95.1 \text{ hr}) = 0.6086.$$

Step 5– The absolute differences, D_i^+ and D_i^-, are calculated for each failure and entered in Columns 6 and 7. For example, for Failure No. 26,

$$
\begin{aligned}
D_{26}^+ &= |Q_E(T_{26}) - Q_O(T_{26}^+)|, \\
&= |0.6086 - 0.5306|, \\
&= 0.0780,
\end{aligned}
$$

and

$$
\begin{aligned}
D_{26}^- &= |Q_E(T_{26}) - Q_O(T_{26}^-)|, \\
&= |0.6086 - 0.5102|, \\
&= 0.0984.
\end{aligned}
$$

Step 6– Failure No. 31 has the maximum differences, $D_{31}^+ = D_{max}^+ = 0.0862$ and $D_{31}^- = D_{max}^- = 0.1067$. Therefore $D_{max} = 0.1067$.

Step 7– From the Kolmogorov-Smirnov tables in Appendix G-1, it may be seen that the maximum allowable, or critical, difference for $N = 49$ failures, at the $\alpha = 0.05$ significance level, is $D_{cr} = 0.19028$.

Step 8– Since the maximum difference, $D_{max} = 0.1067$, is less than the allowable, $D_{cr} = 0.19028$, or

$$D_{max} = 0.1067 < D_{cr} = 0.19028,$$

the exponential distribution is accepted as adequately describing these times-to-failure data at the 0.05 significance level.

EXAMPLE 20–8

The Weibull distribution was found to fit the transistor times-to-failure data presented in Column 2 of Tables 20.4 and 20.8 with reasonable accuracy. According to past experience the parameters are known to be

$$\gamma = 200 \text{ hr}, \quad \beta = 1.39 \text{ and } \eta = 9,916 \text{ hr}.$$

Does this Weibull *pdf* fit these data at the significance level of 0.05?

SOLUTION TO EXAMPLE 20–8

Step 1– The data are arranged in order of increasing times to failure in Column 2 of Table 20.8.

Step 2– The total number of failures is 24, or $N = 24$, from Column 1.

Step 3– The $Q_O(T_i^+)$ and $Q_O(T_i^-)$ for each successive failure are given in Columns 3 and 4. For example, for Failure No. 11 at $T_{11} = 740$ hr,

$$Q_O(T_{11}^+ = 740^+ \text{ hr}) \;=\; \frac{11}{24} = 0.4583,$$

and

$$Q_O(T_{11}^- = 740^- \text{ hr}) \;=\; \frac{11-1}{24} = 0.4167.$$

Step 4– The *expected probability of failure* is calculated, for each time to failure, from

$$Q_E(T_i) = 1 - e^{-(\frac{T_i-\gamma}{\eta})^\beta},$$

and entered in Column 5. For example, for Failure No. 11, $T_{11} = 740$ hr, $\gamma = 200$ hr, $\beta = 1.39$ and $\eta = 9,916$ hr. Then

$$Q_E(T_{11} = 740 \text{ hr}) = 1 - e^{-(\frac{740-200}{9,916})^{1.39}} = 0.0174.$$

Step 5– The absolute differences, D_i^+ and D_i^- are calculated for each failure and entered in Columns 6 and 7. For example, for Failure No. 11,

$$\begin{aligned}
D_{11}^+ \;&=\; |\,Q_E(T_{11}) - Q_O(T_{11}^+)\,|, \\
&=\; |\,0.0174 - 0.4583\,|, \\
&=\; 0.4409,
\end{aligned}$$

TABLE 20.8– Kolmogorov-Smirnov goodness-of-fit test for the Weibull distributed data of Example 20–8.

i^*	T_i, hr	$Q_O(T_i^+)^{**}$	$Q_O(T_i^-)^{***}$	$Q_E(T_i)$	D_i^{+****}	D_i^{-*****}
1	260	0.0417	0.0000	0.0008	0.0409	0.0008
2	350	0.0833	0.0417	0.0029	0.0804	0.0388
3	420	0.1250	0.0833	0.0050	0.1200	0.0783
4	440	0.1667	0.1250	0.0057	0.1610	0.1193
5	480	0.2083	0.1667	0.0070	0.2013	0.1597
6	480	0.2500	0.2083	0.0070	0.2430	0.2013
7	530	0.2917	0.2500	0.0088	0.2829	0.2412
8	580	0.3333	0.2917	0.0107	0.3226	0.2810
9	680	0.3750	0.3333	0.0148	0.3602	0.3185
10	710	0.4167	0.3750	0.0160	0.4007	0.3590
11	740	0.4583	0.4167	0.0174	0.4409	0.3993
12	780	0.5000	0.4583	0.0191	0.4809	0.4392
13	820	0.5417	0.5000	0.0210	0.5207	0.4790
14	840	0.5833	0.5417	0.0219	0.5614	0.5198
15	920	0.6250	0.5833	0.0258	0.5992	0.5575
16	930	0.6667	0.6250	0.0263	0.6404	0.5987
17	1,050	0.7083	0.6667	0.0323	0.6760	0.6344
18	1,060	0.7500	0.7083	0.0329	0.7171	0.6754
19	1,070	0.7917	0.7500	0.0334	0.7583	0.7166
20	1,270	0.8333	0.7917	0.0443	0.7890	0.7474
21	1,340	0.8750	0.8333	0.0482	0.8268	0.7851
22	1,370	0.9167	0.8750	0.0500	0.8667	0.8250
23	1,880	0.9583	0.9167	0.0813	0.8770	0.8354
24	2,130	1.0000	0.9583	0.0977	0.9023†	0.8606‡

* i = failure number.
** $Q_O(T_i^+) = \frac{i}{N}$.
*** $Q_O(T_i^-) = \frac{i-1}{N}$.
**** $D_i^+ = |Q_E(T_i) - Q_O(T_i^+)|$.
***** $D_i^- = |Q_E(T_i) - Q_O(T_i^-)|$.
† Maximum absolute difference, $D_{max}^+ = 0.9023$.
‡ Maximum absolute difference, $D_{max}^- = 0.8606$.

and

$$D_{11}^- = |Q_E(T_{11}) - Q_O(T_{11}^-)|,$$
$$= |0.0174 - 0.4167|,$$
$$= 0.3993.$$

Step 6– Failure No. 24 has the maximum differences, $D_{24}^+ = D_{max}^+ = 0.9023$ and $D_{24}^- = D_{max}^- = 0.8606$. Therefore $D_{max} = 0.9023$.

Step 7– From the Kolmogorov-Smirnov tables in Appendix G-1, it may be seen that the maximum allowable, or critical, difference for $N = 24$ failures, at the $\alpha = 0.05$ significance level, is $D_{cr} = 0.26931$.

Step 8– Since the maximum difference, $D_{max} = 0.9023$, is greater than the allowable difference, $D_{cr} = 0.26931$, or

$$D_{max} = 0.9023 > D_{cr} = 0.26931,$$

the Weibull distribution cannot be accepted as adequately describing these times-to-failure data at the 0.05 significance level.

20.4 MODIFIED *K-S* GOODNESS-OF-FIT TEST FOR CONTINUOUS DISTRIBUTIONS WITH PARAMETERS UNKNOWN

The Kolmogorov-Smirnov D statistic provides a means of testing whether a set of observations are from some completely specified continuous distribution, $Q_E(T)$. When certain parameters of the hypothesized distribution are unknown but must be estimated from the sample, then the standard *K-S test* critical value tables are conservative in the sense that the true significance level is much less than the level associated with the standard critical values. That is to say, the test critical values should be modified for the distributions with unknown parameters. Since the distributions of the *K-S test* statistics depend on the hypothesized distribution family when parameters are estimated, a separate set of tables of critical values must be obtained for each family. Furthermore, it is usually not possible to obtain an analytic expression for the test statistic distributions. Therefore, Monte Carlo methods are often used to obtain critical values for the *modified* tests with estimated parameters.

Lilliefors [2; 3] has generated tables for the modified *K-S test* for the exponential distribution with estimated mean and for the normal distribution with estimated mean and variance, as shown in Appendices I-2 and I-3. The normal critical values can also be used for the lognormal distribution as follows:

Let $T' = \log_e T$; then T' is normally distributed with mean of $\hat{T}'$ and standard deviation of $\sigma_{T'}$ to be estimated from the sample. Thus the problem is reduced to testing that T' values have the normal distribution with estimated $\hat{T}'$ and $\sigma_{T'}$.

A more convenient and compact table which gives the critical values at various significance levels for the *modified K-S* goodness-of-fit test, for the exponential and the normal distributions, is given in Appendix G-4. It is based on the percentage points of the modified D_{max} statistic, D^*_{max}, whose quantifications are also given in this table. The detailed procedure of applying this appendix is given at the bottom thereof.

Woodruff, et al [11] have generated tables for the *modified K-S test* for the Weibull distribution with estimated location and scale parameters and known shape parameter values of 0.5, 1.0, 1.5, 2.0, 2.5, 3.0, 3.5 and 4.0 respectively. These tables are given in Appendix G-5.

Woodruff, et al [12] generated tables for the *modified K-S test* for the gamma distribution with estimated location and scale parameters, and known shape parameter values of 0.5, 1.0, 1.5, 2.0, 2.5, 3.0, 3.5 and 4.0 respectively. These tables are given in Appendix G-6.

Chandra, et al [10] generated tables for the *modified K-S test* for the extreme-value distribution for the following three cases:

Case 1– The location parameter, γ, is estimated, but the scale parameter, η, is known.

Case 2– The scale parameter, η, is estimated, but the location parameter, γ, is known.

Case 3– Both the location parameter, γ, and the scale parameter, η, are estimated.

These tables are given in Appendix G-7. They can also be used for the two-parameter Weibull distribution

$$Q(T) = 1 - e^{-(\frac{T}{\eta})^\beta}.$$

When one or both η and β are unknown, but are estimated from the sample, the test is reduced to a *modified K-S test* for the extreme-value distribution as follows:
Let

$$y = -\log_e T;$$

then, y is extreme-value distributed, or

$$Q(y) = e^{-e^{-\frac{y-\gamma}{\eta'}}}, \quad -\infty < y < +\infty,$$

where

$$\gamma = -\log_e \eta,$$

and

$$\eta' = \frac{1}{\beta}.$$

Thus the problem is reduced to testing that the y values have the extreme-value distribution, with γ and η', or both, unknown but estimated from the sample.

Littell, et al [9] and Mann, et al [4] generated tables for the *modified K-S test* for the two-parameter Weibull distribution with estimated shape and scale parameters. These tables are given in Appendix G-8.

The whole procedure for the *modified K-S test* is identical to the *basic* and the *refined K-S tests*, except that the *modified critical K-S value* tables should be applied in the final decision making step instead of using the *standard K-S critical value* tables as we did before for the completely specified distributions.

EXAMPLE 20–9

Assume that the times-to-failure data presented in Column 2 of Tables 20.1 and 20.5 may be described by a normal distribution. The mean and standard deviation have been estimated from the given sample as $\hat{\bar{T}} = 2,722.26$ hr and $\hat{\sigma}_T = 21.92$ hr, because they were not known a priori. Does this normal *pdf* fit the data at the significance level of 0.05?

SOLUTION TO EXAMPLE 20–9

From Example 20-5 it has been found that the maximum difference between the expected and the observed *cdf's* is

$$D_{max} = 0.0515.$$

From the *modified K-S test* tables for the normal distribution in Appendix G-3, it may be found that the maximum allowable, or critical, difference for $N = 19$ failures, at the $\alpha = 0.05$ significance level, is

$$D_{cr} = 0.195.$$

Since

$$D_{max} = 0.0515 < D_{cr} = 0.195,$$

the normal distribution with $\hat{T} = 2,722.26$ hr and $\hat{\sigma}_T = 21.92$ hr is accepted as adequately describing the times-to-failure data at the 0.05 significance level.

EXAMPLE 20–10

Assume the times-to-failure data presented in Column 2 of Tables 20.2 and 20.6 may be described by a lognormal distribution. The log mean and log standard deviation have been estimated from the given sample as $\hat{T}' = 4.3553$ and $\hat{\sigma}_{T'} = 0.67670$, because they were not known a priori. Does this lognormal *pdf* fit the data at the significance level of 0.05?

SOLUTION TO EXAMPLE 20–10

Let

$$T' = \log_e T;$$

then T' is normally distributed with

$$
\begin{aligned}
Q_E(T') &= \Phi(\frac{T' - 4.3553}{0.67670}), \\
&= \Phi(\frac{\log_e T - 4.3553}{0.67670}), \\
&= Q_E(T).
\end{aligned}
$$

Similarly

$$Q_O(T_i'^-) = Q_O(T_i^-) = \frac{i-1}{N},$$

and

$$Q_O(T_i'^+) = Q_O(T_i^+) = \frac{i}{N}.$$

Therefore,

$$D'_{max} = D_{max}$$

From Example 20-6 it has been found that the maximum difference between the observed and the expected distributions is

$$D_{max} = 0.1072.$$

From the *modified K-S test* tables for the normal distribution in Appendix G-3, it may be found that the maximum allowable, or critical, difference for $N = 9$ failures, at the $\alpha = 0.05$ significance level, is

$$D_{cr} = 0.271.$$

Since

$$D'_{max} = 0.1072 < D_{cr} = 0.271,$$

the lognormal distribution with $\overline{T}' = 4.3553$ hr and $\hat{\sigma}_{T'} = 0.67670$ hr is accepted as adequately describing the times-to-failure data at the 0.05 significance level.

EXAMPLE 20–11

Assume the times-to-failure data presented in Column 2 of Tables 20.3 and 20.7 may be described by an exponential distribution. The mean time between failures has been estimated as $MTBF = \hat{m} = 101.4$ hr, because it was not known a priori. Does this exponential *pdf* fit these data at the significance level of 0.05 ?

SOLUTION TO EXAMPLE 20–11

From Example 20-7 it has been found that the maximum difference between the observed and the expected distributions is

$$D_{max} = 0.1067.$$

From the modified *K-S test* tables for the exponential distribution in Appendix G-2, it may be found that the maximum allowable, or critical, difference for $N = 49$ failures, at the $\alpha = 0.05$ significance level, is

$$D_{cr} = \frac{1.06}{\sqrt{49}} = \frac{1.06}{7} = 0.15143.$$

Since

$$D_{max} = 0.1067 < D_{cr} = 0.15143,$$

the exponential distribution with $MTBF = \hat{m} = 101.4$ hr is accepted as adequately describing the times-to-failure data at the 0.05 significance level.

EXAMPLE 20–12

Assume the transistor times-to-failure data presented in Column 2 of Tables 20.4 and 20.8 are Weibull distributed with known β, but unknown η and γ. According to the past experience the shape parameter is known to be

$$\beta = 1.39.$$

By the probability plotting method the scale and location parameters are found to be

$$\hat{\eta} = 9,916 \text{ hr} \text{ and } \hat{\gamma} = 200 \text{ hr},$$

because they were not known a priori. Does this Weibull *pdf* fit these data at the significance level of 0.05?

SOLUTION TO EXAMPLE 20–12

From Example 20-8 it has been found that the maximum difference between the observed and the expected distributions is

$$D_{max} = 0.9023.$$

From the *modified K-S test* tables for the Weibull distribution, with known shape parameter but unknown scale and location parameters, in Appendix G-5, it may be found that the maximum allowable, or critical, difference for $N = 24$ failures, at the $\alpha = 0.05$ significance level, is (by a double interpolation)

$$D_{cr} = 0.2084.$$

Since

$$D_{max} = 0.9023 > D_{cr} = 0.2084,$$

the Weibull distribution with $\beta = 1.39$, $\hat{\eta} = 9,916$ hr and $\hat{\gamma} = 200$ hr cannot be accepted as adequately describing these times-to-failure data at the 0.05 significance level.

20.5 COMMENTS ON THE KOLMOGOROV-SMIRNOV GOODNESS-OF-FIT TEST

1. Both the *standard* and the *modified K-S tests* are more powerful than the *chi-squared test* for any sample size [2; 3].

2. The *K-S test* can be used with *small sample sizes*, where the validity of the *chi-squared test* would be questionable.

3. Strictly speaking, the *modified K-S test* should be used for distributions with estimated parameters. However, in engineering practice, if the modified critical value tables for a particular distribution are not available, the *standard K-S* tables can still be used satisfactorily. Even though the *standard K-S* tables are conservative, they are often good enough for engineering accuracy.

PROBLEMS

20-1. Using the data of Table 4.3, apply the Kolmogorov-Smirnov goodness-of-fit test, using all three methods, when fitting the normal *pdf* to these data, and discuss the results at the 5% and the 10% levels of significance.

20-2. Using the data of Table 4.2, apply the Kolmogorov-Smirnov goodness-of-fit test, using all three methods, when fitting the exponential *pdf* to these data, and discuss the results at the 5% and the 10% levels of significance.

20-3. Using the data of Table 4.3, apply the Kolmogorov-Smirnov goodness-of-fit test, using all three methods, when fitting the Weibull *pdf* to these data, and discuss the results at the 5% and the 10% levels of significance.

20-4. Using the data of Table 19.1, apply the Kolmogorov-Smirnov goodness-of-fit test, using all three methods, when fitting the normal *pdf* to these data, and discuss the results at the 5% and the 10% levels of significance.

20-5. Using the data of Table 19.1, apply the Kolmogorov-Smirnov goodness-of-fit test, using all three methods, when fitting the exponential *pdf* to these data, and discuss the results at the 5% and the 10% levels of significance.

20-6. Using the data of Table 19.1, apply the Kolmogorov-Smirnov goodness-of-fit test, using all three methods, when fitting the Weibull *pdf* to these data, and discuss the results at the 5% and the 10% levels of significance.

20-7. Using the data of Table 19.2, apply the Kolmogorov-Smirnov goodness-of-fit test, using all three methods, when fitting the normal *pdf* to these data, and discuss the results at the 5% and the 10% levels of significance.

20-8. Using the data of Table 19.2, apply the Kolmogorov-Smirnov goodness-of-fit test, using all three methods, when fitting the exponential *pdf* to these data, and discuss the results at the 5% and the 10% levels of significance.

TABLE 20.9 – Field data for a wear-out failure characteristic of Problem 20–10.

Life in 1,000 hr, starting at 750 hr	Number of failures
0 – 1	40
1 – 2	210
2 – 3	300
3 – 4	250
4 – 5	80
5 – 6	20

Total = 900

20-9. Using the data of Table 19.2, apply the Kolmogorov-Smirnov goodness-of-fit test, using all three methods, when fitting the Weibull *pdf* to these data, and discuss the results at the 5% and the 10% levels of significance.

20-10. Using the data in Table 20.9 do the following:

(1) Apply the chi-squared goodness-of-fit test using:
1.1 The fitted normal distribution.
1.2 The fitted Weibull distribution.
1.3 Comparatively discuss the results of the previous two cases at the 5% and the 10% levels of significance.

(2) Apply the Kolmogorov-Smirnov goodness-of-fit test using all three methods to:
2.1 The fitted normal distribution.
2.2 The fitted Weibull distribution.
2.3 Comparatively discuss the results of the previous two cases at the 5% and the 10% levels of significance.

(3) Comparatively discuss the chi-squared and *K-S* goodness-of-fit test results.

20-11. Using the data in Table 20.10 do the following:

(1) Apply the chi-squared goodness-of-fit test using:
1.1 The fitted normal distribution.
1.2 The fitted Weibull distribution.
1.3 Comparatively discuss the results of the previous two cases at the 5% and the 10% levels of significance.

TABLE 20.10 – **Field data for a wear-out failure characteristic of Problem 20–11.**

Life in 1,000 hr	Number of failures
0 – 1	4
1 – 2	21
2 – 3	30
3 – 4	25
4 – 5	8
5 – 6	2
Total =	90

(2) Apply the Kolmogorov-Smirnov goodness-of-fit test using all three methods to:

2.1 The fitted normal distribution.

2.2 The fitted Weibull distribution.

2.3 Comparatively discuss the results of the previous two cases at the 5% and the 10% levels of significance.

(3) Comparatively discuss the chi-squared and K-S goodness-of-fit test results.

20-12. Using the data in Table 20.11 do the following:

(1) Apply the chi-squared goodness-of-fit test using:

1.1 The fitted exponential distribution.

1.2 The fitted Weibull distribution.

1.3 Comparatively discuss the results of the previous two cases at the 5% and the 10% levels of significance.

(2) Apply the Kolmogorov-Smirnov goodness-of-fit test using all three methods to:

2.1 The fitted exponential distribution.

2.2 The fitted Weibull distribution.

2.3 Comparatively discuss the results of the previous two cases at the 5% and the 10% levels of significance.

(3) Comparatively discuss the chi-squared and K-S goodness-of-fit test results.

20-13. Using the data in Table 20.12 do the following:

TABLE 20.11 – Field data for a sample of identical units of Problem 20–12.

Life in 100 hr, starting at 200 hr	Number of failures
0 – 10	300
10 – 20	200
20 – 30	140
30 – 40	90
40 – 50	60
50 – 60	40
> 60	70

Total = 900

(1) Apply the chi-squared goodness-of-fit test using:

1.1 The fitted exponential distribution.

1.2 The fitted Weibull distribution.

1.3 Comparatively discuss the results of the previous two cases at the 5% and the 10% levels of significance.

(2) Apply the Kolmogorov-Smirnov goodness-of-fit test using all three methods to:

2.1 The fitted exponential distribution.

2.2 The fitted Weibull distribution.

2.3 Comparatively discuss the results of the previous two cases at the 5% and the 10% levels of significance.

(3) Comparatively discuss the chi-squared and K-S goodness-of-fit test results.

20-14. Using the data in Table 20.13 do the following:

(1) Apply the chi-squared goodness-of-fit test using:

1.1 The fitted exponential distribution.

1.2 The fitted Weibull distribution.

1.3 Comparatively discuss the results of the previous two cases at the 5% and the 10% levels of significance.

(2) Apply the Kolmogorov-Smirnov goodness-of-fit test using all three methods to:

2.1 The fitted exponential distribution.

2.2 The fitted Weibull distribution.

TABLE 20.12 – Field data for a sample of identical units of Problem 20–13.

Life in 1,000 hr	Number of failures
0 – 1	30
1 – 2	20
2 – 3	14
3 – 4	9
4 – 5	6
5 – 6	4
> 6	7
Total =	90

2.3 Comparatively discuss the results of the previous two cases at the 5% and the 10% levels of significance.

(3) Comparatively discuss the chi-squared and K-S goodness-of-fit test results.

20-15. The cycles-to-failure data given in Table 20.14 have been obtained from fatigue tests on 35 identical specimens. Do the following:

(1) Fit the normal distribution to these data by the calculation of its parameters.

(2) Fit the lognormal distribution to these data by the calculation of its parameters.

(3) Calculate the chi-squared goodness-of-fit value for the fitted normal *pdf*.

(4) Calculate the Kolmogorov-Smirnov goodness-of-fit D_{max} values for the fitted normal *pdf* using the first two methods.

(5) Repeat Case 3 for the fitted lognormal *pdf*.

(6) Repeat Case 4 for the fitted lognormal *pdf*.

(7) Calculate the coefficient of skewness for the raw data and for the logarithms of the data.

(8) Calculate the coefficient of kurtosis for the raw data and for the logarithms of the data.

TABLE 20.13 – Field data for a sample of identical units of Problem 20–14.

Life starting at 300 hr	Number of failures
0 – 100	75
100 – 200	50
200 – 300	35
300 – 400	22
400 – 500	15
500 – 600	10
> 600	18

Total = 225

TABLE 20.14 – The cycles-to-failure data from fatigue tests on 35 identical specimens of Problem 20–15.

61,667.0	62,882.0	68,394.0	69,342.0	69,372.0
71,179.0	71,268.0	71,624.0	71,713.0	72,039.0
72,305.0	76,602.0	77,817.0	78,262.0	80,010.0
84,218.0	84,544.0	85,433.0	85,788.0	87,152.0
88,337.0	89,759.0	89,789.0	90,767.0	90,797.0
91,923.0	96,131.0	98,679.0	99,953.0	100,961.0
102,620.0	110,799.0	112,577.0	126,031.0	127,690.0

(9) Tabulate the chi-squared, D_{max}, α_3 and α_4 values found in Cases 3 through 8 in one table for the normal and log-normal *pdf*'s. Discuss the results and justify which one of the two distributions should be chosen at the 5% and the 10% significance levels.

20-16. Rework Problem 19-6 applying the Kolmogorov-Smirnov goodness-of-fit test using all three methods and discuss the results at the 5% and the 10% levels of significance.

20-17. Rework Problem 19-7 applying the Kolmogorov-Smirnov goodness-of-fit test using all three methods and discuss the results at the 5% and the 10% levels of significance.

20-18. Rework Problem 19-8 applying the Kolmogorov-Smirnov goodness-of-fit test using all three methods and discuss the results at the 5% and the 10% levels of significance.

20-19. Rework Problem 19-9 applying the Kolmogorov-Smirnov goodness-of-fit test using all three methods and discuss the results at the 5% and the 10% levels of significance.

20-20. Rework Problem 19-10 applying the Kolmogorov-Smirnov goodness-of-fit test using all three methods and discuss the results at the 5% and the 10% levels of significance.

20-21. Rework Problem 19-11 applying the Kolmogorov-Smirnov goodness-of-fit test using all three methods and discuss the results at the 5% and the 10% levels of significance.

20-22. Rework Problem 19-12 applying the Kolmogorov-Smirnov goodness-of-fit test using all three methods and discuss the results at the 5% and the 10% levels of significance.

20-23. Rework Problem 19-13 applying the Kolmogorov-Smirnov goodness-of-fit test using all three methods and discuss the results at the 5% and the 10% levels of significance.

20-24. Rework Problem 19-14 applying the Kolmogorov-Smirnov goodness-of-fit test using all three methods and discuss the results at the 5% and the 10% levels of significance.

20-25. Rework Problem 19-15 applying the Kolmogorov-Smirnov goodness-of-fit test using all three methods and discuss the results at the 5% and the 10% levels of significance.

20-26. Rework Problem 19-16 applying the Kolmogorov-Smirnov goodness-of-fit test using all three methods and discuss the results at the 5% and the 10% levels of significance.

20-27. Rework Problem 19-17 applying the Kolmogorov-Smirnov goodness-of-fit test using all three methods and discuss the results at the 5% and the 10% levels of significance.

20-28. Rework Problem 19-18 applying the Kolmogorov-Smirnov goodness-of-fit test using all three methods and discuss the results at the 5% and the 10% levels of significance.

20-29. Rework Problem 19-19 applying the Kolmogorov-Smirnov goodness-of-fit test using all three methods and discuss the results at the 5% and the 10% levels of significance.

20-30. Rework Problem 19-20 applying the Kolmogorov-Smirnov goodness-of-fit test using all three methods and discuss the results at the 5% and the 10% levels of significance.

REFERENCES

1. Massey, F. J., "The Kolmogorov-Smirnov Test for Goodness of Fit," *Journal of the American Statistical Association*, Vol. 46, pp. 68-78, March 1951.

2. Lilliefors, H. W., "On the Kolmogorov-Smirnov Test for Normality with Mean and Variance Unknown," *Jour. Am. Stat. Assn.*, pp. 399-402, June 1967.

3. Lilliefors, H. W., "On the Kolmogorov-Smirnov Test for Exponential Distribution with Mean Unknown," *Jour. Am. Stat. Assn.*, pp. 387-389, March 1969.

4. Mann, N. R., Scheuer, E. M. and Fertig, K. W., "A New Goodness-of-fit Test for the Two-parameter Weibull or Extreme-value Distribution with Unknown Parameters," *Commun. Statist.*, Vol. 2, pp. 383-400, 1973.

5. Stephens, M. A., "EDF Statistics for Goodness of Fit and Some Comparisons," *Jour. Am. Stat. Assn.*, Vol. 69, No. 347, pp. 730-737, September 1974.

6. Durbin, J., "Kolmogorov-Smirnov Tests when Parameters are Estimated with Applications to Tests of Exponentiality and Tests on Spacing," *Biometrika*, Vol. 62, No. 1, pp. 5-21, 1975.

7. Green, J. R. and Hegazy, J. R., "Powerful Modified-EDF Goodness-of-Fit Tests," *Jour. Am. Stat. Assn.*, Vol. 71, No. 353, pp. 204-209, March 1976.

8. Stephens, M. A., "Goodness of Fit for the Extreme Value Distribution," *Biometrika*, Vol. 64, pp. 583-588, 1977.

9. Littell, R. C., McClave, J. T. and Offen, W. W., "Goodness-of-fit Tests for the Two-parameter Weibull Distribution," *Commun. Statist.*, Vol. B8, No. 3, pp. 257-269, 1979.

10. Chandra, M., Singpurwalla, N. D. and Stephens, M. A., "Kolmogorov Statistics for Tests of Fit for the Extreme-value and Weibull Distributions," *Jour. Am. Stat. Assn.*, Vol. 76, No. 375, pp. 729-731, September 1981.

11. Woodruff, B. W., Moore, A. H., Dunne, E. J. and Cortes, R., "A Modified Kolmogorov-Smirnov Test for Weibull Distribution with Unknown Location and Scale Parameters," *IEEE Transactions on Reliability*, Vol. R-32, No. 2, pp. 209-213, June 1983.

12. Woodruff, B. W., Viviano, P. J., Moore, A. H. and Dunne, E. J., "Modified Goodness-of-Fit Test for Gamma Distributions with Unknown Location and Scale Parameters," *IEEE Transactions on Reliability*, Vol. R-33, No. 3, pp. 241-245, August 1984.

13. Guilbaud, O., "Exact Kolmogorov-Type Tests for Left-Truncated and/or Right-Censored Data," *Jour. Am. Stat. Assn.*, Vol. 83, No. 401, pp. 213-221, June 1988.

14. Lawless, J. F., *Statistical Models and Methods for Lifetime Data*, John Wiley & Sons, Inc., New York, 580 pp., 1982.

Chapter 21

ANDERSON-DARLING AND CRAMER-VON MISES GOODNESS-OF-FIT TESTS

21.1   INTRODUCTION

In addition to Pearson's χ^2 goodness-of-fit test [1; 11], and the *Kolmogorov-Smirnov* goodness-of-fit test which is based on the empirical distribution function (EDF), there are other goodness-of-fit tests. Two such good tests, which are also based on the EDF, are the *Anderson-Darling (AD)* [1; 2] and the *Cramer-von Mises (CvM)* [1; 3] goodness-of-fit tests. The difference among these three EDF-based tests is mainly on their test statistics. Therefore, the corresponding critical values are different. But theoretically they should yield the same conclusion for the same sample data in terms of rejecting or accepting a hypothesized distribution, though they may have different powers of discrimination for a particular distribution. The step-by-step application of these tests to the normal, lognormal, exponential and Weibull distributions are presented in this chapter and illustrated by examples.

The standard *Anderson-Darling* and the *Cramer-von Mises* tests are only valid for the continuous and completely specified distributions. They cannot be used for discrete distributions and ungrouped data. However, they can be modified to handle the censored sample and the unknown parameter cases [1; 5; 6; 7; 8; 9; 10].

21.2 THE STANDARD ANDERSON-DARLING AND CRAMER-VON MISES GOODNESS-OF-FIT TESTS FOR CONTINUOUS DISTRIBUTIONS WITH PARAMETERS KNOWN

Let $T_1 \leq T_2 \leq ... \leq T_N$ be an ordered random sample of size N from a population with unknown $Q(T)$. Let $Q_E(T)$ be a completely specified cumulative distribution function. Our goal is to test

$$H_0: \quad Q(T) \quad = \quad Q_E(T) \quad \text{for all } T,$$

versus

$$H_1: \quad Q(T) \quad \neq \quad Q_E(T). \tag{21.1}$$

Anderson and Darling [1; 2] suggested computing the empirical, or observed, cumulative distribution function, $Q_O(T)$, based on the data $T_1, T_2, ..., T_N$, and then considering the statistic

$$A_N^2 = N \int_{-\infty}^{+\infty} \frac{[Q_O(T) - Q_E(T)]^2}{Q_E(T)[1 - Q_E(T)]} dQ_E(T) \tag{21.2}$$

as a measure of the "distance" or "discrepancy" between the empirical and the specified (assumed) distributions; i.e., $Q_O(T)$ and $Q_E(T)$.

Cramer and Mises [1; 3] suggested computing the empirical, or observed, cumulative distribution function, $Q_O(T)$, based on the data $T_1, T_2, ..., T_N$, and then considering the statistic

$$W_N^2 = N \int_{-\infty}^{+\infty} [Q_O(T) - Q_E(T)]^2 \, dQ_E(T) \tag{21.3}$$

as a measure of the "distance" or "discrepancy" between the empirical and the specified (assumed) distributions; i.e., $Q_O(T)$ and $Q_E(T)$. Here $Q_O(T)$ is defined by

$$Q_O(T) = \begin{cases} 0, & T < T_1, \\ f_i/N, & T_i \leq T < T_{i+1}, \quad i = 1, 2, ..., N-1, \\ 1, & T \geq T_N, \end{cases} \tag{21.4}$$

where

$$f_i \quad = \quad \text{total number of observations up to } T_i.$$

Substituting Eq. (21.4) into Eqs. (21.2) and (??) yields the following alternate expressions:

$$A_N^2 = -\sum_{i=1}^{N} \frac{2i-1}{N} \{\log_e[Q_E(T_i)] + \log_e[1 - Q_E(T_{N+1-i}]\} - N,$$

(21.5)

and

$$W_N^2 = \sum_{i=1}^{N}[Q_E(T_i) - \frac{i-0.5}{N}]^2 + \frac{1}{12N},$$

(21.6)

which will be used, instead of Eqs. (??) and (??), to implement these goodness-of-fit tests.

The asymptotic distributions of these two statistics are known, but Stephens [4; 5] and [10, in which Stephens' work of 1970 is discussed] has been able to show that appropriate modifications of the finite-sample statistics, defined by Eqs. (21.7) and (21.8), may be referred without serious error to their asymptotic distributions. This ingenious empirical procedure is based partly on theory and partly on Monte Carlo simulation. The *modified Anderson-Darling* test statistic is

$$A_N^{2*} = A_N^2, \quad \text{for } N \geq 5.$$

(21.7)

The *modified Cramer-von Mises* test statistic is

$$W_N^{2*} = (W_N^2 - 0.4/N + 0.6/N^2)(1.0 + 1.0/N).$$

(21.8)

The percentage points for these two modified statistics are given in Table 21.1. With Eqs. (21.7) and (??), and Table 21.1, we can conduct the *Anderson-Darling* and the *Cramer-von Mises* goodness-of-fit tests without extensive tables.

Then for a given level of significance, α, the decision can be made on rejecting, or not rejecting, hypothesis H_0 by applying the following rule:

Reject, if $A_N^{2*} \geq A_{cr}^{2*}$, or $A_N^2 \geq A_{cr}^2$ if $N \geq 5$,
Do not reject, if $A_N^{2*} < A_{cr}^{2*}$, or $A_N^2 < A_{cr}^2$ if $N \geq 5$,

or by applying the following rule:

Reject, if $W_N^{2*} \geq W_{cr}^{2*}$,
Do not reject, if $W_N^{2*} < W_{cr}^{2*}$,

TABLE 21.1– Percentage points for the modified Anderson-Darling and Cramer-von Mises test statistics for completely specified distributions.

Statistic	Modified form	Percentage points, A_{cr}^{2*} and W_{cr}^{2*}				
	α	0.15	0.10	0.05	0.025	0.01
A_N^2	A_N^{2*} †	1.610	1.933	2.492	3.070	3.857
W_N^2	W_N^{2*} ‡	0.284	0.347	0.461	0.581	0.743

† $A_N^{2*} = A_N^2$, for $N \geq 5$.
‡ $W_N^{2*} = (W_N^2 - 0.4/N + 0.6/N^2)(1.0 + 1.0/N)$.

where A_{cr}^2 and W_{cr}^2 are the critical, or allowable, values for A_N^2 and W_N^2 corresponding to the specified significance level, α; and A_{cr}^{2*} and W_{cr}^{2*} are the critical, or allowable, values for A_N^{2*} and W_N^{2*} corresponding to the specified significance level, α. The step-by-step procedure for applying these tests to different distributions, is discussed and illustrated next.

21.3 STEP-BY-STEP METHOD FOR THE STANDARD ANDERSON-DARLING AND CRAMER-VON MISES GOODNESS-OF-FIT TESTS WITH PARAMETERS KNOWN

21.3.1 ANDERSON-DARLING GOODNESS-OF-FIT TEST

Step 1– Test the whole sample of size N to failure. Arrange all times to failure in increasing value, or rank them as $T_1 \leq T_2 \leq ... \leq T_N$.

Step 2– Determine the total number of failures, which equals N.

Step 3– After each successive failure, determine the total number of failures already observed, $f_i = i$, at T_i. Calculate the quantity $(2i - 1)/N$.

Step 4– Calculate in the same order, using the equation of the selected distribution, the theoretical, or expected, probability of failure, $Q_E(T_i)$, for each failure that has occurred.

Step 5– Calculate in the same order, the quantities $\log_e[Q_E(T_i)]$, $\log_e[1 - Q_E(T_{N+1-i})]$ and

$$A_i = \frac{2i - 1}{N}\{\log_e[Q_E(T_i)] + \log_e[1 - Q_E(T_{N+1-i})]\},$$

$$(21.9)$$

for each failure that has occurred.

Step 6– Calculate the *Anderson-Darling* goodness-of-fit test statistic from

$$
\begin{aligned}
A_N^2 &= -\sum_{i=1}^{N} \frac{2i - 1}{N}\{\log_e[Q_E(T_i)] \\
&\quad + \log_e[1 - Q_E(T_{N+1-i})]\} - N, \\
&= -\sum_{i=1}^{N} A_i - N,
\end{aligned}
\qquad (21.5')
$$

for the given sample.

Step 7– In the *Anderson-Darling* goodness-of-fit test modified critical value table, given in Table 21.1, look up the allowable, or critical, statistic percentile value, A_{cr}^2, for the sample size N, at the desired, or required, level of significance, usually 0.05.

Step 8– If the calculated test statistic value is less than the allowable critical statistical percentile value from the table, or if

$$A_N^2 < A_{cr}^2,$$

then the selected distribution is accepted as representing the actual test, or operational, data at the indicated level of significance. If the calculated test statistic value is equal to or greater than the allowable critical test statistic percentile value, or if

$$A_N^2 \geq A_{cr}^2,$$

then the selected distribution is not accepted, as representing the data, at the indicated level of significance.

21.3.2   CRAMER-VON MISES GOODNESS-OF-FIT TEST

Step 1– Test the whole sample of size N to failure. Arrange all times to failure in increasing value, or rank them as $T_1 \leq T_2 \leq \ldots \leq T_N$.

Step 2– Determine the total number of failures, which equals N.

Step 3– After each successive failure, determine the total number of failures already observed, f_i, at T_i. Calculate the quantity $(i - 0.5)/N$.

Step 4– Calculate in the same order, using the equation of the selected distribution, the theoretical, or expected, probability of failure, $Q_E(T_i)$, for each failure that has occurred.

Step 5– Calculate in the same order, the quantity $[Q_E(T_i) - \frac{i-0.5}{N}]^2$ for each failure that has occurred.

Step 6– Calculate the *Cramer-von Mises* goodness-of-fit test statistic from

$$W_N^2 = \sum_{i=1}^{N}[Q_E(T_i) - \frac{i - 0.5}{N}]^2 + \frac{1}{12N}, \qquad (21.10)$$

for the given sample.

Step 7– Calculate the *modified Cramer-von Mises* goodness-of-fit test statistic from

$$W_N^{2*} = (W_N^2 - 0.4/N + 0.6/N^2)(1.0 + 1.0/N), (21.11)$$

for the given sample.

Step 8– In the *Cramer-von Mises* goodness-of-fit test critical value table, given in Table 21.1, look up the allowable, or critical, statistic percentile value, W_{cr}^{2*}, for the sample size N, at the desired, or required, level of significance, usually 0.05.

Step 9– If the calculated (modified) test statistic value is less than the allowable critical statistical percentile value from the table, or if

$$W_N^{2*} < W_{cr}^{2*},$$

then the selected distribution is accepted as representing the actual test, or operational, data at the indicated level of significance. If the calculated (modified) test statistic

value is equal to or greater than the allowable critical test statistic percentile value, or if

$$W_N^{2*} \geq W_{cr}^{2*},$$

then the selected distribution is not accepted, as representing the data, at the indicated level of significance.

EXAMPLE 21–1

Rework Example 20–1 using the *Anderson-Darling* and the *Cramer-von Mises* goodness-of-fit tests.

SOLUTION TO EXAMPLE 21–1

1. Anderson-Darling Goodness-of-fit Test

Step 1– The data are arranged in the order of increasing times to failure in Column 2 of Table 21.1.1.

Step 2– The total number of failures is 19, or $N = 19$, from Column 1 of Table 21.1.1.

Step 3– The quantity $(2i - 1)/N$ is calculated for each failure and is given in Column 3 of Table 21.1.1. For example, for Failure No. 5 at $T_5 = 2,706$ hr,

$$\frac{2i - 1}{N} = \frac{2 \times 5 - 1}{19} = 0.473684.$$

Step 4– The *expected probability of failure* is calculated, for each time to failure, from

$$Q_E(T_i) = \Phi\left(\frac{T_i - \overline{T}}{\sigma_T}\right),$$

and entered in Column 4 of Table 21.1.1. For example, for Failure No. 5,

$T_5 = 2,706$ hr, $\overline{T} = 2,722.26$ hr and $\sigma_T = 21.92$ hr. Then,

$$\begin{aligned}
Q_E(T_5 = 2,706 \text{ hr}) &= \Phi(\frac{2,706 - 2,722.26}{21.92}), \\
&= \Phi(-0.74),
\end{aligned}$$

or, from Appendix B,

$$Q_E(T_5 = 2,706 \text{ hr}) = 0.2296.$$

TABLE 21.1.1– Anderson-Darling goodness-of-fit test for the normally distributed data of Example 21-1.

Failure number, i	Time to failure, T_i, hr	$\frac{2i-1}{N}$	Expected unreliability, $Q_E(T_i)$	$\log_e[Q_E(T_i)]$	$\log_e[1-Q_E(T_{N+1-i})]$	$A_i^{\dagger}$
1	2,681	0.052632	0.0301	-3.503230	-3.460947	-0.366536
2	2,691	0.157895	0.0764	-2.571773	-2.610470	-0.818249
3	2,697	0.263158	0.1251	-2.078642	-2.046394	-1.085536
4	2,702	0.368421	0.1788	-1.721487	-1.825730	-1.306870
5	2,706	0.473684	0.2296	-1.471417	-1.497897	-1.406517
6	2,709	0.578947	0.2743	-1.293533	-1.330670	-1.519276
7	2,712	0.684211	0.3192	-1.141937	-1.012802	-1.474295
8	2,716	0.789474	0.3859	-0.952177	-0.922812	-1.480254
9	2,720	0.894737	0.4602	-0.776094	-0.838173	-1.444344
10	2,722	1.000000	0.4960	-0.701179	-0.685179	-1.386358
11	2,726	1.105263	0.5675	-0.566515	-0.616557	-1.307605
12	2,728	1.210526	0.6026	-0.506502	-0.487597	-1.203383
13	2,730	1.315789	0.6842	-0.451300	-0.384487	-1.099719
14	2,736	1.421053	0.7357	-0.306933	-0.320619	-0.891783
15	2,739	1.526316	0.7764	-0.253087	-0.260845	-0.784424
16	2,744	1.631579	0.8389	-0.175664	-0.196989	-0.608012
17	2,747	1.736842	0.8708	-0.138343	-0.133646	-0.472401
18	2,754	1.842105	0.9265	-0.076341	-0.079476	-0.287032
19	2,763	1.947368	0.9686	-0.031904	-0.030562	-0.121644

$$\sum = -19.064238$$

$\dagger\ A_i = \frac{2i-1}{N}\{\log_e[Q_E(T_i)] + \log_e[1 - Q_E(T_{N+1-i})]\}.$

For another example, for Failure No. (19+1-5) or No. 15,

$T_{15} = 2,739$ hr, $\overline{T} = 2,722.26$ hr and $\sigma_T = 21.92$ hr. Then,

$$
\begin{aligned}
Q_E(T_{15} = 2,739 \text{ hr}) &= \Phi(\frac{2,739 - 2,722.26}{21.92}), \\
&= \Phi(-0.7637),
\end{aligned}
$$

or, from Appendix B,

$$Q_E(T_{15} = 2,739 \text{ hr}) = 0.7764.$$

Step 5— Then, the quantities $\log_e[Q_E(T_i)]$, $\log_e[1 - Q_E(T_{N+1-i})]$ and

$$A_i = \frac{2i - 1}{N}\{\log_e[Q_E(T_i)] + \log_e[1 - Q_E(T_{N+1-i})]\},$$

are calculated for each failure and entered in Columns 5, 6 and 7 of Table 21.1.1. For example, for Failure No. 5 at $T_5 = 2,706$ hr,

$$
\begin{aligned}
\log_e[Q_E(T_5)] &= \log_e[Q_E(2,706 \text{ hr})], \\
&= \log_e[0.2296],
\end{aligned}
$$

or

$$\log_e[Q_E(T_5)] = -1.471417.$$

$$
\begin{aligned}
\log_e[1 - Q_E(T_{19+1-5})] &= \log_e[1 - Q_E(T_{15})], \\
&= \log_e[1 - Q_E(2,739 \text{ hr})], \\
&= \log_e[1 - 0.7764],
\end{aligned}
$$

or

$$\log_e[1 - Q_E(T_{15})] = -1.497897.$$

Then, from Eq. (21.9),

$$
\begin{aligned}
A_5 &= \frac{2 \times 5 - 1}{19}\{\log_e[Q_E(T_5)] \\
&\quad + \log_e[1 - Q_E(T_{19+1-5})]\}, \\
&= \frac{2 \times 5 - 1}{19}\{\log_e[Q_E(T_5)] \\
&\quad + \log_e[1 - Q_E(T_{15})]\}, \\
&= 0.473684 \times \{-1.471417 - 1.497897\},
\end{aligned}
$$

or

$$A_5 = -1.406517.$$

Step 6– The *Anderson-Darling* goodness-of-fit test statistic for the given sample, from Eq. (21.5′), is

$$
\begin{aligned}
A_N^2 &= -\sum_{i=1}^{19} \frac{2i-1}{19} \{\log_e[Q_E(T_i)] \\
&\quad + \log_e[1 - Q_E(T_{19+1-i}]\} - 19, \\
&= -\sum_{i=1}^{19} A_i - 19, \\
&= -(-19.064238) - 19,
\end{aligned}
$$

or

$$
A_N^2 = 0.064238.
$$

Step 7– From the *Anderson-Darling* goodness-of-fit test critical value table, given in Table 21.1, it may be seen that the allowable, or critical, statistic value for $N = 19 > 5$ failures, at the $\alpha = 0.05$ significance level, is $A_{cr}^2 = A_{cr}^{2*} = 2.492$.

Step 8– Since the calculated statistic value, $A_N^2 = 0.064238$, is less than the allowable, or critical value, $A_{cr}^2 = 2.492$, or

$$
A_N^2 = 0.064238 < A_{cr}^2 = 2.492,
$$

the normal distribution is accepted as adequately describing the times-to-failure data at the 0.05 significance level.

2. Cramer-von Mises Goodness-of-fit Test

Step 1– The data are arranged in the order of increasing times to failure in Column 2 of Table 21.1.2.

Step 2– The total number of failures is 19, or $N = 19$, from Column 1 of Table 21.1.2.

Step 3– The quantity $(i-0.5)/N$ for each failure is calculated and is given in Column 3 of Table 21.1.2. For example, for Failure No. 5 at $T_5 = 2,706$ hr,

$$
\frac{i - 0.5}{N} = \frac{5 - 0.5}{19} = 0.236842.
$$

TABLE 21.1.2– Cramer-von Mises goodness-of-fit test for the normally distributed data of Example 21–1.

Failure number, i	Time to failure, T_i, hr	$\frac{i-0.5}{N}$	Expected unreliability, $Q_E(T_i)$	$[Q_E(T_i) - \frac{i-0.5}{N}]^2$
1	2,681	0.026316	0.0301	0.000014
2	2,691	0.078947	0.0764	0.000006
3	2,697	0.131579	0.1251	0.000042
4	2,702	0.184211	0.1788	0.000029
5	2,706	0.236842	0.2296	0.000052
6	2,709	0.289474	0.2743	0.000230
7	2,712	0.342105	0.3192	0.000525
8	2,716	0.394737	0.3859	0.000078
9	2,720	0.447368	0.4602	0.000165
10	2,722	0.500000	0.4960	0.000016
11	2,726	0.552632	0.5675	0.000221
12	2,728	0.605263	0.6026	0.000007
13	2,730	0.657895	0.6368	0.000445
14	2,736	0.710526	0.7357	0.000630
15	2,739	0.763158	0.7764	0.000175
16	2,744	0.815789	0.8389	0.000534
17	2,747	0.868421	0.8708	0.000006
18	2,754	0.921053	0.9265	0.000030
19	2,763	0.973684	0.9686	0.000026

$$\sum = 0.003236$$

Step 4–   The *expected probability of failure* is calculated, for each time to failure, from

$$Q_E(T_i) = \Phi\left(\frac{T_i - \overline{T}}{\sigma_T}\right),$$

and entered in Column 4 of Table 21.1.2, which is the same as that given in Column 4 of Table 21.1.1, or

$$Q_E(T_5 = 2,706 \text{ hr}) = 0.2296.$$

Step 5–   Then, the quantity $[Q_E(T_i)] - \frac{i-0.5}{N}]^2$ is calculated for each failure and entered in Column 5 of Table 21.1.2. For example, for Failure No. 5 at $T_5 = 2,706$ hr,

$$[Q_E(T_5)] - \frac{5 - 0.5}{19}]^2 \;=\; [0.2296 - 0.236842]^2,$$
$$=\; 0.000052.$$

Step 6–   The *Cramer-von Mises* goodness-of-fit test statistic, calculated for the given sample from Eq. (21.10), is

$$W_N^2 \;=\; \sum_{i=1}^{19}[Q_E(T_i) - \frac{i - 0.5}{19}]^2 + \frac{1}{12 \times 19},$$
$$=\; 0.003236 + 0.004386,$$

or

$$W_N^2 \;=\; 0.007622.$$

Step 7–   The *modified Cramer-von Mises* goodness-of-fit test statistic, calculated for the given sample from Eq. (21.11), is

$$W_N^{2*} \;=\; (W_N^2 - 0.4/N + 0.6/N^2)(1.0 + 1.0/N),$$
$$=\; (0.007622 - 0.4/19 + 0.6/19^2)(1.0 + 1.0/19),$$

or

$$W_N^{2*} \;=\; -0.012388.$$

Step 8–   From the *Cramer-von Mises* goodness-of-fit test critical value table, given in Table 21.1, it may be seen that the allowable, or critical, statistic value for $N = 19$ failures, at the $\alpha = 0.05$ significance level, is $W_{cr}^{2*} = 0.461$.

Step 9– Since the calculated statistic value, $W_N^{2*} = -0.012388$, is less than the allowable, or critical value, $W_{cr}^{2*} = 0.461$, or

$$W_N^{2*} = -0.012388 < W_{cr}^{2*} = 0.461,$$

the normal distribution is accepted as adequately describing the times-to-failure data at the 0.05 significance level.

EXAMPLE 21–2

Rework Example 20–2 using the *Anderson-Darling* and the *Cramer-von Mises* goodness-of-fit tests.

SOLUTION TO EXAMPLE 21–2

1. Anderson-Darling Goodness-of-fit Test

Step 1– The data are arranged in the order of increasing times to failure in Column 2 of Table 21.2.1.

Step 2– The total number of failures is nine, or $N = 9$, from Column 1 of Table 21.2.1.

Step 3– The quantity $(2i-1)/N$ for each failure is calculated and is given in Column 3 of Table 21.2.1. For example, for Failure No. 3 at $T_3 = 53.3$ hr,

$$\frac{2i-1}{N} = \frac{2 \times 3 - 1}{9} = 0.555556.$$

Step 4– The *expected probability of failure* is calculated, for each time to failure, from

$$Q_E(T_i) = \Phi\left(\frac{\log_e T_i - \overline{T}'}{\sigma_{T'}}\right),$$

and entered in Column 4 of Table 21.2.1. For example, for Failure No. 3,

$$T_3 = 53.3 \text{ hr},$$

$\log_e T_3 = 3.97593$, $\overline{T}' = 4.3553$ and $\sigma_{T'} = 0.67670$. Then,

$$Q_E(T_3 = 53.3 \text{ hr}) = \Phi\left(\frac{3.97593 - 4.3553}{0.67670}\right),$$
$$= \Phi(-0.56),$$

TABLE 21.2.1– Anderson-Darling goodness-of-fit test for the lognormally distributed data of Example 21–2.

Failure number, i	Time to failure, T_i, hr	$\frac{2i-1}{N}$	Expected unreliability, $Q_E(T_i)$	$\log_e[Q_E(T_i)]$	$\log_e[1 - Q_E(T_{N+1-i})]$	A_i[†]
1	30.4	0.111111	0.0823	-2.497384	-3.259798	-0.639676
2	36.7	0.333333	0.1335	-2.013654	-1.649219	-1.220958
3	53.3	0.555556	0.2877	-1.245837	-1.257725	-1.390868
4	58.5	0.777778	0.3372	-1.087079	-1.023319	-1.641421
5	74.0	1.000000	0.4681	-0.759073	-0.631300	-1.390373
6	99.3	1.222222	0.6406	-0.445350	-0.411282	-1.046995
7	114.3	1.444444	0.7157	-0.334494	-0.339256	-0.973195
8	140.1	1.666667	0.8078	-0.213441	-0.143293	-0.594557
9	257.9	1.888889	0.9616	-0.039157	-0.085885	-0.236189

$$\sum = -9.134232$$

† $A_i = \frac{2i-1}{N}\{\log_e[Q_E(T_i)] + \log_e[1 - Q_E(T_{N+1-i})]\}$.

752

or, from Appendix B,

$$Q_E(T_3 = 53.3 \text{ hr}) = 0.2877.$$

For another example, for Failure No. (9+1-3) or No. 7,

$T_7 = 114.3$ hr, $\overline{T}' = 4.3553$ and $\sigma_{T'} = 0.67670$. Then,

$$Q_E(T_7 = 114.3 \text{ hr}) = \Phi\left(\frac{4.7388 - 4.3553}{0.67670}\right),$$
$$= \Phi(0.5668),$$

or, from Appendix B,

$$Q_E(T_7 = 114.3 \text{ hr}) = 0.7157.$$

Step 5— Then, the quantities $\log_e[Q_E(T_i)]$, $\log_e[1-Q_E(T_{N+1-i})]$ and

$$A_i = \frac{2i-1}{N}\{\log_e[Q_E(T_i)] + \log_e[1 - Q_E(T_{N+1-i})]\},$$

are calculated for each failure and entered in Columns 5, 6 and 7 of Table 21.2.1. For example, for Failure No. 3 at $T_5 = 53.3$ hr,

$$\log_e[Q_E(T_3)] = \log_e[Q_E(53.3 \text{ hr})],$$
$$= \log_e[0.2877],$$

or

$$\log_e[Q_E(T_3)] = -1.245837.$$

$$\log_e[1 - Q_E(T_{9+1-3})] = \log_e[1 - Q_E(T_7)],$$
$$= \log_e[1 - Q_E(114.3 \text{ hr})],$$
$$= \log_e[1 - 0.7157],$$

or

$$\log_e[1 - Q_E(T_7)] = -1.257725.$$

Then, from Eq. (21.9),

$$A_3 = \frac{2 \times 3 - 1}{9}\{\log_e[Q_E(T_3)]$$
$$+ \log_e[1 - Q_E(T_{9+1-3})]\},$$
$$= \frac{2 \times 3 - 1}{9}\{\log_e[Q_E(T_3)]$$
$$+ \log_e[1 - Q_E(T_7)]\},$$
$$= 0.555556 \times \{-1.245837 - 1.257725\},$$

or

$$A_3 = -1.390868.$$

Step 6– The *Anderson-Darling* goodness-of-fit test statistic, for the given sample, from Eq. (21.5'), is

$$A_N^2 = -\sum_{i=1}^{9} \frac{2i-1}{9}\{\log_e[Q_E(T_i)]$$
$$+ \log_e[1 - Q_E(T_{9+1-i}]\} - 9,$$
$$= -\sum_{i=1}^{9} A_i - 9,$$
$$= -(-9.134232) - 9,$$

or

$$A_N^2 = 0.134232.$$

Step 7– From the *Anderson-Darling* goodness-of-fit test critical value table, given in Table 21.1, it may be seen that the allowable, or critical, statistic value for $N = 9 > 5$ failures, at the $\alpha = 0.05$ significance level, is $A_{cr}^2 = A_{cr}^{2*} = 2.492$.

Step 8– Since the calculated statistic value, $A_N^2 = 0.134232$, is less than the allowable, or critical value, $A_{cr}^2 = 2.492$, or

$$A_N^2 = 0.134232 < A_{cr}^2 = 2.492,$$

the lognormal distribution is accepted as adequately describing the times-to-failure data at the 0.05 significance level.

2. Cramer-von Mises Goodness-of-fit Test

Step 1– The data are arranged in the order of increasing times to failure in Column 2 of Table 21.2.2.

Step 2– The total number of failures is 9, or $N = 9$, from Column 1 of Table 21.2.2.

Step 3– The quantity $(i-0.5)/N$ for each failure is calculated and is given in Column 3 of Table 21.2.2. For example, for Failure No. 3 at $T_3 = 53.3$ hr,

$$\frac{i-0.5}{N} = \frac{3-0.5}{9} = 0.277778.$$

TABLE 21.2.2– Cramer-von Mises goodness-of-fit test for the lognormally distributed data of Example 21–2.

Failure number, i	Time to failure, T_i, hr	$\frac{i-0.5}{N}$	Expected unreliability, $Q_E(T_i)$	$[Q_E(T_i) - \frac{i-0.5}{N}]^2$
1	30.4	0.055556	0.0823	0.000715
2	36.7	0.166667	0.1335	0.001100
3	53.3	0.277778	0.2877	0.000098
4	58.5	0.388889	0.3372	0.002672
5	74.0	0.500000	0.4681	0.001018
6	99.3	0.611111	0.6406	0.000870
7	114.3	0.722222	0.7157	0.000043
8	140.1	0.833333	0.8078	0.000652
9	257.9	0.944444	0.9616	0.000294
				$\sum = 0.007461$

Step 4– The *expected probability of failure* is calculated, for each time to failure, from

$$Q_E(T_i) = \Phi\left(\frac{\log_e T_i - \overline{T}'}{\sigma_{T'}}\right),$$

and entered in Column 4 of Table 21.2.2, which is the same as that given in Column 4 of Table 21.2.1.

Step 5– Then the quantity $[Q_E(T_i)] - \frac{i-0.5}{N}]^2$ is calculated for each failure and entered in Column 5 of Table 21.2.2. For example, for Failure No. 3 at $T_3 = 53.3$ hr,

$$[Q_E(T_3)] - \frac{3-0.5}{9}]^2 = [0.2877 - 0.277778]^2,$$
$$= 0.000098.$$

Step 6– The *Cramer-von Mises* goodness-of-fit test statistic, calculated for the given sample from Eq. (21.10), is

$$W_N^2 = \sum_{i=1}^{9}[Q_E(T_i) - \frac{i-0.5}{9}]^2 + \frac{1}{12 \times 9},$$
$$= 0.007461 + 0.009259,$$

or

$$W_N^2 \; = \; 0.016721.$$

Step 7– The *modified Cramer-von Mises* goodness-of-fit test statistic, calculated for the given sample from Eq. (21.11), is

$$W_N^{2*} \; = \; (W_N^2 - 0.4/N + 0.6/N^2)(1.0 + 1.0/N),$$
$$= \; (0.016721 - 0.4/9 + 0.6/9^2)(1.0 + 1.0/9),$$

or

$$W_N^{2*} \; = \; -0.022573.$$

Step 8– From the *Cramer-von Mises* goodness-of-fit test critical value table, given in Table 21.1, it may be seen that the allowable, or critical, statistic value for $N = 9$ failures, at the $\alpha = 0.05$ significance level, is $W_{cr}^{2*} = 0.461$.

Step 9– Since the calculated (modified) statistic value, $W_N^{2*} = -0.022573$, is less than the allowable, $W_{cr}^{2*} = 0.461$, or

$$W_N^{2*} = -0.022573 < W_{cr}^{2*} = 0.461,$$

the lognormal distribution is accepted as adequately describing the times-to-failure data at the 0.05 significance level.

EXAMPLE 21–3

Rework Example 20–3 using the *Anderson-Darling* and the *Cramer-von Mises* goodness-of-fit tests.

SOLUTION TO EXAMPLE 21–3

1. Anderson-Darling Goodness-of-fit Test

Step 1– The data are arranged in the order of increasing times to failure in Column 2 of Table 21.3.1.

Step 2– The total number of failures is 49, or $N = 49$, from Column 1 of Table 21.3.1.

Step 3– The quantity $(2i - 1)/N$ for each failure is calculated and is given in Column 3 of Table 21.3.1. For example, for Failure No. 26 at $T_{26} = 95.1$ hr,

$$\frac{2i - 1}{N} = \frac{2 \times 26 - 1}{49} = 1.040816.$$

TABLE 21.3.1– Anderson-Darling goodness-of-fit test for the exponentially distributed data of Example 21–3.

Failure number, i	Time to failure, T_i, hr	$\frac{2i-1}{N}$	Expected unreliability, $Q_E(T_i)$	$\log_e[Q_E(T_i)]$	$\log_e[1 - Q_E(T_{N+1-i})]$	A_i†
1	1.2	0.020408	0.0117	-4.448166	-3.001750	-0.152039
2	1.5	0.061214	0.0147	-4.219908	-2.877061	-0.434508
3	2.8	0.102041	0.0272	-3.604538	-2.505926	-0.623517
4	4.9	0.142857	0.0472	-3.053361	-2.598593	-0.764565
5	6.8	0.183673	0.0648	-2.736450	-2.263365	-0.918333
6	7.0	0.224490	0.0667	-2.707550	-2.014403	-1.060030
7	12.1	0.265306	0.1126	-2.183913	-2.001740	-1.110479
8	13.7	0.306122	0.1264	-2.068304	-1.845160	-1.197999
9	15.1	0.346939	0.1384	-1.977607	-1.826351	-1.319741
10	15.2	0.387755	0.1392	-1.971843	-1.760843	-1.447368
11	23.9	0.428571	0.2100	-1.560648	-1.644547	-1.373655
12	24.3	0.469388	0.2131	-1.545994	-1.618984	-1.485602
13	25.1	0.510204	0.2193	-1.517315	-1.504627	-1.541807
14	35.8	0.551020	0.2974	-1.212677	-1.494771	-1.491859
15	38.9	0.591837	0.3186	-1.143819	-1.485011	-1.555838
16	47.9	0.632653	0.3764	-0.977103	-1.455430	-1.538950

† $A_i = \frac{2i-1}{N}\{\log_e[Q_E(T_i)] + \log_e[1 - Q_E(T_{N+1-i})]\}$.

TABLE 21.3.1– Continued.

Failure number, i	Time to failure, T_i, hr	$\frac{2i-1}{N}$	Expected unreliability, $Q_E(T_i)$	$\log_e[Q_E(T_i)]$	$\log_e[1 - Q_E(T_{N+1-i})]$	$A_i^\dagger$
17	48.9	0.673469	0.3826	-0.960765	-1.420886	-1.603969
18	49.3	0.714286	0.3850	-0.954512	-1.317515	-1.622877
19	53.2	0.755102	0.4082	-0.895998	-1.269045	-1.634828
20	55.6	0.795918	0.4420	-0.816445	-1.069733	-1.501244
21	62.7	0.836735	0.4611	-0.774140	-1.013628	-1.495888
22	72.4	0.877551	0.5103	-0.672757	-0.982164	-1.452177
23	73.6	0.918367	0.5160	-0.661649	-0.965218	-1.494061
24	76.8	0.959184	0.5310	-0.632993	-0.938025	-1.506895
25	83.8	1.000000	0.5623	-0.575720	-0.826222	-1.401941
26	95.1	1.040816	0.6086	-0.496594	-0.757153	-1.304920
27	97.9	1.081633	0.6191	-0.479489	-0.725670	-1.303539
28	99.6	1.122449	0.6255	-0.469204	-0.713962	-1.328044
29	102.8	1.163265	0.6371	-0.450829	-0.618225	-1.243593
30	108.5	1.204082	0.6569	-0.420224	-0.583396	-1.208440
31	128.7	1.244898	0.7189	-0.330033	-0.524586	-1.063914
32	133.6	1.285724	0.7322	-0.311702	-0.486133	-1.025787
33	144.1	1.326531	0.7585	-0.276413	-0.482238	-1.006373

$\dagger$ $A_i = \frac{2i-1}{N}\{\log_e[Q_E(T_i)] + \log_e[1 - Q_E(T_{N+1-i})]\}$.

TABLE 21.3.1– Continued.

Failure number, i	Time to failure, T_i, hr	$\frac{2i-1}{N}$	Expected unreliability, $Q_E(T_i)$	$\log_e[Q_E(T_i)]$	$\log_e[1 - Q_E(T_{N+1-i})]$	A_i†
34	147.6	1.367347	0.7667	-0.265660	-0.472246	-1.008973
35	150.6	1.408163	0.7735	-0.256830	-0.383606	-0.901838
36	151.6	1.448980	0.7757	-0.253889	-0.352968	-0.879468
37	152.6	1.489796	0.7779	-0.251157	-0.247564	-0.742993
38	164.2	1.530612	0.8019	-0.220771	-0.239654	-0.704733
39	166.8	1.571429	0.8069	-0.214556	-0.235722	-0.707579
40	178.6	1.612245	0.8281	-0.188621	-0.149893	-0.545768
41	185.21	1.653061	0.8390	-0.175545	-0.148964	-0.536433
42	187.1	1.693878	0.8420	-0.171975	-0.135133	-0.520203
43	203.0	1.734694	0.8649	-0.145141	-0.119459	-0.459001
44	204.3	1.775510	0.8666	-0.143178	-0.069029	-0.376775
45	229.5	1.816326	0.8960	-0.109815	-0.066995	-0.321144
46	233.1	1.857143	0.8996	-0.105805	-0.048350	-0.286288
47	254.1	1.897959	0.9184	-0.085122	-0.027577	-0.213898
48	291.7	1.938776	0.9437	-0.057947	-0.014809	-0.141058
49	304.4	1.799592	0.9503	-0.050978	-0.011769	-0.124213

$$\sum = -49.685249$$

† $A_i = \frac{2i-1}{N}\{\log_e[Q_E(T_i)] + \log_e[1 - Q_E(T_{N+1-i})]\}$.

759

Step 4—   The *expected probability of failure* is calculated, for each time to failure, from

$$Q_E(T_i) = 1 - e^{-\frac{T_i}{m}},$$

and entered in Column 4 of Table 21.3.1. For example, for Failure No. 26,

$T_{26} = 95.1$ hr and $m = 101.4$ hr. Then,

$$Q_E(T_{26} = 95.1 \text{ hr}) = 1 - e^{-\frac{95.1}{101.4}},$$

or

$$Q_E(T_{26} = 95.1 \text{ hr}) = 0.6086.$$

For another example, for Failure No. (49+1-26) or No. 24,

$T_{24} = 76.8$ hr, and $m = 101.4$ hr. Then,

$$Q_E(T_{24} = 76.8 \text{ hr}) = 1 - e^{-\frac{76.8}{101.4}},$$

or

$$Q_E(T_{24} = 76.8 \text{ hr}) = 0.5310.$$

Step 5—   Then, the quantities $\log_e[Q_E(T_i)]$, $\log_e[1-Q_E(T_{N+1-i})]$ and

$$A_i = \frac{2i - 1}{N} \{\log_e[Q_E(T_i)] + \log_e[1 - Q_E(T_{N+1-i})]\},$$

are calculated for each failure and entered in Columns 5, 6 and 7 of Table 21.3.1. For example, for Failure No. 26 at $T_{26} = 95.1$ hr,

$$\begin{aligned}
\log_e[Q_E(T_{26})] &= \log_e[Q_E(95.1 \text{ hr})], \\
&= \log_e[0.6086],
\end{aligned}$$

or

$$\log_e[Q_E(T_{26})] = -0.496594.$$

$$\begin{aligned}
\log_e[1 - Q_E(T_{49+1-26})] &= \log_e[1 - Q_E(T_{24})], \\
&= \log_e[1 - Q_E(76.8 \text{ hr})], \\
&= \log_e[1 - 0.5310],
\end{aligned}$$

or

$$\log_e[1 - Q_E(T_{24})] = -0.757153.$$

Then, from Eq. (21.9),

$$
\begin{aligned}
A_{26} &= \frac{2 \times 26 - 1}{49}\{\log_e[Q_E(T_{26})] \\
&\quad + \log_e[1 - Q_E(T_{49+1-26})]\}, \\
&= \frac{2 \times 26 - 1}{49}\{\log_e[Q_E(T_{26})] \\
&\quad + \log_e[1 - Q_E(T_{24})]\}, \\
&= 1.040816 \times \{-0.496594 - 0.757153\}, \\
&= -1.304920.
\end{aligned}
$$

Step 6– The *Anderson-Darling* goodness-of-fit test statistic, for the given sample, from Eq. (21.5′), is

$$
\begin{aligned}
A_N^2 &= -\sum_{i=1}^{49} \frac{2i-1}{49}\{\log_e[Q_E(T_i)] \\
&\quad + \log_e[1 - Q_E(T_{49+1-i})]\} - 49, \\
&= -\sum_{i=1}^{49} A_i - 49, \\
&= -(-49.685249) - 49,
\end{aligned}
$$

or

$$
A_N^2 = 0.685249.
$$

Step 7– From the *Anderson-Darling* goodness-of-fit test critical value table, given in Table 21.1, it may be seen that the allowable, or critical, statistic value for $N = 49 > 5$ failures, at the $\alpha = 0.05$ significance level, is $A_{cr}^2 = A_{cr}^{2*} = 2.492$.

Step 8– Since the calculated statistic value, $A_N^2 = 0.685249$, is less than the allowable, or critical, value $A_{cr}^2 = 2.492$, or

$$
A_N^2 = 0.685249 < A_{cr}^2 = 2.492,
$$

the exponential distribution is accepted as adequately describing the times-to-failure data at the 0.05 significance level.

2. Cramer-von Mises Goodness-of-fit Test

Step 1– The data are arranged in the order of increasing times to failure in Column 2 of Table 21.3.2.

TABLE 21.3.2– Cramer-von Mises goodness-of-fit test for
the exponentially distributed data of Example 21–3.

Failure number, i	Time to failure, T_i, hr	$\frac{i-0.5}{N}$	Expected unreliability, $Q_E(T_i)$	$[Q_E(T_i) - \frac{i-0.5}{N}]^2$
1	1.2	0.010204	0.0117	0.000002
2	1.5	0.030612	0.0147	0.000253
3	2.8	0.051020	0.0272	0.000567
4	4.9	0.071429	0.0472	0.000587
5	6.8	0.091837	0.0648	0.000731
6	7.0	0.112245	0.0667	0.002074
7	12.1	0.132653	0.1126	0.000402
8	13.7	0.153061	0.1264	0.000711
9	15.1	0.173469	0.1384	0.001230
10	15.2	0.193878	0.1392	0.002990
11	23.9	0.214286	0.2100	0.000018
12	24.3	0.234696	0.2131	0.000466
13	25.1	0.255102	0.2193	0.001282
14	35.8	0.275510	0.2974	0.000479
15	38.9	0.295918	0.3186	0.000514
16	47.9	0.316327	0.3764	0.003609
17	48.9	0.336735	0.3826	0.002104
18	49.3	0.357143	0.3850	0.000776
19	53.2	0.377551	0.4082	0.000939
20	55.6	0.397959	0.4420	0.001940
21	62.7	0.418367	0.4611	0.001826
22	72.4	0.438776	0.5103	0.005116
23	73.6	0.459184	0.5160	0.003228
24	76.8	0.479592	0.5310	0.002643
25	83.8	0.500000	0.5623	0.003881

TABLE 21.3.2– Continued.

Failure number, i	Time to failure, T_i, hr	$\frac{i-0.5}{N}$	Expected unreliability, $Q_E(T_i)$	$[Q_E(T_i) - \frac{i-0.5}{N}]^2$
26	95.1	0.520408	0.6086	0.007778
27	97.9	0.540816	0.6191	0.006128
28	99.6	0.561224	0.6255	0.004131
29	102.8	0.581633	0.6371	0.003077
30	108.5	0.602041	0.6569	0.003010
31	128.7	0.622449	0.7189	0.009303
32	133.6	0.642857	0.7322	0.007982
33	144.1	0.663265	0.7585	0.009070
34	147.6	0.683673	0.7667	0.006893
35	150.6	0.704082	0.7735	0.004819
36	151.6	0.724490	0.7757	0.002622
37	152.6	0.744898	0.7779	0.001089
38	164.2	0.765306	0.8019	0.001339
39	166.8	0.785714	0.8069	0.000449
40	178.6	0.806122	0.8281	0.000483
41	185.21	0.826531	0.8390	0.000155
42	187.1	0.846939	0.8420	0.000024
43	203.0	0.867347	0.8649	0.000006
44	204.3	0.887755	0.8666	0.000448
45	229.5	0.908163	0.8960	0.000148
46	233.1	0.928571	0.8996	0.000839
47	254.1	0.948980	0.9184	0.000935
48	291.7	0.969388	0.9437	0.000660
49	304.4	0.989796	0.9503	0.001560

$$\sum = 0.111318$$

Step 2– The total number of failures is 49, or $N = 49$, from Column 1 of Table 21.3.2.

Step 3– The quantity $(i-0.5)/N$ for each failure is calculated and is given in Column 3 of Table 21.3.2. For example, for Failure No. 26 at $T_{26} = 95.1$ hr,

$$\frac{i - 0.5}{49} = \frac{26 - 0.5}{49} = 0.520408.$$

Step 4– The *expected probability of failure* is calculated, for each time to failure, from

$$Q_E(T_i) = 1 - e^{-\frac{T_i}{m}},$$

and entered in Column 4, which is the same as that given in Column 4 of Table 21.3.1.

Step 5– Then, the quantity $[Q_E(T_i)] - \frac{i-0.5}{N}]^2$ is calculated for each failure and entered in Column 5 of Table 21.3.2. For example, for Failure No. 26 at $T_{26} = 95.1$ hr,

$$[Q_E(T_{26})] - \frac{26 - 0.5}{49}]^2 = [0.6086 - 0.520408]^2,$$
$$= 0.007778.$$

Step 6– The *Cramer-von Mises* goodness-of-fit test statistic, calculated for the given sample from Eq. (21.10), is

$$W_N^{2*} = \sum_{i=1}^{49}[Q_E(T_i) - \frac{i - 0.5}{49}]^2 + \frac{1}{12 \times 49},$$
$$= 0.111318 + 0.001701,$$

or

$$W_N^{2*} = 0.113019.$$

Step 7– The *modified Cramer-von Mises* goodness-of-fit test statistic, calculated for the given sample from Eq. (21.11), is

$$W_N^{2*} = (W_N^2 - 0.4/N + 0.6/N^2)(1.0 + 1.0/N),$$
$$= (0.113019 - 0.4/49 + 0.6/49^2)(1.0 + 1.0/49),$$

or

$$W_N^{2*} = 0.107251.$$

Step 8– From the *Cramer-von Mises* goodness-of-fit test critical value table, given in Table 21.1, it may be seen that the allowable, or critical, statistic value for $N = 49$ failures, at the $\alpha = 0.05$ significance level, is $W_{cr}^{2*} = 0.461$.

Step 9– Since the calculated (*modified*) statistic value, $W_N^{2*} = 0.107251$, is less than the allowable, $W_{cr}^{2*} = 0.461$, or

$$W_N^{2*} = 0.107251 < W_{cr}^{2*} = 0.461,$$

the exponential distribution is accepted as adequately describing these times-to-failure data at the 0.05 significance level.

EXAMPLE 21–4

Rework Example 20–4 using the *Anderson-Darling* and the *Cramer-von Mises* goodness-of-fit tests.

SOLUTION TO EXAMPLE 21–4

1. Anderson-Darling Goodness-of-fit Test

Step 1– The data are arranged in the order of increasing times to failure in Column 2 of Table 21.4.1.

Step 2– The total number of failures is 24, or $N = 24$, from Column 1 of Table 21.4.1.

Step 3– The quantity $(2i - 1)/N$ for each failure is calculated and is given in Column 3 of Table 21.4.1. For example, for Failure No. 11 at $T_{11} = 740$ hr,

$$\frac{2i - 1}{N} = \frac{2 \times 11 - 1}{24} = 0.875000.$$

Step 4– The *expected probability of failure* is calculated, for each time to failure, from

$$Q_E(T_i) = 1 - e^{-(\frac{T_i - \gamma}{\eta})^\beta},$$

and entered in Column 4 of Table 21.4.1. For example, for Failure No. 11,

$T_{11} = 740$ hr, $\gamma = 200$ hr, $\beta = 1.39$ and $\eta = 9,916$ hr. Then,

$$Q_E(T_{11} = 740 \text{ hr}) = 1 - e^{-(\frac{740-200}{9,916})^{1.39}},$$
$$= 0.0174.$$

TABLE 21.4.1– Anderson-Darling goodness-of-fit test for the Weibull distributed data of Example 21–4.

Failure number, i	Time to failure, T_i, hr	$\frac{2i-1}{N}$	Expected unreliability, $Q_E(T_i)$	$\log_e[Q_E(T_i)]$	$\log_e[1-Q_E(T_{N+1-i})]$	A_i†
1	260	0.041667	0.0008	-7.130899	-0.102808	-0.301404
2	350	0.125000	0.0029	-5.843045	-0.084796	-0.740980
3	420	0.208333	0.0050	-5.298317	-0.051293	-1.114502
4	440	0.291667	0.0057	-5.167289	-0.049400	-1.521534
5	480	0.375000	0.0070	-4.961845	-0.045311	-1.877684
6	480	0.458333	0.0070	-4.961845	-0.033971	-2.289749
7	530	0.541667	0.0088	-4.733004	-0.033453	-2.581831
8	580	0.625000	0.0107	-4.537511	-0.032833	-2.856465
9	680	0.708333	0.0148	-4.213128	-0.026652	-3.003177
10	710	0.791667	0.0160	-4.135167	-0.026139	-3.294367
11	740	0.875000	0.0174	-4.051285	-0.022143	-3.564250
12	780	0.958333	0.0191	-3.958067	-0.021224	-3.813487

† $A_i = \frac{2i-1}{N}\{\log_e[Q_E(T_i)] + \log_e[1-Q_E(T_{N+1-i})]\}$.

766

TABLE 21.4.1– Continued.

Failure number, i	Time to failure, T_i, hr	$\frac{2i-1}{N}$	Expected unreliability, $Q_E(T_i)$	$\log_e[Q_E(T_i)]$	$\log_e[1 - Q_E(T_{N+1-i})]$	$A_i^\dagger$
12	780	0.958333	0.0191	-3.958067	-0.021224	-3.813487
13	820	1.041667	0.0210	-3.863233	-0.019285	-4.044289
14	840	1.125000	0.0219	-3.821269	-0.017553	-4.318675
15	920	1.208333	0.0258	-3.657381	-0.016125	-4.438825
16	930	1.291667	0.0263	-3.638186	-0.014911	-4.718584
17	1,050	1.375000	0.0323	-3.432688	-0.010758	-4.734738
18	1,060	1.458333	0.0329	-3.414283	-0.008839	-4.992052
19	1,070	1.541667	0.0334	-3.399199	-0.007025	-5.251262
20	1,270	1.625000	0.0443	-3.116771	-0.007025	-5.076167
21	1,340	1.708333	0.0482	-3.032396	-0.005716	-5.190109
22	1,370	1.791667	0.0500	-2.995732	-0.005013	-5.376334
23	1,880	1.875000	0.0813	-2.509609	-0.002904	-4.710963
24	2,130	1.958333	0.0977	-2.325854	-0.000800	-4.556365

$$\sum = -84.367790$$

$\dagger$ $A_i = \frac{2i-1}{N}\{\log_e[Q_E(T_i)] + \log_e[1 - Q_E(T_{N+1-i})]\}$.

767

For another example, for Failure No. (24+1-11) or No. 14, $T_{14} = 840$ hr, and $m = 101.4$ hr. Then,

$$Q_E(T_{14} = 840 \text{ hr}) = 1 - e^{-(\frac{840-200}{9,916})^{1.39}},$$

or

$$Q_E(T_{14} = 840 \text{ hr}) = 0.0219.$$

Step 5— Then, the quantities $\log_e[Q_E(T_i)]$, $\log_e[1 - Q_E(T_{N+1-i})]$ and

$$A_i = \frac{2i-1}{N}\{\log_e[Q_E(T_i)] + \log_e[1 - Q_E(T_{N+1-i})]\},$$

are calculated for each failure and entered in Columns 5, 6 and 7 of Table 21.4.1. For example, for Failure No. 11 at $T_{26} = 740$ hr,

$$
\begin{aligned}
\log_e[Q_E(T_{11})] &= \log_e[Q_E(740 \text{ hr})], \\
&= \log_e[0.0174],
\end{aligned}
$$

or

$$\log_e[Q_E(T_{11})] = -4.051285.$$

$$
\begin{aligned}
\log_e[1 - Q_E(T_{24+1-11})] &= \log_e[1 - Q_E(T_{14})], \\
&= \log_e[1 - Q_E(840 \text{ hr})], \\
&= \log_e[1 - 0.0219],
\end{aligned}
$$

or

$$\log_e[1 - Q_E(T_{14})] = -0.022143.$$

Then, from Eq. (21.9),

$$
\begin{aligned}
A_{11} &= \frac{2 \times 11 - 1}{24}\{\log_e[Q_E(T_{11})] \\
&\quad + \log_e[1 - Q_E(T_{24+1-11})]\}, \\
&= \frac{2 \times 11 - 1}{24}\{\log_e[Q_E(T_{11})] \\
&\quad + \log_e[1 - Q_E(T_{11})]\}, \\
&= 0.875000 \times \{-4.051285 - 0.022143\},
\end{aligned}
$$

or

$$A_{11} = -3.564250.$$

Step 6– The *Anderson-Darling* goodness-of-fit test statistic, for the given sample, from Eq. (21.5′), is

$$
\begin{aligned}
A_N^2 &= -\sum_{i=1}^{24} \frac{2i-1}{24}\{\log_e[Q_E(T_i)] \\
&\quad + \log_e[1 - Q_E(T_{24+1-i})]\} - 24, \\
&= -\sum_{i=1}^{24} A_i - 24, \\
&= -(-84.367790) - 24,
\end{aligned}
$$

or

$$
A_N^2 = 60.36779.
$$

Step 7– From the *Anderson-Darling* goodness-of-fit test critical value table, given in Table 21.1, it may be seen that the allowable, or critical, statistic value for $N = 24 > 5$ failures, at the $\alpha = 0.05$ significance level, is $A_{cr}^2 = A_{cr}^{2*} = 2.492$.

Step 8– Since the calculated statistic value, $A_N^2 = 60.36779$, is larger than the allowable, or critical, value $A_{cr}^2 = 2.492$, or

$$
A_N^2 = 60.36779 > A_{cr}^2 = 2.492,
$$

the Weibull distribution cannot be accepted as adequately describing these times-to-failure data at the 0.05 significance level.

2. Cramer-von Mises Goodness-of-fit Test

Step 1– The data are arranged in the order of increasing times to failure in Column 2 of Table 21.4.2.

Step 2– The total number of failures is 24, or $N = 24$, from Column 1 of Table 21.4.2.

Step 3– The quantity $(i-0.5)/N$ for each failure is calculated and is given in Column 3 of Table 21.4.2. For example, for Failure No. 11 at $T_{11} = 740$ hr,

$$
\frac{i-0.5}{24} = \frac{11-0.5}{14} = 0.437570.
$$

TABLE 21.4.2– Cramer-von Mises goodness-of-fit test for the Weibull distributed data of Example 21–4.

Failure number, i	Time to failure, T_i, hr	$\frac{i-0.5}{N}$	Expected unreliability, $Q_E(T_i)$	$[Q_E(T_i) - \frac{i-0.5}{N}]^2$
1	260	0.020833	0.0008	0.000401
2	350	0.062500	0.0029	0.003552
3	420	0.104167	0.0050	0.009834
4	440	0.145833	0.0057	0.019637
5	480	0.187500	0.0070	0.032580
6	480	0.229167	0.0070	0.049358
7	530	0.270833	0.0088	0.068661
8	580	0.312500	0.0107	0.091083
9	680	0.354167	0.0148	0.115170
10	710	0.395833	0.0160	0.144273
11	740	0.437570	0.0174	0.176484
12	780	0.479167	0.0191	0.211661
13	820	0.520833	0.0210	0.249833
14	840	0.562500	0.0219	0.292248
15	920	0.604167	0.0258	0.334508
16	930	0.645833	0.0263	0.383822
17	1,050	0.687500	0.0323	0.429287
18	1,060	0.729167	0.0329	0.484787
19	1,070	0.770833	0.0334	0.543808
20	1,270	0.812500	0.0443	0.590131
21	1,340	0.854167	0.0482	0.649582
22	1,370	0.895833	0.0500	0.715434
23	1,880	0.937500	0.0813	0.733078
24	2,130	0.979167	0.0977	0.776983

$$\sum = 7.106201$$

Step 4– The *expected probability of failure* is calculated, for each time to failure, from

$$Q_E(T_i) = 1 - e^{-(\frac{T_i-\gamma}{\eta})^\beta},$$

and entered in Column 4, which is the same as that given in Column 4 of Table 21.4.1.

Step 5– Then the quantity $[Q_E(T_i)] - \frac{i-0.5}{N}]^2$ is calculated for each failure and entered in Column 5 of Table 21.4.2. For example, for Failure No. 11 at $T_{11} = 740$ hr,

$$[Q_E(T_{11})] - \frac{11-0.5}{24}]^2 = [0.0174 - 0.437570]^2,$$
$$= 0.176484.$$

Step 6– The *Cramer-von Mises* goodness-of-fit test statistic, calculated for the given sample from Eq. (21.10), is

$$W_N^{2*} = \sum_{i=1}^{24}[Q_E(T_i) - \frac{i-0.5}{24}]^2 + \frac{1}{12 \times 24},$$
$$= 7.106201 + 0.003472,$$

or

$$W_N^{2*} = 7.109673.$$

Step 7– The *modified Cramer-von Mises* goodness-of-fit test statistic, calculated for the given sample from Eq. (21.11), is

$$W_N^{2*} = (W_N^2 - 0.4/N + 0.6/N^2)(1.0 + 1.0/N),$$
$$= (7.109673 - 0.4/24 + 0.6/24^2)(1.0 + 1.0/24),$$

or

$$W_N^{2*} = 7.389633.$$

Step 8– From the *Cramer-von Mises* goodness-of-fit test critical value table, given in Table 21.1, it may be seen that the allowable, or critical, statistic value for $N = 24$ failures, at the $\alpha = 0.05$ significance level, is $W_{cr}^{2*} = 0.461$.

Step 9– Since the calculated statistic value, $W_N^{2*} = 7.389633$, is greater than the allowable, or critical, value $W_{cr}^{2*} = 0.461$, or

$$W_N^{2*} = 7.389633 > W_{cr}^{2*} = 0.461,$$

the Weibull distribution cannot be accepted as adequately describing these times-to-failure data at the 0.05 significance level.

21.4 MODIFIED AD AND CvM GOODNESS-OF-FIT TESTS FOR CONTINUOUS DISTRIBUTIONS WITH PARAMETERS UNKNOWN

The *Anderson-Darling* A^2 statistic and the *Cramer-von Mises* W^2 statistic provide two more means of testing whether a set of observations is from some completely specified continuous distribution. When certain parameters of the hypothesized distribution are unknown but must be estimated from the sample, then the *standard AD* and the *CvM* test critical value tables are conservative in the sense that the true significance level is much less than the level associated with the standard critical values. To rectify this, the test critical values should be modified for the distributions with unknown parameters. Since the distributions of the *AD* and the *CvM* test statistics depend on the hypothesized distribution family when parameters are estimated, a separate set of tables of critical values must be obtained for each family. To obtain the asymptotic percentage points of the *AD* and the *CvM* test statistics, some theoretical modifications on both test statistics have been made for different distributions, and Monte Carlo methods have also been used to obtain the critical values for the *modified* tests with estimated parameters.

Stephens [4; 5; 10] modified the *standard AD* and *CvM* test statistics and calculated the corresponding asymptotic percentage points theoretically for the *modified AD* and *CvM* tests for the exponential distribution with estimated mean, and for the normal distribution in the following cases:

Case 1– The mean, μ, is estimated, but the variance, σ^2, is known.

Case 2– The variance, σ^2, is estimated, but the mean, μ, is known.

Case 3– Both the mean, μ, and the variance, σ^2, are estimated.

The modified test statistic forms and their corresponding percentage points are summarized in Table 21.5.

Note that the normal critical values can also be used for the log-normal distribution as follows:

Let $T' = \log_e T$; then T' is normally distributed with mean, $\overline{T}'$, and standard deviation, $\sigma_{T'}$, to be estimated from the sample. Thus the problem is reduced to testing that the T' values are normally distributed with estimated $\overline{T}'$ and $\sigma_{T'}$.

Littell, et al [7] have generated the critical value tables by the Monte Carlo simulation method for the *modified AD* and the *CvM* tests for

TABLE 21.5–
Percentage points for the modified Anderson-Darling and Cramer-von Mises test statistics for the exponential and the normal distributions with unknown parameters.

Statistic	Modified form	Percentage points, A_{cr}^{2*} and W_{cr}^{2*}				
	α	0.15	0.10	0.05	0.025	0.01
A_N^2	Exponential distribution with estimated m					
	$A_N^{2*} = A_N^2(1 + 0.6/N)$	0.922	1.078	1.341	1.606	1.957
	Normal distribution, Case 1: estimated μ and known σ^2					
	$A_N^{2*} = A_N^2$	$--$	0.908	1.105	1.304	1.573
	Normal distribution, Case 2: estimated σ^2 and known μ					
	$A_N^{2*} = A_N^2$, for $N \geq 5$	$--$	1.706	2.323	2.904	3.690
	Normal distribution, Case 3: estimated μ and σ^2					
	$A_N^{2*} = A_N^2(1 + 4/N - 25/N^2)$	0.576	0.656	0.787	0.918	1.092
W_N^2	Exponential distribution with estimated m					
	$W_N^{2*} = W_N^2(1 + 0.16/N)$	0.149	0.177	0.224	0.273	0.337
	Normal distribution, Case 1: estimated μ and known σ^2					
	$W_N^{2*} = W_N^2$	$--$	0.135	0.165	0.196	0.237
	Normal distribution, Case 2: estimated σ^2 and known μ					
	$W_N^{2*} = W_N^2$, for $N \geq 5$	$--$	0.329	0.443	0.562	0.723
	Normal distribution, Case 3: estimated μ and σ^2					
	$W_N^{2*} = W_N^2(1 + 0.5/N)$	0.091	0.104	0.126	0.148	0.178

TABLE 21.6– Percentage points for the modified Anderson-Darling and the Cramer-von Mises test statistics for the two-parameter Weibull distribution with unknown shape and scale parameters.

N	Statistic	α				
		0.20	0.15	0.10	0.05	0.01
10	A_N^2	0.505	0.549	0.616	0.730	0.988
	W_N^2	0.081	0.089	0.101	0.120	0.164
15	A_N^2	0.506	0.554	0.622	0.740	0.994
	W_N^2	0.080	0.089	0.101	0.121	0.168
20	A_N^2	0.506	0.556	0.625	0.744	1.012
	W_N^2	0.080	0.089	0.101	0.123	0.172
25	A_N^2	0.507	0.556	0.625	0.745	1.019
	W_N^2	0.080	0.089	0.101	0.122	0.172
30	A_N^2	0.508	0.557	0.626	0.741	1.007
	W_N^2	0.079	0.088	0.101	0.123	0.172
35	A_N^2	0.509	0.559	0.626	0.739	1.003
	W_N^2	0.079	0.088	0.101	0.123	0.173
40	A_N^2	0.509	0.560	0.627	0.739	1.001
	W_N^2	0.079	0.088	0.101	0.123	0.173

the two-parameter Weibull distribution with zero location parameter, and estimated shape and scale parameters. They are summarized in Table 21.6.

Stephens [8] has modified the *standard AD* and the *CvM* test statistics, and calculated the corresponding asymptotic percentage points theoretically for the *modified AD* and *CvM* tests for the extreme-value distribution, for the following three cases:

Case 1– The location parameter, γ, is estimated, but the scale parameter, η, is known.

Case 2– The scale parameter, η, is estimated, but the location parameter, γ, is known.

Case 3– Both the location parameter, γ, and the scale parameter, η, are estimated.

The results were verified by Monte Carlo simulation [8], and are summarized in Table 21.7.

TABLE 21.7– Percentage points for the modified Anderson-Darling and Cramer-von Mises test statistics for the extreme-value distributions with unknown parameters.

Statistic	Modified form	Percentage points, A_{cr}^{2*} and W_{cr}^{2*}				
	α	0.25	0.10	0.05	0.025	0.01
A_N^2	Case 1					
	$A_N^{2*} = A_N^2(1 + 0.3/N)$	0.736	1.062	1.321	1.591	1.959
	Case 2					
	$A_N^{2*} = A_N^2$	1.060	1.725	2.277	2.854	3.640
	Case 3					
	$A_N^{2*} = A_N^2(1 + 0.2/\sqrt{N})$	0.474	0.637	0.757	0.877	1.038
A_N^2	Case 1					
	$W_N^{2*} = W_N^2(1 + 0.16/N)$	0.116	0.175	0.222	0.271	0.338
	Case 2					
	$W_N^{2*} = W_N^2$	0.186	0.320	0.431	0.547	0.705
	Case 3					
	$W_N^{2*} = W_N^2(1 + 0.2/\sqrt{N})$	0.073	0.102	0.124	0.146	0.175

Woodruff, et al [9] have generated tables for the *modified AD* and *CvM* tests for the gamma distribution with estimated location and scale parameters, and known shape parameter values of 0.5, 1.0, 1.5, 2.0, 2.5, 3.0, 3.5 and 4.0, respectively. The results are given in Tables 21.8 and 21.9.

The procedures for the *modified AD* and *CvM* tests are identical to those of the standard *AD* and *CvM* tests, except that the new modified forms of the test statistics and the corresponding critical percentage points should be applied to the decision making process.

EXAMPLE 21–5

Rework Example 20–9 using the *modified AD* and *CvM* goodness-of-fit tests.

SOLUTION TO EXAMPLE 21–5

1. *Modified AD* Test

In Example 21–1 it was found, in Step 6, that the *standard AD* statistic value is

$$A_N^2 = 0.064238.$$

TABLE 21.8– Percentage points for the modified Anderson- Darling test for the gamma distributions with unknown location and scale parameters but known shape parameter.

		Shape parameter, β							
α	N	0.50	1.0	1.5	2.0	2.5	3.0	3.5	4.0
0.20	5	1.072	2.088	0.691	0.611	0.570	0.555	0.547	0.529
	10	0.922	1.509	0.710	0.627	0.589	0.569	0.552	0.548
	15	0.880	1.283	0.706	0.633	0.592	0.576	0.578	0.544
	20	0.870	1.212	0.715	0.642	0.611	0.575	0.567	0.557
	25	0.837	1.083	0.728	0.643	0.599	0.571	0.560	0.556
	30	0.833	1.056	0.730	0.648	0.601	0.585	0.560	0.558
0.15	5	1.168	2.308	0.758	0.667	0.626	0.596	0.597	0.584
	10	1.030	0.660	0.783	0.695	0.650	0.628	0.602	0.600
	15	1.982	1.440	0.794	0.693	0.656	0.638	0.621	0.601
	20	0.976	1.351	0.790	0.715	0.670	0.636	0.626	0.607
	25	0.947	1.215	0.805	0.700	0.659	0.636	0.619	0.612
	30	0.943	1.175	0.811	0.717	0.657	0.649	0.613	0.619
0.10	5	1.314	2.593	0.849	0.742	0.696	0.673	0.659	0.646
	10	1.192	1.881	0.902	0.788	0.741	0.706	0.673	0.672
	15	1.126	1.649	0.901	0.788	0.733	0.725	0.694	0.682
	20	1.135	1.548	0.912	0.805	0.750	0.716	0.706	0.692
	25	1.096	1.385	0.917	0.798	0.746	0.723	0.696	0.700
	30	1.111	1.375	0.927	0.821	0.739	0.736	0.697	0.697
0.05	5	1.624	3.071	1.060	0.867	0.832	0.781	0.775	0.754
	10	1.438	2.298	1.092	0.924	0.874	0.842	0.807	0.803
	15	1.408	2.036	1.107	0.941	0.869	0.862	0.808	0.815
	20	1.483	1.904	1.092	0.969	0.903	0.876	0.843	0.820
	25	1.391	1.712	1.133	0.967	0.902	0.881	0.842	0.843
	30	1.432	1.689	1.119	0.998	0.903	0.883	0.830	0.824
0.01	5	2.590	4.392	1.431	1.176	1.097	1.063	1.017	0.972
	10	2.232	3.491	1.560	1.226	1.208	1.170	1.132	1.102
	15	2.212	2.956	1.583	1.345	1.205	1.165	1.115	1.104
	20	2.199	2.779	1.572	1.344	1.228	1.150	1.150	1.102
	25	2.063	2.568	1.557	1.396	1.233	1.171	1.159	1.209
	30	2.405	2.448	1.613	1.464	1.216	1.260	1.176	1.162

TABLE 21.9– Percentage points for the modified Cramer-von Mises test for the gamma distribution with unknown location and scale parameters but known shape parameter.

α	N	Shape parameter, β							
		0.50	1.0	1.5	2.0	2.5	3.0	3.5	4.0
0.20	5	0.122	0.145	0.109	0.101	0.096	0.093	0.092	0.089
	10	0.133	0.134	0.113	0.105	0.097	0.094	0.091	0.091
	15	0.138	0.131	0.114	0.104	0.098	0.095	0.093	0.092
	20	0.143	0.132	0.120	0.106	0.099	0.094	0.094	0.091
	25	0.145	0.130	0.116	0.105	0.098	0.094	0.092	0.092
	30	0.144	0.132	0.117	0.108	0.098	0.096	0.093	0.093
0.15	5	0.136	0.156	0.121	0.112	0.105	0.101	0.101	0.097
	10	0.153	0.151	0.127	0.116	0.108	0.104	0.101	0.100
	15	0.160	0.149	0.129	0.115	0.110	0.108	0.103	0.101
	20	0.163	0.148	0.135	0.117	0.110	0.106	0.105	0.102
	25	0.165	0.149	0.131	0.117	0.110	0.105	0.103	0.102
	30	0.165	0.150	0.132	0.122	0.109	0.108	0.102	0.102
0.10	5	0.158	0.181	0.139	0.126	0.119	0.115	0.113	0.111
	10	0.180	0.172	0.147	0.134	0.126	0.121	0.115	0.115
	15	0.186	0.174	0.154	0.133	0.126	0.123	0.117	0.115
	20	0.192	0.175	0.155	0.136	0.126	0.121	0.120	0.116
	25	0.195	0.174	0.150	0.136	0.128	0.122	0.119	0.118
	30	0.199	0.176	0.152	0.142	0.127	0.125	0.117	0.118
0.05	5	0.193	0.231	0.174	0.148	0.143	0.137	0.134	0.130
	10	0.226	0.221	0.182	0.162	0.152	0.144	0.140	0.139
	15	0.234	0.219	0.185	0.163	0.153	0.150	0.142	0.140
	20	0.253	0.222	0.192	0.169	0.153	0.151	0.147	0.142
	25	0.251	0.225	0.185	0.170	0.155	0.152	0.145	0.143
	30	0.257	0.223	0.187	0.174	0.154	0.152	0.142	0.143
0.01	5	0.320	0.357	0.236	0.210	0.190	0.190	0.183	0.176
	10	0.357	0.357	0.267	0.223	0.219	0.205	0.201	0.197
	15	0.366	0.333	0.274	0.241	0.221	0.209	0.200	0.197
	20	0.374	0.333	0.290	0.240	0.221	0.206	0.211	0.196
	25	0.377	0.324	0.279	0.248	0.221	0.214	0.212	0.211
	30	0.417	0.342	0.287	0.259	0.224	0.226	0.206	0.205

Referring to Table 21.5, for the normal distribution Case 3, it is found that the *modified AD* statistic is

$$
\begin{aligned}
A_N^{2*} &= A_N^2(1 + 4/N - 25/N^2), \\
&= 0.064238(1 + 4/19 - 25/19^2),
\end{aligned}
$$

or

$$
A_N^{2*} = 0.073313,
$$

and the critical value, at the $\alpha = 0.05$ significance level, is

$$
A_{cr}^{2*} = 0.787.
$$

Since

$$
A_N^{2*} = 0.073313 < A_{cr}^{2*} = 0.787,
$$

the normal distribution, with $\hat{\bar{T}} = 2{,}722.26$ hr and $\hat{\sigma}_T = 21.92$ hr, is accepted as adequately describing the times-to-failure data at the 0.05 significance level.

2. *Modified CvM Test*

In Example 21–1 it was found, in Step 6, that the standard *CvM* statistic value is

$$
W_N^2 = 0.007622.
$$

Referring to Table 21.5, for the normal distribution Case 3, it is found that the *modified CvM* statistic is

$$
\begin{aligned}
W_N^{2*} &= W_N^2(1 + 0.5/N), \\
&= 0.007622(1 + 0.5/19),
\end{aligned}
$$

or

$$
W_N^{2*} = 0.007823,
$$

and the critical value, at the $\alpha = 0.05$ significance level, is

$$
W_{cr}^{2*} = 0.126.
$$

Since

$$
W_N^{2*} = 0.007832 < W_{cr}^{2*} = 0.126,
$$

the normal distribution, with $\hat{\bar{T}} = 2{,}722.26$ hr and $\hat{\sigma}_T = 21.92$ hr, is accepted as adequately describing the times-to-failure data at the 0.05 significance level.

EXAMPLE 21–6

Rework Example 20–10 using the *modified AD* and *CvM* goodness-of-fit tests.

SOLUTION TO EXAMPLE 21–6

Let

$$T' = \log_e T;$$

then T' is normally distributed with

$$
\begin{aligned}
Q_E(T') &= \Phi(\frac{T' - 4.3553}{0.67670}), \\
&= \Phi(\frac{\log_e T - 4.3553}{0.67670}),
\end{aligned}
$$

or

$$Q_E(T') = Q_E(T).$$

Therefore,

$$A_N^{2\,\prime} = A_N^2,$$

and

$$W_N^{2\,\prime} = W_N^2.$$

1. *Modified AD* Test

In Example 21–2 it was found, in Step 6, that the *standard AD* statistic value is

$$A_N^2 = 0.134232.$$

Referring to Table 21.5, for the normal distribution Case 3, it is found that the *modified AD* statistic is

$$
\begin{aligned}
A_N^{2*} &= A_N^2(1 + 4/N - 25/N^2), \\
&= 0.134232(1 + 4/9 - 25/9^2),
\end{aligned}
$$

or

$$A_N^{2*} = 0.152461,$$

and the critical value, at the $\alpha = 0.05$ significance level, is

$$A_{cr}^{2*} = 0.787.$$

Since

$$A_N^{2*} = 0.152461 < A_{cr}^{2*} = 0.787,$$

the lognormal distribution with $\hat{\bar{T}}' = 4.3553$ and $\hat{\sigma}_{T'} = 0.67670$ is accepted as adequately describing the times-to-failure data at the 0.05 significance level.

2. *Modified CvM Test*

In Example 21–2 it was found, in Step 6, that the standard *CvM* statistic value is

$$W_N^2 = 0.016721.$$

Referring to Table 21.5, for the normal distribution Case 3, it is found that the *modified CvM* statistic is

$$
\begin{aligned}
W_N^{2*} &= W_N^2(1 + 0.5/N), \\
&= 0.016721(1 + 0.5/9),
\end{aligned}
$$

or

$$W_N^{2*} = 0.017650,$$

and the critical value, at the $\alpha = 0.05$ significance level, is

$$W_{cr}^{2*} = 0.126.$$

Since

$$W_N^{2*} = 0.017650 < W_{cr}^{2*} = 0.126,$$

the lognormal distribution, with $\hat{\bar{T}}' = 4.3553$ and $\hat{\sigma}_{T'} = 0.67670$, is accepted as adequately describing the times-to-failure data at the 0.05 significance level.

EXAMPLE 21–7

Rework Example 20–11 using the *modified AD* and *CvM* goodness-of-fit tests.

SOLUTION TO EXAMPLE 21–7

1. *Modified AD Test*

In Example 21–3 it was found, in Step 6, that the *standard AD* statistic value is

$$A_N^2 = 0.685249.$$

Referring to Table 21.5, for the exponential distribution, it is found that the *modified AD* statistic is

$$
\begin{aligned}
A_N^{2*} &= A_N^2(1 + 0.6/N), \\
&= 0.685249(1 + 0.6/49),
\end{aligned}
$$

or

$$
A_N^{2*} = 0.693640,
$$

and the critical value, at the $\alpha = 0.05$ significance level, is

$$
A_{cr}^{2*} = 1.341.
$$

Since

$$
A_N^{2*} = 0.693640 < A_{cr}^{2*} = 1.341,
$$

the exponential distribution, with $MTBF = \hat{m} = 101.4$ hr, is accepted as adequately describing the times-to-failure data at the 0.05 significance level.

2. *Modified CvM* Test

In Example 21–3 it was found, in Step 6, that the standard *CvM* statistic value is

$$
W_N^2 = 0.113019.
$$

Referring to Table 21.5, for the exponential distribution, it is found that the *modified CvM* statistic is

$$
\begin{aligned}
W_N^{2*} &= W_N^2(1 + 0.16/N), \\
&= 0.113019(1 + 0.16/49),
\end{aligned}
$$

or

$$
W_N^{2*} = 0.113388,
$$

and the critical value, at the $\alpha = 0.05$ significance level, is

$$
W_{cr}^{2*} = 0.224.
$$

Since

$$
W_N^{2*} = 0.113388 < W_{cr}^{2*} = 0.224,
$$

the exponential distribution with $MTBF = \hat{m} = 101.4$ hr is accepted as adequately describing the times-to-failure data at the 0.05 significance level.

EXAMPLE 21–8

Assume the transistor times-to-failure data presented in Column 2 of Tables 20.4 and 20.8 are Weibull distributed with known $\gamma = 200$ hr, but unknown η and β. By the probability plotting method the scale and location parameters are found to be

$$\hat{\eta} = 9,916 \text{ hr} \quad \text{and} \quad \hat{\beta} = 1.39,$$

because they were not known a priori. Does this Weibull *pdf* fit these data at the significance level of 0.05?

SOLUTION TO EXAMPLE 21–8

1. *Modified AD* Test

In Example 21–4 it was found, in Step 6, that the *standard AD* statistic value is

$$A_N^2 = 60.36779.$$

Referring to Table 21.6, by interpolation, it is found that the critical value, at the $\alpha = 0.05$ significance level, is

$$A_{cr}^2 = 0.7448.$$

Since

$$A_N^2 = 60.36779 > A_{cr}^2 = 0.7448,$$

the Weibull distribution, with $\gamma = 200$ hr, $\hat{\beta} = 1.39$ and $\hat{\eta} = 9,916$ hr, cannot be accepted as adequately describing these times-to-failure data at the 0.05 significance level.

2. *Modified CvM* Test

In Example 21–4 it was found, in Step 6, that the standard *CvM* statistic value is

$$W_N^2 = 7.109673.$$

Referring to Table 21.6, by interpolation, it is found that the critical value, at the $\alpha = 0.05$ significance level, is

$$W_{cr}^2 = 0.1222.$$

Since

$$W_N^2 = 7.109673 > W_{cr}^2 = 0.1222,$$

the Weibull distribution, with $\gamma = 200$ hr, $\hat{\beta} = 1.39$ and $\hat{\eta} = 9,916$ hr, cannot be accepted as adequately describing these times-to-failure data at the 0.05 significance level.

21.5 COMMENTS ON THE ANDERSON-DARLING AND CRAMER-VON MISES GOODNESS-OF-FIT TESTS

1. Like the *K-S test*, both the *standard* and the *modified AD* and *CvM* tests are more powerful than the *chi-squared test* for any sample size [1; 5].

2. The *AD* and *CvM* tests can be used with *small sample sizes*, where the validity of the *chi-squared test* would be questionable.

3. Broadly speaking, the *K-S* and *CvM* tests are more effective at detecting departures in the middle of a distribution, whereas the *AD* test is relatively more effective at detecting tail departures [1].

4. When unknown parameters are present, the *CvM* test, and especially the *AD* test, are often substantially more powerful than the *K-S* test [1].

5. The *CvM* test is the most powerful test for the *gamma* distribution, while the *AD* test is the most powerful test for the gamma distribution if the lognormal distribution is taken as an alternative.[9].

6. Strictly speaking, the *modified AD* and *CvM* tests should be used for distributions with estimated parameters. However, in engineering practice, if the modified critical value tables for a particular distribution are not available, the *standard AD* and *CvM* tables can still be used satisfactorily. Even though the *standard K-S* tables are conservative, they are often good enough for engineering accuracy.

21.6 RECOMMENDATIONS FOR THE SELECTION OF GOODNESS-OF-FIT TESTS

There has been an increasing number of goodness-of-fit tests proposed in the literature for different distributions. Some of them can be used for a broad range of distribution families, but others can only be used for one or more particular distributions. Though their powers may vary from distribution to distribution, they are quite important and useful in reliability and maintainability engineering. Table 21.10

TABLE 21.10– Recommendations for selection of good-ness-of-fit tests.

Distribution	Recommended goodness-of-fit test statistics (in the order of preference)
Exponential	G_N, W_N^2, A_N^2, D_{max}, χ^2
Normal (Lognormal)	W, A_N^2, W_N^2, Z^*, D_{max}, χ^2
Weibull (Extreme)	M, Z^*, A_N^2, W_N^2, D_{max}, χ^2
Gamma	W_N^2, A_N^2, D_{max}, χ^2

summarizes these goodness-of-fit tests and gives a rough recommendation for selection preference.

In Table 21.10 the following additional goodness-of-fit test statistics are introduced:

$$G_N = \text{Gail-Gastwirth test statistic [1]},$$

$$= \sum_{i=1}^{N}\sum_{j=1}^{N}|T_i - T_j|/2N(N-1)\overline{T},$$

$$W = \text{Shapiro-Wilk test statistic [1]},$$

$$= (\sum_{i=1}^{N} a_i T_{(i)})^2 \sum_{i=1}^{N}(T_i - \overline{T})^2,$$

$$Z^* = \text{Tiku-Singh test statistic [1]},$$

$$= (2\sum_{i=1}^{r-2}(r-i-1)l_i)/(r-2)\sum_{i=1}^{r-1}l_i,$$

and

$$M = \text{Mann-Scheuer-Fertig test statistic [1]},$$

$$= [\frac{r}{2}]\sum_{i=[r/2]+1}^{r-1} l_i/[\frac{r-1}{2}]\sum_{i=1}^{[r/2]}l_i,$$

where

$$\overline{T} = \frac{1}{N}\sum_{i=1}^{N}T_i,$$

$$T_{(i)} = ith \text{ smallest observation in the given sample of size } N,$$

$$l_i = \frac{T_{(i+1)} - T_{(i)}}{E(Z_{(i+1)} - Z_{(i)})}, \quad i = 1, 2, ..., r - 1,$$

r = order number of observation at which the sample is censored,

Z_i = *ith* order statistic in a random sample of size N from the standard extreme-value distribution,

a_i = functions of the means, variances and covariances of the order statistics in a sample of size N from the standardized normal distribution,

$[\frac{r}{2}]$ = largest integer less than, or equal to, $\frac{r}{2}$,

and

$[\frac{r-1}{2}]$ = largest integer less than or equal to $\frac{r-1}{2}$.

PROBLEMS

21-1. Rework Problem 20-1 using the standard and modified Anderson-Darling and Cramer-von Mises goodness-of-fit tests.

21-2. Rework Problem 20-2 using the standard and modified Anderson-Darling and Cramer-von Mises goodness-of-fit tests.

21-3. Rework Problem 20-3 using the standard and modified Anderson-Darling and Cramer-von Mises goodness-of-fit tests.

21-4. Rework Problem 20-4 using the standard and modified Anderson-Darling and Cramer-von Mises goodness-of-fit tests.

21-5. Rework Problem 20-5 using the standard and modified Anderson-Darling and Cramer-von Mises goodness-of-fit tests.

21-6. Rework Problem 20-6 using the standard and modified Anderson-Darling and Cramer-von Mises goodness-of-fit tests.

21-7. Rework Problem 20-7 using the standard and modified Anderson-Darling and Cramer-von Mises goodness-of-fit tests.

21-8. Rework Problem 20-8 using the standard and modified Anderson-Darling and Cramer-von Mises goodness-of-fit tests.

21-9. Rework Problem 20-9 using the standard and modified Anderson-Darling and Cramer-von Mises goodness-of-fit tests.

21-10. Rework Problem 20-10 using the standard and modified Anderson-Darling and Cramer-von Mises goodness-of-fit tests.

21-11. Rework Problem 20-11 using the standard and modified Anderson-Darling and Cramer-von Mises goodness-of-fit tests.

21-12. Rework Problem 20-12 using the standard and modified Anderson-Darling and Cramer-von Mises goodness-of-fit tests.

21-13. Rework Problem 20-13 using the standard and modified Anderson-Darling and Cramer-von Mises goodness-of-fit tests.

21-14. Rework Problem 20-14 using the standard and modified Anderson-Darling and Cramer-von Mises goodness-of-fit tests.

21-15. Rework Problem 20-15 using the standard and modified Anderson-Darling and Cramer-von Mises goodness-of-fit tests.

21-16. Rework Problem 20-16 using the standard and modified Anderson-Darling and Cramer-von Mises goodness-of-fit tests.

21-17. Rework Problem 20-17 using the standard and modified Anderson-Darling and Cramer-von Mises goodness-of-fit tests.

21-18. Rework Problem 20-18 using the standard and modified Anderson-Darling and Cramer-von Mises goodness-of-fit tests.

21-19. Rework Problem 20-19 using the standard and modified Anderson-Darling and Cramer-von Mises goodness-of-fit tests.

21-20. Rework Problem 20-20 using the standard and modified Anderson-Darling and Cramer-von Mises goodness-of-fit tests.

21-21. Rework Problem 20-21 using the standard and modified Anderson-Darling and Cramer-von Mises goodness-of-fit tests.

21-22. Rework Problem 20-22 using the standard and modified Anderson-Darling and Cramer-von Mises goodness-of-fit tests.

21-23. Rework Problem 20-23 using the standard and modified Anderson-Darling and Cramer-von Mises goodness-of-fit tests.

21-24. Rework Problem 20-24 using the standard and modified Anderson-Darling and Cramer-von Mises goodness-of-fit tests.

21-25. Rework Problem 20-25 using the standard and modified Anderson-Darling and Cramer-von Mises goodness-of-fit tests.

21-26. Rework Problem 20-26 using the standard and modified Anderson-Darling and Cramer-von Mises goodness-of-fit tests.

21-27. Rework Problem 20-27 using the standard and modified Anderson-Darling and Cramer-von Mises goodness-of-fit tests.

21-28. Rework Problem 20-28 using the standard and modified Anderson-Darling and Cramer-von Mises goodness-of-fit tests.

21-29. Rework Problem 20-29 using the standard and modified Anderson-Darling and Cramer-von Mises goodness-of-fit tests.

21-30. Rework Problem 20-30 using the standard and modified Anderson-Darling and Cramer-von Mises goodness-of-fit tests.

REFERENCES

1. Lawless, J. F., *Statistical Models and Methods for Lifetime Data*, John Wiley & Sons, Inc., New York, 580 pp., 1982.

2. Anderson, T. W. and Darling, D. A., "Asymptotic Theory of Certain Goodness-of-fit Criteria Based on Stochastic Process," *Annals of Mathematical Statistics*, Vol. 23, pp. 193-212, 1952.

3. Darling, D. A., "The Cramer-Smirnov Test in the Parametric Case," *Annals of Mathematical Statistics*, Vol. 26, pp. 1-20, 1955.

4. Stephens, M. A., "Use of the Kolmogorov-Smirnov, Cramer-von Mises and Related Statistics Without Extensive Tables," *Journal of the Royal Statistical Society*, Ser. B, Vol. 32, No. 1, pp. 115-122, 1970.

5. Stephens, M. A., "EDF Statistics for Goodness of Fit and Some Comparisons," *Journal of the American Statistical Association*, Vol. 69, No. 347, pp. 730-737, September 1974.

6. Green, J. R. and Hegazy, J. R., "Powerful Modified-EDF Goodness-of-Fit Tests," *Journal of the American Statistical Association*, Vol. 71, No. 353, pp. 204-209, March 1976.

7. Littell, R. C., McClave, J. T. and Offen, W. W., "Goodness-of-fit Tests for the Two-parameter Weibull Distribution," *Communication in Statistics*, Vol. B8, No. 3, pp. 257-269, 1979.

8. Stephens, M. A., "Goodness of Fit for the Extreme Value Distribution," *Biometrika*, Vol. 64, pp. 583-588, 1977.

9. Woodruff, B. W., Viviano, P. J., Moore, A. H. and Dunne, E. J., "Modified Goodness-of-Fit Test for Gamma Distributions with Unknown Location and Scale Parameters," *IEEE Transactions on Reliability*, Vol. R-33, No. 3, pp. 241-245, August 1984.

10. Pearson, E. S. and Hartley, H. O., *Biometrika Tables for Statisticians*, Vol. 2, Cambridge University Press, 1972.

11. Pearson, K., "On the Criterion that a Given System of Deviation from the Probable in the Case of a Correlated System of Variables is Such that it Can be Reasonably Supposed to Have Arisen from Random Sampling," *Philosophy Magazine Series 5*, Vol. 50, pp. 157-172, 1900.

APPENDICES

APPENDIX A
RANK TABLES

SAMPLE SIZE = 1

ORDER NUMBER	0.1	0.5	1.0	2.5	5.0	10.0	25.0	50.0	75.0	90.0	95.0	97.5	99.0	99.5	99.9	ORDER NUMBER
1	.100	.500	1.000	2.500	5.000	10.000	25.000	50.000	75.000	90.000	95.000	97.500	99.000	99.500	99.900	1
	99.9	99.5	99.0	97.5	95.0	90.0	75.0	50.0	25.0	10.0	5.0	2.5	1.0	0.5	0.1	

SAMPLE SIZE = 2

ORDER NUMBER	0.1	0.5	1.0	2.5	5.0	10.0	25.0	50.0	75.0	90.0	95.0	97.5	99.0	99.5	99.9	ORDER NUMBER
1	.050	.250	.501	1.258	2.632	5.132	13.397	29.289	50.000	68.377	77.639	84.189	90.000	92.929	96.838	2
2	3.162	7.071	10.000	15.811	22.361	31.623	50.000	70.711	86.603	94.868	97.468	98.742	99.499	99.750	99.950	1
	99.9	99.5	99.0	97.5	95.0	90.0	75.0	50.0	25.0	10.0	5.0	2.5	1.0	0.5	0.1	

SAMPLE SIZE = 3

ORDER NUMBER	0.1	0.5	1.0	2.5	5.0	10.0	25.0	50.0	75.0	90.0	95.0	97.5	99.0	99.5	99.9	ORDER NUMBER
1	.033	.167	.334	.840	1.695	3.451	9.144	20.630	37.004	53.584	63.160	70.760	78.456	82.900	90.000	3
2	1.856	4.140	5.891	9.431	13.535	19.579	32.635	50.001	67.365	80.421	86.465	90.569	94.109	95.860	98.164	2
3	10.000	17.100	21.544	29.240	36.840	46.416	62.996	79.370	90.856	96.549	98.305	99.160	99.666	99.833	99.967	1
	99.9	99.5	99.0	97.5	95.0	90.0	75.0	50.0	25.0	10.0	5.0	2.5	1.0	0.5	0.1	

SAMPLE SIZE = 4

ORDER NUMBER	0.1	0.5	1.0	2.5	5.0	10.0	25.0	50.0	75.0	90.0	95.0	97.5	99.0	99.5	99.9	ORDER NUMBER
1	.025	.125	.251	.631	1.274	2.600	6.940	15.910	29.289	43.766	52.713	60.236	68.377	73.409	82.217	4
2	1.302	2.944	4.200	6.759	9.762	14.256	24.302	38.573	54.366	67.954	75.140	80.589	85.914	88.912	93.597	3
3	6.403	11.088	14.086	19.411	24.860	32.046	45.632	61.427	76.698	85.744	90.238	93.241	95.800	97.056	98.698	2
4	17.783	26.591	31.623	39.764	47.287	56.234	70.711	84.090	93.060	97.400	98.726	99.369	99.740	99.875	99.975	1
	99.9	99.5	99.0	97.5	95.0	90.0	75.0	50.0	25.0	10.0	5.0	2.5	1.0	0.5	0.1	

SAMPLE SIZE=5

ORDER NO.	0.1	0.5	1.0	2.5	5.0	10.0	25.0	50.0	75.0	90.0	95.0	97.5	99.0	99.5	99.9	ORDER NO.
1	0.020	0.100	0.201	0.505	1.021	2.085	5.591	12.945	24.214	36.904	45.072	52.182	60.189	66.343	74.881	5
2	1.011	2.288	3.206	5.274	7.644	11.224	19.376	31.380	45.419	58.389	65.740	71.642	77.793	81.490	87.798	4
3	4.755	8.283	10.564	14.663	18.926	24.664	35.944	50.001	64.056	75.336	81.074	85.337	89.436	91.717	95.245	3
4	12.202	18.510	22.207	28.358	34.260	41.611	54.581	68.620	80.624	88.776	92.356	94.726	96.732	97.712	98.989	2
5	25.119	34.057	39.811	47.818	54.928	63.096	75.786	87.055	94.409	97.915	98.979	99.495	99.799	99.900	99.980	1
	99.9	99.5	99.0	97.5	95.0	90.0	75.0	50.0	25.0	10.0	5.0	2.5	1.0	0.5	0.1	ORDER NO.

SAMPLE SIZE=6

ORDER NO.	0.1	0.5	1.0	2.5	5.0	10.0	25.0	50.0	75.0	90.0	95.0	97.5	99.0	99.5	99.9	ORDER NO.
1	0.017	0.084	0.167	0.421	0.851	1.741	4.682	10.910	20.630	31.871	39.304	45.926	53.584	58.648	68.377	6
2	0.826	1.871	2.676	4.327	6.284	9.260	16.116	26.444	38.947	51.032	58.181	64.123	70.568	74.601	81.861	5
3	3.791	6.628	8.472	11.811	15.316	20.090	29.691	42.141	55.320	66.680	72.866	77.723	82.693	85.641	90.604	4
4	9.396	14.359	17.307	22.277	27.134	33.320	44.680	57.859	70.309	79.910	84.684	88.189	91.528	93.372	96.209	3
5	18.139	25.399	29.432	35.877	41.819	48.968	61.053	73.556	83.884	90.740	93.716	95.673	97.324	98.129	99.174	2
6	31.623	41.352	46.416	54.074	60.696	68.129	79.370	89.090	95.318	98.259	99.149	99.579	99.833	99.916	99.983	1
	99.9	99.5	99.0	97.5	95.0	90.0	75.0	50.0	25.0	10.0	5.0	2.5	1.0	0.5	0.1	ORDER NO.

SAMPLE SIZE=7

ORDER NO.	0.1	0.5	1.0	2.5	5.0	10.0	25.0	50.0	75.0	90.0	95.0	97.5	99.0	99.5	99.9	ORDER NO.
1	0.014	0.072	0.143	0.361	0.730	1.494	4.026	9.428	17.966	28.031	34.816	40.962	48.205	53.088	62.724	7
2	0.698	1.585	2.267	3.669	5.338	7.882	13.798	22.849	34.071	45.257	52.070	57.873	64.336	68.491	76.252	6
3	3.156	5.531	7.081	9.899	12.876	16.964	25.307	36.411	46.600	59.618	65.874	70.957	76.368	79.704	85.622	5
4	7.665	11.770	14.227	18.406	22.532	27.860	37.885	49.999	62.115	72.140	77.468	81.594	85.773	88.230	92.335	4
5	14.378	20.296	23.632	29.043	34.126	40.382	51.391	63.589	74.693	83.036	87.124	90.101	92.919	94.469	96.844	3
6	23.748	31.509	35.664	42.127	47.930	54.743	65.929	77.151	86.202	92.118	94.662	96.331	97.733	98.415	99.302	2
7	37.276	46.912	51.795	59.036	65.184	71.969	82.034	90.572	95.974	98.506	99.270	99.639	99.857	99.928	99.986	1
	99.9	99.5	99.0	97.5	95.0	90.0	75.0	50.0	25.0	10.0	5.0	2.5	1.0	0.5	0.1	ORDER NO.

SAMPLE SIZE = 8

ORDER NO.	0.1	0.5	1.0	2.5	5.0	10.0	25.0	50.0	75.0	90.0	95.0	97.5	99.0	99.5	99.9	ORDER NO.
1	0.013	0.063	0.126	0.316	0.639	1.308	3.832	8.300	15.910	25.011	31.234	36.942	43.766	48.433	57.830	8
2	0.605	1.374	1.966	3.185	4.639	6.863	12.063	20.113	30.270	40.624	47.068	52.651	58.994	63.152	71.128	7
3	2.705	4.746	6.084	8.523	11.111	14.686	22.057	32.052	43.320	53.822	59.969	65.086	70.676	74.216	80.730	6
4	6.483	9.086	12.095	15.701	19.291	23.966	32.908	44.016	55.549	66.537	71.076	75.513	80.180	83.030	88.042	5
5	11.958	15.970	19.820	24.487	28.924	34.463	44.451	55.984	67.092	76.034	80.709	84.299	87.905	90.014	93.517	4
6	19.270	25.784	29.524	34.914	40.031	46.178	56.660	67.948	77.943	85.314	88.889	91.477	93.916	95.254	97.295	3
7	28.872	36.848	41.006	47.340	52.932	59.376	69.730	79.887	87.937	93.137	95.361	96.815	98.034	98.626	99.395	2
8	42.170	51.567	56.234	63.058	68.766	74.989	84.090	91.700	96.468	98.692	99.361	99.684	99.874	99.937	99.987	1
	99.9	99.5	99.0	97.5	95.0	90.0	75.0	50.0	25.0	10.0	5.0	2.5	1.0	0.5	0.1	ORDER NO.

SAMPLE SIZE = 9

ORDER NO.	0.1	0.5	1.0	2.5	5.0	10.0	25.0	50.0	75.0	90.0	95.0	97.5	99.0	99.5	99.9	ORDER NO.
1	0.011	0.056	0.112	0.281	0.568	1.164	3.146	7.413	14.276	22.574	28.313	33.627	40.052	44.495	53.584	9
2	0.533	1.212	1.736	2.814	4.102	6.077	10.717	17.962	27.227	36.835	42.913	48.249	54.403	58.497	66.511	8
3	2.356	4.159	5.335	7.485	9.774	12.949	19.650	28.623	39.054	49.007	54.964	60.009	65.632	69.260	76.115	7
4	5.621	8.678	10.526	13.700	16.875	21.040	29.099	39.309	50.199	59.942	65.505	70.071	74.998	78.086	83.714	6
5	10.252	14.605	17.097	21.201	25.137	30.097	39.196	49.999	60.804	69.903	74.863	78.799	82.903	85.395	89.748	5
6	16.286	21.914	25.002	29.929	34.495	40.058	49.801	60.691	70.901	78.960	83.125	86.300	89.474	91.322	94.379	4
7	23.885	30.740	34.368	39.991	45.036	50.993	60.946	71.377	80.450	87.051	90.226	92.515	94.665	95.841	97.634	3
8	33.489	41.503	45.597	51.751	57.087	63.165	72.773	82.038	89.283	93.923	95.898	97.186	98.264	98.788	99.467	2
9	46.416	55.505	59.948	66.373	71.687	77.426	85.724	92.587	96.854	98.836	99.432	99.719	99.888	99.944	99.989	1
	99.9	99.5	99.0	97.5	95.0	90.0	75.0	50.0	25.0	10.0	5.0	2.5	1.0	0.5	0.1	ORDER NO.

SAMPLE SIZE = 10

ORDER NO.	0.1	0.5	1.0	2.5	5.0	10.0	25.0	50.0	75.0	90.0	95.0	97.5	99.0	99.5	99.9	ORDER NO.
1	0.010	0.050	0.100	0.253	0.512	1.048	2.836	6.697	12.945	20.567	25.887	30.850	36.904	41.130	49.881	10
2	0.477	1.086	1.554	2.522	3.677	5.453	9.640	16.227	24.737	33.685	39.416	44.501	50.436	54.429	62.373	9
3	2.103	3.701	4.751	6.673	8.726	11.582	17.558	25.857	35.545	44.961	50.690	55.610	61.175	64.820	71.845	8
4	4.963	7.677	9.321	12.156	15.003	18.757	26.086	35.509	45.770	55.173	60.662	65.246	70.289	73.612	79.636	7
5	8.981	12.830	15.044	18.708	22.243	26.731	35.069	45.170	55.549	64.579	69.646	73.762	78.165	80.909	85.870	6
6	14.130	19.091	21.835	26.238	30.354	35.421	44.461	54.830	64.931	73.269	77.757	81.292	84.956	87.170	91.019	5
7	20.464	26.488	29.711	34.754	39.338	44.827	54.830	64.491	73.914	81.243	84.997	87.844	90.679	92.323	95.037	4
8	28.155	35.180	38.825	44.390	49.310	55.039	64.465	74.143	82.442	88.418	91.274	93.327	95.249	96.299	97.697	3
9	37.627	45.571	49.664	55.490	60.584	66.316	75.263	83.773	90.360	94.547	96.323	97.478	98.446	98.914	99.523	2
10	50.119	58.870	63.096	69.150	74.113	79.433	87.055	93.303	97.164	98.952	99.488	99.747	99.900	99.950	99.990	1
	99.9	99.5	99.0	97.5	95.0	90.0	75.0	50.0	25.0	10.0	5.0	2.5	1.0	0.5	0.1	ORDER NO.

SAMPLE SIZE = 11

ORDER NO.	0.1	0.5	1.0	2.5	5.0	10.0	25.0	50.0	75.0	90.0	95.0	97.5	99.0	99.5	99.9	ORDER NO.
1	0.009	0.046	0.091	0.230	0.465	0.953	2.581	6.107	11.841	18.887	23.840	28.491	34.207	38.225	46.633	11
2	0.433	0.982	1.406	2.283	3.332	4.945	8.761	14.796	22.663	31.025	36.436	41.277	46.981	50.857	58.665	10
3	1.894	3.335	4.282	6.022	7.882	10.477	15.934	23.579	32.609	41.515	47.009	51.775	57.232	60.850	67.931	9
4	4.444	6.884	8.366	10.926	13.508	16.923	23.640	32.380	42.046	51.077	56.437	60.975	66.042	69.328	75.594	8
5	7.995	11.446	13.439	16.749	19.958	24.053	31.734	41.189	51.107	59.948	65.018	69.210	73.780	76.679	82.061	7
6	12.493	16.932	19.398	23.379	27.125	31.773	40.157	50.001	59.843	68.227	72.875	76.621	80.602	83.068	87.507	6
7	17.939	23.321	26.226	30.790	34.982	40.052	48.893	58.811	68.266	75.947	80.042	83.251	86.561	88.554	92.005	5
8	24.406	30.672	33.958	39.025	43.563	48.923	57.954	67.620	76.360	83.077	86.492	89.074	91.634	93.116	95.556	4
9	32.069	39.150	42.768	48.225	52.991	58.485	67.391	76.421	84.066	89.523	92.118	93.978	95.718	96.665	98.106	3
10	41.335	49.143	53.019	58.723	63.564	68.975	77.337	85.204	91.239	95.055	96.668	97.717	98.594	99.018	99.567	2
11	53.367	61.776	65.793	71.809	76.160	81.113	88.159	93.893	97.419	99.047	99.535	99.770	99.909	99.954	99.991	1
	99.9	99.5	99.0	97.5	95.0	90.0	75.0	50.0	25.0	10.0	5.0	2.5	1.0	0.5	0.1	ORDER NO.

SAMPLE SIZE = 12

ORDER NO.	0.1	0.5	1.0	2.5	5.0	10.0	25.0	50.0	75.0	90.0	95.0	97.5	99.0	99.5	99.9	ORDER NO.
1	0.008	0.042	0.084	0.211	0.427	0.874	2.369	5.613	10.910	17.460	22.092	26.465	31.871	35.695	43.766	12
2	0.394	0.896	1.284	2.087	3.046	4.523	8.028	13.598	20.908	28.750	33.868	38.480	43.954	47.703	55.335	11
3	1.722	3.034	3.898	5.486	7.188	9.565	14.585	21.668	30.119	38.552	43.810	48.414	53.735	57.294	64.356	10
4	4.023	6.240	7.589	9.925	12.286	15.418	21.616	29.757	38.877	47.527	52.732	57.186	62.220	65.522	71.920	9
5	7.206	10.336	12.147	15.165	18.102	21.868	28.985	37.853	47.309	55.900	60.914	65.113	69.759	72.752	78.413	8
6	11.202	15.220	17.461	21.095	24.529	28.817	36.653	45.951	55.466	63.772	68.476	72.334	76.511	79.147	84.011	7
7	15.989	20.853	23.469	27.666	31.534	36.228	44.534	54.049	63.367	71.183	75.471	78.905	82.539	84.780	88.798	6
8	21.587	27.248	30.241	34.887	39.086	44.100	52.691	62.147	71.015	78.132	81.898	84.835	87.853	89.664	92.794	5
9	28.080	34.478	37.780	42.814	47.268	52.473	61.123	70.243	78.384	84.582	87.714	90.075	92.411	93.760	95.977	4
10	35.644	42.706	46.265	51.586	56.190	61.448	69.881	78.332	85.415	90.435	92.812	94.514	96.102	96.966	98.278	3
11	44.665	52.297	56.046	61.520	66.132	71.250	79.092	86.402	91.972	95.477	96.954	97.913	98.716	99.104	99.606	2
12	56.234	64.305	68.129	73.535	77.908	82.540	89.090	94.387	97.631	99.126	99.573	99.789	99.916	99.958	99.992	1
	99.9	99.5	99.0	97.5	95.0	90.0	75.0	50.0	25.0	10.0	5.0	2.5	1.0	0.5	0.1	ORDER NO.

SAMPLE SIZE = 13

ORDER NO.	0.1	0.5	1.0	2.5	5.0	10.0	25.0	50.0	75.0	90.0	95.0	97.5	99.0	99.5	99.9	ORDER NO.
1	0.008	0.039	0.077	0.195	0.394	0.807	2.189	5.192	10.115	16.232	20.582	24.705	29.830	33.473	41.220	13
2	0.362	0.825	1.182	1.920	2.805	4.169	7.409	12.579	19.405	26.783	31.634	36.030	41.282	44.903	52.340	12
3	1.579	2.782	3.577	5.038	6.605	8.799	13.448	20.045	27.979	35.978	41.010	45.448	50.617	54.104	61.095	11
4	3.677	5.708	6.945	9.092	11.266	14.161	19.913	27.528	36.140	44.427	49.465	53.812	58.776	62.063	68.514	10
5	6.561	9.423	11.063	13.857	16.566	20.051	26.676	35.017	44.028	52.343	57.262	61.427	66.090	69.128	74.966	9
6	10.157	13.527	15.882	19.224	22.396	26.373	33.681	42.507	51.670	59.824	64.520	68.423	72.712	75.457	80.622	8
7	14.431	18.870	21.288	25.135	28.706	33.086	40.902	49.999	59.098	66.914	71.294	74.865	78.712	81.130	85.569	7
8	19.378	24.543	27.288	31.577	35.480	40.176	48.330	57.493	66.319	73.627	77.604	80.776	84.118	86.173	89.843	6
9	25.034	30.872	33.910	38.573	42.738	47.657	55.972	64.983	73.324	79.949	83.434	86.143	88.917	90.577	93.439	5
10	31.486	37.937	41.224	46.188	50.535	55.573	63.851	72.472	80.087	85.839	88.734	90.908	93.055	94.292	96.323	4
11	38.905	45.896	49.383	54.552	58.990	64.022	72.021	79.955	86.552	91.201	93.395	94.962	96.423	97.218	98.421	3
12	47.660	55.097	58.718	63.970	68.366	73.217	80.595	87.421	92.591	95.831	97.195	98.080	98.818	99.175	99.638	2
13	58.780	66.537	70.170	75.295	79.418	83.768	89.885	94.808	97.811	99.193	99.606	99.805	99.923	99.961	99.992	1
	99.9	99.5	99.0	97.5	95.0	90.0	75.0	50.0	25.0	10.0	5.0	2.5	1.0	0.5	0.1	ORDER NO.

SAMPLE SIZE = 14

ORDER NO.	0.1	0.5	1.0	2.5	5.0	10.0	25.0	50.0	75.0	90.0	95.0	97.5	99.0	99.5	99.9	ORDER NO.
1	0.007	0.036	0.072	0.181	0.366	0.750	2.034	4.830	9.428	15.166	19.264	23.164	28.031	31.508	38.946	14
2	0.336	0.764	1.095	1.780	2.599	3.866	6.879	11.703	18.104	25.068	29.673	33.869	38.909	42.402	49.635	13
3	1.458	2.570	3.306	4.658	6.110	8.147	12.473	18.647	26.122	33.721	38.538	42.812	47.826	51.231	58.117	12
4	3.385	5.259	6.403	8.389	10.404	13.004	18.459	25.608	33.775	41.698	46.566	50.797	55.666	58.918	65.366	11
5	6.022	8.660	10.192	12.760	15.272	18.513	24.709	32.575	41.167	49.107	54.000	58.103	62.743	65.795	71.752	10
6	9.293	12.672	14.568	17.601	20.608	24.316	31.174	39.544	48.351	56.310	60.958	64.861	69.202	72.015	77.384	9
7	13.155	17.240	19.473	23.035	26.559	30.456	37.824	46.514	55.350	63.087	67.496	71.139	75.120	77.657	82.407	8
8	17.593	22.343	24.880	28.861	32.504	36.913	44.650	53.486	62.176	69.544	73.641	76.965	80.527	82.760	86.845	7
9	22.616	27.985	30.798	35.139	39.042	43.690	51.649	60.456	68.826	75.684	79.392	82.399	85.432	87.328	90.707	6
10	28.268	34.205	37.257	41.897	46.000	50.803	58.833	67.425	75.291	81.487	84.728	87.240	89.808	91.340	93.978	5
11	34.634	41.082	44.334	49.203	53.434	58.302	66.225	74.392	81.541	86.996	89.596	91.611	93.597	94.741	96.615	4
12	41.883	48.769	52.174	57.188	61.462	66.279	73.878	81.353	87.525	91.853	93.590	95.342	96.694	97.430	98.542	3
13	50.365	57.598	61.091	66.131	70.337	74.932	81.896	88.297	93.121	96.134	97.401	98.220	98.905	99.236	99.664	2
14	61.054	68.492	71.969	76.836	80.736	84.834	90.572	95.170	97.966	99.250	99.634	99.819	99.928	99.964	99.993	1
	99.9	99.5	99.0	97.5	95.0	90.0	75.0	50.0	25.0	10.0	5.0	2.5	1.0	0.5	0.1	ORDER NO.

SAMPLE SIZE = 15

ORDER NO.	0.1	0.5	1.0	2.5	5.0	10.0	25.0	50.0	75.0	90.0	95.0	97.5	99.0	99.5	99.9	ORDER NO.
1	0.007	0.033	0.067	0.169	0.341	0.700	1.900	4.516	8.828	14.230	18.104	21.802	26.436	29.758	36.904	15
2	0.314	0.712	1.020	1.658	2.422	3.603	6.420	10.940	16.965	23.557	27.940	31.948	36.790	40.159	47.182	14
3	1.354	2.389	3.072	4.331	5.685	7.586	11.634	17.432	24.496	31.728	36.344	40.461	45.316	48.634	55.393	13
4	3.136	4.876	5.939	7.787	9.666	12.177	17.204	23.939	31.692	39.280	43.978	48.089	52.861	56.052	62.460	12
5	5.566	8.012	9.435	11.825	14.167	17.197	23.016	30.453	38.653	46.397	51.075	55.100	59.689	62.731	68.713	11
6	8.556	11.697	13.457	16.337	19.086	22.559	29.017	36.967	45.426	53.170	57.745	61.620	65.971	68.816	74.317	10
7	12.090	15.873	17.947	21.267	24.372	28.219	35.183	43.484	52.039	59.647	64.043	67.713	71.770	74.387	79.359	9
8	16.117	20.515	22.872	26.586	29.998	34.151	41.499	50.001	58.501	65.849	70.002	73.414	77.128	79.485	83.883	8
9	20.641	25.613	28.230	32.287	35.957	40.353	47.961	56.516	64.817	71.781	75.628	78.733	82.053	84.127	87.910	7
10	25.663	31.184	34.029	38.380	42.255	46.830	54.574	63.033	70.983	77.441	80.914	83.663	86.543	88.303	91.434	6
11	31.287	37.209	40.311	44.900	48.925	53.603	61.347	69.547	76.984	82.803	85.833	88.175	90.565	91.988	94.434	5
12	37.540	43.948	47.140	51.911	56.022	60.720	68.308	76.061	82.796	87.823	90.334	92.213	94.061	95.124	96.864	4
13	44.607	51.366	54.684	59.539	63.656	68.272	75.504	82.568	88.366	92.414	94.315	95.669	96.928	97.611	98.646	3
14	52.818	59.841	63.210	68.052	72.060	76.443	83.035	89.060	93.580	96.397	97.578	98.342	98.980	99.288	99.686	2
15	63.096	70.242	73.564	78.198	81.896	85.770	91.172	95.484	98.100	99.300	99.659	99.831	99.933	99.967	99.993	1
	99.9	99.5	99.0	97.5	95.0	90.0	75.0	50.0	25.0	10.0	5.0	2.5	1.0	0.5	0.1	ORDER NO.

SAMPLE SIZE = 16

ORDER NO.	0.1	0.5	1.0	2.5	5.0	10.0	25.0	50.0	75.0	90.0	95.0	97.5	99.0	99.5	99.9	ORDER NO.
1	0.006	0.031	0.063	0.158	0.320	0.656	1.782	4.240	8.300	13.404	17.075	20.591	25.011	28.190	35.062	16
2	0.292	0.666	0.954	1.551	2.266	3.374	6.017	10.270	15.961	22.218	26.395	30.231	34.884	38.136	44.952	15
3	1.264	2.230	2.869	4.047	5.314	7.096	10.899	16.366	23.058	29.957	34.382	38.348	43.049	46.276	52.897	14
4	2.921	4.545	5.538	7.265	9.025	11.381	16.108	22.474	29.848	37.122	41.657	45.646	50.294	53.455	59.772	13
5	5.173	7.455	8.784	11.018	13.212	16.056	21.539	28.588	36.423	43.893	48.440	52.377	56.898	59.913	65.899	12
6	7.944	10.862	12.505	15.199	17.777	21.041	27.140	34.705	42.831	50.352	54.835	58.662	62.995	65.849	71.429	11
7	11.189	14.710	16.647	19.763	22.669	26.292	32.890	40.823	49.093	56.544	60.899	64.565	68.659	71.323	76.449	10
8	14.874	18.969	21.172	24.651	27.860	31.783	38.769	46.941	55.225	62.496	66.662	70.123	73.931	76.377	81.007	9
9	18.993	23.623	26.069	29.877	33.338	37.504	44.775	53.059	61.231	68.217	72.140	75.349	78.828	81.031	85.126	8
10	23.551	28.677	31.341	35.435	39.101	43.456	50.907	59.177	67.110	73.708	77.331	80.247	83.353	85.290	89.811	7
11	28.571	34.181	37.005	41.338	45.165	49.648	57.169	65.295	72.860	78.959	82.223	84.801	87.495	89.158	92.056	6
12	34.101	40.564	43.102	47.623	51.560	56.107	63.577	71.412	78.461	83.944	86.788	88.982	91.216	92.545	94.827	5
13	40.228	46.564	49.706	54.354	58.343	62.878	70.152	77.526	83.892	88.619	90.975	92.735	94.462	95.465	97.079	4
14	47.103	53.724	56.951	61.652	65.618	70.043	76.942	83.634	89.101	92.904	94.686	95.953	97.131	97.770	98.736	3
15	55.048	61.864	64.769	69.769	73.605	77.782	84.039	89.730	93.983	96.626	97.732	98.449	99.046	99.334	99.706	2
16	64.938	71.810	74.989	79.409	82.925	86.596	91.700	95.760	98.218	99.344	99.680	99.842	99.937	99.969	99.994	1
	99.9	99.5	99.0	97.5	95.0	90.0	75.0	50.0	25.0	10.0	5.0	2.5	1.0	0.5	0.1	ORDER NO.

SAMPLE SIZE = 17

ORDER NO.	0.1	0.5	1.0	2.5	5.0	10.0	25.0	50.0	75.0	90.0	95.0	97.5	99.0	99.5	99.9	ORDER NO.
1	0.006	0.029	0.059	0.149	0.301	0.618	1.678	3.995	7.831	12.667	16.157	19.506	23.730	26.777	33.392	17
2	0.275	0.635	0.896	1.458	2.132	3.173	6.663	9.678	15.069	21.021	25.013	28.689	33.163	36.303	42.916	16
3	1.185	2.093	2.692	3.799	4.990	6.667	10.252	15.421	21.781	28.370	32.619	36.440	40.992	44.129	50.603	15
4	2.735	4.256	5.189	6.811	8.465	10.682	15.144	21.178	28.208	35.188	39.564	43.432	47.962	51.040	57.290	14
5	4.833	6.969	8.216	10.314	12.377	15.058	20.240	26.940	34.437	41.639	46.055	49.900	54.339	57.317	63.277	13
6	7.409	10.139	11.681	14.210	16.636	19.717	25.494	32.703	40.613	47.807	52.193	55.958	60.252	63.099	68.718	12
7	10.412	13.708	15.524	18.444	21.191	24.613	30.876	38.466	46.459	53.736	58.030	61.672	65.771	68.459	73.687	11
8	13.813	17.645	19.711	22.984	26.011	29.726	36.381	44.234	52.288	59.440	63.599	67.075	70.939	73.443	78.242	10
9	17.596	21.928	24.224	27.811	31.083	35.040	41.993	50.001	58.007	64.960	68.917	72.189	75.775	78.072	82.404	9
10	21.758	26.557	29.061	32.925	36.401	40.551	47.712	55.766	63.619	70.274	73.989	77.016	80.289	82.355	86.187	8
11	26.313	31.541	34.229	38.328	41.970	46.264	53.541	61.533	69.122	75.387	78.809	81.556	84.476	86.292	89.588	7
12	31.284	36.901	39.748	44.042	47.808	52.103	59.387	67.297	74.506	80.263	83.564	85.790	88.319	89.861	92.691	6
13	36.723	42.683	45.661	50.100	53.945	58.361	65.563	73.060	79.760	84.942	87.623	89.688	91.784	93.031	95.167	5
14	42.710	48.960	52.038	56.568	60.436	64.812	71.792	78.823	84.856	89.318	91.635	93.189	94.811	95.744	97.265	4
15	49.397	55.871	59.008	63.560	67.381	71.630	78.219	84.579	89.748	93.333	95.010	96.201	97.308	97.907	98.815	3
16	57.084	63.697	66.837	71.311	74.987	78.979	84.931	90.322	94.337	96.827	97.868	98.542	99.104	99.376	99.725	2
17	66.608	73.223	76.270	80.494	83.843	87.333	92.169	96.005	98.322	99.382	99.699	99.851	99.941	99.971	99.994	1
	99.9	99.5	99.0	97.5	95.0	90.0	75.0	50.0	25.0	10.0	5.0	2.5	1.0	0.5	0.1	ORDER NO.

SAMPLE SIZE = 18

ORDER NO.	0.1	0.5	1.0	2.5	5.0	10.0	25.0	50.0	75.0	90.0	95.0	97.5	99.0	99.5	99.9	ORDER NO.
1	0.006	0.028	0.056	0.141	0.285	0.584	1.536	3.778	7.413	12.008	15.332	18.530	22.574	25.499	31.871	18
2	0.259	0.590	0.846	1.376	2.010	2.995	6.347	9.160	14.271	19.947	23.766	27.294	31.602	34.655	41.051	17
3	1.116	1.971	2.535	3.579	4.702	6.286	9.676	14.581	20.637	26.942	31.026	34.711	39.118	42.167	48.492	16
4	2.570	4.002	4.879	6.409	7.969	10.064	14.239	20.023	26.737	33.442	37.669	41.418	45.831	48.841	54.990	15
5	4.536	6.544	7.719	9.695	11.643	14.176	19.091	25.471	32.655	39.602	43.888	47.637	51.989	54.923	60.835	14
6	6.940	9.507	10.958	13.343	15.635	18.549	24.035	30.921	38.430	45.502	49.783	53.480	57.720	60.548	66.169	13
7	9.737	12.835	14.544	17.228	19.895	23.139	29.101	36.372	44.088	51.183	55.404	59.008	63.091	65.786	71.075	12
8	12.894	16.494	18.441	21.529	24.397	27.923	34.270	41.832	49.642	56.672	60.784	64.256	68.142	70.683	75.600	11
9	16.393	20.466	22.630	26.019	29.121	32.885	39.539	47.274	55.099	61.980	65.940	69.242	72.898	75.260	79.772	10
10	20.228	24.740	27.102	30.758	34.060	38.020	44.901	52.726	60.461	67.115	70.879	73.981	77.370	79.534	83.607	9
11	24.400	29.317	31.858	35.744	39.216	43.328	50.356	58.174	65.730	72.077	75.603	78.471	81.559	83.606	87.106	8
12	28.925	34.214	36.909	40.992	44.589	48.817	55.912	63.628	70.899	76.861	80.105	82.702	85.105	87.165	90.263	7
13	33.831	39.452	42.280	46.520	50.217	54.498	61.570	69.079	75.965	81.451	84.365	86.657	89.042	90.493	93.060	6
14	39.165	45.077	48.011	52.363	56.112	60.398	67.346	74.529	80.909	85.824	88.357	90.305	92.281	93.436	95.464	5
15	45.010	51.159	54.169	58.582	62.331	66.558	73.263	79.977	85.711	89.936	92.031	93.591	95.121	95.996	97.430	4
16	51.508	57.833	60.882	65.289	68.974	73.058	79.363	85.419	90.324	93.714	95.298	96.421	97.465	98.029	98.884	3
17	58.949	65.365	68.398	72.706	76.234	80.063	85.729	90.850	94.653	97.005	97.990	98.624	99.154	99.410	99.741	2
18	66.129	74.501	77.426	81.470	84.668	87.992	92.667	96.222	98.414	99.416	99.715	99.859	99.944	99.972	99.994	1
	99.9	99.5	99.0	97.5	95.0	90.0	75.0	50.0	25.0	10.0	5.0	2.5	1.0	0.5	0.1	ORDER NO.

SAMPLE SIZE =19

ORDER NO.	0.1	0.5	1.0	2.5	5.0	10.0	25.0	50.0	75.0	90.0	95.0	97.5	99.0	99.5	99.9	ORDER NO.
1	0.005	0.026	0.053	0.133	0.270	0.553	1.503	3.582	7.036	11.413	14.587	17.647	21.524	24.335	30.481	19
2	0.245	0.558	0.800	1.301	1.904	2.834	5.065	8.677	13.554	18.977	22.637	26.028	30.180	33.111	39.338	18
3	1.054	1.861	2.396	3.382	4.447	5.947	9.164	13.827	19.607	25.651	29.580	33.138	37.406	40.368	46.542	17
4	2.424	3.777	4.606	6.052	7.529	9.515	13.526	18.989	25.411	31.860	35.943	39.579	43.873	46.816	52.856	16
5	4.272	6.168	7.278	9.147	10.990	13.393	18.064	24.154	31.046	37.763	41.912	45.565	49.825	52.711	58.556	15
6	6.528	8.950	10.322	12.576	14.747	17.513	22.736	29.322	36.550	43.404	47.579	51.203	55.379	58.179	63.781	14
7	9.147	12.067	13.682	16.288	18.751	21.833	27.517	34.490	41.946	48.856	52.996	56.550	60.601	63.291	68.609	13
8	12.093	15.488	17.327	20.252	22.971	26.327	32.395	39.660	47.247	54.133	58.193	61.642	65.531	68.090	73.087	12
9	15.348	19.189	21.235	24.447	27.395	30.984	37.360	44.830	52.400	59.246	63.189	66.500	70.196	72.602	77.244	11
10	18.903	23.161	25.396	28.864	32.009	35.793	42.408	50.001	57.592	64.207	67.991	71.136	74.604	76.839	81.097	10
11	22.756	27.398	29.804	33.500	36.811	40.754	47.540	55.170	62.640	69.016	72.605	75.553	78.765	80.811	84.652	9
12	26.913	31.910	34.469	38.358	41.807	45.867	52.753	60.340	67.605	73.673	77.029	79.748	82.673	84.512	87.907	8
13	31.391	36.709	39.599	43.450	47.004	51.144	58.054	65.510	72.483	78.167	81.249	83.712	86.318	87.933	90.853	7
14	36.219	41.821	44.621	48.797	52.421	56.596	63.450	70.678	77.264	82.487	85.253	87.424	89.678	91.050	93.472	6
15	41.444	47.289	50.175	54.435	58.088	62.247	68.954	75.846	81.936	86.607	89.010	90.853	92.722	93.832	95.728	5
16	47.144	53.184	56.127	60.421	64.057	68.140	74.589	81.011	86.474	90.485	92.471	93.948	95.394	96.223	97.576	4
17	53.458	59.632	62.594	66.862	70.420	74.349	80.393	86.173	90.836	94.053	95.563	96.618	97.604	98.139	98.946	3
18	60.662	66.889	69.820	73.972	77.363	81.023	86.446	91.323	94.935	97.166	98.096	98.699	99.200	99.442	99.755	2
19	69.519	75.665	78.476	82.353	85.413	88.587	92.964	96.418	98.497	99.447	99.730	99.867	99.947	99.974	99.995	1
	99.9	99.5	99.0	97.5	95.0	90.0	75.0	50.0	25.0	10.0	5.0	2.5	1.0	0.5	0.1	ORDER NO.

SAMPLE SIZE = 20

ORDER NO.	99.9	99.5	99.0	97.5	95.0	90.0	75.0	50.0	25.0	10.0	5.0	2.5	1.0	0.5	0.1	ORDER NO.
1	0.005	0.025	0.050	0.127	0.256	0.525	1.428	3.406	6.697	10.875	13.911	16.843	20.567	23.273	29.205	20
2	0.233	0.530	0.759	1.235	1.806	2.691	4.812	8.251	12.905	18.096	21.610	24.873	28.880	31.715	37.759	19
3	0.999	1.765	2.271	3.207	4.217	5.642	8.701	13.148	18.674	24.476	28.262	31.698	35.833	38.712	44.738	18
4	2.294	3.576	4.362	5.733	7.136	9.022	12.840	18.055	24.210	30.419	34.357	37.893	42.072	44.947	50.874	17
5	4.038	5.833	6.884	8.657	10.409	12.693	17.142	22.967	29.588	36.066	40.102	43.661	47.828	50.661	56.431	16
6	6.164	8.456	9.754	11.893	13.955	16.587	21.571	27.880	34.844	41.489	45.559	49.105	53.211	55.976	61.543	15
7	8.624	11.388	12.917	15.391	17.731	20.666	26.098	32.795	40.000	46.726	50.782	54.279	58.286	60.961	66.283	14
8	11.387	14.599	16.341	19.118	21.706	24.906	30.715	37.710	45.069	51.803	55.804	59.219	63.094	65.656	70.702	13
9	14.431	18.066	20.005	23.088	25.864	29.293	35.410	42.626	50.060	56.733	60.642	63.946	67.658	70.090	74.827	12
10	17.747	21.775	23.896	27.197	30.195	33.817	40.182	47.543	54.975	61.524	65.307	68.472	71.992	74.277	78.674	11
11	21.336	25.723	28.008	31.528	34.603	38.476	45.025	52.457	59.818	66.183	69.805	72.803	76.104	78.225	82.253	10
12	25.173	29.910	32.342	36.054	39.358	43.267	49.940	57.374	64.590	70.707	74.136	76.942	79.995	81.934	85.569	9
13	29.298	34.344	36.906	40.781	44.196	48.197	54.931	62.290	69.285	75.094	78.294	80.882	83.659	85.401	88.613	8
14	33.717	39.039	41.714	45.721	49.218	53.274	60.000	67.205	73.902	79.334	82.269	84.609	87.083	88.613	91.376	7
15	38.457	44.024	46.789	50.898	54.441	58.511	65.156	72.120	78.429	83.413	86.045	88.107	90.246	91.544	93.836	6
16	43.569	49.339	52.172	56.339	59.806	63.934	70.412	77.033	82.868	87.307	89.591	91.343	93.116	94.167	95.962	5
17	49.126	55.053	57.928	62.107	65.633	69.581	75.790	81.945	87.160	90.978	92.864	94.267	95.638	96.424	97.706	4
18	55.262	61.288	64.167	68.302	71.736	75.524	81.328	86.852	91.299	94.358	95.783	96.793	97.729	98.235	99.001	3
19	62.241	68.285	71.120	75.127	78.590	81.904	87.095	91.749	95.188	97.309	98.194	98.765	99.241	99.470	99.767	2
20	70.795	76.727	79.433	83.157	86.089	89.125	93.303	96.594	98.572	99.475	99.744	99.873	99.950	99.975	99.995	1
ORDER NO.	0.1	0.5	1.0	2.5	5.0	10.0	25.0	50.0	75.0	90.0	95.0	97.5	99.0	99.5	99.9	ORDER NO.

SAMPLE SIZE =21

ORDER NO.	99.9	99.5	99.0	97.5	95.0	90.0	75.0	50.0	25.0	10.0	5.0	2.5	1.0	0.5	0.1	ORDER NO.
1	28.031	22.299	19.491	16.110	13.295	10.385	6.388	3.247	1.361	0.500	0.244	0.120	0.048	0.024	0.005	21
2	36.298	30.429	27.664	23.816	20.673	17.294	12.515	7.864	4.583	2.562	1.719	1.175	0.722	0.504	0.221	20
3	43.063	37.185	34.386	30.378	27.055	23.405	17.826	12.631	8.283	5.367	4.009	3.049	2.158	1.676	0.948	19
4	49.026	43.217	40.411	36.343	32.922	29.102	23.118	17.210	12.220	8.578	6.780	5.447	4.142	3.394	2.178	18
5	54.443	48.757	45.979	41.906	38.441	34.522	28.260	21.891	16.311	12.061	9.884	8.218	6.552	5.594	3.828	17
6	59.440	53.924	51.199	47.166	43.697	39.733	33.289	26.574	20.518	15.755	13.245	11.282	9.246	8.012	5.837	16
7	64.091	58.782	56.130	52.175	48.739	44.772	38.226	31.258	24.819	19.619	16.817	14.588	12.235	10.781	8.157	15
8	68.443	63.374	60.816	56.968	53.594	49.660	43.082	35.943	29.202	23.632	20.576	18.107	15.464	13.807	10.758	14
9	72.521	67.722	65.276	61.564	58.280	54.417	47.866	40.629	33.656	27.779	24.499	21.819	18.913	17.068	13.618	13
10	76.347	71.847	69.527	65.980	62.811	59.046	52.583	45.315	38.178	32.052	28.580	25.713	22.567	20.548	16.724	12
11	79.929	75.755	73.579	70.219	67.190	63.557	57.233	50.001	42.767	36.443	32.810	29.781	26.421	24.245	20.071	11
12	83.276	79.452	77.433	74.287	71.420	67.948	61.822	54.685	47.417	40.954	37.189	34.020	30.473	28.153	23.653	10
13	86.382	82.932	81.087	78.181	75.501	72.221	66.344	59.371	52.134	45.583	41.720	38.436	34.724	32.278	27.479	9
14	89.242	86.193	84.536	81.893	79.424	76.368	70.798	64.057	56.918	50.340	46.406	43.032	39.184	36.626	31.557	8
15	91.843	89.219	87.765	85.412	83.183	80.381	75.181	68.742	61.774	55.228	51.261	47.825	43.870	41.218	36.909	7
16	94.163	91.988	90.764	88.718	86.765	84.245	79.462	73.426	66.711	60.267	56.303	52.834	48.801	46.076	40.560	6
17	96.172	94.406	93.408	91.782	90.116	87.939	83.689	78.109	71.740	65.478	61.559	58.094	54.021	51.243	45.557	5
18	97.822	96.606	95.858	94.553	93.220	91.422	87.760	82.790	76.882	70.898	67.078	63.657	59.589	56.783	50.974	4
19	99.052	98.524	97.842	96.981	95.991	94.633	91.717	87.469	82.172	76.595	72.945	69.622	65.614	62.815	56.937	3
20	99.779	99.496	99.278	98.825	98.281	97.438	95.417	92.136	87.685	82.706	79.327	76.184	72.316	69.571	63.702	2
21	99.995	99.976	99.952	99.880	99.756	99.500	98.639	96.753	93.612	89.615	86.705	83.890	80.309	77.701	71.969	1
	0.1	0.5	1.0	2.5	5.0	10.0	25.0	50.0	75.0	90.0	95.0	97.5	99.0	99.5	99.9	ORDER NO.

SAMPLE SIZE =22

ORDER NO.	0.1	0.5	1.0	2.5	5.0	10.0	25.0	50.0	75.0	90.0	95.0	97.5	99.0	99.5	99.9	ORDER NO.
1	0.005	0.023	0.046	0.115	0.233	0.478	1.299	3.102	6.107	9.937	12.731	15.437	18.887	21.403	26.947	22
2	0.211	0.480	0.689	1.121	1.640	2.444	4.374	7.512	11.777	16.559	19.812	22.844	26.584	29.243	34.945	21
3	0.904	1.597	2.056	2.906	3.823	5.117	7.905	11.970	17.052	22.422	25.947	29.162	33.050	35.772	41.506	20
4	2.071	3.231	3.944	5.187	6.460	8.176	11.658	16.439	22.120	27.894	31.591	34.913	38.873	41.611	47.305	19
5	3.640	5.262	6.214	7.821	9.411	11.491	15.537	20.911	27.045	33.105	36.909	40.284	44.263	46.987	52.583	18
6	5.543	7.613	8.790	10.729	12.603	15.002	19.584	25.384	31.866	38.117	41.981	45.370	49.326	52.010	57.467	17
7	7.740	10.236	11.620	13.865	15.993	18.674	23.650	29.859	36.600	42.970	46.850	50.222	54.121	56.744	62.025	16
8	10.197	13.097	14.677	17.197	19.556	22.483	27.631	34.335	41.261	47.684	51.545	54.873	58.685	61.228	66.302	15
9	12.893	16.176	17.934	20.710	23.272	26.417	32.059	38.810	45.853	52.274	56.087	59.342	63.041	65.490	70.327	14
10	15.816	19.456	21.380	24.386	27.131	30.463	36.339	43.285	50.385	56.761	60.485	63.645	67.205	69.544	74.119	13
11	18.967	22.932	25.008	28.220	31.126	34.618	40.726	47.762	54.858	61.120	64.745	67.789	71.188	73.402	77.688	12
12	22.312	26.598	28.812	32.211	35.255	38.880	45.142	52.238	59.274	65.382	68.874	71.780	74.992	77.068	81.043	11
13	25.881	30.456	32.795	36.365	39.518	43.249	49.615	56.715	63.631	69.537	72.869	75.614	78.620	80.544	84.184	10
14	29.673	34.510	36.959	40.668	43.913	47.726	54.147	61.190	67.931	73.583	76.728	79.290	82.066	83.825	87.107	9
15	33.698	38.772	41.315	45.127	48.455	52.316	58.739	65.665	72.169	77.517	80.444	82.803	85.323	86.903	89.803	8
16	37.975	43.256	45.879	49.778	53.180	57.030	63.400	70.141	76.340	81.326	84.007	86.135	88.380	89.764	92.260	7
17	42.533	47.990	50.674	54.630	58.019	61.883	68.134	74.616	80.436	84.998	87.397	89.271	91.210	92.387	94.457	6
18	47.417	53.013	55.737	59.716	63.091	66.895	72.945	79.089	84.443	88.509	90.589	92.179	93.786	94.738	96.360	5
19	52.695	58.389	61.127	65.087	68.409	72.106	77.880	83.561	88.342	91.828	93.540	94.813	96.056	96.769	97.929	4
20	58.494	64.228	66.950	70.838	74.053	77.578	82.948	88.030	92.095	94.883	96.177	97.094	97.944	98.403	99.096	3
21	65.055	70.757	73.416	77.156	80.188	83.441	88.223	92.488	95.626	97.556	98.360	98.879	99.311	99.520	99.789	2
22	73.053	78.597	81.113	84.563	87.269	90.063	93.893	96.898	98.701	99.522	99.767	99.885	99.954	99.977	99.995	1
	99.9	99.5	99.0	97.5	95.0	90.0	75.0	50.0	25.0	10.0	5.0	2.5	1.0	0.5	0.1	

SAMPLE SIZE = 23

ORDER NO.	0.1	0.5	1.0	2.5	5.0	10.0	25.0	50.0	75.0	90.0	95.0	97.5	99.0	99.5	99.9	ORDER NO.
1	0.004	0.022	0.044	0.110	0.223	0.457	1.243	2.969	5.849	9.526	12.212	14.819	18.145	20.575	25.943	23
2	0.202	0.459	0.656	1.071	1.567	2.337	4.184	7.191	11.284	15.884	19.020	21.949	25.567	28.144	33.687	22
3	0.863	1.525	1.965	2.775	3.652	4.890	7.558	11.457	16.343	21.519	24.924	28.037	31.812	34.460	40.055	21
4	1.977	3.083	3.762	4.961	6.167	7.809	11.144	15.734	21.203	26.781	30.364	33.588	37.446	40.118	45.693	20
5	3.468	5.016	5.928	7.461	8.961	10.972	14.869	20.014	25.932	31.797	35.493	38.781	42.667	45.336	50.840	19
6	5.277	7.253	8.375	10.229	12.022	14.318	18.696	24.297	30.560	36.626	40.389	43.703	47.681	50.220	55.610	18
7	7.363	9.743	11.066	13.210	15.247	17.815	22.605	28.580	35.107	41.305	45.098	48.405	52.242	54.633	60.075	17
8	9.692	12.457	13.966	16.376	18.635	21.441	26.583	32.864	39.585	45.855	49.644	52.920	56.687	59.214	64.274	16
9	12.243	15.372	17.052	19.707	22.164	25.182	30.628	37.147	44.001	50.291	54.046	57.265	60.940	63.384	68.240	15
10	15.003	18.475	20.315	23.191	25.825	29.028	34.722	41.431	48.363	54.621	58.315	61.459	65.015	67.365	71.989	14
11	17.963	21.755	23.742	26.820	29.609	32.972	38.874	45.716	52.668	58.852	62.461	65.505	68.923	71.165	75.533	13
12	21.117	25.210	27.329	30.589	33.515	37.011	43.078	50.001	56.922	62.989	66.485	69.411	72.671	74.790	78.883	12
13	24.467	28.835	31.077	34.495	37.539	41.148	47.333	54.284	61.126	67.028	70.391	73.180	76.258	78.245	82.037	11
14	28.011	32.635	34.985	38.541	41.685	45.379	51.636	58.569	65.278	70.972	74.175	76.809	79.685	81.525	84.997	10
15	31.760	36.616	39.060	42.735	45.954	49.709	55.999	62.853	69.375	74.818	77.836	80.293	82.948	84.628	87.757	9
16	35.728	40.786	43.313	47.080	50.356	54.145	60.415	67.136	73.417	78.559	81.365	83.624	86.034	87.543	90.308	8
17	39.925	45.167	47.758	51.595	54.902	58.695	64.893	71.420	77.395	82.185	84.753	86.790	88.934	90.257	92.637	7
18	44.390	49.780	52.419	56.297	59.611	63.374	69.440	75.703	81.304	85.682	87.978	89.771	91.625	92.747	94.723	6
19	49.160	54.664	57.333	61.219	64.507	68.203	74.068	79.986	85.131	89.028	91.019	92.539	94.074	94.984	96.532	5
20	54.307	59.882	62.554	66.412	69.636	73.219	78.797	84.266	88.856	92.191	93.833	95.049	96.238	96.917	98.023	4
21	59.945	65.540	68.168	71.963	75.076	78.481	83.657	88.543	92.442	95.110	96.348	97.225	98.035	98.475	99.137	3
22	66.313	71.856	74.433	78.051	80.980	84.116	88.716	92.809	95.816	97.663	98.433	98.929	99.342	99.541	99.798	2
23	74.057	79.425	81.855	86.181	87.788	90.474	94.151	97.031	98.757	99.543	99.777	99.890	99.956	99.978	99.996	1
	99.9	99.5	99.0	97.5	95.0	90.0	75.0	50.0	25.0	10.0	5.0	2.5	1.0	0.5	0.1	ORDER NO.

SAMPLE SIZE = 24

ORDER NO.	0.1	0.5	1.0	2.5	5.0	10.0	25.0	50.0	75.0	90.0	95.0	97.5	99.0	99.5	99.9	ORDER NO.
1	0.004	0.021	0.042	0.105	0.213	0.438	1.192	2.847	5.613	9.148	11.735	14.247	17.460	19.809	25.011	24
2	0.193	0.440	0.630	1.026	1.601	2.238	4.009	6.893	10.830	13.263	16.280	21.120	24.625	27.125	32.516	23
3	0.826	1.460	1.879	2.656	3.495	4.682	7.241	10.987	15.690	20.688	23.980	26.997	30.663	33.239	38.700	22
4	1.890	2.947	3.597	4.736	5.901	7.473	10.676	15.087	20.361	25.755	29.227	32.362	36.117	38.726	44.186	21
5	3.312	4.794	5.662	7.131	8.588	10.497	14.240	19.192	24.905	30.589	34.180	37.385	41.181	43.796	49.203	20
6	5.038	6.925	7.999	9.772	11.491	13.694	17.901	23.298	29.356	35.247	38.914	42.152	45.953	48.547	53.863	19
7	7.021	9.296	10.561	12.615	14.568	17.034	21.639	27.406	33.730	39.764	43.468	46.711	50.484	53.042	58.233	18
8	9.234	11.878	13.320	15.630	17.796	20.493	25.443	31.513	38.039	44.160	47.872	51.095	54.815	57.317	62.354	17
9	11.655	14.646	16.254	18.800	21.157	24.058	29.305	35.621	42.292	48.449	52.142	55.322	58.966	61.399	66.256	16
10	14.271	17.590	19.350	22.109	24.639	27.721	33.221	39.729	46.493	52.641	56.289	59.406	62.981	65.302	69.956	15
11	17.069	20.696	22.599	25.553	28.236	31.475	37.185	43.858	50.643	56.742	60.322	63.857	66.782	69.039	73.467	14
12	20.048	23.962	25.994	29.124	31.942	35.317	41.196	47.945	54.746	60.755	64.243	67.179	70.466	72.617	76.798	13
13	23.202	27.383	29.534	32.821	35.757	39.245	45.254	52.055	58.804	64.663	68.058	70.876	74.006	76.038	79.952	12
14	26.533	30.901	33.218	36.643	39.678	43.258	49.357	56.162	62.815	68.525	71.764	74.447	77.401	79.304	82.931	11
15	30.044	34.698	37.049	40.594	43.711	47.359	53.507	60.271	66.779	72.279	75.361	77.891	80.650	82.410	85.729	10
16	33.744	38.601	41.035	44.678	47.858	51.551	57.708	64.379	70.695	75.942	78.843	81.200	83.746	85.354	88.345	9
17	37.646	42.683	45.185	48.905	52.128	55.840	61.961	68.487	74.557	79.507	82.204	84.370	86.680	88.122	90.766	8
18	41.767	46.988	49.516	53.289	56.532	60.236	66.270	72.594	78.361	82.966	85.432	87.385	89.439	90.704	92.979	7
19	46.137	51.465	54.047	57.848	61.086	64.765	70.644	76.702	82.099	86.306	88.509	90.228	92.001	93.075	94.962	6
20	50.797	56.205	58.819	62.615	65.820	69.411	75.025	80.808	85.760	89.503	91.412	92.869	94.338	95.206	96.688	5
21	55.814	61.274	63.883	67.638	70.773	74.245	79.639	84.913	89.324	92.627	94.099	95.264	96.403	97.053	98.110	4
22	61.300	66.761	69.337	73.003	76.020	79.314	84.310	89.013	92.759	95.318	96.505	97.344	98.121	98.540	99.174	3
23	67.484	72.875	75.376	78.880	81.711	84.736	89.170	93.106	95.991	97.762	98.499	98.974	99.370	99.560	99.807	2
24	74.989	80.191	82.540	85.753	88.265	90.852	94.337	97.153	98.808	99.562	99.787	99.895	99.958	99.979	99.996	1
	99.9	99.5	99.0	97.5	95.0	90.0	75.0	50.0	25.0	10.0	5.0	2.5	1.0	0.5	0.1	ORDER NO.

SAMPLE SIZE =25

ORDER NO.	99.9	99.5	99.0	97.5	95.0	90.0	75.0	50.0	25.0	10.0	5.0	2.5	1.0	0.5	0.1	ORDER NO.
1	24.142	19.098	16.824	13.719	11.293	8.799	5.394	2.735	1.144	0.421	0.205	0.101	0.040	0.020	0.004	25
2	31.422	26.176	23.749	20.352	17.612	14.687	10.412	6.623	3.849	2.148	1.440	0.984	0.605	0.422	0.185	24
3	37.431	32.101	29.594	26.031	23.104	19.913	15.087	10.554	6.950	4.491	3.352	2.546	1.801	1.398	0.791	23
4	42.771	37.426	34.878	31.219	28.172	24.802	19.581	14.492	10.244	7.166	5.656	4.537	3.448	2.824	1.809	22
5	47.663	42.352	39.793	36.083	32.961	29.467	23.957	18.435	13.662	10.062	8.230	6.831	5.422	4.589	3.170	21
6	52.216	46.976	44.427	40.704	37.540	33.966	28.242	22.379	17.171	13.123	11.005	9.356	7.655	6.625	4.816	20
7	56.493	51.387	48.838	45.129	41.952	38.331	32.456	26.324	20.754	16.317	13.947	12.072	10.102	8.889	6.709	19
8	60.835	55.632	53.036	49.387	46.221	42.582	36.610	30.270	24.398	19.625	17.031	14.950	12.733	11.350	8.817	18
9	64.371	59.524	57.107	53.500	50.364	46.734	40.710	34.216	28.097	23.031	20.238	17.971	15.530	13.987	11.121	17
10	68.018	63.349	61.002	57.479	54.394	50.796	44.761	38.161	31.844	26.530	23.559	21.125	18.476	16.787	13.607	16
11	71.490	67.020	64.756	61.335	58.317	54.772	48.766	42.107	35.638	30.111	26.986	24.403	21.563	19.738	16.264	15
12	74.793	70.545	68.374	65.072	62.138	58.668	52.728	46.053	39.474	33.773	30.512	27.796	24.786	22.836	19.083	14
13	77.935	73.925	71.859	68.694	65.861	62.486	56.647	50.001	43.353	37.514	34.139	31.306	28.141	26.075	22.065	13
14	80.917	77.164	75.214	72.204	69.488	66.227	60.526	53.947	47.272	41.332	37.862	34.928	31.626	29.455	25.207	12
15	83.736	80.262	78.437	75.597	73.014	69.889	64.362	57.893	51.234	45.228	41.683	38.665	35.244	32.980	28.510	11
16	86.393	83.213	81.524	78.875	76.441	73.470	68.156	61.839	55.239	49.204	45.606	42.521	38.998	36.651	31.982	10
17	88.879	86.013	84.470	82.029	79.762	76.969	71.903	65.784	59.290	53.266	49.636	46.500	42.893	40.476	35.629	9
18	91.183	88.680	87.267	85.050	82.969	80.375	75.602	69.730	63.390	57.418	53.779	50.613	46.944	44.468	39.465	8
19	93.291	91.111	89.898	87.928	86.053	83.683	79.246	73.676	67.544	61.669	58.048	54.871	51.162	48.643	43.507	7
20	95.184	93.375	92.345	90.644	88.995	86.877	82.829	77.621	71.758	66.035	62.460	59.296	55.573	53.024	47.784	6
21	96.830	95.411	94.578	93.169	91.770	89.938	86.338	81.565	76.043	70.533	67.039	63.917	60.207	57.648	52.337	5
22	98.191	97.176	96.552	95.463	94.344	92.834	89.756	85.508	80.419	75.198	71.828	68.781	65.122	62.574	57.229	4
23	99.209	98.602	98.199	97.454	96.648	95.509	93.050	89.446	84.913	80.087	76.896	73.969	70.406	67.899	62.569	3
24	99.815	99.578	99.395	99.016	98.560	97.852	96.151	93.377	89.588	85.313	82.388	79.648	76.251	73.824	68.578	2
25	99.996	99.980	99.960	99.899	99.795	99.579	98.856	97.265	94.606	91.201	88.707	86.281	83.176	80.902	75.858	1
ORDER NO.	0.1	0.5	1.0	2.5	5.0	10.0	25.0	50.0	75.0	90.0	95.0	97.5	99.0	99.5	99.9	ORDER NO.

APPENDIX B
STANDARDIZED NORMAL DISTRIBUTION'S AREA TABLES

$$1 - F(z) = \int_z^\infty \frac{1}{\sqrt{2\pi}} e^{-\frac{t^2}{2}} \, dt.$$

USAGE

For $z \geq 3.0$, entries are in abbreviated notation $\times . \times \times \times \times - p$, where

$$1 - F(z) = \times . \times \times \times \times \times 10^{-p}.$$

EXAMPLE 1

X is normally distributed with mean $\mu = 27$ and standard deviation $\sigma = 4$. What is the probability X will exceed 41?

Answer: $z = (41 - 27)/4 = 3.50$,

$$
\begin{aligned}
P(X \geq 41) &= 1 - F(3.50), \\
&= 2.3263 \times 10^{-4} = 0.00023263.
\end{aligned}
$$

EXAMPLE 2

From Example 1, what is the probability that X will be less than 41?

$$
\begin{aligned}
P(X < 41) &= 1 - [1 - F(3.50)], \\
&= 1 - 0.00023263, \\
&= 0.99976737.
\end{aligned}
$$

z	$1 - F(z)$	z	$1 - F(z)$	z	$1 - F(z)$	z	$1 - F(z)$
0.00	0.50000	0.40	0.34458	0.80	0.21186	1.20	0.11507
0.01	0.49601	0.41	0.34090	0.81	0.20897	1.21	0.11314
0.02	0.49202	0.42	0.33724	0.82	0.20611	1.22	0.11123
0.03	0.48803	0.43	0.33360	0.83	0.20327	1.23	0.10935
0.04	0.48405	0.44	0.32997	0.84	0.20045	1.24	0.10749
0.05	0.48006	0.45	0.32636	0.85	0.19766	1.25	0.10565
0.06	0.47608	0.46	0.32276	0.86	0.19489	1.26	0.10383
0.07	0.47210	0.47	0.31918	0.87	0.19215	1.27	0.10204
0.08	0.46812	0.48	0.31561	0.88	0.18943	1.28	0.10027
0.09	0.46414	0.49	0.31207	0.89	0.18673	1.29	0.098525
0.10	0.46017	0.50	0.30854	0.90	0.18406	1.30	0.096800
0.11	0.45620	0.51	0.30503	0.91	0.18141	1.31	0.095093
0.12	0.45224	0.52	0.30153	0.92	0.17879	1.32	0.093418
0.13	0.44828	0.53	0.29806	0.93	0.17619	1.33	0.091759
0.14	0.44433	0.54	0.29460	0.94	0.17361	1.34	0.090123
0.15	0.44038	0.55	0.29116	0.95	0.17106	1.35	0.088508
0.16	0.43644	0.56	0.28774	0.96	0.16853	1.36	0.086915
0.17	0.43251	0.57	0.28434	0.97	0.16602	1.37	0.085343
0.18	0.42858	0.58	0.28096	0.98	0.16354	1.38	0.083793
0.19	0.42465	0.59	0.27760	0.99	0.16109	1.39	0.082264
0.20	0.42074	0.60	0.27425	1.00	0.15866	1.40	0.080757
0.21	0.41683	0.61	0.27093	1.01	0.15625	1.41	0.079270
0.22	0.41294	0.62	0.26763	1.02	0.15386	1.42	0.077804
0.23	0.40905	0.63	0.26435	1.03	0.15151	1.43	0.076359
0.24	0.40517	0.64	0.26109	1.04	0.14917	1.44	0.074934
0.25	0.40129	0.65	0.25785	1.05	0.14686	1.45	0.073529
0.26	0.39743	0.66	0.25463	1.06	0.14457	1.46	0.072145
0.27	0.39358	0.67	0.25143	1.07	0.14231	1.47	0.070781
0.28	0.38974	0.68	0.24825	1.08	0.14007	1.48	0.069437
0.29	0.38591	0.69	0.24510	1.09	0.13786	1.49	0.068112
0.30	0.38209	0.70	0.24196	1.10	0.13567	1.50	0.066807
0.31	0.37828	0.71	0.23885	1.11	0.13350	1.51	0.065522
0.32	0.37448	0.72	0.23576	1.12	0.13136	1.52	0.064255
0.33	0.37070	0.73	0.23270	1.13	0.12924	1.53	0.063008
0.34	0.36693	0.74	0.22965	1.14	0.12714	1.54	0.061780
0.35	0.36317	0.75	0.22663	1.15	0.12507	1.55	0.060571
0.36	0.35942	0.76	0.22363	1.16	0.12302	1.56	0.059380
0.37	0.35569	0.77	0.22065	1.17	0.12100	1.57	0.058208
0.38	0.35197	0.78	0.21770	1.18	0.11900	1.58	0.057053
0.39	0.34827	0.79	0.21476	1.19	0.11702	1.59	0.055917

z	$1 - F(z)$	z	$1 - F(z)$	z	$1 - F(z)$	z	$1 - F(z)$
1.60	0.054799	2.00	0.022750	2.40	0.0081975	2.80	0.0025551
1.61	0.053699	2.01	0.022216	2.41	0.0079763	2.81	0.0024771
1.62	0.052616	2.02	0.021692	2.42	0.0077603	2.82	0.0024012
1.63	0.051551	2.03	0.021178	2.43	0.0075494	2.83	0.0023274
1.64	0.050503	2.04	0.020675	2.44	0.0073436	2.84	0.0022557
1.65	0.049471	2.05	0.020182	2.45	0.0071428	2.85	0.0021860
1.66	0.048457	2.06	0.019699	2.46	0.0069469	2.86	0.0021182
1.67	0.047460	2.07	0.019226	2.47	0.0067557	2.87	0.0020524
1.68	0.046479	2.08	0.018763	2.48	0.0065691	2.88	0.0019884
1.69	0.045514	2.09	0.018309	2.49	0.0063872	2.89	0.0019262
1.70	0.044565	2.10	0.017864	2.50	0.0062097	2.90	0.0018658
1.71	0.043633	2.11	0.017429	2.51	0.0060366	2.91	0.0018071
1.72	0.042716	2.12	0.017003	2.52	0.0058677	2.92	0.0017502
1.73	0.041815	2.13	0.016586	2.53	0.0057031	2.93	0.0016948
1.74	0.040930	2.14	0.016177	2.54	0.0055426	2.94	0.0016411
1.75	0.040059	2.15	0.015778	2.55	0.0053861	2.95	0.0015889
1.76	0.039204	2.16	0.015386	2.56	0.0052336	2.96	0.0015382
1.77	0.038364	2.17	0.015003	2.57	0.0050849	2.97	0.0014890
1.78	0.037538	2.18	0.014629	2.58	0.0049400	2.98	0.0014412
1.79	0.036727	2.19	0.014262	2.59	0.0047988	2.99	0.0013949
1.80	0.035930	2.20	0.013903	2.60	0.0046612	3.00	1.3499 -3
1.81	0.035148	2.21	0.013553	2.61	0.0045271	3.01	1.3062
1.82	0.034380	2.22	0.013209	2.62	0.0043965	3.02	1.2639
1.83	0.033625	2.23	0.012874	2.63	0.0042692	3.03	1.2228
1.84	0.032884	2.24	0.012545	2.64	0.0041453	3.04	1.1829
1.85	0.032157	2.25	0.012224	2.65	0.0040246	3.05	1.1442 -3
1.86	0.031443	2.26	0.011911	2.66	0.0039070	3.06	1.1067
1.87	0.030742	2.27	0.011604	2.67	0.0037926	3.07	1.0703
1.88	0.030054	2.28	0.011304	2.68	0.0036811	3.08	1.0350
1.89	0.029379	2.29	0.011011	2.69	0.0035726	3.09	1.0008
1.90	0.028717	2.30	0.010724	2.70	0.0034670	3.10	9.6760 -4
1.91	0.028067	2.31	0.010444	2.71	0.0033642	3.11	9.3544
1.92	0.027429	2.32	0.010170	2.72	0.0032641	3.12	9.0426
1.93	0.026803	2.33	0.009903	2.73	0.0031667	3.13	8.7403
1.94	0.026190	2.34	0.009642	2.74	0.0030720	3.14	8.4474
1.95	0.025588	2.35	0.0093867	2.75	0.0029798	3.15	8.1635 -4
1.96	0.024998	2.36	0.0091375	2.76	0.0028901	3.16	7.8885
1.97	0.024419	2.37	0.0088940	2.77	0.0028028	3.17	7.6219
1.98	0.023852	2.38	0.0086563	2.78	0.0027179	3.18	7.3638
1.99	0.023295	2.39	0.0084242	2.79	0.0026354	3.19	7.1136

z	$1 - F(z)$	z	$1 - F(z)$	z	$1 - F(z)$	z	$1 - F(z)$
3.20	6.8714 -4	3.60	1.5911 -4	4.00	3.1671 -5	4.40	5.4125 -6
3.21	6.6367	3.61	1.5310	4.01	3.0359	4.41	5.1685
3.22	6.4095	3.62	1.4730	4.02	2.9099	4.42	4.9350
3.23	6.1895	3.63	1.4171	4.03	2.7888	4.43	4.7117
3.24	5.9765	3.64	1.3632	4.04	2.6726	4.44	4.4979
3.25	5.7703 -4	3.65	1.3112 -4	4.05	2.5609 -5	4.45	4.2935 -6
3.26	5.5706	3.66	1.2611	4.06	2.4536	4.46	4.0980
3.27	5.3774	3.67	1.2128	4.07	2.3507	4.47	3.9110
3.28	5.1904	3.68	1.1662	4.08	2.2518	4.48	3.7322
3.29	5.0094	3.69	1.1213	4.09	2.1569	4.49	3.5612
3.30	4.8342 -4	3.70	1.0780 -4	4.10	2.0658 -5	4.50	3.3977 -6
3.31	4.6648	3.71	1.0363	4.11	1.9783	4.51	3.2414
3.32	4.5009	3.72	9.9611 -5	4.12	1.8944	4.52	3.0920
3.33	4.3423	3.73	9.5740	4.13	1.8138	4.53	2.9492
3.34	4.1889	3.74	9.2010	4.14	1.7365	4.54	2.8127
3.35	4.0406 -4	3.75	8.8417 -5	4.15	1.6624 -5	4.55	2.6823 -6
3.36	3.8971	3.76	8.4957	4.16	1.5912	4.56	2.5577
3.37	3.7584	3.77	8.1624	4.17	1.5230	4.57	2.4386
3.38	3.6243	3.78	7.8414	4.18	1.4575	4.58	2.3249
3.39	3.4946	3.79	7.5324	4.19	1.3948	4.59	2.2162
3.40	3.3693 -4	3.80	7.2348 -5	4.20	1.3346 5	4.60	2.1125 -6
3.41	3.2481	3.81	6.9483	4.21	1.2769	4.61	2.0133
3.42	3.1311	3.82	6.6726	4.22	1.2215	4.62	1.9187
3.43	3.0179	3.83	6.4072	4.23	1.1685	4.63	1.8283
3.44	2.9086	3.84	6.1517	4.24	1.1176	4.64	1.7420
3.45	2.8029 -4	3.85	5.9059 -5	4.25	1.0689 -5	4.65	1.6597 -6
3.46	2.7009	3.86	5.6694	4.26	1.0221	4.66	1.5810
3.47	2.6023	3.87	5.4418	4.27	9.7736 -6	4.67	1.5060
3.48	2.5071	3.88	5.2228	4.28	9.3447	4.68	1.4344
3.49	2.4151	3.89	5.0122	4.29	8.9337	4.69	1.3660
3.50	2.3263 -4	3.90	4.8096 -5	4.30	8.5399 -6	4.70	1.3008 -6
3.51	2.2405	3.91	4.6148	4.31	8.1627	4.71	1.2386
3.52	2.1577	3.92	4.4274	4.32	7.8015	4.72	1.1792
3.53	2.0778	3.93	4.2473	4.33	7.4555	4.73	1.1226
3.54	2.0006	3.94	4.0741	4.34	7.1241	4.74	1.0686
3.55	1.9262 -4	3.95	3.9076 -5	4.35	6.8069 -6	4.75	1.0171 -6
3.56	1.8543	3.96	3.7475	4.36	6.5031	4.76	9.6796 -7
3.57	1.7849	3.97	3.5936	4.37	6.2123	4.77	9.2113
3.58	1.7180	3.98	3.4458	4.38	5.9340	4.78	8.7648
3.59	1.6534	3.99	3.3037	4.39	5.6675	4.79	8.3391

z	$1 - F(z)$	z	$1 - F(z)$	z	$1 - F(z)$	z	$1 - F(z)$
4.80	7.9333 -7	5.20	9.9644 -8	5.60	1.0718 -8	6.00	9.8659 -10
4.81	7.5465	5.21	9.4420	5.61	1.0116	6.01	9.2761
4.82	7.1779	5.22	8.9462	5.62	9.5479 -9	6.02	8.7208
4.83	6.8267	5.23	8.4755	5.63	9.0105	6.03	8.1980
4.84	6.4920	5.24	8.0288	5.64	8.5025	6.04	7.7057
4.85	6.1731 -7	5.25	7.6050 -8	5.65	8.0224 -9	6.05	7.2423 -10
4.86	5.8693	5.26	7.2028	5.66	7.5687	6.06	6.8061
4.87	5.5799	5.27	6.8212	5.67	7.1399	6.07	6.3955
4.88	5.3043	5.28	6.4592	5.68	6.7347	6.08	6.0091
4.89	5.0418	5.29	6.1158	5.69	6.3520	6.09	5.6455
4.90	4.7918 -7	5.30	5.7901 -8	5.70	5.9904 -9	6.10	5.3034 -10
4.91	4.5538	5.31	5.4813	5.71	5.6488	6.11	4.9815
4.92	4.3272	5.32	5.1884	5.72	5.3262	6.12	4.6788
4.93	4.1115	5.33	4.9106	5.73	5.0215	6.13	4.3939
4.94	3.9061	5.34	4.6473	5.74	4.7338	6.14	4.1261
4.95	3.7107 -7	5.35	4.3977 -8	5.75	4.4622 -9	6.15	3.8742 -10
4.96	3.5247	5.36	4.1611	5.76	4.2057	6.16	3.6372
4.97	3.3476	5.37	3.9368	5.77	3.9636	6.17	3.4145
4.98	3.1792	5.38	3.7243	5.78	3.7350	6.18	3.2050
4.99	3.0190	5.39	3.5229	5.79	3.5193	6.19	3.0082
5.00	2.8665 -7	5.40	3.3320 -8	5.80	3.3157 -9	6.20	2.8231 -10
5.01	2.7215	5.41	3.1512	5.81	3.1236	6.21	2.6492
5.02	2.5836	5.42	2.9800	5.82	2.9424	6.22	2.4858
5.03	2.4524	5.43	2.8177	5.83	2.7714	6.23	2.3321
5.04	2.3277	5.44	2.6640	5.84	2.6100	6.24	2.1878
5.05	2.2091 -7	5.45	2.5185 -8	5.85	2.4579 -9	6.25	2.0523 -10
5.06	2.0963	5.46	2.3807	5.86	2.3143	6.26	1.9249
5.07	1.9891	5.47	2.2502	5.87	2.1790	6.27	1.8052
5.08	1.8872	5.48	2.1266	5.88	2.0513	6.28	1.6929
5.09	1.7903	5.49	2.0097	5.89	1.9310	6.29	1.5873
5.10	1.6983 -7	5.50	1.8990 -8	5.90	1.8175 -9	6.30	1.4882 -10
5.11	1.6108	5.51	1.7942	5.91	1.7105	6.31	1.3952
5.12	1.5277	5.52	1.6950	5.92	1.6097	6.32	1.3078
5.13	1.4487	5.53	1.6012	5.93	1.5147	6.33	1.2258
5.14	1.3737	5.54	1.5124	5.94	1.4251	6.34	1.1488
5.15	1.3024 -7	5.55	1.4283 -8	5.95	1.3407 -9	6.35	1.0765 -10
5.16	1.2347	5.56	1.3489	5.96	1.2612	6.36	1.0088
5.17	1.1705	5.57	1.2737	5.97	1.1863	6.37	9.4514 -11
5.18	1.1094	5.58	1.2026	5.98	1.1157	6.38	8.8544
5.19	1.0515	5.59	1.1353	5.99	1.0492	6.39	8.2943

z	$1 - F(z)$	z	$1 - F(z)$	z	$1 - F(z)$	z	$1 - F(z)$
6.40	7.7689 -11	6.80	5.2310 -12	7.20	3.0106 -13	7.60	1.4807 -14
6.41	7.2760	6.81	4.8799	7.21	2.7976	7.61	1.3705
6.42	6.8137	6.82	4.5520	7.22	2.5994	7.62	1.2684
6.43	6.3802	6.83	4.2457	7.23	2.4150	7.63	1.1738
6.44	5.9737	6.84	3.9597	7.24	2.2434	7.64	1.0861
6.45	5.5925 -11	6.85	3.6925 -12	7.25	2.0839 2-13	7.65	1.0049 -14
6.46	5.2351	6.86	3.4430	7.26	1.9355	7.66	9.2967
6.47	4.9001	6.87	3.2101	7.27	1.7974	7.67	8.5998
6.48	4.5861	6.88	2.9926	7.28	1.6691	7.68	7.9544
6.49	4.2918	6.89	2.7896	7.29	1.5498	7.69	7.3568
6.50	4.0160 -11	6.90	2.6001 -12	7.30	1.4388 -13	7.70	6.8033 -15
6.51	3.7575	6.91	2.4233	7.31	1.3357	7.71	6.2909
6.52	3.5154	6.92	2.2582	7.32	1.2399	7.72	5.8165
6.53	3.2885	6.93	2.1042	7.33	1.1508	7.73	5.3773
6.54	3.0759	6.94	1.9605	7.34	1.0680	7.74	4.9708
6.55	2.8769 -11	6.95	1.8264 -12	7.35	9.9103 -14	7.75	4.5946 -15
6.56	2.6904	6.96	1.7014	7.36	9.1955	7.76	4.2465
6.57	2.5158	6.97	1.5847	7.37	8.5314	7.77	3.9243
6.58	2.3522	6.98	1.4759	7.38	7.9145	7.78	3.6262
6.59	2.1991	6.99	1.3744	7.39	7.3414	7.79	3.3505
6.60	2.0558 -11	7.00	1.2798 -12	7.40	6.8092 -14	7.80	3.0954 -15
6.61	1.9216	7.01	1.1916	7.41	6.3150	7.81	2.8594
6.62	1.7960	7.02	1.1093	7.42	5.8560	7.82	2.6412
6.63	1.6784	7.03	1.0327	7.43	5.4299	7.83	2.4394
6.64	1.5684	7.04	9.6120 -13	7.44	5.0343	7.84	2.2527
6.65	1.4655 -11	7.05	8.9459 -13	7.45	4.6670 -14	7.85	2.0802 -15
6.66	1.3691	7.06	8.3251	7.46	4.3261	7.86	1.9207
6.67	1.2790	7.07	7.7467	7.47	4.0097	7.87	1.7732
6.68	1.1947	7.08	7.2077	7.48	3.7161	7.88	1.6369
6.69	1.1159	7.09	6.7056	7.49	3.4437	7.89	1.5109
6.70	1.0421 -11	7.10	6.2378 -13	7.50	3.1909 -14	7.90	1.3945 -15
6.71	9.7312 -12	7.11	5.8022	7.51	2.9564	7.91	1.2869
6.72	9.0862	7.12	5.3964	7.52	2.7388	7.92	1.1876
6.73	8.4832	7.13	5.0184	7.53	2.5370	7.93	1.0957
6.74	7.9193	7.14	4.6665	7.54	2.3499	7.94	1.0109
6.75	7.3923 -12	7.15	4.3389 -13	7.55	2.1763 -14	7.95	9.3256 -16
6.76	6.8996	7.16	4.0339	7.56	2.0153	7.96	8.6020
6.77	6.4391	7.17	3.7499	7.57	1.8661	7.97	7.9337
6.78	6.0088	7.18	3.4856	7.58	1.7278	7.98	7.3167
6.79	5.6067	7.19	3.2396	7.59	1.5995	7.99	6.7469

z	$1 - F(z)$	z	$1 - F(z)$	z	$1 - F(z)$	z	$1 - F(z)$
8.00	6.2210 -16	8.40	2.2324 -17	8.80	6.8408 -19	9.20	1.7897 -20
8.01	5.7354	8.41	2.0501	8.81	6.2573	9.21	1.6306
8.02	5.2873	8.42	1.8824	8.82	5.7230	9.22	1.4855
8.03	4.8736	8.43	1.7283	8.83	5.2338	9.23	1.3532
8.04	4.4919	8.44	1.5867	8.84	4.7859	9.24	1.2325
8.05	4.1397 -16	8.45	1.4565 -17	8.85	4.3760 -19	9.25	1.1225 -20
8.06	3.8147	8.46	1.3369	8.86	4.0007	9.26	1.0222
8.07	3.5149	8.47	1.2270	8.87	3.6573	9.27	9.3073 -21
8.08	3.2383	8.48	1.1260	8.88	3.3430	9.28	8.4739
8.09	2.9832	8.49	1.0332	8.89	3.0554	9.29	7.7144
8.10	2.7480 -16	8.50	9.4795 -18	8.90	2.7923 -19	9.30	7.0223 -21
8.11	2.5310	8.51	8.6967	8.91	2.5516	9.31	6.3916
8.12	2.3309	8.52	7.9777	8.92	2.3314	9.32	5.8170
8.13	2.1465	8.53	7.3174	8.93	2.1300	9.33	5.2935
8.14	1.9764	8.54	6.7111	8.94	1.9459	9.34	4.8167
8.15	1.8196 -16	8.55	6.1544 -18	8.95	1.7774 -19	9.35	4.3824 -21
8.16	1.6751	8.56	5.6434	8.96	1.5234	9.36	3.9868
8.17	1.5419	8.57	5.1743	8.97	1.4826	9.37	3.6266
8.18	1.4192	8.58	4.7437	8.98	1.3538	9.38	3.2986
8.19	1.3061	8.59	4.3485	8.99	1.2362	9.39	3.0000
8.20	1.2019 -16	8.60	3.9858 -18	9.00	1.1286 -19	9.40	2.7282 -21
8.21	1.1059	8.61	3.6530	9.01	1.0303	9.41	2.4807
8.22	1.0175	8.62	3.3477	9.02	9.4045 -20	9.42	2.2554
8.23	9.3607 -17	8.63	3.0676	9.03	8.5836	9.43	2.0504
8.24	8.6105	8.64	2.8107	9.04	7.8336	9.44	1.8639
8.25	7.9197 -17	8.65	2.5750 -18	9.05	7.1484 -20	9.45	1.6942 -21
8.26	7.2836	8.66	2.3588	9.06	6.5225	9.46	1.5397
8.27	6.6980	8.67	2.1606	9.07	5.9509	9.47	1.3992
8.28	6.1588	8.68	1.9788	9.08	5.4287	9.48	1.2614
8.29	5.6624	8.69	1.8122	9.09	4.9520	9.49	1.1552
8.30	5.2056 -17	8.70	1.6594 -18	9.10	4.5166 -20	9.50	1.0495 -21
8.31	4.7851	8.71	1.5194	9.11	4.1191	9.51	9.5331 -22
8.32	4.3982	8.72	1.3910	9.12	3.7562	9.52	8.6590
8.33	4.0421	8.73	1.2734	9.13	3.4250	9.53	7.8642
8.34	3.7145	8.74	1.1656	9.14	3.1226	9.54	7.1416
8.35	3.4131 -17	8.75	1.0668 -18	9.15	2.8467 -20	9.55	6.4848 -22
8.36	3.1359	8.76	9.7625 -19	9.16	2.5949	9.56	5.8878
8.37	2.8809	8.77	8.9333	9.17	2.3651	9.57	5.3453
8.38	2.6464	8.78	8.1737	9.18	2.1555	9.58	4.8522
8.39	2.4307	8.79	7.4780	9.19	1.9642	9.59	4.4043

z	$1 - F(z)$	z	$1 - F(z)$	z	$1 - F(z)$	z	$1 - F(z)$
9.60	3.9972 -22	9.70	1.5075 -22	9.80	5.6293 -23	9.90	2.0814 -23
9.61	3.6274	9.71	1.3667	9.81	5.0984	9.91	1.8832
9.62	3.2916	9.72	1.2389	9.82	4.6172	9.92	1.7038
9.63	2.9865	9.73	1.1230	9.83	4.1809	9.93	1.5413
9.64	2.7094	9.74	1.0178	9.84	3.7855	9.94	1.3941
9.65	2.4578 -22	9.75	9.2234 -23	9.85	3.4272 -23	9.95	1.2609 -23
9.66	2.2293	9.76	8.3578	9.86	3.1025	9.96	1.1403
9.67	2.0219	9.77	7.5726	9.87	2.8082	9.97	1.0311
9.68	1.8336	9.78	6.8605	9.88	2.5416	9.98	9.3233 -24
9.69	1.6626	9.79	6.2148	9.89	2.3001	9.99	8.4291
						10.00	7.6199 -24

APPENDIX C
STANDARDIZED NORMAL DISTRIBUTION'S ORDINATE VALUES OR PROBABILITY DENSITIES

$$\phi(z) = \frac{1}{\sqrt{2\pi}} e^{-\frac{1}{2}z^2}, \text{ for } 0 \leq z \leq 4.99, \phi(-z) = \phi(z)$$

z	0.00	0.01	0.02	0.03	0.04	0.05	0.06	0.07	0.08	0.09
0.0	0.3989	0.3989	0.3989	0.3988	0.3986	0.3984	0.3982	0.3980	0.3977	0.3973
0.1	0.3970	0.3965	0.3961	0.3956	0.3951	0.3945	0.3939	0.3932	0.3925	0.3918
0.2	0.3910	0.3902	0.3894	0.3885	0.3876	0.3867	0.3857	0.3847	0.3836	0.3825
0.3	0.3814	0.3802	0.3790	0.3778	0.3765	0.3752	0.3739	0.3725	0.3712	0.3697
0.4	0.3683	0.3668	0.3653	0.3637	0.3621	0.3605	0.3589	0.3572	0.3555	0.3538
0.5	0.3521	0.3503	0.3485	0.3467	0.3448	0.3429	0.3410	0.3391	0.3372	0.3352
0.6	0.3332	0.3312	0.3292	0.3271	0.3251	0.3230	0.3209	0.3187	0.3166	0.3144
0.7	0.3123	0.3101	0.3079	0.3056	0.3034	0.3011	0.2989	0.2966	0.2943	0.2920
0.8	0.2897	0.2874	0.2850	0.2827	0.2803	0.2780	0.2756	0.2732	0.2709	0.2685
0.9	0.2661	0.2637	0.2613	0.2589	0.2565	0.2541	0.2516	0.2492	0.2468	0.2444
1.0	0.2420	0.2396	0.2371	0.2347	0.2323	0.2299	0.2275	0.2251	0.2227	0.2203
1.1	0.2179	0.2155	0.2131	0.2107	0.2083	0.2059	0.2036	0.2012	0.1989	0.1965
1.2	0.1942	0.1919	0.1895	0.1872	0.1849	0.1826	0.1804	0.1781	0.1758	0.1736
1.3	0.1714	0.1691	0.1669	0.1647	0.1626	0.1604	0.1582	0.1561	0.1539	0.1518
1.4	0.1497	0.1476	0.1456	0.1435	0.1415	0.1394	0.1374	0.1354	0.1334	0.1315
1.5	0.1295	0.1276	0.1257	0.1238	0.1219	0.1200	0.1182	0.1163	0.1145	0.1127
1.6	0.1109	0.1092	0.1074	0.1057	0.1040	0.1023	0.1006	0.09893	0.09728	0.09566
1.7	0.09405	0.09246	0.09089	0.08933	0.08780	0.08628	0.08478	0.08329	0.08183	0.08038
1.8	0.07895	0.07754	0.07614	0.07477	0.07341	0.07206	0.07074	0.06943	0.06814	0.06687
1.9	0.06562	0.06438	0.06316	0.06195	0.06077	0.05959	0.05844	0.05730	0.05618	0.05508
2.0	0.05399	0.05292	0.05186	0.05082	0.04980	0.04879	0.04780	0.04682	0.04586	0.04491
2.1	0.04398	0.04307	0.04217	0.04128	0.04041	0.03955	0.03871	0.03788	0.03706	0.03626
2.2	0.03547	0.03470	0.03394	0.03319	0.03246	0.03174	0.03103	0.03034	0.02965	0.02898
2.3	0.02833	0.02768	0.02705	0.02643	0.02582	0.02522	0.02463	0.02406	0.02349	0.02294
2.4	0.02239	0.02186	0.02134	0.02083	0.02033	0.01984	0.01936	0.01888	0.01842	0.01797
2.5	0.01753	0.01709	0.01667	0.01625	0.01585	0.01545	0.01506	0.01468	0.01431	0.01394
2.6	0.01358	0.01323	0.01289	0.01256	0.01223	0.01191	0.01160	0.01130	0.01100	0.01071
2.7	0.010421	0.010143	$0.0^2 9871$	$0.0^2 9606$	$0.0^2 9347$	$0.0^2 9094$	$0.0^2 8846$	$0.0^2 8605$	$0.0^2 8370$	$0.0^2 8140$
2.8	$0.0^2 7915$	$0.0^2 7697$	$0.0^2 7483$	$0.0^2 7274$	$0.0^2 7071$	$0.0^2 6873$	$0.0^2 6679$	$0.0^2 6491$	$0.0^2 6307$	$0.0^2 6127$
2.9	$0.0^2 5953$	$0.0^2 5782$	$0.0^2 5616$	$0.0^2 5454$	$0.0^2 5296$	$0.0^2 5143$	$0.0^2 4993$	$0.0^2 4847$	$0.0^2 4705$	$0.0^2 4567$
3.0	$0.0^2 4432$	$0.0^2 4301$	$0.0^2 4173$	$0.0^2 4049$	$0.0^2 3928$	$0.0^2 3810$	$0.0^2 3695$	$0.0^2 3584$	$0.0^2 3475$	$0.0^2 3370$
3.1	$0.0^2 3267$	$0.0^2 3167$	$0.0^2 3070$	$0.0^2 2975$	$0.0^2 2884$	$0.0^2 2794$	$0.0^2 2707$	$0.0^2 2623$	$0.0^2 2541$	$0.0^2 2461$
3.2	$0.0^2 2384$	$0.0^2 2309$	$0.0^2 2236$	$0.0^2 2165$	$0.0^2 2096$	$0.0^2 2029$	$0.0^2 1964$	$0.0^2 1901$	$0.0^2 1840$	$0.0^2 1780$
3.3	$0.0^2 1723$	$0.0^2 1667$	$0.0^2 1612$	$0.0^2 1560$	$0.0^2 1508$	$0.0^2 1459$	$0.0^2 1411$	$0.0^2 1364$	$0.0^2 1319$	$0.0^2 1275$
3.4	$0.0^2 1232$	$0.0^2 1191$	$0.0^2 1151$	$0.0^2 1112$	$0.0^2 1075$	$0.0^2 1038$	$0.0^2 1003$	$0.0^3 9689$	$0.0^3 9358$	$0.0^3 9037$
3.5	$0.0^3 8727$	$0.0^3 8426$	$0.0^3 8135$	$0.0^3 7853$	$0.0^3 7581$	$0.0^3 7317$	$0.0^3 7061$	$0.0^3 6814$	$0.0^3 6575$	$0.0^3 6343$
3.6	$0.0^3 6119$	$0.0^3 5902$	$0.0^3 5693$	$0.0^3 5490$	$0.0^3 5294$	$0.0^3 5105$	$0.0^3 4921$	$0.0^3 4744$	$0.0^3 4573$	$0.0^3 4408$
3.7	$0.0^3 4248$	$0.0^3 4093$	$0.0^3 3944$	$0.0^3 3800$	$0.0^3 3661$	$0.0^3 3526$	$0.0^3 3396$	$0.0^3 3271$	$0.0^3 3149$	$0.0^3 3032$
3.8	$0.0^3 2919$	$0.0^3 2810$	$0.0^3 2705$	$0.0^3 2604$	$0.0^3 2506$	$0.0^3 2411$	$0.0^3 2320$	$0.0^3 2232$	$0.0^3 2147$	$0.0^3 2065$
3.9	$0.0^3 1987$	$0.0^3 1910$	$0.0^3 1837$	$0.0^3 1766$	$0.0^3 1698$	$0.0^3 1633$	$0.0^3 1569$	$0.0^3 1508$	$0.0^3 1449$	$0.0^3 1393$
4.0	$0.0^3 1338$	$0.0^3 1286$	$0.0^3 1235$	$0.0^3 1186$	$0.0^3 1140$	$0.0^3 1094$	$0.0^3 1051$	$0.0^3 1009$	$0.0^4 9687$	$0.0^4 9299$
4.1	$0.0^4 8926$	$0.0^4 8567$	$0.0^4 8222$	$0.0^4 7890$	$0.0^4 7570$	$0.0^4 7263$	$0.0^4 6967$	$0.0^4 6683$	$0.0^4 6410$	$0.0^4 6147$
4.2	$0.0^4 5894$	$0.0^4 5652$	$0.0^4 5418$	$0.0^4 5194$	$0.0^4 4979$	$0.0^4 4772$	$0.0^4 4573$	$0.0^4 4382$	$0.0^4 4199$	$0.0^4 4023$
4.3	$0.0^4 3854$	$0.0^4 3691$	$0.0^4 3535$	$0.0^4 3386$	$0.0^4 3242$	$0.0^4 3104$	$0.0^4 2972$	$0.0^4 2845$	$0.0^4 2723$	$0.0^4 2606$
4.4	$0.0^4 2494$	$0.0^4 2387$	$0.0^4 2284$	$0.0^4 2185$	$0.0^4 2090$	$0.0^4 1999$	$0.0^4 1912$	$0.0^4 1829$	$0.0^4 1749$	$0.0^4 1672$
4.5	$0.0^4 1598$	$0.0^4 1528$	$0.0^4 1461$	$0.0^4 1396$	$0.0^4 1334$	$0.0^4 1275$	$0.0^4 1218$	$0.0^4 1164$	$0.0^4 1112$	$0.0^4 1062$
4.6	$0.0^4 1014$	$0.0^5 9684$	$0.0^5 9248$	$0.0^5 8830$	$0.0^5 8430$	$0.0^5 8047$	$0.0^5 7681$	$0.0^5 7331$	$0.0^5 6996$	$0.0^5 6676$
4.7	$0.0^5 6370$	$0.0^5 6077$	$0.0^5 5797$	$0.0^5 5530$	$0.0^5 5274$	$0.0^5 5029$	$0.0^5 4796$	$0.0^5 4573$	$0.0^5 4360$	$0.0^5 4156$
4.8	$0.0^5 3961$	$0.0^5 3775$	$0.0^5 3598$	$0.0^5 3428$	$0.0^5 3267$	$0.0^5 3112$	$0.0^5 2965$	$0.0^5 2824$	$0.0^5 2690$	$0.0^5 2561$
4.9	$0.0^5 2439$	$0.0^5 2322$	$0.0^5 2211$	$0.0^5 2105$	$0.0^5 2003$	$0.0^5 1907$	$0.0^5 1814$	$0.0^5 1727$	$0.0^5 1643$	$0.0^5 1563$

APPENDIX D
CHI-SQUARE DISTRIBUTION PERCENTILE VALUES

Tabulation of $P(X^2 > \chi^2_{\delta;\nu}) = \left[2^{\frac{\nu}{2}}\Gamma\left(\frac{\nu}{2}\right)\right]^{-1} \int_{\chi^2_{\delta;\nu}}^{\infty} e^{-t/2}t^{\frac{\nu}{2}-1}dt.$

ν $\diagdown$ δ	0.995	0.99	0.98	0.975	0.95	0.90
1	0.0000	0.0002	0.0006	0.0010	0.0039	0.0158
2	0.0100	0.0201	0.0404	0.0506	0.1026	0.2107
3	0.0717	0.1148	0.1848	0.2158	0.3518	0.5844
4	0.2070	0.2971	0.4294	0.4844	0.7107	1.0636
5	0.4117	0.5543	0.7519	0.8312	1.1455	1.6103
6	0.6757	0.8721	1.1344	1.2373	1.6354	2.2041
7	0.9893	1.2390	1.5643	1.6899	2.1674	2.8331
8	1.3445	1.6465	2.0325	2.1797	2.7326	3.4895
9	1.7349	2.0879	2.5323	2.7004	3.3251	4.1682
10	2.1558	2.5582	3.0591	3.2469	3.9403	4.8652
11	2.6033	3.0535	3.6086	3.8157	4.5748	5.5778
12	3.0737	3.5706	4.1783	4.4038	5.2260	6.3038
13	3.5649	4.1069	4.7654	5.0088	5.8918	7.0415
14	4.0747	4.6604	5.3682	5.6287	6.5706	7.7896
15	4.6011	5.2294	5.9849	6.2621	7.2609	8.5468
16	5.1421	5.8123	6.6143	6.9077	7.9617	9.3123
17	5.6973	6.4077	7.2550	7.5642	8.6718	10.0852
18	6.2646	7.0149	7.9063	8.2307	9.3904	10.8649
19	6.8440	7.6328	8.5670	8.9065	10.1170	11.6509
20	7.4338	8.2605	9.2367	9.5908	10.8508	12.4426
21	8.0337	8.8972	9.9146	10.2830	11.5913	13.2396
22	8.6426	9.5425	10.6001	10.9823	12.3380	14.0415
23	9.2603	10.1958	11.2926	11.6886	13.0905	14.8480
24	9.8862	10.8564	11.9918	12.4011	13.8484	15.6587
25	10.5195	11.5239	12.6973	13.1198	14.6115	16.4734
26	11.1602	12.1982	13.4086	13.8439	15.3792	17.2919
27	11.8076	12.8784	14.1255	14.5734	16.1514	18.1139
28	12.4614	13.5647	14.8474	15.3079	16.9279	18.9393
29	13.1211	14.2563	15.5745	16.0471	17.7084	19.7678
30	13.7866	14.9536	16.3062	16.7908	18.4927	20.5992

APPENDIX D–Continued.

ν \ Q	0.85	0.80	0.75	0.70	0.60	0.50
1	0.0358	0.0642	0.1015	0.1485	0.2750	0.4549
2	0.3250	0.4463	0.5754	0.7133	1.0216	1.3863
3	0.7978	1.0052	1.2125	1.4237	1.8692	2.3660
4	1.3665	1.6488	1.9226	2.1947	2.7528	3.3567
5	1.9938	2.3425	2.6746	2.9999	3.6555	4.3515
6	2.6613	3.0701	3.4546	3.8275	4.5702	5.3481
7	3.3583	3.8223	4.2549	4.6713	5.4932	6.3458
8	4.0782	4.5936	5.0706	5.5274	6.4226	7.3441
9	4.8165	5.3801	5.8988	6.3933	7.3570	8.3428
10	5.5701	6.1791	6.7372	7.2672	8.2955	9.3418
11	6.3364	6.9887	7.5841	8.1479	9.2373	10.3410
12	7.1138	7.8073	8.4384	9.0343	10.1820	11.3403
13	7.9008	8.6339	9.2991	9.9257	11.1292	12.3398
14	8.6963	9.4673	10.1653	10.8215	12.0785	13.3393
15	9.4993	10.3069	11.0365	11.7212	13.0298	14.3389
16	10.3090	11.1521	11.9122	12.6244	13.9827	15.3385
17	11.1249	12.0023	12.7919	13.5307	14.9373	16.3382
18	11.9463	12.8570	13.6753	14.4398	15.8932	17.3379
19	12.7727	13.7158	14.5620	15.3517	16.8504	18.3376
20	13.6039	14.5784	15.4518	16.2659	17.8088	19.3374
21	14.4393	15.4446	16.3444	17.1823	18.7683	20.3372
22	15.2787	16.3140	17.2396	18.1007	19.7288	21.3370
23	16.1219	17.1865	18.1373	19.0211	20.6902	22.3369
24	16.9686	18.0618	19.0373	19.9432	21.6525	23.3367
25	17.8185	18.9398	19.9393	20.8670	22.6156	24.3366
26	18.6714	19.8202	20.8434	21.7924	23.5794	25.3365
27	19.5272	20.7030	21.7494	22.7192	24.5440	26.3363
28	20.3857	21.5880	22.6572	23.6475	25.5092	27.3362
29	21.2468	22.4750	23.5666	24.5770	26.4751	28.3361
30	22.1104	23.3641	24.4776	25.5078	27.4416	29.3360

APPENDIX D–Continued.

ν \ Q	0.40	0.30	0.25	0.20	0.15	0.10
1	0.7083	1.0742	1.3233	1.6423	2.0722	2.7054
2	1.8326	2.4079	2.7726	3.2189	3.7942	4.6052
3	2.9462	3.6649	4.1083	4.6416	5.3170	6.2514
4	4.0446	4.8784	5.3853	5.9886	6.7449	7.7794
5	5.1319	6.0644	6.6257	7.2893	8.1152	9.2364
6	6.2108	7.2311	7.8408	8.5581	9.4461	10.6447
7	7.2832	8.3834	9.0371	9.8033	10.7479	12.0170
8	8.3505	9.5245	10.2188	11.0301	12.0271	13.3616
9	9.4136	10.6564	11.3888	12.2422	13.2880	14.6837
10	10.4732	11.7807	12.5489	13.4420	14.5339	15.9872
11	11.5298	12.8987	13.7007	14.6314	15.7671	17.2750
12	12.5838	14.0111	14.8454	15.8120	16.9893	18.5493
13	13.6356	15.1187	15.9839	16.9848	18.2020	19.8120
14	14.6853	16.2221	17.1169	18.1508	19.4063	21.0641
15	15.7332	17.3217	18.2451	19.3107	20.6030	22.3071
16	16.7795	18.4179	19.3689	20.4651	21.7930	23.5419
17	17.8244	19.5110	20.4887	21.6146	22.9771	24.7690
18	18.8679	20.6013	21.6049	22.7595	24.1555	25.9894
19	19.9102	21.6891	22.7178	23.9004	25.3289	27.2036
20	20.9514	22.7745	23.8277	25.0375	26.4976	28.4120
21	21.9915	23.8578	24.9348	26.1711	27.6620	29.6151
22	23.0307	24.9390	26.0392	27.3015	28.8224	30.8133
23	24.0689	26.0184	27.1414	28.4288	29.9792	32.0069
24	25.1064	27.0959	28.2411	29.5533	31.1324	33.1963
25	26.1430	28.1719	29.3389	30.6752	32.2825	34.3816
26	27.1789	29.2463	30.4346	31.7946	33.4294	35.5632
27	28.2141	30.3193	31.5284	32.9117	34.5736	36.7412
28	29.2486	31.3909	32.6205	34.0266	35.7150	37.9159
29	30.2825	32.4612	33.7109	35.1393	36.8538	39.0875
30	31.3159	33.5302	34.7997	36.2502	37.9902	40.2560

APPENDIX D–Continued.

ν \ Q	0.05	0.025	0.02	0.01	0.005	0.001
1	3.8415	5.0239	5.4119	6.6348	7.8794	10.8281
2	5.9915	7.3777	7.8240	9.2104	10.5967	13.8164
3	7.8147	9.3484	9.8374	11.3447	12.8379	16.2656
4	9.4877	11.1433	11.6678	13.2769	14.8604	18.4688
5	11.0705	12.8325	13.3882	15.0864	16.7500	20.5156
6	12.5916	14.4495	15.0332	16.8120	18.5479	22.4570
7	14.0671	16.0127	16.6223	18.4751	20.2773	24.3203
8	15.5073	17.5345	18.1682	20.0903	21.9551	26.1250
9	16.9189	19.0227	19.6790	21.6660	23.5898	27.8750
10	18.3070	20.4832	21.1606	23.2090	25.1885	29.5859
11	19.6752	21.9202	22.6179	24.7251	26.7568	31.2656
12	21.0261	23.3367	24.0540	26.2168	28.2998	32.9102
13	22.3621	24.7356	25.4714	27.6885	29.8193	34.5273
14	23.6848	26.1189	26.8728	29.1411	31.3193	36.1250
15	24.9958	27.4883	28.2595	30.5781	32.8008	37.6953
16	26.2963	28.8452	29.6333	31.9998	34.2676	39.2500
17	27.5872	30.1909	30.9951	33.4087	35.7188	40.7891
18	28.8693	31.5264	32.3462	34.8052	37.1563	42.3125
19	30.1436	32.8523	33.6875	36.1909	38.5820	43.8203
20	31.4104	34.1697	35.0195	37.5664	39.9971	45.3125
21	32.6707	35.4790	36.3433	38.9321	41.4004	46.7969
22	33.9244	36.7808	37.6597	40.2891	42.7959	48.2656
23	35.1725	38.0757	38.9683	41.6387	44.1816	49.7266
24	36.4150	39.3640	40.2705	42.9795	45.5586	51.1797
25	37.6525	40.6465	41.5659	44.3145	46.9277	52.6172
26	38.8851	41.9231	42.8560	45.6416	48.2900	54.0547
27	40.1133	43.1946	44.1401	46.9629	49.6445	55.4766
28	41.3372	44.4609	45.4189	48.2783	50.9941	56.8906
29	42.5569	45.7222	46.6929	49.5879	52.3359	58.3047
30	43.7729	46.9792	47.9619	50.8926	53.6719	59.7031

APPENDIX E

Student's t distribution percentile values.

Tabulation of the values of α versus $t_{\alpha;\nu}$ for different values of ν.

$$\alpha = P(t > t_{\alpha;\nu}) = \int_{t_{\alpha;\nu}}^{\infty} f(t)dt$$

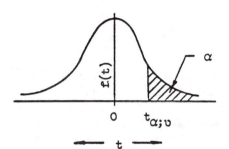

ν \ α	.40	.30	.20	.10	.050	.025	.010	.005	.001	.0005
1	.325	.727	1.376	3.078	6.314	12.71	31.82	63.66	318.3	636.6
2	.289	.617	1.061	1.886	2.920	4.303	6.965	9.925	22.33	31.60
3	.277	.584	.978	1.638	2.353	3.182	4.541	5.841	10.22	12.94
4	.271	.569	.941	1.533	2.132	2.776	3.747	4.604	7.173	8.610
5	.267	.559	.920	1.476	2.015	2.571	3.365	4.032	5.893	6.859
6	.265	.553	.906	1.440	1.943	2.447	3.143	3.707	5.208	5.959
7	.263	.549	.896	1.415	1.895	2.365	2.998	3.499	4.785	5.405
8	.262	.546	.889	1.397	1.860	2.306	2.896	3.355	4.501	5.041
9	.261	.543	.883	1.383	1.833	2.262	2.821	3.250	4.297	4.781
10	.260	.542	.879	1.372	1.812	2.228	2.764	3.169	4.144	4.587
11	.260	.540	.876	1.363	1.796	2.201	2.718	3.106	4.025	4.437
12	.259	.539	.873	1.356	1.782	2.179	2.681	3.055	3.930	4.318
13	.259	.538	.870	1.350	1.771	2.160	2.650	3.012	3.852	4.221
14	.258	.537	.868	1.345	1.761	2.145	2.624	2.977	3.787	4.140
15	.258	.536	.866	1.341	1.753	2.131	2.602	2.947	3.733	4.073
16	.258	.535	.865	1.337	1.746	2.120	2.583	2.921	3.686	4.015
17	.257	.534	.863	1.333	1.740	2.110	2.567	2.898	3.646	3.965
18	.257	.534	.862	1.330	1.734	2.101	2.552	2.878	3.611	3.922
19	.257	.533	.861	1.328	1.729	2.093	2.539	2.861	3.579	3.883
20	.257	.533	.860	1.325	1.725	2.086	2.528	2.845	3.552	3.850
21	.257	.532	.859	1.323	1.721	2.080	2.518	2.831	3.527	3.819
22	.256	.532	.858	1.321	1.717	2.074	2.508	2.819	3.505	3.792
23	.256	.532	.858	1.319	1.714	2.069	2.500	2.807	3.485	3.767
24	.256	.531	.857	1.318	1.711	2.064	2.492	2.797	3.467	3.745
25	.256	.531	.856	1.316	1.708	2.060	2.485	2.787	3.450	3.725
26	.256	.531	.856	1.315	1.706	2.056	2.479	2.779	3.435	3.707
27	.256	.531	.855	1.314	1.703	2.052	2.473	2.771	3.421	3.690
28	.256	.530	.855	1.313	1.701	2.048	2.467	2.763	3.408	3.674
29	.256	.530	.854	1.311	1.699	2.045	2.462	2.756	3.396	3.659
30	.256	.530	.854	1.310	1.697	2.042	2.457	2.750	3.385	3.646
40	.255	.529	.851	1.303	1.684	2.021	2.423	2.704	3.307	3.551
50	.255	.528	.849	1.298	1.676	2.009	2.403	2.678	3.262	3.495
60	.254	.527	.848	1.296	1.671	2.000	2.390	2.660	3.232	3.460
80	.254	.527	.846	1.292	1.664	1.990	2.374	2.639	3.195	3.415
100	.254	.526	.845	1.290	1.660	1.984	2.365	2.626	3.174	3.389
200	.254	.525	.843	1.286	1.653	1.972	2.345	2.601	3.131	3.339
500	.253	.525	.842	1.283	1.648	1.965	2.334	2.586	3.106	3.310
∞	.253	.524	.842	1.282	1.645	1.960	2.326	2.576	3.090	3.291

APPENDIX F

PERCENTAGE POINTS, F–DISTRIBUTION

This talbe gives the values of F such that

$$F(F) = \int_o^F \frac{\Gamma(\frac{m+n}{2})}{\Gamma(\frac{m}{2})\Gamma(\frac{n}{2})} m^{\frac{m}{2}} n^{\frac{n}{2}} x^{\frac{m-2}{2}} (n + mx)^{-\frac{m+n}{2}} dx$$

for selected values of m, the number of degrees of freedom of the numerator of F; and for selected values of n, the number of degrees of freedom of the denominator of F. The table also provides values corresponding to $F(F) = 0.10, 0.05, 0.25, 0.01, 0.005, 0.001$ since $F_{1-\alpha}$ for m and n degrees of freedom is the reciprocal of F_α for n and m of degrees of freedom. Thus

$$F_{0.05}(4, 7) = \frac{1}{F_{0.95}(7, 4)} = \frac{1}{6.09} = 0.164.$$

APPENDIX F.1-F(F)=0.50

ν_2 \ ν_1	1	2	3	4	5	6	7	8	9
1	1.0000	1.5000	1.7092	1.8227	1.8936	1.9420	1.9771	2.0038	2.0247
2	0.6667	1.0000	1.1349	1.2071	1.2519	1.2824	1.3046	1.3213	1.3344
3	0.5851	0.8811	1.0000	1.0632	1.1023	1.1289	1.1481	1.1628	1.1741
4	0.5486	0.8284	0.9405	1.0000	1.0367	1.0616	1.0797	1.0933	1.1040
5	0.5281	0.7988	0.9072	0.9646	1.0000	1.0240	1.0414	1.0545	1.0648
6	0.5149	0.7798	0.8858	0.9419	0.9765	1.0000	1.0170	1.0297	1.0398
7	0.5057	0.7665	0.8710	0.9262	0.9602	0.9833	1.0000	1.0126	1.0224
8	0.4990	0.7568	0.8600	0.9146	0.9483	0.9711	0.9876	1.0000	1.0097
9	0.4938	0.7494	0.8517	0.9058	0.9391	0.9618	0.9780	0.9904	1.0000
10	0.4897	0.7435	0.8451	0.8988	0.9320	0.9543	0.9705	0.9828	0.9923
11	0.4864	0.7387	0.8398	0.8931	0.9261	0.9484	0.9644	0.9766	0.9861
12	0.4837	0.7348	0.8353	0.8885	0.9213	0.9434	0.9594	0.9715	0.9809
13	0.4814	0.7315	0.8316	0.8845	0.9171	0.9392	0.9552	0.9672	0.9766
14	0.4794	0.7286	0.8284	0.8812	0.9137	0.9357	0.9516	0.9636	0.9730
15	0.4778	0.7262	0.8257	0.8783	0.9107	0.9327	0.9485	0.9605	0.9698
16	0.4763	0.7241	0.8233	0.8758	0.9082	0.9300	0.9458	0.9577	0.9671
17	0.4750	0.7222	0.8212	0.8735	0.9059	0.9276	0.9434	0.9553	0.9646
18	0.4738	0.7205	0.8194	0.8717	0.9038	0.9256	0.9413	0.9532	0.9624
19	0.4728	0.7191	0.8177	0.8698	0.9020	0.9238	0.9394	0.9513	0.9606
20	0.4719	0.7177	0.8162	0.8682	0.9004	0.9221	0.9378	0.9496	0.9588
21	0.4711	0.7165	0.8149	0.8669	0.8989	0.9206	0.9362	0.9480	0.9573
22	0.4703	0.7155	0.8137	0.8656	0.8975	0.9192	0.9348	0.9466	0.9559
23	0.4696	0.7145	0.8126	0.8644	0.8963	0.9180	0.9336	0.9454	0.9546
24	0.4690	0.7136	0.8115	0.8633	0.8952	0.9168	0.9324	0.9442	0.9534
25	0.4684	0.7127	0.8106	0.8624	0.8942	0.9159	0.9314	0.9431	0.9524
26	0.4679	0.7120	0.8097	0.8615	0.8933	0.9149	0.9305	0.9421	0.9514
27	0.4674	0.7113	0.8089	0.8606	0.8924	0.9140	0.9295	0.9412	0.9505
28	0.4670	0.7106	0.8082	0.8599	0.8916	0.9132	0.9287	0.9404	0.9496
29	0.4665	0.7100	0.8075	0.8592	0.8909	0.9124	0.9279	0.9396	0.9488
30	0.4662	0.7094	0.8069	0.8584	0.8902	0.9117	0.9272	0.9389	0.9480
40	0.4633	0.7053	0.8023	0.8536	0.8851	0.9065	0.9219	0.9336	0.9427
60	0.4605	0.7012	0.7977	0.8487	0.8802	0.9014	0.9168	0.9283	0.9375
120	0.4577	0.6972	0.7932	0.8439	0.8752	0.8964	0.9116	0.9232	0.9322
∞	0.45494	0.69315	0.78866	0.83917	0.87029	0.89135	0.90654	0.91802	0.92698

APPENDIX F.1-F(F)=0.50-Continued.

ν_2 \ ν_1	10	12	15	20	24	30	40	60	120	∞
1	2.0416	2.0676	2.0922	2.1191	2.1337	2.1450	2.1575	2.1717	2.1848	2.1981
2	1.3450	1.3610	1.3771	1.3933	1.4014	1.4096	1.4178	1.4261	1.4344	1.427
3	1.1833	1.1972	1.2111	1.2252	1.2323	1.2393	1.2465	1.2536	1.2608	1.2680
4	1.1126	1.1255	1.1386	1.1517	1.1583	1.1650	1.1716	1.1782	1.1849	1.1916
5	1.0730	1.0855	1.0981	1.1106	1.1170	1.1233	1.1298	1.1361	1.1426	1.1490
6	1.0478	1.0600	1.0722	1.0845	1.0907	1.0969	1.1032	1.1093	1.1156	1.1219
7	1.0304	1.0423	1.0543	1.0664	1.0724	1.0785	1.0847	1.0907	1.0969	1.1031
8	1.0175	1.0293	1.0411	1.0531	1.0591	1.0651	1.0711	1.0772	1.0832	1.0893
9	1.0077	1.0194	1.0312	1.0430	1.0488	1.0548	1.0607	1.0667	1.0727	1.0788
10	1.0000	1.0116	1.0232	1.0349	1.0408	1.0467	1.0526	1.0585	1.0645	1.0705
11	0.9937	1.0053	1.0168	1.0285	1.0342	1.0401	1.0460	1.0519	1.0578	1.0637
12	0.9886	1.0000	1.0115	1.0231	1.0289	1.0347	1.0405	1.0463	1.0523	1.0582
13	0.9842	0.9956	1.0071	1.0186	1.0243	1.0301	1.0359	1.0418	1.0477	1.0535
14	0.9805	0.9919	1.0033	1.0147	1.0205	1.0263	1.0321	1.0379	1.0436	1.0495
15	0.9773	0.9886	1.0000	1.0114	1.0172	1.0229	1.0287	1.0345	1.0402	1.0461
16	0.9746	0.9858	0.9972	1.0085	1.0143	1.0200	1.0257	1.0316	1.0373	1.0431
17	0.9721	0.9833	0.9947	1.0060	1.0117	1.0174	1.0232	1.0290	1.0347	1.0405
18	0.9699	0.9811	0.9924	1.0038	1.0095	1.0152	1.0209	1.0267	1.0325	1.0382
19	0.9680	0.9792	0.9905	1.0018	1.0075	1.0132	1.0189	1.0246	1.0303	1.0361
20	0.9663	0.9774	0.9887	1.0000	1.0057	1.0114	1.0171	1.0228	1.0286	1.0343
21	0.9647	0.9759	0.9871	0.9984	1.0041	1.0098	1.0154	1.0211	1.0269	1.0326
22	0.9633	0.9745	0.9856	0.9969	1.0026	1.0082	1.0139	1.0197	1.0254	1.0311
23	0.9620	0.9732	0.9843	0.9956	1.0012	1.0069	1.0126	1.0182	1.0240	1.0297
24	0.9608	0.9719	0.9831	0.9944	1.0000	1.0057	1.0113	1.0170	1.0227	1.0284
25	0.9597	0.9709	0.9820	0.9932	0.9989	1.0045	1.0102	1.0159	1.0215	1.0273
26	0.9587	0.9698	0.9810	0.9922	0.9978	1.0035	1.0091	1.0148	1.0205	1.0262
27	0.9578	0.9689	0.9801	0.9912	0.9969	1.0025	1.0081	1.0138	1.0195	1.0252
28	0.9569	0.9680	0.9791	0.9904	0.9960	1.0016	1.0073	1.0129	1.0186	1.0243
29	0.9561	0.9672	0.9783	0.9896	0.9951	1.0008	1.0064	1.0120	1.0178	1.0234
30	0.9554	0.9664	0.9776	0.9887	0.9944	1.0000	1.0056	1.0113	1.0170	1.0226
40	0.9500	0.9611	0.9721	0.9832	0.9888	0.9944	1.0000	1.0056	1.0113	1.0169
60	0.9448	0.9557	0.9667	0.9777	0.9833	0.9889	0.9944	1.0000	1.0056	1.0112
120	0.9394	0.9503	0.9613	0.9722	0.9778	0.9833	0.9888	0.9945	1.0000	1.0056
∞	0.93418	0.94503	0.95592	0.96687	0.97236	0.97787	0.98338	0.98891	0.99445	1.0000

APPENDIX F.2-F(F)=0.75

ν_1 ν_2	1	2	3	4	5	6	7	8	9
1	5.8284	7.5000	8.1999	8.5809	8.8198	8.9833	9.1021	9.1923	9.2631
2	2.5714	3.0000	3.1534	3.2321	3.2799	3.3121	3.3352	3.3526	3.3661
3	2.0239	2.2798	2.3556	2.3902	2.4096	2.4217	2.4303	2.4364	2.4409
4	1.8074	2.0000	2.0467	2.0642	2.0723	2.0766	2.0790	2.0805	2.0814
5	1.6925	1.8528	1.8843	1.8927	1.8946	1.8945	1.8935	1.8923	1.8910
6	1.6214	1.7622	1.7844	1.7872	1.7852	1.7821	1.7789	1.7760	1.7733
7	1.5732	1.7010	1.7170	1.7158	1.7111	1.7059	1.7011	1.6969	1.6931
8	1.5384	1.6569	1.6683	1.6642	1.6575	1.6508	1.6448	1.6395	1.6350
9	1.5121	1.6236	1.6316	1.6253	1.6170	1.6092	1.6021	1.5961	1.5909
10	1.4915	1.5975	1.6029	1.5948	1.5854	1.5765	1.5688	1.5621	1.5563
11	1.4749	1.5767	1.5799	1.5704	1.5598	1.5502	1.5418	1.5346	1.5284
12	1.4613	1.5595	1.5609	1.5503	1.5389	1.5286	1.5197	1.5120	1.5053
13	1.4500	1.5452	1.5451	1.5336	1.5214	1.5106	1.5012	1.4931	1.4861
14	1.4403	1.5331	1.5318	1.5193	1.5066	1.4952	1.4854	1.4770	1.4697
15	1.4321	1.5227	1.5203	1.5072	1.4938	1.4820	1.4719	1.4631	1.4556
16	1.4249	1.5137	1.5102	1.4965	1.4828	1.4706	1.4601	1.4510	1.4433
17	1.4186	1.5057	1.5015	1.4872	1.4730	1.4605	1.4497	1.4405	1.4325
18	1.4130	1.4988	1.4938	1.4790	1.4644	1.4516	1.4405	1.4311	1.4230
19	1.4081	1.4925	1.4870	1.4718	1.4568	1.4437	1.4324	1.4228	1.4145
20	1.4037	1.4870	1.4809	1.4652	1.4499	1.4367	1.4252	1.4154	1.4068
21	1.3997	1.4820	1.4752	1.4593	1.4438	1.4302	1.4186	1.4086	1.4000
22	1.3961	1.4774	1.4702	1.4540	1.4382	1.4245	1.4126	1.4025	1.3937
23	1.3928	1.4733	1.4657	1.4491	1.4331	1.4191	1.4072	1.3969	1.3880
24	1.3898	1.4695	1.4615	1.4447	1.4285	1.4144	1.4022	1.3919	1.3828
25	1.3870	1.4661	1.4578	1.4406	1.4242	1.4099	1.3977	1.3872	1.3781
26	1.3845	1.4629	1.4542	1.4369	1.4202	1.4059	1.3935	1.3828	1.3736
27	1.3821	1.4600	1.4509	1.4334	1.4166	1.4021	1.3895	1.3788	1.3696
28	1.3800	1.4573	1.4479	1.4302	1.4133	1.3985	1.3860	1.3752	1.3658
29	1.3780	1.4547	1.4452	1.4273	1.4102	1.3954	1.3826	1.3718	1.3623
30	1.3761	1.4524	1.4426	1.4245	1.4072	1.3923	1.3795	1.3686	1.3590
40	1.3626	1.4355	1.4239	1.4045	1.3863	1.3705	1.3571	1.3454	1.3354
60	1.3493	1.4188	1.4055	1.3848	1.3657	1.3491	1.3348	1.3226	1.3120
120	1.3362	1.4024	1.3873	1.3654	1.3453	1.3278	1.3128	1.2999	1.2886
∞	1.3233	1.3836	1.3694	1.3463	1.3251	1.3068	1.2910	1.2774	1.2654

APPENDIX F.2-F(F)=0.75-Continued.

ν_2 \ ν_1	10	12	15	20	24	30	40	60	120	∞
1	9.3201	9.4064	9.4934	9.5813	9.6254	9.6698	9.7144	9.7592	9.8041	9.8492
2	3.3770	3.3934	3.4098	3.4263	3.4346	3.4428	3.4511	3.4594	3.4677	3.4761
3	2.4446	2.4498	2.4552	2.4600	2.4629	2.4650	2.4674	2.4697	2.4719	2.4742
4	2.0821	2.0825	2.0828	2.0829	2.0829	2.0825	2.0821	2.0817	2.0812	2.0806
5	1.8898	1.8876	1.8850	1.8821	1.8801	1.8782	1.8762	1.8742	1.8719	1.8694
6	1.7708	1.7668	1.7621	1.7569	1.7540	1.7509	1.7476	1.7443	1.7407	1.7368
7	1.6898	1.6843	1.6781	1.6712	1.6675	1.6636	1.6593	1.6546	1.6501	1.6452
8	1.6310	1.6243	1.6170	1.6087	1.6043	1.5996	1.5946	1.5893	1.5836	1.5777
9	1.5863	1.5789	1.5705	1.5611	1.5561	1.5507	1.5449	1.5389	1.5325	1.5257
10	1.5513	1.5430	1.5338	1.5235	1.5178	1.5119	1.5056	1.4990	1.4919	1.4843
11	1.5229	1.5141	1.5041	1.4930	1.4869	1.4804	1.4736	1.4665	1.4587	1.4504
12	1.4996	1.4902	1.4796	1.4678	1.4613	1.4544	1.4471	1.4393	1.4310	1.4221
13	1.4801	1.4701	1.4590	1.4465	1.4397	1.4324	1.4246	1.4165	1.4076	1.3980
14	1.4634	1.4531	1.4414	1.4283	1.4212	1.4136	1.4055	1.3967	1.3874	1.3772
15	1.4491	1.4383	1.4263	1.4127	1.4053	1.3973	1.3888	1.3797	1.3698	1.3591
16	1.4366	1.4255	1.4131	1.3990	1.3913	1.3830	1.3742	1.3646	1.3544	1.3432
17	1.4256	1.4142	1.4014	1.3869	1.3790	1.3704	1.3613	1.3513	1.3406	1.3290
18	1.4159	1.4042	1.3911	1.3762	1.3680	1.3592	1.3497	1.3396	1.3284	1.3162
19	1.4072	1.3953	1.3819	1.3667	1.3582	1.3491	1.3394	1.3289	1.3174	1.3048
20	1.3995	1.3873	1.3736	1.3580	1.3494	1.3401	1.3302	1.3193	1.3075	1.2943
21	1.3925	1.3801	1.3661	1.3502	1.3414	1.3319	1.3217	1.3105	1.2983	1.2848
22	1.3861	1.3735	1.3593	1.3431	1.3341	1.3244	1.3140	1.3026	1.2901	1.2761
23	1.3803	1.3676	1.3531	1.3366	1.3275	1.3176	1.3069	1.2952	1.2824	1.2681
24	1.3750	1.3621	1.3474	1.3307	1.3214	1.3114	1.3005	1.2885	1.2753	1.2607
25	1.3701	1.3571	1.3422	1.3253	1.3158	1.3056	1.2944	1.2823	1.2689	1.2538
26	1.3656	1.3524	1.3374	1.3202	1.3106	1.3002	1.2889	1.2766	1.2628	1.2474
27	1.3614	1.3481	1.3329	1.3155	1.3058	1.2952	1.2838	1.2712	1.2572	1.2414
28	1.3577	1.3441	1.3287	1.3112	1.3013	1.2906	1.2790	1.2662	1.2519	1.2358
29	1.3540	1.3404	1.3249	1.3071	1.2971	1.2864	1.2745	1.2615	1.2471	1.2306
30	1.3507	1.3370	1.3213	1.3034	1.2933	1.2823	1.2703	1.2571	1.2424	1.2256
40	1.3266	1.3119	1.2952	1.2758	1.2648	1.2529	1.2397	1.2248	1.2081	1.1883
60	1.3025	1.2870	1.2691	1.2481	1.2361	1.2229	1.2081	1.1912	1.1715	1.1474
120	1.2787	1.2621	1.2429	1.2199	1.2068	1.1920	1.1752	1.1555	1.1314	1.0987
∞	1.2549	1.2371	1.2163	1.1914	1.1767	1.1600	1.1404	1.1164	1.0838	1.0000

APPENDIX F.3-F(F)=0.90

ν_2 \ ν_1	1	2	3	4	5	6	7	8	9
1	39.8635	49.5000	53.5932	55.8330	57.2401	58.2044	58.9060	59.4390	59.8576
2	8.5263	9.0000	9.1618	9.2434	9.2926	9.3255	9.3491	9.3668	9.3805
3	5.5383	5.4624	5.3910	5.3426	5.3090	5.2848	5.2657	5.2520	5.2402
4	4.5448	4.3246	4.1910	4.1074	4.0507	4.0099	3.9792	3.9549	3.9359
5	4.0604	3.7797	3.6196	3.5201	3.4529	3.4044	3.3678	3.3392	3.3163
6	3.7759	3.4633	3.2887	3.1808	3.1074	3.0546	3.0144	2.9831	2.9578
7	3.5894	3.2574	3.0740	2.9605	2.8833	2.8273	2.7850	2.7515	2.7246
8	3.4579	3.1131	2.9239	2.8064	2.7264	2.6683	2.6242	2.5893	2.5613
9	3.3603	3.0065	2.8129	2.6926	2.6106	2.5508	2.5053	2.4695	2.4403
10	3.2850	2.9245	2.7277	2.6054	2.5217	2.4606	2.4140	2.3771	2.3473
11	3.2252	2.8595	2.6602	2.5362	2.4512	2.3890	2.3416	2.3040	2.2735
12	3.1765	2.8068	2.6055	2.4801	2.3941	2.3310	2.2828	2.2446	2.2135
13	3.1362	2.7632	2.5602	2.4337	2.3468	2.2830	2.2341	2.1954	2.1638
14	3.1022	2.7265	2.5223	2.3947	2.3070	2.2425	2.1932	2.1539	2.1219
15	3.0732	2.6952	2.4897	2.3614	2.2730	2.2081	2.1582	2.1186	2.0862
16	3.0481	2.5682	2.4619	2.3327	2.2438	2.1784	2.1281	2.0880	2.0553
17	3.0262	2.6446	2.4374	2.3077	2.2182	2.1524	2.1017	2.0613	2.0283
18	3.0070	2.6239	2.4160	2.2857	2.1959	2.1296	2.0785	2.0379	2.0047
19	2.9899	2.6056	2.3970	2.2663	2.1759	2.1094	2.0580	2.0171	1.9836
20	2.9747	2.5893	2.3801	2.2490	2.1582	2.0913	2.0396	1.9986	1.9648
21	2.9610	2.5746	2.3649	2.2333	2.1423	2.0752	2.0233	1.9819	1.9480
22	2.9486	2.5613	2.3512	2.2193	2.1280	2.0605	2.0084	1.9669	1.9328
23	2.9374	2.5493	2.3387	2.2065	2.1150	2.0472	1.9950	1.9531	1.9189
24	2.9271	2.5383	2.3274	2.1948	2.1030	2.0351	1.9826	1.9407	1.9063
25	2.9177	2.5283	2.3170	2.1843	2.0921	2.0241	1.9713	1.9292	1.8946
26	2.9091	2.5191	2.3074	2.1745	2.0822	2.0139	1.9611	1.9188	1.8841
27	2.9012	2.5106	2.2988	2.1654	2.0730	2.0045	1.9515	1.9091	1.8743
28	2.8938	2.5028	2.2906	2.1572	2.0645	1.9959	1.9428	1.9002	1.8651
29	2.8870	2.4955	2.2832	2.1494	2.0567	1.9878	1.9346	1.8918	1.8568
30	2.8807	2.4887	2.2761	2.1423	2.0492	1.9803	1.9269	1.8842	1.8490
40	2.8354	2.4404	2.2261	2.0909	1.9968	1.9268	1.8726	1.8289	1.7930
60	2.7911	2.3933	2.1774	2.0410	1.9457	1.8747	1.8195	1.7747	1.7381
120	2.7478	2.3473	2.1300	1.9923	1.8959	1.8238	1.7675	1.7220	1.6842
∞	2.7055	2.3026	2.0838	1.9449	1.8473	1.7741	1.7167	1.6702	1.6315

APPENDIX F.3-F(F)=0.90-Continued.

ν_2 \ ν_1	10	12	15	20	24	30	40	60	120	∞
1	60.1950	60.7052	61.2203	61.7403	62.0020	62.2650	62.5291	62.7943	63.0606	63.3280
2	9.3916	9.4081	9.4247	9.4413	9.4496	9.4579	9.4662	9.4746	9.4829	9.4912
3	5.2297	5.2159	5.1995	5.1854	5.1755	5.1681	5.1597	5.1512	5.1425	5.1337
4	3.9197	3.8958	3.8702	3.8448	3.8306	3.8179	3.8036	3.7896	3.7753	3.7607
5	3.2972	3.2682	3.2382	3.2069	3.1901	3.1741	3.1577	3.1402	3.1228	3.1050
6	2.9369	2.9046	2.8713	2.8364	2.8185	2.7997	2.7814	2.7620	2.7423	2.7222
7	2.7026	2.6682	2.6321	2.5947	2.5752	2.5553	2.5353	2.5139	2.4928	2.4708
8	2.5380	2.5020	2.4642	2.4247	2.4040	2.3831	2.3613	2.3389	2.3162	2.2926
9	2.4163	2.3788	2.3396	2.2984	2.2768	2.2548	2.2320	2.2083	2.1843	2.1592
10	2.3226	2.2841	2.2434	2.2008	2.1784	2.1555	2.1316	2.1072	2.0818	2.0554
11	2.2483	2.2087	2.1672	2.1231	2.1001	2.0763	2.0515	2.0263	1.9997	1.9721
12	2.1878	2.1474	2.1049	2.0597	2.0359	2.0115	1.9860	1.9599	1.9323	1.9036
13	2.1376	2.0966	2.0532	2.0070	1.9827	1.9577	1.9314	1.9044	1.8757	1.8462
14	2.0955	2.0537	2.0096	1.9624	1.9376	1.9120	1.8851	1.8573	1.8281	1.7973
15	2.0594	2.0171	1.9722	1.9243	1.8991	1.8729	1.8454	1.8167	1.7866	1.7551
16	2.0281	1.9854	1.9399	1.8912	1.8656	1.8388	1.8108	1.7815	1.7509	1.7182
17	2.0009	1.9578	1.9117	1.8624	1.8362	1.8090	1.7806	1.7506	1.7191	1.6856
18	1.9770	1.9333	1.8868	1.8368	1.8103	1.7826	1.7537	1.7232	1.6911	1.6567
19	1.9557	1.9117	1.8647	1.8142	1.7873	1.7593	1.7297	1.6987	1.6658	1.6308
20	1.9367	1.8923	1.8449	1.7938	1.7666	1.7382	1.7083	1.6768	1.6434	1.6074
21	1.9197	1.8750	1.8272	1.7756	1.7481	1.7193	1.6890	1.6569	1.6228	1.5872
22	1.9042	1.8593	1.8110	1.7589	1.7313	1.7021	1.6714	1.6389	1.6042	1.5668
23	1.8902	1.8450	1.7965	1.7439	1.7159	1.6865	1.6554	1.6223	1.5871	1.5490
24	1.8775	1.8319	1.7831	1.7301	1.7019	1.6721	1.6406	1.6073	1.5714	1.5327
25	1.8657	1.8200	1.7708	1.7175	1.6889	1.6590	1.6272	1.5933	1.5569	1.5176
26	1.8550	1.8090	1.7596	1.7059	1.6771	1.6468	1.6147	1.5806	1.5436	1.5036
27	1.8451	1.7989	1.7491	1.6952	1.6662	1.6356	1.6032	1.5686	1.5314	1.4906
28	1.8359	1.7895	1.7395	1.6851	1.6560	1.6251	1.5924	1.5576	1.5198	1.4784
29	1.8274	1.7809	1.7306	1.6759	1.6465	1.6155	1.5825	1.5472	1.5091	1.4670
30	1.8195	1.7727	1.7223	1.6673	1.6377	1.6065	1.5733	1.5376	1.4988	1.4564
40	1.7627	1.7146	1.6624	1.6052	1.5741	1.5411	1.5056	1.4671	1.4248	1.3769
60	1.7070	1.6574	1.6033	1.5435	1.5107	1.4756	1.4374	1.3952	1.3476	1.2915
120	1.6524	1.6012	1.5449	1.4821	1.4473	1.4093	1.3676	1.3203	1.2646	1.1926
∞	1.5987	1.5458	1.4871	1.4206	1.3832	1.3419	1.2951	1.2400	1.1686	1.000

APPENDIX F.4 — F(F) = 0.95

ν_2 \ ν_1	1	2	3	4	5	6	7	8	9
1	161.4476	199.5000	215.7073	224.5832	230.1619	233.9860	236.7684	238.8827	240.5433
2	18.5128	19.0000	19.1643	19.2468	19.2964	19.3295	19.3532	19.3710	19.3848
3	10.1280	9.5521	9.2761	9.1169	9.0129	8.9419	8.8856	8.8450	8.8127
4	7.7086	6.9443	6.5913	6.3881	6.2561	6.1627	6.0939	6.0412	5.9989
5	6.6079	5.7861	5.4097	5.1920	5.0505	4.9504	4.8761	4.8186	4.7724
6	5.9874	5.1433	4.7572	4.5337	4.3872	4.2837	4.2065	4.1468	4.0992
7	5.5914	4.7374	4.3468	4.1204	3.9716	3.8660	3.7870	3.7256	3.6765
8	5.3177	4.4590	4.0662	3.8377	3.6874	3.5806	3.5005	3.4382	3.3881
9	5.1174	4.2565	3.8625	3.6331	3.4817	3.3737	3.2928	3.2295	3.1789
10	4.9646	4.1028	3.7082	3.4781	3.3259	3.2173	3.1354	3.0717	3.0203
11	4.8443	3.9823	3.5875	3.3567	3.2039	3.0946	3.0123	2.9481	2.8963
12	4.7472	3.8853	3.4904	3.2591	3.1058	2.9962	2.9134	2.8485	2.7963
13	4.6672	3.8056	3.4106	3.1791	3.0255	2.9152	2.8320	2.7669	2.7143
14	4.6001	3.7389	3.3440	3.1123	2.9582	2.8478	2.7643	2.6996	2.6458
15	4.5431	3.6823	3.2874	3.0556	2.9012	2.7905	2.7066	2.6407	2.5876
16	4.4940	3.6337	3.2388	3.0069	2.8525	2.7413	2.6572	2.5910	2.5376
17	4.4513	3.5915	3.1967	2.9647	2.8100	2.6987	2.6142	2.5479	2.4942
18	4.4139	3.5546	3.1600	2.9277	2.7728	2.6613	2.5767	2.5101	2.4563
19	4.3807	3.5219	3.1273	2.8951	2.7401	2.6283	2.5435	2.4768	2.4226
20	4.3512	3.4928	3.0985	2.8661	2.7109	2.5989	2.5140	2.4471	2.3929
21	4.3248	3.4668	3.0724	2.8400	2.6848	2.5727	2.4876	2.4204	2.3661
22	4.3009	3.4434	3.0491	2.8167	2.6613	2.5491	2.4638	2.3965	2.3420
23	4.2793	3.4221	3.0280	2.7956	2.6401	2.5276	2.4422	2.3747	2.3201
24	4.2597	3.4028	3.0088	2.7763	2.6207	2.5081	2.4226	2.3551	2.3002
25	4.2417	3.3852	2.9912	2.7586	2.6030	2.4903	2.4047	2.3370	2.2820
26	4.2252	3.3690	2.9752	2.7425	2.5868	2.4741	2.3884	2.3205	2.2654
27	4.2100	3.3541	2.9605	2.7277	2.5719	2.4592	2.3732	2.3053	2.2501
28	4.1960	3.3404	2.9466	2.7141	2.5580	2.4452	2.3592	2.2912	2.2360
29	4.1830	3.3277	2.9339	2.7014	2.5453	2.4325	2.3464	2.2783	2.2229
30	4.1709	3.3158	2.9223	2.6896	2.5336	2.4205	2.3344	2.2662	2.2107
40	4.0847	3.2317	2.8387	2.6060	2.4495	2.3359	2.2490	2.1802	2.1241
60	4.0012	3.1504	2.7581	2.5252	2.3683	2.2541	2.1665	2.0969	2.0400
120	3.9201	3.0718	2.6802	2.4472	2.2899	2.1750	2.0868	2.0164	1.9588
∞	3.8415	2.9957	2.6049	2.3719	2.2141	2.0986	2.0096	1.9384	1.8799

APPENDIX F.4-F(F)=0.95-Continued.

ν_2 \ ν_1	10	12	15	20	24	30	40	60	120	∞
1	241.8817	243.9060	245.9499	248.0131	249.0518	250.0951	251.1432	252.1957	253.2529	254.3100
2	19.3959	19.4125	19.4291	19.4458	19.4541	19.4624	19.4707	19.4791	19.4874	19.4960
3	8.7855	8.7448	8.7013	8.6626	8.6412	8.6166	8.5944	8.5720	8.5494	8.5264
4	5.9650	5.9109	5.8578	5.8028	5.7744	5.7450	5.7170	5.6877	5.6581	5.6281
5	4.7349	4.6774	4.6186	4.5582	4.5267	4.4962	4.4631	4.4314	4.3985	4.3650
6	4.0598	4.0002	3.9383	3.8738	3.8414	3.8077	3.7735	3.7398	3.7047	3.6689
7	3.6366	3.5745	3.5106	3.4448	3.4102	3.3762	3.3403	3.3051	3.2674	3.2298
8	3.3470	3.2838	3.2183	3.1504	3.1153	3.0797	3.0432	3.0048	2.9669	2.9276
9	3.1373	3.0728	3.0060	2.9365	2.9003	2.8637	2.8257	2.7875	2.7475	2.7067
10	2.9783	2.9131	2.8450	2.7740	2.7371	2.6997	2.6609	2.6208	2.5801	2.5379
11	2.8537	2.7876	2.7186	2.6465	2.6090	2.5704	2.5307	2.4899	2.4480	2.4045
12	2.7534	2.6867	2.6168	2.5436	2.5055	2.4664	2.4259	2.3840	2.3410	2.2962
13	2.6710	2.6036	2.5331	2.4588	2.4201	2.3803	2.3392	2.2967	2.2523	2.2064
14	2.6021	2.5342	2.4630	2.3879	2.3487	2.3082	2.2662	2.2230	2.1777	2.1307
15	2.5437	2.4752	2.4035	2.3275	2.2879	2.2468	2.2042	2.1599	2.1142	2.0658
16	2.4935	2.4246	2.3523	2.2756	2.2355	2.1938	2.1507	2.1057	2.0591	2.0096
17	2.4499	2.3807	2.3078	2.2303	2.1898	2.1477	2.1040	2.0584	2.0107	1.9604
18	2.4118	2.3420	2.2687	2.1906	2.1497	2.1071	2.0630	2.0167	1.9679	1.9168
19	2.3779	2.3079	2.2341	2.1554	2.1142	2.0713	2.0265	1.9797	1.9301	1.8780
20	2.3478	2.2776	2.2033	2.1241	2.0824	2.0391	1.9939	1.9464	1.8964	1.8432
21	2.3209	2.2503	2.1757	2.0961	2.0541	2.0103	1.9645	1.9164	1.8659	1.8117
22	2.2967	2.2258	2.1507	2.0707	2.0284	1.9841	1.9380	1.8893	1.8379	1.7831
23	2.2747	2.2037	2.1282	2.0476	2.0050	1.9605	1.9139	1.8648	1.8128	1.7570
24	2.2548	2.1834	2.1076	2.0267	1.9837	1.9390	1.8919	1.8424	1.7897	1.7330
25	2.2364	2.1649	2.0888	2.0075	1.9643	1.9191	1.8719	1.8217	1.7683	1.7110
26	2.2197	2.1479	2.0716	1.9899	1.9464	1.9011	1.8532	1.8027	1.7488	1.6906
27	2.2042	2.1322	2.0558	1.9736	1.9299	1.8842	1.8361	1.7851	1.7308	1.6717
28	2.1900	2.1178	2.0411	1.9585	1.9147	1.8687	1.8203	1.7689	1.7138	1.6541
29	2.1768	2.1045	2.0274	1.9446	1.9005	1.8543	1.8055	1.7537	1.6982	1.6376
30	2.1646	2.0921	2.0148	1.9316	1.8873	1.8409	1.7918	1.7396	1.6836	1.6223
40	2.0772	2.0035	1.9244	1.8389	1.7929	1.7444	1.6928	1.6373	1.5766	1.5089
60	1.9926	1.9174	1.8364	1.7480	1.7001	1.6492	1.5943	1.5343	1.4673	1.3893
120	1.9105	1.8337	1.7505	1.6587	1.6084	1.5543	1.4952	1.4290	1.3519	1.2539
∞	1.8307	1.7522	1.6664	1.5705	1.5173	1.4591	1.3940	1.3180	1.2214	1.0000

APPENDIX F .5-F(F)=0.975

ν_2 \ ν_1	1	2	3	4	5	6	7	8	9
1	647.7890	799.5000	864.1630	899.5833	921.8479	937.1111	948.2169	956.6552	963.2846
2	38.5063	39.0000	39.1655	39.2484	39.2982	39.3315	39.3552	39.3730	39.3869
3	17.4434	16.0441	15.4395	15.1029	14.8840	14.7374	14.6274	14.5421	14.4720
4	12.2179	10.6491	9.9789	9.6037	9.3636	9.1969	9.0742	8.9801	8.9047
5	10.0070	8.4336	7.7637	7.3882	7.1467	6.9781	6.8536	6.7572	6.6806
6	8.8131	7.2599	6.5986	6.2274	5.9876	5.8199	5.6956	5.6000	5.5231
7	8.0727	6.5415	5.8900	5.5228	5.2852	5.1187	4.9952	4.8995	4.8233
8	7.5709	6.0595	5.4158	5.0528	4.8174	4.6505	4.5285	4.4330	4.3571
9	7.2093	5.7147	5.0781	4.7181	4.4843	4.3197	4.1970	4.1019	4.0260
10	6.9367	5.4564	4.8257	4.4682	4.2360	4.0720	3.9497	3.8548	3.7791
11	6.7241	5.2559	4.6299	4.2750	4.0441	3.8806	3.7587	3.6638	3.5878
12	6.5538	5.0959	4.4742	4.1213	3.8911	3.7283	3.6066	3.5118	3.4357
13	6.4143	4.9653	4.3472	3.9960	3.7666	3.6042	3.4828	3.3879	3.3121
14	6.2979	4.8567	4.2416	3.8920	3.6635	3.5013	3.3799	3.2854	3.2092
15	6.1995	4.7650	4.1527	3.8043	3.5765	3.4147	3.2933	3.1988	3.1226
16	6.1151	4.6867	4.0768	3.7293	3.5022	3.3407	3.2194	3.1249	3.0487
17	6.0420	4.6189	4.0111	3.6649	3.4379	3.2766	3.1557	3.0611	2.9848
18	5.9781	4.5597	3.9540	3.6083	3.3819	3.2209	3.0999	3.0053	2.9291
19	5.9216	4.5075	3.9033	3.5586	3.3328	3.1718	3.0509	2.9562	2.8801
20	5.8715	4.4613	3.8588	3.5146	3.2890	3.1283	3.0074	2.9127	2.8366
21	5.8266	4.4199	3.8187	3.4754	3.2500	3.0895	2.9686	2.8740	2.7977
22	5.7863	4.3828	3.7830	3.4402	3.2151	3.0547	2.9338	2.8391	2.7628
23	5.7498	4.3492	3.7505	3.4083	3.1835	3.0232	2.9023	2.8076	2.7313
24	5.7166	4.3187	3.7210	3.3794	3.1549	2.9945	2.8738	2.7791	2.7026
25	5.6864	4.2909	3.6943	3.3530	3.1287	2.9685	2.8479	2.7530	2.6767
26	5.6586	4.2655	3.6698	3.3290	3.1047	2.9448	2.8241	2.7293	2.6527
27	5.6331	4.2421	3.6473	3.3067	3.0827	2.9228	2.8021	2.7074	2.6308
28	5.6096	4.2205	3.6264	3.2863	3.0625	2.9027	2.7819	2.6872	2.6106
29	5.5878	4.2006	3.6072	3.2673	3.0438	2.8841	2.7633	2.6686	2.5919
30	5.5675	4.1821	3.5894	3.2499	3.0265	2.8666	2.7461	2.6513	2.5746
40	5.4239	4.0510	3.4633	3.1261	2.9037	2.7445	2.6238	2.5289	2.4519
60	5.2856	3.9253	3.3425	3.0077	2.7863	2.6274	2.5068	2.4116	2.3344
120	5.1523	3.8046	3.2269	2.8943	2.6740	2.5154	2.3948	2.2994	2.2217
∞	5.0239	3.6889	3.1161	2.7858	2.5665	2.4082	2.2875	2.1918	2.1136

APPENDIX F.5-F(F)=0.975-Continued.

ν_2 \ ν_1	10	12	15	20	24	30	40	60	120	∞
1	968.6274	976.7079	984.8668	993.1028	997.2492	1001.4144	1005.5981	1009.8001	1014.0202	1018.3000
2	39.3980	39.4146	39.4313	39.4479	39.4562	39.4646	39.4729	39.4812	39.4896	39.4890
3	14.4217	14.3330	14.2591	14.1696	14.1344	14.0805	14.0365	13.9921	13.9473	13.9020
4	8.8451	8.7519	8.6566	8.5586	8.5124	8.4630	8.4111	8.3604	8.3092	8.2573
5	6.6196	6.5254	6.4268	6.3286	6.2794	6.2265	6.1741	6.1225	6.0693	6.0153
6	5.4611	5.3666	5.2686	5.1682	5.1174	5.0650	5.0117	4.9589	4.9044	4.8491
7	4.7610	4.6661	4.5681	4.4663	4.4146	4.3622	4.3094	4.2536	4.1989	4.1423
8	4.2954	4.1995	4.1014	3.9995	3.9471	3.8939	3.8398	3.7842	3.7279	3.6702
9	3.9639	3.8680	3.7693	3.6669	3.6143	3.5607	3.5057	3.4489	3.3918	3.3329
10	3.7167	3.6211	3.5217	3.4184	3.3652*	3.3109	3.2557	3.1988	3.1399	3.0798
11	3.5257	3.4295	3.3301	3.2261	3.1727	3.1177	3.0616	3.0032	2.9441	2.8828
12	3.3735	3.2774	3.1772	3.0727	3.0187	2.9632	2.9065	2.8478	2.7874	2.7249
13	3.2498	3.1532	3.0527	2.9477	2.8933	2.8371	2.7796	2.7204	2.6591	2.5955
14	3.1469	3.0502	2.9494	2.8437	2.7889	2.7323	2.6743	2.6140	2.5516	2.4872
15	3.0603	2.9632	2.8620	2.7560	2.7006	2.6437	2.5851	2.5244	2.4613	2.3953
16	2.9862	2.8891	2.7875	2.6808	2.6252	2.5679	2.5087	2.4473	2.3830	2.3163
17	2.9222	2.8250	2.7230	2.6159	2.5597	2.5021	2.4421	2.3802	2.3155	2.2474
18	2.8664	2.7688	2.6667	2.5591	2.5026	2.4445	2.3842	2.3213	2.2556	2.1869
19	2.8172	2.7195	2.6170	2.5089	2.4523	2.3938	2.3329	2.2695	2.2035	2.1333
20	2.7736	2.6759	2.5731	2.4645	2.4075	2.3487	2.2872	2.2233	2.1561	2.0853
21	2.7348	2.6367	2.5338	2.4248	2.3676	2.3081	2.2465	2.1819	2.1140	2.0422
22	2.6998	2.6016	2.4984	2.3889	2.3316	2.2718	2.2096	2.1447	2.0762	2.0032
23	2.6682	2.5700	2.4664	2.3566	2.2990	2.2390	2.1763	2.1108	2.0416	1.9677
24	2.6395	2.5411	2.4374	2.3273	2.2692	2.2089	2.1460	2.0800	2.0098	1.9353
25	2.6135	2.5149	2.4110	2.3004	2.2422	2.1816	2.1182	2.0517	1.9809	1.9055
26	2.5896	2.4908	2.3867	2.2759	2.2174	2.1565	2.0929	2.0258	1.9543	1.8781
27	2.5676	2.4688	2.3645	2.2533	2.1947	2.1334	2.0693	2.0018	1.9300	1.8527
28	2.5473	2.4484	2.3438	2.2325	2.1735	2.1121	2.0476	1.9797	1.9073	1.8291
29	2.5287	2.4295	2.3248	2.2131	2.1541	2.0924	2.0276	1.9592	1.8861	1.8072
30	2.5112	2.4121	2.3071	2.1952	2.1359	2.0740	2.0089	1.9400	1.8664	1.7867
40	2.3882	2.2881	2.1819	2.0677	2.0069	1.9429	1.8752	1.8028	1.7242	1.6371
60	2.2701	2.1692	2.0613	1.9444	1.8817	1.8152	1.7441	1.6668	1.5810	1.4821
120	2.1570	2.0548	1.9450	1.8250	1.7597	1.6899	1.6142	1.5299	1.4327	1.3104
∞	2.0483	1.9447	1.8326	1.7085	1.6402	1.5660	1.4835	1.3883	1.2684	1.0000

APPENDIX F.6-F(F)=0.99

ν_2 / ν_1	1	2	3	4	5	6	7	8	9
1	4052.1807	4999.5000	5403.3520	5624.5833	5763.6496	5858.9861	5928.3357	5981.0703	6022.4732
2	98.5025	99.0000	99.1662	99.2494	99.2993	99.3326	99.3564	99.3742	99.3881
3	34.1162	30.8165	29.4606	28.7088	28.2387	27.9075	27.6723	27.5048	27.3365
4	21.1977	18.0000	16.6927	15.9760	15.5202	15.2064	14.9773	14.8014	14.6591
5	16.2582	13.2739	12.0596	11.3910	10.9668	10.6727	10.4566	10.2904	10.1577
6	13.7450	10.9248	9.7791	9.1488	8.7454	8.4657	8.2597	8.1022	7.9755
7	12.2464	9.5466	8.4511	7.8469	7.4601	7.1918	6.9927	6.8403	6.7187
8	11.2586	8.6491	7.5909	7.0062	6.6316	6.3706	6.1776	6.0291	5.9106
9	10.5614	8.0215	6.9920	6.4221	6.0567	5.8018	5.6127	5.4670	5.3513
10	10.0443	7.5594	6.5521	5.9942	5.6362	5.3859	5.2003	5.0566	4.9427
11	9.6460	7.2057	6.2167	5.6683	5.3162	5.0692	4.8859	4.7445	4.6315
12	9.3302	6.9266	5.9525	5.4119	5.0644	4.8205	4.6397	4.4995	4.3874
13	9.0738	6.7010	5.7395	5.2053	4.8615	4.6204	4.4411	4.3021	4.1911
14	8.8616	6.5149	5.5638	5.0354	4.6949	4.4559	4.2778	4.1400	4.0297
15	8.6831	6.3589	5.4169	4.8933	4.5556	4.3183	4.1415	4.0046	3.8949
16	8.5310	6.2262	5.2923	4.7726	4.4375	4.2017	4.0259	3.8897	3.7804
17	8.3997	6.1121	5.1850	4.6690	4.3359	4.1016	3.9267	3.7909	3.6824
18	8.2854	6.0129	5.0920	4.5791	4.2480	4.0146	3.8407	3.7054	3.5972
19	8.1849	5.9259	5.0104	4.5003	4.1708	3.9385	3.7653	3.6305	3.5225
20	8.0960	5.8489	4.9382	4.4306	4.1027	3.8715	3.6987	3.5644	3.4566
21	8.0166	5.7804	4.8742	4.3689	4.0422	3.8118	3.6396	3.5055	3.3981
22	7.9454	5.7190	4.8166	4.3133	3.9880	3.7583	3.5868	3.4530	3.3457
23	7.8811	5.6637	4.7650	4.2635	3.9393	3.7103	3.5391	3.4056	3.2987
24	7.8229	5.6136	4.7182	4.2183	3.8950	3.6667	3.4959	3.3628	3.2560
25	7.7698	5.5680	4.6755	4.1773	3.8549	3.6271	3.4568	3.3239	3.2171
26	7.7213	5.5263	4.6367	4.1399	3.8183	3.5910	3.4211	3.2884	3.1819
27	7.6767	5.4881	4.6009	4.1055	3.7848	3.5580	3.3883	3.2559	3.1495
28	7.6356	5.4529	4.5681	4.0741	3.7538	3.5275	3.3581	3.2259	3.1196
29	7.5977	5.4204	4.5379	4.0449	3.7255	3.4994	3.3302	3.1982	3.0921
30	7.5625	5.3903	4.5097	4.0179	3.6991	3.4734	3.3044	3.1727	3.0664
40	7.3141	5.1785	4.3126	3.8283	3.5138	3.2910	3.1238	2.9931	2.8876
60	7.0771	4.9774	4.1259	3.6490	3.3389	3.1187	2.9552	2.8233	2.7184
120	6.8509	4.7865	3.9491	3.4795	3.1735	2.9559	2.7918	2.6629	2.5586
∞	6.6349	4.6052	3.7816	3.3192	3.0173	2.8020	2.6393	2.5113	2.4073

APPENDIX F.6-F(F)=0.99-Continued.

ν_2＼ν_1	10	12	15	20	24	30	40	60	120	∞
1	6055.8467	6106.3207	6157.2846	6208.7302	6234.6309	6260.6486	6286.7821	6313.0301	6339.3913	6365.9000
2	99.3992	99.4159	99.4325	99.4492	99.4575	99.4658	99.4742	99.4825	99.4908	99.4990
3	27.2169	27.0339	26.8530	26.6902	26.6026	26.5045	26.4108	26.3164	26.2211	26.1250
4	14.5498	14.3724	14.1945	14.0238	13.9362	13.8365	13.7454	13.6522	13.5581	13.4630
5	10.0516	9.8897	9.7231	9.5520	9.4646	9.3769	9.2965	9.2020	9.1118	9.0204
6	7.8736	7.7177	7.5594	7.3965	7.3124	7.2283	7.1453	7.0567	6.9690	6.8800
7	6.6195	6.4694	6.3138	6.1562	6.0746	5.9917	5.9076	5.8219	5.7373	5.6495
8	5.8144	5.6667	5.5154	5.3395	5.2788	5.1989	5.1162	5.0326	4.9461	4.8588
9	5.2565	5.1115	4.9621	4.8083	4.7285	4.6480	4.5664	4.4837	4.3978	4.3105
10	4.8491	4.7058	4.5579	4.4058	4.3270	4.2469	4.1656	4.0826	3.9965	3.9090
11	4.5393	4.3976	4.2508	4.0988	4.0212	3.9409	3.8600	3.7761	3.6904	3.6024
12	4.2959	4.1552	4.0098	3.8585	3.7803	3.7008	3.6192	3.5358	3.4494	3.3608
13	4.1002	3.9603	3.8152	3.6644	3.5868	3.5069	3.4252	3.3417	3.2541	3.1654
14	3.9394	3.8003	3.6559	3.5051	3.4275	3.3474	3.2659	3.1811	3.0939	3.0040
15	3.8050	3.6661	3.5221	3.3719	3.2941	3.2141	3.1321	3.0469	2.9599	2.8684
16	3.6909	3.5528	3.4088	3.2587	3.1809	3.1006	3.0185	2.9332	2.8444	2.7528
17	3.5931	3.4553	3.3117	3.1615	3.0835	3.0033	2.9205	2.8345	2.7457	2.6530
18	3.5081	3.3705	3.2272	3.0771	2.9991	2.9186	2.8354	2.7494	2.6599	2.5660
19	3.4337	3.2966	3.1532	3.0031	2.9250	2.8442	2.7608	2.6743	2.5842	2.4893
20	3.3682	3.2311	3.0880	2.9376	2.8594	2.7786	2.6949	2.6078	2.5167	2.4212
21	3.3099	3.1730	3.0299	2.8796	2.8010	2.7201	2.6358	2.5485	2.4570	2.3603
22	3.2575	3.1210	2.9779	2.8274	2.7488	2.6676	2.5831	2.4952	2.4030	2.3055
23	3.2105	3.0741	2.9311	2.7806	2.7018	2.6201	2.5356	2.4472	2.3544	2.2558
24	3.1682	3.0317	2.8887	2.7380	2.6591	2.5774	2.4924	2.4035	2.3098	2.2107
25	3.1294	2.9931	2.8501	2.6994	2.6202	2.5384	2.4530	2.3636	2.2694	2.1694
26	3.0941	2.9579	2.8150	2.6640	2.5847	2.5026	2.4169	2.3272	2.2326	2.1315
27	3.0618	2.9256	2.7827	2.6316	2.5523	2.4699	2.3839	2.2938	2.1985	2.0965
28	3.0320	2.8960	2.7529	2.6018	2.5223	2.4396	2.3536	2.2629	2.1671	2.0642
29	3.0046	2.8685	2.7256	2.5741	2.4946	2.4118	2.3253	2.2343	2.1380	2.0342
30	2.9791	2.8432	2.7002	2.5487	2.4690	2.3859	2.2992	2.2078	2.1106	2.0062
40	2.8005	2.6648	2.5216	2.3688	2.2880	2.2034	2.1142	2.0194	1.9173	1.8047
60	2.6318	2.4961	2.3523	2.1978	2.1154	2.0285	1.9361	1.8363	1.7263	1.6006
120	2.4721	2.3363	2.1915	2.0346	1.9500	1.8600	1.7628	1.6557	1.5330	1.3805
∞	2.3209	2.1847	2.0385	1.8783	1.7908	1.6964	1.5923	1.4730	1.3346	1.0000

APPENDIX F.7 - F(F) = 0.995

ν_2 \ ν_1	1	2	3	4	5	6	7	8	9
1	16210.7227	19999.5000	21614.7414	22499.5833	23055.7982	23437.1111	23714.5658	23925.4062	24091.0041
2	198.5013	199.0000	199.1664	199.2497	199.2996	199.3330	199.3568	199.3746	199.3885
3	55.5520	49.7993	47.4555	46.1732	45.4171	44.8536	44.4028	44.1064	43.6433
4	31.3328	26.2843	24.2579	23.1518	22.4552	21.9769	21.6207	21.3526	21.1392
5	22.7848	18.3138	16.5302	15.5558	14.9397	14.5140	14.2009	13.9603	13.7702
6	18.6350	14.5441	12.9174	12.0284	11.4629	11.0726	10.7863	10.5662	10.3929
7	16.2356	12.4040	10.8828	10.0510	9.5220	9.1551	8.8855	8.6784	8.5144
8	14.6882	11.0424	9.5961	8.8052	8.3020	7.9522	7.6942	7.4963	7.3385
9	13.6136	10.1067	8.7172	7.9556	7.4714	7.1341	6.8850	6.6930	6.5411
10	12.8265	9.4270	8.0810	7.3429	6.8721	6.5449	6.3025	6.1159	5.9679
11	12.2263	8.9122	7.6004	6.8807	6.4219	6.1018	5.8646	5.6824	5.5369
12	11.7542	8.5096	7.2260	6.5212	6.0711	5.7569	5.5244	5.3450	5.2020
13	11.3735	8.1865	6.9256	6.2334	5.7911	5.4821	5.2531	5.0760	4.9351
14	11.0603	7.9216	6.6805	5.9986	5.5624	5.2573	5.0312	4.8566	4.7174
15	10.7980	7.7008	6.4760	5.8028	5.3722	5.0708	4.8472	4.6742	4.5364
16	10.5755	7.5138	6.3034	5.6378	5.2117	4.9134	4.6921	4.5205	4.3839
17	10.3842	7.3536	6.1557	5.4967	5.0744	4.7791	4.5593	4.3893	4.2535
18	10.2181	7.2148	6.0277	5.3747	4.9561	4.6628	4.4449	4.2759	4.1411
19	10.0725	7.0935	5.9160	5.2681	4.8527	4.5613	4.3447	4.1771	4.0427
20	9.9439	6.9865	5.8176	5.1743	4.7616	4.4721	4.2569	4.0898	3.9565
21	9.8295	6.8914	5.7306	5.0911	4.6809	4.3931	4.1790	4.0129	3.8799
22	9.7271	6.8064	5.6524	5.0168	4.6088	4.3224	4.1092	3.9439	3.8115
23	9.6348	6.7300	5.5822	4.9501	4.5440	4.2592	4.0469	3.8823	3.7503
24	9.5513	6.6609	5.5190	4.8898	4.4856	4.2019	3.9904	3.8263	3.6949
25	9.4753	6.5982	5.4615	4.8351	4.4328	4.1500	3.9394	3.7759	3.6448
26	9.4059	6.5409	5.4091	4.7851	4.3844	4.1028	3.8927	3.7296	3.5990
27	9.3423	6.4885	5.3613	4.7396	4.3402	4.0595	3.8502	3.6874	3.5570
28	9.2838	6.4403	5.3169	4.6976	4.2996	4.0197	3.8111	3.6487	3.5187
29	9.2297	6.3958	5.2763	4.6592	4.2620	3.9830	3.7750	3.6130	3.4832
30	9.1797	6.3547	5.2388	4.6233	4.2276	3.9492	3.7415	3.5801	3.4504
40	8.8279	6.0664	4.9758	4.3738	3.9861	3.7130	3.5088	3.3497	3.2219
60	8.4946	5.7950	4.7290	4.1399	3.7599	3.4918	3.2912	3.1345	3.0083
120	8.1788	5.5393	4.4972	3.9207	3.5482	3.2849	3.0874	2.9330	2.8083
∞	7.8794	5.2983	4.2794	3.7151	3.499	3.0913	2.8968	2.7444	2.6210

APPENDIX F.7-F(F)=0.995-Continued.

ν_2 \ ν_1	10	12	15	20	24	30	40	60	120	∞
1	24224.4868	24426.3662	24630.2051	24835.9709	24939.5653	25043.6277	25148.1532	25253.1369	25358.5735	25464.0000
2	199.3996	199.4163	199.4329	199.4496	199.4579	199.4663	199.4746	199.4829	199.4912	199.5000
3	43.7331	43.3825	43.1296	42.6840	42.6531	42.4658	42.3082	42.1494	41.9895	41.8280
4	20.9733	20.7021	20.4276	20.1686	20.0418	19.8853	19.7518	19.6107	19.4684	19.3250
5	13.6211	13.3850	13.1473	12.9046	12.7763	12.6609	12.5267	12.4024	12.2737	12.1440
6	10.2488	10.0346	9.8141	9.5873	9.4763	9.3568	9.2436	9.1219	9.0015	8.8793
7	8.3797	8.1758	7.9667	7.7537	7.6447	7.5349	7.4227	7.3101	7.1953	7.0760
8	7.2111	7.0152	6.8150	6.6083	6.5029	6.3973	6.2871	6.1781	6.0649	5.9506
9	6.4175	6.2274	6.0326	5.8321	5.7286	5.6246	5.5182	5.4118	5.3001	5.1875
10	5.8470	5.6617	5.4710	5.2736	5.1727	5.0706	4.9653	4.8587	4.7501	4.6385
11	5.4183	5.2363	5.0490	4.8552	4.7560	4.6547	4.5509	4.4458	4.3367	4.2255
12	5.0857	4.9060	4.7214	4.5297	4.4316	4.3306	4.2286	4.1222	4.0149	3.9039
13	4.8198	4.6430	4.4602	4.2704	4.1723	4.0726	3.9705	3.8654	3.7576	3.6465
14	4.6032	4.4279	4.2470	4.0586	3.9616	3.8619	3.7603	3.6552	3.5478	3.4359
15	4.4235	4.2496	4.0699	3.8824	3.7857	3.6868	3.5847	3.4800	3.3721	3.2602
16	4.2719	4.0993	3.9203	3.7340	3.6376	3.5387	3.4373	3.3326	3.2243	3.1115
17	4.1424	3.9710	3.7928	3.6073	3.5112	3.4122	3.3109	3.2060	3.0972	2.9839
18	4.0305	3.8599	3.6827	3.4976	3.4015	3.3031	3.2015	3.0959	2.9872	2.8732
19	3.9329	3.7631	3.5866	3.4021	3.3060	3.2075	3.1058	3.0004	2.8904	2.7762
20	3.8471	3.6780	3.5020	3.3177	3.2222	3.1235	3.0214	2.9158	2.8059	2.6904
21	3.7708	3.6024	3.4271	3.2432	3.1473	3.0487	2.9467	2.8409	2.7303	2.6140
22	3.7031	3.5349	3.3599	3.1764	3.0807	2.9820	2.8797	2.7735	2.6628	2.5455
23	3.6420	3.4745	3.2998	3.1165	3.0208	2.9222	2.8197	2.7130	2.6019	2.4837
24	3.5870	3.4199	3.2457	3.0624	2.9667	2.8680	2.7653	2.6583	2.5464	2.4276
25	3.5370	3.3705	3.1964	3.0132	2.9175	2.8186	2.7161	2.6088	2.4961	2.3765
26	3.4915	3.3252	3.1516	2.9685	2.8729	2.7739	2.6709	2.5634	2.4502	2.3297
27	3.4499	3.2840	3.1104	2.9275	2.8317	2.7327	2.6296	2.5216	2.4079	2.2867
28	3.4116	3.2461	3.0726	2.8899	2.7942	2.6949	2.5915	2.4833	2.3689	2.2470
29	3.3765	3.2111	3.0378	2.8552	2.7594	2.6599	2.5564	2.4478	2.3330	2.2102
30	3.3439	3.1787	3.0057	2.8231	2.7272	2.6277	2.5241	2.4152	2.2998	2.1760
40	3.1168	2.9532	2.7811	2.5985	2.5021	2.4015	2.2958	2.1838	2.0634	1.9318
60	2.9042	2.7419	2.5704	2.3872	2.2897	2.1874	2.0789	1.9621	1.8342	1.6885
120	2.7052	2.5439	2.3728	2.1882	2.0890	1.9840	1.8709	1.7469	1.6055	1.4311
∞	2.5188	2.3583	2.1868	1.9998	1.8983	1.7891	1.6691	1.5325	1.3637	1.0000

APPENDIX F.8 $F(F)=0.9975$

ν_2 \ ν_1	1	2	3	4	5	6	7	8	9
1	64844.8909	79999.5000	86460.2989	89999.5833	92224.3929	93749.6111	94859.4061	95702.7500	96365.1276
2	398.5006	399.0000	399.1666	399.2498	399.2998	399.3331	399.3570	399.3748	399.3887
3	89.5843	79.9325	76.0558	73.8924	72.5565	71.7558	71.1303	70.5514	70.0220
4	45.6740	38.0000	34.9621	33.3031	32.2677	31.5417	31.0179	30.6039	30.2966
5	31.4067	24.9640	22.4266	21.0503	20.1782	19.5791	19.1421	18.8011	18.5317
6	24.8073	19.1042	16.8661	15.6515	14.8841	14.3334	13.9625	13.6662	13.4294
7	21.1107	15.8871	13.8441	12.7336	12.0305	11.5457	11.1871	10.9133	10.6986
8	18.7797	13.8885	11.9780	10.9409	10.2839	9.8286	9.4923	9.2360	9.0326
9	17.1876	12.5392	10.7263	9.7407	9.1161	8.6832	8.3645	8.1193	7.9244
10	16.0363	11.5723	9.8332	8.8877	8.2878	7.8707	7.5637	7.3276	7.1397
11	15.1674	10.8480	9.1667	8.2524	7.6709	7.2679	6.9697	6.7406	6.5586
12	14.4896	10.2865	8.6514	7.7615	7.1962	6.8030	6.5130	6.2892	6.1113
13	13.9468	9.8392	8.2423	7.3728	6.8200	6.4352	6.1509	5.9319	5.7573
14	13.5026	9.4748	7.9095	7.0569	6.5150	6.1369	5.8578	5.6423	5.4710
15	13.1328	9.1726	7.6344	6.7961	6.2628	5.8911	5.6157	5.4035	5.2348
16	12.8201	8.9179	7.4028	6.5769	6.0510	5.6842	5.4129	5.2035	5.0366
17	12.5525	8.7006	7.2051	6.3900	5.8709	5.5085	5.2404	5.0331	4.8680
18	12.3208	8.5130	7.0353	6.2293	5.7159	5.3572	5.0919	4.8868	4.7228
19	12.1184	8.3494	6.8871	6.0892	5.5808	5.2257	4.9629	4.7594	4.5968
20	11.9401	8.2056	6.7570	5.9663	5.4625	5.1104	4.8496	4.6476	4.4865
21	11.7817	8.0782	6.6418	5.8580	5.3578	5.0085	4.7495	4.5491	4.3890
22	11.6403	7.9646	6.5389	5.7610	5.2646	4.9177	4.6604	4.4611	4.3021
23	11.5131	7.8626	6.4471	5.6745	5.1813	4.8365	4.5807	4.3825	4.2244
24	11.3982	7.7706	6.3639	5.5964	5.1061	4.7634	4.5088	4.3117	4.1543
25	11.2938	7.6871	6.2889	5.5254	5.0381	4.6970	4.4438	4.2476	4.0909
26	11.1987	7.6111	6.2204	5.4610	4.9759	4.6368	4.3847	4.1893	4.0333
27	11.1115	7.5416	6.1577	5.4021	4.9196	4.5816	4.3308	4.1361	3.9807
28	11.0315	7.4778	6.1006	5.3481	4.8677	4.5312	4.2812	4.0873	3.9323
29	10.9576	7.4190	6.0476	5.2987	4.8198	4.4848	4.2358	4.0426	3.8878
30	10.8803	7.3646	5.9987	5.2527	4.7758	4.4420	4.1937	4.0011	3.8468
40	10.4111	6.9857	5.6589	4.9336	4.4695	4.1440	3.9017	3.7134	3.5625
60	9.9616	6.6317	5.3425	4.6373	4.1854	3.8681	3.6314	3.4471	3.2992
120	9.5387	6.3008	5.0479	4.3619	3.9218	3.6122	3.3809	3.2005	3.0553
∞	9.1406	5.9915	4.7734	4.1060	3.6771	3.3749	3.1486	2.9718	2.8292

APPENDIX F

APPENDIX F.8-F(F)=0.9975-Continued.

ν_2 \ ν_1	10	12	15	20	24	30	40	60	120	∞
1	96899.0474	97706.5481	98521.8873	99344.9336	99759.3027	100175.5440	100593.6379	101013.5643	101435.3021	101860.0000
2	399.3998	399.4165	399.4331	399.4498	399.4581	399.4665	399.4748	399.4831	399.4915	399.5000
3	69.7920	69.3211	68.9673	68.3545	67.8584	67.8009	67.5420	67.2811	67.0184	66.7540
4	30.0288	29.6534	29.2790	28.8303	28.6911	28.4694	28.2433	28.0346	27.8240	27.6120
5	18.3268	17.9903	17.6538	17.3200	17.1513	16.9669	16.7832	16.6220	16.4425	16.2610
6	13.2383	12.9488	12.6517	12.3477	12.1951	12.0326	11.8750	11.7201	11.5583	11.3950
7	10.5205	10.2542	9.9803	9.6993	9.5536	9.4114	9.2605	9.1119	8.9641	8.8109
8	8.8670	8.6144	8.3548	8.0877	7.9524	7.8148	7.6746	7.5351	7.3912	7.2447
9	7.7656	7.5241	7.2761	7.0209	6.8906	6.7583	6.6255	6.4854	6.3483	6.2063
10	6.9878	6.7534	6.5140	6.2668	6.1403	6.0124	5.8806	5.7487	5.6114	5.4723
11	6.4097	6.1831	5.9486	5.7083	5.5848	5.4580	5.3299	5.1985	5.0652	4.9280
12	5.9660	5.7441	5.5148	5.2790	5.1570	5.0336	4.9075	4.7771	4.6452	4.5093
13	5.6150	5.3972	5.1721	4.9392	4.8195	4.6970	4.5719	4.4442	4.3138	4.1775
14	5.3313	5.1161	4.8940	4.6646	4.5457	4.4247	4.3012	4.1741	4.0443	3.9084
15	5.0965	4.8843	4.6650	4.4378	4.3204	4.2003	4.0776	3.9505	3.8201	3.6857
16	4.8998	4.6903	4.4732	4.2473	4.1311	4.0116	3.8894	3.7634	3.6339	3.4985
17	4.7327	4.5251	4.3098	4.0860	3.9701	3.8515	3.7293	3.6041	3.4744	3.3388
18	4.5891	4.3828	4.1694	3.9467	3.8314	3.7132	3.5918	3.4661	3.3366	3.2009
19	4.4642	4.2595	4.0473	3.8261	3.7115	3.5936	3.4722	3.3470	3.2167	3.0807
20	4.3545	4.1512	3.9403	3.7201	3.6059	3.4882	3.3671	3.2417	3.1110	2.9749
21	4.2579	4.0557	3.8457	3.6262	3.5123	3.3951	3.2741	3.1485	3.0183	2.8811
22	4.1719	3.9706	3.7616	3.5430	3.4292	3.3123	3.1912	3.0659	2.9346	2.7972
23	4.0947	3.8945	3.6862	3.4682	3.3548	3.2380	3.1171	2.9913	2.8597	2.7218
24	4.0252	3.8258	3.6182	3.4010	3.2877	3.1708	3.0502	2.9244	2.7929	2.6536
25	3.9622	3.7638	3.5569	3.3400	3.2270	3.1103	2.9894	2.8636	2.7313	2.5917
26	3.9052	3.7073	3.5009	3.2845	3.1717	3.0552	2.9341	2.8080	2.6756	2.5351
27	3.8530	3.6557	3.4500	3.2340	3.1211	3.0047	2.8834	2.7571	2.6241	2.4831
28	3.8051	3.6083	3.4033	3.1874	3.0748	2.9583	2.8372	2.7107	2.5774	2.4353
29	3.7612	3.5650	3.3602	3.1448	3.0322	2.9158	2.7944	2.6677	2.5337	2.3912
30	3.7204	3.5247	3.3205	3.1054	2.9929	2.8764	2.7550	2.6281	2.4937	2.3502
40	3.4385	3.2461	3.0444	2.8310	2.7188	2.6020	2.4794	2.3499	2.2114	2.0600
60	3.1773	2.9878	2.7886	2.5760	2.4636	2.3457	2.2209	2.0872	1.9411	1.7756
120	2.9355	2.7486	2.5509	2.3387	2.2252	2.1051	1.9763	1.8356	1.6760	1.4802
∞	2.7112	2.5265	2.3300	2.1168	2.0014	1.8777	1.7425	1.5891	1.4007	1.0000

APPENDIX F.9-F(F)=0.999

ν_2 \ ν_1	1	2	3	4	5	6	7	8	9
1	405284.0679	499999.5000	540379.2017	562499.5833	576404.5558	585937.1111	592873.2879	598144.1563	602283.9916
2	998.5003	999.0000	999.1666	999.2499	999.2999	999.3333	999.3571	999.3749	999.3888
3	167.0292	148.5000	141.0065	137.1238	134.7583	132.9745	131.7449	130.6970	130.0866
4	74.1373	61.2456	56.1733	53.4545	51.7076	50.5233	49.6622	48.9611	48.4676
5	47.1808	37.1223	33.2022	31.0847	29.7464	28.8398	28.1549	27.6525	27.2481
6	35.5075	27.0000	23.7029	21.9252	20.8012	20.0287	19.4625	19.0335	18.6869
7	29.2452	21.6890	18.7718	17.1989	16.2054	15.5219	15.0176	14.6339	14.3319
8	25.4148	18.4937	15.8297	14.3912	13.4842	12.8577	12.3975	12.0459	11.7654
9	22.8571	16.3871	13.9016	12.5600	11.7134	11.1282	10.6976	10.3672	10.1069
10	21.0396	14.9054	12.5524	11.2825	10.4813	9.9254	9.5170	9.2036	8.9556
11	19.6868	13.8116	11.5609	10.3462	9.5779	9.0463	8.6550	8.3547	8.1161
12	18.6433	12.9737	10.8041	9.6326	8.8924	8.3791	8.0006	7.7101	7.4802
13	17.8154	12.3127	10.2090	9.0727	8.3539	7.8559	7.4882	7.2059	6.9817
14	17.1434	11.7789	9.7293	8.6225	7.9219	7.4356	7.0773	6.8015	6.5828
15	16.5874	11.3391	9.3351	8.2526	7.5672	7.0917	6.7406	6.4708	6.2561
16	16.1202	10.9710	9.0059	7.9441	7.2719	6.8048	6.4603	6.1948	5.9840
17	15.7222	10.6584	8.7267	7.6832	7.0218	6.5625	6.2235	5.9621	5.7541
18	15.3793	10.3899	8.4876	7.4594	6.8078	6.3549	6.0207	5.7628	5.5574
19	15.0808	10.1568	8.2797	7.2657	6.6223	6.1755	5.8450	5.5904	5.3875
20	14.8188	9.9526	8.0984	7.0962	6.4604	6.0185	5.6919	5.4401	5.2392
21	14.5869	9.7723	7.9384	6.9467	6.3180	5.8805	5.5573	5.3077	5.1088
22	14.3803	9.6120	7.7962	6.8140	6.1915	5.7581	5.4376	5.1903	4.9930
23	14.1950	9.4685	7.6688	6.6955	6.0783	5.6487	5.3309	5.0853	4.8895
24	14.0280	9.3394	7.5545	6.5892	5.9769	5.5504	5.2350	4.9914	4.7969
25	13.8767	9.2225	7.4511	6.4933	5.8851	5.4617	5.1482	4.9062	4.7130
26	13.7390	9.1163	7.3573	6.4057	5.8018	5.3812	5.0698	4.8290	4.6370
27	13.6131	9.0194	7.2717	6.3260	5.7258	5.3077	4.9983	4.7590	4.5681
28	13.4976	8.9305	7.1931	6.2532	5.6565	5.2407	4.9327	4.6948	4.5048
29	13.3912	8.8488	7.1207	6.1862	5.5926	5.1791	4.8726	4.6358	4.4466
30	13.2930	8.7734	7.0545	6.1244	5.5338	5.1222	4.8173	4.5816	4.3931
40	12.6094	8.2508	6.5945	5.6981	5.1284	4.7306	4.4355	4.2070	4.0243
60	11.9730	7.7678	6.1712	5.3067	4.7665	4.3721	4.0864	3.8648	3.6872
120	11.3802	7.3211	5.7814	4.9472	4.4157	4.0437	3.7670	3.5519	3.3792
∞	10.8280	6.9078	5.4221	4.6167	4.1030	3.7430	3.4746	3.2656	3.0975

APPENDIX F.9-F(F)=0.999-Continued.

ν_2	10	12	15	20	24	30	40	60	120	∞
1	605620.9712	610667.8213	615763.6620	620007.6727	623497.4649	626098.9585	628712.0309	631336.5558	633972.4030	636620.0000
2	999.3999	999.4166	999.4333	999.4499	999.4583	999.4666	999.4749	999.4833	999.4916	999.5000
3	129.4743	128.2520	127.6751	126.6939	126.8828	125.4486	124.9590	124.4658	123.9692	123.4700
4	48.0107	47.4160	46.7758	46.0535	45.8235	45.4966	45.0886	44.7457	44.3998	44.0510
5	26.9095	26.4204	25.9073	25.4101	25.1460	24.8566	24.5869	24.3326	24.0605	23.7850
6	18.4097	17.9869	17.5612	17.1221	16.8970	16.6798	16.4414	16.2143	15.9812	15.7450
7	14.0818	13.7053	13.3252	12.9279	12.7337	12.5257	12.3251	12.1265	11.9090	11.6960
8	11.5405	11.1945	10.8426	10.4796	10.2944	10.1081	9.9175	9.7236	9.5321	9.33337
9	9.8937	9.5706	9.2390	8.8989	8.7255	8.5463	8.3705	8.1844	8.0014	7.8128
10	8.7546	8.4460	8.1286	7.8031	7.6372	7.4700	7.2985	7.1224	6.9443	6.7625
11	7.9219	7.6258	7.3216	7.0074	6.8474	6.6833	6.5169	6.3483	6.1753	5.9983
12	7.2923	7.0044	6.7097	6.4048	6.2486	6.0895	5.9287	5.7619	5.5931	5.4195
13	6.7989	6.5193	6.2308	5.9335	5.7810	5.6264	5.4668	5.3052	5.1371	4.9671
14	6.4042	6.1301	5.8483	5.5564	5.4073	5.2547	5.0977	4.9372	4.7736	4.6042
15	6.0808	5.8118	5.5350	5.2482	5.1009	4.9500	4.7962	4.6371	4.4760	4.3070
16	5.8117	5.5474	5.2742	4.9916	4.8460	4.6971	4.5447	4.3881	4.2259	4.0592
17	5.5844	5.3235	5.0544	4.7754	4.6308	4.4836	4.3325	4.1765	4.0152	3.8496
18	5.3901	5.1525	4.8665	4.5901	4.4471	4.3008	4.1505	3.9960	3.8356	3.6698
19	5.2221	4.9673	4.7055	4.4299	4.2881	4.1427	3.9935	3.8393	3.6799	3.5141
20	5.0752	4.8228	4.5618	4.2901	4.1494	4.0049	3.8566	3.7028	3.5441	3.3778
21	4.9463	4.6959	4.4367	4.1668	4.0272	3.8836	3.7356	3.5827	3.4242	3.2575
22	4.8316	4.5834	4.3263	4.0579	3.9189	3.7757	3.6284	3.4760	3.3170	3.1505
23	4.7296	4.4831	4.2275	3.9605	3.8223	3.6799	3.5328	3.3806	3.2212	3.0548
24	4.6379	4.3930	4.1389	3.8733	3.7354	3.5934	3.4469	3.2945	3.1356	2.9685
25	4.5552	4.3115	4.0588	3.7945	3.6571	3.5154	3.3693	3.2172	3.0584	2.8904
26	4.4801	4.2377	3.9861	3.7230	3.5859	3.4449	3.2985	3.1467	2.9875	2.8193
27	4.4116	4.1705	3.9199	3.6577	3.5212	3.3801	3.2344	3.0825	2.9234	2.7543
28	4.3491	4.1090	3.8594	3.5980	3.4617	3.3213	3.1754	3.0236	2.8644	2.6947
29	4.2918	4.0527	3.8039	3.5432	3.4074	3.2671	3.1215	2.9696	2.8097	2.6397
30	4.2387	4.0007	3.7527	3.4927	3.3573	3.2171	3.0716	2.9196	2.7593	2.5889
40	3.8744	3.6426	3.4003	3.1450	3.0110	2.8720	2.7269	2.5736	2.4102	2.2326
60	3.5414	3.3154	3.0782	2.8265	2.6938	2.5550	2.4086	2.2523	2.0821	1.8905
120	3.2372	3.0162	2.7833	2.5344	2.4019	2.2622	2.1129	1.9502	1.7667	1.5433
∞	2.9558	2.7425	2.5132	2.2657	2.1324	1.9901	1.8350	1.6601	1.4468	1.0000

APPENDIX G-1

Critical values for the Kolmogorov-Smirnov (K-S) goodness-of-fit test.

Sample size N	Level of significance				
	0.20	0.15	0.10	0.05	0.01
1	0.900	0.925	0.950	0.975	0.995
2	0.684	0.726	0.776	0.842	0.929
3	0.565	0.597	0.642	0.708	0.828
4	0.494	0.575	0.564	0.624	0.733
5	0.446	0.424	0.510	0.454	0.669
6	0.410	0.436	0.470	0.521	0.618
7	0.381	0.405	0.438	0.486	0.577
8	0.358	0.381	0.411	0.457	0.543
9	0.339	0.360	0.388	0.432	0.514
10	0.322	0.342	0.368	0.410	0.490
11	0.307	0.326	0.452	0.391	0.468
12	0.295	0.313	0.338	0.375	0.405
13	0.284	0.302	0.325	0.361	0.433
14	0.274	0.292	0.314	0.349	0.418
15	0.266	0.293	0.304	0.338	0.404
16	0.258	0.274	0.295	0.328	0.392
17	0.250	0.266	0.286	0.318	0.381
18	0.244	0.259	0.278	0.309	0.371
19	0.237	0.252	0.272	0.301	0.363
20	0.231	0.246	0.264	0.294	0.356
25	0.21	0.22	0.24	0.27	0.32
30	0.19	0.20	0.22	0.24	0.29
35	0.18	0.19	0.21	0.23	0.27
> 35	$\dfrac{1.07}{\sqrt{N}}$	$\dfrac{1.14}{\sqrt{N}}$	$\dfrac{1.22}{\sqrt{N}}$	$\dfrac{1.36}{\sqrt{N}}$	$\dfrac{1.63}{\sqrt{N}}$

APPENDIX G-2

Critical values for the K-S test [3*, pp 387-389] for the exponential distribution with estimated mean. *See Reference 3 in Chapter 20.

Sample Size N	Level of Significance for $D = \text{Max}\|F^\bullet(X) - S_N(X)\|$				
	.20	.15	.10	.05	.01
3	.451	.479	.511	.551	.600
4	.396	.422	.449	.487	.548
5	.359	.382	.406	.442	.504
6	.331	.351	.375	.408	.470
7	.309	.327	.350	.382	.442
8	.291	.308	.329	.360	.419
9	.277	.291	.311	.341	.399
10	.263	.277	.295	.325	.380
11	.251	.264	.283	.311	.365
12	.241	.254	.271	.298	.351
13	.232	.245	.261	.287	.338
14	.224	.237	.252	.277	.326
15	.217	.229	.244	.269	.315
16	.211	.222	.236	.261	.306
17	.204	.215	.229	.253	.297
18	.199	.210	.223	.246	.289
19	.193	.204	.218	.239	.283
20	.188	.199	.212	.234	.278
25	.170	.180	.191	.210	.247
30	.155	.164	.174	.192	.226
Over 30	$\dfrac{.86}{\sqrt{N}}$	$\dfrac{.91}{\sqrt{N}}$	$\dfrac{.96}{\sqrt{N}}$	$\dfrac{1.06}{\sqrt{N}}$	$\dfrac{1.25}{\sqrt{N}}$

APPENDIX G-3

Critical values for the K-S test [2*, pp 399-402] for the normal distribution with estimated mean and standard deviation. *See Reference 2 in Chapter 20.

Sample Size N	Level of Significance for $D = \text{Max} \, |F^*(X) - S_N(X)|$				
	.20	.15	.10	.05	.01
4	.300	.319	.352	.381	.417
5	.285	.299	.315	.337	.405
6	.265	.277	.294	.319	.364
7	.247	.258	.276	.300	.348
8	.233	.244	.261	.285	.331
9	.223	.233	.249	.271	.311
10	.215	.224	.239	.258	.294
11	.206	.217	.230	.249	.284
12	.199	.212	.223	.242	.275
13	.190	.202	.214	.234	.268
14	.183	.194	.207	.227	.261
15	.177	.187	.201	.220	.257
16	.173	.182	.195	.213	.250
17	.169	.177	.189	.206	.245
18	.166	.173	.184	.200	.239
19	.163	.169	.179	.195	.235
20	.160	.166	.174	.190	.231
25	.149	.153	.165	.180	.203
30	.131	.136	.144	.161	.187
Over 30	$\dfrac{.736}{\sqrt{N}}$	$\dfrac{.768}{\sqrt{N}}$	$\dfrac{.805}{\sqrt{N}}$	$\dfrac{.886}{\sqrt{N}}$	$\dfrac{1.031}{\sqrt{N}}$

APPENDIX G-4 †

Percentage points of the modified K-S test statistics [5††, pp. 730-737] for the exponential and the normal distributions with estimated parameters.

Statistic	Modified form	Percentage points, D_{cr}^*				
	α	0.15	0.10	0.05	0.025	0.01
D_{max}	Exponential distribution with estimated m					
	$D_{max}^* = (D_{max} - 0.2/N)(\sqrt{N} + 0.26 + 0.5/\sqrt{N})$	0.926	0.990	1.094	1.190	1.308
	Normal distribution with estimated μ and σ^2					
	$D_{max}^* = D_{max}(\sqrt{N} - 0.01 + 0.85/\sqrt{N})$	0.775	0.819	0.895	0.955	1.035

† This table can be used as a substitute for APPENDICES G-2 and G-3, respectively, as follows: Substitute the calculated sample D_{max} value into the corresponding modified statistic equation given in the second column of this table and calculate the D_{max}^* value; compare this modified statistic value, D_{max}^*, with the asymptotic critical value, D_{cr}^*, corresponding to the specified significance level, α· Reject the hypothesis, if $D_{max}^* \geq D_{cr}^*$; Do not reject, if $D_{max}^* < D_{cr}^*$.

†† See Reference 5 in Chapter 20.

APPENDIX G-5

Critical values for the K-S test [11*, pp.209-213] for the Weibull distribution with known shape parameters but estimated scale and location parameters. *See Reference 11 in Chapter 20.

APPENDIX G-5.1
Critical Values for Shape Parameter = 0.5

Sample Size N	Level of s-significance				
	0.20	0.15	0.10	0.05	0.01
5	.335	.350	.369	.399	.491
6	.312	.326	.346	.377	.437
7	.296	.313	.333	.361	.413
8	.278	.292	.313	.340	.408
9	.269	.283	.302	.329	.386
10	.252	.265	.282	.310	.370
11	.243	.258	.276	.301	.354
12	.233	.247	.265	.290	.346
13	.228	.240	.255	.280	.331
14	.219	.231	.247	.274	.322
15	.214	.225	.241	.264	.312
20	.186	.197	.211	.233	.276
25	.169	.178	.190	.208	.246
30	.154	.162	.174	.192	.228

APPENDIX G-5.2
Critical Values for Shape Parameter = 1.0

Sample Size N	Level of s-significance				
	0.20	0.15	0.10	0.05	0.01
5	.321	.336	.358	.385	.445
6	.302	.316	.331	.369	.437
7	.284	.299	.322	.358	.413
8	.275	.298	.309	.338	.394
9	.261	.273	.292	.320	.384
10	.249	.262	.279	.308	.353
11	.241	.253	.271	.301	.350
12	.230	.243	.259	.284	.334
13	.222	.234	.250	.277	.327
14	.218	.230	.246	.270	.315
15	.210	.222	.237	.259	.303
20	.187	.197	.211	.231	.273
25	.167	.176	.188	.207	.242
30	.154	.162	.172	.190	.225

APPENDIX G-5.3
Critical Values for Shape Parameter = 1.5

Sample Size N	Level of s-significance				
	0.20	0.15	0.10	0.05	0.01
5	.338	.357	.381	.414	.468
6	.314	.328	.349	.378	.445
7	.298	.311	.329	.364	.420
8	.278	.292	.314	.342	.399
9	.266	.281	.300	.326	.384
10	.253	.267	.284	.313	.367
11	.243	.256	.271	.298	.351
12	.233	.246	.261	.287	.336
13	.225	.237	.252	.279	.325
14	.218	.229	.244	.267	.307
15	.211	.221	.237	.259	.304
20	.182	.193	.205	.225	.265
25	.167	.175	.186	.203	.236
30	.153	.161	.173	.186	.226

APPENDIX G-5.4
Critical Values for Shape Parameter = 2.0

Sample Size N	Level of s-significance				
	0.20	0.15	0.10	0.05	0.01
5	.328	.345	.366	.394	.461
6	.306	.320	.338	.368	.423
7	.285	.298	.315	.343	.391
8	.271	.283	.301	.329	.383
9	.254	.266	.282	.305	.359
10	.244	.256	.271	.297	.345
11	.232	.244	.258	.284	.332
12	.223	.235	.249	.270	.321
13	.216	.227	.242	.263	.306
14	.210	.221	.236	.259	.297
15	.202	.212	.225	.246	.288
20	.177	.185	.197	.216	.254
25	.159	.167	.178	.195	.227
30	.148	.155	.165	.180	.216

APPENDIX G-5.5
Critical Values for Shape Parameter = 2.5

Sample Size N	Level of s-significance				
	0.20	0.15	0.10	0.05	0.01
5	.320	.336	.355	.380	.443
6	.298	.313	.329	.359	.411
7	.279	.292	.310	.339	.393
8	.262	.274	.290	.315	.363
9	.250	.262	.277	.302	.348
10	.237	.250	.265	.287	.336
11	.228	.239	.254	.278	.332
12	.219	.230	.244	.265	.311
13	.211	.222	.234	.254	.299
14	.206	.216	.229	.250	.294
15	.196	.207	.219	.240	.279
20	.174	.188	.193	.211	.247
25	.156	.164	.174	.190	.226
30	.143	.151	.161	.175	.203

APPENDIX G-5.6
Critical Values for Shape Parameter = 3.0

Sample Size N	Level of s-significance				
	0.20	0.15	0.10	0.05	0.01
5	.316	.331	.349	.377	.437
6	.292	.306	.323	.352	.406
7	.274	.288	.304	.328	.383
8	.259	.272	.286	.308	.356
9	.243	.255	.270	.291	.337
10	.233	.244	.259	.281	.325
11	.224	.234	.247	.269	.312
12	.216	.226	.240	.262	.308
13	.208	.220	.233	.252	.304
14	.202	.212	.225	.246	.287
15	.193	.204	.217	.237	.278
20	.170	.177	.188	.206	.242
25	.154	.161	.171	.187	.218
30	.141	.148	.157	.172	.202

APPENDIX G-5.7
Critical Values for Shape Parameter = 3.5

Sample Size N	Level of s-significance				
	0.20	0.15	0.10	0.05	0.01
5	.314	.330	.346	.371	.433
6	.287	.300	.320	.348	.396
7	.268	.282	.298	.324	.373
8	.256	.268	.284	.310	.355
9	.242	.254	.270	.293	.341
10	.232	.243	.257	.280	.319
11	.223	.233	.248	.270	.306
12	.213	.223	.235	.257	.304
13	.204	.215	.230	.250	.291
14	.200	.209	.221	.241	.279
15	.192	.201	.214	.233	.269
20	.169	.177	.188	.204	.243
25	.150	.157	.167	.182	.212
30	.139	.145	.154	.169	.205

APPENDIX G-5.8
Critical Values for Shape Parameter = 4.0

Sample Size N	Level of s-significance				
	0.20	0.15	0.10	0.05	0.01
5	.314	.329	.347	.371	.419
6	.288	.301	.318	.345	.400
7	.269	.283	.299	.326	.374
8	.255	.268	.283	.306	.351
9	.241	.252	.265	.287	.331
10	.229	.240	.255	.278	.325
11	.221	.230	.244	.266	.308
12	.211	.221	.235	.258	.296
13	.203	.214	.225	.246	.286
14	.196	.205	.218	.237	.275
15	.192	.201	.213	.231	.268
20	.168	.176	.187	.203	.239
25	.151	.159	.168	.183	.214
30	.137	.143	.152	.166	.197

APPENDIX G-6

Critical values for the K-S test[12*, pp 241-245] for the gamma distribution with known shape parameter but estimated scale and location parameters. *See Reference 12 in Chapter 20.

α	n	β=.5	1.0	1.5	2.0	2.5	3.0	3.5	4.0
.20	5	.353	.370	.333	.326	.320	.318	.316	.313
	10	.264	.265	.250	.243	.237	.234	.231	.231
	15	.225	.218	.208	.201	.196	.194	.192	.190
	20	.197	.192	.184	.175	.172	.171	.170	.167
	25	.176	.169	.162	.159	.155	.152	.151	.151
	30	.164	.156	.151	.146	.142	.141	.138	.138
.15	5	.368	.385	.351	.342	.337	.334	.332	.329
	10	.280	.279	.262	.257	.248	.245	.242	.242
	15	.237	.231	.218	.211	.207	.204	.200	.199
	20	.208	.204	.194	.184	.180	.179	.178	.175
	25	.187	.179	.171	.167	.164	.160	.159	.157
	30	.174	.165	.158	.154	.148	.148	.145	.145
.10	5	.391	.399	.374	.364	.358	.352	.351	.347
	10	.301	.296	.280	.273	.263	.260	.256	.257
	15	.251	.245	.234	.224	.219	.217	.213	.212
	20	.223	.218	.206	.198	.191	.190	.188	.186
	25	.201	.191	.181	.177	.174	.170	.169	.167
	30	.184	.176	.168	.164	.158	.157	.154	.154
.05	5	.433	.433	.415	.393	.390	.380	.377	.373
	10	.333	.327	.305	.297	.288	.283	.281	.279
	15	.280	.271	.255	.246	.241	.235	.233	.231
	20	.246	.239	.224	.216	.210	.207	.206	.200
	25	.223	.211	.198	.194	.189	.186	.183	.182
	30	.204	.196	.184	.179	.173	.171	.168	.169
.01	5	.525	.527	.473	.448	.440	.437	.431	.427
	10	.386	.393	.360	.338	.338	.328	.323	.327
	15	.332	.321	.300	.285	.281	.275	.272	.267
	20	.294	.277	.266	.249	.246	.240	.240	.231
	25	.262	.252	.232	.226	.221	.213	.213	.214
	30	.245	.230	.222	.210	.202	.203	.199	.197

Two-sided tolerance factors for the normal distribution.

Factors K such that the Probability is γ that at least a Proportion $1 - \alpha$ of the Distribution Will Be Included between $\bar{X} \pm KS$, where $\bar{X}$ and S Are Estimators of the Mean and the Standard Deviation Computed from a Random Sample of Size n.

n	$\gamma = 0.75$					$\gamma = 0.90$				
	0.25	0.10	0.05	0.01	0.001	0.25	0.10	0.05	0.01	0.001
2	4.498	6.301	7.414	9.531	11.920	11.407	15.978	18.800	24.167	30.227
3	2.501	3.538	4.187	5.431	6.844	4.132	5.847	6.919	8.974	11.309
4	2.035	2.892	3.431	4.471	5.657	2.932	4.166	4.943	6.440	8.149
5	1.825	2.599	3.088	4.033	5.117	2.454	3.494	4.152	5.423	6.879
6	1.704	2.429	2.889	3.779	4.802	2.196	3.131	3.723	4.870	6.188
7	1.624	2.318	2.757	3.611	4.593	2.034	2.902	3.452	4.521	5.750
8	1.568	2.238	2.663	3.491	4.444	1.921	2.743	3.264	4.278	5.446
9	1.525	2.178	2.593	3.400	4.330	1.839	2.626	3.125	4.098	5.220
10	1.492	2.131	2.537	3.328	4.241	1.775	2.535	3.018	3.959	5.046
11	1.465	2.093	2.493	3.271	4.169	1.724	2.463	2.933	3.849	4.906
12	1.443	2.062	2.456	3.223	4.110	1.683	2.404	2.863	3.758	4.792
13	1.425	2.036	2.424	3.183	4.059	1.648	2.355	2.805	3.682	4.697
14	1.409	2.013	2.398	3.148	4.016	1.619	2.314	2.756	3.618	4.615
15	1.395	1.994	2.375	3.118	3.979	1.594	2.278	2.713	3.562	4.545
16	1.383	1.977	2.355	3.092	3.946	1.572	2.246	2.676	3.514	4.484
17	1.372	1.962	2.337	3.069	3.917	1.552	2.219	2.643	3.471	4.430
18	1.363	1.948	2.321	3.048	3.891	1.535	2.194	2.614	3.433	4.382
19	1.355	1.936	2.307	3.030	3.867	1.520	2.172	2.588	3.399	4.339
20	1.347	1.925	2.294	3.013	3.846	1.506	2.152	2.564	3.368	4.300
21	1.340	1.915	2.282	2.998	3.827	1.493	2.135	2.543	3.340	4.264
22	1.334	1.906	2.271	2.984	3.809	1.482	2.118	2.524	3.315	4.232
23	1.328	1.898	2.261	2.971	3.793	1.471	2.103	2.506	3.292	4.203
24	1.322	1.891	2.252	2.959	3.778	1.462	2.089	2.489	3.270	4.176
25	1.317	1.883	2.244	2.948	3.764	1.453	2.077	2.474	3.251	4.151
26	1.313	1.877	2.236	2.938	3.751	1.444	2.065	2.460	3.232	4.127
27	1.309	1.871	2.229	2.929	3.740	1.437	2.054	2.447	3.215	4.106
28	1.305	1.865	2.222	2.920	3.728	1.430	2.044	2.435	3.199	4.085
29	1.301	1.860	2.216	2.911	3.718	1.423	2.034	2.424	3.184	4.066
30	1.297	1.855	2.210	2.904	3.708	1.417	2.025	2.413	3.170	4.049
31	1.294	1.850	2.204	2.896	3.699	1.411	2.017	2.403	3.157	4.032
32	1.291	1.846	2.199	2.890	3.690	1.405	2.009	2.393	3.145	4.016
33	1.288	1.842	2.194	2.883	3.682	1.400	2.001	2.385	3.133	4.001
34	1.285	1.838	2.189	2.877	3.674	1.395	1.994	2.376	3.122	3.987
35	1.233	1.834	2.185	2.871	3.667	1.390	1.988	2.368	3.112	3.974
36	1.280	1.830	2.181	2.866	3.660	1.386	1.981	2.361	3.102	3.961
37	1.278	1.827	2.177	2.860	3.653	1.381	1.975	2.353	3.092	3.949
38	1.275	1.824	2.173	2.855	3.647	1.377	1.969	2.346	3.083	3.938
39	1.273	1.821	2.169	2.850	3.641	1.374	1.964	2.340	3.075	3.927
40	1.271	1.818	2.166	2.846	3.635	1.370	1.959	2.334	3.066	3.917
41	1.269	1.815	2.162	2.841	3.629	1.366	1.954	2.328	3.059	3.907
42	1.267	1.812	2.159	2.837	3.624	1.363	1.949	2.322	3.051	3.897
43	1.266	1.810	2.156	2.833	3.619	1.360	1.944	2.316	3.044	3.888
44	1.264	1.807	2.153	2.829	3.614	1.357	1.940	2.311	3.037	3.879
45	1.262	1.805	2.150	2.826	3.609	1.354	1.935	2.306	3.030	3.871
46	1.261	1.802	2.148	2.822	3.605	1.351	1.931	2.301	3.024	3.863
47	1.259	1.800	2.145	2.819	3.600	1.348	1.927	2.297	3.018	3.855
48	1.258	1.798	2.143	2.815	3.596	1.345	1.924	2.292	3.012	3.847
49	1.256	1.796	2.140	2.812	3.592	1.343	1.920	2.288	3.006	3.840
50	1.255	1.794	2.138	2.809	3.588	1.340	1.916	2.284	3.001	3.833

[1] Reproduced from C. Eisenhart, M. W. Hastay, and W. A. Wallis, *Techniques of Statistical Analysis*, Chapter 2, McGraw-Hill Book Company, Inc., New York, 1947.

n	α		$\gamma = 0.95$					$\gamma = 0.99$		
	0.25	0.10	0.05	0.01	0.001	0.25	0.10	0.05	0.01	0.001
2	22.858	32.019	37.674	48.430	60.573	114.363	160.193	188.491	242.300	303.054
3	5.922	8.380	9.916	12.861	16.208	13.378	18.930	22.401	29.055	36.616
4	3.779	5.369	6.370	8.299	10.502	6.614	9.398	11.150	14.527	18.383
5	3.002	4.275	5.079	6.634	8.415	4.643	6.612	7.855	10.260	13.015
6	2.604	3.712	4.414	5.775	7.337	3.743	5.337	6.345	8.301	10.548
7	2.361	3.369	4.007	5.248	6.676	3.233	4.613	5.488	7.187	9.142
8	2.197	3.136	3.732	4.891	6.226	2.905	4.147	4.936	6.468	8.234
9	2.078	2.967	3.532	4.631	5.899	2.677	3.822	4.550	5.966	7.600
10	1.987	2.839	3.379	4.433	5.649	2.508	3.582	4.265	5.594	7.129
11	1.916	2.737	3.259	4.277	5.452	2.378	3.397	4.045	5.308	6.766
12	1.858	2.655	3.162	4.150	5.291	2.274	3.250	3.870	5.079	6.477
13	1.810	2.587	3.081	4.044	5.158	2.190	3.130	3.727	4.893	6.240
14	1.770	2.529	3.012	3.955	5.045	2.120	3.029	3.608	4.737	6.043
15	1.735	2.480	2.954	3.878	4.949	2.060	2.945	3.507	4.605	5.876
16	1.705	2.437	2.903	3.812	4.865	2.009	2.872	3.421	4.492	5.732
17	1.679	2.400	2.858	3.754	4.791	1.965	2.808	3.345	4.393	5.607
18	1.655	2.366	2.819	3.702	4.725	1.926	2.753	3.279	4.307	5.497
19	1.635	2.337	2.784	3.656	4.667	1.891	2.703	3.221	4.230	5.399
20	1.616	2.310	2.752	3.615	4.614	1.860	2.659	3.168	4.161	5.312
21	1.599	2.286	2.723	3.577	4.567	1.833	2.620	3.121	4.100	5.234
22	1.584	2.264	2.697	3.543	4.523	1.808	2.584	3.078	4.044	5.163
23	1.570	2.244	2.673	3.512	4.484	1.785	2.551	3.040	3.993	5.098
24	1.557	2.225	2.651	3.483	4.447	1.764	2.522	3.004	3.947	5.039
25	1.545	2.208	2.631	3.457	4.413	1.745	2.494	2.972	3.904	4.985
26	1.534	2.193	2.612	3.432	4.382	1.727	2.469	2.941	3.865	4.935
27	1.523	2.178	2.595	3.409	4.353	1.711	2.446	2.914	3.828	4.888
28	1.514	2.164	2.579	3.388	4.326	1.695	2.424	2.888	3.794	4.845
29	1.505	2.152	2.554	3.368	4.301	1.681	2.404	2.864	3.763	4.805
30	1.497	2.140	2.549	3.350	4.278	1.668	2.385	2.841	3.733	4.768
31	1.489	2.129	2.536	3.332	4.256	1.656	2.367	2.820	3.706	4.732
32	1.481	2.118	2.524	3.316	4.235	1.644	2.351	2.801	3.680	4.699
33	1.475	2.108	2.512	3.300	4.215	1.633	2.335	2.782	3.655	4.668
34	1.468	2.099	2.501	3.286	4.197	1.623	2.320	2.764	3.632	4.639
35	1.462	2.090	2.490	3.272	4.179	1.613	2.306	2.748	3.611	4.611
36	1.455	2.081	2.479	3.258	4.161	1.604	2.293	2.732	3.590	4.585
37	1.450	2.073	2.470	3.246	4.146	1.595	2.281	2.717	3.571	4.560
38	1.446	2.068	2.464	3.237	4.134	1.587	2.269	2.703	3.552	4.537
39	1.441	2.060	2.455	3.226	4.120	1.579	2.257	2.690	3.534	4.514
40	1.435	2.052	2.445	3.213	4.104	1.571	2.247	2.677	3.518	4.493
41	1.430	2.045	2.437	3.202	4.090	1.564	2.236	2.665	3.502	4.472
42	1.426	2.039	2.429	3.192	4.077	1.557	2.227	2.653	3.486	4.453
43	1.422	2.033	2.422	3.183	4.065	1.551	2.217	2.642	3.472	4.434
44	1.418	2.027	2.415	3.173	4.053	1.545	2.208	2.631	3.458	4.416
45	1.414	2.021	2.408	3.165	4.042	1.539	2.200	2.621	3.444	4.399
46	1.410	2.016	2.402	3.156	4.031	1.533	2.192	2.611	3.431	4.383
47	1.406	2.011	2.396	3.148	4.021	1.527	2.184	2.602	3.419	4.367
48	1.403	2.006	2.390	3.140	4.011	1.522	2.176	2.593	3.407	4.352
49	1.399	2.001	2.384	3.133	4.002	1.517	2.169	2.584	3.396	4.337
50	1.396	1.969	2.379	3.126	3.993	1.512	2.162	2.576	3.385	4.323

APPENDIX I

One-sided tolerance factors for the normal distribution.

n	$\gamma = 0.75$					$\gamma = 0.90$				
α	0.25	0.10	0.05	0.01	0.001	0.25	0.10	0.05	0.01	0.001
3	1.464	2.501	3.152	4.396	5.805	2.602	4.258	5.310	7.340	9.651
4	1.256	2.134	2.680	3.726	4.910	1.972	3.187	3.957	5.437	7.128
5	1.152	1.961	2.463	3.421	4.507	1.698	2.742	3.400	4.666	6.112
6	1.087	1.860	2.336	3.243	4.273	1.540	2.494	3.091	4.242	5.556
7	1.043	1.791	2.250	3.126	4.118	1.435	2.333	2.894	3.972	5.201
8	1.010	1.740	2.190	3.042	4.008	1.360	2.219	2.755	3.783	4.955
9	0.984	1.702	2.141	2.977	3.924	1.302	2.133	2.649	3.641	4.772
10	0.964	1.671	2.103	2.927	3.858	1.257	2.065	2.568	3.532	4.629
11	0.947	1.646	2.073	2.885	3.804	1.219	2.012	2.503	3.444	4.515
12	0.933	1.624	2.048	2.851	3.760	1.188	1.966	2.448	3.371	4.420
13	0.919	1.606	2.026	2.822	3.722	1.162	1.928	2.403	3.310	4.341
14	0.909	1.591	2.007	2.796	3.690	1.139	1.895	2.363	3.257	4.274
15	0.899	1.577	1.991	2.776	3.661	1.119	1.866	2.329	3.212	4.215
16	0.891	1.566	1.977	2.756	3.637	1.101	1.842	2.299	3.172	4.164
17	0.883	1.554	1.964	2.739	3.615	1.085	1.820	2.272	3.136	4.118
18	0.876	1.544	1.951	2.723	3.595	1.071	1.800	2.249	3.106	4.078
19	0.870	1.536	1.942	2.710	3.577	1.058	1.781	2.228	3.078	4.041
20	0.865	1.528	1.933	2.697	3.561	1.046	1.765	2.208	3.052	4.009
21	0.859	1.520	1.923	2.686	3.545	1.035	1.750	2.190	3.028	3.979
22	0.854	1.514	1.916	2.675	3.532	1.025	1.736	2.174	3.007	3.952
23	0.849	1.508	1.907	2.665	3.520	1.016	1.724	2.159	2.987	3.927
24	0.845	1.502	1.901	2.656	3.509	1.007	1.712	2.145	2.969	3.904
25	0.842	1.496	1.895	2.647	3.497	0.999	1.702	2.132	2.952	3.882
30	0.825	1.475	1.869	2.613	3.454	0.966	1.657	2.080	2.884	3.794
35	0.812	1.458	1.849	2.588	3.421	0.942	1.623	2.041	2.833	3.730
40	0.803	1.445	1.834	2.568	3.395	0.923	1.598	2.010	2.793	3.679
45	0.795	1.435	1.821	2.552	3.375	0.908	1.577	1.986	2.762	3.638
50	0.788	1.426	1.8 11	2.538	3.358	0.894	1.560	1.965	2.735	3.604

n	$\gamma = 0.95$					$\gamma = 0.99$				
α	0.25	0.10	0.05	0.01	0.001	0.25	0.10	0.05	0.01	0.001
3	3.804	6.158	7.655	10.552	13.857					
4	2.619	4.163	5.145	7.042	9.215					
5	2.149	3.407	4.202	5.741	7.501					
6	1.895	3.006	3.707	5.062	6.612	2.849	4.408	5.409	7.334	9.540
7	1.732	2.755	3.399	4.641	6.061	2.490	3.856	4.730	6.411	8.348
8	1.617	2.582	3.188	4.353	5.686	2.252	3.496	4.287	5.811	7.566
9	1.532	2.454	3.031	4.143	5.414	2.085	3.242	3.971	5.389	7.014
10	1.465	2.355	2.911	3.981	5.203	1.954	3.048	3.739	5.075	6.603
11	1.411	2.275	2.815	3.852	5.036	1.854	2.897	3.557	4.828	6.284
12	1.366	2.210	2.736	3.747	4.900	1.771	2.773	3.410	4.633	6.032
13	1.329	2.155	2.670	3.659	4.787	1.702	2.677	3.290	4.472	5.826
14	1.296	2.108	2.614	3.585	4.690	1.645	2.592	3.189	4.336	5.651
15	1.268	2.068	2.566	3.520	4.607	1.596	2.521	3.102	4.224	5.507
16	1.242	2.032	2.523	3.463	4.534	1.553	2.458	3.028	4.124	5.374
17	1.220	2.001	2.486	3.415	4.471	1.514	2.405	2.962	4.038	5.268
18	1.200	1.974	2.453	3.370	4.415	1.481	2.357	2.906	3.961	5.167
19	1.183	1.949	2.423	3.331	4.364	1.450	2.315	2.855	3.893	5.078
20	1.167	1.926	2.396	3.295	4.319	1.424	2.275	2.807	3.832	5.003
21	1.152	1.905	2.371	3.262	4.276	1.397	2.241	2.768	3.776	4.932
22	1.138	1.887	2.350	3.233	4.238	1.376	2.208	2.729	3.727	4.866
23	1.126	1.869	2.329	3.206	4.204	1.355	2.179	2.693	3.680	4.806
24	1.114	1.853	2.309	3.181	4.171	1.336	2.154	2.663	3.638	4.755
25	1.103	1.838	2.292	3.158	4.143	1.319	2.129	2.632	3.601	4.706
30	1.059	1.778	2.220	3.064	4.022	1.249	2.029	2.516	3.446	4.508
35	1.025	1.732	2.166	2.994	3.934	1.195	1.957	2.431	3.334	4.364
40	0.999	1.697	2.126	2.941	3.866	1.154	1.902	2.365	3.250	4.255
45	0.978	1.669	2.092	2.897	3.811	1.122	1.857	2.313	3.181	4.168
50	0.961	1.646	2.065	2.863	3.766	1.096	1.821	2.296	3.124	4.096

One-Sided Tolerance Factors for Normal Population. Probability of Inclusion= 0.50 = γ. Proportion of the Normal Population Included= $1 - \alpha = R_{L1}(T)$.

Sample Size N	$1-\alpha$								
	0.85	0.90	0.95	0.990	0.995	0.999	0.9995	0.99995	0.999995
5	1.1145	1.3818	1.7793	2.5257	2.7989	3.3622	3.5814	4.2381	4.8139
10	1.0697	1.3240	1.7016	2.4103	2.6698	3.2047	3.4130	4.0367	4.5841
15	1.0575	1.3085	1.6809	2.3795	2.6352	3.1626	3.3681	3.9831	4.5231
20	1.0519	1.3012	1.6712	2.3652	2.6193	3.1432	3.3473	3.9582	4.4948
25	1.0486	1.2971	1.6656	2.3570	2.6100	3.1320	3.3351	3.9439	4.4784
30	1.0465	1.2944	1.6620	2.3516	2.6040	3.1246	3.3274	3.9347	4.4678
1000	1.0364	1.2815	1.6449	2.3264	2.5758	3.0903	3.2906	3.8927	4.4175

N	0.9_65	0.9_75	0.9_85	0.9_95	$0.9_{10}5$	$0.9_{11}5$	$0.9_{12}5$	$0.9_{13}5$	$0.9_{14}5$
5	5.3329	5.8075	6.2498	6.6632	7.0547	7.4259	7.7783	8.1193	8.4461
10	5.0776	5.5289	5.9490	6.3430	6.7148	7.0674	7.4046	7.7270	8.0375
15	5.0094	5.4551	5.8692	6.2577	6.6237	6.9718	7.3028	7.6224	7.9286
20	4.9781	5.4205	5.8322	6.2169	6.5817	6.9278	7.2568	7.5742	7.8770
25	4.9597	5.4002	5.8103	6.1942	6.5571	6.9009	7.2308	7.5459	7.8473
30	4.9476	5.3876	5.7960	6.1794	6.5417	6.8859	7.2132	7.5176	7.8176
1000	4.8919	5.3265	5.7306	6.1093	6.4669	6.8065	7.1305	7.4409	7.7392

APPENDIX K

Coefficients of the BLUEs for the extreme value distribution with parameters μ and σ.

From John S. White, *Industrial Mathematics*, Vol. 14, Part 1, pp. 21-60, 1964. By permission of J.S. White, who developed the table and R. Schmidt, editor of *Industrial Mathematics*, where the table first appeared.

	i	=	failure order number.
Where:	r	=	number of test failures.
	n	=	test sample size.

i	r	N	b(i;N;r)	c(i;N;r)
1	2	2	0.0836269	-0.7213475
2	2	2	0.9163731	0.7213475
1	2	3	-0.3777001	-0.8221011
2	2	3	1.3777000	0.8221011
1	3	3	0.0879664	-0.3747251
2	3	3	0.2557135	-0.2558163
3	3	3	0.6563201	0.6305411
1	2	4	-0.7063194	-0.8690149
2	2	4	1.7063194	0.8690149
1	3	4	-0.0801058	-0.4143997
2	3	4	0.0604318	-0.3258576
3	3	4	1.0196739	0.7402573
1	4	4	0.0713800	-0.2487965
2	4	4	0.1536799	-0.2239193
3	4	4	0.2639426	-0.0859033
4	4	4	0.5109975	0.5586191
1	2	5	-0.9598627	-0.8962840
2	2	5	1.9598626	0.8962840
1	3	5	-0.2101147	-0.4343423
2	3	5	-0.0860216	-0.3642452
3	3	5	1.2961362	0.7985875
1	4	5	-0.0153827	-0.2730535
2	4	5	0.0519625	-0.2499443
3	4	5	0.1520766	-0.1491092
4	4	5	0.8113436	0.6720870
1	5	5	0.0583501	-0.1844827
2	5	5	0.1088237	-0.1816557
3	5	5	0.1676093	-0.1304549
4	5	5	0.2462824	-0.0065338
5	5	5	0.4189344	0.5031271
1	2	6	-1.1655650	-0.9141358
2	2	6	2.1655650	0.9141358
1	3	6	-0.3153967	-0.4466058
2	3	6	-0.2034317	-0.3886493
3	3	6	1.5188283	0.8352511
1	4	6	-0.0865380	-0.2858649
2	4	6	-0.0280567	-0.2654763
3	4	6	0.0649474	-0.1858689
4	4	6	1.0496472	0.7372101

i	r	N	b(i;N;r)	c(i;N;r)
1	5	6	0.0057312	-0.2015427
2	5	6	0.0465760	-0.1972715
3	5	6	0.1002434	-0.1536128
4	5	6	0.1722854	-0.0645867
5	5	6	0.6751639	0.6170138
1	6	6	0.0488670	-0.1458073
2	6	6	0.0835217	-0.1495343
3	6	6	0.1210537	-0.1267241
4	6	6	0.1656198	-0.0731993
5	6	6	0.2254881	0.0359918
6	6	6	0.3554496	0.4592732
1	2	7	-1.3382740	-0.9267370
2	2	7	2.3332740	0.9267370
1	3	7	-0.4036103	-0.4549547
2	3	7	-0.3012125	-0.4055742
3	3	7	1.7048228	0.8605289
1	4	7	-0.1463256	-0.2940405
2	4	7	-0.0940681	-0.2760193
3	4	7	-0.0070967	-0.2101609
4	4	7	1.2474903	0.7802208
1	5	7	-0.0392570	-0.2110152
2	5	7	-0.0043641	-0.2064593
3	5	7	0.0458324	-0.1691176
4	5	7	0.1134204	-0.0991823
5	5	7	0.8843683	0.6857749
1	6	7	0.0137303	-0.1586850
2	6	7	0.0417973	-0.1608703
3	6	7	0.0756807	-0.1396394
4	6	7	0.1175849	-0.0950700
5	6	7	0.1721153	-0.0176462
6	6	7	0.5790915	0.5719108
1	7	7	0.0418411	-0.1201405
2	7	7	0.0673314	-0.1258588
3	7	7	0.0957470	-0.1148675
4	7	7	0.1232231	-0.0873391
5	7	7	0.1585900	-0.0361916
6	7	7	0.2062600	0.0606980
7	7	7	0.3090075	0.4236996
1	2	8	-1.4869219	-0.9361095

i	r	N	b(i;N;r)	c(i;N;r)
2	2	8	2.4869219	0.9361095
1	3	8	-0.4793981	-0.4610279
2	3	8	-0.3848212	-0.4180145
3	3	8	1.8642193	0.8790424
1	4	8	-0.1977185	-0.2997588
2	4	8	-0.1502017	-0.2836886
3	4	8	-0.0684910	-0.2274858
4	4	8	1.4164112	0.8109332
1	5	8	-0.0781440	-0.2172447
2	5	8	-0.0474155	-0.2127596
3	5	8	-0.0000857	-0.1802817
4	5	8	0.0637119	-0.1225159
5	5	8	1.0619333	0.7328019
1	6	8	-0.0172425	-0.1661332
2	6	8	0.0065283	-0.1674873
3	6	8	0.0380245	-0.1482977
4	6	8	0.0779914	-0.1105318
5	6	8	0.1292040	-0.0499901
6	6	8	0.7654942	0.6424402
1	7	8	0.0168081	-0.1302929
2	7	8	0.0375935	-0.1347891
3	7	8	0.0612335	-0.1238688
4	7	8	0.0888741	-0.0990771
5	7	8	0.1224279	-0.0571224
6	7	8	0.1654593	0.0108656
7	7	8	0.5076035	0.5342847
1	8	8	0.0364852	-0.1019365
2	8	8	0.0561317	-0.1080740
3	8	8	0.0759037	-0.1027278
4	8	8	0.0971419	-0.0871624
5	8	8	0.1211747	-0.0589284
6	8	8	0.1501999	-0.0111245
7	8	8	0.1894276	0.0757666
8	8	8	0.2735352	0.3941872
1	2	9	-1.6172796	-0.9433541
2	2	9	2.6172795	0.9433541
1	3	9	-0.5457604	-0.4656486
2	3	9	-0.4577309	-0.4275495
3	3	9	2.0034912	0.8931981

i	r	N	b(i;N;r)	c(i;N;r)
1	4	9	-0.2427102	-0.3140002
2	4	9	-0.9898152	-0.2895314
3	4	9	-0.1218960	-0.2404933
4	4	9	1.5635877	0.8340249
1	5	9	-0.1122653	-0.2217028
2	5	9	-0.0846628	-0.2174080
3	5	9	-0.0398298	-0.1887179
4	5	9	0.0206119	-0.1394352
5	5	9	1.2161461	0.7672640
1	6	9	-0.0446124	-0.1711693
2	6	9	-0.0239090	-0.1720278
3	6	9	0.0057259	-0.1546900
4	6	9	0.0439637	-0.1219925
5	6	9	0.0924579	-0.0720775
6	6	9	0.9263739	0.6919572
1	7	9	-0.0057741	-0.1363684
2	7	9	0.0117905	-0.1400394
3	7	9	0.0335898	-0.1297226
4	7	9	0.0599713	-0.1076490
5	7	9	0.0921887	-0.0723188
6	7	9	0.1325194	-0.0193719
7	7	9	0.6757144	0.6054702
1	8	9	0.0177962	-0.1101884
2	8	9	0.0339652	-0.1154096
3	8	9	0.0515769	-0.1097441
4	8	9	0.0713590	-0.0950005
5	8	9	0.0942696	-0.0700075
6	8	9	0.1218432	-0.0312302
7	8	9	0.1569098	0.0292244
8	8	9	0.4522801	0.5023558
1	9	9	0.0322910	-0.0883912
2	9	9	0.0479566	-0.0945693
3	9	9	0.0633995	-0.0919651
4	9	9	0.0795686	-0.0826548
5	9	9	0.0972179	-0.0655738
6	9	9	0.1173565	-0.0379772
7	9	9	0.1417891	0.0064858
8	9	9	0.1748819	0.0852032
9	9	9	0.2455388	0.3692424

i	r	N	b(i;N;r)	c(i;N;r)
1	2	10	-1.7332827	-0.9491221
2	2	10	2.7332827	0.9491221
1	3	10	-0.6047400	-0.4692847
2	3	10	-0.5222903	-0.4350929
3	3	10	2.1270303	0.9043776
1	4	10	-0.2826778	-0.3072783
2	4	10	-0.2420734	-0.2941358
3	4	10	-0.1630791	-0.2506307
4	4	10	1.6933303	0.8520448
1	5	10	-0.1426067	-0.2250678
2	5	10	-0.1174612	-0.2209984
3	5	10	-0.0748846	-0.1953226
4	5	10	-0.0174033	-0.1523120
5	5	10	1.3523158	0.7937008
1	6	10	-0.0690324	-0.1748478
2	6	10	-0.0506448	-0.1753912
3	6	10	-0.0225702	-0.1596414
4	5	10	0.0140717	-0.1308280
5	6	10	0.0601887	-0.0882729
6	6	10	1.0679869	0.7289813
1	7	10	-0.0260821	-0.1405755
2	7	10	-0.0108783	-0.1436594
3	7	10	0.0095210	-0.1340341
4	7	10	0.0348369	-0.1142583
5	7	10	0.0658392	-0.0837641
6	7	10	0.1040540	-0.0401921
7	7	10	0.8227093	0.6564835
1	8	10	0.0006317	-0.1152690
2	8	10	0.0143203	-0.1197883
3	8	10	0.0304554	-0.1142025
4	8	10	0.0492565	-0.1005933
5	8	10	0.0713719	-0.0785228
6	8	10	0.0978952	-0.0460265
7	8	10	0.1306530	0.0008844
8	8	10	0.6054161	0.5735230
1	9	10	0.0178349	-0.0952545
2	9	10	0.0308561	-0.1005502
3	9	10	0.0445900	-0.0977581
4	9	10	0.0595653	-0.0836049

i	r	N	b(i;N;r)	c(i;N;r)
5	9	10	0.0763356	-0.0727481
6	9	10	0.0956763	-0.0486079
7	9	10	0.1188411	-0.0128578
8	9	10	0.1481415	0.0415222
9	9	10	0.4081592	0.4748591
1	10	10	0.0289290	-0.0779399
2	10	10	0.0417478	-0.0835515
3	10	10	0.0541930	-0.0827706
4	10	10	0.0669876	-0.0770208
5	10	10	0.0806178	-0.0660648
6	10	10	0.0956359	-0.0486710
7	10	10	0.1128684	-0.0221794
8	10	10	0.1338452	0.0192100
9	10	10	0.1623082	0.0911583
10	10	10	0.2228670	0.3478297
1	2	11	-1.8377307	-0.9538235
2	2	11	2.8377307	0.9538235
1	3	11	-0.6577871	-0.4722220
2	3	11	-0.5801607	-0.4412107
3	3	11	2.2379478	0.9134327
1	4	11	-0.3186053	-0.3098909
2	4	11	-0.2806371	-0.2978600
3	4	11	-0.2112831	-0.2587592
4	4	11	1.8105255	0.8665101
1	5	11	-0.1698921	-0.2277048
2	5	11	-0.1467420	-0.2238631
3	5	11	-0.1061141	-0.2006363
4	5	11	-0.0513661	-0.1624608
5	5	11	1.4741115	0.8146650
1	6	11	-0.0910322	-0.1776687
2	6	11	-0.0744627	-0.1780023
3	6	11	-0.0477433	-0.1636020
4	6	11	-0.0125799	-0.1378512
5	6	11	0.0314153	-0.1007180
6	6	11	1.1944028	0.7578422
1	7	11	-0.0444511	-0.1437042
2	7	11	-0.0310560	-0.1463524
3	7	11	-0.0117979	-0.1373925
4	7	11	0.0125571	-0.1195226

i	r	N	b(i;N;r)	c(i;N;r)
5	7	11	0.0424182	-0.0926953
6	7	11	0.0788225	-0.0555804
7	7	11	0.9535072	0.6952474
1	8	11	-0.0150243	-0.1188596
2	8	11	-0.0032013	-0.1228351
3	8	11	0.0118120	-0.1174590
4	8	11	0.0298233	-0.1049450
5	8	11	0.0512298	-0.0852557
6	8	11	0.0768116	-0.0572781
7	8	11	0.1078325	-0.0187425
8	8	11	0.7407162	0.6253750
1	9	11	0.0043670	-0.0995823
2	9	11	0.0154366	-0.1043068
3	9	11	0.0279638	-0.1014022
4	9	11	0.0421240	-0.0927166
5	9	11	0.0582942	-0.0782329
6	9	11	0.0770601	-0.0570311
7	9	11	0.0993329	-0.0271922
8	9	11	0.1266006	0.0148710
9	9	11	0.5488207	0.5455931
1	10	11	0.0174317	-0.0837528
2	10	11	0.0281914	-0.0888527
3	10	11	0.0392646	-0.0877099
4	10	11	0.0510690	-0.0818786
5	10	11	0.0639625	-0.0713651
6	10	11	0.0783871	-0.0554233
7	10	11	0.0949853	-0.0324598
8	10	11	0.1148044	0.0005783
9	10	11	0.1397781	0.0499838
10	10	11	0.3721260	0.4508800
1	11	11	0.0261799	-0.0596435
2	11	11	0.0368858	-0.0748304
3	11	11	0.0471592	-0.0749774
4	11	11	0.0575780	-0.0713809
5	11	11	0.0684852	-0.0640708
6	11	11	0.0802218	-0.0524642
7	11	11	0.0932342	-0.0352839
8	11	11	0.1082257	-0.0100318
9	11	11	0.1265219	0.0286042
10	11	11	0.1513848	0.0948691
11	11	11	0.2041233	0.3292097
1	2	12	-1.9326838	-0.9577292
2	2	12	2.9326838	0.9577292
1	3	12	-0.7059671	-0.4746448
2	3	12	-0.6325588	-0.4462730
3	3	12	2.3385259	0.9209178
1	4	12	-0.3512178	-0.3120236
2	4	12	-0.3155127	-0.3009354
3	4	12	-0.2494171	-0.2654253
4	4	12	1.9161476	0.8783843
1	5	12	-0.1946622	-0.2298303
2	5	12	-0.1731727	-0.2262054
3	5	12	-0.1343335	-0.2050051
4	5	12	-0.0820210	-0.1706756
5	5	12	1.5841895	0.8317164
1	6	12	-0.1110243	-0.1799074
2	6	12	-0.0959246	-0.1800965
3	6	12	-0.0704054	-0.1668469
4	6	12	-0.0366109	-0.1435706
5	6	12	0.0054727	-0.1106083
6	6	12	1.3084924	0.7810298
1	7	12	-0.0611825	-0.1461377
2	7	12	-0.0492184	-0.1484513
3	7	12	-0.0309288	-0.1401000
4	7	12	-0.0074549	-0.1238187
5	7	12	0.0213293	-0.0998649
6	7	12	0.0561330	-0.0674908
7	7	12	1.0713258	0.7258634
1	8	12	-0.0273475	-0.1215721
2	8	12	-0.0189783	-0.1251165
3	8	12	-0.0048692	-0.1199910
4	8	12	0.0124567	-0.1084511
5	8	12	0.0331956	-0.0907082
6	8	12	0.0578972	-0.0661294
7	8	12	0.0874668	-0.0333351
8	8	12	0.8621786	0.6653035
1	9	12	-0.0080605	-0.1026948
2	9	12	0.0015064	-0.1069507
3	9	12	0.0131077	-0.1040492
4	9	12	0.0266268	-0.0958851
5	9	12	0.0422903	-0.0826430
6	9	12	0.0605311	-0.0637936
7	9	12	0.0820263	-0.0381598
8	9	12	0.1078161	-0.0036644
9	9	12	0.6741558	0.5978406
1	10	12	0.0065957	-0.0874933
2	10	12	0.0157917	-0.0921338
3	10	12	0.0258664	-0.0908158
4	10	12	0.0369782	-0.0851487
5	10	12	0.0493695	-0.0753004
6	10	12	0.0633846	-0.0608340
7	10	12	0.0795237	-0.0407555
8	10	12	0.0985502	-0.0132751
9	10	12	0.1217001	0.0248324
10	10	12	0.5022398	0.5209243
1	11	12	0.0168287	-0.0746398
2	11	12	0.0258986	-0.0794387
3	11	12	0.0350553	-0.0792736
4	11	12	0.0446466	-0.0755164
5	11	12	0.0549243	-0.0683231
6	11	12	0.0661658	-0.0573406
7	11	12	0.0787381	-0.0417422
8	11	12	0.0931854	-0.0200137
9	11	12	0.1104001	0.0106385
10	11	12	0.1320313	0.0559065
11	11	12	0.3421256	0.4297430
1	12	12	0.0238938	-0.0629056
2	12	12	0.0329842	-0.0676705
3	12	12	0.0416281	-0.0683572
4	12	12	0.0503027	-0.0661223
5	12	12	0.0592658	-0.0611125
6	12	12	0.0687474	-0.0530529
7	12	12	0.0790177	-0.0412780
8	12	12	0.0904553	-0.0245480
9	12	12	0.1036731	-0.0005340
10	12	12	0.1198380	0.0355552
11	12	12	0.1418326	0.0970860
12	12	12	0.1883611	0.3128398

APPENDIX L

Weights for obtaining the best linear invariant estimates of the parameters of the Weibull distribution.

From Mann, N.R., *Results on Location and Scale Parameter Estimation to the Extreme-Value Distribution*, Aerospace Research Laboratories, Wright-Patterson Air Force Base, Ohio, ARL 67-0023, Contract No. AF 33(615)-2818, February, 1967.

	N	r	i	α_i	c_i		N	r	i	α_i	c_i
	2	2	1	0.110731	−0.421383		4	3	1	−0.044975	−0.297651
			2	0.889269	0.421383				2	0.088057	−0.234054
E(LU)[b]				0.65712995					3	0.956918	0.531705
E(CP)				0.03757418		E(LU)				0.42315147	
E(LB)					0.41583918	E(CP)				0.08477554	
						E(LB)					0.28172930
	3	2	1	−0.166001	−0.452110						
			2	1.166001	0.452110		4	4	1	0.064336	−0.203052
E(LU)				0.79546061					2	0.147340	−0.182749
E(CP)				0.25750956					3	0.261510	−0.070109
E(LB)					0.45005549				4	0.526813	0.455910
						E(LU)				0.29247651	
	3	3	1	0.081063	−0.278666	E(CP)				−0.02831210	
			2	0.251001	−0.190239	E(LB)					0.18386193
			3	0.667936	0.468904						
E(LU)				0.40240741			5	2	1	−0.481434	−0.472962
E(CP)				−0.01842169					2	1.481434	0.472962
E(LB)					0.25634620	E(LU)				1.24921018	
						E(CP)				0.53379141	
	4	2	1	−0.346974	−0.465455	E(LB)					0.47230837
			2	1.346974	0.465455						
E(LU)				1.01477788			5	3	1	−0.137958	−0.306562
E(CP)				0.41350875					2	−0.025510	−0.257087
E(LB)					0.46438768				3	1.163468	0.563650

[b] E(LU) = Expected loss for estimate of u.
E(CP) = Expected cross product.
E(LB) = Expected loss for estimate of b.

	N	r	i	α_i	c_i
E(LU)				0.49029288	
E(CP)				0.16612899	
E(LB)					0.29419192
	5	4	1	−0.006983	−0.217766
			2	0.059652	−0.199351
			3	0.156664	−0.118927
			4	0.790668	0.536044
E(LU)				0.29062766	
E(CP)				0.03076329	
E(LB)					0.20241894
	5	5	1	0.052975	−0.158131
			2	0.103531	−0.155707
			3	0.163808	−0.111820
			4	0.246092	−0.005600
			5	0.433593	0.431259
E(LU)				0.23040495	
E(CP)				−0.02913523	
E(LB)					0.14284288
	6	2	1	−0.588298	−0.477782
			2	1.588298	0.477782
E(LU)				1.48102383	
E(CP)				0.63148980	
E(LB)					0.47734078
	6	3	1	−0.211474	−0.311847
			2	−0.112994	−0.271381
			3	1.324468	0.583229

	N	r	i	α_i	c_i
E(LU)				0.57539484	
E(CP)				0.23269670	
E(LB)					0.30173252
	6	4	1	−0.063569	−0.225141
			2	−0.006726	−0.209083
			3	0.079882	−0.146386
			4	0.990412	0.580610
E(LU)				0.31552097	
E(CP)				0.08035062	
E(LB)					0.21242254
	6	5	1	0.007521	−0.169920
			2	0.048328	−0.166319
			3	0.101608	−0.129510
			4	0.172859	−0.054453
			5	0.669685	0.520201
E(LU)				0.22351297	
E(CP)				0.00888019	
E(LB)					0.15690540
	6	6	1	0.044826	−0.128810
			2	0.079377	−0.132102
			3	0.117541	−0.111951
			4	0.163591	−0.064666
			5	0.226486	0.031796
			6	0.368179	0.405733
E(LU)				0.19030430	
E(CP)				−0.02771574	
E(LB)					0.11657671

	N	r	i	α_i	c_i		N	r	i	α_i	c_i
	7	2	1	−0.676894	−0.481140		7	6	1	0.013524	−0.138436
			2	1.676894	0.481140				2	0.041588	−0.140342
E(LU)				1.70468001					3	0.075499	−0.121821
E(CP)					0.71366553				4	0.117461	−0.082938
E(LB)					0.48082310				5	0.172092	−0.015394
									6	0.579835	0.498931
	7	3	1	−0.272195	−0.315369	E(LU)				0.18269947	
			2	−0.184061	−0.281139	E(CP)					−0.00130057
			3	1.456255	0.596507	E(LB)					0.12760617
E(LU)				0.66758707							
E(CP)					0.28885432	7	7	1	0.038743	−0.108323	
E(LB)					0.30681307				2	0.064086	−0.113479
									3	0.090785	−0.103569
	7	4	1	−0.110274	−0.229691				4	0.120971	−0.078748
			2	−0.060226	−0.215613				5	0.157657	−0.032632
			3	0.018671	−0.164168				6	0.207825	0.054727
			4	1.151829	0.609472				7	0.319934	0.382022
E(LU)				0.35340223		E(LU)				0.16219070	
E(CP)					0.12260834	E(CP)					−0.02578937
E(LB)					0.21884662	E(LB)					0.09836496
	7	5	1	−0.030368	−0.176203	8	2	1	−0.752513	−0.483616	
			2	0.004333	−0.172399				2	1.752513	0.483616
			3	0.052957	−0.141218	E(LU)				1.91861540	
			4	0.117599	−0.082820	E(CP)					0.78453314
			5	0.855480	0.572640	E(LB)					0.48337662
E(LU)				0.23316740							
E(CP)					0.04212562	8	3	1	−0.323875	−0.317890	
E(LB)					0.16497315				2	−0.243808	−0.288231
									3	1.567683	0.606120

	N	r	i	α_i	c_i		N	r	i	α_i	c_i
E(LU)				0.76198737			8	7	1	0.015973	−0.116317
E(CP)					0.33734068				2	0.036729	−0.120331
E(LB)					0.31047652				3	0.060439	−0.110582
									4	0.088239	−0.088450
	8	4	1	−0.149973	−0.232805				5	0.122062	−0.050995
			2	−0.105015	−0.220324				6	0.165529	0.009700
			3	−0.032257	−0.176675				7	0.511030	0.476975
			4	1.287245	0.629805	E(LU)				0.15505149	
E(LU)				0.39805551		E(CP)				−0.00641304	
E(CP)					0.15928131	E(LB)					0.10726405
E(LB)					0.22335819						
							8	8	1	0.034052	−0.093270
	8	5	1	−0.062656	−0.180231				2	0.053552	−0.098886
			2	−0.032248	−0.176510				3	0.073452	−0.093994
			3	0.012767	−0.149566				4	0.095062	−0.079752
			4	0.072446	−0.101642				5	0.119768	−0.053918
			5	1.009691	0.607948				6	0.149934	−0.010179
E(LU)				0.25192092					7	0.191236	0.069325
E(CP)					0.07129172				8	0.282943	0.360675
E(LB)					0.17037848	E(LU)				0.14136026	
						E(CP)				−0.02386561	
	8	6	1	−0.013509	−0.143834	E(LB)					0.08501680
			2	0.010292	−0.145006						
			3	0.041357	−0.128393		9	2	1	−0.818444	−0.485517
			4	0.080475	−0.095696				2	1.818444	0.485517
			5	0.130327	−0.043280	E(LU)				2.12272209	
			6	0.751058	0.556209	E(CP)					0.84680378
E(LU)				0.18599844		E(LB)					0.48532951
E(CP)					0.02247163						
E(LB)					0.13422386						

	N	r	i	α_i	c_i		N	r	i	α_i	c_i
	9	3	1	−0.368833	−0.319786		9	7	1	−0.004220	−0.120988
			2	−0.295280	−0.293621				2	0.013386	−0.124245
			3	1.664113	0.613407				3	0.035068	−0.115091
E(LU)				0.85621748					4	0.061198	−0.095508
E(CP)					0.37995861				5	0.093013	−0.064162
E(LB)					0.31324611				6	0.132740	−0.017187
									7	0.668815	0.537180
	9	4	1	−0.184461	−0.235080	E(LU)				0.15547192	
			2	−0.143505	−0.223891	E(CP)					0.01139509
			3	−0.075815	−0.185970	E(LB)					0.11278822
			4	1.403781	0.644941						
E(LU)				0.44625568			9	8	1	0.016797	−0.100011
E(CP)					0.19160927				2	0.032919	−0.104750
E(LB)					0.22671251				3	0.050582	−0.099608
									4	0.070497	−0.086226
	9	5	1	−0.090726	−0.183061				5	0.093635	−0.063541
			2	−0.063541	−0.179515				6	0.121560	−0.028346
			3	−0.021495	−0.155825				7	0.157175	0.026525
			4	0.034159	−0.115133				8	0.456836	0.455956
			5	1.141604	0.633534	E(LU)				0.13496842	
E(LU)				0.27605014		E(CP)				−0.00906894	
E(CP)					0.09715351	E(LB)					0.09236358
E(LB)					0.17429417						
							9	9	1	0.030338	−0.081777
	9	6	1	−0.037118	−0.147411				2	0.045872	−0.087308
			2	−0.016377	−0.148150				3	0.061368	−0.085084
			3	0.012499	−0.133219				4	0.077742	−0.076470
			4	0.049305	−0.105060				5	0.095769	−0.060667
			5	0.095614	−0.062073				6	0.116517	−0.035136
			6	0.896078	0.595913				7	0.141932	0.006001
E(LU)				0.19579592					8	0.176764	0.078828
E(CP)					0.04378261				9	0.253697	0.341614
E(LB)					0.13880129						

	N	r	i	α_i	c_i		N	r	i	α_i	c_i
E(LU)				0.12529518			10	6	1	−0.058017	−0.149985
E(CP)					−0.02209438				2	−0.039595	−0.150451
E(LB)					0.07482425				3	−0.012513	−0.136941
									4	0.022314	−0.112224
	10	2	1	−0.876869	−0.487022				5	0.065750	−0.075721
			2	1.876869	0.487022				6	1.022062	0.625321
E(LU)				2.31744054		E(LU)				0.20973843	
E(CP)					0.90232208	E(CP)					0.06299841
E(LB)					0.48687150	E(LB)					0.14219828
	10	3	1	−0.408602	−0.321265		10	7	1	−0.022198	−0.124170
			2	−0.340443	−0.297858				2	−0.006909	−0.126894
			3	1.749045	0.619124				3	0.013224	−0.118392
E(LU)				0.94907551					4	0.037994	−0.100924
E(CP)					0.41795081				5	0.068153	−0.073988
E(LB)					0.31541467				6	0.105164	−0.035501
									7	0.804572	0.579868
	10	4	1	−0.214930	−0.236817	E(LU)				0.16066059	
			2	−0.177223	−0.226688	E(CP)					0.02762724
			3	−0.113820	−0.193159	E(LB)					0.11670571
			4	1.505973	0.656663						
E(LU)				0.49619736			10	8	1	0.001179	−0.104082
E(CP)					0.22047816				2	0.014889	−0.108163
E(LB)					0.22930885				3	0.030998	−0.103119
									4	0.049734	−0.090835
	10	5	1	−0.115524	−0.185169				5	0.071745	−0.070902
			2	−0.090868	−0.181821				6	0.098114	−0.041560
			3	−0.051341	−0.160697				7	0.130649	0.000799
			4	0.000925	−0.125311				8	0.602692	0.517864
			5	1.256809	0.652997	E(LU)				0.13403554	
E(LU)				0.30344549		E(CP)					0.00474963
E(CP)					0.12033056	E(LB)					0.09704810
E(LB)					0.17727542						

N	r	i	α_i	c_i		N	r	i	α_i	c_i
10	9	1	0.016841	−0.087538		11	3	1	−0.444245	−0.322452
		2	0.029807	−0.092405				2	−0.380642	−0.301277
		3	0.043570	−0.089839				3	1.824887	0.623729
		4	0.058640	−0.081428	E(LU)				1.03995578	
		5	0.075576	−0.066855	E(CP)					0.45220741
		6	0.095169	−0.044670	E(LB)					0.31715930
		7	0.118707	−0.011816						
		8	0.148575	0.038159		11	4	1	−0.242206	−0.238188
		9	0.413116	0.436394				2	−0.207204	−0.228941
E(LU)			0.11965747					3	−0.147490	−0.198888
E(CP)			−0.01043859					4	1.596900	0.666017
E(LB)				0.08100409	E(LU)				0.54681985	
					E(CP)					0.24653583
10	10	1	0.027331	−0.072734	E(LB)					0.23138012
		2	0.040034	−0.077971						
		3	0.052496	−0.077242		11	5	1	−0.137718	−0.186803
		4	0.065408	−0.071876				2	−0.115110	−0.183651
		5	0.079263	−0.061652				3	−0.077762	−0.164597
		6	0.094638	−0.045420				4	−0.028411	−0.133278
		7	0.112414	−0.020698				5	1.359000	0.668329
		8	0.134239	0.017927	E(LU)				0.33282848	
		9	0.164178	0.085070	E(CP)					0.14129911
		10	0.230001	0.324597	E(LB)					0.17962678
E(LU)			0.11252220							
E(CP)			−0.02050852			11	6	1	−0.076739	−0.151936
E(LB)				0.06679250				2	−0.060142	−0.152221
								3	−0.034581	−0.139907
11	2	1	−0.929310	−0.488243				4	−0.001490	−0.117886
		2	1.929310	0.488243				5	0.039518	−0.086131
E(LU)			2.50340024					6	1.133434	0.648081
E(CP)			0.95239887		E(LU)				0.22640907	
E(LB)				0.48812000	E(CP)					0.08045010
					E(LB)					0.14483423

N	r	i	α_i	c_i	N	r	i	α_i	c_i
11	7	1	−0.038349	−0.126507	E(LU)			0.11809425	
		2	−0.024842	−0.128838	E(CP)			0.00058414	
		3	−0.005964	−0.120951	E(LB)				0.08503131
		4	0.017632	−0.105219					
		5	0.046354	−0.081602	11	10	1	0.016502	−0.077717
		6	0.081182	−0.048929			2	0.027205	−0.082449
		7	0.923987	0.612047			3	0.038291	−0.081388
E(LU)			0.16905710				4	0.050160	−0.075977
E(CP)			0.04246025				5	0.063170	−0.066222
E(LB)				0.11966982			6	0.077772	−0.051429
							7	0.094625	−0.030120
11	8	1	−0.012943	−0.106922			8	0.114811	0.000537
		2	−0.001050	−0.110498			9	0.140333	0.046381
		3	0.013869	−0.105662			10	0.377130	0.418384
		4	0.031661	−0.094405	E(LU)			0.10756449	
		5	0.052723	−0.076693	E(CP)			−0.01109747	
		6	0.077815	−0.051525	E(LB)				0.07207183
		7	0.108161	−0.016860					
		8	0.729765	0.562564	11	11	1	0.024850	−0.065444
E(LU)			0.13669382				2	0.035456	−0.070318
E(CP)			0.01751192				3	0.045727	−0.070456
E(LB)				0.19043756			4	0.056215	−0.067076
							5	0.067261	−0.060207
11	9	1	0.004425	−0.091115			6	0.079220	−0.049300
		2	0.015498	−0.095437			7	0.092560	−0.033156
		3	0.028023	−0.092780			8	0.108034	−0.009427
		4	0.042178	−0.084833			9	0.127068	0.026879
		5	0.058340	−0.071581			10	0.153197	0.089148
		6	0.077093	−0.052182			11	0.210412	0.309357
		7	0.099349	−0.024880	E(LU)			0.10212039	
		8	0.126592	0.013606	E(CP)			−0.01910164	
		9	0.548502	0.499201	E(LB)				0.06030372

N	r	i	α_i	c_i		N	r	i	α_i	c_i
12	2	1	−0.976872	−0.489254		12	6	1	−0.093679	−0.153471
		2	1.976872	0.489254				2	−0.078561	−0.153632
E(LU)			2.68127021					3	−0.054320	−0.142329
E(CP)				0.99799849				4	−0.022769	−0.122474
E(LB)				0.48915157				5	0.016136	−0.094355
								6	1.233193	0.666261
12	3	1	−0.476530	−0.323426	E(LU)			0.24490094	
		2	−0.416836	−0.304093	E(CP)				0.09641022
		3	1.893367	0.627519	E(LB)				0.14694548
E(LU)			1.12857097							
E(CP)				0.48338667		12	7	1	−0.052987	−0.128308
E(LB)				0.31859354				2	−0.040893	−0.130339
								3	−0.023072	−0.123007
12	4	1	−0.266888	−0.239300				4	−0.000515	−0.108712
		2	−0.234180	−0.230796				5	0.026930	−0.087681
		3	−0.177681	−0.203562				6	0.059918	−0.059256
		4	1.678749	0.673657				7	1.030620	0.637304
E(LU)			0.59748043		E(LU)			0.17967935	
E(CP)				0.27026774	E(CP)				0.05607919
E(LB)				0.23307201	E(LB)				0.12200601
12	5	1	−0.157792	−0.188109		12	8	1	−0.025785	−0.109045
		2	−0.136884	−0.185142				2	−0.015312	−0.112224
		3	−0.101445	−0.167790				3	−0.001353	−0.107627
		4	−0.054640	−0.139693				4	0.015634	−0.097276
		5	1.450761	0.680734				5	0.035853	−0.081361
E(LU)			0.36338878					6	0.059835	−0.059315
E(CP)				0.16042600				7	0.088444	−0.029900
E(LB)				0.18153147				8	0.842684	0.596748

N	r	i	α_i	c_i	N	r	i	α_i	c_i
E(LU)			0.14186580		12	11	1	0.015982	−0.069798
E(CP)			0.02930146				2	0.024997	−0.074285
E(LB)				0.10304331			3	0.034156	−0.074131
							4	0.043790	−0.070617
	12	9	1	−0.006944	−0.093658		5	0.054149	−0.063891
			2	0.002669	−0.097540		6	0.065515	−0.053621
			3	0.014239	−0.094893		7	0.078264	−0.039034
			4	0.027669	−0.087448		8	0.092958	−0.018715
			5	0.043189	−0.075371		9	0.110521	0.009948
			6	0.061225	−0.058180		10	0.132666	0.052280
			7	0.082441	−0.034802		11	0.347003	0.401864
			8	0.107856	−0.003342	E(LU)			0.09775217	
			9	0.667655	0.545234	E(CP)			−0.01134890	
E(LU)			0.11929957		E(LB)				0.06487266
E(CP)			0.01087297						
E(LB)				0.08799386	12	12	1	0.022771	−0.059449
							2	0.031776	−0.063952
	12	10	1	0.006411	−0.080881		3	0.040408	−0.064601
			2	0.015598	−0.085171		4	0.049122	−0.062489
			3	0.025675	−0.083952		5	0.058175	−0.057754
			4	0.036799	−0.078714		6	0.067800	−0.050137
			5	0.049211	−0.069610		7	0.078281	−0.039010
			6	0.063256	−0.056237		8	0.090017	−0.023199
			7	0.079438	−0.037675		9	0.103664	−0.000505
			8	0.098522	−0.012272		10	0.120475	0.033696
			9	0.121752	0.022956		11	0.143566	0.091751
			10	0.503338	0.481555		12	0.193947	0.295648
E(LU)			0.10573191		E(LU)			0.09348388	
E(CP)			−0.00210755		E(CP)			−0.01785537	
E(LB)				0.07557509	E(LB)				0.05495436

	N	r	i	α_i	c_i		N	r	i	α_i	c_i
	13	2	1	−1.020378	−0.490105		13	6	1	−0.109140	−0.154711
			2	2.020377	0.490105				2	−0.095246	−0.154785
E(LU)				2.85169694					3	−0.072165	−0.144347
E(CP)				1.03985071					4	−0.041997	−0.126268
E(LB)					0.49001823				5	−0.004940	−0.101028
									6	1.323488	0.681140
	13	3	1	−0.506031	−0.324239	E(LU)				0.26460952	
			2	−0.449735	−0.306454	E(CP)				0.11109896	
			3	1.955765	0.630694	E(LB)					0.14867755
E(LU)				1.21480934							
E(CP)				0.51198847			13	7	1	−0.066358	−0.129743
E(LB)					0.31979363				2	−0.055414	−0.131538
									3	−0.038503	−0.124701
	13	4	1	−0.289420	−0.240219				4	−0.016879	−0.111609
			2	−0.258687	−0.232349				5	0.009416	−0.092649
			3	−0.205024	−0.207450				6	0.040810	−0.067475
			4	1.753131	0.680018				7	1.126930	0.657714
E(LU)				0.64778295		E(LU)				0.19187273	
E(CP)				0.29204583		E(CP)				0.06864731	
E(LB)					0.23448055	E(LB)					0.12390133
	13	5	1	−0.176109	−0.189177		13	8	1	−0.037540	−0.110704
			2	−0.156637	−0.186381				2	−0.028206	−0.113563
			3	−0.122893	−0.170454				3	−0.015049	−0.109206
			4	−0.078337	−0.144971				4	0.001231	−0.099644
			5	1.533976	0.690983				5	0.020686	−0.085204
E(LU)				0.39459617					6	0.043677	−0.065581
E(CP)				0.17799724					7	0.070830	−0.039995
E(LB)					0.18310709				8	0.944372	0.623896

	N	r	i	α_i	c_i		N	r	i	α_i	c_i
E(LU)				0.14885020		13	11	1	0.007628	-0.072617
E(CP)				0.04022462				2	0.015408	-0.076746
E(LB)					0.10512398			3	0.023732	-0.076418
								4	0.032743	-0.072938
	13	9	1	-0.017389	-0.095590			5	0.042611	-0.066531
			2	-0.008934	-0.099109			6	0.053556	-0.057014
			3	0.001863	-0.096521			7	0.065876	-0.043886
			4	0.014684	-0.089554			8	0.080005	-0.026244
			5	0.029637	-0.078490			9	0.096594	-0.002552
			6	0.047027	-0.063068			10	0.116703	0.029910
			7	0.067346	-0.042607			11	0.465143	0.465037
			8	0.091328	-0.015928	E(LU)			0.09583611	
			9	0.774437	0.580865	E(CP)			-0.00388188	
E(LU)				0.12250342		E(LB)				0.06795140
E(CP)				0.02046326						
E(LB)					0.09030201	13	12	1	0.015382	--0.063288
								2	0.023100	-0.067492
	13	10	1	-0.002927	-0.083170			3	0.030818	-0.067892
			2	0.005067	-0.087085			4	0.038824	-0.065622
			3	0.014356	-0.085792			5	0.047302	-0.060887
			4	0.024891	-0.080789			6	0.056444	-0.053540
			5	0.036816	-0.072325			7	0.066482	-0.043158
			6	0.050389	-0.060181			8	0.077739	-0.028970
			7	0.065995	-0.043768			9	0.090699	-0.009644
			8	0.084201	-0.022048			10	0.106166	0.017233
			9	0.105863	0.006715			11	0.125627	0.056547
			10	0.615348	0.528441			12	0.321416	0.386713
E(LU)				0.10607774		E(LU)			0.08961947	
E(CP)				0.00635741		E(CP)			-0.01136145	
E(LB)					0.07818835	E(LB)				0.05895232

N	r	i	α_i	c_i		N	r	i	α_i	c_i
13	13	1	0.021005	−0.054436	E(LU)				0.69748231	
		2	0.028757	−0.058585	E(CP)				0.31216081	
		3	0.036127	−0.059535	E(LB)					0.23567174
		4	0.043501	−0.058259						
		5	0.051078	−0.054942		14	5	1	−0.192947	−0.190068
		6	0.059028	−0.049472				2	−0.174709	−0.187427
		7	0.067533	−0.041504				3	−0.142478	−0.172710
		8	0.076831	−0.030398				4	−0.099930	−0.149393
		9	0.087274	−0.015037				5	1.610065	0.699598
		10	0.099441	0.006644	E(LU)				0.42609561	
		11	0.114446	0.038943	E(CP)				0.19423903	
		12	0.135068	0.093324	E(LB)					0.18443288
		13	0.179913	0.283257						
E(LU)			0.08619744			14	6	1	−0.123352	−0.155736
E(CP)			−0.01674914					2	−0.110490	−0.155747
E(LB)				0.05046988				3	−0.088443	−0.146054
								4	−0.059523	−0.129460
14	2	1	−1.060461	−0.490831				5	−0.024111	−0.106556
		2	2.060461	0.490831				6	1.405919	0.693553
E(LU)			3.01527998		E(LU)				0.28511973	
E(CP)			1.07852097		E(CP)				0.12469427	
E(LB)				0.49075663	E(LB)					0.15012578
14	3	1	−0.533185	−0.324929		14	7	1	−0.078656	−0.130915
		2	−0.479874	−0.308462				2	−0.068666	−0.132521
		3	2.013059	0.633391				3	−0.052554	−0.126123
E(LU)			1.29865775					4	−0.031776	−0.114051
E(CP)			0.53840104					5	−0.006522	−0.096788
E(LB)				0.32081269				6	0.023467	−0.074184
								7	1.214708	0.674581
14	4	1	−0.310144	−0.240992	E(LU)				0.20518434	
		2	−0.281132	−0.233670	E(CP)				0.08030259	
		3	−0.229990	−0.210735	E(LB)					0.12547311
		4	1.821266	0.685397						

N	r	i	α_i	c_i	N	r	i	α_i	c_i
14	8	1	−0.048365	−0.112041	14	10	1	−0.011580	−0.084931
		2	−0.039964	−0.114637			2	−0.004548	−0.088528
		3	−0.027495	−0.110509			3	0.004100	−0.087207
		4	−0.011849	−0.101635			4	0.014144	−0.082451
		5	0.006905	−0.088422			5	0.025647	−0.074573
		6	0.029002	−0.070735			6	0.038794	−0.063473
		7	0.054897	−0.048074			7	0.053879	−0.048768
		8	1.036868	0.646052			8	0.071335	−0.029776
E(LU)			0.15716466				9	0.091783	−0.005398
E(CP)				0.05038249			10	0.716445	0.565105
E(LB)				0.10683049	E(LU)			0.10803536	
14	9	1	−0.027030	−0.097117	E(CP)				0.01430729
		2	−0.019516	−0.100334	E(LB)				0.08024763
		3	−0.009363	−0.097827					
		4	0.002928	−0.091298	14	11	1	−0.000170	−0.074686
		5	0.017368	−0.081103			2	0.006622	−0.078499
		6	0.034165	−0.067124			3	0.014283	−0.078064
		7	0.053685	−0.048921			4	0.022800	−0.074680
		8	0.076476	−0.025720			5	0.032273	−0.068624
		9	0.871287	0.609445			6	0.042866	−0.059816
E(LU)			0.12719148				7	0.054817	−0.047926
E(CP)				0.02941694			8	0.068463	−0.032355
E(LB)				0.09216556			9	0.084290	−0.012126
							10	0.103025	0.014349
							11	0.570731	0.512429

N	r	i	α_i	c_i	N	r	i	α_i	c_i
E(LU)			0.09566494		E(LU)			0.08276211	
E(CP)			0.00320055		E(CP)			−0.01123278	
E(LB)				0.07027548	E(LB)				0.05400148
	14	12	1	0.008361	−0.065816	14	14	1	0.019487	−0.050186
			2	0.015058	−0.069728			2	0.026238	−0.054008
			3	0.022076	−0.069962			3	0.032614	−0.055130
			4	0.029552	−0.067659			4	0.038947	−0.054419
			5	0.037615	−0.063070			5	0.045399	−0.052075
			6	0.046411	−0.056130			6	0.052097	−0.048066
			7	0.056132	−0.046558			7	0.059168	−0.042197
			8	0.067039	−0.033834			8	0.066767	−0.034099
			9	0.079506	−0.017101			9	0.075102	−0.023149
			10	0.094096	0.005064			10	0.084482	−0.008285
			11	0.111723	0.035156			11	0.095428	0.012430
			12	0.432431	0.449638			12	0.108942	0.043015
E(LU)			0.08771669				13	0.127523	0.094166
E(CP)			−0.00506397				14	0.167807	0.272004
E(LB)				0.06168210	E(LU)			0.07996685	
	14	13	1	0.014760	−0.057849	E(CP)			−0.01576372	
			2	0.021453	−0.061764	E(LB)				0.04665712
			3	0.028064	−0.062506	15	2	1	−1.097617	−0.491458
			4	0.034842	−0.061074			2	2.097617	0.491458
			5	0.041933	−0.057693	E(LU)			3.17256460	
			6	0.049474	−0.052317	E(CP)			1.11445612	
			7	0.057619	−0.044707	E(LB)				0.49139327
			8	0.066569	−0.034420					
			9	0.076605	−0.020713	15	3	1	−0.558336	−0.325521
			10	0.088151	−0.002338			2	−0.507671	−0.310191
			11	0.101914	0.022943			3	2.066007	0.635712
			12	0.119200	0.059643	E(LU)			1.38015851	
			13	0.299416	0.372795	E(CP)			0.56293169	
						E(LB)				0.32168886

	N	r	i	α_i	c_i		N	r	i	α_i	c_i
	15	4	1	−0.329324	−0.241651		15	7	1	−0.090036	−0.131891
			2	−0.301829	−0.234806				2	−0.080850	−0.133342
			3	−0.252948	−0.213548				3	−0.065446	−0.127335
			4	1.884101	0.690005				4	−0.045441	−0.116138
E(LU)				0.74642859					5	−0.021137	−0.100291
E(CP)				0.33084387					6	0.007597	−0.079774
E(LB)					0.23669248				7	1.295312	0.688771
						E(LU)				0.21929214	
	15	5	1	−0.208525	−0.190823	E(CP)				0.09116039	
			2	−0.191357	−0.188323	E(LB)					0.12679942
			3	−0.160491	−0.174645						
			4	−0.119748	−0.153153		15	8	1	−0.058390	−0.113143
			5	1.680121	0.706944				2	−0.050767	−0.115520
E(LU)				0.45764555					3	−0.038897	−0.111607
E(CP)				0.20933279					4	−0.023825	−0.103332
E(LB)					0.18556433				5	−0.005717	−0.091156
									6	0.015565	−0.075053
	15	6	1	−0.136498	−0.156597				7	0.040351	−0.054703
			2	−0.124518	−0.156563				8	1.121680	0.664514
			3	−0.103401	−0.147517	E(LU)				0.16646559	
			4	−0.075614	−0.132182	E(CP)				0.05986446	
			5	−0.041680	−0.111215	E(LB)					0.10825884
			6	1.481712	0.704074						
E(LU)				0.30614004							
E(CP)				0.13734100							
E(LB)					0.15135556						

	N	r	i	α_i	c_i		N	r	i	α_i	c_i
	15	9	1	−0.035972	−0.098361		15	11	1	−0.007450	−0.076297
			2	−0.029235	−0.101322				2	−0.001467	−0.079835
			3	−0.019633	−0.098904				3	0.005652	−0.079332
			4	−0.007812	−0.092773				4	0.013759	−0.076068
			5	0.006156	−0.083327				5	0.022893	−0.070355
			6	0.022403	−0.070544				6	0.033174	−0.062181
			7	0.041203	−0.054142				7	0.044787	−0.051331
			8	0.062969	−0.033595				8	0.057997	−0.037396
			9	0.959920	0.632967				9	0.073180	−0.019723
E(LU)				0.13300106					10	0.090865	0.002701
E(CP)					0.03779810				11	0.666610	0.549817
E(LB)					0.09370837	E(LU)				0.09681113	
	15	10	1	−0.019626	−0.086339	E(CP)					0.00989471
			2	−0.013383	−0.089664	E(LB)					0.07212492
			3	−0.005271	−0.088341						
			4	0.004351	−0.083828		15	12	1	0.001756	−0.067695
			5	0.015475	−0.076474				2	0.007624	−0.071342
			6	0.028227	−0.066261				3	0.014079	−0.071459
			7	0.042832	−0.052943				4	0.021133	−0.069178
			8	0.059624	−0.036054				5	0.028861	−0.064779
			9	0.079072	−0.014863				6	0.037374	−0.058256
			10	0.808700	0.594768				7	0.046827	−0.049425
E(LU)				0.11121862					8	0.057431	−0.037926
E(CP)					0.02177795				9	0.069479	−0.023180
E(LB)					0.08192616				10	0.083393	−0.004280
									11	0.099799	0.020236
									12	0.532243	0.497284
						E(LU)				0.08723346	
						E(CP)					0.00094612
						E(LB)					0.06376409

N	r	i	α_i	c_i
15	13	1	0.008779	−0.060130
		2	0.014620	−0.063805
		3	0.020637	−0.064394
		4	0.026961	−0.062900
		5	0.033693	−0.059574
		6	0.040939	−0.054417
		7	0.048828	−0.047269
		8	0.057528	−0.037821
		9	0.067265	−0.025565
		10	0.078368	−0.009694
		11	0.091330	0.011113
		12	0.106947	0.039155
		13	0.404106	0.435302
E(LU)			0.08092217	
E(CP)			−0.00585240	
E(LB)				0.05644073
15	14	1	0.014143	−0.053241
		2	0.020013	−0.056879
		3	0.025750	−0.057827
		4	0.031576	−0.056973
		5	0.037611	−0.054542
		6	0.043958	−0.050539
		7	0.050725	−0.044833
		8	0.058045	−0.037157
		9	0.066092	−0.027072
		10	0.075114	−0.013872
		11	0.085490	0.003612
		12	0.097844	0.027465
		13	0.113340	0.061879
		14	0.280298	0.359980

N	r	i	α_i	c_i
E(LU)			0.07689745	
E(CP)			−0.01102126	
E(LB)				0.04980248
15	15	1	0.018170	−0.046538
		2	0.024108	−0.050064
		3	0.029685	−0.051279
		4	0.035191	−0.050957
		5	0.040762	−0.049298
		6	0.046496	−0.046315
		7	0.052488	−0.041899
		8	0.058844	−0.035827
		9	0.065696	−0.027731
		10	0.073230	−0.017008
		11	0.081725	−0.002653
		12	0.091651	0.017156
		13	0.103914	0.04619†
		14	0.120784	0.094483
		15	0.157255	0.261738
E(LU)			0.07457775	
E(CP)			−0.01488220	
E(LB)				0.04337628
16	2	1	−1.132243	−0.492005
		2	2.132243	0.492005
E(LU)			3.32404220	
E(CP)			1.14801534	
E(LB)				0.49194784

N	r	i	α_i	c_i
16	3	1	−0.581757	−0.326035
		2	−0.533457	−0.311694
		3	2.115214	0.637730
E(LU)			1.45938438	
E(CP)			0.58582769	
E(LB)				0.32245028
16	4	1	−0.347172	−0.242220
		2	−0.321026	−0.235794
		3	−0.274186	−0.215984
		4	1.942384	0.693998
E(LU)			0.79453329	
E(CP)			0.34828173	
E(LB)				0.23757701
16	5	1	−0.223015	−0.191470
		2	−0.206788	−0.189099
		3	−0.177158	−0.176323
		4	−0.138048	−0.156390
		5	1.745009	0.713282
E(LU)			0.48908000	
E(CP)			0.22342597	
E(LB)				0.18654151
16	6	1	−0.148725	−0.157331
		2	−0.137508	−0.157263
		3	−0.117232	−0.148785
		4	−0.090481	−0.134532
		5	−0.057883	−0.115196
		6	1.551828	0.713108
E(LU)			0.32746210	
E(CP)			0.14915808	
E(LB)				0.15241337

N	r	i	α_i	c_i
16	7	1	−0.100621	−0.132718
		2	−0.092121	−0.134040
		3	−0.077354	−0.128381
		4	−0.058057	−0.117942
		5	−0.034624	−0.103296
		6	−0.007020	−0.084506
		7	1.369798	0.700883
E(LU)			0.23396225	
E(CP)			0.10131710	
E(LB)				0.12793461
16	8	1	−0.067719	−0.114069
		2	−0.060754	−0.116260
		3	−0.049415	−0.112545
		4	−0.034868	−0.104798
		5	−0.017357	−0.093508
		6	0.003178	−0.078726
		7	0.026973	−0.060251
		8	1.199963	0.680158
E(LU)			0.17650200	
E(CP)			0.06874770	
E(LB)				0.10947376
16	9	1	−0.044303	−0.099396
		2	−0.038218	−0.102138
		3	−0.029094	−0.099811
		4	−0.017697	−0.094037
		5	−0.004166	−0.085242
		6	0.011570	−0.073467
		7	0.029712	−0.058535
		8	0.050576	−0.040084
		9	1.041619	0.652711
E(LU)			0.13966768	
E(CP)			0.04566615	
E(LB)				0.09501012

N	r	i	α_i	c_i	N	r	i	α_i	c_i
16	10	1	−0.027135	−0.087496	16	12	1	−0.004450	−0.069172
		2	−0.021550	−0.090585			2	0.000732	−0.072584
		3	−0.013895	−0.089277			3	0.006721	−0.072615
		4	−0.004646	−0.084992			4	0.013424	−0.070383
		5	0.006132	−0.078105			5	0.020868	−0.066184
		6	0.018515	−0.068653			6	0.029134	−0.060054
		7	0.032675	−0.056482			7	0.038344	−0.051876
		8	0.048869	−0.041268			8	0.048668	−0.041398
		9	0.067459	−0.022503			9	0.060342	−0.028216
		10	0.893576	0.619360			10	0.073692	−0.011716
E(LU)			0.11534960				11	0.089173	0.009035
E(CP)			0.02881067				12	0.623351	0.535164
E(LB)				0.03352716	E(LU)			0.08784015	
					E(CP)			0.00665801	
16	11	1	−0.014263	−0.077597	E(LB)				0.06543511
		2	−0.008950	−0.080895					
		3	−0.002286	−0.080340	16	13	1	0.003118	−0.061843
		4	0.005469	−0.077313			2	0.008256	−0.065297
		5	0.014303	−0.071820			3	0.013789	−0.065770
		6	0.024297	−0.064207			4	0.019747	−0.064259
		7	0.035593	−0.054237			5	0.026189	−0.061031
		8	0.048404	−0.041625			6	0.033196	−0.056120
		9	0.063020	−0.025917			7	0.040872	−0.049427
		10	0.079847	−0.006432			8	0.049357	−0.040731
		11	0.754566	0.580293			9	0.058836	−0.029675
E(LU)			0.09897866				10	0.069568	−0.015710
E(CP)			0.01622073				11	0.081920	0.002010
E(LB)				0.07364497			12	0.096438	0.024833
							13	0.498713	0.483018

	N	r	i	α_i	c_i		N	r	i	α_i	c_i
E(LU)				0.08025299			16	15	1	0.013547	−0.049291
E(CP)					−0.00069037				2	0.018743	−0.052670
E(LB)					0.05831799				3	0.023778	−0.053739
									4	0.028849	−0.053290
	16	14	1	0.008992	−0.055309				5	0.034060	−0.051538
			2	0.014141	−0.058750				6	0.039489	−0.048520
			3	0.019370	−0.059563				7	0.045218	−0.044164
			4	0.024804	−0.058635				8	0.051338	−0.038307
			5	0.030525	−0.056208				9	0.057965	−0.030678
			6	0.036615	−0.052317				10	0.065253	−0.020850
			7	0.043164	−0.046878				11	0.073425	−0.008156
			8	0.050284	−0.039699				12	0.082818	0.008503
			9	0.058124	−0.030467				13	0.093994	0.031075
			10	0.066884	−0.018695				14	0.107995	0.063476
			11	0.076854	−0.003625				15	0.263528	0.348149
			12	0.088469	0.015969	E(LU)				0.07182155	
			13	0.102433	0.042224	E(CP)					−0.01076262
			14	0.379341	0.421953	E(LB)					0.04619787
E(LU)				0.07514429							
E(CP)					−0.00637294						
E(LB)					0.05199709						

N	r	i	α_i	c_i
16	16	1	0.017016	−0.043375
		2	0.022284	−0.046633
		3	0.027208	−0.047890
		4	0.032046	−0.047839
		5	0.036912	−0.046675
		6	0.041887	−0.044432
		7	0.047042	−0.041053
		8	0.052455	−0.036402
		9	0.058216	−0.030249
		10	0.064444	−0.022230
		11	0.071304	−0.011772
		12	0.079051	0.002079
		13	0.088111	0.021044
		14	0.099315	0.048675
		15	0.114733	0.094419
		16	0.147977	0.252333
E(LU)			0.06987019	
E(CP)			−0.01409012	
E(LB)				0.04052374
17	2	1	−1.164659	−0.492486
		2	2.164659	0.492486
E(LU)			3.47015408	
E(CP)			1.17949167	
E(LB)				0.49243526
17	3	1	−0.603668	−0.326486
		2	−0.557497	−0.313014
		3	2.161166	0.639500

N	r	i	α_i	c_i
E(LU)			1.53642388	
E(CP)			0.60729095	
E(LB)				0.32311812
17	4	1	−0.363861	−0.242716
		2	−0.338922	−0.236662
		3	−0.293934	−0.218114
		4	1.996717	0.697492
E(LU)			0.84174810	
E(CP)			0.36462724	
E(LB)				0.23835098
17	5	1	−0.236557	−0.192031
		2	−0.221164	−0.189778
		3	−0.192661	−0.177793
		4	−0.155037	−0.159206
		5	1.805419	0.718809
E(LU)			0.52028442	
E(CP)			0.23663986	
E(LB)				0.18739415
17	6	1	−0.160149	−0.157965
		2	−0.149601	−0.157871
		3	−0.130090	−0.149896
		4	−0.104290	−0.136581
		5	−0.072907	−0.118639
		6	1.617037	0.720952
E(LU)			0.34893506	
E(CP)			0.16024410	
E(LB)				0.15333326

N	r	i	α_i	c_i	N	r	i	α_i	c_i
17	7	1	−0.110512	−0.133428	E(LU)			0.14699387	
		2	−0.102606	−0.134640	E(CP)			0.05307401	
		3	−0.088415	−0.129294	E(LB)				0.09612512
		4	−0.069771	−0.119517					
		5	−0.047139	−0.105901	17	10	1	−0.034167	−0.088465
		6	−0.020560	−0.088568			2	−0.029139	−0.091350
		7	1.439003	0.711349			3	−0.021881	−0.090064
E(LU)			0.24902198				4	−0.012965	−0.085992
E(CP)			0.11085361				5	−0.002507	−0.079521
E(LB)				0.12891783			6	0.009531	−0.070728
							7	0.023273	−0.059520
17	8	1	−0.076441	−0.114859			8	0.038922	−0.045671
		2	−0.070039	−0.116891			9	0.056761	−0.028822
		3	−0.059173	−0.113357			10	0.972172	0.640135
		4	−0.045110	−0.106076	E(LU)			0.12022174	
		5	−0.028154	−0.095554	E(CP)			0.03544569	
		6	−0.008307	−0.081890	E(LB)				0.08451762
		7	0.014595	−0.064968					
		8	1.272628	0.693595	17	11	1	−0.020654	−0.078673
E(LU)			0.18708688				2	−0.015906	−0.081761
E(CP)			0.07709833				3	−0.009632	−0.081188
E(LB)				0.11052085			4	−0.002186	−0.078180
							5	0.006378	−0.073083
17	9	1	−0.052096	−0.100271			6	0.016104	−0.065964
		2	−0.046565	−0.102825			7	0.027102	−0.056744
		3	−0.037862	−0.100587			8	0.039540	−0.045224
		4	−0.026851	−0.095136			9	0.053648	−0.031078
		5	−0.013728	−0.086910			10	0.069744	−0.013827
		6	0.001531	−0.075995			11	0.835861	0.605723
		7	0.019069	−0.062288	E(LU)			0.10195092	
		8	0.039129	−0.045535	E(CP)			0.02220540	
		9	1.117373	0.669546	E(LB)				0.07492279

N	r	i	α_i	c_i		N	r	i	α_i	c_i
17	12	1	−0.010288	−0.070375	E(LU)				0.08049558	
		2	−0.005683	−0.073577	E(CP)				0.00423893	
		3	−0.000086	−0.073546	E(LB)					0.05983608
		4	0.006316	−0.071375						
		5	0.013511	−0.067372		17	14	1	0.004088	−0.056878
		6	0.021553	−0.061602				2	0.008636	−0.060131
		7	0.030535	−0.053996				3	0.013446	−0.060836
		8	0.040597	−0.044377				4	0.018560	−0.059871
		9	0.051928	−0.032455				5	0.024028	−0.057487
		10	0.064785	−0.017797				6	0.029909	−0.053742
		11	0.079517	0.000228				7	0.036278	−0.048586
		12	0.707314	0.566244				8	0.043231	−0.041881
E(LU)			0.08930564					9	0.050892	−0.033402
E(CP)			0.01203216					10	0.059426	−0.022799
E(LB)				0.06681858				11	0.069060	−0.009558
								12	0.080119	0.007112
17	13	1	−0.002231	−0.063202				13	0.093083	0.028459
		2	0.002313	−0.066454				14	0.469244	0.469601
		3	0.007448	−0.066839	E(LU)				0.07436842	
		4	0.013101	−0.065333	E(CP)				−0.00189289	
		5	0.019298	−0.062220	E(LB)					0.05369960
		6	0.026098	−0.057556						
		7	0.033584	−0.051282						
		8	0.041872	−0.043242						
		9	0.051113	−0.033181						
		10	0.061516	−0.020708						
		11	0.073364	−0.005250						
		12	0.087058	0.014056						
		13	0.585461	0.521211						

N	r	i	α_i	c_i
17	15	1	0.009066	−0.051176
		2	0.013648	−0.054390
		3	0.018244	−0.055341
		4	0.022974	−0.054815
		5	0.027908	−0.053042
		6	0.033111	−0.050075
		7	0.038648	−0.045871
		8	0.044600	−0.040314
		9	0.051065	−0.033203
		10	0.058176	−0.024231
		11	0.066111	−0.012936
		12	0.075128	0.001394
		13	0.085616	0.019905
		14	0.098200	0.044587
		15	0.357506	0.409507

E(LU) 0.07016498
E(CP) −0.00670775
E(LB) 0.04818440

N	r	i	α_i	c_i
17	16	1	0.012979	−0.045870
		2	0.017617	−0.049009
		3	0.022076	−0.050145
		4	0.026538	−0.049982
		5	0.031091	−0.048727
		6	0.035799	−0.046430
		7	0.040724	−0.043057
		8	0.045932	−0.038508
		9	0.051504	−0.032609
		10	0.057542	−0.025090
		11	0.064186	−0.015545
		12	0.071635	−0.003341
		13	0.080195	0.012556
		14	0.090373	0.033974
		15	0.103110	0.064588
		16	0.248699	0.337194

E(LU) 0.06738336
E(CP) −0.01047916
E(LB) 0.04307100

N	r	i	α_i	c_i
17	17	1	0.015998	−0.040607
		2	0.020706	−0.043624
		3	0.025089	−0.044891
		4	0.029378	−0.045031
		5	0.033671	−0.044229
		6	0.038035	−0.042531
		7	0.042527	−0.039913
		8	0.047204	−0.036289
		9	0.052133	−0.031512
		10	0.057392	−0.025352
		11	0.063089	−0.017458
		12	0.069375	−0.007282
		13	0.076482	0.006082
		14	0.084803	0.024262
		15	0.095098	0.050618
		16	0.109270	0.094076
		17	0.139752	0.243681

E(LU) 0.06572241
E(CP) −0.01337530
E(LB) 0.03802109

N	r	i	α_i	c_i
18	2	1	−1.195128	−0.492912
		2	2.195128	0.492912

E(I.U) 3.61129585
E(CP) 1.20912723
E(LB) 0.49286703

N	r	i	α_i	c_i
18	3	1	−0.624252	−0.326884
		2	−0.580008	−0.314183
		3	2.204260	0.641066

	N	r	i	α_i	c_i
E(LU)				1.61137253	
E(CP)					0.62748837
E(LB)					0.32370865
	18	4	1	−0.379529	−0.243153
			2	−0.355679	−0.237429
			3	−0.312382	−0.219992
			4	2.047590	0.700574
E(LU)				0.88805128	
E(CP)					0.38000703
E(LB)					0.23903395
	18	5	1	−0.249266	−0.192523
			2	−0.234618	−0.190376
			3	−0.207148	−0.179091
			4	−0.170883	−0.161679
			5	1.861914	0.723670
E(LU)				0.55118001	
E(CP)					0.24907530
E(LB)					0.18814472
	18	6	1	−0.170868	−0.158518
			2	−0.160910	−0.158405
			3	−0.142100	−0.150876
			4	−0.117175	−0.138383
			5	−0.086906	−0.121647
			6	1.677960	0.727829

	N	r	i	α_i	c_i
E(LU)				0.37044855	
E(CP)					0.17068152
E(LB)					0.15414076
	18	7	1	−0.119793	−0.134044
			2	−0.112406	−0.135163
			3	−0.098738	−0.130098
			4	−0.080698	−0.120904
			5	−0.058807	−0.108183
			6	−0.033165	−0.092095
			7	1.503605	0.720486
E(LU)				0.26434202	
E(CP)					0.11983820
E(LB)					0.12977806
	18	8	1	−0.084626	−0.115541
			2	−0.078711	−0.117434
			3	−0.068272	−0.114068
			4	−0.054656	−0.107202
			5	−0.038217	−0.097349
			6	−0.019006	−0.084645
			7	0.003084	−0.069031
			8	1.340405	0.705270
E(LU)				0.19807869	
E(CP)					0.08497296
E(LB)					0.11143330

N	r	i	α_i	c_i	N	r	i	α_i	c_i
18	9	1	−0.059414	−0.101022	18	11	1	−0.026669	−0.079582
		2	−0.054359	−0.103411			2	−0.022402	−0.082484
		3	−0.046030	−0.101260			3	−0.016466	−0.081896
		4	−0.035375	−0.096099			4	−0.009294	−0.079012
		5	−0.022631	−0.088374			5	−0.000979	−0.074183
		6	−0.007819	−0.078203			6	0.008496	−0.067503
		7	0.009161	−0.065532			7	0.019212	−0.058930
		8	0.028495	−0.050186			8	0.031300	−0.048324
		9	1.187973	0.684087			9	0.044947	−0.035451
E(LU)			0.15482946				10	0.060404	−0.019962
E(CP)				0.06006815			11	0.911449	0.627325
E(LB)				0.09709201	E(LU)			0.10556433	
					E(CP)				0.02787638
18	10	1	−0.040776	−0.089291	E(LB)				0.07601539
		2	−0.036223	−0.091997					
		3	−0.029314	−0.090739					
		4	−0.020701	−0.086863	18	12	1	−0.015793	−0.071378
		5	−0.010540	−0.080764			2	−0.011677	−0.074393
		6	0.001172	−0.072544			3	−0.006416	−0.074315
		7	0.014523	−0.062157			4	−0.000278	−0.072211
		8	0.029671	−0.049445			5	0.006695	−0.068395
		9	0.046841	−0.034147			6	0.014529	−0.062952
		10	1.045347	0.657947			7	0.023297	−0.055848
E(LU)			0.12567798				8	0.033110	−0.046959
E(CP)				0.04172006			9	0.044122	−0.036073
E(LB)				0.08554362			10	0.056540	−0.022877
							11	0.070637	−0.006924
							12	0.785235	0.592326
					E(LU)			0.09145851	
					E(CP)				0.01723593
					E(LB)				0.06798899

	N	r	i	α_i	c_i		N	r	i	α_i	c_i
	18	13	1	−0.007289	−0.064317	E(LU)				0.07436294	
			2	−0.003238	−0.067387	E(CP)					0.00240300
			3	0.001550	−0.067701	E(LB)						0.05508562
			4	0.006940	−0.066218						
			5	0.012925	−0.063222		18	15	1	0.004780	−0.052617
			6	0.019540	−0.058792				2	0.008843	−0.055674
			7	0.026851	−0.052898				3	0.013074	−0.056526
			8	0.034951	−0.045430				4	0.017522	−0.055953
			9	0.043969	−0.036200				5	0.022232	−0.054191
			10	0.054072	−0.024926				6	0.027249	−0.051307
			11	0.065486	−0.011201				7	0.032630	−0.047281
			12	0.078516	0.005561				8	0.038443	−0.042029
			13	0.665728	0.552731				9	0.044772	−0.035403
E(LU)				0.08146655					10	0.051728	−0.027176
E(CP)					0.00893995				11	0.059460	−0.017018
E(LB)					0.06110111				12	0.068169	−0.004442
									13	0.078145	0.011289
	18	14	1	−0.000568	−0.058133				14	0.089813	0.031340
			2	0.003471	−0.061213				15	0.443142	0.456986
			3	0.007930	−0.061830						
			4	0.012775	−0.060849	E(LU)				0.06933298	
			5	0.018027	−0.058527	E(CP)				−0.00278409	
			6	0.023730	−0.054936	E(LB)					0.04973648
			7	0.029942	−0.050053						
			8	0.036744	−0.043781						
			9	0.044239	−0.035952						
			10	0.052564	−0.026314						
			11	0.061904	−0.014497						
			12	0.072509	0.000034						
			13	0.084730	0.018080						
			14	0.552004	0.507970						

N	r	i	α_i	c_i	N	r	i	α_i	c_i
18	16	1	0.009048	−0.047594	18	17	1	0.012444	−0.042879
		2	0.013157	−0.050597			2	0.016611	−0.045800
		3	0.017235	−0.051629			3	0.020593	−0.046965
		4	0.021397	−0.051393			4	0.024555	−0.047008
		5	0.025706	−0.050102			5	0.028573	−0.046121
		6	0.030212	−0.047820			6	0.032702	−0.044362
		7	0.034966	−0.044532			7	0.036990	−0.041722
		8	0.040027	−0.040165			8	0.041487	−0.038137
		9	0.045465	−0.034587			9	0.046252	−0.033494
		10	0.051368	−0.027599			10	0.051355	−0.027618
		11	0.057855	−0.018906			11	0.056889	−0.020248
		12	0.065087	−0.008069			12	0.062980	−0.010994
		13	0.073294	0.005581			13	0.069810	0.000742
		14	0.082827	0.023119			14	0.077655	0.015937
		15	0.094248	0.046408			15	0.086979	0.036313
		16	0.338109	0.397887			16	0.098636	0.065331
E(LU)			0.06582537				17	0.235490	0.327023
E(CP)			−0.00691185		E(LU)			0.06346845	
E(LB)				0.04487895	E(CP)			−0.01018489	
					E(LB)				0.04033369

N	r	i	α_i	c_i
18	18	1	0.015092	−0.038165
		2	0.019328	−0.040965
		3	0.023258	−0.042221
		4	0.027089	−0.042497
		5	0.030909	−0.041963
		6	0.034773	−0.040676
		7	0.038728	−0.038627
		8	0.042820	−0.035765
		9	0.047095	−0.031992
		10	0.051612	−0.027160
		11	0.056443	−0.021041
		12	0.061685	−0.013300
		13	0.067477	−0.003410
		14	0.074032	0.009488
		15	0.081713	0.026940
		16	0.091221	0.052132
		17	0.104314	0.093529
		18	0.132411	0.235693
E(LU)			0.06204005	
E(CP)			−0.01272745	
E(LB)				0.03580789
19	2	1	−1.223869	−0.493292
		2	2.223869	0.493292
E(LU)			3.74782267	
E(CP)			1.23712437	
E(LB)				0.49325215
19	3	1	−0.643659	−0.327238
		2	−0.601169	−0.315224
		3	2.244827	0.642462

N	r	i	α_i	c_i
E(LU)			1.68432765	
E(CP)			0.64655945	
E(LB)				0.32423458
19	4	1	−0.394294	−0.243540
		2	−0.371431	−0.238113
		3	−0.329685	−0.221662
		4	2.095409	0.703314
E(LU)			0.93343886	
E(CP)			0.39452713	
E(LB)				0.23964110
19	5	1	−0.261237	−0.192958
		2	−0.247259	−0.190909
		3	−0.220739	−0.180245
		4	−0.185724	−0.163869
		5	1.914959	0.727980
E(LU)			0.58171310	
E(CP)			0.26081700	
E(LB)				0.18881059
19	6	1	−0.180964	−0.159004
		2	−0.171530	−0.158877
		3	−0.153365	−0.151748
		4	−0.129250	−0.139981
		5	−0.100003	−0.124297
		6	1.735111	0.733908
E(LU)			0.39192137	
E(CP)			0.18053993	
E(LB)				0.15485543

N	r	i	α_i	c_i	N	r	i	α_i	c_i
19	7	1	−0.128533	−0.134583	E(LU)			0.16305851	
		2	−0.121602	−0.135622	E(CP)			0.06668914	
		3	−0.108414	−0.130811	E(LB)				0.09793914
		4	−0.090935	−0.122136					
		5	−0.069730	−0.110197	19	10	1	−0.047007	−0.090004
		6	−0.044949	−0.095185			2	−0.042865	−0.092551
		7	1.564162	0.728535			3	−0.036265	−0.091324
E(LU)			0.27982455				4	−0.027929	−0.087629
E(CP)			0.12832890				5	−0.018045	−0.081863
E(LB)				0.13053726			6	−0.006641	−0.074147
							7	0.006343	−0.064468
19	8	1	−0.092336	−0.116135			8	0.021029	−0.052718
		2	−0.086846	−0.117908			9	0.037595	−0.038703
		3	−0.076795	−0.114696			10	1.113786	0.673407
		4	−0.063593	−0.108201	E(LU)			0.13159684	
		5	−0.047637	−0.098937	E(CP)			0.04766707	
		6	−0.029016	−0.087065	E(LB)				0.08643819
		7	−0.007670	−0.072570					
		8	1.403893	0.715513	19	11	1	−0.032345	−0.080360
E(LU)			0.20936888				2	−0.028492	−0.083097
E(CP)			0.09242024				3	−0.022852	−0.082502
E(LB)				0.11223593			4	−0.015928	−0.079736
							5	−0.007843	−0.075152
19	9	1	−0.066309	−0.101674			6	0.001395	−0.068861
		2	−0.061667	−0.103918			7	0.011842	−0.060851
		3	−0.053675	−0.101850			8	0.023603	−0.051025
		4	−0.043347	−0.096952			9	0.036827	−0.039209
		5	−0.030960	−0.089671			10	0.051716	−0.025147
		6	−0.016565	−0.080147			11	0.982076	0.645940
		7	−0.000103	−0.068365	E(LU)			0.10969257	
		8	0.018570	−0.054204	E(CP)			0.03326000	
		9	1.254056	0.696782	E(LB)				0.07696225

N	r	i	α_i	c_i		N	r	i	α_i	c_i
19	12	1	−0.020995	−0.072230	E(LU)				0.08302831	
		2	−0.017298	−0.075078	E(CP)					0.01342381
		3	−0.012331	−0.074965	E(LB)					0.06217738
		4	−0.006425	−0.072929						
		5	0.000344	−0.069288		19	14	1	−0.004989	−0.059169
		6	0.007984	−0.064141				2	−0.001384	−0.062091
		7	0.016548	−0.057480				3	0.002773	−0.062636
		8	0.026124	−0.049218				4	0.007387	−0.061652
		9	0.036839	−0.039200				5	0.012453	−0.059399
		10	0.048859	−0.027194				6	0.017997	−0.055961
		11	0.062404	−0.012872				7	0.024066	−0.051334
		12	0.857947	0.614595				8	0.030726	−0.045450
E(LU)			0.09416748					9	0.038064	−0.038185
E(CP)				0.02213853				10	0.046194	−0.029351
E(LB)				0.06899532				11	0.055267	−0.018676
								12	0.065482	−0.005781
19	13	1	−0.012077	−0.065253				13	0.077109	0.009882
		2	−0.008453	−0.068158				14	0.628854	0.539802
		3	−0.003960	−0.068416	E(LU)				0.07498023	
		4	0.001201	−0.066962	E(CP)					0.00651543
		5	0.006996	−0.064084	E(LB)					0.05624730
		6	0.013442	−0.059872						
		7	0.020588	−0.054320						
		8	0.028511	−0.047351						
		9	0.037316	−0.038827						
		10	0.047141	−0.028538						
		11	0.058169	−0.016185						
		12	0.070640	−0.001353						
		13	0.740487	0.579318						

N	r	i	α_i	c_i	N	r	i	α_i	c_i
19	15	1	0.000692	−0.053779	19	16	1	0.005271	−0.048924
		2	0.004313	−0.056686			2	0.008929	−0.051792
		3	0.008234	−0.057456			3	0.012687	−0.052735
		4	0.012444	−0.056855			4	0.016601	−0.052449
		5	0.016963	−0.055121			5	0.020708	−0.051151
		6	0.021823	−0.052332			6	0.025048	−0.048913
		7	0.027068	−0.048486			7	0.029663	−0.045735
		8	0.032757	−0.043523			8	0.034603	−0.041566
		9	0.038961	−0.037334			9	0.039929	−0.036308
		10	0.045774	−0.029749			10	0.045717	−0.029809
		11	0.053320	−0.020523			11	0.052068	−0.021850
		12	0.061762	−0.009310			12	0.059114	−0.012118
		13	0.071323	0.004394			13	0.067036	−0.000151
		14	0.082316	0.021334			14	0.076094	0.014738
		15	0.522250	0.495426			15	0.086669	0.033642

E(LU)   0.06915784
E(CP)   0.00099208
E(LB)   0.05100764

| | | 16 | 0.419861 | 0.445119 |

E(LU)   0.06496979
E(CP)   −0.00344791
E(LB)   0.04630055

N	r	i	α_i	c_i	N	r	i	α_i	c_i
19	17	1	0.008968	−0.044464	19	18	1	0.011941	−0.040244
		2	0.012677	−0.047270			2	0.015709	−0.042966
		3	0.016326	−0.048344			3	0.019289	−0.044136
		4	0.020023	−0.048319			4	0.022835	−0.044327
		5	0.023825	−0.047390			5	0.026412	−0.043716
		6	0.027772	−0.045626			6	0.030068	−0.042366
		7	0.031906	−0.043028			7	0.033841	−0.040281
		8	0.036272	−0.039552			8	0.037772	−0.037423
		9	0.040920	−0.035112			9	0.041903	−0.033716
		10	0.045913	−0.029573			10	0.046286	−0.029044
		11	0.051331	−0.022740			11	0.050984	−0.023232
		12	0.057280	−0.014330			12	0.056083	−0.016030
		13	0.063906	−0.003928			13	0.061695	−0.007067
		14	0.071419	0.009098			14	0.067989	0.004227
		15	0.080136	0.025759			15	0.075216	0.018774
		16	0.090563	0.047806			16	0.083802	0.038205
		17	0.320763	0.387016			17	0.094528	0.065788
							18	0.223648	0.317554

E(LU) 0.06200679
E(CP) −0.00702291
E(LB) 0.04198714

E(LU) 0.05998848
E(CP) −0.00988868
E(LB) 0.03791809

	N	r	i	α_i	c_i		N	r	i	α_i	c_i
	19	19	1	0.014282	−0.035995	E(LU)				1.75538518	
			2	0.018115	−0.038600	E(CP)				0.66462201	
			3	0.021661	−0.039833	E(LB)					0.32470597
			4	0.025107	−0.040204						
			5	0.028531	−0.039873		20	4	1	−0.408252	−0.243885
			6	0.031980	−0.038897				2	−0.386289	−0.238726
			7	0.035494	−0.037282				3	−0.345972	−0.223154
			8	0.039109	−0.034997				4	2.140513	0.705766
			9	0.042861	−0.031977	E(LU)				0.97791855	
			10	0.046794	−0.028121	E(CP)				0.40827717	
			11	0.050958	−0.023280	E(LB)					0.24018443
			12	0.055419	−0.017234						
			13	0.060267	−0.009660		20	5	1	−0.272551	−0.193344
			14	0.065629	−0.000055				2	−0.259179	−0.191385
			15	0.071704	0.012400				3	−0.233536	−0.181278
			16	0.078826	0.029177				4	−0.199675	−0.165821
			17	0.087648	0.053305				5	1.964941	0.731828
			18	0.099799	0.092832	E(LU)				0.61184794	
			19	0.125817	0.228292	E(CP)				0.27193675	
E(LU)				0.05874886		E(LB)					0.18940540
E(CP)				−0.01213794							
E(LB)					0.03383684		20	6	1	−0.190502	−0.159435
									2	−0.181539	−0.159298
	20	2	1	−1.251068	−0.493634				3	−0.163969	−0.152528
			2	2.251068	0.493634				4	−0.140605	−0.141408
E(LU)				3.88005370					5	−0.112303	−0.126651
E(CP)				1.26365389					6	1.788917	0.739321
E(LB)					0.49359782	E(LU)				0.41329354	
						E(CP)				0.18987852	
	20	3	1	−0.662014	−0.327555	E(LB)					0.15549251
			2	−0.621129	−0.316157						
			3	2.283144	0.643713						

N	r	i	α_i	c_i	N	r	i	α_i	c_i
20	7	1	−0.136790	−0.135060	E(LU)			0.17159045	
		2	−0.130264	−0.136029	E(CP)			0.07297238	
		3	−0.117518	−0.131448	E(LB)				0.09868793
		4	−0.100561	−0.123236					
		5	−0.079995	−0.111990	20	10	1	−0.052900	−0.090626
		6	−0.056007	−0.097918			2	−0.049115	−0.093031
		7	1.621135	0.735681			3	−0.042792	−0.091837
E(LU)			0.29539488				4	−0.034710	−0.088309
E(CP)			0.13637533				5	−0.025087	−0.082842
E(LB)				0.13121241			6	−0.013973	−0.075573
							7	−0.001335	−0.066511
20	8	1	−0.099621	−0.116659			8	0.012921	−0.055584
		2	−0.094504	−0.118326			9	0.028939	−0.042651
		3	−0.084808	−0.115255			10	1.178052	0.686964
		4	−0.071993	−0.109093	E(LU)			0.13788300	
		5	−0.056488	−0.100352	E(CP)			0.05331635	
		6	−0.038416	−0.089210	E(LB)				0.08722579
		7	−0.017755	−0.075681					
		8	1.463585	0.724575	20	11	1	−0.037715	−0.081036
E(LU)			0.22087332				2	−0.034222	−0.083625
E(CP)			0.09948206				3	−0.028845	−0.083028
E(LB)				0.11294771			4	−0.022146	−0.080373
							5	−0.014276	−0.076014
20	9	1	−0.072826	−0.102246			6	−0.005262	−0.070070
		2	−0.068544	−0.104362			7	0.004930	−0.062554
		3	−0.060858	−0.102371			8	0.016382	−0.053398
		4	−0.050834	−0.097711			9	0.029216	−0.042476
		5	−0.038781	−0.090828			10	0.043593	−0.029594
		6	−0.024779	−0.081874			11	1.048347	0.662168
		7	−0.008798	−0.070863	E(LU)			0.11423656	
		8	0.009270	−0.057714	E(CP)			0.03838053	
		9	1.316151	0.707969	E(LB)				0.07779186

	N	r	i	α_i	c_i		N	r	i	α_i	c_i
	20	12	1	-0.025922	-0.072964		20	14	1	-0.009191	-0.060043
			2	-0.022589	-0.075662				2	-0.005961	-0.062821
			3	-0.017879	-0.075522				3	-0.002065	-0.063307
			4	-0.012183	-0.073554				4	0.002348	-0.062329
			5	-0.005600	-0.070076				5	0.007248	-0.060148
			6	0.001858	-0.065197				6	0.012649	-0.056856
			7	0.010227	-0.058928				7	0.018585	-0.052465
			8	0.019578	-0.051211				8	0.025109	-0.046929
			9	0.030012	-0.041931				9	0.032297	-0.040154
			10	0.041668	-0.030912				10	0.040241	-0.031999
			11	0.054724	-0.017911				11	0.049069	-0.022261
			12	0.926107	0.633868				12	0.058941	-0.010659
E(LU)				0.09732994					13	0.070069	0.003196
E(CP)					0.02680890				14	0.700660	0.566775
E(LB)					0.06987174	E(LU)				0.07610847	
	20	13	1	-0.016619	-0.066052	E(CP)					0.01045130
			2	-0.013364	-0.068809	E(LB)					0.05724068
			3	-0.009129	-0.069021						
			4	-0.004170	-0.067601		20	15	1	-0.003203	-0.054744
			5	0.001453	-0.064835				2	0.000035	-0.057513
			6	0.007742	-0.060825				3	0.003690	-0.058213
			7	0.014732	-0.055581				4	0.007695	-0.057596
			8	0.022485	-0.049051				5	0.012048	-0.055899
			9	0.031087	-0.041132				6	0.016769	-0.053209
			10	0.040654	-0.031666				7	0.021892	-0.049537
			11	0.051333	-0.020429				8	0.027467	-0.044842
			12	0.063321	-0.007116				9	0.033552	-0.039043
			13	0.810474	0.602120				10	0.040230	-0.032010
E(LU)				0.08507430					11	0.047602	-0.023560
E(CP)					0.01770390				12	0.055804	-0.013436
E(LB)					0.06310742				13	0.065012	-0.001280
									14	0.075467	0.013416
									15	0.595940	0.527466

	N	r	i	α_i	c_i		N	r	i	α_i	c_i
E(LU)				0.06951916			20	17	1	0.005617	−0.045695
E(CP)					0.00461910				2	0.008931	−0.048385
E(LB)					0.05207861				3	0.012297	−0.049380
									4	0.015773	−0.049304
	20	16	1	0.001656	−0.050002				5	0.019394	−0.048357
			2	0.004926	−0.052742				6	0.023192	−0.046613
			3	0.008410	−0.053608				7	0.027201	−0.044083
			4	0.012111	−0.053287				8	0.031459	−0.040736
			5	0.016048	−0.051997				9	0.036010	−0.036510
			6	0.020247	−0.049816				10	0.040910	−0.031298
			7	0.024742	−0.046757				11	0.046228	−0.024954
			8	0.029575	−0.042785				12	0.052055	−0.017265
			9	0.034801	−0.037825				13	0.058509	−0.007934
			10	0.040482	−0.031763				14	0.065756	0.003474
			11	0.046707	−0.024432				15	0.074029	0.017606
			12	0.053585	−0.015601				16	0.083671	0.035486
			13	0.061262	−0.004939				17	0.398968	0.433947
			14	0.069938	0.008023	E(LU)				0.06114864	
			15	0.079893	0.023984	E(CP)					−0.00394307
			16	0.495616	0.483548	E(LB)					0.04329477
E(LU)				0.06467887							
E(CP)					−0.00010333						
E(LB)					0.04747116						

N	r	i	α_i	c_i	N	r	i	α_i	c_i
20	18	1	0.008847	−0.041706	20	19	1	0.011469	−0.037905
		2	0.012215	−0.044331			2	0.014895	−0.040446
		3	0.015502	−0.045422			3	0.018135	−0.041607
		4	0.018813	−0.045550			4	0.021329	−0.041903
		5	0.022197	−0.044896			5	0.024538	−0.041503
		6	0.025690	−0.043529			6	0.027802	−0.040467
		7	0.029324	−0.041460			7	0.031153	−0.038810
		8	0.033136	−0.038666			8	0.034524	−0.036509
		9	0.037162	−0.035086			9	0.038247	−0.033514
		10	0.041450	−0.030632			10	0.042061	−0.029746
		11	0.046055	−0.025168			11	0.046112	−0.025085
		12	0.051050	−0.018506			12	0.050458	−0.019364
		13	0.056532	−0.010374			13	0.055176	−0.012340
		14	0.062634	−0.000381			14	0.060372	−0.003660
		15	0.069547	0.012071			15	0.066198	0.007217
		16	0.077558	0.027938			16	0.072887	0.021167
		17	0.087131	0.048871			17	0.080830	0.039737
		18	0.305157	0.376826			18	0.090746	0.066024
E(LU)			0.05861867				19	0.212971	0.308714
E(CP)			−0.00706723		E(LU)			0.05687410	
E(LB)				0.03943688	E(CP)			−0.00959610	
					E(LB)				0.03577112

N	r	i	α_i	c_i		N	r	i	α_i	c_i
20	20	1	0.013553	−0.034055	E(LU)				0.05578958	
		2	0.017039	−0.036484	E(CP)				−0.01159947	
		3	0.020257	−0.037686	E(LB)					0.03207039
		4	0.023376	−0.038123						
		5	0.026464	−0.037945						
		6	0.029565	−0.037211						
		7	0.032711	−0.035932						
		8	0.035932	−0.034091						
		9	0.039258	−0.031646						
		10	0.042720	−0.028527						
		11	0.046357	−0.024632						
		12	0.050215	−0.019814						
		13	0.054354	−0.013860						
		14	0.058856	−0.006460						
		15	0.063842	0.002866						
		16	0.069496	0.014902						
		17	0.076128	0.031052						
		18	0.084346	0.054203						
		19	0.095669	0.092028						
		20	0.119862	0.221415						

APPENDIX M

Binomial or attribute reliability testing curves for confidence levels of 50%, 80%, 90%, 95% and 99%, where N=number of entries and F=number of failures observed.
From David K. Lloyd and Myron Lipow, *Reliability: Management, Methods and Mathematics*, 1962. By permission of Prentice-Hall, Inc., Englewood Cliffs, New Jersey.

APPENDIX M.1
$$\gamma = 0.50$$

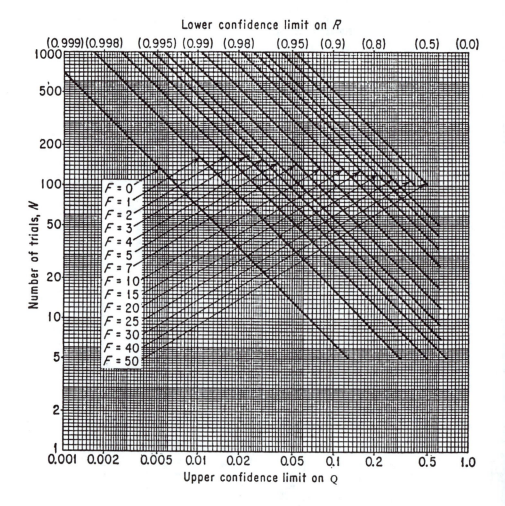

APPENDIX M.2
$\gamma = 0.80$

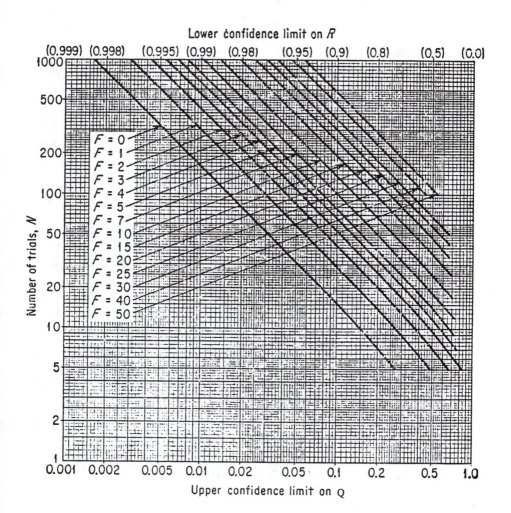

Lower confidence limit on R

APPENDIX M.3
$$\gamma = 0.90$$

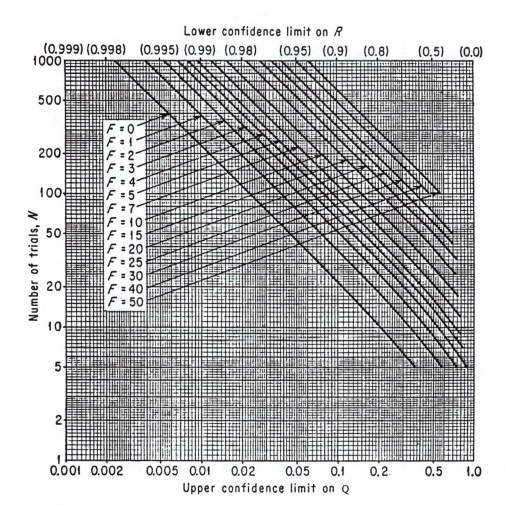

APPENDIX M.4
$\gamma = 0.95$

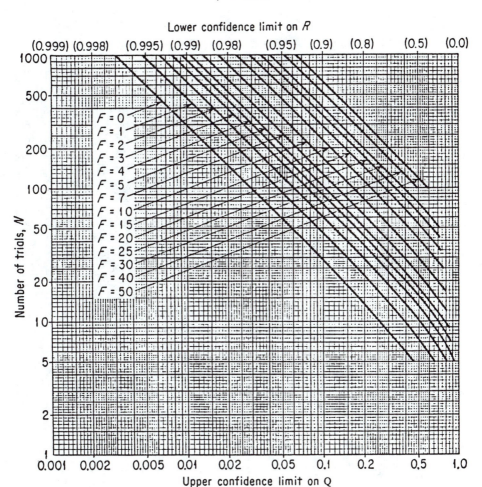

Lower confidence limit on R

(0.999) (0.998) (0.995) (0.99) (0.98) (0.95) (0.9) (0.8) (0.5) (0.0)

$F = 0$
$F = 1$
$F = 2$
$F = 3$
$F = 4$
$F = 5$
$F = 7$
$F = 10$
$F = 15$
$F = 20$
$F = 25$
$F = 30$
$F = 40$
$F = 50$

Number of trials, N

Upper confidence limit on Q

APPENDIX M.5
$\gamma = 0.99$

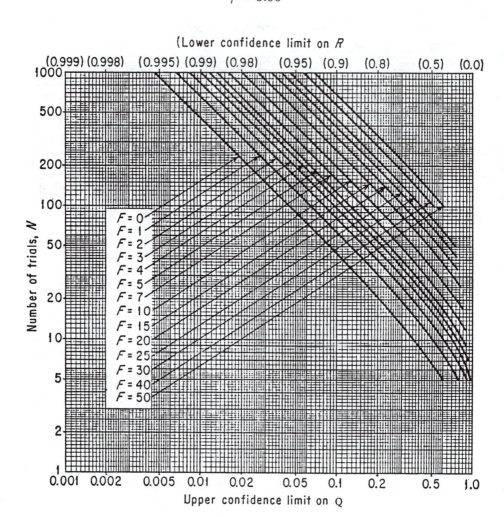

(Lower confidence limit on R

Index

ABOUT THE AUTHOR

This book is based on the following extensive experience of the author in Reliability Engineering an Life Testing:

1. He initiated and was the Director of the Corporate Reliability Engineering Program at the Allis-Chalmers Manufacturing Co., Milwaukee, Wisconsin, from 1960 to 1963.

2. He started the Reliability Engineering Instructional Program at The University of Arizona in 1963, which now has more than ten courses in it. A Master's Degree with a Reliability Engineering Option is currently being offered in the Aerospace and Mechanical Engineering Department at The University of Arizona. This option started in 1969. A Master's Degree in Reliability Engineering is also being offered now in the Systems and Industrial Engineering Department at The University of Arizona. This degree started in January 1987.

3. He conceived and directed the first two Summer Institutes for College Teachers in reliability engineering ever to be supported by the National Science Foundation. The first was in 1965 and the second in 1966, for 30 college and university faculty each summer. These faculty started teaching reliability engineering courses at their respective universities and/or incorporating reliability engineering concepts into their courses.

4. He helped initiate The Professional Certificate Award in Reliability and Quality Engineering program at The University of Arizona in 1991. This is a 15-unit program. The certificate's requirements are met via videotapes of the VIDEOCAMPUS organization. No participant need to be on the campus of The University of Arizona to get this certificate.

5. In 1963 he conceived, initiated, and has directed since then the now internationally famous *Annual Reliability Engineering and Management Institute* at The University of Arizona.

6. In 1975 he conceived, initiated, and has directed since then the now internationally famous *Annual Reliability Testing Institute* at The University of Arizona.

914

7. He has lectured extensively and conducted over 285 training courses, short courses and seminars worldwide and has exposed over 10,000 reliability, maintainability, test, design, and product assurance engineers to the concepts in this book.

8. He has been the principal investigator of mechanical reliability research for the NASA-Lewis Research Center, the Office of Naval Research, and the Naval Weapons Engineering Support Activity for ten years.

9. He has been consulted extensively by over 82 industries and government agencies worldwide on reliability engineering, reliability and life testing, maintainability engineering, and mechanical reliability matters.

10. He has been active in the Reliability and Maintainability Symposia and Conferences dealing with reliability engineering since 1963.

11. He founded the Tucson Chapter of the Society of Reliability Engineers in 1974 and was its first President. He also founded the first and very active Student Chapter of the Society of Reliability Engineers at The University of Arizona.

12. He has authored or co-authored over 120 papers and articles, of which over 107 are in all areas of reliability engineering.

13. In addition to this book, he authored or contributed to the following books:

    (1) *Bibliography on Plasticity - Theory and Applications,* Dimitri B. Kececioglu, published by the American Society of Mechanical Engineers, New York, 191 pp., 1950.

    (2) *Manufacturing, Planning and Estimating Handbook,* Dimitri B. Kececioglu and Lawrence Karvonen contributed part of Chapter 19, pp. 19-1 to 19-12, published by McGraw-Hill Book Co., Inc., New York, 864 pp., 1963.

    (3) *Introduction to Probabilistic Design for Reliability,* Dimitri B. Kececioglu, published by the United States Army Management Engineering Training Agency, Rock Island, Illinois, contributed Chapter 7 of 109 pp., and Chapter 8 of 137 pp., May 1974.

(4) *Manual of Product Assurance Films on Reliability Engineering and Management, Reliability Testing, Maintainability, and Quality Control,* Dimitri Kececioglu, printed by Dr. Dimitri B. Kececioglu, 7340 N. La Oesta Avenue, Tucson, Arizona 85704, 178 pp., 1976.

(5) *Manual of Product Assurance Films and Videotapes,* Dimitri Kececioglu, printed by Dimitri B. Kececioglu, 7340 N. La Oesta Avenue, Tucson, Arizona 85704, 327 pp., 1980.

(6) *Reliability Engineering Handbook,* Dimitri B. Kececioglu, Prentice Hall Inc., Englewood Cliffs, New Jersey 07632, Vol. 1, 720 pp. and Vol. 2, 568 pp., 1991

(7) *Maintainability, Availability and Operational Readiness Engineering Handbook,* Dimitri B. Kececioglu, will be published in 1993 by Prentice Hall.

14. He has received over 30 prestigious awards and has been recognized for his valuable contributions to the field of reliability engineering. Among these are the following: (1) Fulbright Scholar in 1971. (2) Ralph Teetor Award of the Society of Automotive Engineers as "Outstanding Engineering Educator" in 1977. (3) Certificate of Excellence by the Society of Reliability Engineers for his "personal contributions made toward the advancement of the philosophy and principles of reliability engineering" in 1978. (4) ASQC-Reliability Division, Reliability Education Advancement Award for his "outstanding contributions to the development and presentation of meritorious reliability educational programs" in 1980. (5) ASQC Allen Chop Award for his "outstanding contributions to Reliability Science and Technology" in 1981. (6) The University of Arizona College of Engineering Anderson Prize for "engineering the Master's Degree program in the Reliability Engineering Option" in 1983. (7) Designation of "Senior Extension Teacher" by Dr. Leonard Freeman, Dean, Continuing Education and University Extension, the University of California, Los Angeles in 1983. (8) Honorary Member, Golden Key National Honor Society in 1984. (9) Honorary Professor, Shanghai University of Technology in 1984. (10) Honorary Professor, Phi Kappa Phi Honor Society in 1988. (11) The American Hellenic Educational Progressive Association "Academy of Achievement Award in Education" in 1992.

15. He conceived and established *The Dr. Dimitri Basil Kececioglu Reliability Engineering Research Fellowships Endowment Fund* in 1987. The cosponsors of his institutes, mentioned in Items 5 and 6, have contributed generously to this fund which has now crossed the $320,000 mark.

16. He received his BSME from Robert College, Istanbul, Turkey in 1942, and his M.S. in Industrial Engineering in 1948 and his Ph.D. in Engineering Mechanics in 1953, both from Purdue University, Lafayette, Indiana.

17. He has also been granted five patents.